Computer Aided Analysis and Optimization of Mechanical System Dynamics

NATO ASI Series

Advanced Science Institutes Series

A series presenting the results of activities sponsored by the NATO Science Committee, which aims at the dissemination of advanced scientific and technological knowledge, with a view to strengthening links between scientific communities.

The Series is published by an international board of publishers in conjunction with the NATO Scientific Affairs Division

A Life Sciences **B Physics**	Plenum Publishing Corporation London and New York
C Mathematical and Physical Sciences	D. Reidel Publishing Company Dordrecht, Boston and Lancaster
D Behavioural and Social Sciences **E Applied Sciences**	Martinus Nijhoff Publishers Boston, The Hague, Dordrecht and Lancaster
F Computer and Systems Sciences **G Ecological Sciences**	Springer-Verlag Berlin Heidelberg New York Tokyo

Series F Computer and Systems Sciences Vol. 9

Computer Aided Analysis and Optimization of Mechanical System Dynamics

Edited by

Edward J. Haug

Center for Computer Aided Design, College of Engineering
University of Iowa, Iowa City, IA 52242, USA

Springer-Verlag Berlin Heidelberg GmbH 1984

Proceedings of the NATO Advanced Study Institute on Computer Aided Analysis and Optimization of Mechanical System Dynamics held at Iowa City/USA, August 1–12, 1983

ISBN 978-3-642-52467-7 ISBN 978-3-642-52465-3 (eBook)
DOI 10.1007/978-3-642-52465-3

Library of Congress Cataloging in Publication Data. NATO Advanced Study Institute on Computer Aided Analysis and Optimization of Mechanical System Dynamics (1983 : Iowa City, Iowa) Computer aided analysis and optimization of mechanical system dynamics. (NATO ASI series. Series F, Computer and systems sciences ; vol. 9) "Proceedings of the NATO Advanced Study Institute on Computer Aided Analysis and Optimization of Mechanical System Dynamics held at Iowa City/USA, August 1–12, 1983"--Includes bibliographical references. 1. Machinery, Dynamics of--Data processing--Congresses. 2. Dynamics–Data processing--Congresses. I. Haug, Edward J. II. Title. III. Series: NATO ASI series. Series F, Computer and system sciences ; no. 9. TJ173.N38 1983 621.8'11 84-10482 ISBN 0-387-12887-5 (U.S.)

Originally published by Springer-Verlag Berlin Heidelberg New York Tokyo in 1984

2145/3140-543210

NATO-NSF-ARO ADVANCED STUDY INSTITUTE ON

COMPUTER AIDED ANALYSIS AND

OPTIMIZATION OF MECHANICAL SYSTEM DYNAMICS

IOWA CITY, IOWA, UNITED STATES, AUGUST 1-12, 1983

Director: Edward J. Haug
Center for Computer Aided Design
College of Engineering
University of Iowa
Iowa City, Iowa

Scientific Content of the Advanced Study Institute

The Advanced Study Institute was organized to bring together engineers, numerical analysts, and applied mathematicians working in the field of mechanical system dynamic analysis and optimization. The principal focus of the Institute was on dynamic analysis and optimization of mechanical systems that are comprised of multiple bodies connected by kinematic joints and compliant elements. Specialists working in this area from throughout North America and Western Europe presented alternative approaches to computer generation and solution of the equations of system dynamics. Numerical analysis considerations such as sparse matrix methods, loop closure topological analysis methods, symbolic computation methods, and computer graphics were explored and applied to system dynamic analysis and design. This forum provided ample opportunity for expression of divergent views and spirited discussion of alternatives and their pros and cons. Emerging developments in dynamics of systems with flexible bodies, feedback control, intermittent motion, and other interdisciplinary effects were presented and illustrated. Animated graphics was shown to be a valuable tool in visualization of system dynamics, as illustrated through applications in mechanism and vehicle dynamics. Recently developed methods of kinematic synthesis, kinematic and dynamic design sensitivity analysis, and iterative optimization of mechanisms and machines were presented and illustrated.

Scientific Program of the Advanced Study Institute

The scientific program began with a review (Haug) of alternative approaches that are possible and trade-offs that must be made in selecting an efficient, unified method of system dynamic analysis.

Fundamental analytical methods in machine dynamics were reviewed (Paul and Wittenburg) and computational applications discussed. Theoretical methods for kinematic definition of system state were discussed (Wittenburg and Wehage). Lagrangian formulations of equations of mechanical system dynamics, using symbolic computation and a minimal set of generalized coordinates, were presented and applied to study vehicle dynamics (Schiehlen). An alternative formulation, using a maximal set of Cartesian generalized coordinates and the resulting simplified form of sparse equations, were presented and illustrated (Chace and Nikravesh). The potential for application of general purpose symbolic computation languages for support of dynamic analysis was considered and test problems illustrated (Noble). A comprehensive review of numerical methods that are available for solving differential equations of motion, regardless of how derived, was presented (Enright) and computer software that is available for applications was discussed. Special numerical analysis problems associated with mixed differential-algebraic equations and numerical methods for treating systems with both high frequency and low frequency content were discussed and the state of the art evaluated (Gear). Application of numerical integration methods to various formulations of equations of motion were discussed and use of high speed computer graphics to create an animation as output of dynamic analysis was illustrated (Nikravesh). Formulations for dynamic analysis of mechanisms and machines with flexible components were presented and their relationship with finite element structural analysis codes discussed (van der Werff). Systematic incorporation of feedback control and hydraulic effects in large scale mechanical system dynamics were discussed and illustrated (Vanderploeg). Methods of kinematic synthesis were presented and their application using microcomputers illustrated (Rankers). Methods for design sensitivity analysis and optimization of large scale kinematically and dynamically driven systems were presented and illustrated (Haug). Iterative optimization methods that are suitable for application in kinematic and dynamic system synthesis were reviewed and their pros and cons discussed (Fleury and Gill).

LECTURERS

M. Chace, Mechanical Dynamics, Inc., Ann Arbor, Michigan, U.S.A.
W. Enright, University of Toronto, Toronto, Ontario, CANADA
W. Gear, University of Illinois, Champaign-Urbana, Illinois, U.S.A.
E. Haug, The University of Iowa, Iowa City, Iowa, U.S.A.
M. Hussain, General Electric Corporation, Schenectady, New York, U.S.A.
P. Nikravesh, The University of Iowa, Iowa City, Iowa, U.S.A.
B. Noble, University of Wisconsin, Madison, Wisconsin, U.S.A.
B. Paul, University of Pennsylvania, Philadelphia, Pennsylvania, U.S.A.
H. Rankers, Delft University of Technology, Delft, The NETHERLANDS
W. Schiehlen, University of Stuttgart, Stuttgart, GERMANY
K. van der Werff, Delft University of Technology, Delft, The NETHERLANDS
M. Vanderploeg, The University of Iowa, Iowa City, Iowa, U.S.A.
R. Wehage, U.S. Army Tank Automotive R & D Command, Warren, Michigan, U.S.A.
J. Wittenburg, Karlsruhe University, Karlsruhe, GERMANY

PARTICIPANTS

R. Albrecht, Technische Universitat Braunschweig, Braunschweig, WEST GERMANY
G. Andrews, University of Waterloo, Waterloo, Ontario, CANADA
J. Avery, University of Idaho, Moscow, Idaho, U.S.A.
P. Bourassa, University of Sherbrooke, Sherbrooke, Quebec, CANADA
S. Chang, Chung-cheng Institute of Technology, Taishi, Taoyen, TAIWAN ROC
S. Desa, Stanford University, Stanford, California, U.S.A.
A. Dilpare, Jacksonville University, Jacksonville, Florida, U.S.A.
M. Dixon, Clemson University, Clemson, South Carolina, U.S.A.
S. Dwivedi, University of North Carolina at Charlotte, Charlotte, North Carolina, U.S.A.
C. Fleury, University of Liege, Liege, BELGIUM
M. Geradin, University of Liege, Liege, BELGIUM
J. Granda, California State University, Sacramento, California, U.S.A.

P. Grewal, McMaster University at Hamilton, Hamilton, Ontario, CANADA

M. Guns, Katholieke University, Leuven, BELGIUM

J. Hansen, Technical University of Denmark, Lyngby, DENMARK

J. Inoue, Kyushu Institute of Technology, Kitakyshu-shi, JAPAN

Y. Inoue, Kobe Steel, Ltd., Kobe, JAPAN

D. Johnson, University Catholique of Louvain, Louvain, BELGIUM

A. Jones, University of California, Davis, California, U.S.A.

S. Kim, The University of Iowa, Iowa City, Iowa, U.S.A.

T. Knudsen, Technical Universtiy of Denmark, Lyngby, DENMARK

E. Kreuzer, University of Stuttgart, Stuttgart, WEST GERMANY

M. Lawo, Essen University, Essen, WEST GERMANY

S. Lukowski, University of Akron, Arkon, Ohio, U.S.A.

M. Magi, Chalmers University of Technology, Gotteborg, SWEDEN

N. Mani, The University of Iowa, Iowa City, Iowa, U.S.A.

K. Matsui, Tokyo Denki University, Saitama, JAPAN

Y. Morel, FRAMATOME, Saint Marcel, FRANCE

U. Neumann, Technische Hockschule Darmstadt, Darmstadt, GERMANY

G. Ostermeyer, Technische Universitat Braunschweig, Braunschweig, WEST GERMANY

A. Pasrija, Drexel University, Philadelphia, Pennsylvania, U.S.A.

E. Pennestri, University of Rome, Rome, ITALY

M. Pereira, Instituto Superior Tecnico, Lisboa Codex, PORTUGAL

N. Petersmann, University of Hannover, Hannover, GERMANY

J. Pinto, San Diego State University, San Diego, California, U.S.A.

S. Rao, San Diego State University, San Diego, California, U.S.A.

P. Reinhall, University of Washington, Seattle, Washington, U.S.A.

W. Schramm, Technische Hockschule Darmstadt, Darmstadt, GERMANY

A. Schwab, Delft University of Technology, Delft, The NETHERLANDS

A. Shabana, University of Illinois at Chicago, Chicago, Illinois, U.S.A.

G. Shiflett, University of Southern California, Los Angeles, California, U.S.A.

K. Singhal, University of Waterloo, Waterloo, Ontario, CANADA

A. Soni, Oklahoma State University, Stillwater, Oklahoma, U.S.A.

M. Srinivasan, The University of Iowa, Iowa City, Iowa, U.S.A.

W. Teschner, Technische Hockschule Darmstadt, Darmstadt, GERMANY

M. Thomas, University of Florida, Gainesville, Florida, U.S.A.

C. Vibet, Universite Paris Val de Marne, Evry-Cedex, FRANCE

D. Vojin, University of Stuttgart, Stuttgart, WEST GERMANY

O. Wallrapp, DFVLR (German Aerospace Research Establishment), Wessling, WEST GERMANY

J. Whitesell, University of Michigan, Ann Arbor, Michigan, U.S.A.

T. Wielenga, University of Michigan, Ann Arbor, Michigan, U.S.A.

J. Wiley, John Deere Technical Center, Moline, Illinois, U.S.A.

U. Wolz, University of Karlsruhe, Karlsruhe, GERMANY

J. Wong, University of Louisville, Louisville, Kentucky, U.S.A.

W. Yoo, The University of Iowa, Iowa City, Iowa, U.S.A.

Y. Yoo, The Korea Advanced Institute of Science and Technology, Seoul, KOREA

Y. Youm, Catholic University of America, Washington, D.C., U.S.A.

PREFACE

These proceedings contain lectures presented at the NATO-NSF-ARO sponsored Advanced Study Institute on "Computer Aided Analysis and Optimization of Mechanical System Dynamics" held in Iowa City, Iowa, 1-12 August, 1983. Lectures were presented by free world leaders in the field of machine dynamics and optimization. Participants in the Institute were specialists from throughout NATO, many of whom presented contributed papers during the Institute and all of whom participated actively in discussions on technical aspects of the subject.

The proceedings are organized into five parts, each addressing a technical aspect of the field of computational methods in dynamic analysis and design of mechanical systems. The introductory paper presented first in the text outlines some of the numerous technical considerations that must be given to organizing effective and efficient computational methods and computer codes to serve engineers in dynamic analysis and design of mechanical systems. Two substantially different approaches to the field are identified in this introduction and are given attention throughout the text. The first and most classical approach uses a minimal set of Lagrangian generalized coordinates to formulate equations of motion with a small number of constraints. The second method uses a maximal set of cartesian coordinates and leads to a large number of differential and algebraic constraint equations of rather simple form. These fundamentally different approaches and associated methods of symbolic computation, numerical integration, and use of computer graphics are addressed throughout the proceedings. At the conclusion of the Institute, participants agreed that a tabulation of available software should be prepared, to include a summary of capabilities and availability. A survey was carried out following the Institute to provide information on software that is available. Results of this survey are included in the introductory paper.

Basic analytical methods of formulating governing equations of mechanical system dynamics are presented in Part 1 of the proceedings. Implications of selection of alternative formulations of the equations of classical mechanics are identified and discussed, with attention to their suitability for computer implementation. Algebraic and analytical properties of alternative generalized coordinate sets are discussed in some detail.

Part 2 of the proceedings focuses on methods of computer generation of the equations of dynamics for large scale, constrained dynamic systems. Both the loop closure Lagrangian generalized coordinate approach for formulating a minimal system of governing equations of motion and the cartesian coordinate approach that leads to a maximal set of loosely coupled equations are presented and illustrated. Use of symbolic computation techniques is presented as an integral part of the Lagrangian coordinate approach and as an independent method for analytical studies in system dynamics.

Numerical methods of solving systems of ordinary differential equations and mixed systems of differential-algebraic equations are treated extensively in Part 3. Theoretical properties of numerical integration algorithms are reviewed and their favorable and unfavorable attributes for application to system dynamics analyzed. A review of available computer codes for use in solution of equations of dynamics is presented. Applications of integration techniques and high speed computer graphics to aid in solution of dynamic equations and in interpretation of results are presented and illustrated.

Two important interdisciplinary aspects of machine dynamics are presented in Part 4. Methods of including the effects of flexible bodies in machine dynamics applications, based primarily on finite element structural analysis models, are presented and illustrated. A method for incorporating feedback control subsystems into modern mechanical system dynamic analysis formulations is presented and examples that illustrate first order coupling between control and phyical dynamic effects are illustrated.

Part 5 of the proceedings focuses on synthesis and optimization of kinematic and dynamic systems. An extensive treatment of methods of type and parameter synthesis of mechanisms and machines is presented and illustrated through applications on a microcomputer. Methods of design sensitivity analysis and optimization of both kinematically and dynamically driven systems, using large scale computer codes for formulation and solution of dynamic and design sensitivity equations, are presented. Finally, surveys are presented on leading iterative optimization methods that are available and applicable for design optimization of mechanical system dynamics.

The extent and variety of the lectures presented in these proceedings illustrate the contribution of numerous individuals in preparation and conduct of the Institute. The Institute Director wishes to thank all the contributors to these proceedings and participants in the Institute, who refused to be passive listeners and participated actively in discussions and contributed presentations. Special thanks go to C. Flack, S. Lustig, and R. Huff for their efforts in administrative planning and support of the Institute. Finally, without the financial support* of the NATO Office of Scientific Affairs, the U.S. National Science Foundation, and the U.S. Army Research Office, the Institute and these proceedings would not have been possible. Their support is gratefully acknowledged by all concerned with the Institute.

February, 1984
E. J. Haug

* The views, opinions, and/or findings contained in these proceedings are those of the authors and should not be construed as an official position, policy, or decision of the sponsors, unless so designated by other documentation.

TABLE OF CONTENTS

Part 4

INTERDISCIPLINARY PROBLEMS

Part 5

SYNTHESIS AND OPTIMIZATON

Introduction

ELEMENTS AND METHODS OF COMPUTATIONAL DYNAMICS

Edward J. Haug
Center for Computer Aided Design
University of Iowa
Iowa City, Iowa 52242

Abstract. The impact of the digital computer on all fields of science and engineering is already significant and will become dominant within the decade. Well developed computer software for analysis and design has, in fact, revolutionized the fields of structural and electronic circuit analysis and design. The situation, however, is somewhat different in mechanical system dynamic analysis and optimization. While the potential for use of computational techniques in this field is at least as great as for structures and circuits, development has lagged behind these fields. The purpose of the Advanced Study Institute "Computer Aided Analysis and Optimization of Mechanical System Dynamics" and these proceedings is to investigate basic methods for computer formulation and solution of the equations of kinematics and dynamics of mechanical systems. A second objective is to investigate methods for optimizing design of such systems. This paper serves as an introduction to the proceedings and presents a summary of computer software in the field, based on information provided by participants in the Institute.

1. THE SCOPE OF MECHANICAL SYSTEM DYNAMICS

For the purposes of this paper a dynamic system is defined as a collection of interconnected bodies (rigid or flexible) that can move relative to one another, as defined by joints or constraints that limit the relative motion that may occur. Motion of a mechanical system may be determined by defining the time history of the position of one or more of the bodies, or by application of external forces, in which case the motion of the body is determined by laws of physics.

NATO ASI Series, Vol. F9
Computer Aided Analysis and Optimization of Mechanical System Dynamics
Edited by E. J. Haug

Dynamics of such systems is characterized by large amplitude motion, leading to geometric nonlinearity that is reflected in the differential equations of motion and algebraic equations of constraint. Three basically different types of analysis of such systems arise in design of mechanical systems.

Types of Dynamic Analysis

Kinematic analysis of a mechanical system concerns the motion of the system, irrespective of the forces that produce the motion. Typically, the time history of position (or relative position) of one or more bodies of the system is prescribed. The time history of position, velocity, and acceleration of the remaining bodies is then determined by solving systems of nonlinear algebraic equations for position, and linear algebraic equations for velocity and acceleration.

Dynamic analysis of a mechanical system involves determining the time history of position, velocity, and acceleration of the system due to the action of external and internal forces. A special case of dynamic analysis is determination of an equilibrium position of the system, under the action of forces that are independent of time. The motion of the system, under the action of specified forces, is required to be consistent with the kinematic relations imposed on the system by joints that connect bodies in the system. The equations of dynamics are differential or differential-algebraic equations.

Inverse dynamic analysis is a hybrid form of kinematic and dynamic analysis, in which the time history of the positions or relative positions of one or more bodies in the system is prescribed, leading to complete determination of position, velocity, and acceleration of the system from the equations of kinematics. The equations of motion of the system are then solved as algebraic equations to determine the forces that are required to generate the prescribed motion.

Forces Acting on a Mechanical System

An important consideration that serves to classify mechanical systems concerns the source of forces that act on the system. This is particularly important in modern mechanical systems, in which some form of control is exerted. Force effects due to electrical and hydraulic feedback control subsystems play a crucial role in dynamics of mechanical systems. The scope of mechanical system dynamic considerations is, therefore, heavily dependent on a definition of the classes of force systems that are permitted to act in the system.

The most elementary form of force acting on a system is gravitational force, which is normally taken as constant and acting perpendicular to the surface of the earth. Other relatively simple forces that act on bodies making up a system, due to interaction with their environment, include aerodynamic forces due to air or fluid drag associated with motion of the system, and friction and damping forces that act due to relative motion between components that make up the mechanical system.

An important class of forces that act in a mechanical system are associated with compliant elements such as coil springs, leaf springs, tires, plastic hinges, and a multitude of other deformable components that have reaction forces and moments associated with them. Forces associated with compliant elements act between bodies in the system, as a function of relative position and velocity. A special form of force of interaction is associated with impact between bodies, leading to essentially impulsive forces that are characterized by discontinuities in velocity of bodies. Motion with discontinuous velocity is called intermittent motion. Such impact forces are generally associated with local deformation of bodies and occur over a very short period of time. Modeling such force effects may be done by accounting for the micromechanics of force-deformation, or by defining an impulsive force and accounting for jump discontinuities in velocity.

Of utmost importance in modern mechanical systems are force effects due to control systems. Controllers and regulators in most modern mechanical systems sense position and velocity of components of the system and exert corrective effects through hydraulic or electrical actuators, in order to adjust motion of the system to some desired result. Such feedback control systems may be under manual control of an operator or under control of an analog or digital computer that impliments a control law and determines the actuator force that is exerted between bodies in the system. Such feedback controllers often involve time delay that is associated with operator reaction time or time required for computation, prior to implementation of an actuator force. Such forces are often of dominant importance in determining the time history of system response and must be carefully incorporated in a mechanical system simulation.

Typical Mechanical Systems

The conventional concept of a mechanical system has focused on mechanisms and linkages that are used to control position and, in some cases, velocity and acceleration of components of a system. Four-bar linkages, such as those shown in Fig. 1 are commonly encountered in mechanisms and machines. The linkage in Fig. 1(a) may be used to control the angular position ψ of body 3 by specifying the input angle θ of body 1. In the case of the slider-crank shown in Fig. 1(b), the crank angle θ may be given as a function of time to control the position of a slider relative to ground. Such a slider-crank is employed in numerous machine tool applications and is the linkage used in internal combustion engines. Much more complex linkages and machine subsystems are encountered in mechanical systems.

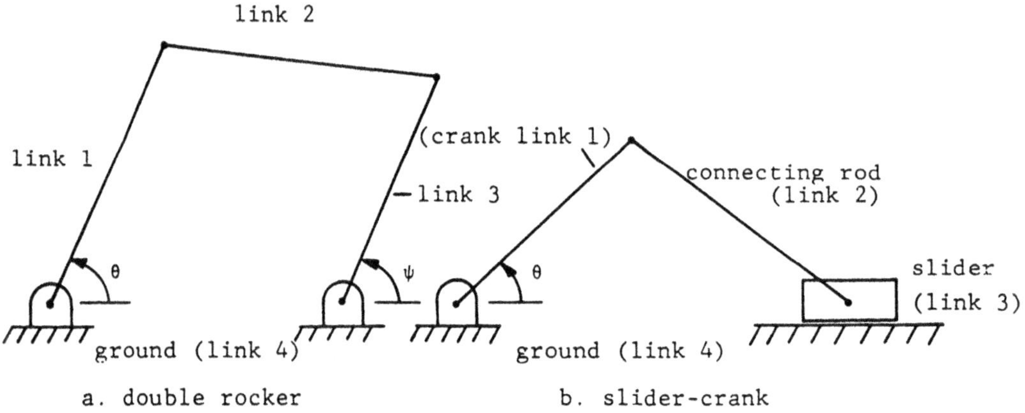

Figure 1. Four-Bar Linkages

A more complex class of mechanical systems concerns complete vehicle systems, whose dynamic performance is of critical importance. The three dimensional tractor-trailer model shown in Fig. 2 reflects an intricate interrelationship of bodies that make up the suspension subsystems of a vehicular system. While the suspension kinematics of such a vehicle are more complicated than elementary four-bar linkages, they define similiar kinematic relationships among bodies that limit motion of the vehicle suspension. Suspension springs and dampers are used in this application to control motion of the vehicle as it traverses a road surface or off-road terrain.

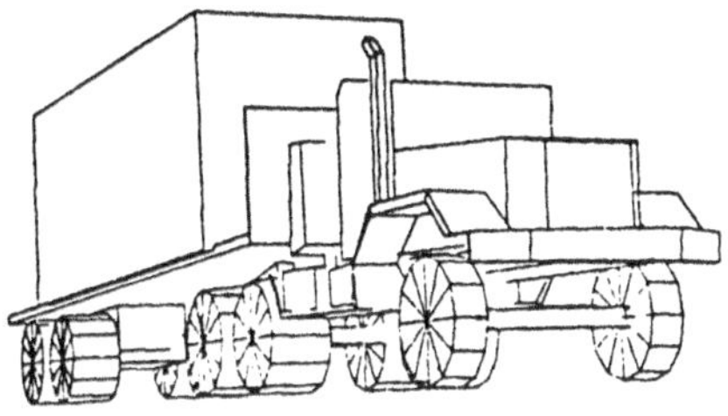

a. Tractor - Trailer

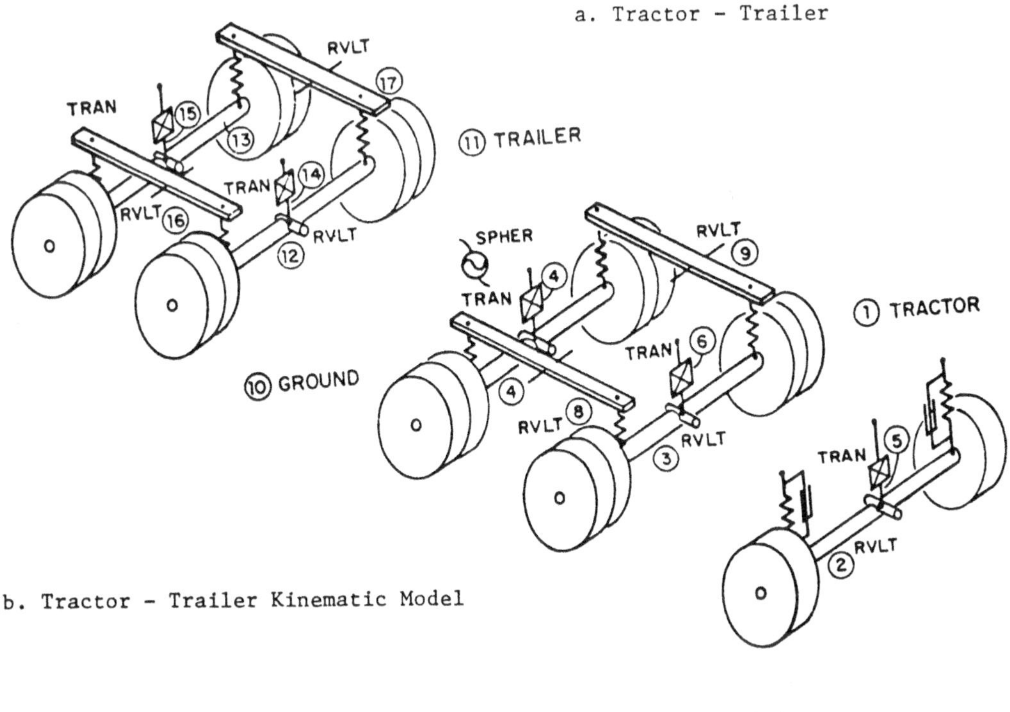

b. Tractor - Trailer Kinematic Model

Figure 2. Tractor-Trailer Vehicle Model

Robotic devices are becoming increasingly important in manufacturing, material manipulation in hazardous environments, and in numerous fields of application. The robotic device shown in Fig. 3 represents an elementary manipulator with six degrees of freedom and associated actuators at each of the rotational joints. Some form of feedback control law is used to determine actuator forces at each of the joints, to control the end effector position, orientation, and velocity for welding, painting, parts positioning, and other applications.

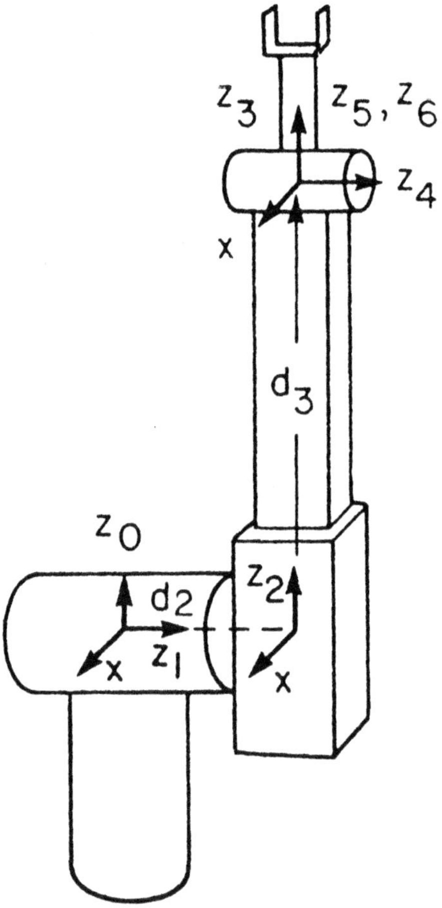

Figure 3. Robotic Manipulator

The preceding examples represent typical machines that are encountered in mechanical system dynamic analysis and design. The breadth of such systems is extensive and includes a multitude of mechanical engineering analysis and design applications.

2. CONVENTIONAL METHODS OF DYNAMIC ANALYSIS

Owing to the nonlinear nature of large displacement kinematic analysis, the mechanism designer has traditionally resorted to graphical techniques and physical models for the study of kinematics of mechanical systems [1,2]. As might be expected, such methods are limited in generality and rely heavily on the designer's intuition. For more modern treatments of mechanism and machine dynamics, Refs. 3-6 may be consulted.

The conventional approach to dynamic analysis of mechanical systems is to use Lagrangian methods of formulating the system equations of motion in terms of a set of displacement generalized coordinates. Excellent classical texts on dynamics [7-9] provide fundamentals of dynamics that are needed for mechanical system dynamic analysis.

Models of kinematic and dynamic systems with several degrees of freedom have traditionally been characterized by "clever" formulations that take advantage of the specific problem to obtain a simplified form of equations of kinematics and dynamics. Specifically, ingenious selection of position coordinates can often lead to unconstrained formulations with independent variables that allow manual generation of equations of motion, with only moderate effort. More often, analysis of systems with up to 10 degrees-of-freedom leads to massive algebraic manipulation and derivative calculation that is required in constructing equations of motion. This ad-hoc approach is, therefore, limited to mechanical systems of only moderate complexity. Some extension has been achieved using symbolic computation [10], in which the computer is used to carry out differentiation and algebraic manipulation, creating terms that are required in the equations of motion.

After the governing equations of motion have been derived, by manual or symbollic computation methods, the engineer or analyst is still faced with the problem of obtaining a solution of the differential equations and initial conditions. Since these equations are highly nonlinear, the prospect of obtaining closed form solutions is remote, except in very simple cases. With the advent of analog and digital computers, engineers began to use the computer and available numerical integration methods to solve their equations of motion. This, however, involves a substantial amount of manpower for patching circuit boards for analog computation or writing FORTRAN programs that define the differential equations that are to be solved by numerical integration techniques.

In contrast to the traditional ad-hoc approach that has been employed in mechanical system dynamics, a massive literature has evolved in finite element structural analysis [11,12] and analysis of electronic circuits [13,14]. Developments in these two areas are characterized by the same technical approach. Rather than capitalizing on "clever" formulation, a systematic approach is typically taken when digital computers are used for organization of calculations and implementation of iterative numerical solution of equations. Through systematic formulation and selection of numerical techniques, user oriented computer codes have been developed that are capable of handling a broad range of structures and circuits. The overwhelming success of finite element structural and electronic circuit analysis codes indicates that systematic formulations are possible in these areas, which can provide a guide for development of a similar set of techniques and computer codes for mechanical system dynamics.

3. THE OBJECTIVE OF COMPUTATIONAL DYNAMICS

The objectives of computational methods in dynamics are to (1) create a formulation and digital computer software that allows the engineer to input data that define the mechanical system of interest and automatically formulate the governing equations of kinematics and dynamics, (2) develop and implement numerical algorithms that automatically solve the nonlinear equations for dynamic response of the system, and (3) provide computer graphic output of results of the simulation to communicate results to the designer or analyst. The essence of these objectives is to make maximum use of the power of the digital computer for rapid and accurate data manipulation and numerical computation to relieve the engineer of routine calculations that heretofore have been carried out manually.

As suggested by advancements in the fields of finite element structural analysis and electrical circuit theory, both of which are more advanced than the field of mechanical system dynamics, a systematic approach to the formulation and solution of kinematics and dynamics problems is required to reduce computations to computer programs. Great care must be taken to consider the many alternatives available in selecting a formulation and numerical methods to achieve this objective. Basic ingredients that must be considered in computational dynamics are reviewed in the following section.

Several general purpose computer programs for kinematic and dynamic analysis have been developed, along the lines suggested above, in the late 1960's and early 1970's [15-17]. These programs are satisfactory for many applications. An alternate method of formulating system constraints and equations of motion, in terms of a maximal set of generalized coordinates, was introduced in the later 1970's [18-20], bypassing topological analysis and making it easier for the user to supply constraints and forcing functions. This approach leads to a general computer program, with practically no limitation on mechanism or machine type. The penalty, however, is a larger system of equations that must be solved.

4. INGREDIENTS OF COMPUTATIONAL DYNAMICS

In order to be specific concerning some of the alternatives and tradeoffs that exist in the field of computational dynamics, an elementary example is employed in this section to discuss mathematical properties of system dynamics and to identify alternatives that will be considered in detail in individual papers of these proceedings.

An Elementary Example

To illustrate concepts of system dynamics, consider a simplified model of the slider crank mechanism shown in Fig. 1, idealized to include the slider mass at the right end of the connecting rod (body 2). The center of mass of the connecting rod has been adjusted to reflect incorporation of the mass of the slider as a point mass at right end, which must move along the x-axis.

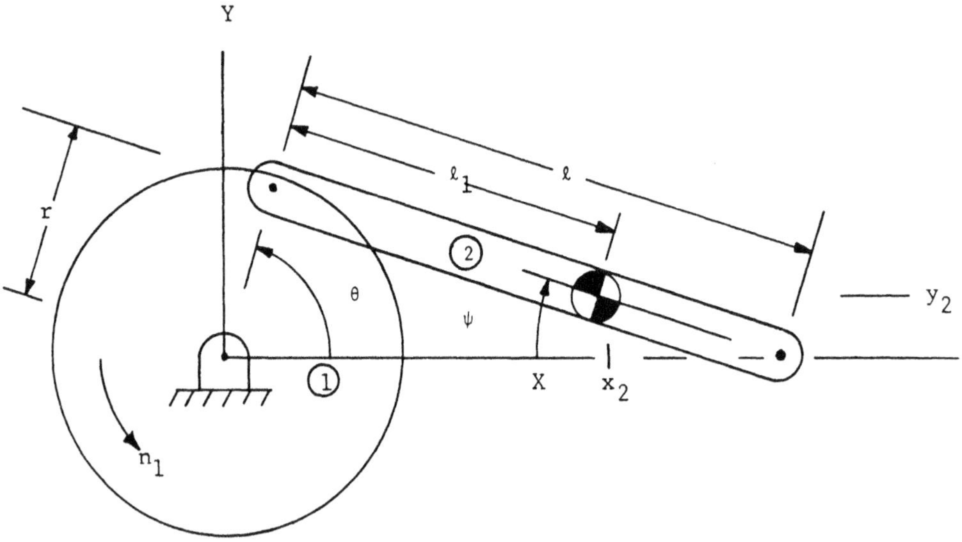

Figure 4. Elementary Slider-Crank Model

It is clear from simple trigonometry that once the angle θ (called a Lagrangian generalized coordinate) is fixed, as long as $\ell > r$, the angle ψ is uniquely determined by simple trigonometry as follows:

$$\sin \psi = (r\sin \theta)/\ell \tag{1}$$

or

$$\psi = \text{Arcsin}\,(\frac{r}{\ell} \sin \theta) \tag{2}$$

The global coordinates of the center of mass of the connecting rod are

$$x_2 = r\cos \theta + \left[\ell_1{}^2 - \frac{r^2 \ell_1^2}{\ell^2} \sin^2 \theta\right]^{1/2} \tag{3}$$

$$y_2 = \frac{r(\ell - \ell_1)}{\ell} \sin \theta \tag{4}$$

It is possible in this simple example to completely define the location and orientation of the connecting rod in terms of the crank angle θ, which will be taken as an input angle.

Presuming the input angle θ is known as a function of time; i.e., $\theta = \theta(t)$, differentiating Eqs. 2-4 with respect to time, using the chain rule of differentiation, yields

$$\dot{\psi} = \left[1 - \frac{r^2}{\ell^2}\sin^2\theta\right]^{-1/2} \left(\frac{r}{\ell}\cos\theta\right)\dot{\theta} \tag{5}$$

$$\dot{x}_2 = \left\{-r\sin\theta - \left[\ell_1^2 - \frac{r^2\ell_1^2}{\ell^2}\sin^2\theta\right]^{-1/2} \frac{r^2\ell_1^2}{\ell^2}\sin\theta\cos\theta\right\}\dot{\theta} \tag{6}$$

$$\dot{y}_2 = \left[r\left(1 - \frac{\ell_1}{\ell}\right)\cos\theta\right]\dot{\theta} \tag{7}$$

Thus, given the input position θ and angular velocity $\dot{\theta}$, Eqs. 2-7 determine the position and velocity of the connecting rod.

In preparation for writing Lagrange's equations of motion for the system, its kinetic energy, using the velocity relations of Eqs. 5-7, can be written as

$$T = \tfrac{1}{2} J_1 \dot{\theta}^2 + \tfrac{1}{2} J_2 \dot{\psi}^2 + \tfrac{1}{2} m_2 \left(\dot{x}_2^2 + \dot{y}_2^2\right) \tag{8}$$

$$= \tfrac{1}{2}\left\{J_1 + J_2\left[\frac{r^2\cos^2\theta}{\ell^2 - r^2\sin^2\theta}\right] + m_2 r^2\left[1 - \frac{\ell_1}{\ell}\right]^2 \cos^2\theta \right.$$
$$\left. + m_2\left(r + \left[\ell_1^2 - \frac{r^2\ell_1^2}{\ell^2}\sin^2\theta\right]^{-1/2} \frac{r^2\ell_1^2}{\ell^2}\cos\theta\right)^2 \sin^2\theta\right\}\dot{\theta}^2$$

Note that the kinetic energy has been written totally in terms of the input angle θ and input angle rate $\dot{\theta}$.

In order to write Lagrange's equations of motion, the virtual work of the torque n_1 acting on the crank link and the gravitational force acting on the connecting rod must be calculated. Using the differential of y_2 from Eq. 4, the virtual work can be written as

$$\delta W = n_1\delta\theta - m_2 g\delta y_2 = \left[n_1 - m_2 g r\left(1 - \frac{\ell_1}{\ell}\right)\cos\theta\right]\delta\theta \equiv g_\theta \delta\theta \tag{9}$$

where the coefficient of the virtual displacement $\delta\theta$ on the right is defined to be the generalized force acting on the system.

Lagrange's equation of motion [7] for this system is

$$\frac{d}{dt}\left(\frac{\partial T}{\partial \dot{\theta}}\right) - \frac{\partial T}{\partial \theta} = g_\theta \tag{10}$$

Noting that the kinetic energy in Eq. 8 depends on $\dot{\theta}$ in a simple way, but on θ in a very complex way, it becomes clear that the expansion that is required to calculate terms on the left side of Eq. 10 will be extensive. The reader who is interested in seeing just how complicated this calculation can be is invited to carry out the expansion and write the explicit second order differential equation for θ in Eq. 10, to obtain

$$
\begin{aligned}
&\ddot{\theta}\Bigg\{J_1 + J_2\left[\frac{r^2\cos^2\theta}{\ell^2 - r^2\sin^2\theta}\right] + m_2 r^2\left[\frac{1-\ell_1^2}{\ell^2}\right]^2\cos^2\theta \\
&+ m_2\left[r + \left(\ell_1^2 - \frac{r^2\ell_1^2}{\ell^2}\sin^2\theta\right)^{-1/2}\frac{r^2\ell_1^2}{\ell^2}\cos\theta\right]^2\sin^2\theta\Bigg\} \\
&+ \dot{\theta}\Bigg\{J_1 + J_2\left[\frac{(\ell^2 - r^2\sin^2\theta)(r^2 2\cos\theta)(-\sin\theta)\dot{\theta} + (r^2\cos^2\theta)(-r^2 2\sin\theta\cos\theta)\dot{\theta}}{(\ell^2 - r^2\sin^2\theta)^2}\right] \\
&+ m_2\left[r^2\left[\frac{\ell-\ell_1}{\ell}\right]^2(2\cos\theta)(-\sin\theta)\dot{\theta} + (2\sin\theta\cos\theta)\dot{\theta}\left[r + \left(\ell_1^2 - \frac{r^2\ell_1^2}{\ell^2}\sin^2\theta\right)^{-1/2}\frac{r^2\ell_1^2}{\ell^2}\cos\theta\right]^2\right] \\
&+ m_2\sin^2\theta\, 2\left[r + \left(\ell_1^2 - \frac{r^2\ell_1^2}{\ell^2}\sin^2\theta\right)^{-1/2}\frac{r^2\ell_1^2}{\ell^2}\cos\theta\right] \\
&\times\left[\frac{\left(\ell_1^2 - \frac{r^2\ell_1^2}{\ell^2}\sin^2\theta\right)^{-1/2}\frac{r^2\ell_1^2}{\ell^2}(-\sin\dot{\theta})\theta - \frac{r^2\ell_1^2}{\ell^2}(\cos\theta)\frac{1}{2}\left(\ell_1^2 - \frac{r^2\ell_1^2}{\ell^2}\sin^2\theta\right)^{-3/2} - \frac{r^2\ell_1^2}{\ell^2}(2\sin\theta\cos\theta)\dot{\theta}}{\left(\ell_1^2 - \frac{r^2}{\ell^2}\sin^2\theta\right)}\right] \\
&- \frac{\dot{\theta}^2}{2}\Bigg\{J_1 + J_2\left[\frac{(\ell^2 - r^2\sin^2\theta)(r^2 2\cos\theta)(-\sin\theta) - (r^2\cos^2\theta)(-r^2 2\sin\theta\cos\theta)}{(\ell^2 - r^2\sin^2\theta)^2}\right] \\
&+ m_2\left[r^2\left[1 - \frac{\ell_1}{\ell}\right]^2 2\cos\theta(-\sin\theta) + 2\sin\theta\cos\theta\left[r + \left(\ell^2 - \frac{r^2\ell_1^2}{\ell^2}\sin^2\theta\right)^{-1/2}\frac{r^2\ell_1^2}{\ell^2}\cos\theta\right]^2\right] \\
&+ 2\, m_2\sin^2\theta\left[r + \left(\ell_1^2 - \frac{r^2\ell_1^2}{\ell^2}\sin^2\theta\right)^{-1/2}\frac{r^2\ell_1^2}{\ell^2}\cos\theta\right] \\
&\times\left[\frac{\left(\ell_1^2 - \frac{r^2\ell_1^2}{\ell^2}\sin^2\theta\right)^{-1/2}\frac{r^2\ell_1^2}{\ell^2}(-\sin\theta) - \frac{r^2\ell_1^2}{\ell^2}(\cos\theta)\frac{1}{2}\left(\ell^2 - \frac{r^2\ell_1^2}{\ell^2}\sin^2\theta\right)^{-3/2}\left(-\frac{r_1^2\ell_2^2}{\ell^2}\sin\theta\cos\theta\right)}{\left(\ell_1^2 - \frac{r^2}{\ell^2}\sin^2\theta\right)}\right]\Bigg\} \\
&= n_1 - m_2 g r\left(1 - \frac{\ell_1}{\ell}\right)\cos\theta
\end{aligned}
\tag{11}
$$

The unnerving complexity of writing Lagranges equations for even this trivial machine dynamics problem serves to illustrate technical difficulties associated with writing equations of motion in terms of a minimal set of position coordinates. While such calculations can in theory be carried out for complex systems, the level of technical difficulty grows rapidly. Even if symbolic computation [10] is used to carry out the expansions on the left side of Eq. 10, a complicated set of FORTRAN expressions results.

To illustrate a more systematic approach to development of equations of motion, consider again the simplified slider crank of Fig. 1. In order to develop the required expressions, first consider the bodies making up the system as being disconnected, as shown in Fig. 5. In this formulation, the angular orientation ϕ_1 of the crank link and the coordinates x_2 and y_2 of the center of mass of body 2 and its angular orientation ϕ_2 are taken as position coordinates that are called Cartesian generalized coordinates. In order to assemble the linkage, however, these four variables must satisfy kinematic relations. Specifically, points A_1 and A_2 must coincide, in order to have a rotational joint, leading to the algebraic relations

$$\left.\begin{aligned} \Phi_1 &\equiv r\cos\phi_1 - (x_2 - \ell_1\cos\phi_2) = 0 \\ \Phi_2 &\equiv r\sin\phi_1 - (y_2 - \ell_1\sin\phi_2) = 0 \end{aligned}\right\} \qquad (12)$$

Similarly, in order that point B_2 on the connecting rod slide in a horizontal slot situated along the X-axis, it is necessary that its y coordinate be zero, leading to the condition

$$\Phi_3 \equiv y_2 + (\ell - \ell_1)\sin\phi_2 = 0 \qquad (13)$$

These three algebraic equations comprise three constraints among the four position coordinates. Since these equations are nonlinear, a closed form solution for three of the variables in terms of the remaining variable is difficult to obtain.

In order to obtain velocity relationships among the four position coordinates, one may differentiate Eqs. 1 and 2 with respect to time to obtain

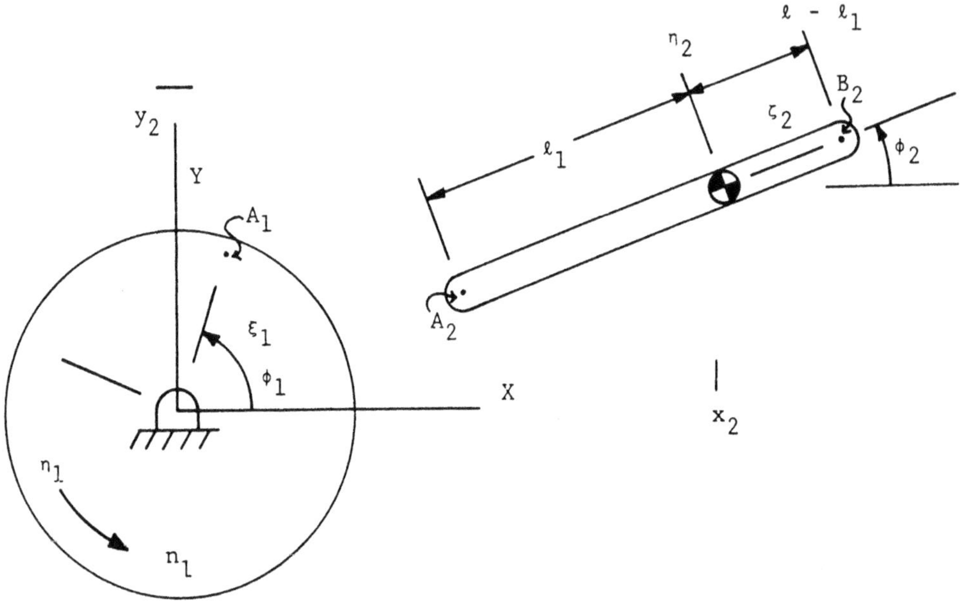

Figure 5. Cartesian Coordinates for Slider-Crank

$$\begin{bmatrix} -r\sin\phi_1 & -1 & 0 & -\ell_1\sin\phi_2 \\ r\cos\phi_1 & 0 & -1 & \ell_1\cos\phi_2 \\ 0 & 0 & 1 & (\ell-\ell_1)\cos\phi_2 \end{bmatrix} \begin{bmatrix} \dot\phi_1 \\ \dot x_2 \\ \dot y_2 \\ \dot\phi_2 \end{bmatrix} = \begin{bmatrix} 0 \\ 0 \\ 0 \end{bmatrix} \qquad (14)$$

Note that these equations are linear in the four velocity variables. These algebraic equations may be rewritten, taking terms depending on $\dot\phi_1$ to the right side, to obtain

$$\begin{bmatrix} -1 & 0 & -\ell_1\sin\phi_2 \\ 0 & -1 & \ell_1\cos\phi_2 \\ 0 & 1 & (\ell-\ell_1)\cos\phi_2 \end{bmatrix} \begin{bmatrix} \dot x_2 \\ \dot y_2 \\ \dot\phi_2 \end{bmatrix} = \begin{bmatrix} r\sin\phi_1 \\ -r\cos\phi_1 \\ 0 \end{bmatrix} \dot\phi_1 \qquad (15)$$

Recall that in the preceding formulation it was possible, for $\ell > r$, to solve for the positions and velocities of the connecting rod in terms of the input angle, which in this case is denoted as ϕ_1. To see that this can be done in the present formulation, it must be shown that the determinant of the coefficient matrix on the left side of Eq. 4 is nonsingular. Calculating its determinant,

$$\det = (\ell - \ell_1) \cos \phi_2 + \ell_1 \cos \phi_2 = \ell \cos \phi_2 \tag{16}$$

which is nonzero if $\ell > r$, since in this case $-\pi/2 < \phi_2 < \pi/2$. Note that the Jacobian matrix of Eqs. 1 and 2 with respect to x_2, y_2, and ϕ_2 is the coefficient matrix on the left side of Eq. 4. By the implicit function theorem, if the linkage can initially be assembled, then a unique solution of Eqs. 1 and 2 is guaranteed for x_2, y_2, and ϕ_2 as differentiable functions of ϕ_1. This theoretical result serves as the foundation for one of the modern methods of formulating and solving equations of kinematics of such systems.

In order to write the equations of motion for this example in terms of Cartesian coordinates, the kinetic energy of the system is first written as

$$T = \tfrac{1}{2} J_1 \dot{\phi}_1^{\,2} + \tfrac{1}{2} J_2 \dot{\phi}_2^2 + \tfrac{1}{2} m_2 (\dot{x}_2^2 + \dot{y}_2^2) \tag{17}$$

Similarly, the generalized forces acting on this system are

$$\left.\begin{aligned} g_{\phi_1} &= n_1 \\ g_{y_2} &= -m_2 g \end{aligned}\right\} \tag{18}$$

The Lagrange multiplier form of Lagrange's equations of motion [7] are now written in the form

$$\frac{d}{dt} \left(\frac{\partial T}{\partial \dot{q}_i}\right) - \frac{\partial T}{\partial q_i} + \sum_{j=1}^{3} \lambda_j \frac{\partial \Phi_j}{\partial q_i} = g_{q_i} \quad , \; i=1, \ldots, 4 \tag{19}$$

In the present example, this yields the following system of differential equations of motion:

$$j_1 \ddot{\phi}_1 - \lambda_1 r\sin \phi_1 + \lambda_2 r\cos \phi_1 = n_1 \tag{20}$$

$$m_2\ddot{x}_2 - \lambda_1 = 0 \tag{21}$$

$$m_2\ddot{y}_2 - \lambda_2 + \lambda_3 = -m_2 g \tag{22}$$

$$j_2\ddot{\phi}_2 - \lambda_1\ \ell_1 \sin\ \phi_2 + \lambda_2\ \ell_1 \cos\ \phi_2 + \lambda_3(\ell - \ell_1)\cos\ \phi_2 = 0 \tag{23}$$

Note that the system of equations comprises four second order differential equations (Eqs. 20-23) and three algebraic equations of constraint (Eqs. 12-13) for four Cartesian generalized coordinates (ϕ_1, x_2, y_2, and ϕ_2) and three algebraic variables (the Lagrange multipliers λ_1, λ_2, and λ_3). This is a mixed system of differential-algebraic equations that must be solved to determine motion of the system.

This example is presented to illustrate that there are alternative methods of formulating equations of motion that lead to very different sets of governing equations, even though the solutions are identical. In the first formulation, with independent Lagrangian coordinates, a single second order differential equation of motion is obtained that is highly nonlinear and complex. In the second formulation, with Cartesian coordinates, the form of the governing equations is much simplified, but a mixed system of differential-algebraic equations results.

Library of Elements

In order to systematically formulate models of mechanical systems, it is important to define a library of components or elements that can be used in assembling a model. The most common element is a rigid body, several of which can be used to represent components in a mechanical system. There are several ways of representing the position of a rigid body in space. One approach that can be used to locate a rigid body in the plane or in three dimensional space is Cartesian generalized coordinates, such as those identified for the connecting rod in the slider crank mechanism in Fig. 5. Using Cartesian coordinates, each body is located independently. Kinematic constraints between bodies are then defined as algebraic equations among the Cartesian generalized coordinates; e.g., Eqs. 12-13 for the slider-crank. Standard joints between bodies, such as rotational joints, translational joints, point follower joints, etc. can be defined and a set of equations that represent standard constraints between bodies can be derived and put into a constraint definition library. In this way, the user can call as many of these joints as needed to connect the bodies in the mechanism. Thus, in addition to a library of bodies, a library of kinematic joints that connect bodies is needed.

Algebraic Equation Solving

As indicated in the Cartesian coordinate approach, a substantial number of nonlinear kinematic constraint equations are obtained and must be solved to determine the position of the system. In the case of the elementary slider crank model, Eqs. 12 and 13 represent a system of three nonlinear equations in x_2, y_2, and ϕ_2, presuming an input angle ϕ_1 is given. Since these equations are nonlinear, iterative numerical methods must be used to obtain solutions. The most common method is Newton iteration [21], which begins with an estimate of the solution and iteratively improves it, with excellent convergence properties. It is interesting to note that the sequence of matrix equations that must be solved in Newton's method has the same coefficient matrix as appears on the left side of the velocity equation of Eq. 15. Thus, once the coefficient matrix of the constraint system is formulated, it is used over and over in iteratively solving for positions and for velocities in Eq. 15. Note, furthermore, that the form of the matrix in Eq. 15 is very simple, containing primarily zeros and constants, with only a few terms involving the position coordinates. This is common in large scale mechanical systems. In fact, as the complexity of the mechanical system increases, the form of this coefficient matrix becomes even more sparse (meaning that most entries in the matrix are zeros), with a moderate number (perhaps 5%) being simple nonzero constants or algebraic expressions. Thus, the special form of this matrix equation, which must be solved repetitively, suggests that matrix methods that take advantage of sparsity may enhance efficiency.

Differential Equation Formulation

Just as equations of constraint can be organized in a systematic way, using a library of standard kinematic constraint elements, the governing differential equations of motion of the mechanical system can be assembled with the aid of a digital computer. As indicated in Eqs. 8-10, a manual derivation of the equations of motion may lead to exceptionally complex algebraic expressions. This difficulty may be somewhat overcome through use of modern symbolic computation techniques [10] that use a table of differential calculus formulas that allow the computer to carry out the differentiations called for in Eq. 10, using a kinetic energy expression similar to that presented in Eq. 8. While the result of symbollic computation may be extra long statements, it is possible for the computer to generate such terms and provide all expressions that are needed in creating the governing differential equations of motion. This technique has been explored in the field of mechanical system dynamics [10] and holds potential for systematic application in systems with Lagrangian coordinates.

Derivation of governing equations of motion, using Lagrangian coordinates, has been investigated and developed to a high degree in Refs. 4 and 17, using loop closure techniques. The resulting internal representation of equations of motion is rather complex, but is transparent to the user. The advantage of this approach is that a small number of governing equations of motion is obtained, which may be integrated by standard numerical integration techniques. The disadvantage of this approach is that the internal representation of the equations of motion is complex, so to represent nonstandard effects within a system requires a great deal of ad-hoc work to modify the basic code.

As illustrated by the Cartesian coordinate approach for the slider crank mechanism, algebraic equations of constraint (Eqs. 12 and 13) and second order differential equations of motion (Eqs. 20-23) involving the Cartesian coordinates and Lagrange multipliers may be written. As is clear from development of these equations for the slider crank mechanism, equations of constraint and equations of motion can be assembled in a systematic way. Thus, a computer code that begins with a library of rigid bodies and kinematic constraints, along with compliant elements such as springs and dampers, can be used to assemble the governing system of differential and algebraic equations of motion for broad classes of machines.

While it is easy to formulate large systems of mixed differential and algebraic equations of motion using the Cartesian coordinate approach, the challenge of solving such mixed differential-algebraic equations remains. As will be noted in the following subsection, there are technical approaches that make solution of such systems possible, given that they can be formulated in an automatic way.

Numerical Integration Methods

If the system of equations of motion can be reduced to only differential equations, as might be the case in simple problems that use Lagrangian coordinates, then a massive literature on numerical integration of ordinary differential equations is available. Specifically, powerful predictor-corrector algorithms [22,23] are available with codes that implement the numerical integration algorithms and require little or no ad-hoc programming by the user. Such codes are readily available for general use and are employed in integrating even the most complex equations of motion that arise in a Lagrangian formulation.

The theory of numerical methods for mixed systems of differential-algebraic equations, such as arise in the Cartesian coordinate approach, is not as complete. It is becoming well known [24,25] that such systems of differential-algebraic equations, taken as a single system, are stiff, in the sense that high frequency damped components exist and cause problems for predictor-corrector numerical integration algorithms. To resolve this difficulty, a stiff integration technique, such as the Gear stiff integration algorithm [23], may be employed to solve the system of equations. This approach, however, requires that an implicit integration algorithm be employed. Thus, the algorithm is required to solve a very large system of nonlinear equations.

Alternate techniques are presented in Refs. 19 and 26. One method uses matrix computational techniques to identify a set of independent generalized coordinates and implicitly reduces the differential-algebraic system to differential equations in only the independent generalized coordinates. A second method employs constraint correction terms that allow use of standard predictor-corrector numerical integration algorithms and avoids the stiffness problem.

As might be expected, there is a massive literature on numerical integration theory that implements the techniques outlined above. There are also hybrid techniques that show promise for improved performance in mechanical system dynamics. It is clear, however, that the type of formulation selected for mechanical system dynamics determines the set of numerical integration algorithms that is feasible for the class of applications being considered.

Interdisciplinary Effects

As noted in Section 1, virtually all modern mechanical systems involve technical considerations beyond kinematics and dynamics of rigid bodies. Many applications require that flexibility of bodies that make up a mechanical system be considered in dynamic analysis. This, of course, requires that deformation of bodies be characterized in some rational way, perhaps by use of finite element modeling methods that have been highly developed for structural analysis. Only a modest amount of work has been done to incorporate flexibility effects into large scale machine dynamics codes. The area is, however, receiving considerable attention and is developing rapidly.

Feedback control effects have become increasingly important in modern mechanical systems and are dominant in many system applications. It is important to provide a convenient means for incorporating differential equations of control, to include the capability for automatic generation of controller equations, in order to include control system modeling in the same framework that is used in creating mechanical system models. Very little has been done in this direction.

Intimately related to the question of feedback control is representation of electrial or hydraulic actuator forces and torques that act on components of the mechanical system. In many cases, it is inadequate to simply represent the output of the control device as a force. Rather, a command is given to an electrical or hydraulic actuator that responds according to its own internal dynamics. Particularly in the case of hydraulic devices, dynamics of internal flow and compressibility of fluid can have a significant influence on system dynamics.

While a long discussion of interdisciplinary effects associated with mechanical systems could be given, the purpose of this subsection is served by simply noting that there are numerous such important effects that must be accounted for in dynamic analysis of real mechanical systems. It is therefore imperative that the formulation selected for representing mechanical system dynamics allow for easy insertion of components or modules that represent special effects that might not have been envisioned when the formulation was initially adopted.

Code Organization for Generality

As stated in Section 1, the basic objective of computer aided analysis of mechanical systems is to create a formulation, numerical methods, and a computer code to allow the digital computer to carry out both detailed computations associated with formulating governing equations of motion, solving them, and displaying results in an interpretable fashion. To meet this objective and to retain generality that is required for future expansion to represent nonstandard effects, careful consideration must be given to organization of the computer code that formulates, solves, and displays results of system dynamic analysis.

Three forms of digital computer code might be considered for such applications. First is a rigidly structured computer code that is written to carry out a single class of applications, using a well defined set of components. Such a computer code can be made quite efficient and transparent to the user, but will most likely be limited in its capability for expansion. This approach generally leads to a computer code of maximum efficiency but minimal flexibility.

An alternate approach to computer code organization is to employ a formulation that defines a rigidly structured form of system equations and constructs contributing equations using standard components from a library in formulating dynamics models that fall within the originally envisioned scope. This approach allows the user to derive and incorporate equations governing nonstandard effects and to enter FORTRAN descriptions of such equations and array dimension modifications that are required to assemble and solve the system equations of motion. This approach has the desirable feature of allowing a high degree of efficiency, but at the same time provides a capability for the sophisticated user to incorporate nonstandard effects for special purpose subsystem descriptions, as they are encountered in applications. The drawback of such an approach is that a relatively high level of sophistication on the part of the user is required in creating user supplied routines that are required for special purpose applications.

A third alternative is to adopt a formulation and construct a computer code that is modular in nature, with certain classes of functions carried out in standard and nonstandard modules in the digital computer code. This approach maximizes flexibility of the software system for expansion of capability to incorporate interdisciplinary effects that may not have been envisioned when the code was originally constructed. Further, basic functions within the code can be organized in such a way that the level of sophistication that is required of the user in creating a new module to represent some interdisciplinary effect or a special purpose subsystem is minimal.

Trade-Offs

As is suggested in each of the foregoing subsections, there are numerous alternative methods of describing a physical system, methods for assembling equations of motion, methods for solving equations of motion, methods for incorporating nonstandard effects, and methods for computer code organization. If all possible combinations of decisions associated with these considerations are considered, a very large number of combinations would be found. It is also clear that decisions regarding one aspect of formulation and code development will influence decisions regarding other aspects. Independent decisions regarding all of these aspects cannot be made if an effective formulation and computer code is to be created. Optimization of capability of software developed in this area is predicated on careful consideration of all of the factors involved and selection of approaches that are in harmony and lead to an effective computational framework and computer code.

5. A SURVEY OF DYNAMICS SOFTWARE

A topic agreed upon by participants in this Institute during a concluding discussion was the need for identification of computer codes that are available for analysis and design optimization of dynamic systems. The author agreed to prepare a survey form and distribute it to participants and other interested parties, following the Institute, to collect information for a survey to be published with the proceedings. This section documents the results of that survey.

Due to time restrictions, it was possible in this survey only to communicate once with individuals who completed the survey form. As a result, some questions were imperfectly formulated and some questions were left unasked. The author has attempted to extract pertinent information from the survey forms provided by participants and colleagues, to present a summary of the capabilities offered by software in the area of system dynamics and to provide addresses where more detailed information may be obtained.

Adequate information was obtained on twelve general purpose dynamic analysis codes and ten special purpose codes in the areas of kinematics and dynamics to allow their inclusion in this survey. Summary information on capabilities and sources of further information for codes in each catagory are presented in the following subsections. If special rates are available to Universities, they are indicated. Otherwise, commercial rates apply.

General Purpose Analysis Codes

Of the twelve general purpose dynamic analysis codes identified in this section, several are large scale software packages that run on mainframe computers and are capable of efficiently analyzing large classes of complex mechanical systems. Other codes are more limited, but are available on microcomputers that can be run for very low cost. The following codes are included in the survey:

ADAMS (Automated Dynamic Analysis of Mechanical Systems)
Developed by N. Orlandea, J. Angell, and R. Rampalli.
Commercially available from Mechanical Dynamics Incorporated, 55 South Forest, Ann Arbor, Michigan 48104. Special terms available for academic institutions.

DADS (Dynamic Analysis and Design System)
Developed by E. J. Haug, G. M. Lance, P. E. Nikravesh, M. J. Vanderploeg, and R. A. Wehage.
Commercially available January 1985, from Computer Aided Design Software Incorporated, P.O. Box 203, Oakdale, Iowa 52319. Available to Universities for instructional use at cost of documentation.

DYSPAM (Dynamics of Spatial Mechanisms)
Developed by B. Paul and R. Schaffa.
Commercially available in late 1984 from Professor B. Paul, Department of Mechanical Engineering, University of Pennsylvania, Philadelphia, Pennsylvania 19104.

IMP-84 (Integrated Mechanisms Program)
Developed by J. J. Uicker and P. Sheth.
Commercially available from Professor J. J. Uicker, Department of Mechanical Engineering, University of Wisconsin, Madison, Wisconsin.

MESA VERDE (Mechanism, Satellite, Vehicle, and Robot Dynamics)
Developed by J. Wittenburg and U. Wolz.
Commercially available from Professor J. Wittenburg, Institut Fur Mechanik, Universitat Karlsruhe, D - 7500 Karsruhe 1, FRG.

NEWEUL
Developed by W. O. Schiehlen and E. J. Kreuzer.
Commercially available from Professor W. O. Schiehlen, Institute B. of Mechanics, University of Stuttgarte, Paffenwaldring 9, 7000 Stuttgarte 80, FRG.

SPACAR
Developed by Van der Werff, A. Schwab, and J. B. Jonker.
Not yet commercially available. Inquiries should be directed to CADOM Group, Delft University of Technology, Mekelweg 2, Delft, Netherlands.

DRAM (Dynamic Response of Articulated Machinery)
Developed by D. A. Smith, J. Angell, M. A. Chace, M. Korybalski, and A. Rubens.
Commercially available from Mechanical Dynamics Incorporated, 555 South Forest, Ann Arbor, Michigan 48104. Special terms available to academic institutions.

DYMAC-G (Dynamics of Machinery-General Version)
Developed by B. Paul and A. Amin.
Commercially available from Professor B. Paul, Department of Mechanical Engineering, University of Pennsylvania, Philadelphia, Pennsylvania 19104.

KAVM (Kinematic and Kinetostatic Analysis of Vector Method)
Developed by B. Pauwels.
Commercially available form I. C. Systems n.v., Gouverneuer Verwilghensingel 4, B 3500 Hasselt, BELGIUM.

MICRO-MECH (Micro Computer Based Mechanism Analysis)
Developed by R. J. Williams.
Commercially available from Ham Lake Software Company, 631 Harriet Avenue, Shoreview, Minnesota 55112.

SINDYS (Simulation Program for Nonlinear Dynamic Systems)
Developed by Y. Inoue, E. Imanishi, and T. Fujikawa.
Under consideration for commercial distribution. Inquiries should be directed to Y. Inoue, Mechanical Engineering Research Laboratory, Kobe Steel Limited, 3-18, 1-Chrome, Wakinohama-Cho, Chou Ku, Kobe, Japan. Available to academic institutions for the cost of documentation.

A compact summary of capabilities of the twelve general purpose software packages surveyed are presented in Table 1. The dimension and mode of analysis are self explanatory. A Code is indicated as using Lagragian coordinates if any generalized coordinates are defined in a moving reference frame. Otherwise, the code is designated as having Cartesian generalized coordinates. Use of Lagrange multipliers is identified, since Lagrange multipliers define reaction forces in joints associated with constraints, providing needed force information for component design. A distinction is made between codes that identify a minimal set of independent variables for the purpose of integration and, if so, whether the user must define these coordinates or whether they are automatically defined by the code. Use of sparse matrices is identified as an indicator of those Cartesian coordinate codes that allow large systems to be treated.

Interdisciplinary effects are becoming more common in modern dynamic system analysis, some of which are identified in the tabulation of properties in Table 1. A code is designated as treating flexible bodies only if deformation generalized coordinates are defined for each flexible body in the system. Codes that model flexibility by lumping mass and stiffness into discrete bodies are not interpreted, for the purposes of this survey, as having a flexible body capability. Codes that treat feedback control are identified. A distinction is made between codes that require the user to write his controller equations and define their effect and codes that provide for automatic formulation of control equations, parralleling the capability for mechanical system analysis. Codes treating impact between bodies, either by providing user definition of reaction force

Table 1 CHARACTERISTICS OF GENERAL PURPOSE ANALYSIS CODES

CHARACTERISTIC	ADAMS	DADS	DYSPAM	IMP-84	MESA VERDE	NEWEUL	SPACAR	DRAM	DYMAC -G	KAVM	MICRO-MECH	SINDYS
Dimension: Planar (P), Spatial (S)	S	S,P	S	S	S	S	S	P	P	P	P	P
Modes of Analysis: Kinematic(K), Static (S), Dynamic (D), Inverse Dynamic (I)	K,S D,I	K,S D,I	K,S D,I	K,S D,I	K,S D,I	K D,I	K,S D,I	K,S D,I	K,S D,I	K,S D,I	K,S I	K,S D
Generalized Coordinates: Lagrangian (L), Cartesian (C) C C	L	L	L	L	C	L	L	C	L	C		
Lagrange Multipliers: Yes (Y), No (N)	Y	Y	N	N	N	N	Y	Y	Y	N	N	N
Independent Variables: No (N), Yes (Y)/ User Defined (U), Automatic (A)	N	Y/A	Y/A	Y/A	Y/A	Y/U	Y/U	Y/A	Y/U	Y/U	Y/U	Y
Sparse Matrices: Yes (Y), No (N)	Y	Y	N	N	N	N	N	N	N	N	N	N
Flexible Bodies: Yes (Y), No (N)	Y	Y	N	N	N	N	Y	N	N	N	N	Y
Feedback Control: No (N), Yes (Y)/ User Supplied (U), Automatic (A)	Y/U	Y/A	Y/U	Y/U	Y/U	Y/U	N	Y/U	Y/U	N	N	Y/U
Impact: No (N), Yes (Y)/User Supplied (U), Automatic (A)	Y/U	Y/A	Y/U	Y/U	Y/A	N	N	Y/U	Y/U	N	N	Y/U
Friction: No (N), Yes (Y)/ User Supplied (U), Automatic (A)	Y/U	Y/A	Y	Y/A	N	Y/U	Y/U	Y/U	Y/U	Y/A	Y/U	Y/U
Computers: IBM (I), VAX (V), CDC (C), PRIME (P), H.P.(HP), Harris (H), Micro(M)	I,V C,P	I,V C,P	I,V	V,H	I	I,V C,P		I,V C,P	I,V	HP M	I,M	I
Pre-Processing: Interactive (I), Graphic (G) I,G I,G		I,G	I,G	I		I,G		I,G	I	I		
Post-Processing: Plots (P), Graphic (G), Animation (A) A	P,G A	P,G	A	P,G			A	P,G	P,G	G,A	P,G	

or by providing automatic formulation, are identified. Similarly, codes allowing user supplied representation of friction versus those providing an automated formulation of friction effects are identified.

The final items denoted under characteristics are associated with computer implementation. Principal computers on which codes are available are identified. Codes that have interactive preprocessing with graphic aids are identified. Postprocessing capabilities for creating plots, graphic displays, and animations are also identified.

Special Purpose Analysis Codes

In addition to the twelve general purpose dynamic analysis codes, ten special purpose codes were identified in the survey and are included in this subsection. The distinction between general purpose and special purpose was based on the respondent's identification of special structure or limited applicability for their code. In addition to a tabulation of characteristic properties of these ten codes in Table 2, a brief summary of the special nature of the code is included with the code title and availability. The following ten software packages were identified in this category:

CAM-NCCAM (Computer Aided Manufacturing of Cams and Numerical Control Machining of Cams)

Analysis of kinematic and dynamic behavior of cam systems, generating instructions for manufacturing cams.

Developed by J. De Fraine.

Commercially available from J. De Fraine, WTCM-CRIF, Celestijnenlaan, 300 C-B-3030 Heverlee, BELGIUM.

CAMDYN (Cam Dynamics)

Analysis and design optimization of roller or flat faced followers to generate standard or user supplied motion schedules. Provides dynamic contact force, stress, and torque on shaft. Optimizes Cam dimensions to bring contact stresses below maximum allowable level.

Developed by B. Paul and C. Van Dyke.

Commercially available from Professor B. Paul, Department of Mechanical Engineering, University of Pennsylvania, Philadelphia, Pennsylvania 19104.

CMLAW (Motion Laws for Cams and Dynamic Behavior)

Analysis of Cam mechanisms with one degree of freedom. Defines classical and general motion laws as input to kinematic analysis programs or Cam design programs.

Developed by B. Pauwels.

Commercially available from IC Systems n.v., Gouverneur Verwilghensingel 4, B3500 Hasselt, BELGIUM.

DYREC-MC (Dynamics of Reciprocating Machinery-Multiple Cylinders)

Dynamic analysis of reciprocating (slider-crank) machines with multiple sliders.

Developer B. Paul.

Commercially available from Professor B. Paul, Department of Mechanical Engineering, University of Pennsylvania, Philadelphia, Pennsylvania 19104.

KINMAC (Kinematics of Machinery)

Analysis of planar machines for position, velocity, and acceleration, when all input motions are specified.

Developed by B. Paul and A. Amin.

Commercially available from Professor B. Paul, Professor of Mechanical Engineering, University of Pennsylvania, Philadelphia, Pennsylvania 19104.

MEDUSA (Multi-body System Dynamics and Analysis)

Analysis of flexible, multi-body systems with small motions relative to a large reference motion. Special modules for vehicle/guideway interaction and wheel/rail contact.

Developed by O. Wallrapp.

Commercially available from Dr. W. Kortum, DFVLR (German Aerospace Research Establishment), D-8031 Wessling FRG. Special rates available to universities.

MULTIBODY

Dynamic analysis of multibody systems with tree configurations.

Developed by R. Schwertassek and R. E. Robberson.

Commercially available from R. Schwertassek, DFVLR (German Aerospace Research Establishment), D-8031 Wessling, FRG. Available to academic institutions in exchange for comparable software.

NEPTUNIX 2

General purpose simulation code for nuclear simulation, electrical networks, and large scale systems having up to 3,000 differential-algebraic equations.

Developed by M. Nakhle.

Commercially available from M. Nakhle, CISI, 35 Boulevard Brune, 75015, Paris, FRANCE. Available to academic institutions for cost of documentation.

ROBSIM (Robotic Simulations Package)

Analysis of Robotic systems for development of control systems, structures and algorithms, motion programming and planning algorithms, and robot system design.

Developed by D. C. Haley, B. Almand, M. Thomas, L. Kruze, and K. Gremban, Martin Marietta Denver Aerospace, Denver, Colorado.

Not commercially available.

STATMAC (Statics of Machinery)

Static analysis of planar machines for configurations when forces are specified, or for forces when configuration is specified.

Developed by B. Paul and A. Amin.

Commercially available from Professor B. Paul, Department of Mechanical Engineering, University of Pennsylvania, Philadelphia, Pennsylvania 19104.

A tabulation of characteristics of each of these ten codes, using the same format and definition of terms employed in the preceding subsection, may be found in Table 2.

6. DESIGN SYNTHESIS AND OPTIMIZATION

Considerations thus far in this paper have been limited to analysis of dynamic performance of a mechanical system, presuming that the design is specified and applied loads or kinematic inputs are given. While development of an analytical formulation and computer code to carry out even this analysis function is a demanding task, it is only a part of the engineering design process in which an optimized design is sought. The synthesis or design optimization process may be viewed as an inverse analysis problem, in which a design is sought to cause desired performance of the system. To appreciate the complexity

Table 2 CHARACTERISTICS OF SPECIAL PURPOSE ANALYSIS CODES

CHARACTERISTIC	CAM-NCCAM	CAM-DYN	CMLAW	DYREC-MC	KIN-MAC	MED-USA	MULTI-BODY	NEPTUNIX2	ROB-SIM	STAT-MAC
Dimension: Planar (P), Spatial (S)	P	P	P	P	P	S	S	P,S	S	P
Modes of Analysis: Kinematic (K), Static (S), Dynamic (D), Inverse Dynamic (I)	K,D	K,I	K,D	D	K	K,S D	D	D,I	K,S D,I	S
Generalized Coordinates: Lagrangian (L), Cartesian (C)				L	L	L	L	L	L	L
Lagrange Multipliers: Yes (Y), No (N)				N	N	N	N		N	N
Independent Variables: No (N), Yes (Y)/ User Defined (U), Automatic (A)				Y/A	Y	Y/A	Y/A		Y	Y/U
Sparse Matrices: Yes (Y), No (N)				N	N	Y	Y		N	N
Flexible Bodies: Yes (Y), No (N)	N	N	N	N	N	Y	N	N	N	N
Feedback Control: No (N), Yes (Y)/ User Supplied (U), Automatic (A)	N	N	N	N	N	Y/U	Y/U	Y/U	Y/A	
Impact: No (N), Yes (Y)/User Supplied (U), Automatic (A)	Y/U	Y/A	N	N	N	N	N	N	N	
Friction: No (N), Yes (Y)/ User Supplied (U), Automatic (A)	N	N	Y/A	Y/U		Y	N	N	Y/A	Y/U
Computers: IBM (I), VAX (V), CDC (C), PRIME (P), H.P.(HP), Harris (H), Micro (M)	V,C HP	I,V	HP	I,M	I,V M	I,V C	I	I	V	I,V
Pre-Processing: Interactive (I), Graphic (G)	I			I			I,G			
Post-Processing: Plots (P), Graphic (G), Animation (A) P,G	P,G			P	P		P,G			

of the design-performance functional relationship, consider that evaluation of performance for a single design requires one complete cycle of dynamic performance analysis. Analytical solution of this inverse problem is generally impossible.

Considerations in Design Synthesis

In spite of the grave difficulty in analytical prediction of the effect of design variables on system performance, the engineer must in some way create a design that exhibits acceptable, hopefully optimized, dynamic performance. The conventional approach to this process involves designers with many years of experience, who have developed an intuitive understanding of the effect of design variations on system performance. Trial designs are created, based on rules of thumb or rough calculations, and their performance is estimated by either fabrication and test or dynamic analysis. The objective of organized design synthesis and optimization is to provide the designer with tools that give him information needed to make a change in design to improve performance. The ideal, of course, is automated synthesis techniques that create optimized designs.

Part V of these proceedings addresses selected aspects of the design synthesis and optimization process. Since this segment of the proceedings is rather compact, no attempt is made here to summarize details of methods, approach, or examples. Only a few comments are offered on basic approaches to such problems.

The extremely difficult problem of configuration or type synthesis of mechanisms, to carry out desired kinematic functions, is addressed in Ref. 27. The creation of linkage configurations within certain classes of mechanisms is carried out using precision point and Fourrier analysis techniques that make use of computer software to identify the form of mechanisms that may be acceptable. Once the configuration or mechanism type is selected, dimension synthesis is carried out to proportion each of the candidate mechanisms to best perform desired kinematic and load carrying functions. Each phase of type and dimension synthesis is carried out with the assistance of computer subroutines that provide information to the designer. It should be noted that this type of machine synthesis is essentially limited to kinematically driven systems; i.e., systems in which prescribed time histories of certain input variables uniquely determine the time histories of all other variables, without consideration of forces that act on the system. The problem of type synthesis of dynamically driven systems; i.e., systems in which free degrees of freedom exist and applied forces determine motion, appears to be beyond the current state of the arc.

Design sensitivity analysis methods for dimension synthesis of large scale kinematically and dynamically driven systems are presented in Refs. 28 and 29. The foundation for these methods is a body of techniques for calculating derivatives of kinematic and dynamic performance measures with respect to design parameters, called design sensitivity analysis. Armed with such sensitivity information, the designer may maintain control of the design process, or control may be turned over to an optimization algorithm to iteratively arrive at an optimized design, considering both kinematic and strength aspects of system performance. Techniques and examples presented in Refs. 28 and 29 indicate feasibility of dynamic design optimization, but also show that considerable development in algorithms and software is required.

Finally, two survey papers on modern iterative optimization techniques are presented in Refs. 30 and 31. Analysis of the properties of available iterative optimization methods is given, as regards their potential for fruitful application in kinematic and dynamic optimization of mechanical systems. These and the other papers in Part V of these proceedings represent a beginning step toward development of dynamic system synthesis techniques that will require considerable attention in the future to accrue the potential benefits that exist.

Design Synthesis Codes

The complexity of design synthesis requires codes that implement synthesis theory be more specialized and focused than the general purpose analysis codes discussed in Section 5 of this paper. Information is presented on just three kinematic and dynamic system synthesis codes in this section. The first code addresses both planar and spatial systems and concerns synthesis for both kinematic and dynamic performance. The latter two codes forcus more heavily on kinematic synthesis, using a higher degree of automation.

Owing to the diversity of technical approaches and capability sought, tabulation of capabilities for the three synthesis codes considered here is not practical. Therefore, a narrative discussion of the capability of each code is presented. All three codes are commercially available. Further information may be obtained from the proponent organizations, whose addresses are given in the following:

CADOM (Computer Aided Design of Mechanisms)

CADOM consists of a substantial number of general purpose and special purpose modules that carry out a range of analysis and synthesis functions. The code is based on a formalism that is presented in some detail in the paper by Rankers in these proceedings [27].

CADOM continues to be development by a task group in the Delft University of Technology consisting of H. Rankers, K. Vander Werff, A. J. Kline-Breteler, A. Van Dyk, B. Van den Berg, and H. Drent. CADOM treats kinematic, static, and dynamic performance of both planar and spatial systems. Goal functions, in the form of desired motion or path, are defined by the user. Precision point and Fourrier transform techniques are employed to synthesize characteristics of designs, which are then defined from a catalog of available designs. The software is under interactive control of the user and runs on a variety of computers, including a microcomputer. Substantial use is made of graphics for both preprocessing and postprocessing of information.

CADOM is available from the CADOM Task Group, Department of Mechanical Engineering, Delft University of Technology, Netherlands. The code is available to academic institutions for the cost of documentation.

KINSYN

KINSYN VII and Micro KINSYN are kinematic synthesis codes for design of planar linkages. The codes use a variety of closed form interactive and heuristic techniques to create and graphically evaluate kinematic performance of linkages. Graphics based preprocessing of information and graphic presentation of design configuration and performance are imbedded in the software to enhance user interaction. The software is available on both super minicomputers and microprocessors.

KINSYN was developed by R. Kaufman, M. R. Dandu, and D. L. Kaufman. The software is available from KINTECH Incorporated, 1441

Springvale Avenue, MacClean, Virginia 22101. Academic discounts are available.

LINCAGES (Linkages, Computer Analysis, and Graphically Enhanced Synthesis Package)

The LINCAGES code addresses synthesis of planar linkages for combinations of motion and path and function generation for three, four, and five prescribed positions. The code has an embedded interactive graphics capability for specification of desired characteristics and postprocessing graphics for evaluation of performance prediction, in the form of animation for four bar mechanisms. Advanced versions of the software are under development for n-bar linkages, including gears and sliders.

LINCAGES was developed by A. G. Erdman and D. R. Riley. It is available commercially from them at the Department of Mechanical Engineering, University of Minnesota, Minneapolis, Minnesota. Academic discounts are available.

REFERENCES

1. Beyer, R., The Kinematic Synthesis of Mechanisms, McGraw-Hill, N.Y., 1963.

2. Hirschhorn, J., Kinematics and Dynamics of Plane Mechanisms; McGraw-Hill, N.Y., 1962.

3. Paul, B., Kinematics and Dynamics of Planar Machinery, Prentice-Hall, Englewood Cliffs, N.J., 1979.

4. Wittenburg, J., Dynamics of Systems of Rigid Bodies, Teubner, Stuttgart, 1977.

5. Soni, A.H., Mechanism Synthesis and Analysis, McGraw-Hill, N.Y., 1974.

6. Suh, C.H. and Radcliffe, Kinematics and Mechanisms Design, Wiley, N.Y., 1978.

7. Greenwood, D.T., Principles of Dynamics, Prentice-Hall, Englewood Cliffs, New Jersey, 1965.

8. Kane, T.R., Dynamics, Holt, Rinehart, and Winston, N.Y., 1968.

9. Goldstein, H., Classical Mechanics, Addison-Wesley, Reading, Massachusetts, 1980.

10. Noble, B. and Hussain, M.A., "Applications of MACSYMA to Calculations in Dynamics," Computer Aided Analysis and Optimization of Mechanical System Dynamics, (ed. E.J. Haug), Springer-Verlag, Heidelberg, 1984.

11. Zienkiewiez, O., The Finite Element Method, McGraw-Hill, N.Y., 1977.

12. Gallagher, R.H., Finite Element Analysis: Fundamentals, Prentice-Hall, Englewood Cliffs, N.J., 1975.

13. Chua, L.O. and Lin, P-M., Computer Aided Analysis of Electronic Circuits, Prentice-Hall, Englewood Cliffs, N.J., 1975.

14. Calahan, D.A., Computer Aided Network Design, McGraw-Hill, N.Y., 1972.

15. Paul, B. and Krajcinovic, D., "Computer Analysis of Machines With Planar Motion - Part I: Kinematic; Part II: Dynamics," Journal of Applied Mechanics, Vol. 37, pp. 697-712, 1970.

16. Chace, M.A. and Smith, D.A., "DAMN-A Digital Computer Program for the Dynamic Analysis of Generalized Mechanical Systems," SAE paper 710244, January 1971.

17. Sheth, P.N. and Uicker, J.J., Jr., "IMP (Integrated Mechanisms Program), A Computer Aided Design Analysis System for Mechanisms and Linkages," Journal of Engineering for Industry, Vol. 94, pp. 454-464, 1972.

18. Orlandea, N., Chace, M.A., and Calahan, D.A., "A Sparsity-Oriented Approach to the Dynamic Analysis and Design of Mechanical Systems, Parts I and II," Journal of Engineering for Industry, Vol. 99, pp. 773-784, 1977.

19. Wehage, R.A. and Haug, E.J., "Generalized Coordinate Partitioning for Dimension Reduction in Analysis of Constrained Dynamic Systems," Journal of Mechanical Design, Vol. 104, No. 1, pp. 247-255, 1982.

20. Nikravesh, P.E. and Chung, I.S., "Application of Euler Parameters to the Dynamic Analysis of Three Dimensional Constrained Mechanical Systems," Journal of Mechanical Design, Vol. 104, No. 4, pp. 785-791, 1982.

21. Atkinson, K.E., An Introduction to Numerical Analysis, Wiley, New York, 1978.

22. Shampine, L.F. and Gordon, M.K., Computer Solution of Ordinary Differential Equations: The Initial Value Problem, Freeman, San Fransisco, CA, 1975.

23. Gear, C.W., Numerical Initial Value Problems in Ordinary Differential Equations, Prentice-Hall, Englewood Cliffs, N.J., 1971.

24. Petzold, L.D., Differential/Algebraic Equations Are Not ODES, Rept. SAND 81-8668, Sandia National Laboratories, Livermore, CA, 1981.

25. Gear, C.W., "Differential - Algebraic Equations," Computer Aided Analysis and Optimization of Mechanical System Dynamics (ed. E.J. Haug), Springer-Verlag, Heidelberg, 1984.

26. Nikravesh, P.E., "Some Methods for Dynamic Analysis of Constrained Mechanical Systems: A Survey," Computer Aided Analysis and Optimization of Mechanical System Dynamics (ed. E.J. Haug), Springer-Verlag, Heidelberg, 1984.

27. Rankers, I.H., "Synthesis of Mechanisms," Computer Aided Analysis and Optimization of Mechanical System Dynamics (ed. E.J. Haug), Springer-Verlag, Heidelberg, 1984.

28. Haug, E.J. and Sohoni, V.N., "Design Sensitivity Analysis and Optimization of Kinematically Driven Systems," Computer Aided Analysis and Optimization of Mechanical System Dynamics (ed. E.J. Haug), Springer-Verlag, Heidelberg, 1984.

29. Haug, E.J., Mani, N.K., and Krishnaswami, P., "Design Sensitivity Analysis and Optimization of Dynamically Driven Systems," Computer Aided Analysis and Optimization of Mechanical System Dynamics (ed. E.J. Haug), Springer-Verlag, Heidelberg, 1984.

30. Fleury, C. and Braibant, V., "Optimization Methods," Computer Aided Analysis and Optimization of Mechanical System Dynamics (ed. E.J. Haug), Springer-Verlag, Heidelberg, 1984.

31. Gill, P.E., Murray, W., Saunders, M.A., and Wright, M.H., "Sequential Quadratic Programming Methods for Nonlinear Programming," Computer Aided Analysis and Optimization of Mechanical System Dynamics (ed. E.J. Haug), Springer-Verlag, Heidelberg, 1984.

Part 1

ANALYTICAL METHODS

COMPUTER ORIENTED ANALYTICAL DYNAMICS OF MACHINERY

Burton Paul
Asa Whitney Professor of Dynamical Engineering
Department of Mechanical Engineering and Applied Mechanics
University of Pennsylvania
Philadelphia, Penna. 19104

Abstract. These lectures consist of a survey of the theory of multi-body machine systems based on analytical mechanics formulated in such a way as to facilitate solutions by digital computers. The material is divided into the following four subdivisions:

1. Analytical Kinematics

The relationship between Lagrangian Coordinates, and Generalized Coordinates will be introduced, and used to clarify the concept of Degree of Freedom. The determination of displacements, velocities, and accelerations of individual links, or of points of interest anywhere in the system, is expressed in a uniform notation associated with the Lagrangian Coordinates. Special consideration is given to the determination of appropriate velocity and acceleration coefficients for both plane and spatial motions. Existing computer programs for kinematic analysis of mechanical systems are briefly reviewed.

2. Statics of Machine Systems

The advantages and disadvantages of Vectorial Mechanics (the direct use of Newton's Laws) versus Analytical Mechanics are discussed. The basic terminology needed for Analytical Mechanics is introduced in describing the Principle of Virtual Work for ideal mechanical systems. The use of Lagrange Multipliers is demonstrated, and the special case of conservative systems is covered. The determination of reaction forces and moments at joints of the system, and the effects of Coulomb friction is also treated.

3. Kinetics of Machine Systems

It is shown how the governing system of differential equations for any mechanical system is conveniently formulated through the use of Lagrange's form of d'Alembert's Principle (a generalization of the Principle of Virtual Work to Dynamics).

NATO ASI Series, Vol. F9
Computer Aided Analysis and Optimization of Mechanical System Dynamics
Edited by E. J. Haug

This formulation is contrasted to the more common use of Lagrange's Equations (of second type). The important special case of single-freedom systems is described. Problems of kinetostatics, including the determination of joint reactions and other kinetostatic considerations, are covered. A brief discussion is given of balancing of machinery, the effects of flexibility, and of joint clearances.

4. Numerical Methods of Solving the Differential Equations of Motion

Prior to utilizing any numerical method of integrating the differential equations of motion, it is necessary to establish the appropriate initial values of all Lagrangian coordinates and all Lagrangian velocities. It is shown how this may be achieved, thereby paving the way for use of any of the differential equation algorithms discussed elsewhere in this Symposium. A brief survey of existing general-purpose computer programs for the dynamic analysis of machine systems is presented.

1. INTRODUCTION

The following material is a survey of the theory of Machine Systems based on analytical mechanics and digital computation. Much of the material to be presented is described in greater detail in [1,36,64-66], but some of it is an extension of previously published works. For a discussion of the subject based on the more traditional approach of vector dynamics and graphical constructions see [51,52,83].

The organization of this chapter follows the classification of Kelvin[93]which divides Mechanics into kinematics (the study of motion irrespective of its cause) and dynamics (the study of motion due to forces); dynamics is then subdivided into the categories statics (forces in equilibrium) and kinetics (forces not in equilibrium).

The systems which we shall consider are comprised of resistant bodies interconnected in such a way that specified input forces and motions are transformed in a predictable way to produce desired output forces and motions. The term "resistant" body includes both rigid bodies and components such as cables or fluid columns which momentarily serve the same function as rigid bodies. This description of a mechanical system is essentially the definition of the term machine given by Reuleaux [75]. When discussing the purely kinematic aspects of such a system, the assemblage of links (i.e. resistant bodies) is frequently referred to as a kinematic chain; when one of the links of a kinematic chain is fixed, the chain is said to form a mechanism. Various types of kinematic pairs or interconnections (e.g. revolutes, sliders, cams, rolling pairs, etc.) between pairs of linke are described in Sec. 5. We wish to point out here that two or more such kinematic pairs may be superposed at an idealized point or joint of the system.

Notation

Boldface type is used to represent ordinary vectors, matrices, and column vectors, with their respective scalar components indicated as in the following examples:

$$\mathbf{r} = x\mathbf{i} + y\mathbf{j} + z\mathbf{k} \equiv (x,y,z)$$

$$\mathbf{C} = [c_{ij}] = \begin{bmatrix} c_{11} & \cdots & c_{1N} \\ \cdots & & \cdots \\ c_{M1} & & c_{MN} \end{bmatrix}$$

$$\mathbf{x} \equiv \{x_1, x_2, \ldots x_N\} \equiv [x_1, x_2, \ldots x_N]^T$$

We shall occasionally refer to a matrix such as $\mathbf{C}$ as "an (MxN) matrix." Note the use of a superscript T to indicate the transpose of a matrix. In what follows, we shall make frequent use of the rule for transposition of a matrix product; i.e. $(\mathbf{AB})^T = \mathbf{B}^T\mathbf{A}^T$.

If $F(\psi) = F(\psi_1, \psi_2, \ldots \psi_M)$ we will use subscript commas to denote the partial derivatives of F with respect to its arguments, as follows:

$$F_{,i1} = \frac{\partial F(\psi)}{\partial \psi_i}; \quad F_{,ij} = \frac{\partial^2 F(\psi)}{\partial \psi_i \partial \psi_j}$$

2. ANALYTICAL KINEMATICS

(a) Lagrangian Coordinates

The cartesian coordinates (x^i, y^i, z^i) of a particle P^i with respect to a fixed right-handed reference frame will be called the global coordinates of the particle. Occasionally, we shall represent the global coordinates by the alternative notation $(x_1^{\ i}, x_2^{\ i}, x_3^{\ i})$, abbreviated as

$$\mathbf{x}^i \equiv (x_1^{\ i}, x_2^{\ i}, x_3^{\ i}) = (x^i, y^i, z^i) \tag{1}$$

When all particles move parallel to the global x-y plane, the motion is said to be planar.

Since the number of particles in a mechanical system is usually infinite, it is convenient to represent the global position vector

$\mathbf{x}^k \equiv (x^k, y^k, z^k)$ of an arbitrary particle P^k in terms of a set of M variables $(\psi_1, \psi_2, \ldots \psi_M)$ called Lagrangian coordinates. By introducing a suitable number of Lagrangian coordinates, it is always possible to express the global coordinates of every particle in the system via transformation equations of the form

$$\mathbf{x}^k = \mathbf{x}^k(\psi_1, \ldots \psi_M) \tag{2-a}$$

or

$$x_i^{\ k} = x_i^{\ k}(\psi_1, \ldots \psi_M), \quad (i = 1,2,3) \tag{2-b}$$

where the time t does not appear explicitly. For example, in Fig. 1 the global (x,y) coordinates of the midpoint P^4 of bar BC are given by

$$\begin{aligned} x_1^{\ 4} &= x^4 = \frac{1}{2} a_4 \cos \psi_4 \\ x_2^{\ 4} &= y^4 = - \psi_1 + \frac{1}{2} a_4 \sin \psi \end{aligned} \tag{3}$$

In general, the Lagrangian coordinates ψ_i need not be independent, but may be related by equations of constraint of the form

$$f_i(\psi_1, \psi_2, \ldots \psi_M) = 0, \quad (i = 1,2,\ldots N_s) \tag{4}$$

or

$$g_k(\psi_1, \psi_2, \ldots \psi_M, t) = 0, \quad (k = 1,2,\ldots,N_t) \tag{5}$$

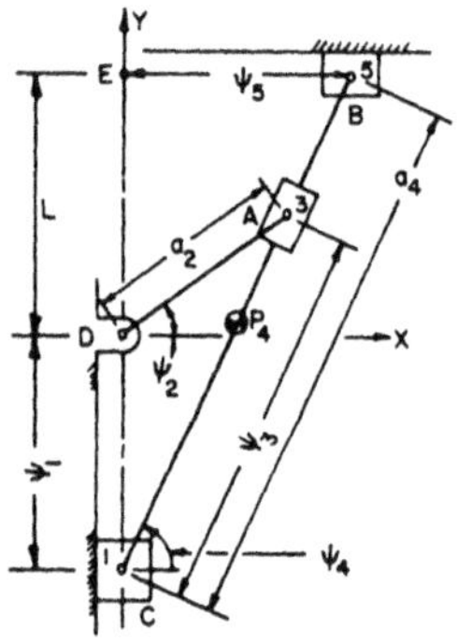

Fig. 1. Example of multiloop planar mechanism

Equations (4) represent spatial (or scleronomic) constraints because only the space variables ψ_i appear as arguments; there are N_s spatial constraints. On the other hand, time t does appear explicitly in Eqs. (5) which are therefore said to represent temporal (or rheonomic) constraints; there are N_t such constraints. For example, closure of the two <u>independent</u>* loops ECBE and DCAD requires that the following <u>spatial</u> constraints be satisfied:

$$\begin{aligned} f_1(\psi) &= a_4 \sin \psi_4 - \psi_5 = 0 \\ f_2(\psi) &= a_4 \sin \psi_4 - L - \psi_1 = 0 \\ f_3(\psi) &= \psi_3 \cos \psi_4 - a_2 \cos \psi_2 = 0 \\ f_4(\psi) &= \psi_3 \sin \psi_4 - a_2 \sin \psi_2 - \psi_3 = 0. \end{aligned} \tag{6}$$

Other equations of spatial constraint described by Eq. (4) might arise from the action of gears (see Sec. 7.30 of [66]) or from cams (see Sec. 7.40 of [66]), or from kinematic pairs in three-dimensional mechanisms (see p. 752 of [62].

As an example of a temporal constraint consider the function

$$g(\psi,t) = \psi_1 - \beta - \omega t - \alpha t^2/2 = 0 \tag{7}$$

where α, β, and ω are constants. This constraint specifies that the variable ψ_1, increases with constant acceleration ($\ddot{\psi}_1=\alpha$), starting from a reference state (at t = 0) where its initial value was β and its initial velocity was $\dot{\psi}_1 = \omega$. The special case where $\alpha = 0$ corresponds to a link being driven by a constant speed motor.

A second example is given by the constraint

$$g(\psi,t) = \psi_1+\psi_2+\psi_3 - 3\beta - 3\omega t = 0 \tag{8}$$

which requires that the average velocity $\frac{1}{3}(\dot{\psi}_1+\dot{\psi}_2+\dot{\psi}_3)$ remains fixed at a constant value ω. Such a constraint could conceivably be imposed by a feedback control system.

If the total number of constraints, given by

$$N_c = N_s + N_t \tag{9}$$

just equals the number M of Lagrangian variables, the M equations (4) and (5), which can be written in the form

$$F_i(\psi_1,\psi_2,\ldots\psi_M,t) = 0, \quad (i = 1,2,\ldots M) \tag{10}$$

can be solved for the M unknown values of ψ_j (except for certain <u>singular</u> states; see Sec. 8.2 of [66]). Because Eqs. (10) are usually highly nonlinear, it is usually necessary to solve them by a numerical

---

*The equations of constraint corresponding to loop closure will be independent if each loop used is a <u>simple closed circuit</u>; i.e. it encloses exactly one polygon with non-intersecting sides. See Chap. 8 of [66] for criteria of loop independence.

procedure, such as the Newton-Raphson algorithm. Details of the technique are given in Chapter 9 of [66].

To find the Lagrangian velocities it is necessary to differentiate Eqs. (10) with respect to time t, thereby obtaining

$$\sum_{j=1}^{M} \frac{\partial F_i}{\partial \psi_j} \dot{\psi}_j + \frac{\partial F_i}{\partial t} = 0, \quad (i = 1,\ldots M) \tag{11}$$

In matrix terms this may be expressed as

$$\mathbf{G}\dot{\boldsymbol{\psi}} = \boldsymbol{\mu} \tag{12}$$

where

$$\mathbf{G} = [\partial F_i/\partial \psi_j], \ \boldsymbol{\mu} = -\{\partial F_i/\partial t\} \tag{13}$$

In the nonsingular case, Eq. 11) can be solved for the $\dot{\psi}_j$.

To find Lagrangian accelerations, we may differentiate Eqs. (11) to obtain

$$\sum_{j=1}^{M} (F_{i,j}\ddot{\psi}_j + \sum_{k=1}^{M} F_{i,jk}\dot{\psi}_k\dot{\psi}_j) = \sum_{j=1}^{M} (\frac{\partial F_i}{\partial t})_{,j} \dot{\psi}_j + \frac{\partial^2 F_i}{\partial t^2} \tag{14}$$

where use has been made of the comma notation for denoting partial derivatives.

In matrix notation Eq. (14) can be expressed as

$$\mathbf{G}\ddot{\boldsymbol{\psi}} = \mathbf{b} \tag{15}$$

where $\mathbf{G}$ has been previously defined and

$$\mathbf{b} = \dot{\boldsymbol{\mu}} - \dot{\mathbf{G}}\dot{\boldsymbol{\psi}} \tag{16}$$

or

$$b_i = -\frac{\partial^2 F_i}{\partial t^2} - 2\sum_{j=1}^{M} (\frac{\partial F_i}{\partial t})_{,j} \dot{\psi}_j - \sum_{j=1}^{M}\sum_{k=1}^{M} F_{i,jk}\dot{\psi}_k\dot{\psi}_j \tag{17}$$

To find velocities of the point of interest $\mathbf{x}^k$, we differentiate Eq. (2-b) to obtain

$$\dot{x}_i^k = \sum_{j=1}^{M} x_{i,j}^k \dot{\psi}_j \tag{18}$$

or

$$\dot{\mathbf{x}}^k = \mathbf{P}^k \dot{\boldsymbol{\psi}} \tag{19}$$

where the (3xM) matrix $\mathbf{P}^k$ is defined by

$$\mathbf{P}^k = [x_{i,j}^k] \tag{20}$$

The corresponding acceleration is given by

$$\ddot{x}_i^{\,k} = \sum_{j=1}^{M} x_{i,j}^{k}\,\ddot{\psi}_j + \sum_{j=1}^{M}\sum_{r=1}^{M} x_{i,jr}^{k}\,\dot{\psi}_r\dot{\psi}_j \tag{21}$$

or

$$\ddot{\mathbf{x}}^k = \mathbf{P}^k\ddot{\boldsymbol{\psi}} + \mathbf{p}^k \tag{22}$$

where the components of the vector $\mathbf{p}^k$ are given by

$$p_i^{\,k} = \sum_{j=1}^{M}\sum_{r=1}^{M} x_{i,jr}^{k}\,\dot{\psi}_r\dot{\psi}_j \tag{23}$$

(b) <u>Degree of Freedom (DOF)</u>

For a system described by M Lagrangian coordinates, the most general form for the equations of spatial constraint is

$$f_i(\psi_1,\ldots\psi_M) = 0,\quad (i = 1,\ldots N_s) \tag{24}$$

The "rate form" of these equations is

$$\sum_{j=1}^{M} \frac{\partial f_i}{\partial \psi_j}\,\dot{\psi}_j = 0,\quad \text{or}\quad \mathbf{D}\dot{\boldsymbol{\psi}} = \mathbf{0} \tag{25}$$

where

$$\mathbf{D} = [D_{ij}] = [\partial f_i/\partial \psi_j] \tag{26}$$

is an $(N_s \times M)$ matrix. If the rank of matrix **D** is r, we know from a theorem of linear algebra [9] that we may arbitrarily assign any values to (M-r) of the $\dot{\psi}_j$ and the others will be uniquely determined by Eq. (25). In kinematics terms, we say that the degree of freedom or mobility* of the system is

$$F = M - r \tag{27}$$

This result is due to Freudenstein [24]

In the nonsingular case, $r = N_s$ and the DOF is

$$F = M - N_s \tag{28}$$

For example, consider the mechanism of Fig. 1, where M = 5, and $N_s = 4$ [see Eqs. (6)]; this mechanism has 1 DOF.

If any temporal constraints are also present, we should refer to F as the <u>unreduced DOF</u>, because the imposition of N_t temporal constraints will lower the degree of freedom still further to the reduced

*See Chapter 8 of [66] for a discussion of mobility criteria.

DOF

$$F_R = F - N_t = M - N_s - N_t \tag{29}$$

For example, if we drive link AD of Fig. 1 with uniform angular velocity ω, we are in essence imposing the single temporal constraint

$$g_1(\psi,t) = \psi_2 - \omega t = 0 \tag{30}$$

Hence the reduced degree of freedom is

$$F_R = 1 - 1 = 0$$

In short, the system is kinematically determinate. No matter what external forces are applied, the motion is always the same. Of course, the internal forces at the joints of the mechanism are dependent on the applied forces, as will be discussed in the following sections on statics and kinetics of machine systems.

(c) Generalized Coordinates

For purposes of kinematics, it is sufficient to work directly with the Lagrangian coordinates ψ_i. However, in problems of dynamics it is useful to express the terms ψ_i in terms of a smaller number of variables q_r called generalized coordinates (also called primary coordinates). In problems of kinematics, we may think of the q_r as "driving variables." For example, if a motor drives link DA of Fig. 2.1, $\psi_2 = q_1$ is a good choice for this problem.

In a system with F (unreduced) DOF, we are free to choose any subset of F Lagrangian coordinates to serve as generalized coordinates. This choice may be expressed by the matrix relationship

$$\mathbf{T}\boldsymbol{\psi} = \mathbf{q} \tag{31}$$

where $\mathbf{T}$ is an (F x M) matrix whose elements are defined as follows:

$$T_{ij} = 1 \text{ if } q_i = \psi_j \tag{32-a}$$

$$T_{ij} = 0 \text{ otherwise} \tag{32-b}$$

For example, if M = 5 and F = 2, we might select ψ_2 and ψ_5 as the primary variables q_1 and q_2. Then Eq. (31) becomes

$$\begin{bmatrix} 0 & 1 & 0 & 0 & 0 \\ 0 & 0 & 0 & 0 & 1 \end{bmatrix} \begin{bmatrix} \psi_1 \\ \psi_2 \\ \psi_3 \\ \psi_4 \\ \psi_5 \end{bmatrix} = \begin{bmatrix} q_1 \\ q_2 \end{bmatrix}$$

If desired, the matrix $\mathbf{T}$ may be chosen so that any q_r is some desired linear combination of the ψ_i, but the simpler choice embodied in Eqs. (32-a,b) should suffice for most purposes.

The F equations (31), together with the (M-F) spatial constraint equations

$$f_i(\psi_1,\ldots\psi_M) = 0, \quad (i = 1,\ldots N_s) \tag{33}$$

comprise a set of M equations in the M variables $\dot{\psi}_k$. We can represent the differentiated form of these equations by

$$\sum_{j=1}^{M} f_{i,j}\,\dot{\psi}_j = 0, \quad (i = 1,\ldots,N_s) \tag{34}$$

$$\sum_{j=1}^{M} T_{kj}\,\dot{\psi}_j = \dot{q}_k, \quad (k = 1,\ldots,F) \tag{35}$$

or

$$\mathbf{A}\dot{\boldsymbol{\psi}} = \mathbf{B}\dot{\mathbf{q}} \tag{36}$$

where

$$A_{ij} = f_{i,j};\; B_{ij} = 0 \qquad ;\ (i=1,\ldots N_s) \tag{37-a}$$

$$\left.\begin{aligned} A_{ij} = T_{ij};\quad B_{ij} &= 1, \text{ if } i = j + N_s \\ B_{ij} &= 0, \text{ if } i \neq j + N_s \end{aligned}\right\} ;\ (i=N_s+1,\ldots M) \tag{37-b}$$

In general, the square matrix $\mathbf{A}$ will be nonsingular for some choice of driving variables; i.e. for some arrangement of the ones and zeroes in the matrix $\mathbf{T}$. It may be necessary to redefine the initial choice of the $\mathbf{T}$ matrix during the course of time in order for $\mathbf{A}$ to remain nonsingular at certain critical configurations. For example, in a slider-crank mechanism, we could not use the displacement of the slider as a generalized coordinate when the slider is at either of its extreme positions (i.e. in the so-called dead-center positions).

When $\mathbf{A}$ is nonsingular, we may assign independent numerical values to all the q_i, and solve the (MxM) system of nonlinear algebraic equations (33) and (31) for $(\psi_1,\ldots\psi_M)$ by means of a numerical technique such as the Newton-Raphson method. In order for real solutions to exist, it is necessary that the q_i lie within geometrically meaningful ranges. Then one may find the Lagrangian velocities $\dot{\psi}_k$ from Eq. (36) in the form

$$\dot{\boldsymbol{\psi}} = \mathbf{A}^{-1}\mathbf{B}\,\dot{\mathbf{q}} \equiv \mathbf{C}\dot{\mathbf{q}}, \text{ or } \dot{\psi}_j = \sum_{r=1}^{F} C_{jr}\,\dot{q}_r \tag{38}$$

where

$$\mathbf{C} \equiv \mathbf{A}^{-1}\mathbf{B} \tag{39}$$

To find the accelerations associated with given values of q_r, it is only necessary to solve the differentiated form of Eq. (36); i.e.

$$\mathbf{A}\ddot{\psi} = \mathbf{B}\ddot{\mathbf{q}} + \mathbf{v} \tag{40}$$

where

$$\mathbf{v} \equiv -\dot{\mathbf{A}}\dot{\psi} \tag{41}$$

and

$$A_{ij} \equiv \sum_{k=1} f_{i,jk}\, \dot{\psi}_k, \quad (i = 1,\ldots N_s) \tag{42}$$

$$A_{ij} \equiv 0, \qquad (i = N_s+1,\ldots M) \tag{43}$$

Note that $\dot{\mathbf{B}} \equiv \mathbf{0}$, since all B_{ij} are constants by definition.

From Eq. (40), it follows that

$$\ddot{\psi} = \mathbf{C}\ddot{\mathbf{q}} + \mathbf{e} \tag{44}$$

where

$$\mathbf{e} = \mathbf{A}^{-1}\mathbf{v} \tag{45}$$

From previous definitions, it may be verified that each component of the vector **e** is a quadratic form in the generalized velocities $\dot{q}_r$. Hence, we see from Eq. (44) that any component of Lagrangian acceleration is expressible as a linear combination of the generalized accelerations $\ddot{q}_r$ plus a quadratic form in the generalized velocities $\dot{q}_r$. The quadratic velocity terms are manifestations of centripetal and Coriolis accelerations (also called compound centripetal accelerations). We will accordingly refer to them as <u>centripetal</u> terms for brevity.

(d) Points of Interest

To express the velocity of point $\mathbf{x}^k$ in terms of $\dot{\mathbf{q}}$ we merely substitute Eqs. (38) into Eq. (19) and find

$$\dot{\mathbf{x}}^k = \mathbf{U}^k\dot{\mathbf{q}}, \text{ or } \dot{x}_i^{\,k} = \sum_{j=1}^{M} U_{ij}^k\, \dot{q}_j \tag{46}$$

where

$$\mathbf{U}^k = \mathbf{P}^k\mathbf{C}\;, \text{ or } U_{ij}^k = \sum_{n=1}^{M} P_{in}^k\, C_{nj} \tag{47}$$

Similarly, the acceleration for the same point of interest is found by substituting Eq. (44) into Eq. (22) to yield

$$\ddot{\mathbf{x}}^k = \mathbf{U}^k\ddot{\mathbf{q}} + \mathbf{v}^k \tag{48}$$

where

$$\mathbf{v}^k \equiv \mathbf{P}^k \mathbf{e} + \mathbf{p}^k \tag{49}$$

From previous definitions, it is readily verified that each component of the vector $\mathbf{v}^k$ is a quadratic form in $\dot{q}_i$.

Upon multiplying each term in Eq. (46) by an infinitesimal time increment δt we find that any set of displacements which satisfy the spatial constraint Eqs. (33) must satisfy the differential relation

$$\delta x_i{}^k = \sum_{j=1}^{F} U_{ij}^k \, \delta q_j \tag{50}$$

This result will be useful in our later discussion of virtual displacements.

(e) Angular Position, Velocity and Acceleration

Let (ξ,η,ζ) be a set of right-handed orthogonal axes (body axes) fixed in a rigid body, and let (x,y,z) be a right-handed orthogonal set of axes parallel to directions fixed in inertial space (space axes). Both sets of axes have a common origin. The angular orientation of the body relative to fixed space is defined by the three Euler angles Θ,Ψ,Φ shown in Fig. 2.

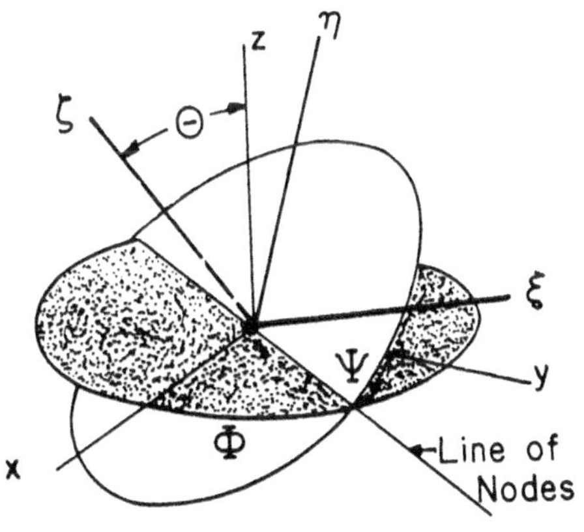

Fig. 2. Euler Angles

With this definition of the Euler angles [15,29,53,61], the body is brought from its initial position (where ξ,η,ζ coincide with x,y,z) to its final position by the following sequence:

1) rotation by Φ (precession angle) about the z-axis, thereby carrying the ξ axis into the position marked "line of nodes";
2) rotation by Θ (nutation angle) about the line of nodes, thereby carrying the ζ axis into its final position shown;
3) rotation by Ψ (spin angle) about the ζ axis thereby carrying

the ξ and η axes into their final positions as shown.

It may be shown [15] that the angular velocity components $(\omega_x, \omega_y, \omega_z)$ relative to the fixed axes (x,y,z) are given by

$$\begin{bmatrix} \omega_x \\ \omega_y \\ \omega_z \end{bmatrix} = \begin{bmatrix} \cos\Phi & \sin\Theta\sin\Phi & 0 \\ \sin\Phi & -\sin\Theta\cos\Phi & 0 \\ 0 & \cos\Theta & 1 \end{bmatrix} \begin{bmatrix} \dot{\Theta} \\ \dot{\Psi} \\ \dot{\Phi} \end{bmatrix} \tag{51}$$

$$\omega = \mathbf{R}\{\dot{\Theta}, \dot{\Psi}, \dot{\Phi}\} \tag{52}$$

where $\mathbf{R}$ is the square matrix shown in Eq. (51).

Now the Euler angles for body k can be defined as a subset of the Lagrangian coordinates as follows:

$$\{\Theta, \Psi, \Phi\}^k \equiv \mathbf{E}^k \psi \tag{53}$$

where the components of the (3xM) matrices $\mathbf{E}^k$ are all zeroes or ones.

Upon noting from Eq. (38) that $\dot{\psi} = \mathbf{C}\dot{\mathbf{q}}$, we can write Eq. (52) in the form

$$\omega^k = \mathbf{R}^k\,\mathbf{E}^k\mathbf{C}\dot{\mathbf{q}} = \mathbf{W}^k\dot{\mathbf{q}} \tag{54}$$

where

$$\mathbf{W}^k \equiv \mathbf{R}^k\,\mathbf{E}^k\mathbf{C} \tag{55}$$

Later we will be interested in studying infinitesimal rotations $(\delta\theta_1, \delta\theta_2, \delta\theta_3)^k$ of body k about the (x,y,z) axes. From Eq. (54) it follows that

$$\delta\theta^k = \mathbf{W}^k\delta\mathbf{q} \quad \text{or} \quad \delta\theta_i^k = \sum_{r=1}^{F} W_{ir}^k\,\delta q_r \tag{56}$$

If desired, the angular accelerations are readily found by differentation of Eqs. (51) or (54).

It is worth noting that use of Euler angles can lead to singular states when Θ is zero or π, because in such cases there is no clear distinction between the angles Φ and Ψ (see Fig. 2). For this reason, it is sometimes convenient to work with other measures of angular position, such as the four (dependent) Euler parameters [29,41,57,63, 78].

(f) Kinematics of Systems with One Degree of Freedom

The special case of F = 1 is sufficiently important to introduce a notational simplification which is possible because the vector of generalized coordinates $\mathbf{q}$ has only the single component $q_1 \equiv q$. Similarly, we may drop the second subscript in all the matrices which operate on q. For example, (Eq. 36) becomes

$$\mathbf{A}\dot{\psi} = \mathbf{b}\dot{q} \tag{57}$$

where

$$b_i \equiv B_{i1} \tag{58}$$

In like manner, we may rewrite Eqs. (38) and (44) in the form

$$\dot{\psi}_i = c_i\dot{q} \tag{59}$$

$$\ddot{\psi}_i = c_i\ddot{q} + e_i = c_i\ddot{q} + \dot{c}_i'\dot{q}^2 \tag{60}$$

where

$$c_i' \equiv -\sum_{j=1}^{M} \frac{\partial c_i}{\partial \psi_j} c_j \tag{61}$$

Similarly, the velocities and accelerations, for point of interest k, can be written in the compact form

$$\dot{\mathbf{x}}^k = \mathbf{u}^k\dot{q} \tag{62}$$

$$\ddot{\mathbf{x}}^k = \mathbf{u}^k\ddot{q} + \mathbf{v}^k = \mathbf{u}^k\ddot{q} + \mathbf{u}'^k\dot{q}^2 \tag{63}$$

where

$$\mathbf{v}^k \equiv \dot{q} \sum_{j=1}^{M} \frac{\partial \mathbf{u}^k}{\partial \psi_j} \dot{\psi}_j = \dot{q}^2 \sum_{j=1}^{M} \frac{\partial \mathbf{u}^k}{\partial \psi_j} c_j$$

$$\mathbf{u}'^k \equiv \sum_{j=1}^{M} \frac{\partial \mathbf{u}^k}{\partial \psi_j} c_j$$

Finally, we note that the components of the angular velocity of any rigid-body member k of the system are given by Eq. (54) in the simplified form

$$\omega_i^k = w_i^k\dot{q} \tag{64}$$

and the corresponding angular accelerations are given by

$$\dot{\omega}_i^k = w_i^k\ddot{q} + \dot{w}_i^k\dot{q} \tag{65}$$

where

$$w_i^k \equiv W_{i1}^k \tag{66}$$

(g) Computer Programs for Kinematics Analysis

The foregoing analysis forms the basis of a digital computer program called KINMAC (KINematics of MAChinery) developed by the author and his colleagues [1,36,69]. This FORTRAN program calculates displacements, velocities and accelerations for up to 30 points of interest in planar mechanisms with up to 10 DOF, up to 10 loops, and up to 10 user-supplied auxiliary constraint relations (in addition to the automatically formulated loop closure constraint relations). The program also contains built in subroutines for use in the modelling of cams and temporal constraint relations. A less sophisticated student-oriented program for kinematics analysis, called KINAL (KINematics anALysis) is described and listed in full in [66].

In [89], Suh and Radcliffe give listings for a package of FORTRAN subroutines called LINCPAC which can be used to solve a variety of fundamental problems in planar kinematics. They also give listings for a package called SPAPAC to be similarly used for problems of spatial kinematics.

Other programs have been developed primarily for dynamics analysis, but they can function in a so-called kinematic mode; among these are IMP and DRAM which will be discussed further in connection with programs for dynamics analysis. Computer programs which have been developed primarily for problems of kinematic synthesis [42],but have some application to analysis and simulation, include R. Kaufman's KINSYN [80], A. Erdman's LINCAGES [22]; and H. Ranker's CADOM [129].

3. STATICS OF MACHINE SYSTEMS

(a) Direct Use of Newton's Laws (Vectorial Mechanics)

The most direct way of formulating the equations of statics for a machine is to write equations of equilibrium for each body of the system. For example, Fig. 3 shows the set of five free body diagrams associated with the mechanism of Fig. 1. The internal reaction forces

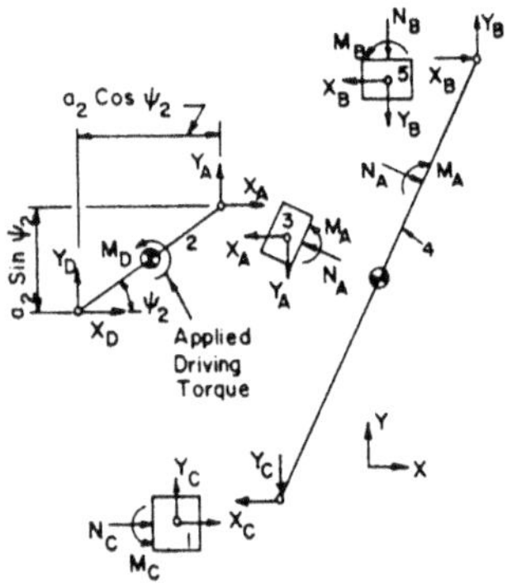

Fig. 3. Free-body diagrams for the mechanism of Fig. 1.

at pin joints (e.g. X_A, Y_A) and the normal forces and moments at smooth slider contacts (e.g. N_A, M_A) appear as external loadings on the various bodies of the system. A driving torque M_D, acting on link 2, is assumed to be a known function of position. Thus, there exists a total of fourteen unknown internal force components, namely: X_A, Y_A, N_A $M_A, X_B, Y_B, N_B, M_B, X_C, Y_C, N_C, M_C, X_D, Y_C$. In addition, there exist five unknown position variables ($\psi_1, \ldots \psi_5$). In order to find the nineteen unknown quantities, we can write three equations of equilibrium for each of the five "free bodies." A typical set of such equations, for link 2, is

$$X_D + X_A = 0 \tag{67-a}$$

$$Y_D + Y_A = 0 \tag{67-b}$$

$$M_D + (X_D - X_A) \frac{a_2}{2} \sin \psi_2 + (Y_A - Y_D) \frac{a_2}{2} \cos \psi_2 = 0 \tag{67-c}$$

In all, we can write a total of fifteen such equations of equilibrium for the five links. We can eliminate the fourteen internal reactions from these equations to yield one master equation involving the five position variables ($\psi_1, \ldots \psi_5$). In addition, we have the four equations of constraint in the form of Eqs. (6). Thus all of the equations mentioned above may be boiled down to five equations for the five unknown Lagrangian coordinates ψ_i. These equations are nonlinear in the ψ_i, but a numerical solution is always possible (see Sec. 9.22 of [66]). Having found the ψ_i, one may solve for the fourteen unknown reactions from the equilibrium equations which are linear in the reactions.

If one only wishes to find the configuration of a mechanism under the influence of a set of static external forces, the above procedure is unnecessarily cumbersome compared to the method of virtual work (discussed below).

However, if the configuration of the system is known (all ψ_i given) then the fifteen equations of equilibrium are convenient for finding the fourteen internal reaction components plus the required applied torque M_D. The details of such a solution are given in Sec. 11.52 of [66] for arbitrary loads on all the links of the example mechanism.

(b) Terminology of Analytical Mechanics

We now define some key terms and concepts from the subject of Analytical Mechanics. For a more detailed discussion of the subject see any of the standard works, e.g. [29,30,46,61,102] or [53,64,66].

(i) Virtual Displacements are defined to be any set of infinitesimal displacements which satisfy all instantaneous constraints.

This means that if there exist F degrees of freedom, the displacements of a typical point P^k must satisfy the differential relationship.

$$\delta x_i^k = \sum_{j=1}^{F} U_{ij}^k \, \delta q_j \tag{68}$$

given by Eq. (50).

If, in addition, there exist N_t temporal constraints of the form

$$g_k(\psi_1, \ldots \psi_M, t) = 0, \quad (k = 1,2,\ldots N_t) \tag{69}$$

They can be expressed in the "rate" form

$$\sum_{j=1}^{M} g_{k,j} \, \dot{\psi}_j + \frac{\partial g_k}{\partial t} = 0 \tag{70}$$

However, from Eq. (38)

$$\dot{\psi}_j = \sum_{r=1}^{F} C_{jr} \, \dot{q}_r$$

Therefore

$$\sum_{r=1}^{F} \alpha_{kr} \, \dot{q}_r + \beta_k = 0, \quad (k = 1,\ldots N_t) \tag{71}$$

where

$$\alpha_{kr} = \sum_{j=1}^{M} g_{k,j} \, C_{jr} \tag{72-a}$$

$$\beta_k = \partial g_k / \partial t \tag{72-b}$$

Since $\dot{q}_r = dq_r/dt$, Eq. (71) can be expressed in the form

$$\sum_{r=1}^{F} \alpha_{kr} \, dq_r + \beta_k \, dt = 0 \tag{73}$$

The virtual displacements δq_r, however, must, by definition, satisfy the equations

$$\sum_{r=1}^{F} \alpha_{kr} \, \delta q_r = 0, \quad (k = 1,\ldots N_t) \tag{74}$$

We say that Eq. (73) represents the actual constraints, and Eq. (74) represents the instantaneous constraints. Any set of infinitesimals dq_r which satisfy Eq. (73) are called possible displacements, in distinction to the virtual displacements which satisfy Eqs. (74).

In problems of statics, such temporal constraints are not normally present, but they could easily arise in problems of dynamics. Equations of constraint of the form (71) sometimes arise which cannot be integrated to give a function of the form $g_k(\psi,t) = 0$. Such non-integrable equations of constraint are said to be nonholonomic.

(ii) Virtual work (δW) is the work done by specified forces on virtual displacements.

(iii) Active forces are those forces which produce nonzero net virtual work.

(iv) Ideal mechanical systems are systems in which constraints are maintained by forces which do no virtual work (e.g. frictionless contacts).

(v) Generalized forces (Q_r) are terms which multiply the virtual generalized displacements (δq_r) in the expression for virtual work of the form $\delta W = \sum Q_r \delta q_r$.

(vi) Equilibrium is a state in which the resultant force on each particle of the system vanishes.

(vii) Generalized equilibrium is a state where all the generalized forces vanish.

(c) The Principle of Virtual Work

The principle of virtual work may be viewed as a basic postulate from which all the well-known laws of vector statics may be deduced, or it may be proved directly from the laws of vector statics. The following two-part statement of the principle is so-proved in Sec. 11.30 of [66].

1. If an ideal mechanical system is in equilibrium, the net virtual work of all the active forces vanishes for every set of virtual displacements.

(2) If the net virtual work of all the active forces vanishes, for every set of virtual displacements, an ideal mechanical system is in a state of generalized equilibrium.

It is pointed out in [66] that no distinction need be made between generalized equilibrium and equilibrium for systems which are at rest (relative to an inertial frame) before the active forces are applied.

For a system which consists of n rigid bodies, we may express the resultant external force acting on a typical body by $\mathbf{X}^P$ acting through a reference point in the body which momentarily occupies the point $\mathbf{x}^P$ measured in the global coordinate frame. Similarly the active torque $\mathbf{M}^P$ is assumed to act on the same body. The virtual displacements of the reference points are denoted by $\delta\mathbf{x}^p$ and the corresponding rotations of each rigid body may be expressed by the infinitesimal rotation vector

$$\delta\theta^p = (\delta\theta_1{}^p, \delta\theta_2{}^p, \delta\theta_3{}^p) \tag{75}$$

where $\delta\theta_i{}^p$ is the rotation component about an axis, through the point $\mathbf{x}^p$, parallel to the global x_i axis. The virtual work of all the active forces is therefore given by

$$\delta W = \sum_{p=1}^{n} \sum_{i=1}^{3} (X_i{}^p \, \delta x_i{}^p + M_i{}^p \, \delta\theta_i{}^p) \tag{76}$$

From Eqs. (50) and (56), we know that

$$\delta x_i{}^p = \sum_{j=1}^{F} U_{ij}^p \, \delta q_j \tag{77}$$

$$\delta\theta_i{}^p = \sum_{j=1}^{F} W_{ij}^p \, \delta q_j \tag{78}$$

Upon substitution of Eqs. (77) and (78) into Eq. (76) we find that

$$\delta W = \sum_{j=1}^{F} Q_j \, \delta q_j \tag{79}$$

where

$$Q_j \equiv \sum_{p=1}^{n} \sum_{i=1}^{3} (U_{ij}^p \, X_i{}^p + W_{ij}^p \, M_i{}^p), \quad (j=1,\ldots F) \tag{80}$$

Equation (80) is an explicit formula for the calculation of generalized forces, although they can usually be calculated in more direct, less formal, ways as described in Sec. 11.22 of [66].

From part 1 of the statement of the principle of virtual work, it follows that if an ideal system is in equilibrium then for all possible choice of δq_i

$$\delta W = \sum_{j=1}^{F} Q_j \, \delta q_j = 0 \tag{81}$$

In the absence of temporal constraints, the δq_j are all independent and it follows from Eq. (81) that for such cases

$$Q_j = 0 \,, \quad (j = 1,2,\ldots F) \tag{82}$$

Thus the vanishing of all generalized forces is a necessary condition for equilibrium and a sufficient condition (by definition) for generalized equilibrium.

For examples of the method of virtual work applied to the statics of machine systems, see Sec. 11.30 of [66].

For those cases where the generalized coordinates q_r are not independent, but are related by constraint relations of the form (71) (arising either from temporal or nonholonomic constraints) we cannot reduce eq. (81) to the simple form of Eq. (82). For such case we can use the method of Lagrange multipliers.

(d) Lagrange Multipliers

It is sometimes convenient to choose as primary coordinates, a subset, $q_1,\ldots q_f$, of the Lagrangian coordinates which are not independent, but are related by equations of constraint of the type (74), i.e.

$$\sum_{j=1}^{F} \alpha_{ij}\, \delta q_j = 0\ , \quad (i = 1,\ldots N_t) \tag{83}$$

we say that such primary coordinates form a set of redundant or superfluous generalized coordinates.

However, we may choose to think of all the q_i as independent coordinates if we conceptually remove the physical constraints (e.g., hinge pins, slider guides, etc.) and consider the reactions produced by the constraints as external forces. In other terms, we must supplement the generalized forces Q_i of the active loads by a set of forces Q_i', associated with the reactions produced by the (now deleted) constraints. Since the reactions in an ideal system, by definition, produce no work on any displacement compatible with the constraints it is necessary that the supplemental forces Q_i' satisfy the condition

$$\sum_{j=1}^{F} Q_i'\, \delta q_j = 0 \tag{84}$$

where the displacements δq_j satisfy the constraint Eqs. (83).

It may be shown [100, p. 562] that in order for Eqs. (83) and (84) to be satisfied for arbitrary choices of the δq_i, it is necessary that

$$Q_j' = \sum_{i=1}^{N_t} \lambda_i \alpha_{ij} \tag{85}$$

where λ_i are unknown quantities called "Lagrangian Multipliers." Accordingly, Eq. (85) provides the desired supplements to the generalized forces, and the equations of equilibrium assume the more general form

$$Q_j + Q_j' = Q_j + \sum_{i=1}^{N_t} \lambda_i \alpha_{ij} = 0, \quad (j = 1,\ldots F) \tag{86}$$

If we consider that Q_j and α_{ij} are known functions of the q_r then Eqs. (86) and (69) provide (for holonomic constraints) $F+N_t$ equations for determining the unknowns $(q_1,\ldots q_F)$ and $(\lambda_1,\ldots\lambda_{N_t})$. However, in practice, it is likely that Q_j and α_j are expressed as functions of $(\psi_1,\ldots\psi_M)$. Then it is necessary to utilize the M additional equations consisting of Eqs. (24) and (31). For nonholonomic constraints, Eqs. (69) are not available and one must utilize instead differential equations of the form (71). Such problems must usually be solved by a process of numerical integration. Numerical integration will be discussed in Sec. 5.

(e) Conservative Systems

If the work done on an ideal mechanical system by a set of forces is a function $\bar{W}(q_1,\ldots q_F)$ of the generalized coordinates only, the system is said to be conservative. Such forces could arise from elastic springs, gravity, and other field effects.

The increment in $\bar{W}$ due to a virtual displacement δq_i is

$$\delta\bar{W} = \frac{\partial\bar{W}}{\partial q_1}\delta q_1 + \ldots \frac{\partial\bar{W}}{\partial q_F}\delta q_F \tag{87}$$

By tradition, one defines the <u>potential energy</u> or <u>potential</u> of the system as

$$V(q) \equiv -\bar{W}(q)$$

Therefore

$$\delta\bar{W} = -\frac{\partial V}{\partial q_1}\delta q_1 \ldots - \frac{\partial V}{\partial q_F}\delta q_F \tag{88}$$

If, in addition, a set of active forces (not related to $\bar{W}$) act on the system and produce the generalized forces Q_r^*, the net work done during a virtual displacement is

$$\delta W = (Q_1^* - \frac{\partial V}{\partial q_1})\delta q_1 \ldots - (Q_F^* - \frac{\partial V}{\partial q_F})\delta q_F = 0 \tag{89}$$

where the right hand zero is a consequence of the principle of virtual work. From the definition of Q_r (Sec. 3-b) it follows that

$$Q_r = Q_r^* - \frac{\partial V}{\partial q_r} \tag{90}$$

If all of the δq_r are independent (i.e., no temporal or non-holonomic constraints exist) then Eq. (89) predicts that

$$Q_r \equiv Q_r^* - \frac{\partial V}{\partial q_r} = 0, \quad (r = 1,\ldots F) \tag{91}$$

If the δq_r are related, as in Eq. (83)

$$\sum_{j=1}^{F} \alpha_{ij}\delta q_j = 0, \quad (i = 1,2,\ldots N_t) \tag{92}$$

it follows from the discussion leading to Eq. (70) that

$$Q_j^* - \frac{\partial V}{\partial q_i} + \sum_{i=1}^{N_t} \lambda_i \alpha_{ij} = 0, \quad (j = 1,\ldots F) \tag{93}$$

It is only necessary to define V within an additive constant, since only the derivatives of V enter into the equilibrium equations.

Some examples of the form of V for some common cases follow:

1. A linear spring with free length s_f, current length s, and stiffness K has potential

$$V(s) = \frac{1}{2} K (s-s_f)^2 \tag{94}$$

2. A system of mass particles close to the surface of the earth (where the acceleration of gravity g is essentially constant), has potential

$$V(\bar{z}) = W\bar{z} \tag{95}$$

where $\bar{z}$ is the elevation of the center of gravity of the system vertically above an arbitrary datum plane, and W is the total weight of all the particles.

3. A mass particle is located at a distance r from the center of the earth. If its weight is W_e when the particle is at a distance r_e from the earth's center (e.g. at the surface of the earth), its potential in general is

$$V(r) = - W_e r_e^2/r \tag{96}$$

In all of these examples, V has been expressed in terms of a Lagrangian variable. More generally, V is of the form

$$V = V(\psi_1,\ldots\psi_M) \tag{97}$$

The required derivatives of V are of the form

$$\frac{\partial V(q)}{\partial q_r} = \sum_{j=1}^{M} \frac{\partial V(\psi)}{\partial \psi_j} \frac{\partial \psi_j}{\partial q_r} = \sum_{j=1}^{M} \frac{\partial V(\psi)}{\partial \psi_r} C_{jr} \tag{98}$$

and C_{jr} are components of the matrix **C** defined by Eq. (39). Illustrations of the use of potential functions in problems of statics of machines are given in Sec. 11.42 of [66].

(f) Reactions at Joints

By vector statics - Perhaps the most straightforward way to compute reactions is to establish the three equations of equilibrium for each link, as exemplified by Eqs. (67). The equations are linear in the desired reactions and are readily solved by the same linear equation subroutine that is needed for other purposes in a numerical analysis. This method has been described in depth for planar mechanisms in Sec. 11.50-52 of [66], and has been used in the author's computer programs STATMAC [70], and DYMAC [68]. It was also used in [72] and in the computer programs MEDUSA [18] and VECNET [2].

By principal of virtual work - In order for the internal reactions at a specific joint to appear in the expression for virtual work it is necessary to imagine that the joint is "broken" and then to introduce, as active forces, the unknown joint reactions needed to maintain closure. This technique is briefly described in [64], and in more detail by Denavit et al [16], and by Uicker [97], who combines the method of virtual work with Lagrange's Equations of second type (see Sec. 4-b).

By Lagrange Multipliers - As shown by Chace et al [10] and Smith [85], it is possible to find certain crucial joint reactions, simultaneously with the solution of the general dynamics problem, by the method of Lagrange multiplers. A brief description of this method is also given in [64].

(g) Effects of Friction in Joints

Up to this point, it has been assumed that all internal reactions are workless. If the friction at the joints is velocity-dependent, we can place a damper (not necessarily linear) at the joints; the associated pairs of velocity-dependent force or torque reactions are then included in the active forces (X^p, M^p) as defined in Section 3-c. These forces, which appear linearly in the generalized forces Q [see Eq. (90)], result in velocity-dependent generalized forces.

However, when Coulomb friction exists at the joints, the generalized forces become functions of the unknown joint reactions. These forces complicate the solution of the equations, but they can be accounted for in a number of approximate ways. A detailed treatment of frictional effects in statics of planar mechanisms is given in Secs. 11.60-63 of [66]. Frictional effects in the dynamics of planar machines are discussed in Sec. 12.70 of [66]. Greenwood [30] (p.271) utilizes Lagrange Multipliers to solve a problem of frictional effects between three bodies moving in a plane with translational motion only.

4. KINETICS OF MACHINE SYSTEMS

(a) Lagrange's Form of d'Alembert's Principle

We now assume that active forces $X^i \equiv (X_1,X_2,X_3)^i$ applied to the mass center $\mathbf{x}^i \equiv (x_1,x_2,x_3)^i$ of link i and an active torque $\mathbf{M}^i \equiv (M_1,M_2,M_3)^i$ is exerted about the mass center. Any set of active forces which does not pass through the mass center may be replaced by a statically equipollent set of forces which does. In general, X^i and $\mathbf{M}^i$ may depend explicitly upon position, velocity and time. Massless springs may be modelled by pairs of equal, but opposite, collinear forces which act on the links at the terminal points of the springs. The spring forces may be any specified linear or nonlinear functions of the distance between the spring's terminals. Dampers may be modelled by similar pairs of equal and opposite collinear forces which are dependent on the rate of relative extension of the line joining the damper terminals. Examples of such springs and dampers are given in Sec. 14.43 of [66].

A typical link is subjected to inertia forces $-m\ddot{\mathbf{x}}$ as well as the real forces X. Thus the virtual work due to <u>both</u> of these "forces" is given by

$$\delta W_F = \mathbf{x}^T(X-m\ddot{\mathbf{x}}) \tag{99}$$

From Eqs. (48) and (50) we observe that

$$\ddot{\mathbf{x}} = \mathbf{U}\ddot{\mathbf{q}} + \mathbf{v} \tag{100-a}$$

$$\delta\mathbf{x} = \mathbf{U}\ \delta\mathbf{q}, \quad \delta\mathbf{x}^T = \delta\mathbf{q}^T\ \mathbf{U}^T \tag{100-b}$$

where a superscript T denotes a matrix transpose. Therefore the virtual work due to "forces" on the rigid body is

$$\delta W_F = \delta\mathbf{q}^T[\mathbf{U}^T\mathbf{X}-m\mathbf{U}^T\mathbf{v}-m\mathbf{U}^T\mathbf{U}\ddot{\mathbf{q}}] \tag{101}$$

To this must be added the virtual work due to the real and inertial torques which act about the mass center. To help calculate these, we introduce the following notation:

G:	the mass center of a rigid body
(ξ_1,ξ_2,ξ_3):	local orthogonal reference axes through G aligned with the principal axes of inertia of the body
$\bar{J}_1,\bar{J}_2,\bar{J}_3$:	principal moments of inertia of the body, referred to axes ξ_1,ξ_2,ξ_3

$\bar{M}_1, \bar{M}_2, \bar{M}_3$: components of external torque in directions of ξ_1, ξ_2, ξ_3

$\bar{\omega}_1, \bar{\omega}_2, \bar{\omega}_3$: components of angular velocity in directions ξ_1, ξ_2, ξ_3

$\delta\bar{\theta}_1, \delta\bar{\theta}_2, \delta\bar{\theta}_3$ infinitesimal rotation components about axes ξ_1, ξ_2, ξ_3

D_{ij}: direction cosine between (local) axis ξ_i and (global) axis x_j

Note that the overbar is used on vector components referred to the local (ξ) axes. Corresponding symbols without overbars refer to global (x) axes. The two systems are related by matrix relationships of the type

$$\bar{\mathbf{M}} = \mathbf{DM} \tag{102}$$

$$\bar{\omega} = \mathbf{D}\omega \tag{103}$$

$$\delta\bar{\boldsymbol{\theta}} = \mathbf{D}\ \delta\boldsymbol{\theta} \tag{104}$$

It may be shown [29,53] that the direction cosines are related to the Euler angles as follows:

$$\mathbf{D} \equiv [D_{ij}] = \begin{bmatrix} (C\Psi\ C\Phi - C\Theta\ S\Phi\ S\Psi) & (C\Psi\ S\Phi + C\Theta\ C\Phi\ S\Psi) & (S\Psi\ S\Theta) \\ (-S\Psi\ C\Phi - C\Theta\ S\Phi\ C\Psi) & (-S\Psi\ S\Phi + C\Theta\ C\Phi\ C\Psi) & (C\Psi\ S\Theta) \\ (S\Theta\ S\Phi) & (-S\Theta\ C\Phi) & (C\Theta) \end{bmatrix} \tag{105}$$

where $C\Psi = \cos\Psi$, $S\Psi = \sin\Psi$, etc. (106)

and that $\mathbf{D}$ is an orthogonal matrix, i.e.

$$\mathbf{D}^T = \mathbf{D}^{-1} \tag{107}$$

We now take note of Euler's equation of motion for rigid bodies expressed in the form [53,94]

$$\bar{M}_i + \bar{T}_i = 0, \quad (i = 1,2,3) \tag{108}$$

where the <u>inertia torque</u> components $\bar{T}_i$ are given by

$$\bar{T}_1 = -\bar{J}_1\dot{\bar{\omega}}_1 + (\bar{J}_2-\bar{J}_3)\ \bar{\omega}_2\bar{\omega}_3 \tag{109-a}$$

$$\bar{T}_2 = -\bar{J}_2\dot{\bar{\omega}}_2 + (\bar{J}_3-\bar{J}_1)\ \bar{\omega}_3\bar{\omega}_1 \tag{109-b}$$

$$\bar{T}_3 = -\bar{J}_3\dot{\bar{\omega}}_3 + (\bar{J}_1-\bar{J}_2)\ \bar{\omega}_1\bar{\omega}_2 \tag{109-c}$$

In matrix form, we can write

$$\bar{T} = -\bar{J}\dot{\bar{\omega}} + \tau \tag{110}$$

where

$$\bar{J} = \begin{bmatrix} \bar{J}_1 & 0 & 0 \\ 0 & \bar{J}_2 & \\ 0 & 0 & \bar{J}_3 \end{bmatrix} \tag{111}$$

$$\tau = \begin{Bmatrix} (\bar{J}_2-\bar{J}_3) & \bar{\omega}_2\bar{\omega}_3 \\ (\bar{J}_3-\bar{J}_1) & \bar{\omega}_3\bar{\omega}_1 \\ (\bar{J}_1-\bar{J}_2) & \bar{\omega}_1\bar{\omega}_2 \end{Bmatrix} \tag{112}$$

From Eq. (54) we recall that

$$\omega = \mathbf{W}\dot{\mathbf{q}} \tag{113}$$

Hence, the local velocity components are given by Eqs. (103) in the form

$$\bar{\omega} = \mathbf{D}\omega = \mathbf{D}\mathbf{W}\dot{\mathbf{q}} \tag{114}$$

The corresponding acceleration components are therefore

$$\dot{\bar{\omega}} = \bar{\mathbf{W}}\ddot{\mathbf{q}} + \bar{\mathbf{V}} \tag{115}$$

where

$$\bar{\mathbf{W}} = \mathbf{D}\mathbf{W} \tag{116}$$

$$\bar{\mathbf{V}} = (\mathbf{D}\dot{\mathbf{W}}+\dot{\mathbf{D}}\mathbf{W})\dot{\mathbf{q}} \tag{117}$$

Thus the inertia torque given by Eqs. (110) can be expressed as

$$\bar{\mathbf{T}} = -\bar{\mathbf{J}}\bar{\mathbf{W}}\ddot{\mathbf{q}} - \bar{\mathbf{J}}\bar{\mathbf{V}} + \tau \tag{118}$$

In accordance with d'Alembert's principle, we may therefore express the virtual work of the active and inertia torques in the form

$$\delta W_M = \sum_{i=1}^{3} \delta\bar{\theta}_i(\bar{M}_i+\bar{T}_i) = \delta\bar{\theta}^T(\bar{M}+\bar{T}) \tag{119}$$

Upon utilization of Eqs. (104) and (118) we find

$$\delta W_M = (\mathbf{D}\delta\theta)^T(\bar{M}+\tau+\bar{\mathbf{J}}\bar{\mathbf{V}}-\bar{\mathbf{J}}\bar{\mathbf{W}}\ddot{\mathbf{q}}) \tag{120}$$

Since $\delta\theta = \mathbf{W}\ \delta q$, and $\bar{\mathbf{W}} = \mathbf{D}\mathbf{W}$, we can write

$$(\mathbf{D}\delta\theta)^T = (\mathbf{D}\mathbf{W}\delta\mathbf{q})^T = (\bar{W}\delta q)^T = \delta q^T\bar{W}^T \tag{121}$$

Now we can write Eq. (120) in the form

$$\delta W_M = \mathbf{q}^T[\bar{W}^T(\bar{M}+\tau-\bar{J}\bar{V}) - \bar{W}^T J W \ddot{\mathbf{q}}] \tag{122}$$

Adding δW_M to δW_F given by Eq. (101), the virtual work of all the real and inertial forces and torques is seen to take the form

$$\delta W_M+\delta W_F=\delta\mathbf{q}^T[(U^T X+\bar{W}^T\bar{M})-mU^T v+\bar{W}^T(\tau-\bar{J}\bar{V})-(mU^T U+\bar{W}^T\bar{J}\bar{W}\ddot{\mathbf{q}}] \tag{123}$$

If we use a superscript k to denote the virtual work done on the kth rigid body in the system, we can express the virtual work of all the real and inertial forces of the complete system by

$$\delta W = \sum_{k=1}^{B} (\delta W_M + \delta W_F)^k = \delta\mathbf{q}^T(Q+Q^{\dagger} - \mathbf{I}\ddot{\mathbf{q}}) = 0 \tag{124}$$

where we have utilized the following definitions.

$$\mathbf{I} \equiv \sum_{k=1}^{B} (mU^T U+\bar{W}^T\bar{J}\bar{W})^k \tag{125}$$

$$Q^{\dagger} = \sum_{k=1}^{B} [-mU^T v+\bar{W}^T(\tau-\bar{J}\bar{V})]^k \tag{126}$$

$$Q = \sum_{k=1}^{B} (U^T X+\bar{W}^T\bar{M})^k = \sum_{k=1}^{B} (U^T X+W^T M)^k \tag{127}$$

B = number of rigid bodies present

It is worth noting that the definition of **I** in Eq. (125) implies that **I** is symmetric, and that the right-most expression just given for Q follows from Eqs. (102), (107) and (116).

We have set $\delta W = 0$, in Eq. (124) in accordance with the generalized version of the principle of virtual work which includes the inertial forces. This form of the principle is referred to as Lagrange's form of d'Alembert's principle [30,64].

When there are no constraints relating the generalized coordinates q_i, the rightmost factor in Eq. (124) must vanish, and we have the equations of motion in their final form

$$\mathbf{I}\ddot{\mathbf{q}} = Q + Q^{\dagger} \tag{128-a}$$

or

$$\sum_{j=1}^{F} I_{ij}\ddot{q}_j = Q_i + Q_i^{\dagger} \quad (i = 1,\dots F) \tag{128-b}$$

Now consider the case where the generalized velocities are related by N_t constraint equations of the form

$$\sum_{j=1} \alpha_{ij}\cdot\dot{q}_j + \beta_i = 0 \quad (i = 1,2,\dots N_t) \tag{129}$$

where α_{ij} and β_i may all be functions of the Lagrangian coordinates $_r$ and of the time t. These relationships may be non-integrable (nonholonomic) or derivable from holonomic constraints of the form

$$g_i(\psi_1,\ldots\psi_M,t) = 0 \tag{130}$$

In either case, the corresponding virtual displacements must satisfy (by definition) the equations

$$\sum_{j=1}^{F} \alpha_{ij}\ \delta q_j = 0 \tag{131}$$

Using the same argument which led to Eq. (85), we conclude that the generalized forces must be supplemented by the "constraint" forces

$$Q_j' = \sum_{i=1}^{N_t} \lambda_i\ \alpha_{ij} \qquad j = 1,2,\ldots F \tag{132-a}$$

or

$$Q' = \alpha^T\lambda \tag{132-b}$$

where the Lagrange multipliers $(\lambda_1,\ldots\lambda N_t)$ are unknown functions of time to be determined. Therefore the equations of motion (128) may be written in the more general form

$$Iq = Q+Q^{\dagger} + \alpha^T\lambda \tag{133-a}$$

or

$$\sum_{j=1}^{F} I_{ij}\ddot{q}_j = Q_i+Q_i^{\dagger} + \sum_{j=1}^{N_t} \alpha_{ji}\lambda_j, \quad (i = 1,\ldots F) \tag{133-b}$$

These F equations together with the N_t equations (129) suffice to find the unknowns $(q_1,..q_F, \lambda_1,\ldots,\lambda N_t)$.

From previous definitions, it may be verified that the terms $Q_i^{\dagger}$ are all quadratic forms in the generalized velocities $\dot{q}_r$, and therefore reflect the effects of the so-called centrifugal and Coriolis (inertial) forces. However, no use is made of this fact in the numerical methods discussed below.

The equations (133) form the basis of the author's general-purpose computer program DYMAC (DYnamics of MAChinery) [1,66 ,68], and its 3-Dimensional Version DYMAC-3D [127,128].

(b) Equations of Motion from Lagrange's Equations

The kinetic energy T of a system of B rigid bodies is given by

$$T = \frac{1}{2}\sum_{k=1}^{B} (m^k\ \dot{\mathbf{x}}^T\dot{\mathbf{x}} + \bar{\omega}^T\ \bar{J}\bar{\omega}) \tag{134}$$

where all terms have been defined above. Upon recalling that $\mathbf{x} \equiv \mathbf{U}\dot{\mathbf{q}}$ and $\bar{\omega}=\bar{\mathbf{W}}\dot{\mathbf{q}}$, it follows that

$$T = \frac{1}{2}\,\dot{q}^T I^k \dot{q} = \frac{1}{2}\sum_{i,j=1}^{F} I_{ij}\,\dot{q}_i\,\dot{q}_j \tag{135}$$

where the generalized inertia coefficients I_{ij} are precisely those defined by Eq. (125). Since the I_{ij} are functions of position only, they may be considered implicit functions of the generalized coordinates q_r; i.e.

$$I_{ij} = I_{ij}[\boldsymbol{\psi}(\mathbf{q})] \tag{136}$$

Until further notice we will assume that I_{ij} is known explicitly in terms of the generalized coordinates; i.e. $I_{ij} = I_{ij}(q)$, and that the q_r are all independent.

Now we may make use of the well known Lagrange Equations (of second type) for independent generalized coordinates*

$$\frac{d}{dt}\frac{\partial T}{\partial \dot{q}_r} - \frac{\partial T}{\partial q_r} = Q_r = Q_r^* - \frac{\partial V}{\partial q_r} \quad , \tag{137}$$

where $\mathbf{Q}_r$ is the generalized force corresponding to $\mathbf{q}_r$ and $\mathbf{Q^*}$ and $\mathbf{V}$ are as defined in Sec. 3-e.

Making use of Eq. (135) and the fact that $I_{ij} = I_{ji}$, we find that:

$$\frac{\partial T}{\partial \dot{q}_r} = \sum_{j=1}^{F} I_{rj}\,\dot{q}_j \tag{138}$$

and

$$\frac{\partial T}{\partial \dot{q}_r} = \sum_{i,j=1}^{F} \frac{1}{2}\, I_r^{ij}\,\dot{q}_i\dot{q}_j \tag{139}$$

where

$$I_r^{ij} \equiv \frac{\partial I_{ij}(q)}{\partial q_r} \tag{140}$$

Equation (138) leads to the further result

$$\frac{d}{dt}\left(\frac{\partial T}{\partial \dot{q}_r}\right) = \sum_{j=1}^{F} I_{rj}\ddot{q}_j + \sum_{j=1}^{F} \dot{I}_{rj}\dot{q}_j \tag{141}$$

---

*These equations may be found in any text on Analytical Mechanics or Variational Mechanics [29,30,40,46,61,102].

We may represent $\dot{I}_{rj}$ in the form

$$\dot{I}_{rj} = \sum_{i=1}^{F} \frac{\partial I_{rj}(q)}{\partial q_i} \dot{q}_i = \sum_{i=1}^{F} I_i^{rj} \dot{q}_i \tag{142}$$

Upon substitution of Eqs. (138)-(142) into Lagrange's Equation (137) we find

$$\sum_{j=1}^{F} I_{rj}\ddot{q}_j + \sum_{i=1}^{F} \sum_{j=1}^{F} G_r^{ij} \dot{q}_i\dot{q}_j = Q_r \quad (r=1,2,\ldots,F) \tag{143}$$

where

$$G_r^{ij} \equiv I_i^{rj} - \frac{1}{2} I_r^{ij} . \tag{144}$$

Equations (143) are a set of second-order differential equations of motion in the generalized coordinates. However, they can be cast in a somewhat more symmetric form by a rearrangement of the coefficients G_r^{ij}. Towards this end, we note that an interchange of the dummy subscripts i and j on the left hand side of the equation

$$\sum_{i,j=1}^{F} I_i^{rj} \dot{q}_i\dot{q}_j \equiv \sum_{i=1}^{F} \sum_{j=1}^{F} I_j^{ri} \dot{q}_j\dot{q}_i \tag{145}$$

results identically in the right hand side. Hence, the double sum in Eq. (143) can be expressed in the form

$$\sum_{i=1}^{F} \sum_{j=1}^{F} G_r^{ij} \dot{q}_i\dot{q}_j = \sum_{i=1}^{F} \sum_{j=1}^{F} [{}^{ij}_{r}] \dot{q}_i\dot{q}_j \tag{146}$$

where

$$[{}^{ij}_{r}] \equiv \frac{1}{2}(I_i^{rj} + I_j^{ri} - I_r^{ij}) \tag{147}$$

is the <u>Christoffel Symbol of the first kind</u> associated with the matrix of inertia coefficients I_{ij}.

Upon substitution of Eq. (146) into Eq. (143) we find

$$\sum_{j=1}^{F} I_{rj}\ddot{q}_j + \sum_{i=1}^{F} \sum_{j=1}^{F} [{}^{ij}_{r}] \dot{q}_i\dot{q}_j = Q_r . \tag{148}$$

These equations are called the "explicit form of Lagrange's equations" by Whitaker [102].

Although they are a particularly good starting point for numerical integration of the equations of motion, these "explicit" equations have not been widely used in the engineering literature. They are mentioned by Beyer [6] who found it necessary to evaluate the coefficients I_{rj} semi-graphically, and to evaluate coefficients of the type I_r^{ij} by an approximate finite difference formula.

Upon comparing Eqs. (148) and (128), we see that the terms $Q_r^\dagger$ defined by Eq. (126) play precisely the same role, in the equations of motion, as the double sum in Eq. (148). The great advantage of the form involving $Q_r^\dagger$ is that these terms can be found without knowing explicit expressions for I_{ij} in terms of q_i, as would be required for the calculation of the Christoffel symbols via Eq. (140). In order to integrate Eqs. (148) numerically, it is necessary to solve them explicitly for the accelerations $\ddot{q}_i$. This may be done by any elimination procedure or by inversion of the matrix $[I_{rs}]$.

If we denote by J_{pr} the elements of the matrix $[I_{pr}]^{-1}$, we have

$$\sum_{r=1}^{F} J_{pr} I_{rj} \equiv \delta_{pj} \tag{149}$$

where δ_{pj} is 1 if p = j and is 0 otherwise. Upon multiplying Eq. (148) through by J_{pr} and summing over r we find [64]

$$\ddot{q}_r = \sum_{r=1}^{F} J_{pr} Q_r - \sum_{i=1}^{F} \sum_{j=1}^{F} \{{}^{p}_{ij}\} \dot{q}_i \dot{q}_j \tag{150}$$

where

$$\{{}^{p}_{ij}\} \equiv \sum_{r=1}^{F} J_{pr} [{}^{ij}_{r}] \tag{151}$$

is the Christoffel symbol of second kind [90].

When the generalized velocities are related by constraints of the form of Eq. (129) Lagrange multipliers may be introduced to generate the supplemental forces of Eq. (132). Then it is only necessary to add the terms

$$Q'_r = \sum_{i=1}^{N_t} \lambda_i \alpha_{ir} \tag{152}$$

to the right hand side of Eq. (148).

Lagrange's Equations (with multipliers) are used in several general-purpose computer programs, such as DRAM [12,85], ADAMS (58,60], DADS [32,101], AAPD [47] and IMP [81,82], IMPUM [81]. Lagrange multipliers are not necessary when treating open-loop systems such as robots [34,73,98] or other systems with tree-like topology as covered by the program UCIN [37,38].

(c) Kinetics of Single-Freedom Systems

For systems with one DOF there is a single generalized coordinate and we may write, for brevity,

$$q_1 \equiv q, \; Q_1 \equiv Q; \; Q_1^{\dagger} \equiv Q^{\dagger}; \quad I_{11} \equiv I \tag{153}$$

Then the equations of motion (128) reduce to the single differential equation

$$I\ddot{q} = Q + Q^{\dagger} \tag{154}$$

This result also follows directly from the principle of power balance, which states that the power of the active forces $(Q\dot{q})$ equals the rate of increase of the kinetic energy (T). From Eq. (135) we see that

$$T \equiv \frac{1}{2} I\dot{q}^2 \tag{155}$$

Therefore, the power balance principle is

$$\frac{dT}{dt} = \frac{d}{dt} \frac{1}{2}(I\dot{q}^2) = Q\dot{q} \tag{156}$$

$$I\dot{q}\ddot{q} + \frac{1}{2} \frac{dI}{dq} \frac{dq}{dt} \dot{q}^2 = Q\dot{q} \tag{157}$$

Hence,

$$\boxed{I\ddot{q} = Q - C\dot{q}^2} \tag{158}$$

where

$$C \equiv \frac{1}{2} \frac{dI}{dq} \tag{159}$$

Equation (158) is clearly of the form (154) with

$$Q^{\dagger} \equiv -C\dot{q}^2 \tag{160}$$

Using the simplified notation of Sec. 2-f, we can express I, via Eq. (125); in the form

$$I = \sum_{k=1}^{B} \sum_{i=1}^{3} (mu_i{}^2 + \bar{J}_i \, \bar{w}_i{}^2)^k \tag{161}$$

Therefore

$$C = \frac{1}{2} \frac{dI}{dq} = \sum_{k=1}^{B} \sum_{i=1}^{3} (mu_i u_i' + \bar{J}_i \bar{w}_i \bar{w}_i')^k \tag{162}$$

where

$$u_i' = \sum_{j=1}^{M} \frac{\partial u_i(\psi)}{\partial \psi_j} \frac{\partial \psi_j}{\partial q} = \sum_{j=1}^{M} \frac{\partial u_i(\psi)}{\partial \psi_j} c_j \tag{163}$$

$$\bar{w}_i' = \sum_{j=1}^{M} \frac{\partial \bar{w}_i(\psi)}{\partial \Psi_j} c_j \tag{164}$$

Similarly, the generalized force given by Eq. (127) can be expressed in the form

$$Q = \sum_{k=1}^{B} \sum_{i=1}^{3} (u_i X_i + w_i M_i)^k \tag{165}$$

For <u>planar motions</u> in which all particles move in paths parallel to the global (x,y) axes, two of the Euler angles shown in Fig. 2 vanish (say $\Theta = \Phi = 0$). It then follows from Eq. (105) that **D** is a unit matrix, and from Eq. (116) that there is no distinction between the angular velocity coefficients w_i and $\bar{w}_i$.

Numerous examples of the application of these equations to planar single-freedom systems are given in Chapter 12 of [66]. For related approaches to the use of the generalized equation of motion (158), see [21] and [84].

For conservative systems (see Sec. 3-e) with one DOF, where all generalized forces are derived from a potential function V(q), the generalized forces can be expressed in the form $Q = -dV/dq$. Therefore, the power balance theorem can be expressed in the form

$$\int_{T_o}^{T} dT = \int_{q_o}^{q} Q\, dq = - \int_{V_o}^{V} dV \tag{166}$$

where q_o is some reference state where the potential energy is V_o and the kinetic energy is T_o.

The integrated form of this equation is

$$T - T_o = -(V - V_o) \tag{167}$$

or

$$T + V = T_o + V_o \equiv E \tag{168}$$

where E (a constant) is the total energy of the system. Equation (168) is called the <u>energy integral</u> or the <u>first integral</u> of the equation of motion.

Recalling Eq. (155), we can write Eq. (168) in the form

$$\frac{dq}{dt} = \dot{q} = \pm \left[2 \frac{T}{I}\right]^{1/2} = \pm \left\{ \frac{2[E - V(q)]}{I(q)} \right\}^{1/2} \tag{169}$$

Finally, a relation can be found between t and q by means of a simple quadrature (e.g. by Simpson's rule) of Eq. (169) written in the form $dt = dq/\dot{q}$; i.e.

$$t - t_o = \int_{q_o}^{q} \frac{dq}{\dot{q}} \int_{q_o}^{q} \left[\frac{I(q)}{2[E - V(q)]} \right]^{1/2} dq \tag{170}$$

where t_o is the time at which $q = q_o$. The correct choice of + or - in Eqs. (168) and (170) is that which makes $\dot{q}$ consistent with the given "initial" velocity at $t = t_o$; thereafter a reversal of sign occurs whenever $\dot{q}$ becomes zero. Examples of the use of these equations are given in Sec. 12.50 of [66], where it is shown, among other results, that a stable oscillatory motion can occur in the neighborhood of equilibrium points (where $dV/dq = 0$) only if V is a local minimum at this point.

Many slightly different versions of the method of power balance have been published [3,21,28,33,74]. A brief review of the differences among these versions is given in [64].

(d) Kinetostatics and Internal Reactions

If all of the external forces acting on the links of a machine are given, the consequent motion can be found by integrating the generalized differential equations of motion (133), or any of their equivalent forms. This is the general problem of kinetics.

However, one is frequently required to find those forces which must act on a system or subsystem (e.g. a single link) in order to produce a given motion. This problem can be solved, as we shall see, by using the methods of statics, and is therefore called a problem of kinetostatics.

Suppose, for example, that the general problem of kinetics has already been solved for a given machine (typically by a numerical solution of the differential equations as explained in Sec. 5. This means that the positions, velocities, and accelerations of all points are known at a given instant of time, and the corresponding inertia force $-m\ddot{\mathbf{x}}$ is known. Similarly, one may now calculate the inertia torque $\bar{\mathbf{T}}$, whose components, referred to principal central axes, are given by Eqs. (109) or (118) for each link. We may then use the matrix $\mathbf{D}$ [defined by Eq. (105)] to express the components of inertia torque T_i along the global (x_i) axes as follows:

$$\mathbf{T} = \mathbf{D}^T\bar{\mathbf{T}} \tag{171}$$

Furthermore, let $\mathbf{X}'$ and $\mathbf{M}'$ be the net force and moment (about the mass center) exerted on the link by all the other links in contact with the subject link. For example, if the subject link (k) has its mass center at $\mathbf{x}^k$ and it is in contact with link n which exerts the force $\mathbf{X}'^{kn}$ at point $\mathbf{x}^{kn}$, we can write

$$\mathbf{X}'^{k} = \sum_n \mathbf{X}'^{kn} \tag{172}$$

$$\mathbf{M}'^{k} = \sum_n [\mathbf{M}'^{kn} + (\mathbf{x}^n - \mathbf{x}^k) \times \mathbf{X}'^{kn}] \tag{173}$$

where the summation is made over all links which contact link k. Using this notation, we may write, for each link k, the following vector equations of "dynamic equilibrium" for forces and moments respectively.

$$\mathbf{X}^k - m^k \ddot{\mathbf{x}}^k + \sum_n \mathbf{X}'^{kn} = \mathbf{0} \tag{174-a}$$

$$\mathbf{M}^k + \mathbf{T}^k + \sum_n [\mathbf{M}'^{kn} + (\mathbf{x}^n - \mathbf{x}^k) \times \mathbf{X}'^{kn}] = \mathbf{0} \tag{174-b}$$

Two such vector equations of kinetostatics can be written for each body in the system (k = 1,2,...B), leading to a set of 6B scalar equations which are linear in the unknown reaction components $X_i'^{kn}$ and $M_i'^{kn}$. If one expresses Newton's law of action and reaction in the form

$$X_i'^{kn} = - X_i'^{nk}; \quad M_i'^{kn} = - M_i'^{nk}$$

there will be a sufficient number of linear equations to solve for all the unknown reaction force and moment components [1,64].

When temporal constraints are present, the applied forces and moments ($\mathbf{X}^k$ and $\mathbf{M}^k$) are not all known a-priori, and some of the scalar quantities X_i^k, M_i^k (or some linear combination of them) must be treated as unknowns in Eqs. (174-a,b). For example, if a motor exerts an unknown torque on link number 7 about the local axis ξ_2^7 the required control torque $\bar{T}_2^7$ is unknown a-priori and the following terms will appear in Eqs. (174-b) as unknown contributions to M^7:

$$\begin{vmatrix} \Delta M_1^7 \\ \Delta M_2^7 \\ \Delta M_3^7 \end{vmatrix} = \begin{vmatrix} D_{11}^7 & D_{21}^7 & D_{31}^7 \\ D_{12}^7 & D_{22}^7 & D_{32}^7 \\ D_{13}^7 & D_{23}^7 & D_{33}^7 \end{vmatrix} \begin{vmatrix} 0 \\ \bar{T}_2^7 \\ 0 \end{vmatrix} = \begin{vmatrix} D_{21}\bar{T}_2^7 \\ D_{22}\bar{T}_2^7 \\ D_{23}\bar{T}_2^7 \end{vmatrix}$$

A systematic procedure for the automatic generation and solution of the appropriate linear equations, when such control forces must be accounted for, is given in [1].

(e) Balancing of Machinery

From elementary mechanics, it is known that the net force acting on a system of particles with fixed total mass equals the total mass times the acceleration of the center of mass. If this force is provided by the foundation of a machine system, the equal but opposite force felt by the foundation is called the shaking force. For a system of rigid bodies, the shaking force is precisely the resultant inertia force $\mathbf{F} = -\Sigma m^k \ddot{\mathbf{x}}^k$.

Similarly, a net shaking couple $\mathbf{T}$ is transmitted to the foundation where $\mathbf{T}$ is the resultant of all the inertia torques defined by Eqs. (109) for an individual link.

The problem of machine balancing is usually posed as one of kinetostatics, wherein the acceleration of every mass particle in the system is given (usually in the steady state condition of a machine), and one seeks to determine the size and location of balancing weights which will reduce $|\mathbf{F}|$ and $|\mathbf{T}|$ to acceptable levels. The subject is usually discussed under the two categories of rotor balancing and linkage balancing. Because of the large body of literature on these subjects, we shall only present a brief qualitative discussion of each, which follows closely that in [67].

Rotor Balancing [8,44,45,48,54,66,67,104] A rotating shaft supported by coaxial bearings, together with any attached mass (e.g. a turbine disk or motor armature), is called a rotor. If the center of mass (CM) of a rotor is not located exactly on the bearing axis, a net centrifugal force will be transmitted, via the bearings, to the foundation. The horizontal and vertical components of this periodic shaking force can travel through the foundation to create serious vibration problems in neighboring components. The magnitude of the shaking force is $F = me\omega^2$, where m = rotor mass, ω = angular speed, e = distance (called eccentricity) of the CM from the rotation axis. The product me, called the unbalance, depends only on the mass distribution, and is usually nonzero because of unavoidable manufacturing tolerances, thermal distortion, etc. When the CM lies exactly on the axis of rotation ($e=0$), no net shaking force occurs and the rotor is said to be in static balance. Static balance is achieved in practice by adding or subtracting balancing weights at any convenient radius until the rotor shows no tendency to turn about its axis, starting from any given initial orientation. Static balancing is adequate for relatively thin disks or short rotors (e.g. automobile wheels). However, a long rotor (e.g. in a turbogenerator) may be in static balance and still exert considerable forces upon individual bearing supports. For example, suppose that a long shaft is supported on bearings which are coaxial with the ζ axis, where ξ,η,ζ are body-fixed axes through the CM of a rotor, and ζ coincides with the fixed (global) z axis. If the body rotates with constant angular speed $|\omega|\equiv\omega_3$ about the ζ axis, the Euler equations (see [94] or Sec. 13.10 of [66]) show that the inertia torque about the rotating ξ and η axes are

$$\bar{T}_1 = \bar{I}_{\zeta\eta}\omega^2$$

$$\bar{T}_2 = -\bar{I}_{\zeta\xi}\omega^2$$

$$\bar{T}_3 = 0$$

where $\bar{I}_{\zeta\eta}$ and $\bar{I}_{\zeta\xi}$ are the (constant) products of inertia referred to the (ξ,η,ζ) axes. From these equations it is clear that the resultant inertia torque rotates with the shaft and produces harmonically fluctuating shaking forces on the foundation. However, if $\bar{I}_{\zeta\eta}$ and $\bar{I}_{\zeta\xi}$ can be made to vanish, the shaking couple will likewise vanish. It is shown in the references cited that any <u>rigid</u> shaft may be <u>dynamically balanced</u> (i.e. the net shaking force <u>and</u> the shaking couple can be simultaneously eliminated) by adding or subtracting definite amounts of mass at any convenient radius in each of two arbitrary transverse cross-sections of the rotor. The so-called <u>balancing planes</u> selected for this purpose are usually located near the ends of the rotor where suitable shoulders or <u>balancing rings</u> have been machined to permit the convenient addition of mass (e.g. lead weights, calibrated bolts, etc.) or the removal of mass (e.g. by drilling or grinding). Long rotors, running at high speeds, may undergo appreciable deformations. For such <u>flexible rotors</u> it is necessary to utilize more than two balancing planes.

<u>Balancing Machines and Procedures</u>. The most common types of rotor <u>balancing machines</u> consist of bearings held in pedestals which support the rotor, which is spun at constant speed (e.g. by a belt drive, or compressed air stream). Electromechanical transducers sense the unbalance forces (or associated vibrations) transmitted to the pedestals and electric circuits automatically perform the calculations necessary to predict the location and amount of balancing weight to be added or subtracted in preselected balancing planes.

For very large rotors, or for rotors which drive several auxiliary devices, commercial balancing machies may not be convenient. The <u>field balancing</u> procedures for such installations may involve the use of accelerometers on the bearing housings, along with vibration meters and phase discriminators (possibly utilizing stroboscopy) to determine the proper location and amount of balance weight. For an example of modern techniques in field balancing of flexible shafts, see [26].

Committees of the International Standards Organization (ISO) and the American National Standards Institute (ANSI) have formulated recommendations for the allowable quality grade $G = e\omega$ for various classes of machines[54].

Balancing of Linkage Machinery. The literature in this field pertaining to the slider-crank mechanism, is huge, because of its relevance to the ubiquitous internal combustion engine[8,14,17,35, 39,43,66,79,91,94]. For this relatively simple mechanism (even with multiple cranks attached to the same crankshaft), the theory is well understood and readily applied in practice. However, for more general linkages, the theory and reduction to practice, has not yet been as thoroughly developed. The state of the field up to the nineteen sixties has been reviewed periodically by Lowen, Tepper and Berkof [4,49,50], and practical techniques for balancing the shaking forces in four-bar and six-bar mechanisms have been described [5,66, 88,95,96,99].

It should be noted that the addition of the balancing weights needed to achieve net force balance will tend to increase the shaking moment, bearing forces, and required input torque. The size of the balancing weights can be reduced if one is willing to accept a partial force balance [92,102]. Because of the many variables involved in the linkage balancing problem, a number of investigations of the problem have been made from the point of view of optimization theory [13,96,103].

(f) Effects of Flexibility and Joint Clearance

In the above analysis, it has been assumed that all of the links are perfectly rigid, and that no clearance (backlash) exists at the joints. A great deal of literature has appeared on the influence of both of these departures from the commonly assumed ideal conditions. In view of the extent of this literature, we will point specifically only to surveys of the topics and to recent representative papers which contain additional references.

Publications on link flexibility (up to 1971) have been reviewed by Erdman and Sandor [23], and more recent work in this area has been described in the papers of Bhagat and Wilmert [7], and

Nath and Ghosh [55,56], on the finite element method of analyzing link flexibility. A straightforward, practical, and accurate approach to this problem has been described by Amin [1], who showed how to model flexible links so that they may be properly included in a dynamic situation by the general-purpose computer program DYMAC [68].

A number of papers dealing with the effects of looseness and compliance of joints have been published by Dubowsky, et al [19], and the entire field has been surveyed by Haines [31]. A later discussion of the topic was given by Ehle and Haug [20].

(g) Survey of General-Purpose Computer Programs for Dynamics of Machinery

Several general-purpose computer programs for the dynamic analysis of mechanisms have been described in the literature. These programs, which automatically generate and numerically integrate the equations of motion, are based on different, but related, analytical and numerical principles. The author's paper [64] reviews the various principles and techniques available for formulating the equations of motion, for integrating them numerically, and for solving the associated kinetostatic problem for the determination of bearing reactions. The relative advantages of vector methods, d'Alembert's principle, Lagrange's equations with and without multipliers, Hamilton's equations, virtual work, and energy methods are discussed. Particular emphasis is placed on how well suited the various methods are to the automatic generation of the equations of motion and to the form and order of the corresponding systems of differential equations. Methods for solving both the general dynamics problem and the kinetostatic problem are reviewed, and the particular methods of implementation used in some available general-purpose computer programs and in other recent literature are described.

The major general-purpose programs available at the time of writing--and sources of information on each--are given in the following list (also see the list of Kaufman [42]).

IMP (Integrated Mechanisms Program) is described by Sheth and Uicker [82] and Sheth [81]. A version, called IMP-UM, was developed by Sheth at the University of Michigan (3D).*

DRAM (Dynamic Response of Articulated Machinery), described in general terms by Chace and Sheth [11], is an updated version of DAMN, reported upon by Chace and Smith [12], and Smith et al [87] (2D).

*"3D" indicates the capability of analyzing three-dimensional mechanisms. "2D" denotes a restriction to planar mechanisms.

MEDUSA (Machine Dynamics Universal System Analyzer) is described by Dix and Lehman [18].

DYMAC (Dynamics of Machinery) [68], DYMAC-3D[127] developed by B. Paul [64,65,66], G. Hud [36], A. Amin [1], and R. Schaffa [128] at the University of Pennsylvania, (2D) and (3D).

STATMAC [70] (Statics of Machinery) is based upon the analysis given in Section 3, and uses essentially the same input format as DYMAC. It will solve for unknown configuration variables when forces are given (special case of the general dynamics problem with vanishing accelerations) or for unknown forces required for equilibrium when the configuration is given (special case of kinetostatics).

VECNET (Vector Network) has been described by Andrews and Kesavan [2] as "a research tool" for mass particles and spherical joints. It has spawned a second-generation program PLANET, for plane motion of rigid bodies; see Rogers and Andrews [76,77] (2D).

ADAMS (Automatic Dynamic Analysis of Mechanical Systems), developed at the University of Michigan by Orlandea et al [58,60] is described in Orlandea and Chace [58] (3D).

DADS (Dynamic Analysis and Design System) was developed at the University of Iowa by Haug et al [32], [57], [101] (3D).

The above-mentioned programs are all applicable to closed-loop systems, with sliding and rotating pairs, that typify linkage machinery. They can also be used for special cases where sliding pairs are not present or only open loops occur (e.g., in modeling human or animal bodies). Special-purpose programs suitable for such cases include the following:

AAPD (Automated Assembly Program--Displacement Analysis) a program for three-dimensional mechanisms, is described by Langrana and Bartel [47]. Only revolute and spherical pairs are allowed (3D).

UCIN, developed by Huston et al [37,38] at the University of Cincinnati, is for strictly open loop (chain-like) systems, e.g., the human body (3D).

5. NUMERICAL METHODS FOR SOLVING THE DIFFERENTIAL EQUATIONS OF MOTION

A system of n ordinary differential equations is said to be in standard form if the derivatives of the unknown functions $w_1, w_2, \ldots w_n$, with respect to the independent variable t, are given in the form

$$dw_i/dt = f_i(w_1, w_2, \ldots w_n, t) \quad (i = 1, \ldots, n). \tag{175}$$

If the initial values $w_1(0)$, $w_2(0)$,... are all given, we are faced with an initial value problem, in standard form, of order n.

Fortunately, a number of well-tested and efficient computer subprograms are available for the solution of the problem posed. One major class of such programs is based on the so-called "one-step" or "direct" methods such as the Runge-Kutta methods. These methods have the advantage of being self-starting, are free of iterative processes, and readily permit arbitrary spacing Δt between successive times where results are needed.

A second major class of numerical integration routines belong to the so-called "predictor-corrector" or "multi-step" methods. These are iterative procedures whose main advantage over the one-step procedures is that they permit more direct control over the size of the numerical errors developed. Most predictor-corrector methods suffer however from the drawback of starting difficulties, and an inability to change the step size Δt once started. However, "almost automatic" starting procedures for multistep methods have been devised [27], which use variable order integration rules for starting or changing interval size. For a review of the theory, and a FORTRAN listing, of the Runge-Kutta algorithm, see App. G of [66]. For examples of the use of other types of differential equation solvers, see [10] and [101].

(a) Determining Initial Values of Lagrangian Coordinates

We will assume that we are dealing with a system described by M Lagrangian coordinates $(\psi_1,\ldots\psi_M)$ which are related by spatial and temporal constraint equations of the form

$$f_i(\psi_1,\ldots\psi_M) = 0, \quad (i = 1,2,\ldots N_s) \tag{176}$$

$$g_k(\psi_1,\ldots\psi_M,t) = 0, (k = 1,2,\ldots N_t) \tag{177}$$

In order for this system of $(N_s + N_t)$ equations in M unknowns to have a solution at time $t = 0$, it is necessary that F_R of the ψ_i be given, where the reduced degree of freedom

$$F_R \equiv M - N_s - N_t$$

was previously defined in Equation (29). Assuming that F_R of the ψ_i are given at $t = 0$, we may solve the $N_s + N_t$ equations for the unknown initial coordinates, using a numerical method, such as the Newton-Raphson algorithm (see Chap. 9 and Appendix F of [66]). At this point, all values of ψ_i will be known.

(b) Initial Values of Lagrangian Velocities

When there are no temporal constraints ($N_t = 0$), we must specify the F independent initial velocities ($\dot{q}_1, \ldots \dot{q}_F$) at time t = 0. Then all of the Lagrangian velocities at t = 0, follow from Eq. (38) in the form

$$\dot{\psi}_j = \sum_{k=1}^{F} C_{jk} \, \dot{q}_k \qquad (j = 1, \ldots M) \tag{178}$$

When there exist N_t temporal relations of the form (171), we can express them in their rate form [see Eq. (71)].

$$\sum_{k=1}^{F} (\alpha_{jk} \, \dot{q}_k + \dot{\alpha}_{jk} q_k) + \dot{\beta}_j = 0 \qquad (j=1, \ldots N_t) \tag{179}$$

These temporal constraints permit us to specify only F_R ($=F-N_t$) generalized velocities, rather than F of them. With N_t values of $\dot{q}_k$ left unspecified, we may solve Eq. (178) and (179) for all of the ($M+N_t$) unknown values of $\dot{\psi}_i$ and $\dot{q}_j$. Thus, all values of both ψ_i and $\dot{\psi}_i$ are known at time t = 0, and the numerical integration can proceed.

(c) Numerical Integration

Let us introduce the notation

$$\{\psi_1, \psi_2, \ldots \psi_M, \dot{q}_1, \ldots \dot{q}_F\} \equiv \{w_1, w_2, \ldots w_M, w_{M+1}, \ldots w_{M+F}\} \tag{180}$$

Now we may rewrite the governing equations (178), (133), and (179), respectively as

$$\dot{w}_j = \sum_{k=1}^{F} C_{jk} \, w_{M+k}; \qquad (j = 1, 2, \ldots M) \tag{181}$$

$$\sum_{k=1}^{F} I_{jk} \, \dot{w}_{M+k} = Q_j + Q_j^{\dagger} + \sum_{m=1}^{N_t} \alpha_{mj} \, \lambda_m; \quad (j = 1, \ldots F) \tag{182}$$

$$\sum_{k=1}^{F} \alpha_{jk} \, \dot{w}_{M+k} = \nu_j; \quad (j = 1, \ldots N_t) \tag{183}$$

where

$$\nu_j \equiv -(\dot{\beta}_j + \sum_k \dot{\alpha}_{jk} \, \dot{q}_k)$$

$$\dot{\beta}_j = \frac{\partial \beta_j}{\partial t} + \sum_{k=1}^{M} \frac{\partial \beta_j}{\partial \psi_k} \dot{\psi}_k; \quad \dot{\alpha}_{jk} = \frac{\partial \alpha_{jk}}{\partial t} + \sum_{n=1}^{M} \frac{\partial \alpha_{jk}}{\partial \psi_n} \dot{\psi}_n$$

We now note that Q_j, $Q_j^{\dagger}$ and β_j are all known functions of ($w_1, \ldots w_{M+F}$), whenever the state ($\psi, \dot{\psi}$) is known (e.g. at t = 0). Then it is clear that we may solve the ($F + N_t$) Eqs. (182) and (183) for the unknowns ($\dot{w}_{M+1}, \ldots \dot{w}_{M+F}, \lambda_1, \ldots \lambda_{N_t}$). Then we may solve Eqs. (181)

for the M remaining velocities $(\dot{w}_1, \dot{w}_2, \ldots \dot{w}_M)$. In short, Eqs. (181) through (183) enable us to calculate all the M + F derivatives $\dot{w}_1$ from a knowledge of the w_i. That is, they constitute a system of M + F first order differential equations in standard form. Any of the well-known subroutines for such initial value problems may be utilized to match out the solution for all the ψ_i and $\dot{\psi}_i$, at user-specified time steps. The Lagrangian accelerations may be computed at any time step from Eq. (44), and the displacements, velocities, and accelerations for any point of interest may be found from Eqs. (2), (36) and (48).

REFERENCES

1 Amin, A., Automatic Formulation and Solution Techniques in Dynamics of Machinery, Ph.D. Dissertation, University of Pennsylvania, Philadelphia, PA, 1979.

2 Andrews, G. C. and Kesavan, H. K., "The Vector Network Model: A New Approach to Vector Dynamics," Mechanism and Machine Theory, vol. 10, 1975, pp. 57-75.

3 Benedict, C. E. and Tesar, D., "Dynamic Response Analysis of Quasi-rigid Mechanical Systems Using Kinematic Influence Coefficients" J. Mechanisms, vol. 6, 1971, pp. 383-403.

4 Berkof, R. S., Lowen, G. and Tepper, F. R., "Balancing of Linkages", Shock & Vibration Dig., vol. 9, no. 6, 1977, pp. 3-10.

5 Berkof, R. S. and Lowen, G., "A New Method for Completely Force Balancing Simple Linkages," J. Eng. Ind., Trans ASME, Ser. B, vol. 91, 1969, pp. 21-26.

6 Beyer, R., Kinematisch-getriebedynamisches Praktikum, Springer, Berlin, 1960.

7 Bhagat, B. M. and Wilmert, K. D., "Finite Element Vibration Analysis of Planar Mechanisms," Mechanism and Machine Theory, vol. 11, 1976, pp. 47-71.

8 Biezeno, C. B. and Grammel, R., Engineering Dynamics, Vol. IV, Internal-Combustion Engines, (trans. by M. P. White), Blackie, London, 1954.

9 Bocher, M., Introduction to Higher Algebra, Macmillan, New York, 1907, pp. 46,47.

10 Chace, M. A. and Bayazitoglu, Y. O.,"Development and Application of a Generalized d'Alembert Force for Multi-freedom Mechanical Systems," J. Eng. Ind., Trans ASME, Ser. B. vol. 93, 1971, pp. 317-327.

11 Chace, M. A. and Sheth, P. N., "Adaptation of Computer Techniques to the Design of Mechanical Dynamic Machinery," ASME Paper 73-DET-58, ASME, New York, 1973.

12 Chace, M. A. and Smith, D. A., "DAMN--Digital Computer Program for the Dynamic Analysis of Generalized Mechanical Systems," SAE Trans., vol. 80, 1971, pp. 969-983.

13 Conte, F. L., George, G. R., Mayne, R. W., and Sadler, J.P., "Optimum Mechanism Design Combining Kinematic and Dynamic Force Considerations," J. Eng. Ind., Trans. ASME, Ser. B., vol. 97, 1975, pp. 662-670.

14 Crandall, S. H., "Rotating and Reciprocating Machines", in Handbook of Engineering Mechanics, ed. by W. Flugge, McGraw-Hill, New York, 1962, Chap. 58.

15 Crandall, S. H., Karnopp, D. C., Kurtz, E.F.,Jr. and Pridmore-Brown, D. C., Dynamics of Mechanical and Electromechanical Systems, McGraw-Hill Publishing Co., New York, 1968.

16 Denavit, J., Hartenberg, R. S., Razi, R., and Uicker, J. J., Jr., "Velocity, Acceleration and Static-Force Analysis of Spatial Linkages, J. Appl. Mech., 32, Trans. ASME, Ser. E., vol. 87, 1965, pp. 903-910.

17 Den Hartog, J. P., Mechanical Vibrations, McGraw-Hill, New York, 1956.

18 Dix, R. C. and Lehman, T. J., "Simulation of the Dynamics of Machinery," J. Eng. Ind., Trans. ASME, Ser. B., vol. 94, 1972, pp. 433-438.

19 Dubowsky, S. and Gardner, T. N., Design and Analysis of Multilink Flexible Mechanisms with Multiple Clearance Connections, J. Eng. Ind., Trans. ASME, Ser. B., vol. 99, 1977, pp. 88-96.

20 Ehle, P. E. and Haug, E. J., "A Logical Function Method for Dynamic Design Sensitivity Analysis of Mechanical Systems with Intermittent Motion," J. Mechanical Design, Trans ASME, vol. 104, 1982, pp. 90-100.

21 Eksergian, R., "Dynamical Analysis of Machines", a series of 15 installments, appearing in J. Franklin Inst., vols. 209, 210, 211, 1930-1931.

22 Erdman, A. G. and Gustavson, J. E., "LINCAGES: Linkage Interaction Computer Analysis and Graphically Enhanced Synthesis Package," ASME Paper 77-DET-5, 1977.

23 Erdman, A. G. and Sandor, G. N., "Kineto-Elastodynamics--A Review of the Art and Trends," Mechanism and Machine Theory, vol. 7, 1972, pp. 19-33.

24 Freudenstein, F., "On the Variety of Motions Generated by Mechanisms," J. Eng. Ind., Trans. ASME, Ser. B, vol. 84, 1962, pp. 156-160.

25 Freudenstein, F. and Sandor, G. N., Kinematics of Mechanisms, Sec. 5 of "Mechanical Design and Systems Handbook," 1st ed., Ed. by H. A. Rothbart, McGraw-Hill Book Co., New York,1964.

26 Fujisaw, F. et al., "Experimental Investigation of Multi-Span Rotor Balancing Using Least Squares Method," J. Mech. Design, Trans. ASME, vol. 102, 1980, pp. 589-596.

27 Gear, C. W., Numerical Initial Value Problems in Ordinary Differential Equations, Prentice-Hall, Englewood Cliffs, NJ, 1971.

28 Givens, E. J. and Wolford, J. C., "Dynamic Characteristics of Spatial Mechanisms," J. Eng. Ind., Trans.ASME, Ser. B, vol. 91, 1969, pp. 228-234.

29 Goldstein, H., Classical Mechanics, 2nd Ed., Addison-Wesley Publishing Co., Reading, MA 1980.

30 Greenwood, D. T., Classical Dynamics, Prentice-Hall, Inc. Englewood Cliffs, NJ, 1977.

31 Haines, R. S., "Survey: 2-Dimensional Motion and Impact at Revolute Joints," Mechanism and Machine Theory, vol. 15, 1980, pp. 361-370.

32 Haug, E. J., Wehage, R. A. and Barman, N. C., "Dynamic Analysis and Design of Constrained Mechanical Systems," J. of Mech. Design, Trans. ASME, Vol. 104, 1982, 778-784.

33 Hirschhorn, J., Kinematics and Dynamics of Plane Mechanisms, McGraw-Hill, New York, 1962.

34 Hollerbach, J. M., "A Recursive Lagrangian Formulation of Manipulator Dynamics and a Comparative Study of Dynamics Formulation," IEEE Trans. on Systems, Man and Cybernetics, SMC-10,11, Nov. 1980, pp. 730-736.

35 Holowenko, A. R., Dynamics of Machinery, Wiley, New York, 1955.

36 Hud, G. C., Dynamics of Inertia Variant Machinery, Ph.D. Dissertation, University of Pennsylvania, Philadelphia, PA, 1976.

37 Huston, R. L., "Multibody Dynamics Including the Effects of Flexibility and Compliance,"Computers and Structures, vol. 14, 1981, pp. 443-451.

38 Huston, R. L., Passerello, C.E. and Harlow,M.W.,"Dynamics of Multirigid-Body Systems," J. of Appl. Mech., vol. 45, 1978, pp. 889-894.

39 Judge, A. W., Automobile and Aircraft Engines, Vol. 1, The Mechanics of Petrol and Diesel Engines, Pitman, New York, 1947.

40 Kane, T. R., Dynamics, Holt, Rinehart and Winston, NY, 1968.

41 Kane, T. R. and Wang, C. F.,"On the Derivation of Equations of Motion,"J.Soc. for Ind. and Appl. Math., vol. 13, 1965, pp. 487-492.

42 Kaufman, R. E., "Kinematic and Dynamic Design of Mechanisms," in Shock and Vibration Computer Programs Reviews and Summaries, ed by W. & B. Pilkey, Shock and Vibration Information Center, Naval Research Lab., Code 8404, Washington, DC, 1975.

43 Kožešnik, J., Dynamics of Machines, Noordhoff, Groningen, Netherlands, 1962.

44 Kroon, R. P., "Balancing of Rotating Apparatus--I," J. Appl. Mech, Trans. ASME, no. 10, vol. 65, 1943, pp. A225-A228.

45 Kroon, R. P., "Balancing of Rotating Apparatus--II," J.Appl. Mech, Trans. ASME, no. 11, vol. 66, 1944, pp. A47-A50.

46 Lanczos, C., The Variational Principles of Mechanics, 3rd ed. Univ. of Toronto Press, Toronto, 1966.

47 Langrana, N. A. and Bartel, D. L., "An Automated Method for Dynamic Analysis of Spatial Linkages for Biomechanical Applications," J. Eng. Ind., Trans. ASME, Ser. B, vol. 97, 1975, pp. 566-574.

48 Loewy, R. G. and Piarulli, V. J., Dynamics of Rotating Shafts, Shock and Vibration Information Center, NRL, Washington, DC, 1969.

49 Lowen, G. G. and Berkof, R. S., "Survey of Investigations into the Balancing of Linkages," J. Mechanisms, vol. 3, 1968, pp. 221-231.

50 Lowen, G. G., Tepper, F. R. and Berkof, R. S., "Balancing of Linkages--An Update," (to be published).

51 Mabie, H. H. and Ocvirk, F. W., Mechanisms and Dynamics of Machinery, Wiley, New York, 1975.

52 Martin, G. H., Kinematics and Dynamics of Machines, McGraw-Hill Book Co.. Inc., New York, 1982.

53 Mitchell, T., Mechanics, Sec. 4, "Mechanical Design and Systems Handbook," 1st ed., Ed. by H. A. Rothbart, McGraw-Hill Book Co., New York, 1964.

54 Muster, D. and Stadelbauer, D. G.,"Balancing of Rotating Machinery,"in Shock and Vibration Handbook, 2nd ed., by C. M. Harris and C. E. Crede, McGraw Hill, New York, 1976, Chap. 39.

55 Nath, P. K. and Ghosh, A., "Kineto-Elastodynamic Analysis of Mechanisms by Finite Element Method," Mechanism and Machine Theory, vol. 15, 1980, pp. 179-197.

56 Nath, P. K. and Ghosh, A., "Steady State Response of Mechanisms with Elastic Links by Finite Element Methods," Mechanisms and Machine Theory, vol. 15, 1980, pp. 199-211.

57 Nikravash, P. E. and Chung, I. S., "Application of Euler Parameters to the Dynamic Analysis of Three-Dimensional Constrained Systems," J. of Mechanical Design, Trans. ASME, vol. 104, 1982, 785-791.

58 Orlandea, N., Node-Analogous, Sparsity-Oriented Methods for Simulation of Mechanical Systems, Ph.D. Dissertation, Univ. of Michigan, Ann Arbor, MI, 1973.

59 Orlandea, N. and Chace, M. A., "Simulation of a Vehicle Suspension with the ADAMS Computer Program," SAE Paper No. 770053, Soc. of Autom. Engrs. Warrendale, PA, 1977.

60 Orlandea, N., Chace, M. A. and Calahan, D. A., "A Sparsity-Oriented Approach to the Dynamic Analysis and Design of Mechanical Systems--Parts I and II," J. Eng. Ind., Trans ASME, Ser. B, vol. 99, 1977, pp. 733-779, 780-784.

61 Pars, L. A., A Treatise on Analytical Dynamics, Wiley, New York, 1968.

62 Paul, B., "A Unified Criterion for the Degree of Constraint of Plane Kinematic Chains," J. Appl. Mech., vol. 27, Trans ASME, Vol. 82, Ser. E, 1960, pp.196-200. (See discussion same volume, pp.751-752).

63 Paul, B., "On the Composition of Finite Rotations," Amer. Math. Monthly, vol. 70, no. 8, 1963, pp. 859-862.

64 Paul, B., "Analytical Dynamics of Mechanisms--A Computer-Oriented Overview, Mechanism and Machine Theory, vol. 10, 1975, pp. 481-507.

65 Paul, B., "Dynamic Analysis of Machinery Via Program DYMAC", SAE Paper 770049, Soc. of Autom. Engrs., Warrendale, PA, 1977.

66 Paul, B., Kinematics and Dynamics of Planar Machinery, Prentice-Hall, Englewood Cliffs, NJ, 1979.

67 Paul, B., "Shaft Balancing," in Encyclopedia of Science and Technology, McGraw-Hill Book Co., New York, to be published.

68 Paul, B. and Amin, A., "User's Manual for Program DYMAC-G4 (DYnamics of MAChinery-General Version 4)," Available from B. Paul, University of Pennsylvania, Phila, PA, MEAM Dept., 1979.

69 Paul, B. and Amin, A., "User's Manual for Program KINMAC (KINematics of MAChinery)," Available from B. Paul, University of Pennsylvania, Phila., MEAM Dept., 1979.

70 Paul, B. and Amin, A., "User's Manual for Program STATMAC (STATics of MAChinery)," Available from B. Paul, University of Pennsylvania, Phila, PA, MEAM Dept., 1979.

71 Paul, B. and Krajcinovic, D., "Computer Analysis of Machines with Planar Motion--Part 1: Kinematics," J. Appl. Mech. 37, Trans ASME, Ser. E., vol. 92, 1970 , pp. 697-702.

72 Paul, B. and Krajcinovic, D., "Computer Analysis of Machines with Planar Motion--Part 2: Dynamics," J. Appl. Mech. 37, Trans ASME, Ser. E, vol. 92, 1970 , pp. 703-712.

73 Paul, R. P., Robot Manipulators: Mathematics, Programming, and Control, MIT Press, Cambridge, MA, 1981.

74 Quinn, B., "Energy Method for Determining Dynamic Characteristics of Mechanisms," J. Appl. Mech. 16, Trans ASME, vol. 71, 1949, pp. 283-288.

75 Reuleaux, F., The Kinematics of Machinery, (trans. and annotated by A.B.W. Kennedy), reprinted by Dover, NY 1963, original date, 1876.

76 Rogers, R. J. and Andrews, G. C., "Simulating Planar Systems Using a Simplified Vector-Network Method," Mechanism and Machine Theory, vol. 10, 1975, pp. 509-519.

77 Rogers, R. J. and Andrews, G. C., "Dynamic Simulation of Planar Mechanical Systems with Lubricated Bearing Clearances Using Vector Network Methods," J. Eng. Ind., Trans. ASME, Ser. B, 99, 1977, pp. 131-137.

78 Rongved, L. and Fletcher, H. J., "Rotational Coordinates," J. Franklin Institute, vol. 277, 1964, pp. 414-421.

79 Root, R. E., Dynamics of Engine and Shaft, Wiley, New York, 1932.

80 Rubel, A. J. and Kaufman, R. E., "KINSYN III: A New Human Engineered System for Interactive Computer-Aided Design of Planar Linkages," ASME Paper No. 76-DET-48, 1976.

81 Sheth, P. N., A Digital Computer Based Simulation Procedure for Multiple Degree of Freedom Mechanical Systems with Geometric Constraints, Ph.D. Thesis, Univ. of Wisconsin, Madison, Wisc., 1972.

82 Sheth, P. N. and Uicker, J. J. Jr., "IMP (Integrated Mechanisms Program), A Computer-Aided Design Analysis System for Mechanisms and Linkage," J. Eng. Ind., Trans. ASME, Ser. B. vol. 94, 1972, pp. 454-464.

83 Shigley, J. E. and Uicker, J. J., Jr., Theory of Machines and Mechanisms, McGraw-Hill Book Co., NY, 1980.

84 Skreiner, M., "Dynamic Analysis Used to Complete the Design of a Mechanism," J. Mechanisms, vol. 5, 1970, pp. 105-109.

85 Smith, D. A., Reaction Forces and Impact in Generalized Two-Dimensional Mechanical Dynamic Systems, Ph.D. Dissertation, Mech. Eng., Univ. of Michigan, Ann Arbor, MI, 1971.

86 Smith, D. A., "Reaction Force Analysis in Generalized Machine Systems," J. Eng. Ind., Trans. ASME, Ser. B., vol. 95, 1973, pp. 617-623.

87 Smith, D. A., Chace, M. A., and Rubens, A. C., "The Automatic Generation of a Mathematical Model for Machinery Systems," ASME Paper 72-Mech-31, ASME, New York, 1972.

88 Stevensen, E. N., Jr., "Balancing of Machines," J. Eng. Ind., Trans ASME, vol. 95, Ser. B, 1973, pp. 650-656.

89 Suh, C. H. and Radcliffe, C. W., Kinematics and Mechanisms Design, J. Wiley and Sons, New York, 1978.

90 Synge, J. L. and Schild, A., Tensor Analysis, Univ. of Toronto Press, Toronto, 1952.

91 Taylor, C. F., The Internal Combustion Engine in Theory and Practice, MIT Press, Cambridge, MA, Vol. I, 1966; Vol. II, 1968.

92 Tepper, F. R. and Lowen, G. G., "Shaking Force Optimization of Four-Bar Linkages with Adjustable Constraints on Ground Bearing Forces," J. Eng. Ind., Trans. ASME, Ser. B, vol. 97, 1975, pp. 643-651.

93 Thomson, W. (Lord Kelvin) and Tait, P. G., Treatise on Natural Philosophy, Part I, Cambridge Univ. Press, Cambridge, 1879, p. vi.

94 Timoshenko, S. P. and Young, D. H., Advanced Dynamics, McGraw-Hill, New York, 1948.

95 Tricamo, S. J. and Lowen, G. G., "A New Concept for Force Balancing Machines for Planar Linkages, Part I: Theory" J. Mech. Design, Trans. ASME, vol. 103, 1981, pp. 637-642.

96 Tricamo, S. J. and Lowen, G. G., "A New Concept for Force Balancing Machines for Planar Linkages, Part 2: Application to Four-Bar Linkages and Experiment," J. Mechanical Design, Trans. ASME, vol. 103, 1981, pp. 784-792.

97 Uicker, J. J. Jr., "Dynamic Force Analysis of Spatial Linkages," J. Appl. Mech, 34, Trans. ASME, Ser. E, 89, 1967, pp. 418-424.

98 Vukobratovic, M., Legged Locomotion Robots and Anthropomorphic Mechanics, Mihailo Pupin Institute, Belgrade, 1975.

99 Walker, M. J. and Oldham, K., "A General Theory of Force Balancing Using Counterweights," Mechanism and Machine Theory, vol. 13, 1978, pp. 175-185.

100 Webster, A. G., Dynamics of Particles and of Rigid Elastic and Fluid Bodies, Dover Publication reprint, New York, 1956.

101 Wehage, R. A. and Haug, E. J., "Generalized Coordinate Partitioning for Dimension Reduction in Analysis of Constrained Dynamic System," J. Mechanical Design, Trans. ASME, vol. 104, 1982, pp. 247-255.

102 Whittaker, E. T., A Treatise on the Analytical Dynamics of Particles and Rigid Bodies, 4th Ed., Dover Publications, New York, 1944.

103 Wiederrich, J. L. and Roth, B., "Momentum Balancing of Four-Bar Linkages," J. Eng. Ind., Trans. ASME, Ser. B, vol. 98, 1976, pp. 1289-1295.

104 Wilcox, J. B., Dynamic Balancing of Rotating Machinery, Pitman, London, 1967.

Added in Proof:

125 Benson, D., "Simulating Anelastic Mechanism Using Vector Processors," Comp. in Mech. Eng., vol. 1, no. 3, 1983, pp. 59-64.

126 Sunada, W. H. and Dubowsky, S., "The Application of Finite Element Methods to the Dynamic Analysis of Flexible Spatial and Co-Planar Linkage Systems," J. Mechanical Design, Trans. ASME, Vol. 103, 1981, 643-651.

127 Paul, B. and Schaffa, R., "User's Manual for DYMAC-3D", Available from B, Paul, MEAM Dept., Univ. of Penna., Phila, PA, 19104.

128 Schaffa, R., "Dynamics of Spatial Mechanisms and Robots", Ph.D. Dissertation, Univ. of Pa., Phila., PA 19104, To be published, 1984.

129 Rankers, H. "Synthesis of Mechanisms," Proceedings of NATO Advanced Study Institute on Computer-Aided Analysis and Optimization of Mechanical System Dynamics, University of Iowa. To be published.

ANALYTICAL METHODS IN MECHANICAL SYSTEM DYNAMICS

Jens Wittenburg
Institute of Mechanics
University at Karlsruhe
D-7500 Karlsruhe, FRG

Abstract. Exact nonlinear differential equations are developed for large motions of articulated rigid body systems such as mechanisms, vehicles, robots etc.. Parameters entering the equations are, among other quantities, the number of bodies, the number and location of joints interconnecting the bodies and the kinematical characteristics of the individual joints. The number of equations equals the total number of degrees of freedom of the system. The equations are developed in an explicit standard form. A computer program for the symbolic (i.e. non-numerical) generation of the equations is described. The user of this program has a free choice of generalized coordinates. The only input data required is a standard set of system parameters which includes the chosen coordinates.

1. INTRODUCTION

The subject of this lecture is the dynamics of articulated rigid body systems, briefly called multi body systems. Engineers are confronted with a large variety of such systems, e.g. vehicles, machine mechanisms, robots etc.. For dynamics simulations of large motions exact nonlinear differential equations must be formulated. The number of physical system parameters entering the equations is very large. The total number of degrees of freedom ranges from one (typical for many machine mechanisms) to rather large numbers. The equations will, therefore, be very complicated. For this reason efficient, systematic methods are required for generating the equations. The call for efficiency rules out that for every mechanical system to be investigated equations of motion are individually deduced from basic principles of mechanics. What is needed is a set of equations which is valid for a large class of systems. In spite of this generality the equations must be appli-

NATO ASI Series, Vol. F9
Computer Aided Analysis and Optimization of Mechanical System Dynamics
Edited by E. J. Haug

cable to any particular system with a minimum of preparatory work. From the user's point of view it is desirable to have a computer program which automatically generates the equations once the user has specified all system parameters. The term parameter must obviously be understood in a broad sense. Not only the masses, moments of inertia and dimensions of individual bodies in a system are parameters but also the interconnection structure of a system - which bodies are connected by which joints? - the location of joints on bodies, the kinematic constraints in joints etc..

It is indeed possible to formulate exact nonlinear equations of motion in such a programmable fashion. Various approaches can be found in the literature. Some of them are described in the present volume. Textbooks on the subject are scarce [1,2,3,4,5,6]. The present lecture introduces simplifications as well as generalizations not found in ref.[2] bei Wittenburg. For an extensive bibliography on multi body dynamics (methods and applications) see Wittenburg [7].

Specification of the Class of Systems Under Consideration

It has already been pointed out that the individual bodies of a multi body system are assumed to be rigid. This restriction rules out the application of the formalism to dynamics simulations where deformations of individual bodies are essential as is the case with highly flexible spacecraft, for example. However, not every system with deformable bodies is automatically excluded. To give an example, the human body in Fig.1 may well be considered as composed of rigid links if only its gross motion is of interest. To be sure, non-rigid elements such as nonlinear springs and dampers may be attached to the joints coupling individual bodies. Also actuators may be attached to joints which apply specified control torques or which prescribe the velocity of two coupled bodies relative to one another. No restrictions are imposed on the number of bodies, on the number of joints, on the number of degrees of freedom in individual joints as well as on the total number of degrees of freedom. This implies that systems with closed kinematic chains (Fig.1a) as well as systems with tree structure (Fig.1b,c) are considered. A system may be kinematically constrained to an external body whose location and angular orientation in space is prescribed as a function of time. Body 0 in Fig.1a,b, for example, can represent inertial space just as well as an accelerating vehicle. Such kinematical constraints to external bodies may also be absent (Fig.1c).

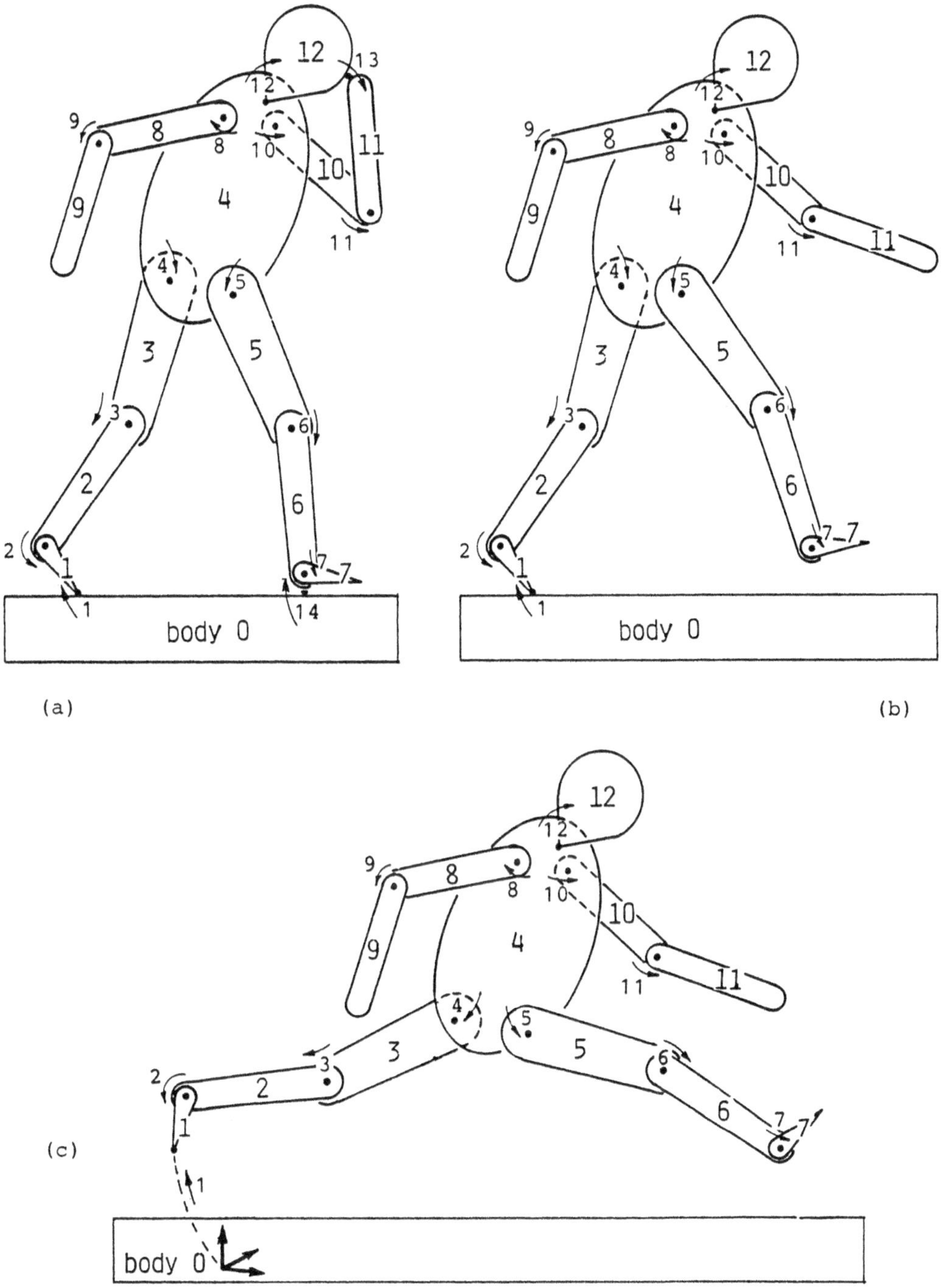

Fig.1: Multi body systems with closed kinematic chains (a) and with tree structure (b,c) which are either constrained (a,b) or not constrained (c) to an external moving body 0

The general case of spatial motion is considered and the dots representing joints in Fig.1 do not indicate that these joints are of some special type. The only restriction concerning joints is that all kinematic constraints are assumed to be ideal in the usual sense of analytical mechanics. This rules out Coulomb friction.

2. D'ALEMBERT'S PRINCIPLE

Starting point is d'Alembert's principle in the Lagrangian form

$$\int_m \delta\vec{r}\cdot(\ddot{\vec{r}}dm - d\vec{F}) = 0$$

(radius vector $\vec{r}$ of mass element dm, force $d\vec{F}$ on dm). It is useful to distinguish between external forces $d\vec{F}_e$ and internal forces. The latter ones are associated with springs, dampers and actuators in the joints between bodies. If δW represents the total virtual work of all internal forces then d'Alembert's principle for a system of n rigid bodies with masses m_i $(i=1...n)$ reads

$$\sum_{i=1}^{n} \int_{m_i} \delta\vec{r}\cdot(\ddot{\vec{r}}dm - d\vec{F}_e) - \delta W = 0.$$

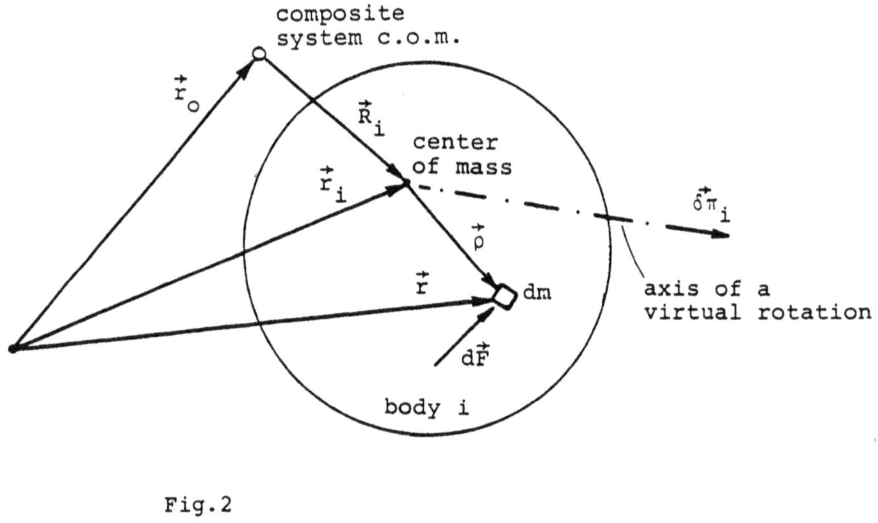

Fig.2

In Fig.2 $\vec{r}_i$ points to the body i center of mass and $\vec{\rho}$ is a vector fixed on body i. It follows that

$$\ddot{\vec{r}} = \ddot{\vec{r}}_i + \ddot{\vec{\rho}}, \qquad \delta\vec{r} = \delta\vec{r}_i + \delta\vec{\pi}_i \times \vec{\rho}$$

where $\delta\vec{\pi}_i$ is an infinitesimal rotation vector along the axis of a virtual rotation of body i. The absolute value of $\delta\vec{\pi}_i$ is equal to the infinitesimal rotation angle about this axis. Taking into account the relationship $\int_{m_i} \vec{\rho}\,dm = 0$ we obtain

$$\int_{m_i} \delta\vec{r}\cdot(\ddot{\vec{r}}dm - d\vec{F}_e) = \delta\vec{r}_i\cdot \int_{m_i}(\ddot{\vec{r}}_i dm - d\vec{F}_e) + \delta\vec{\pi}_i\cdot\int_{m_i} \vec{\rho}\times(\ddot{\vec{\rho}}dm - d\vec{F}_e)$$

$$= \delta\vec{r}_i\cdot(m_i\ddot{\vec{r}}_i - \vec{F}_i) + \delta\vec{\pi}_i\cdot(\mathrm{J}_i\cdot\dot{\vec{\omega}}_i + \vec{\omega}_i\times\mathrm{J}_i\cdot\vec{\omega}_i - \vec{M}_i)$$

and hence for the complete system

$$\sum_{i=1}^{n}\left[\delta\vec{r}_i\cdot(m_i\ddot{\vec{r}}_i - \vec{F}_i) + \delta\vec{\pi}_i\cdot(\mathrm{J}_i\cdot\dot{\vec{\omega}}_i + \vec{\omega}_i\times\mathrm{J}_i\cdot\vec{\omega}_i - \vec{M}_i)\right] - \delta W = 0. \tag{1}$$

Here, $\vec{F}_i$ and $\vec{M}_i$ represent the resultant external force and torque, respectively, on body i and J_i is the central inertia tensor.

If a system has no kinematical constraints to external bodies (c.f.Fig.1c) then it is possible to separate the equation of motion for the composite system center of mass thereby reducing the number of the remaining equations. For this purpose we write in Eq.(1)

$$\ddot{\vec{r}}_i = \ddot{\vec{r}}_o + \ddot{\vec{R}}_i, \qquad \delta\vec{r}_i = \delta\vec{r}_o + \delta\vec{R}_i \qquad i=1...n.$$

According to Fig.2 $\vec{r}_o$ is the radius vector to the system center of of mass and $\vec{R}_i$ is the vector leading from there to the body i center of mass. Taking into account that $\sum_{i=1}^{n} m_i\vec{R}_i = \vec{0}$ we obtain with $M = \sum_{j=1}^{n} m_j$

$$\delta\vec{r}_o\cdot(M\ddot{\vec{r}}_o - \sum_{j=1}^{n}\vec{F}_j) +$$

$$+\sum_{i=1}^{n}\left[\delta\vec{R}_i\cdot(m_i\ddot{\vec{R}}_i + m_i\ddot{\vec{r}}_o - \vec{F}_i) + \delta\vec{\pi}_i\cdot(\mathrm{J}_i\cdot\dot{\vec{\omega}}_i + \vec{\omega}_i\times\mathrm{J}_i\cdot\vec{\omega}_i - \vec{M}_i)\right] - \delta W = 0.$$

By assumption $\delta\vec{r}_o$ is independent so that the associated factor must be zero. This yields Newton's law for the composite system center of mass

$$\ddot{\vec{r}}_o = \frac{1}{M}\sum_{j=1}^{n}\vec{F}_j .$$

With this expression the remaining parts of the equation become

$$\sum_{i=1}^{n}\left[\delta\vec{R}_i\cdot(m_i\ddot{\vec{R}}_i - \sum_{j=1}^{n}\mu_{ij}\vec{F}_j) + \delta\vec{\pi}_i\cdot(\mathrm{J}_i\cdot\dot{\vec{\omega}}_i + \vec{\omega}_i\times\mathrm{J}_i\cdot\vec{\omega}_i - \vec{M}_i)\right] \quad \delta W = 0. \quad (2)$$

$$\mu_{ij} = \delta_{ij} - m_i/M$$

For subsequent reformulations Eqs.(1) and (2) are expressed in matrix form:

$$\delta\underline{\vec{r}}^T\cdot(\underline{m}\ddot{\underline{\vec{r}}} - \underline{\vec{F}}) + \underline{\delta\vec{\pi}}^T\cdot(\underline{\mathrm{J}}\cdot\dot{\underline{\vec{\omega}}} + \underline{\vec{V}} - \underline{\vec{M}}) - \delta W = 0 \quad (3)$$

$$\delta\underline{\vec{R}}^T\cdot(\underline{m}\ddot{\underline{\vec{R}}} - \underline{\mu}\underline{\vec{F}}) + \underline{\delta\vec{\pi}}^T\cdot(\underline{\mathrm{J}}\cdot\dot{\underline{\vec{\omega}}} + \underline{\vec{V}} - \underline{\vec{M}}) - \delta W = 0\ . \quad (4)$$

Here, $\underline{m}$ and $\underline{\mathrm{J}}$ are diagonal matrices with scalar elements $m_i\delta_{ij}$ and tensor elements $\mathrm{J}_i\delta_{ij}$, respectively. $\delta\underline{\vec{r}}$, $\delta\underline{\vec{R}}$, $\ddot{\underline{\vec{r}}}$, $\ddot{\underline{\vec{R}}}$, $\underline{\delta\vec{\pi}}$, $\dot{\underline{\vec{\omega}}}$, $\underline{\vec{F}}$, $\underline{\vec{M}}$ and $\underline{\vec{V}}$ are column matrices with vectors as elements, for example $\underline{\vec{F}} = [\vec{F}_1\vec{F}_2\ldots\vec{F}_n]^T$. The i-th element of $\underline{\vec{V}}$ is $\vec{\omega}_i\times\mathrm{J}_i\cdot\vec{\omega}_i$ and $\underline{\mu}$ is the matrix of the elements μ_{ij} defined by Eq.(2). The exponent T indicates transposition.

Vectors and tensors will be handled in symbolic form throughout the entire development. This explains the multiplication symbols $\cdot$ and $\times$ between matrices. The scalar decomposition of vectors and tensors will be carried out at the very end in a reference frame which is chosen in some optimal sense. The formal similarity of the two matrix equations is reflected in the relationship

$$\underline{\vec{R}} = \underline{\mu}^T\underline{\vec{r}}. \quad (5)$$

The dynamics part of the problem is now complete. What remains to be done is an analysis of systems kinematics.

Illustrative Example: The two-body system shown in Fig.3 serves to demonstrate how simple it is to develop from either Eq.(1) or (2) scalar differential equations of motion for a specific example. The bodies labeled 1 and 2 are coupled by a massless crank- and-slider mechanism whose only purpose is to provide the kinematical constraints of a joint (the system does not have a closed kinematic chain since the links of the mechanism are not physical bodies). There are no kinematical constraints to external bodies so that Eq.(1) as well as (2) is applicable. Both equations will be used although it is clear from the outset that Eq.(2) is more appropriate.

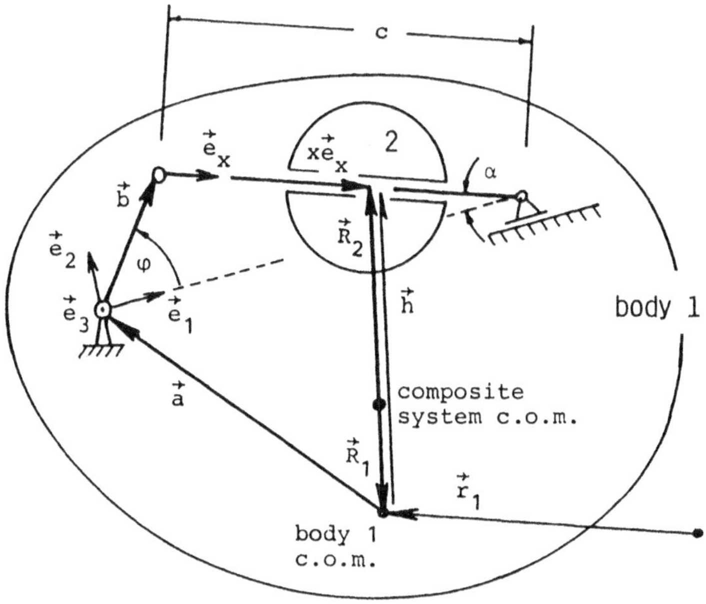

Fig.3: Two bodies 1 and 2 coupled by a massless crank-and-slider mechanism. $\vec{e}_{1,2,3}$ and $\vec{e}_x$ are unit vectors

As variables we use the scalar coordinates φ and x locating body 2 relative to body 1 and the vectors $\vec{r}_1$ and $\vec{\omega}_1$ for body 1. Simple trigonometry yields

$$\sin\alpha = \frac{b}{c}\sin\varphi, \qquad \dot{\alpha} = \dot{\varphi}\lambda \qquad \text{with} \qquad \lambda = \frac{b}{c}\cos\varphi\Big/\sqrt{1-\left(\frac{b}{c}\right)^2\sin^2\varphi} \;.$$

Using the vectors indicated in the figure we can write

$$\begin{aligned} \vec{\omega}_2 &= \vec{\omega}_1 - \dot{\alpha}\vec{e}_3 \\ \dot{\vec{\omega}}_2 &= \dot{\vec{\omega}}_1 - \ddot{\varphi}\lambda\vec{e}_3 + \ldots\text{(lower order terms)} \\ \delta\vec{\pi}_2 &= \delta\vec{\pi}_1 - \delta\varphi\lambda\vec{e}_3 \end{aligned} \tag{6}$$

and with $\vec{h}$ and $\vec{p}$ defined as

$$\left.\begin{aligned} \vec{h} &= \vec{a}+\vec{b}+x\vec{e}_x, \qquad \vec{p} = \vec{b}-\lambda x\vec{e}_x \\ \vec{r}_2 &= \vec{r}_1+\vec{h} \\ &\vdots \end{aligned}\right\} \tag{7}$$

$$\left.\begin{aligned}\ddot{\vec{r}}_2 &= \ddot{\vec{r}}_1 + \dot{\vec{\omega}}_1\times\vec{h} + \ddot{\varphi}\vec{e}_3\times\vec{p} + \ddot{x}\vec{e}_x + \ldots \\ \delta\vec{r}_2 &= \delta\vec{r}_1 + \delta\vec{\pi}_1\times\vec{h} + \delta\varphi\vec{e}_3\times\vec{p} + \delta x\vec{e}_x .\end{aligned}\right\}$$

In the figure the composite system center of mass is indicated from which the vectors $\vec{R}_1$ and $\vec{R}_2$ are measured. Obviously

$$\vec{R}_1 = \frac{-m_2}{m_1+m_2}\vec{h}, \qquad \vec{R}_2 = \frac{m_1}{m_1+m_2}\vec{h}. \tag{8}$$

The expressions for $\ddot{\vec{h}}$ and $\delta\vec{h}$ are copied from Eq.(7). These few expressions constitute the entire kinematics section. First, we substitute Eqs.(6) and (7) into Eq.(1). This leads to a linear combination of the form

$$\delta\vec{r}_1\cdot(\ldots) + \delta\vec{\pi}_1\cdot(\ldots) + \delta\varphi(\ldots) + \delta x(\ldots) = 0.$$

All four expressions in brackets - two vectors and two scalars - must be equal to zero. This results in two vector and two scalar differential equations of motion, i.e. in altogether eight scalar equations of motion for as many degrees of freedom. It is a simple matter to arrange the highest-order derivatives in matrix form so as to obtain the equations in the form ($\mathbb{E}$ is the unit dyadic)

$$\begin{bmatrix} m_1+m_2 & -m_2\vec{h}\times & m_2\vec{e}_3\times\vec{p} & m_2\vec{e}_x \\ m_2\vec{h}\times & [\mathbf{J}_1+\mathbf{J}_2+m_2(\vec{h}^2\mathbb{E}-\vec{h}\vec{h})]\cdot & m_2\vec{h}\times(\vec{e}_3\times\vec{p})-\lambda\mathbf{J}_2\cdot\vec{e}_3 & m_2\vec{h}\times\vec{e}_x \\ m_2\vec{e}_3\times\vec{p}\cdot & [m_2\vec{h}\times(\vec{e}_3\times\vec{p})-\lambda\vec{e}_3\cdot\mathbf{J}_2]\cdot & m_2(\vec{e}_3\times\vec{p})^2+\lambda^2\vec{e}_3\cdot\mathbf{J}_2\cdot\vec{e}_3 & m_2\vec{e}_x\cdot\vec{e}_3\times\vec{p} \\ m_2\vec{e}_x\cdot & m_2\vec{h}\times\vec{e}_x\cdot & m_2\vec{e}_x\cdot\vec{e}_3\times\vec{p} & m_2 \end{bmatrix} \begin{bmatrix} \ddot{\vec{r}}_1 \\ \dot{\vec{\omega}}_1 \\ \ddot{\varphi} \\ \ddot{x} \end{bmatrix} = \begin{bmatrix} \cdot \\ \cdot \\ \cdot \\ \cdot \\ \cdot \\ \cdot \\ \cdot \\ \cdot \end{bmatrix} \tag{9}$$

The lower-order terms are indicated by dots. Once all vectors and tensors are decomposed in the reference base $\vec{e}_{1,2,3}$ on body 1 the resulting 8×8 coefficient matrix is symmetric.

Next, Eq.(2) is developed by substituting Eqs.(6) and (8). This results in equations for $\vec{\omega}$, φ and x. The coefficient matrix in front of $\dot{\vec{\omega}}$, $\ddot{\varphi}$ and $\ddot{x}$ turns out to be the same submatrix as in the previous case (shown by the dashed lines) provided m_2 is replaced everywhere by the reduced mass $m_1m_2/(m_1+m_2)$. Scalar decomposition yields a 5×5 coefficient matrix.

The equations of motion would be much more complicated if $\vec{\omega}_1$ had been expressed in terms of generalized coordinates. We learn from this example that absolute or relative angular velocities which are not kinematically constrained should always be kept as variables in the equations of motion. For each such angular velocity vector an additional set of kinematics differential equations must be formulated from which the angular orientation of the body is determined. For these equations Euler angles are a possible choice of coordinates. But they have the disadvantage that the kinematics differential equations become singular if the intermediate angle is zero. No such problems arise if Euler-Rodrigues parameters q_o, q_1, q_2 and q_3 are used. For their definition the reader is referred to Wittenburg [2] and to other lectures in the present volume. The angular orientation of body 1 in inertial space is specified by the direction cosine matrix $\underline{G}_1$ relating the body 1 reference base to an inertial reference base. In terms of $q_{o,1,2,3}$ this matrix reads

$$\underline{G}_1 = \begin{bmatrix} q_o^2+q_1^2-q_2^2-q_3^2 & 2(q_1q_2+q_oq_3) & 2(q_1q_3-q_oq_2) \\ 2(q_1q_2-q_oq_3) & q_o^2-q_1^2+q_2^2-q_3^2 & 2(q_2q_3+q_oq_1) \\ 2(q_1q_3+q_oq_2) & 2(q_2q_3-q_oq_1) & q_o^2-q_1^2-q_2^2+q_3^2 \end{bmatrix} \tag{10}$$

and the kinematics differential equations are

$$\begin{bmatrix} \dot{q}_o \\ \dot{q}_1 \\ \dot{q}_2 \\ \dot{q}_3 \end{bmatrix} = \frac{1}{2} \begin{bmatrix} 0 & -\omega_1 & -\omega_2 & -\omega_3 \\ \omega_1 & 0 & \omega_3 & -\omega_2 \\ \omega_2 & -\omega_3 & 0 & \omega_1 \\ \omega_3 & \omega_2 & -\omega_1 & 0 \end{bmatrix} \begin{bmatrix} q_o \\ q_1 \\ q_2 \\ q_3 \end{bmatrix}. \tag{11}$$

These equations have to be integrated simultaneously with Eq.(9) if vectors are involved whose components are given in inertial space.

Eq.(11) requires an additional comment. By definition the Euler-Rodrigues parameters satisfy the constraint equation

$$q_o^2+q_1^2+q_2^2+q_3^2 = 1.$$

Numerical solutions of Eq.(11) will not exactly satisfy this constraint. Therefore, at each time step it is necessary to update the numerical values so as to satisfy the constraint before $\underline{G}_1$ is calculated. Optimal updating procedures are discussed by Wittenburg [8].

3. SYSTEM KINEMATICS

Outline of the General Strategy

Suppose that for a multi body system composed of n bodies and with a total number N of degrees of freedom N generalized coordinates $q_1 \ldots q_N$ have been chosen so that for a given set of values $q_1 \ldots q_N$ and at a given time t the position of any point in the system is uniquely determined (the time appears explicitly if the motion of body 0 in Fig.1a,b is prescribed as a function of time and also if motors built into joints introduce prescribed functions of time). Among the quantities determined by $q_1 \ldots q_N$ and t are, in particular, the radius vectors $\vec{r}_i$ of all body centers of mass, the absolute angular velocities $\vec{\omega}_i$ of the bodies and for each joint j a direction cosine matrix $\underline{G}_j$ relating two frames of reference, each fixed on one of the two coupled bodies:

$$\vec{r}_i(q_1 \ldots q_N, t), \qquad \vec{\omega}_i(q_1 \ldots q_N, t), \qquad \underline{G}_j(q_1 \ldots q_N, t).$$

In the dynamics Eq.(3) the quantities $\delta\vec{\underline{r}}$ and $\ddot{\vec{\underline{r}}}$ must be expressed in terms of the coordinates. Differentiating $\vec{r}_i$ once and twice we find

$$\delta\vec{\underline{r}} = \vec{\underline{\alpha}}\delta\underline{q}, \qquad \ddot{\vec{\underline{r}}} = \vec{\underline{\alpha}}\ddot{\underline{q}}+\vec{\underline{u}} \tag{12}$$

where $\vec{\underline{\alpha}}$ is some as yet unknown n×N matrix with vectors as elements and $\vec{\underline{u}}$ is a column matrix of N vectors containing zero and first-order derivatives of $q_1 \ldots q_N$. For Eq.(4) corresponding expressions are required for $\delta\vec{\underline{R}}$ and $\ddot{\vec{\underline{R}}}$. They follow immediately from Eq.(5),

$$\delta\vec{\underline{R}} = \underline{\mu}^T\delta\vec{\underline{r}} = \underline{\mu}^T\vec{\underline{\alpha}}\delta\underline{q}, \qquad \ddot{\vec{\underline{R}}} = \underline{\mu}^T\ddot{\vec{\underline{r}}} = \underline{\mu}^T(\vec{\underline{\alpha}}\delta\underline{q}+\vec{\underline{u}}). \tag{13}$$

Similarly, $\delta\vec{\underline{\pi}}$ and $\dot{\vec{\underline{\omega}}}$ can be expressed in the form

$$\delta\vec{\underline{\pi}} = \vec{\underline{\beta}}\delta\underline{q} \qquad \dot{\vec{\underline{\omega}}} = \vec{\underline{\beta}}\ddot{\underline{q}}+\vec{\underline{v}} \tag{14}$$

with as yet unknown matrices $\vec{\underline{\beta}}$ and $\vec{\underline{v}}$. When the Eqs.(12) and (14) are substituted into Eq.(3) we obtain

$$\delta\underline{q}^T\Big[\underbrace{\left(\vec{\underline{\alpha}}^T\cdot\underline{m}\vec{\underline{\alpha}}+\vec{\underline{\beta}}^T\cdot\underline{\underline{J}}\cdot\vec{\underline{\beta}}\right)}_{\underline{A}}\ddot{\underline{q}} + \underbrace{\left(\vec{\underline{\alpha}}^T\cdot(\underline{m}\vec{\underline{u}}-\vec{\underline{F}}) + \vec{\underline{\beta}}^T\cdot(\underline{\underline{J}}\cdot\vec{\underline{v}}+\vec{\underline{V}}-\vec{\underline{M}})-\underline{Q}\right)}_{-\underline{B}}\Big] = 0 \tag{15}$$

and since the $\delta\underline{q}$ are independent

$$\underline{A}\ddot{\underline{q}} = \underline{B}. \tag{16}$$

These are the desired differential equations of motion. Equations of the same type are obtained for systems without external constraints by substituting Eqs.(13) and (14) into Eq.(4). The N×N matrix $\underline{A}$ has scalar elements. It is symmetric and positive definite. The column matrix $\underline{B}$ consists of N scalar elements which combine all lower order terms including the generalized forces $\underline{Q}$ contributed by the expression $\delta W = \delta \underline{q}^T \underline{Q}$. For carrying out the dot products in the expressions for $\underline{A}$ and $\underline{B}$ all vectors and tensors are decomposed in one common reference frame by means of the direction cosine matrices $\underline{G}_j$. This concludes the outline of the general strategy for formulating equations of motion.

Kinematics of Individual Joints

First the kinematics of a single joint is formulated. Let Fig.4 represent two bodies labeled k and s which are coupled by joint j. In what follows the rest of the system is thought to be non-existent. To each body a cartesian reference frame is rigidly attached. For unambiguity it must be decided which of the two bodies serves as reference body for the other body. Let this decision be indicated by the arrow in the figure. The body from which the arrow is pointing away is understood to be the reference body. The index of the reference body for joint j will be called $i^+(j)$ and the index of the other body $i^-(j)$. Thus, i^+ and i^- are the names of two integer functions with integer argument. In the example we have $i^+(j) = k$ and $i^-(j) = s$.

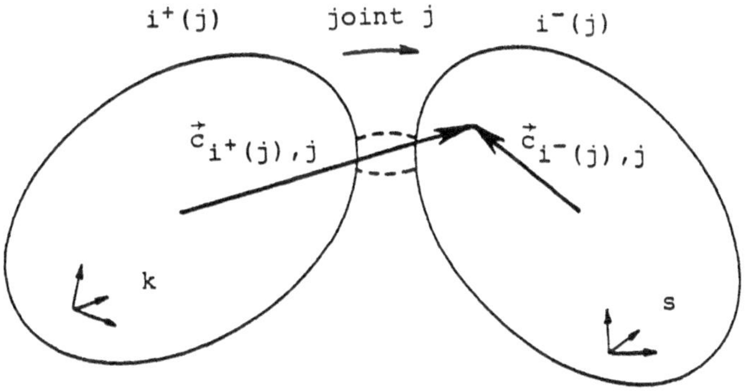

Fig.4: Two bodies k and s coupled by a joint j with unspecified kinematical characteristics

It is a problem of elementary mechanics to describe in terms of appropriately chosen variables the location, the angular orientation, the translational velocity and acceleration and, finally, the angular velocity and acceleration of body $i^-(j)$ relative to body $i^+(j)$.

Choice of variables: Let $1 \leq N_j \leq 6$ be the number of degrees of freedom in joint j. In Fig.5 four joints are shown which have 1, 2, 2 and 4 degrees of freedom, respectively. It is assumed that all constraints are scleronomic so that the time t does not appear explicitly in the expressions to be developed. Practically this means that built-in motors which prescribe relative positions as functions of time are excluded. Such constraints will be introduced later. Let $q_{j\ell}$ $(\ell=1\ldots N_j)$ be a set of generalized coordinates for joint j. The choice is completely arbitrary though, for most practical purposes it suffices to think of $q_{j\ell}$ as either angles of rotation about certain axes or line coordinates such as cartesian coordinates, arc length etc.. Examples are shown in the figure.

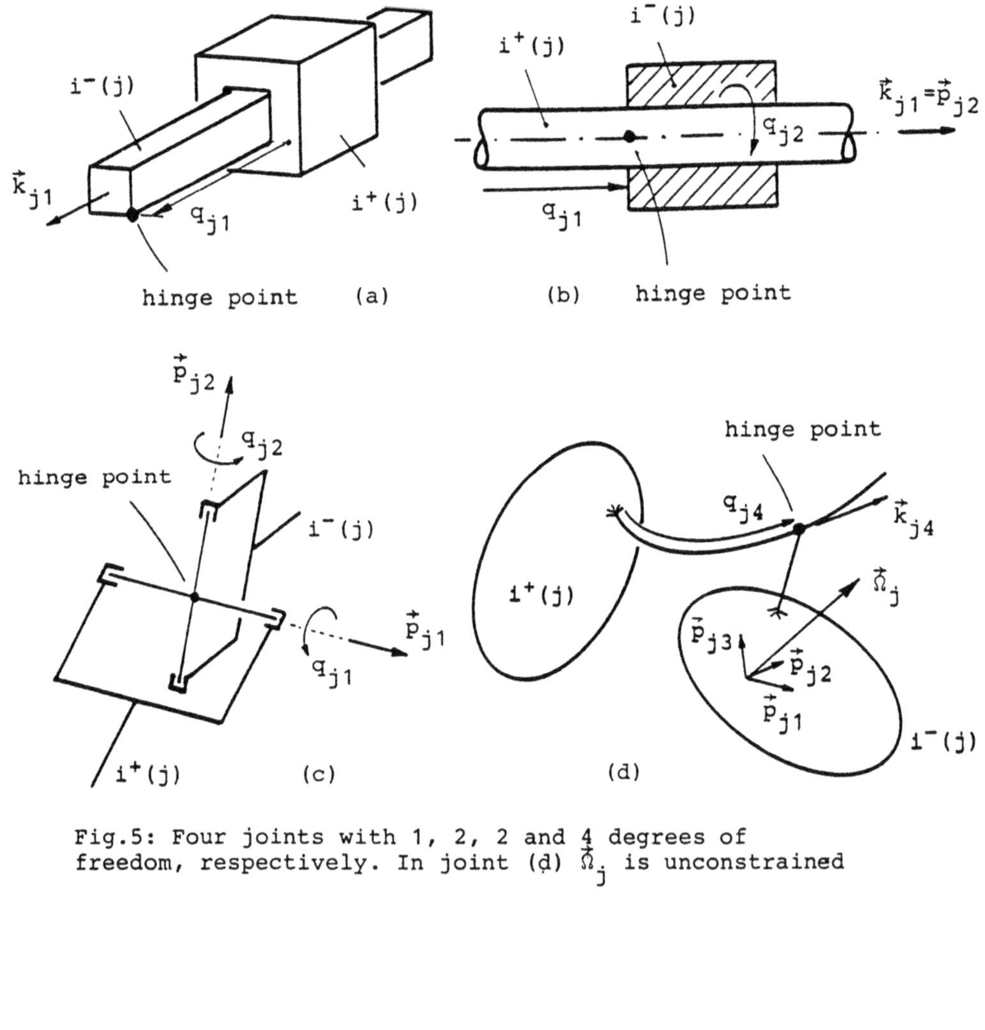

Fig.5: Four joints with 1, 2, 2 and 4 degrees of freedom, respectively. In joint (d) $\vec{\Omega}_j$ is unconstrained

Choice of hinge points: In order to specify the location and the translational velocity and acceleration of body $i^-(a)$ it suffices to do so for a single point fixed on this body. This point will be referred to as the *hinge point*. Appropriate choices of hinge points are demonstrated in Fig.5. The vector from the body $i^-(j)$ center of mass to the hinge point j is denoted by $\vec{c}_{i^-(j),j}$. It is fixed on this body.

The *location* of the hinge point in the reference base on body $i^+(j)$ is specified by the vector $\vec{c}_{i^+(j),j}$ starting from the body center of mass (Fig.4). Its coordinates in the reference base will, in general, be functions of the $q_{j\ell}$. Exceptions are revolute joints, Hooke's joints and ball joints where the hinge point can be chosen such that not only $\vec{c}_{i^-(j),j}$ but also $\vec{c}_{i^+(j),j}$ is fixed on the respective body (see Fig.5c).

The *velocity* of the hinge point relative to the body $i^+(j)$ reference base is the local time derivative $\overset{\circ}{\vec{c}}_{i^+(j),j}$ (the symbol ∘ indicates that it is not the time derivative in inertial space). It has the form

$$\overset{\circ}{\vec{c}}_{i^+(j),j} = \sum_{\ell=1}^{N_j} \vec{k}_{j\ell}\dot{q}_{j\ell} \qquad j=1\ldots n \; . \tag{17}$$

The vector $\vec{k}_{j\ell}$ is a unit vector if $q_{j\ell}$ is a line coordinate. In Fig.5 all $\vec{k}_{j\ell}$ associated with rotation variables are zero but this is not a general statement as is shown by polar coordinates r and φ where the velocity has components $\overset{\circ}{r}$ and $r\dot{\varphi}$.

The *acceleration* of the hinge point relative to the reference base on body $i^+(j)$ is

$$\overset{\circ\circ}{\vec{c}}_{i^+(j),j} = \sum_{\ell=1}^{N_j} \vec{k}_{j\ell}\ddot{q}_{j\ell} + \vec{s}_j \qquad j=1\ldots n \tag{18}$$

where $\vec{s}_j$ is a lower order term. It is non-zero if at least one $\vec{k}_{j\ell}$ is not fixed on body $i^+(j)$. Fig.5d shows such a case.

The *angular orientation* of body $i^-(j)$ is specified by the direction cosine matrix $\underline{G}_j$ relating the two bases fixed on bodies $i^+(j)$ and $i^-(j)$, respectively. It is a function of the $q_{j\ell}$. Special cases are joints with unconstrained rotational motion such as in Fig.5d. For such joints we prefer to express $\underline{G}_j$ by Euler-Rodrigues parameters in the form of Eq.(10).

The *angular velocity* of body $i^-(j)$ relative to body $i^+(j)$ is called $\vec{\Omega}_j$. If $\vec{\Omega}_j$ is unconstrained as in Fig.5d then we decompose it in the body $i^-(j)$ frame of reference. With unit base vectors $\vec{p}_{j\ell}$ this yields

$$\vec{\Omega}_j = \sum_{\ell=1}^{3} \vec{p}_{j\ell}\Omega_{j\ell}.$$

In all other cases $\vec{\Omega}_j$ is expressed in terms of generalized coordinates in the form

$$\vec{\Omega}_j = \sum_{\ell=1}^{N_j} \vec{p}_{j\ell}\dot{q}_{j\ell} \qquad j=1\ldots n$$

where $\vec{p}_{j\ell}$ is an axial unit vector if $q_{j\ell}$ is a rotation angle and zero if $q_{j\ell}$ is a line coordinate. In order to combine the two presentations we introduce $\dot{\pi}_{j\ell}$ as a new name comprising generalized velocities $\dot{q}_{j\ell}$ as well as quasi velocities $\Omega_{j\ell}$. Then for all joints the formula reads

$$\vec{\Omega}_j = \sum_{\ell=1}^{N_j} \vec{p}_{j\ell}\dot{\pi}_{j\ell} \qquad j=1\ldots n. \tag{19}$$

Note that $\delta\pi_{j\ell}$ always exists whereas $\pi_{j\ell}$ exists only if $\dot{\pi}_{j\ell}$ represents a generalized velocity $\dot{q}_{j\ell}$.

A final remark on angular velocities: The components $\Omega_{j\ell}$ of $\vec{\Omega}_j$ in the base of body $i^-(j)$ are related to the direction cosine matrix $\underline{G}_j$ through Poissons's kinematics differential equations

$$\begin{bmatrix} 0 & -\Omega_{j_3} & \Omega_{j_2} \\ \Omega_{j_3} & 0 & -\Omega_{j_1} \\ -\Omega_{j_2} & \Omega_{j_1} & 0 \end{bmatrix} = -\dot{\underline{G}}_j\underline{G}_j^T \qquad j=1\ldots n.$$

This relationship can be used for calculating the $\Omega_{j\ell}$ from a given matrix $\underline{G}_j$ if it is not preferred to identify the vectors $\vec{p}_{j\ell}$ directly by inspection.

The *angular acceleration* of body $i^-(j)$ relative to body $i^+(j)$ is the local time derivative of $\vec{\Omega}_j$. It does not make a difference whether the derivative is taken in the base of body $i^-(j)$ or of body $i^+(j)$ since the two differ by $\vec{\Omega}_j \times \vec{\Omega}_j = \vec{0}$. Choosing the base of body $i^-(j)$ we get from Eq.(19) the formula

$$\mathring{\vec{\Omega}}_j = \sum_{\ell=1}^{N_j} \vec{p}_{j\ell}\ddot{\pi}_{j\ell} + \vec{w}_j \qquad j=1\ldots n. \tag{20}$$

The lower-order term $\vec{w}_j$ is non-zero if at least one vector $\vec{p}_{j\ell}$ is not fixed on body $i^-(j)$. Fig.5c shows such a case. This concludes the section on kinematics of individual joints.

System Structure

With reference to Fig.1 we have distinguished systems with tree structure from those without. We have also distinguished between systems having kinematical constraints to external bodies and systems without such constraints. The latter distinction was shown to be important in the section on dynamics. However, from the point of view of kinematics and systems structure it is immaterial. This can be seen by a comparison of the systems in Fig.1b and 1c which differ only by the presence and absence, respectively, of this kind of constraint. In order to specify for the system in Fig.1c the position relative to the reference frame labeled 0 it is necessary to specify for each joint the relative position of the two contiguous bodies and to specify, in addition, the position of one arbitrarily chosen body relative to the reference frame. From this we conclude that for a complete description of system kinematics a joint with six degrees of freedom must be introduced between the one arbitrarily chosen body and the reference frame 0. In Fig.1c this joint is indicated by a dashed line. The reference frame 0 does not necessarily represent inertial space. If the motion of the system relative to a vehicle with prescribed acceleration is investigated then the reference frame 0 is fixed on this vehicle. With the six-degree-of-freedom joint and with the reference frame 0 the system has the same structure as the one in Fig.1b.

Let the bodies of any multi body system be labeled from 0 to n. The labeling order is arbitrary except for the indices 0 and 1. The index 0 is given to the body with prescribed motion and the index 1 to a body directly coupled to body 0 (see Fig.1a,b,c). Also the joints are labeled. A tree-structured system with bodies 0...n has n joints. In systems with closed chains the number is larger, say $n+\hat{n}$. By cutting $\hat{n}$ joints a system with tree structure can be produced, a so-called *spanning tree* of the original system. Fig.1b, for example, shows a spanning tree of the system in Fig.1a. If more than one spanning tree can be produced from a given system then one of them is chosen arbitrarily. Let the joints be labeled from 1 to $n+\hat{n}$. The labeling order is arbitrary except for the indices 1 and $n+1,\ldots,\hat{n}$. The indices $n+1,\ldots,\hat{n}$ are given to the joints which are chosen to be cut when a spanning tree is produced and the index 1 is given to the joint connecting the bodies 0 and 1 (see Fig.1a,b).

As one result of the kinematics description of the individual joints we have two integer functions $i^+(j)$ and $i^-(j)$ for $j=1\ldots\hat{n}$. These functions uniquely specify the interconnection structure of the system. They also provide full information about the choice of reference bodies for the individual joints. If the arrows in Fig.1a are interpreted the same way as in Fig.4 then the integer functions for this system read

j	1	2	3	4	5	6	7	8	9	10	11	12	13	14
$i^+(j)$	0	2	3	4	4	5	6	4	8	4	10	4	12	0
$i^-(j)$	1	1	2	3	5	6	7	8	9	10	11	12	11	6

Based on the two functions we define the so-called *incidence matrix* $\underline{S}$ of the system. Its elements are

$$S_{ij} = \begin{cases} +1 & \text{if } i=i^+(j) \\ -1 & \text{if } i=i^-(j) \\ 0 & \text{else} \end{cases} \qquad \begin{array}{l} i=1\ldots n \\ j=1\ldots\hat{n}. \end{array}$$

For the system in Fig.1a the matrix has the form

$$\underline{S} = \begin{array}{c} \\ 1\\2\\3\\4\\5\\6\\7\\8\\9\\10\\11\\12 \end{array} \begin{array}{c} \begin{array}{cccccccccccccc} 1 & 2 & 3 & 4 & 5 & 6 & 7 & 8 & 9 & 10 & 11 & 12 & 13 & 14 \end{array} \\ \left[\begin{array}{cccccccccccc:cc}
-1 & -1 & & & & & & & & & & & & \\
 & 1 & -1 & & & & & & & & & & & \\
 & & 1 & -1 & & & & & & & & & & \\
 & & & 1 & 1 & & & 1 & & 1 & & 1 & & \\
 & & & & -1 & 1 & & & & & & & & \\
 & & & & & -1 & 1 & & & & & & & -1 \\
 & & & & & & -1 & & & & & & & \\
 & & & & & & & -1 & 1 & & & & & \\
 & & & & & & & & -1 & & & & & \\
 & & & & & & & & & -1 & 1 & & & \\
 & & & & & & & & & & -1 & & -1 & \\
 & & & & & & & & & & & -1 & 1 &
\end{array}\right] \end{array}$$

Its 12×12 submatrix to the left of the dashed line represents the incidence matrix of the spanning tree shown in Fig.1b.

Another important matrix called $\underline{\vec{C}}$ is closely related to $\underline{S}$. It represents a weighted incidence matrix. Its elements are

$$\vec{C}_{ij} = S_{ij}\,\vec{c}_{ij} \qquad i=1\ldots n, \qquad j=1\ldots\hat{n}. \tag{21}$$

By the definition of S_{ij} only vectors $\vec{c}_{ij}$ with either $i=i^+(j)$ or $i=i^-(j)$ play a role in the matrix $\underline{\vec{C}}$. These are the vectors explained in Fig.4.

Another important matrix is the so-called *path matrix* $\underline{T}$. It can be defined only for tree-structured systems. Its rows and columns correspond to the joints and bodies, respectively (notice that in the matrices $\underline{S}$ and $\underline{\vec{C}}$ it is the inverse correspondence). The elements of $\underline{T}$ are

$$T_{ji} = \begin{cases} +1 & \text{if joint } j \text{ is located along the direct path from body 0 to body } i \text{ and if the arrow of joint } j \text{ is pointing toward body 0} \\ -1 & \text{if joint } j \text{ is located along the direct path from body 0 to body } i \text{ and if the arrow of joint } j \text{ is pointing away from body 0} \\ 0 & \text{else} \qquad\qquad i,j=1\ldots n. \end{cases}$$

In each column i of $\underline{T}$ only the elements in those rows are non-zero which correspond to joints along the direct path from body 0 to body i. This property explains the term path matrix. For the systems of Fig.1b,c the matrix has the form

$$\underline{T} = \begin{array}{r} \\ 1\\2\\3\\4\\5\\6\\7\\8\\9\\10\\11\\12 \end{array} \begin{array}{c} \begin{array}{cccccccccccc} 1&2&3&4&5&6&7&8&9&10&11&12 \end{array} \\ \left[\begin{array}{cccccccccccc}
-1&-1&-1&-1&-1&-1&-1&-1&-1&-1&-1&-1\\
&1&1&1&1&1&1&1&1&1&1&1\\
&&1&1&1&1&1&1&1&1&1&1\\
&&&1&1&1&1&1&1&1&1&1\\
&&&&-1&-1&-1&&&&&\\
&&&&&-1&-1&&&&&\\
&&&&&&-1&&&&&\\
&&&&&&&-1&-1&&&\\
&&&&&&&&-1&&&\\
&&&&&&&&&-1&-1&\\
&&&&&&&&&&-1&\\
&&&&&&&&&&&-1
\end{array}\right] \end{array}$$

An important relationship exists between the incidence matrix $\underline{S}$ of a system with closed chains on the one hand and the path matrix $\underline{T}$ for any spanning tree of the system on the other. The product $\underline{T}\underline{\underline{S}}$ represents what is known as the *fundamental cutset matrix*. For tree-structured systems, in particular, the product $\underline{\underline{T}}\,\underline{\underline{S}}$ is the unit matrix. An example: The matrix $\underline{\underline{S}}$ for the system in Fig.1a and the matrix $\underline{\underline{T}}$ for the spanning tree in Fig.1b yield the fundamental cutset matrix

$$\underline{\underline{T}}\,\underline{\underline{S}} = \begin{array}{c} \\ 1\\2\\3\\4\\5\\6\\7\\8\\9\\10\\11\\12 \end{array} \begin{array}{c} \begin{array}{cccccccccccc|cc} 1&2&3&4&5&6&7&8&9&10&11&12&13&14 \end{array} \\ \left[\begin{array}{cccccccccccc|cc}
1 & & & & & & & & & & & & & 1 \\
 & 1 & & & & & & & & & & & & -1 \\
 & & 1 & & & & & & & & & & & -1 \\
 & & & 1 & & & & & & & & & & -1 \\
 & & & & 1 & & & & & & & & & 1 \\
 & & & & & 1 & & & & & & & & 1 \\
 & & & & & & 1 & & & & & & & \\
 & & & & & & & 1 & & & & & & \\
 & & & & & & & & 1 & & & & & \\
 & & & & & & & & & 1 & & & 1 & \\
 & & & & & & & & & & 1 & & 1 & \\
 & & & & & & & & & & & 1 & -1 &
\end{array}\right] \end{array}$$

In rows 13 and 14 - these are the indices of the cut joints - only the elements in those rows are non-zero which represent the joints in the corresponding closed chains. In the mathematical literature on graph theory the incidence matrix as well as the fundamental cutset matrix are known as important parameters describing a graph. However, the path matrix $\underline{T}$ had never been considered. Its important role was discovered in research on electrical networks by Branin [9] and independently by Roberson and the present author [10]. The matrix $\underline{T}$ will turn out to be one of the essential system parameters describing multi body systems. The matrix is defined once the integer functions $i^+(j)$ and $i^-(j)$ are given for j=1...n. Roberson [11] describes an algorithm by which $\underline{T}$ can be calculated directly from $i^+(j)$ and $i^-(j)$ without having to invert the incidence matrix of the spanning tree.

Formulation of the Matrices $\underline{A}$ and $\underline{B}$

In the previous section tree-structured systems were shown to be special in the sense that only for them a path matrix $\underline{T}$ can be defined. Another special property of equal importance for our kinematics analysis is the following. Between the number N of degrees of freedom that an n-body system has as a whole and the numbers N_j for its individual joints a simple relationship exists only for tree-structured systems. It has the form

$$N = \sum_{j=1}^{n} N_j .$$

In n-body systems with closed kinematic chains the $\hat{n}$ joints in excess of n cause kinematical constraints. Their number ν depends not only on the numbers N_j $(j=1\ldots n+\hat{n})$ but also on geometric parameters of all the bodies forming closed chains. The determination of the number ν and the formulation of ν indepent constraint equations pose mathematical problems which must be solved for every closed kinematic chain individually. Except for closed chains with simple geometry such as plane crank-and-slider mechanisms or plane four-bar-mechanisms the mathematical difficulties are considerable. Use should be made as much as possible of the literature on kinematics of mechanisms where constraint equations can be found for many technically important closed chains. For holonomic systems the constraint equations have the general form

$$f_i(\underline{q},t) = 0 \qquad i=1\ldots\nu \tag{22}$$

where $\underline{q}$ denotes a set of N generalized coordinates for the spanning tree.

Because of the special properties of tree-structured systems we will formulate equations of motion for systems with closed kinematic chains in two steps. In the first step a spanning tree will be produced. For this tree with N degrees of freedom Eq.(15) will be formulated, i.e.

$$\delta\underline{q}^T(\underline{A}\ddot{\underline{q}} - \underline{B}) = 0. \tag{23}$$

It will be seen that for tree-structured systems the matrices $\underline{A}$ and $\underline{B}$ are particularly simple which is one reason for investigating a spanning tree first. The second step consists in incorporating into Eq.(23) the ν constraint Eqs.(22) which are expressed in terms of the same N variables $\underline{q}$. Depending on the mathematical formulation of the constraint equations the incorporation is possible either in explicit form or only numerically. The fact that eventually only numerical methods are available is another reason for investigating a spanning tree first. Independent of whether the incorporation into Eq.(23) is done explicitly or numerically a final set of only $N-\nu$ differential equations for as many independent variables is obtained for the system with closed kinematic chains. These equations benefit from the simplicity of the expressions for $\underline{A}$ and $\underline{B}$.

Systems With Tree Structure: The development of explicit expressions for the matrices $\underline{A}$ and $\underline{B}$ must start from expressions for the radius vectors $\vec{r}_i$ and for the angular velocities $\vec{\omega}_i$ of the bodies $i=1\ldots n$ (cf.Eqs.(12) to (14)). Fig.6 demonstrates that in either expression the direct path from body 0 to body i plays an essential role.

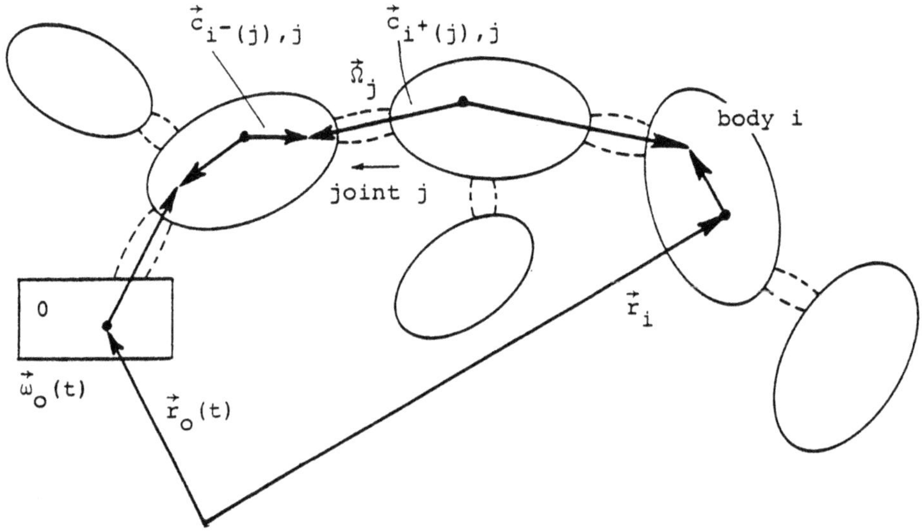

Fig.6: Vectors along the direct path from body 0 to body i

For $\vec{\omega}_i$ we need the sum of all vectors $\vec{\Omega}_j$ (some positive and some negative) along the path and for $\vec{r}_i$ we need the sum of all vectors $\vec{c}_{i^+(j),j}$ and $\vec{c}_{i^-(j),j}$ (some positive and some negative) along the path. The formula for $\vec{\omega}_i$ is

$$\vec{\omega}_i = -\sum_{j=1}^{n} T_{ji}\vec{\Omega}_j + \vec{\omega}_o(t).$$

The matrix elements T_{ji} sort out the direct path to body i and they also provide the correct signs. From this equation follows

$$\underline{\vec{\omega}} = -\underline{T}^T\underline{\vec{\Omega}} + \vec{\omega}_o\underline{1}_n \tag{24}$$

where $\underline{\vec{\omega}}$, $\underline{\vec{\Omega}}$ and $\underline{1}_n$ are the column matrices

$$\underline{\vec{\omega}} = [\vec{\omega}_1\vec{\omega}_2\ldots\vec{\omega}_n]^T, \qquad \underline{\vec{\Omega}} = [\vec{\Omega}_1\vec{\Omega}_2\ldots\vec{\Omega}_n]^T, \qquad \underline{1}_n = [1\ 1\ldots1]^T.$$

The corresponding expressions for $\vec{r}_i$ and for the column matrix $\underline{\vec{r}} = [\vec{r}_1\vec{r}_2\ldots\vec{r}_n]^T$ are

$$\vec{r}_i = -\sum_{j=1}^{n} T_{ji}\left(\vec{c}_{i^+(j),j} - \vec{c}_{i^-(j),j}\right) + \vec{r}_o(t).$$

$$\underline{\vec{r}} = -\underline{T}^T(\underline{\vec{c}}^+ - \underline{\vec{c}}^-) + \vec{r}_o\underline{1}_n \tag{25}$$

where $\underline{\vec{c}}^+$ and $\underline{\vec{c}}^-$ are column matrices which list all vectors $\vec{c}_{i^+(j),j}$ and $\vec{c}_{i^-(j),j}$, respectively, in the order j=1...n.

On the basis of Eqs.(24) and (25) we can now proceed to explicit expressions for $\dot{\underline{\vec{\omega}}}$ and for $\ddot{\underline{\vec{r}}}$. First, $\dot{\underline{\vec{\omega}}}$ is constructed. From Eq.(19) follows

$$\underline{\vec{\Omega}} = \underline{\vec{p}}^T\dot{\underline{\pi}}$$

where $\dot{\underline{\pi}}$ is the column matrix

$$\dot{\underline{\pi}} = [\dot{\pi}_{11}\ldots\dot{\pi}_{1N_1}\ \dot{\pi}_{21}\ldots\dot{\pi}_{2N_2}\ldots\ldots\ldots\dot{\pi}_{n1}\ldots\dot{\pi}_{nN_n}]^T$$

of all generalized velocities and quasi velocities defined for the joints and $\underline{\vec{p}}^T$ is an n×N matrix composed of the vectors $\vec{p}_{j\ell}$. To give an example: For a system containing the four joints shown in Fig.5 the matrices $\underline{\vec{p}}^T$ and $\dot{\underline{\pi}}$ are

$$\underline{\vec{p}}^T = \begin{bmatrix} \vec{\delta} & & & & & & & & \\ & \vec{\delta} & \vec{p}_{22} & & & & & & \\ & & & \vec{p}_{31} & \vec{p}_{32} & & & & \\ & & & & & \vec{p}_{41} & \vec{p}_{42} & \vec{p}_{43} & \vec{\delta} \end{bmatrix}, \quad \dot{\underline{\pi}} = [\dot{q}_{11}\ \dot{q}_{21}\ \dot{q}_{22}\ \dot{q}_{31}\ \dot{q}_{32}\ \Omega_{41}\ \Omega_{42}\ \Omega_{43}\ \dot{q}_{44}]^T.$$

When the expression for $\underline{\vec{\Omega}}$ is substituted into Eq.(24) we get

$$\underline{\vec{\omega}} = -(\underline{\vec{p}}\,\underline{T})^T\dot{\underline{\pi}} + \vec{\omega}_o\underline{1}_n\ .$$

The absolute angular accelerations are

$$\dot{\underline{\vec{\omega}}} = -\underline{T}^T\dot{\underline{\vec{\Omega}}} + \dot{\vec{\omega}}_o\underline{1}_n$$

where with $\mathring{\vec{\Omega}}_j$ from Eq.(20)

$$\dot{\vec{\Omega}}_j = \mathring{\vec{\Omega}}_j + \vec{\omega}_{i^-(j)} \times \vec{\Omega}_j .$$

Combining these expressions we have

$$\dot{\underline{\vec{\omega}}} = \underline{\vec{\beta}}\, \ddot{\underline{\pi}} + \underline{\vec{v}} \tag{26}$$

with

$$\underline{\vec{\beta}} = -(\underline{\vec{p}}\,\underline{T})^T , \qquad \underline{\vec{v}} = -\underline{T}^T \underline{\vec{f}} + \dot{\vec{\omega}}_0 \underline{1}_n . \tag{27}$$

and with a column matrix $\underline{\vec{f}}$ whose elements are

$$\vec{f}_j = \vec{w}_j + \vec{\omega}_{i^-(j)} \times \vec{\Omega}_j \qquad j=1\ldots n.$$

The expression for $\dot{\underline{\vec{\omega}}}$ has the form predicted in Eq.(14). The change from $\ddot{\underline{q}}$ to $\ddot{\underline{\pi}}$ is a consequence of our decision to include quasi velocities as variables.

In a similar manner we find the precise formulation of $\ddot{\underline{\vec{r}}}$ from the expression for $\dot{\underline{\vec{r}}}$. The second time derivative is

$$\ddot{\underline{\vec{r}}} = -\underline{T}^T (\ddot{\underline{\vec{c}}}^+ - \ddot{\underline{\vec{c}}}^-) + \ddot{\vec{r}}_0 \underline{1}_n . \tag{28}$$

The vectors $\vec{c}_{i^+(j),j}$ and $\vec{c}_{i^-(j),j}$ have the derivatives (see Eq.(18))

$$\ddot{\vec{c}}_{ij} = \begin{cases} \sum\limits_{\ell=1}^{N_j} \vec{k}_{j\ell}\ddot{\pi}_{j\ell} + \vec{s}_j - \vec{c}_{ij}\times\dot{\vec{\omega}}_i + \vec{\omega}_i\times(\vec{\omega}_i\times\vec{c}_{ij}) + 2\vec{\omega}_i\times\mathring{\vec{c}}_{ij} & i=i^+(j) \\[2ex] \qquad\qquad - \vec{c}_{ij}\times\dot{\vec{\omega}}_i + \vec{\omega}_i\times(\vec{\omega}_i\times\vec{c}_{ij}) + 2\vec{\omega}_i\times\mathring{\vec{c}}_{ij} & i=i^-(j). \end{cases}$$

The term involving $\dot{\vec{\omega}}_i$ contributes to $\ddot{\underline{\vec{r}}}$ the expression $(\underline{\vec{C}}\,\underline{T})^T \times \dot{\underline{\vec{\omega}}} - \dot{\vec{\omega}}_0 \times \vec{c}_{01}\underline{1}_n$ where $\underline{\vec{C}}$ is the matrix defined by Eq.(21). This is easily verified by multiplying out $\underline{\vec{C}}^T \times \dot{\underline{\vec{\omega}}}$. The contribution to $\ddot{\underline{\vec{r}}}$ of the term involving $\ddot{\pi}_{j\ell}$ is $-\underline{T}^T\underline{\vec{k}}^T\ddot{\underline{\pi}} = -(\underline{\vec{k}}\,\underline{T})^T\ddot{\underline{\pi}}$ where $\underline{\vec{k}}^T$ is an n×N matrix composed of the vectors $\vec{k}_{j\ell}$ in the same manner as $\underline{\vec{p}}^T$ is composed of the vectors $\vec{p}_{j\ell}$. To give an example: For a system containing the four joints of Fig.5 the matrix reads

$$\underline{\vec{k}}^T = \begin{bmatrix} \vec{k}_{11} & & & \\ & \vec{k}_{21} & \vec{0} & \\ & & \vec{0} & \vec{0} \\ & & & \vec{0} & \vec{0} & \vec{0} & \vec{k}_{44} \end{bmatrix}$$

With these expressions we can rewrite Eq.(28) in the form

$$\ddot{\underline{\vec{r}}} = (\underline{\vec{C}}\underline{T})^T \times \dot{\underline{\vec{\omega}}} - (\underline{\vec{k}}\underline{T})^T \ddot{\underline{\pi}} - \underline{T}^T(\underline{\vec{g}}+\underline{\vec{h}}) + \left[\ddot{\vec{r}}_o + \dot{\vec{\omega}}_o \times \vec{c}_{o1} + \vec{\omega}_o \times (\vec{\omega}_o \times \vec{c}_{o1}) + 2\vec{\omega}_o \times \overset{\circ}{\vec{c}}_{o1}\right]\underline{1}_n$$

where $\underline{\vec{g}}$ and $\underline{\vec{h}}$ are column matrices with the elements

$$\left.\begin{aligned} \vec{g}_j &= \vec{\omega}_{i^+(j)} \times \left(\vec{\omega}_{i^+(j)} \times \vec{c}_{i^+(j),j}\right) - \vec{\omega}_{i^-(j)} \times \left(\vec{\omega}_{i^-(j)} \times \vec{c}_{i^-(j),j}\right) \\ \vec{h}_j &= \vec{s}_j + 2\vec{\omega}_{i^+(j)} \times \overset{\circ}{\vec{c}}_{i^+(j),j} \end{aligned}\right\} j=1\ldots n.$$

When for $\dot{\underline{\vec{\omega}}}$ the expression from Eq.(26) is substituted $\ddot{\underline{\vec{r}}}$, finally, takes the form predicted in Eq.(12)

$$\ddot{\underline{\vec{r}}} = \underline{\vec{\alpha}}\,\ddot{\underline{\pi}} + \underline{\vec{u}}$$

with

$$\underline{\vec{\alpha}} = (\underline{\vec{p}}\underline{T} \times \underline{\vec{C}}\underline{T} - \underline{\vec{k}}\underline{T})^T$$

and (29)

$$\underline{\vec{u}} = (\underline{\vec{C}}\underline{T})^T \times \underline{\vec{v}} - \underline{T}^T(\underline{\vec{g}}+\underline{\vec{h}}) + \left[\ddot{\vec{r}}_o + \dot{\vec{\omega}}_o \times \vec{c}_{o1} + \vec{\omega}_o \times (\vec{\omega}_o \times \vec{c}_{o1}) + 2\vec{\omega}_o \times \overset{\circ}{\vec{c}}_{o1}\right]\underline{1}_n .$$

For systems without constraints to an external body $\ddot{\underline{\vec{R}}}$ instead of $\ddot{\underline{\vec{r}}}$ is needed. Eq.(13) yields the matrices valid for this case. They are denoted by an asterisk:

$$\underline{\vec{\alpha}}^* = \underline{\mu}^T \underline{\vec{\alpha}} = (\underline{\vec{p}}\underline{T} \times \underline{\vec{C}}\underline{T}\underline{\mu} - \underline{\vec{k}}\underline{T}\underline{\mu})^T$$

$$\underline{\vec{u}}^* = \underline{\mu}^T \underline{\vec{u}} = (\underline{\vec{C}}\underline{T}\underline{\mu})^T \times \underline{\vec{v}} - (\underline{T}\underline{\mu})^T(\underline{\vec{g}}+\underline{\vec{h}}) .$$

The simple form of $\underline{\vec{u}}^*$ is a consequence of the identity $\underline{\mu}^T \underline{1}_n = \underline{0}$.

We are now in a position to write the equations of motion for tree-structured systems in an explicit form. According to Eq.(16), systems with constraints to an external body 0 are governed by the equation

$$\underline{A}\ddot{\underline{\pi}} = \underline{B}$$

where with $\underline{\vec{\alpha}}$, $\underline{\vec{u}}$, $\underline{\vec{\beta}}$ and $\underline{\vec{v}}$ from Eqs.(29) and (27)

$$\underline{A} = \underline{\vec{\alpha}}^T \cdot \underline{m}\underline{\vec{\alpha}} + \underline{\vec{\beta}}^T \cdot \underline{J} \cdot \underline{\vec{\beta}} \tag{30}$$

$$\underline{B} = \underline{\vec{\alpha}}^T \cdot (\underline{\vec{F}} - \underline{m}\underline{\vec{u}}) + \underline{\vec{\beta}}^T \cdot (\underline{\vec{M}} - \underline{J} \cdot \underline{\vec{v}} - \underline{\vec{V}}) + \underline{Q}. \tag{31}$$

For systems without constraints to an external body 0 the equations are analogously

$$\underline{A}^* \ddot{\underline{\pi}} = \underline{B}^* \tag{32}$$

where $\underline{A}^*$ and $\underline{B}^*$ are given by the same formulas except that $\underline{\vec{\alpha}}$, $\underline{\vec{u}}$ and $\underline{\vec{F}}$ are replaced by $\underline{\vec{\alpha}}^*$ $\underline{\vec{u}}^*$ and $\underline{\mu}\underline{\vec{F}}$, respectively.

The equations above are complete if all variables represent generalized coordinates. If, however, quasi velocities $\Omega_{j\ell}$ are among the variables then the equations must be supplemented by kinematic differential equations. These have the form of Eq.(11). For each set of quasi velocities Ω_{j1}, Ω_{j2}, Ω_{j3} such a set of equations has to be formulated. The quantities ω_{11}, ω_{12} and ω_{13} have to be replaced by Ω_{j1}, Ω_{j2} and Ω_{j3}, respectively, and the Euler-Rodrigues parameters are the same which were used in formulating the direction cosine matrix $\underline{G}_j$ (see the section on kinematics of individual joints).

The formulation of the equations of motion thus far developed is sufficiently explicit for numerical integration. However, as a consequence of the special forms of $\underline{\vec{\alpha}}$ and $\underline{\vec{\beta}}$ it is possible to develop the matrices $\underline{A}$ and $\underline{B}$ still further. The resulting simplifications provide a deeper theoretical insight. They also provide the basis for more efficient computer programs.

Further Development of the Matrices $\underline{A}$ and $\underline{B}$: We begin by an investigation of the matrix $\underline{A}^*$. With the explicit expressions for $\underline{\vec{\alpha}}^*$ and $\underline{\vec{\beta}}^*$ it reads

$$\begin{aligned}\underline{A}^* = {} & (\underline{\vec{p}}\underline{T}\times\underline{\vec{C}}\underline{T}\underline{\mu})\cdot\underline{m}(\underline{\vec{p}}\underline{T}\times\underline{\vec{C}}\underline{T}\underline{\mu})^T + (\underline{\vec{p}}\underline{T})\cdot\underline{J}\cdot(\underline{\vec{p}}\underline{T})^T \\ & -\left[\underline{\vec{p}}\underline{T}\times\underline{\vec{C}}\cdot(\underline{T}\underline{\mu}\underline{m}\underline{\mu}^T\underline{T}^T)\underline{\vec{k}}^T\right] - \left[\underline{\vec{p}}\underline{T}\times\underline{\vec{C}}\cdot(\underline{T}\underline{\mu}\underline{m}\underline{\mu}^T\underline{T}^T)\underline{\vec{k}}^T\right]^T + \underline{\vec{k}}\cdot(\underline{T}\underline{\mu}\underline{m}\underline{\mu}^T\underline{T}^T)\underline{\vec{k}}^T.\end{aligned} \tag{33}$$

The symmetric matrix $\underline{T}\underline{\mu}\underline{m}\underline{\mu}^T\underline{T}^T$ is constant. The identity $\underline{\mu}\underline{m}\underline{\mu}^T \equiv \underline{\mu}\underline{m}$ is worth mentioning. From vector algebra the identity is known

$$(\vec{a}\times\vec{b})\cdot(\vec{c}\times\vec{d}) = (\vec{a}\cdot\vec{c})(\vec{b}\cdot\vec{d}) - (\vec{b}\cdot\vec{c})(\vec{a}\cdot\vec{d}) = \vec{a}\cdot(\vec{c}\vec{b}-\vec{c}\cdot\vec{b}\mathbb{E})\cdot\vec{d}$$

($\mathbb{E}$ is the unit dyadic). This suggests that $\underline{A}^*$ can be written in the form

$$\underline{A}^* = (\vec{\underline{p}}\underline{T})\cdot\underline{\mathbb{K}}^*\cdot(\vec{\underline{p}}\underline{T})^T - \left[\vec{\underline{p}}\underline{T}\times\vec{\underline{C}}\cdot(\underline{T}\underline{\mu}\underline{m}\underline{\mu}^T\underline{T}^T)\vec{\underline{k}}^T\right] - \left[\vec{\underline{p}}\underline{T}\times\vec{\underline{C}}\cdot(\underline{T}\underline{\mu}\underline{m}\underline{\mu}^T\underline{T}^T)\vec{\underline{k}}^T\right]^T + \vec{\underline{k}}\cdot(\underline{T}\underline{\mu}\underline{m}\underline{\mu}^T\underline{T}^T)\vec{\underline{k}}^T. \quad (34)$$

In order to interpret the physical significance of the tensor elements of $\underline{\mathbb{K}}^*$ we must first find out the physical interpretation of the vectors $(\vec{\underline{C}}\underline{T}\underline{\mu})_{ij}$ for i,j=1...n. For what follows the reader is referred to ref.[2]. Associated with each body i=1...n of the system a so-called augmented body is constructed. The method is illustrated in Fig.7 for the body i=4 of Fig.1. In the hinge point of every joint on body i a point mass is concentrated which equals the sum of the masses of all bodies coupled to body i via the respective hinge point. The extended body i together with these point masses represents the augmented body i. The augmented body has the total system mass M.

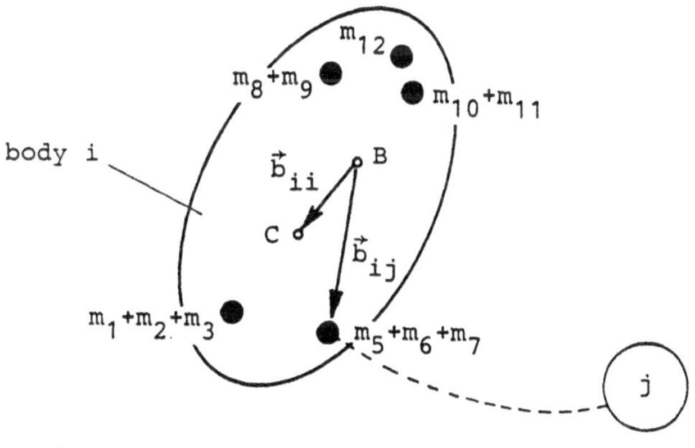

Fig.7: The augmented body i=4 in the system of Fig.1, its center of mass B and the vectors $\vec{b}_{ij}$ (j=1...n) on this body

In general, it is not a rigid body since the hinge points may not be fixed on it. Still, we can define the augmented body i center of mass B and vectors $\vec{b}_{ij}$ as shown in the figure. The vector $\vec{b}_{ii}$ points from

B to the original body i center of mass C and $\vec{b}_{ij}$ ($j \neq i$) points to the hinge point which leads either directly or indirectly to body j. This definition implies that there are fewer different vectors $\vec{b}_{ij}$ then combinations of indices i and j. In Fig.1, for example, we have the identities $\vec{b}_{41}=\vec{b}_{42}=\vec{b}_{43}$. From the definition follows that

$$\sum_{j=1}^{n} m_j \vec{b}_{ij} = \vec{0} \qquad i=1\ldots n. \tag{35}$$

In ref.[2] it is shown that the elements of the matrix $\underline{\vec{C}T\mu}$ are

$$(\underline{\vec{C}T\mu})_{ij} = -\vec{b}_{ij} \qquad i,j=1\ldots n. \tag{36}$$

With this result and knowing the special properties of the vectors $\vec{b}_{ij}$ we investigate the leading term of $\underline{A}^*$ in Eq.(33). In order to simplify matters it is recalled that the left hand side term $(\underline{\vec{p}T})$ originally had the factor $\delta\underline{\pi}^T$ in front and that the right hand side term $(\underline{\vec{p}T})^T$ still has the factor $\ddot{\underline{\pi}}$. We add both factors and use as temporary abbreviations

$$\underline{\vec{a}}^T = \delta\underline{\pi}^T(\underline{\vec{p}T}), \qquad \underline{\vec{d}} = (\underline{\vec{p}T})^T\ddot{\underline{\pi}}.$$

Our task is then to investigate the matrix $\underline{\mathbb{I}}^*$ in the equation

$$-\left[\underline{\vec{a}}^T\times(\underline{\vec{C}T\mu})\right]\cdot\underline{m}\left[(\underline{\vec{C}T\mu})^T\times\underline{\vec{d}}\right] = \underline{\vec{a}}^T\cdot\underline{\mathbb{I}}^*\cdot\underline{\vec{d}}.$$

The left hand side expression is the scalar

$$-\sum_i\sum_j\sum_k m_k(\vec{a}_i\times\vec{b}_{ik})\cdot(\vec{b}_{jk}\times\vec{d}_j) = -\sum_i\sum_j \vec{a}_i\cdot\underbrace{\sum_k m_k(\vec{b}_{jk}\vec{b}_{ik}-\vec{b}_{jk}\cdot\vec{b}_{ik}\mathbb{E})}_{-\mathbb{I}^*_{ij}}\cdot\vec{d}_j \tag{37}$$

In what follows $\mathbb{I}^*_{ij}$ for $j \neq i$ will be considered. Let the bodies of a system be devided into two groups in such a way that according to Fig.8 the division line is drawn somewhere across the direct path between bodies i and j. Let the set of indices of all bodies of the group containing body i be denoted by I and the set of indices of all other bodies by II. Then we know that for all indices k belonging to I (abbreviated $k \in I$) the identity $\vec{b}_{jk}=\vec{b}_{ji}$ holds and for all $k \in II$ the identity $\vec{b}_{ik}=\vec{b}_{ij}$.

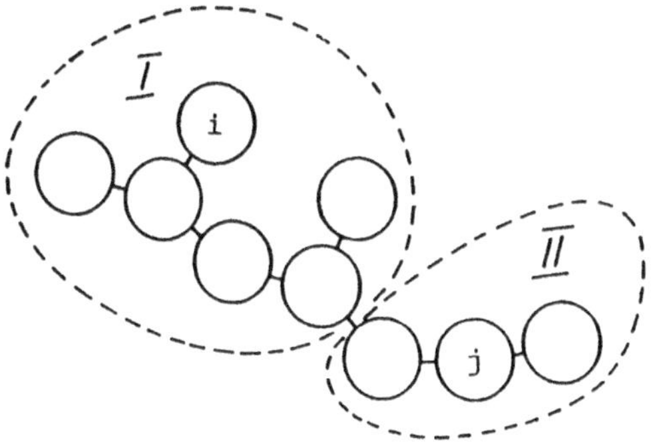

Fig.8

Using repeatedly these identities and making use of Eq.(35) we can write

$$\sum_k m_k \vec{b}_{jk}\vec{b}_{ik} = \vec{b}_{ji}\sum_{k\in I} m_k\vec{b}_{ik} + \Big(\sum_{k\in II} m_k\vec{b}_{jk}\Big)\vec{b}_{ij}$$

$$= \vec{b}_{ji}\Big(\sum_{k=1}^{n} m_k\vec{b}_{ik} - \vec{b}_{ij}\sum_{k\in II} m_k\Big) + \Big(\sum_{k=1}^{n} m_k\vec{b}_{jk} - \vec{b}_{ji}\sum_{k\in I} m_k\Big)\vec{b}_{ij}$$

$$= -\vec{b}_{ji}\Big(\sum_{k\in II} m_k + \sum_{k\in I} m_k\Big)\vec{b}_{ij}$$

$$= -M\vec{b}_{ji}\vec{b}_{ij}. \tag{38}$$

Along the same line of arguments also the second term of $\mathbb{J}^*_{ij}$ can be simplified. The final result reads

$$\mathbb{J}^*_{ij} = -M\big(\vec{b}_{ji}\cdot\vec{b}_{ij}\mathbb{E} - \vec{b}_{ji}\vec{b}_{ij}\big) \qquad i\neq j.$$

The expression for the diagonal elements $\mathbb{J}^*_{ii}$ is left in the form of Eq.(37). Thus, the matrix $\underline{\mathbb{K}}^*$ in Eq.(34) has the elements

$$\mathbb{K}^*_{ij} = \begin{cases} \mathbb{J}_i + \sum_{k=1}^{n} m_k(\vec{b}^2_{ik}\mathbb{E} - \vec{b}_{ik}\vec{b}_{ik}) & i=j \\ -M(\vec{b}_{ji}\cdot\vec{b}_{ij}\mathbb{E} - \vec{b}_{ji}\vec{b}_{ij}) & i\neq j \end{cases} \qquad i,j=1\ldots n.$$

From Fig.7 it is seen that $\mathbb{K}^*_{ii}$ represents the inertia tensor of the augmented body i with respect to its center of mass B. Thus, all elements of the matrix $\underline{\mathbb{K}}^*$ in Eq.(34) have surprisingly simple physical interpretations. $\underline{A}^*$ takes its simplest form if all joints of a system have only rotational degrees of freedom (revolute joints, Hooke's joints and ball joints). In this case $\vec{\underline{k}}$ is zero so that only the term with $\underline{\mathbb{K}}^*$ is left. Moreover, all vectors $\vec{b}_{ij}$ have constant coordinates on the bodies on which they are defined.

The line of arguments just presented for the matrix $\underline{A}^*$ can in an analogous form be repeated for the matrix $\underline{A}$ of Eq.(30). This results in (see ref.[2])

$$\begin{aligned}\underline{A} = \; & (\vec{\underline{p}}\underline{T})\cdot\underline{\mathbb{K}}\cdot(\vec{\underline{p}}\underline{T})^T \\ & -\left[\vec{\underline{p}}\underline{T}\times\vec{\underline{C}}\cdot(\underline{T}\underline{m}\underline{T}^T)\vec{\underline{k}}^T\right]-\left[\vec{\underline{p}}\underline{T}\times\vec{\underline{C}}\cdot(\underline{T}\underline{m}\underline{T}^T)\vec{\underline{k}}^T\right]^T+\vec{\underline{k}}\cdot(\underline{T}\underline{m}\underline{T}^T)\vec{\underline{k}}^T\end{aligned} \tag{39}$$

where $\underline{\mathbb{K}}$ has the elements

$$\mathbb{K}_{ij} = \begin{cases} \mathbb{J}_i + \sum_{k=1}^{n} m_k(\vec{d}_{ik}^{\,2}\mathbb{E} - \vec{d}_{ik}\vec{d}_{ik}) & i=j \\ M(\vec{b}_{jo}\cdot\vec{d}_{ij}\mathbb{E} - \vec{b}_{jo}\vec{d}_{ij}) & \text{body } i \text{ is located along the direct path from body 0 to body } j \\ M(\vec{d}_{ji}\cdot\vec{b}_{io}\mathbb{E} - \vec{d}_{ji}\vec{b}_{io}) & \text{body } j \text{ is located along the direct path from body 0 to body } i \\ 0 & \text{else} \qquad i,j=1\ldots n. \end{cases} \tag{40}$$

The vectors $\vec{b}_{io}$ and $\vec{d}_{ij}$ are defined

$$\vec{b}_{io} = \begin{cases} \vec{b}_{11}+\vec{c}_{11} & i=1 \\ \vec{b}_{i1} & i=2\ldots n \end{cases}$$

$$\vec{d}_{ij} = (\vec{\underline{C}}\underline{T})_{ij} \qquad i,j=1\ldots n.$$

Fig.9 illustrates the physical significance of $\vec{d}_{ij}$. The vector $\vec{d}_{ij}$ is located on body i (first index). It terminates at the hinge point leading toward body 0. Starting point is the body center of mass in the case i=j and the hinge point leading toward body j in the case i≠j. From this follows that there are fewer different vectors $\vec{d}_{ij}$ than there are index combinations i,j. Obviously, $\vec{d}_{ij}$ is zero if the hinge point leading toward body 0 also leads toward body j. This implies that out of any pair $\vec{d}_{ij}$ and $\vec{d}_{ji}$ for i≠j at least one vector is zero.

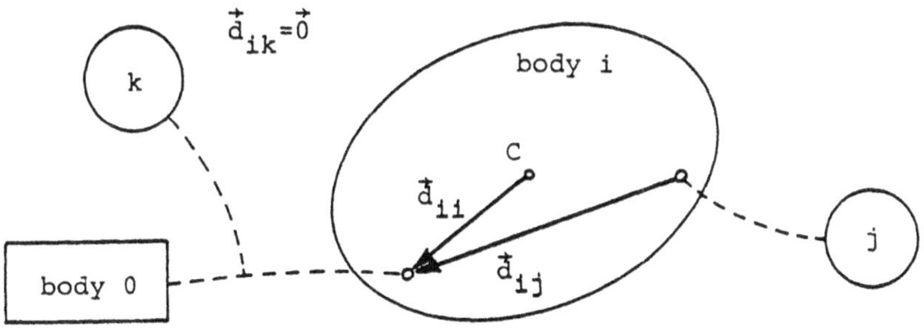

Fig.9: The vectors $\vec{d}_{ij}$ (j=1...n) on body i

The diagonal element $\mathbb{K}_{ii}$ is seen to represent the inertia tensor of the augmented body i with respect to the hinge point on body i which leads toward body 0.

Next, the matrix $\underline{B}^*$ on the right hand side of Eq.(32) is investigated. In explicit form it reads

$$\underline{B}^* = (\vec{\underline{p}}\underline{T}\times\vec{\underline{C}}\underline{T}\underline{\mu} - \vec{\underline{k}}\underline{T}\underline{\mu})\cdot\left\{\underline{\mu}\vec{\underline{F}} - \underline{m}\left[(\vec{\underline{C}}\underline{T}\underline{\mu})^T\times\vec{\underline{v}} - (\underline{T}\underline{\mu})^T(\vec{\underline{g}}+\vec{\underline{h}})\right]\right\} - \vec{\underline{p}}\underline{T}\cdot(\vec{\underline{M}} - \underline{\mathbb{J}}\cdot\vec{\underline{v}} - \vec{\underline{V}}) + \underline{Q}\,.$$

Several by now familiar expressions are recognized. The following reformulations are possible

$$-(\vec{\underline{p}}\underline{T})\times(\vec{\underline{C}}\underline{T}\underline{\mu})\cdot\underline{m}(\vec{\underline{C}}\underline{T}\underline{\mu})^T\times\vec{\underline{v}} + (\vec{\underline{p}}\underline{T})\cdot\underline{\mathbb{J}}\cdot\vec{\underline{v}} = (\vec{\underline{p}}\underline{T})\cdot\underline{\mathbb{K}}^*\cdot\vec{\underline{v}}$$

$$-(\vec{\underline{p}}\underline{T})\times(\vec{\underline{C}}\underline{T}\underline{\mu})\cdot\underline{m}(\underline{T}\underline{\mu})^T\vec{\underline{g}} + (\vec{\underline{p}}\underline{T})\cdot\vec{\underline{V}} = (\vec{\underline{p}}\underline{T})\cdot\vec{\underline{G}}^*$$

where $\vec{\underline{G}}^*$ is a column matrix with the elements

$$\vec{G}_i^* = -M\sum_{\substack{j=1\\ \neq i}}^{n}\vec{b}_{ij}\times[\vec{\omega}_j\times(\vec{\omega}_j\times\vec{b}_{ji})] + \vec{\omega}_i\times\mathbb{K}_{ii}^*\cdot\vec{\omega}_i \qquad i=1\dots n.$$

With these expressions and with the identities $\underline{\mu}\underline{m}\underline{\mu}^T \equiv \underline{\mu}\underline{m}$, $\underline{\mu}\underline{\mu} \equiv \underline{\mu}$ $\underline{B}^*$ takes the final form

$$\underline{B}^* = (\vec{\underline{p}}\underline{T})\cdot\left[\underline{\mathbb{K}}^*\cdot\vec{\underline{v}} + \vec{\underline{G}}^* - \vec{\underline{M}} + (\vec{\underline{C}}\underline{T}\underline{\mu})\times(\vec{\underline{F}}+\underline{m}\underline{T}^T\vec{\underline{h}})\right]$$

$$-(\vec{\underline{k}}\underline{T}\underline{\mu})\cdot\left[\vec{\underline{F}} + \underline{m}\underline{T}^T(\vec{\underline{g}}+\vec{\underline{h}}) - \underline{m}(\vec{\underline{C}}\underline{T})^T\times\vec{\underline{v}}\right] + \underline{Q}.$$

For systems with purely rotational joints the matrices $\vec{\underline{k}}$ and $\vec{\underline{h}}$ are zero.

For the matrix $\underline{B}$ in Eq.(31) analogous arguments result in the formula

$$\underline{B} = (\vec{\underline{p}}\underline{T})\circ\left[\mathbb{K}\cdot\underline{\vec{v}} + \underline{\vec{G}} - \underline{\vec{M}} + (\vec{\underline{C}}\underline{T})\times(\underline{\vec{F}} + \underline{m}\underline{T}^T\underline{\vec{h}} - \ddot{\vec{r}}_{o1}\underline{m}\underline{1}_n)\right]$$

$$-(\vec{\underline{k}}\underline{T})\circ\left[\underline{\vec{F}} + \underline{m}\underline{T}^T(\underline{\vec{g}}+\underline{\vec{h}}) - \underline{m}(\vec{\underline{C}}\underline{T})^T\times\underline{\vec{v}} - \ddot{\vec{r}}_{o1}\underline{m}\underline{1}_n\right] + \underline{Q}. \qquad (41)$$

with $\mathbb{K}$ from Eq. (40) and with a column matrix $\underline{\vec{G}}$ whose elements are

$$\vec{G}_i = M\left\{\sum_{j\neq i}\vec{d}_{ij}\times[\vec{\omega}_j\times(\vec{\omega}_j\times\vec{b}_{jo})] + \vec{b}_{io}\times\sum_{j\neq i}\vec{\omega}_j\times(\vec{\omega}_j\times\vec{d}_{ji})\right\} + \vec{\omega}_i\times\mathbb{K}_{ii}\cdot\vec{\omega}_i$$

$$i=1\ldots n.$$

For an appreciation of the summation ranges in the two sums involving $\vec{d}_{ij}$ and $\vec{d}_{ji}$ the reader is reminded of Fig.9. The symbol $\ddot{\vec{r}}_{o1}$ in the formula for $\underline{B}$ is used as an abbreviation for

$$\ddot{\vec{r}}_{o1} = \ddot{\vec{r}}_o + \dot{\vec{\omega}}_o\times\vec{c}_{o1} + \vec{\omega}_o\times(\vec{\omega}_o\times\vec{c}_{o1}) + 2\vec{\omega}_o\times\mathring{\vec{c}}_{o1}.$$

This concludes the discussion of the matrices $\underline{A}$ and $\underline{B}$.

The generalized forces $\underline{Q}$: The term $\underline{Q}$ in the equations $\underline{A}\ddot{\underline{\pi}} = \underline{B}$ is composed of N generalized forces each of which is associated with one variable,

$$\underline{Q} = [Q_{11}\ldots Q_{1N_1}\quad Q_{21}\ldots Q_{2N_2}\ldots\ldots Q_{n1}\ldots Q_{nN_n}]^T.$$

The generalized forces are caused by springs, dampers and actuators mounted in the joints. They are determined from the virtual work expression

$$\delta W = \delta\underline{q}^T\underline{Q}.$$

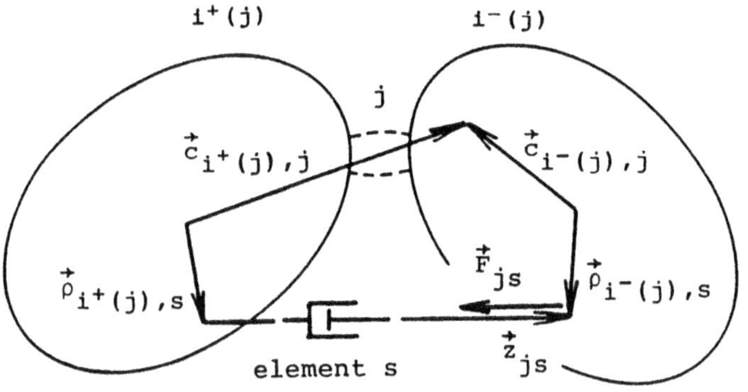

Fig.10: Vectors associated with a passive force element which is attached to joint j

In order to find the generalized forces associated with the variables $q_{j1} \dots q_{jN_j}$ of a single joint j only this joint has to be considered. The following programmable formalism is restricted to passive force elements, i.e. to (linear or nonlinear) springs and dampers. Let Fig.10 represent the same joint j that was shown in Fig.4. The body-fixed vectors $\vec{\rho}_{i^+(j),s}$ and $\vec{\rho}_{i^-(j),s}$ locate the attachment points of the passive element s. To be calculated are the following quantities:

(i) The vector $\vec{z}_{js}$ joining the attachment points:

$$\vec{z}_{js} = \vec{c}_{i^+(j),j} - \vec{c}_{i^-(j),j} - \vec{\rho}_{i^+(j),s} + \vec{\rho}_{i^-(j),s}$$

(ii) The rate of change v_{js} of the distance between the attachment points. It is calculated as

$$v_{js} = \frac{\vec{z}_{js} \cdot \overset{\circ}{\vec{z}}_{js}}{|\vec{z}_{js}|}$$

where $\overset{\circ}{\vec{z}}_{js}$ is the local derivative of $\vec{z}_{js}$ in the body $i^+(j)$ reference frame. Using Eqs.(17) and (19) we get

$$\overset{\circ}{\vec{z}}_{js} = \overset{\circ}{\vec{c}}_{i^+(j),j} + \vec{\Omega}_j \times \left(\vec{\rho}_{i^-(j),s} - \vec{c}_{i^-(j),j}\right)$$

$$= \sum_{\ell=1}^{N_j} \left[\vec{k}_{j\ell} + \vec{p}_{j\ell} \times \left(\vec{\rho}_{i^-(j),s} - \vec{c}_{i^-(j),j}\right)\right] \dot{\pi}_{j\ell}$$

(iii) The virtual change $\delta\vec{z}_{js}$ of $\vec{z}_{js}$ in the body $i^+(j)$ reference frame. It is

$$\delta\vec{z}_{js} = \sum_{\ell=1}^{N_j} \left[\vec{k}_{j\ell} + \vec{p}_{j\ell} \times \left(\vec{\rho}_{i^-(j),s} - \vec{c}_{i^-(j),j}\right)\right] \delta\pi_{j\ell}.$$

The force $\vec{F}_{js}$ acting on body $i^-(j)$ has the direction of $-\vec{z}_{js}$ and a magnitude which is some known function f_{js} of $|\vec{z}_{js}|$ and of v_{js}. Hence

$$\vec{F}_{js} = -\frac{\vec{z}_{js}}{|\vec{z}_{js}|} f_{js}\left(|\vec{z}_{js}|, v_{js}\right). \tag{42}$$

Using the above expressions we can now calculate the virtual work $\vec{F}_{js} \cdot \delta\vec{z}_{js}$. The coefficient of $\delta\pi_{j\ell}$ in this work expression represents the desired contribution of $\vec{F}_{js}$ to the generalized force $Q_{j\ell}$. If there are ν_j force elements in joint j then

$$Q_{j\ell} = \sum_{s=1}^{\nu_j} \vec{F}_{js} \cdot \left[\vec{k}_{j\ell} + \vec{p}_{j\ell} \times \left(\vec{\rho}_{i^-(j),s} - \vec{c}_{i^-(j),j}\right)\right].$$

With these formulas the generalized forces can be calculated once the following minimum amount of data is provided for each joint $j=1\ldots n$:

- the number ν_j of passive force elements in the joint,
- for $s=1\ldots\nu_j$ the coordinates of the attachment points i.e. of the vectors $\vec{\rho}_{i^+(j),s}$ and $\vec{\rho}_{i^-(j),s}$ in the respective body frames of reference,
- for $s=1\ldots\nu_j$ the characteristic function f_{js} (it has to be specified in such a way that the minus sign in Eq.(42) is correct).

Springs and dampers can also be mounted between bodies which are not directly coupled by a joint. With some arbitrariness the forces exerted by such elements were listed among the external forces. It should be clear, however, that the mathematical treatment is not conceptually different from the previous case. Also in this case two attachment points and a characteristic function must be specified and a vector analogous to $\vec{z}_{js}$ must be formed. This time the radius vectors $\vec{r}_i$ leading to the centers of mass of the two bodies come into play. The details are not worked out here.

Scalar Decomposition of Vectors: The matrices $\underline{A}$ and $\underline{B}$ in the equations of motion are sums of products of other matrices whose elements are vectors. For every vector the components are given (as constants or as functions of variables) in certain body-fixed frames of reference. Sums and products of vectors can be evaluated only after the components of all vectors have been transformed into one *master reference frame*. The transformation into the master reference frame of three vector components given on some body k requires a chain of matrix multiplications, each multiplication carrying accross one joint j along the direct path from body k to the master body. The matrix associated with joint j is the direction cosine matrix $\underline{G}_j$ introduced in the section on kinematics of individual joints. It is a given function of variables. Actually, for some joints $\underline{G}_j$ and for others $\underline{G}_j^T$ has to be multiplied depending on the directions of the joint associated arrows. To give an example, in Fig.1a the matrix $\underline{G}_1^T\underline{G}_2\underline{G}_3\underline{G}_4\underline{G}_8^T$ transforms from the body 8 into the body 0 reference frame.

It is important to select the master reference frame optimally. The optimality criterion depends on whether a computer program is written for the numerical evaluation of $\underline{A}$ and $\underline{B}$ or for the symbolic generation of formulas for $\underline{A}$ and $\underline{B}$. Here, the latter is of interest.

The master reference frame should then be selected such that the formulas for all elements of $\underline{A}$ and $\underline{B}$ when pieced together yield the minimum total length. This criterion is not practical. Keeping in mind that roughly speaking each body carries an equal number of vectors to be transformed we might adopt the following criterion. The total number of multiplications by matrices $\underline{G}_j$ necessary to transform one vector from each body must be a minimum. In Fig.1 this criterion yields the reference frame on body 4 as master reference frame. In graph theory the so-called median of a graph is defined by an analogous criterion. We can refine the criterion still further in order to account for the fact that the elements of $\underline{G}_j$ are of different complexity depending on whether joint j has one or two or three degrees of freedom of rotational motion. Accordingly, weighting factors 1, 2 and 3, respectively, can be given to the joints. The statement that all bodies carry equal numbers of vectors to be transformed is a rough approximation. It is no problem to count for each body exactly how many vectors $\vec{b}_{ij}$, $\vec{k}_{j\ell}$, $\vec{p}_{j\ell}$, $\vec{s}_j$ etc. have to be transformed and to define weighting factors for the bodies accordingly. Also these weighting factors can be taken into account in the criterion for the selection of a master reference frame.

Rheonomic Constraints: Consider a tree-structured system with N degrees of freedom which is governed by the equations of motion

$$\underline{A}\,\ddot{\underline{\pi}} = \underline{B}. \tag{43}$$

Suppose that a motor mounted in one of the joints causes one of the generalized coordinates, say π_k, to be a prescribed function of time thereby reducing the number of degrees of freedom to N-1. In the simplest case of practical importance a relative angular velocity $\dot{\pi}_k$ is forced to be constant. It is desired to write a new system of differential equations for the reduced set of N-1 variables. In addition, an expression is desired for the generalized motor force necessary to produce the prescribed function of time.

In order to achieve these goals Eq.(43) is written in the more detailed fashion

$$\begin{bmatrix} A_{11}\cdot\cdot A_{k1}\cdot\cdot\cdot A_{1N} & & \\ \vdots & & \\ A_{k1} & A_{kk} & A_{kN} \\ \vdots & & \\ A_{N1} & A_{Nk} & A_{NN} \end{bmatrix} \begin{bmatrix} \ddot{\pi}_1 \\ \vdots \\ \ddot{\pi}_k(t) \\ \vdots \\ \ddot{\pi}_N \end{bmatrix} = \begin{bmatrix} B_1 \\ \vdots \\ B_k'-Q_k \\ \vdots \\ B_N \end{bmatrix}.$$

Together with $\pi_k(t)$ also $\dot{\pi}_k(t)$ and $\ddot{\pi}_k(t)$ are given functions of time. The element B_k' of $\underline{B}$ has been split into the generalized force $-Q_k$ and the rest B_k. One part of Q_k is the desired generalized motor force Q_k^{mot}. If the motor is counteracted or assisted by additional forces Q_k^* (caused by springs and dampers in the same joint) then

$$Q_k = Q_k^{mot} + Q_k^*.$$

From the matrix equation the k-th equation is extracted and solved for Q_k^{mot}:

$$Q_k^{mot} = B_k' - Q_k^* - \left[A_{k1} \ldots A_{kk} \ldots A_{kN}\right]\ddot{\underline{\pi}}.$$

In the remaining N-1 equations all terms involving $\ddot{\pi}_k(t)$ are shifted to the right hand side. This produces for the remaining N-1 variables the matrix equation

$$\begin{bmatrix} A_{11} \ldots & A_{1,k-1} & A_{1,k+1} \ldots & A_{1N} \\ \vdots & & & \\ A_{k-1,1} & & & \\ A_{k+1,1} & & & \\ \vdots & & & \\ A_{N1} & & & A_{NN} \end{bmatrix} \begin{bmatrix} \ddot{\pi}_1 \\ \vdots \\ \ddot{\pi}_{k-1} \\ \ddot{\pi}_{k+1} \\ \vdots \\ \ddot{\pi}_N \end{bmatrix} = \begin{bmatrix} B_1 \\ \vdots \\ B_{k-1} \\ B_{k+1} \\ \vdots \\ B_N \end{bmatrix} - \ddot{\pi}_k(t) \begin{bmatrix} A_{1k} \\ \\ A_{k-1,k} \\ A_{k+1,k} \\ \\ A_{Nk} \end{bmatrix}$$

This equation has the same standard form as Eq.(43). Its smaller coefficient matrix is symmetric, again. This matrix is now an explicit function of time since it depends on all variables including $\pi_k(t)$. Once numerical solutions have been determined from the equation for the reduced set of variables also Q_k^{mot} is known. The method just described is easily generalized to the case with more than one variable prescribed as a function of time. In the extreme all N variables are prescribed. In this case Eq.(43) can be solved directly for the generalized forces $\underline{Q}$.

Systems With Closed Kinematic Chains

The general method of establishing equations of motion for systems with closed kinematic chains has been explained already. For a spanning tree of the system the equation

$$\delta\underline{\pi}^T(\underline{A}\ddot{\underline{\pi}} - \underline{B}) = 0 \tag{44}$$

is formulated. The variables in this equation are subject to ν independent constraint equations which, in the case of holonomic constraints, have the general form

$$f_i(\underline{\pi}, t) = 0 \qquad i=1\ldots\nu < N \; . \tag{45}$$

It should be noticed that by writing the equations in this form it is assumed that no quasi coordinates are used for the joints involved.

Two cases must be distinguished. It is either possible or not possible to solve the constraint equations explicitly for ν of the variables $\underline{\pi}$. Suppose it is possible so that we can write (after suitably renumbering the variables)

$$\pi_i = \varphi_i(\pi_1 \ldots \pi_{N-\nu}, t) \qquad i=N-\nu+1,\ldots,N. \tag{46}$$

From this follows

$$\dot{\pi}_i = \sum_{j=1}^{N-\nu} J^*_{ij}\dot{\pi}_j \qquad i=N-\nu+1,\ldots,N \tag{47}$$

$$\delta\pi_i = \begin{cases} \delta\pi_i & i=1\ldots N-\nu \\ \sum_{j=1}^{N-\nu} J^*_{ij}\delta\pi_j & i=N-\nu+1,\ldots,N \end{cases} \tag{48}$$

$$\ddot{\pi}_i = \begin{cases} \ddot{\pi}_i & i=1\ldots N-\nu \\ \sum_{j=1}^{N-\nu} J^*_{ij}\ddot{\pi}_j + H^*_i & i=N-\nu+1,\ldots,N \end{cases} \tag{49}$$

where J^*_{ij} and H^*_i are defined as

$$J^*_{ij} = \frac{\partial\varphi_i}{\partial\pi_j} \qquad i=N-\nu+1,\ldots,N;\; j=1\ldots N-\nu$$

$$H^*_i = \sum_{k=1}^{N-\nu} \frac{\partial J^*_{ij}}{\partial\pi_k}\dot{\pi}_j\dot{\pi}_k \qquad i=N-\nu+1,\ldots,N.$$

The two sets of Eqs.(48) and (49) are written in matrix form,

$$\begin{aligned} \delta\underline{\pi} &= \underline{J}\delta\underline{\pi}^* \\ \ddot{\underline{\pi}} &= \underline{J}\ddot{\underline{\pi}}^* + \underline{H}. \end{aligned} \tag{50}$$

Here, $\ddot{\underline{\pi}}$ is the same as in Eq.(44) whereas $\ddot{\underline{\pi}}^*$ is the smaller set of independent accelerations

$$\underline{\ddot{\pi}}^* = [\ddot{\pi}_1 \ldots \ddot{\pi}_{N-\nu}]^T.$$

The Jacobian matrix $\underline{J}$ of size $N\times(N-\nu)$ and the column matrix $\underline{H}$ are composed of zero and unit elements and of the elements J^*_{ij} and H^*_i as follows

$$\underline{J} = \left[\begin{array}{ccc} 1 & & \\ & 1 \ddots & \\ & & 1 \\ \hline & \underline{J}^* & \end{array}\right] \begin{array}{l} \left.\vphantom{\begin{array}{c}1\\1\\1\end{array}}\right\} N-\nu \\ \left.\vphantom{\begin{array}{c}1\end{array}}\right\} \nu \end{array} \qquad \underline{H} = \left[\begin{array}{c} \underline{0} \\ \hline \underline{H}^* \end{array}\right] \tag{51}$$

The Eqs.(50) are now substituted into Eq.(44). This yields

$$\delta\underline{\pi}^{*T}\underline{J}^T[\underline{A}(\underline{J}\underline{\ddot{\pi}}^* + \underline{H}) - \underline{B}] = 0$$

whence follows

$$\underline{J}^T\underline{A}\,\underline{J}\,\underline{\ddot{\pi}}^* = \underline{J}^T(\underline{B} - \underline{A}\underline{H}). \tag{52}$$

This represents the desired equations of motion for the smaller set of $N-\nu$ variables $\pi_1 \ldots \pi_{N-\nu}$. In the matrices $\underline{A}$ and $\underline{B}$ the dependent variables $\pi_{N-\nu+1} \ldots \pi_N$ as well as their first derivatives must, of course, be eliminated by means of Eqs.(46) and (47).

If the constraint Eqs.(45) cannot be written in the explicit form of Eq.(46) then the equations of motion can still be constructed in the form of Eq.(52) but only numerically. For given values of $\pi_1 \ldots \pi_{N-\nu}$ the implicit constraint equations must be solved numerically for $\pi_{N-\nu+1} \ldots \pi_N$. This is not difficult since at each integration step the solutions for the previous integration step represent close approximations. Numerical values for $\underline{J}^*$ and $\underline{H}^*$ are easily obtained from implicit first and second time derivatives of Eq.(45).

4. A COMPUTER PROGRAM FOR THE SYMBOLIC GENERATION OF THE EQUATIONS

At the Institute of Mechanics, Karlsruhe University a computer program was developed by Udo Wolz which generates the equations of motion in symbolic form. The program is written in portable standard PASCAL. It is fully structured in the sense that no GOTO statements are used. Adaptation to different types of computers is simple. Starting from a standard set of system parameters provided as input by the user the program automatically generates the elements of the matrices $\underline{A}$ and $\underline{B}$ in symbolic form. The elements are stored for subsequent use in numerical calculations. Numbers can then be assigned to parameters and variables as often as is desired. In order to economize on speed and storage requirements the program has been designed for the specific purpose without making use of existing general purpose programs. The program is composed of modules. Every module is capable of executing a special mathematical operation such as differentiating a function with respect to a specified variable or vector-cross multiplying two matrices with vector elements. At any given time during a computer run the central processing unit is loaded with only a small part of the program code and with a minute section of the generated formulas. Therefore, it is possible to use a small computer and yet to generate equations for large systems with many degrees of freedom. The formulas stored on external files are optimally located for quick access. Intermediately generated formulas are eliminated as soon as they become obsolete.

A small number of input parameters has to be specified in numerical form, for example the number n of bodies, the number N of variables and the two lists of integers $i^+(j)$ and $i^-(j)$ for $j=1...N$. All other input data can be specified either numerically or in symbolic form as a string of characters of the users own choice. To give an example, the masses of bodies 1 and 2 might be given as MASS1 and as 100., respectively. For every vector which is part of the input data three components must be given in a body-fixed frame of reference. In addition, the index of the body must be specified. To give an example, the vector $\vec{c}_{i^+(j),j}$ shown in Fig.4 has three components in the base fixed on body $i^+(j)$ which are known functions of the generalized coordinates. The input for the first coordinate could be given in the form A52 + B52 * X2 with unspecified parameters A52 and B52 and with X2 being the name of the cartesian variable along the joint axis. It should be mentioned that the vectors $\vec{k}_{j\ell}$ and $\vec{s}_j$ in Eq.(18) are generated automatically from $\vec{c}_{i^+(j),j}$ by the program module for differentiation.

In the course of formula generation various complicated expressions are generated which subsequently occur many times. For such expressions new names are defined as substitution variables in order to improve speed and storage requirements. Of course, not every quantity occuring more than once in subsequent calculations is given a new name because this would destroy one of the main advantages achieved by the symbolic generation of formulas. The advantage lies in the fact that a numerical evaluation of $\underline{A}$ and $\underline{B}$ based on symbolic expressions requires the computation of only a small number of substitution variables and of the final expressions as a whole whereas in a fully numerical program an extraordinary large number of array elements has to be computed. The savings are the more significant the larger the mechanical system is. Some comparative tests showed increases in speed by a factor of ten and more (in the fully numerical computations optimal use was made of recursion formulas for calculations along chains of bodies and also of the sparsity of the matrices involved).

Another point of view regarding substitution variables is worth mentioning. An efficient program for symbolic formula generation is capable of collecting terms which appear more than once. This is not possible if two such terms are hidden in two different substitution variables. Here, another advantage of the expressions developed for $\underline{A}$ and $\underline{B}$ becomes apparent. The elements of the matrix $\underline{\mathbb{K}}$ occur very many times in the final expressions for $\underline{A}$ and $\underline{B}$. They can be declared as substitution variables without running the risk of suppressing any collecting of terms. Prior to Eq.(38) the declaration of substitution variables would have done harm because with this equation a massive cancellation of terms was achieved (note that according to Eq.(36) the individual vectors $\vec{b}_{ij}$ are complicated weighted sums of other vectors).

References

1 Fischer, O., "Theoretische Grundlagen der Mechanik lebender Mechanismen (Theoretical Foundation for the Mechanics of Living Mechanisms)", Teubner, Leipzig, 1906

2 Wittenburg, J., "Dynamics of Systems of Rigid Bodies", Teubner, Stuttgart, 1977, Russian translation Moscow 1980, Chinese translation Peking 1983

3 Popov, E.P., Vereschtschagin, A.V., Senkevic, S.L., "Manipulazionnye Roboty, Dynamika i Algoritmy", Moscow, 1978

4 Paul, B., "Kinematics and Dynamics of Planar Machinary", Prentice Hall, 1979

5 Vukobratovic, M., Potkonjak, V., "Scientific Fundamentals of Robotics 1: Dynamics of Manipulation Robots", Springer, Berlin, 1982

6 Kane, T., Likins, P., Levinson, D., "Spacecraft Dynamics", McGraw-Hill, New York, 1983

7 Wittenburg, J., "Dynamics of Multibody Systems", Proc. XVth IUTAM/ICTAM Congr., Toronto, 1980

8 Wittenburg, J., "A New Correction Formula for Euler-Rodrigues Parameters", ZAMM, Vol.62, 1982, pp.495-497

9 Branin, F.H., "The Relation Between Kron's Method and the Classical Methods of Network Analysis", Matrix and Tensor Quart., Vol.12, 1962, pp.69-105

10 Roberson, R.E., Wittenburg, J., "A Dynamical Formalism for an Arbitrary Number of Interconnected Rigid Bodies. With Reference to the Problem of Satellite Attitude Control", Proc. 3rd IFAC Congr., London, 1966, 46D.2-46D.9

11 Roberson, R.E., "A Path Matrix, its Construction and its Use in Evaluating Certain Products", Comp. Meth. in Appl. Mech., to appear in Vol.151

DUAL QUATERNIONS IN THE KINEMATICS OF SPATIAL MECHANISMS

Jens Wittenburg
Institute of Mechanics
University at Karlsruhe
D-7500 Karlsruhe, FRG

Abstract. Dual quaternions comprise as special cases real numbers, vectors, dual numbers, line vectors and quaternions. All of these mathematical concepts find applications in the kinematics of large displacements of rigid bodies. Dual quaternions are particularly useful in describing the multiply constrained displacements of the individual links of spatial, single-degree-of-freedom mechanisms. An elementary introduction to the theory is presented with special emphasis on simple geometrical interpretations of the mathematical apparatus.

1. INTRODUCTION

In spatial mechanisms of general nature the individual bodies translate and rotate without having fixed points or fixed directions of angular velocity. The kinematical analysis of large displacements of such general nature requires special mathematical methods. Subject of the present paper is a method which is based on Hamilton's quaternions, on Clifford's dual numbers and dual quaternions [1] and on Study's transference principle [2]. Blaschke [3, 4] was the first to recognize the importance of these concepts for technical kinematics. The first engineering publications were due to Keler [5], Yang [6] and Yang/Freudenstein [7]. Dimentberg wrote a textbook on the subject [8]. The theory is applicable to mechanisms with revolute and prismatic joints and with joints which represent combinations of these two basic types (cylindrical joints, Hooke's joints, screw joints, ball joints).

The concepts of quaternion, dual number, line vector and dual quaternion are unfamiliar to most engineers. When these concepts are introduced in an abstract mathematical fashion it is somewhat difficult to recognize how nicely they fit the specific needs of kine-

NATO ASI Series, Vol. F9
Computer Aided Analysis and Optimization of Mechanical System Dynamics
Edited by E. J. Haug

matics. It is the purpose of the present paper to give an introduction to the theory with special emphasis on simple geometrical interpretations. For illustrative purposes the mechanism shown in Fig.1 will be used which was studied in detail by Yang [6].

2. Kinematical Parameters and Variables

Fig.1 shows an R-C-C-C mechanism, i.e. a mechanism having one revolute and three cylindrical joints. The joint axes are labeled 1 to 4 and each body is identified by the digits of the two joint axes located on it. The body labeled 41 is considered to be at rest. For arbitrary body dimensions the system has a single degree of freedom.

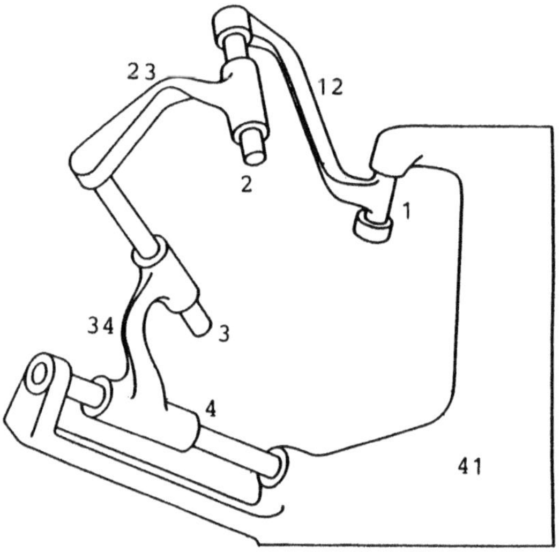

Fig.1

This is seen as follows. By removing joint 1 a system with 3x2=6 degrees of freedom is produced. When the one-degree-of-freedom joint 1 is restored 6-1=5 constraints are introduced so that the total number of degrees of freedom is 3x2-5=1. Obviously, this argument yields only a lower bound for the total number of degrees of freedom. The actual number is larger if the added constraints are not independent. Dependency of constraints occurs only under special conditions regarding the kinematical (i.e. geometrical) parameters of a system. It is one of the goals of kinematics of mechanisms to formulate such conditions. In what follows the system of Fig.1 is assumed to be of general nature and, hence, to have a single degree of freedom. A complete kinematic analysis must yield expressions for the location and angular orientation of all bodies in terms of a single input variable and of system parameters. Fig.2 explains which parameters characterize the system and how the variables should be chosen for specifying the location and angular orientation of the bodies. Shown are two representative bodies ij and jk with their joint axes i, j and k. Obviously, it is of no importance for the motion of the two bodies relative to one another whether body ij has the shape shown in solid lines or the one shown in dashed lines (Fig.a). The two shapes

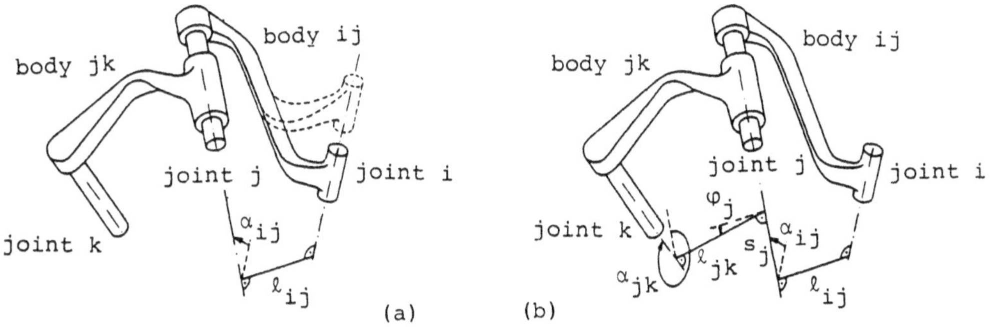

Fig.2: Parameters α_{ij} and ℓ_{ij} of body ij (a) and variables φ_j and s_j in joint j (b)

are kinematically equivalent. All that counts is the relative location of the two joint axes i and j. This can be specified by two parameters, namely by the length ℓ_{ij} of the common perpendicular and by the projected angle α_{ij} between the two axes. In Fig.2b also the respective parameters ℓ_{jk} and α_{jk} for body jk are indicated. From this figure it is seen that the joint axis j in turn represents the common perpendicular of the two common perpendiculars just described. Its pertinent section has a length s_j and in the projection along the joint axis the angle φ_j appears between the perpendiculars of lengths ℓ_{ij} and ℓ_{jk}. For the cylindrical joint j both quantities s_j and φ_j are independent variables. In a revolute joint s_j is constant and only φ_j is variable, in a prismatic joint φ_j is constant while s_j is variable and in a screw joint the two variables are related through $s_j = p_j\varphi_j + s_{jo}$ with p_j being the pitch.

In Fig.3 all the common perpendiculars of the system in Fig.1 are shown. They form a moving spatial polygon with a right angle at every corner. Instead of the mechanism only this polygon has to be investigated. It contains the nine parameters s_1, ℓ_{12}, α_{12}, ℓ_{23}, α_{23}, ℓ_{34}, α_{34}, ℓ_{41} and α_{41} and the seven variables φ_1, s_2, φ_2, s_3, φ_3, s_4 and φ_4. Since the system has a single degree of freedom only one of the variables, say φ_1, can be chosen as independent variable. The remaining ones are functions of φ_1 and of the nine parameters. It is these functions which we want to determine. By simple geometrical arguments it can be predicted that the angular variables φ_2, φ_3 and φ_4 will not depend on s_1, ℓ_{12}, ℓ_{34} and ℓ_{41}. The similarity of the roles

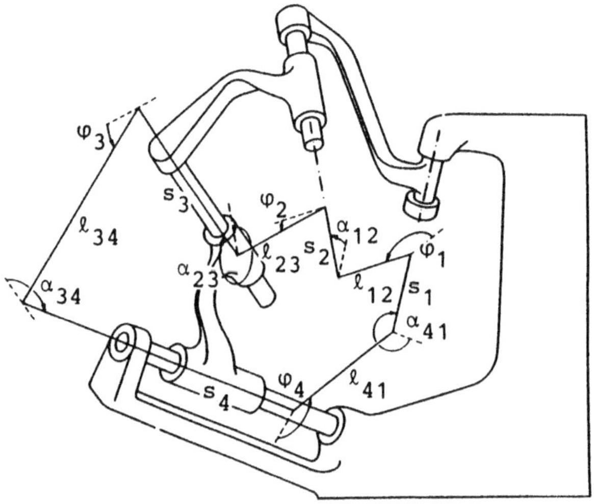

Fig.3: The spatial polygon with its nine parameters and seven variables

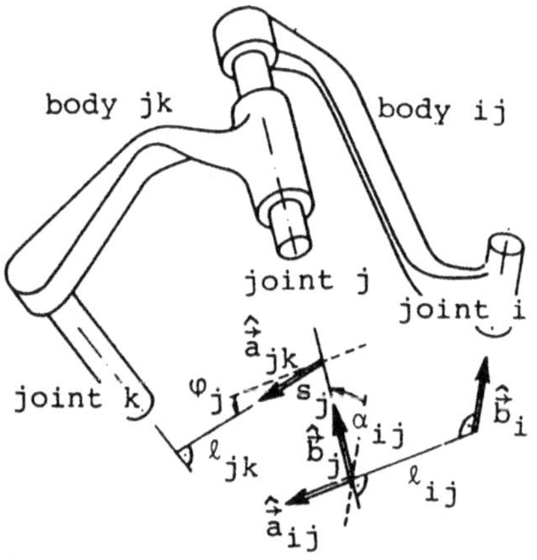

Fig.4: Unit line vectors $\hat{\vec{b}}_i$, $\hat{\vec{b}}_j$, $\hat{\vec{a}}_{ij}$ and $\hat{\vec{a}}_{jk}$ on the spatial polygon

played by the variables and by the parameters in the polygon suggests that the variables are well chosen.

Closure Conditions

With a uniform sense of direction around the spatial polygon we assign to each of its sides a so-called line vector of unit length. In contrast to the ordinary freely translating vectors a line vector is only free to translate along the line to which it is assigned. The symbol ^ will be used for indicating this particular character. In Fig.4 the notation of the line vectors is indicated. The ones assigned to the joint axes i, j etc. are denoted $\hat{\vec{b}}_i$, $\hat{\vec{b}}_j$ etc. and those assigned to the polygon sides of constant length ℓ_{ij}, ℓ_{jk} etc. are denoted $\hat{\vec{a}}_{ij}$, $\hat{\vec{a}}_{jk}$ etc.. It is seen that $\hat{\vec{b}}_j$ can be produced by applying to $\hat{\vec{b}}_i$ a screw displacement with a translational component ℓ_{ij} and a rotational component α_{ij} along the screw axis $\hat{\vec{a}}_{ij}$. The basic idea of the formalism to be developed is to construct from $\hat{\vec{a}}_{ij}$, ℓ_{ij} and α_{ij} a *screw operator* $\hat{q}_{ij}$ such that $\hat{\vec{b}}_j$ can be written as the product

$$\hat{\vec{b}}_j = \hat{q}_{ij}\hat{\vec{b}}_i \qquad \text{with} \qquad \hat{q}_{ij} = \hat{q}_{ij}(\hat{\vec{a}}_{ij}, \ell_{ij}, \alpha_{ij}). \tag{1}$$

The same kind of equation relates the vectors $\hat{\vec{a}}_{ij}$ and $\hat{\vec{a}}_{jk}$. In this case the screw operator is constructed from $\hat{\vec{b}}_j$, s_j and φ_j and it is called $\hat{q}_j$:

$$\hat{\vec{a}}_{jk} = \hat{q}_j\hat{\vec{a}}_{ij} \qquad \text{with} \qquad \hat{q}_j = \hat{q}_j(\hat{\vec{b}}_j, s_j, \varphi_j). \tag{2}$$

For the vectors $\hat{\vec{b}}_j$ of Fig.3 with j=4, 3 and 2 Eq.(1) reads

$$\hat{\vec{b}}_4 = \hat{q}_{34}\hat{\vec{b}}_3, \qquad \hat{\vec{b}}_3 = \hat{q}_{23}\hat{\vec{b}}_2, \qquad \hat{\vec{b}}_2 = \hat{q}_{12}\hat{\vec{b}}_1 \tag{3}$$

and also

$$\hat{\vec{b}}_4 = \hat{q}_{14}\hat{\vec{b}}_1. \tag{4}$$

In the last equation $\hat{q}_{14}$ is the operator of a screw displacement with translational and rotational components ℓ_{41} and α_{41}, respectively, along the axis of $-\hat{\vec{a}}_{41}$. If the associative law holds for the multiplicative rule, i.e. if

$$\hat{q}_{jk}\underbrace{(\hat{q}_{ij}\hat{\vec{b}}_i)}_{\hat{\vec{b}}_j} = (\hat{q}_{jk}\hat{q}_{ij})\hat{\vec{b}}_i \tag{5}$$

then Eq.(3) yields

$$\hat{\vec{b}}_4 = \hat{q}_{34}\hat{q}_{23}\hat{q}_{12}\hat{\vec{b}}_1 \tag{6}$$

and from a comparison with Eq.(4) follows

$$\hat{q}_{34}\hat{q}_{23}\hat{q}_{12} = \hat{q}_{14} \ . \tag{7}$$

This equation represents a closure condition for the mechanism which must yield all the desired relationships. The equation can be given several equivalent forms. Premultiplication by $\hat{q}_{43}$ produces the form

$$\hat{q}_{23}\hat{q}_{12} = \hat{q}_{43}\hat{q}_{14}$$

since it follows from Eqs.(1) and (5) that $\hat{q}_{43}$ is the inverse of $\hat{q}_{34}$. To give another formulation three more pre- and postmultiplications result in

$$\hat{q}_{12}\hat{q}_{41}\hat{q}_{34} = \hat{q}_{32}. \tag{8}$$

It will be seen later that it is useful to have different formulations.

Let it be clear that so far we have not yet established the validity of any of the above equations. The symbol $\hat{\vec{b}}_i$ denoting a line vector is still undefined. We do not know either how to construct a screw operator from its three arguments and how to multiply an operator and a line vector. Finally, the validity of the associative law remains to be shown. All this will be established in the following two sections.

3. The Special Case of Pure Rotation: The Rotation Operator

The mathematical formulation of Eq.(1) is simple if the length ℓ_{ij} of the common perpendicular in Fig.4 is equal to zero. In this case the vector $\hat{\vec{b}}_j$ is obtained from $\hat{\vec{b}}_i$ by a simple rotation through α_{ij}. In what follows ordinary vectors can be used instead of line vectors. In Fig.5 this is indicated by omitting the symbol ^. In this figure $\vec{a}$ and α have been written instead of $\vec{a}_{ij}$ and α_{ij}, respectively, in order to simplify the notation. It is seen that

$$\vec{b}_j = \cos\alpha\vec{b}_i + \sin\alpha\vec{a}\times\vec{b}_i.$$

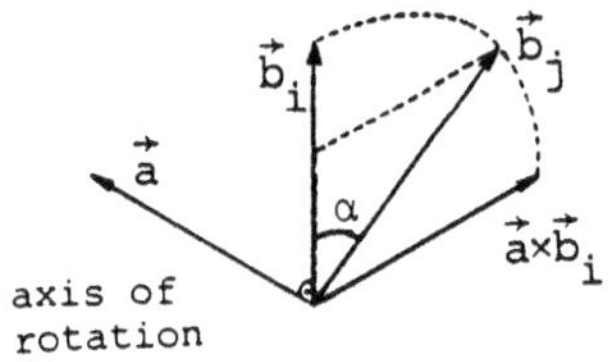

Fig.5

Defining the *rotation operator*

$$q_{ij} = \cos\alpha + \vec{a}\sin\alpha \tag{9}$$

we can express this in the form of Eq.(1) if the multiplication is defined accordingly:

$$\vec{b}_j = (\cos\alpha + \vec{a}\sin\alpha)\vec{b}_i = \cos\alpha\vec{b}_i + \sin\alpha\vec{a}\times\vec{b}_i. \tag{10}$$

The operator q_{ij} consists of a scalar and a vector part. The plus sign between the two parts must not be interpreted as addition. It merely reflects the plus sign on the right hand side of Eq.(10). Instead of + we could just as well write a semicolon.

According to a well-known theorem by Euler two successive rotations can be replaced by a single rotation. The theorem establishes the validity of Eq.(5) - the associative law - in the case of pure rotations. It states that the product $q_{jk}q_{ij}$ of two rotation operators is itself a rotation operator. How the product has to be carried out will now be established. For the sake of simplicity we write the two operators in the short forms

$$q_{ij} = u_1+\vec{v}_1, \qquad q_{jk} = u_2+\vec{v}_2.$$

For the unknown operator $q_{jk}q_{ij}$ the ansatz is made

$$q_{jk}q_{ij} = A+B_1\vec{v}_1+B_2\vec{v}_2+C\vec{v}_2\times\vec{v}_1 \tag{11}$$

with unknowns A, B_1, B_2 and C. Since $\vec{v}_1$, $\vec{v}_2$ and $\vec{v}_2\times\vec{v}_1$ span the entire space this is the most general form possible. Using the multiplication rule defined by Eq.(10) we evaluate the left hand side of Eq.(5):

$$\vec{b}_j = q_{ij}\vec{b}_i = (u_1+\vec{v}_1)\vec{b}_i = u_1\vec{b}_i+\vec{v}_1\times\vec{b}_i$$

$$q_{jk}(q_{ij}\vec{b}_i) = (u_2+\vec{v}_2)\vec{b}_j = u_2\vec{b}_j+\vec{v}_2\times\vec{b}_j.$$

From this follows

$$\begin{aligned} q_{jk}(q_{ij}\vec{b}_i) &= u_2u_1\vec{b}_1 + u_2\vec{v}_1\times\vec{b}_i + u_1\vec{v}_2\times\vec{b}_i + \vec{v}_2\times(\vec{v}_1\times\vec{b}_i) \\ &= (u_2u_1-\vec{v}_2\cdot\vec{v}_1)\vec{b}_i + u_2\vec{v}_1\times\vec{b}_i + u_1\vec{v}_2\times\vec{b}_i + \vec{v}_2\cdot\vec{b}_i\vec{v}_1. \end{aligned} \tag{12}$$

With Eq.(11) the right hand side of Eq.(5) takes the form

$$q_{jk}q_{ij}\vec{b}_i = A\vec{b}_i+B_1\vec{v}_1\times\vec{b}_i + B_2\vec{v}_2\times\vec{b}_i + C(\vec{v}_2\times\vec{v}_1)\times\vec{b}_i.$$

When the double cross product is multiplied out it must be remembered that $\vec{v}_1$ is an abbreviation for $\vec{a}\sin\alpha$ and that $\vec{a}$ is orthogonal to $\vec{b}_i$. Hence

$$q_{jk}q_{ij}\vec{b}_i = A\vec{b}_i + B_1\vec{v}_1\times\vec{b}_i + B_2\vec{v}_2\times\vec{b}_i + C\vec{v}_2\cdot\vec{b}_i\vec{v}_1.$$

Comparison with Eq.(12) yields

$$A = u_2u_1 - \vec{v}_2\cdot\vec{v}_1, \quad B_1 = u_2, \quad B_2 = u_1, \quad C = 1.$$

With these results Eq.(11) yields for the product of two rotation operators the formula

$$(u_2+\vec{v}_2)(u_1+\vec{v}_1) = \underbrace{(u_2u_1-\vec{v}_2\cdot\vec{v}_1)}_{\text{scalar part}} + \underbrace{u_2\vec{v}_1 + u_1\vec{v}_2 + \vec{v}_2\times\vec{v}_1}_{\text{vector part}} \quad . \tag{13}$$

A quantity which is composed of a scalar and a vector and for which this multiplication rule is valid is called a quaternion. The name is derived from the latin word "four" with reference to the number of scalar components of which a quaternion is composed. Quaternions represent numbers which comprise as special cases scalars as well as vectors. According to Eq.(13) the quaternion product of two scalars u_2 and u_1 is u_2u_1 and the quaternion product of two vectors $\vec{v}_2$ and $\vec{v}_1$ is the quaternion $-\vec{v}_2\cdot\vec{v}_1 + \vec{v}_2\times\vec{v}_1$. Also Eq.(10) is a special case of Eq.(13), namely the one with $u_1=0$ and $\vec{v}_2\perp\vec{v}_1$. It should be noticed that rotation operators are special quaternions in that their norm $u^2+\vec{v}^2$ equals one. This follows from Eq.(9) where $\vec{a}$ is a unit vector.

4. The Screw Operator

We begin by a discussion of numbers of the form $x+\varepsilon y$ where x and y are real. For $\varepsilon^2=+1$ and $\varepsilon^2=-1$ we obtain real and complex numbers, respectively. Clifford [1] assumed $\varepsilon^2=0$ and called $x+\varepsilon y$ in this case a dual number with primary part x and dual part y. It must be understood that $\varepsilon^2=0$ does not imply $\varepsilon=0$. The element $\varepsilon = \sqrt{0}$ is the unit of the dual part, just as $i=\sqrt{-1}$ is the imaginary unit. Clifford postulated that the commutative, the distributive and the associative laws are valid for sums and products of dual numbers. This allows to multiply out products term by term and to change the order of terms in sums and in products. One consequence is that together with ε^2 also ε^n for $n>2$ is equal to zero. Clifford showed that his postulates do not have any contradicting consequences. There is only one operation which must be ruled out as undefined, namely division by a dual number $x+\varepsilon y$ with $x=0$. The reason is that

$$\frac{1}{x+\varepsilon y} = \frac{x-\varepsilon y}{(x+\varepsilon y)(x-\varepsilon y)} = \frac{x-\varepsilon y}{x^2-\varepsilon^2 y^2} = \frac{x-\varepsilon y}{x^2} \; .$$

It is straight-forward to explain the meaning of a function $f(x+\varepsilon y)$ of a dual number. A Taylor series expansion about the point $y=0$ consists of only two terms because all second and higher order terms of ε are zero:

$$f(x+\varepsilon y) = f(x)+\varepsilon y \left.\frac{\partial f}{\partial x}\right|_{y=0}. \tag{14}$$

The expression on the right-hand side is a dual number in its standard form. This equation not only defines $f(x+\varepsilon y)$. It also provides a simple method of evaluating the primary and the dual part of $f(x+\varepsilon y)$.

We are now ready to proceed with kinematics. It was Study's idea [2] to define for two non-intersecting straight lines with a common perpendicular of length ℓ and with a projected angle α (Fig.6a) the *dual angle*

$$\hat{\alpha}=\alpha+\varepsilon\ell. \tag{15}$$

Eq.(14) provides the means for calculating functions of $\hat{\alpha}$. Examples are

$$\cos\hat{\alpha} = \cos(\alpha+\varepsilon\ell) = \cos\alpha-\varepsilon\ell\sin\alpha, \quad \sin\hat{\alpha} = \sin\alpha+\varepsilon\ell\cos\alpha. \tag{16}$$

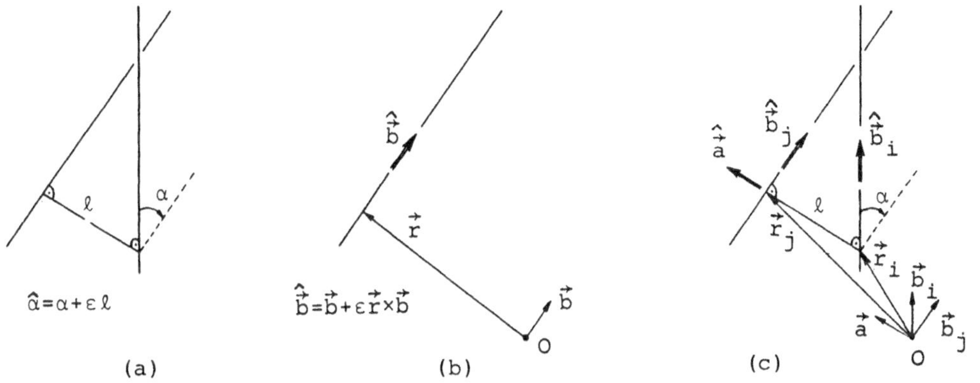

Fig.6a: Components α and ℓ of the dual angle formed by two non-intersecting straight lines.

Fig.6b: The unit line vector $\hat{\vec{b}}$ is specified by two ordinary vectors $\vec{b}$ and $\vec{r}$.

Fig.6c: Unit line vectors $\hat{\vec{b}}_i$, $\hat{\vec{b}}_j$ and $\hat{\vec{a}}$ along two non-intersecting lines and along the common perpendicular, respectively.

Next, we turn to the problem of describing line vectors in mathematical form. Fig.6b shows a unit line vector $\hat{\vec{b}}$. To be specified are the location of its line in space relative to some reference point O and the sense of direction of $\hat{\vec{b}}$. This can be done by means of two ordinary vectors $\vec{r}$ and $\vec{b}$. The former connects O with an arbitrary point on the line and the latter is a unit vector parallel to $\hat{\vec{b}}$. The two vectors uniquely specify $\hat{\vec{b}}$ but they do not define it. At this point dual numbers are introduced once more. The unit line vector is defined as the *dual vector*

$$\hat{\vec{b}} = \vec{b}+\varepsilon\vec{r}\times\vec{b} \ . \tag{17}$$

That the dual part is written as $\vec{r}\times\vec{b}$ and not simply as $\vec{r}$ has the advantage that $\vec{r}\times\vec{b}$ is the same vector no matter to which point of the line $\vec{r}$ is directed. However, how ingenious the definition really is can only be seen when relationships are established between the line vectors and angles in Fig.6c. The figure shows two non-intersecting lines with the projected angle α and with the common perpendicular of length ℓ. Unit line vectors $\hat{\vec{b}}_i$, $\hat{\vec{b}}_j$ and $\hat{\vec{a}}$ are located on the three lines. The situation is the same as in Fig.4 except that the indices ij have been omitted from $\hat{\vec{a}}_{ij}$, ℓ_{ij} and α_{ij}. The essential difference to the situation in Fig.5 is the translational displacement $\ell \neq 0$. In Fig.6c also the ordinary vectors $\vec{r}_i$, $\vec{r}_j$, $\vec{a}$, $\vec{b}_i$ and $\vec{b}_j$ are shown which define the line vectors as

$$\hat{\vec{a}} = \vec{a}+\varepsilon\vec{r}_i\times\vec{a}. \quad \hat{\vec{b}}_i = \vec{b}_i+\varepsilon\vec{r}_i\times\vec{b}_i, \quad \hat{\vec{b}}_j = \vec{b}_j+\varepsilon\vec{r}_j\times\vec{b}_j. \tag{18}$$

Between the ordinary vectors we have the relationships

$$\begin{aligned} &\vec{b}_i\cdot\vec{b}_j = \cos\alpha, \qquad &&\vec{b}_i\times\vec{b}_j = \vec{a}\sin\alpha\,, \qquad &&\vec{r}_j-\vec{r}_i = \vec{a}\ell, \\ &\vec{b}_i\cdot\vec{r}_j = \vec{b}_i\cdot\vec{r}_i, \qquad &&(\vec{r}_j-\vec{r}_i)\cdot(\vec{b}_i\times\vec{b}_j) = \ell\sin\alpha\,, \qquad &&\vec{a}\cdot\vec{b}_i = 0. \end{aligned} \tag{19}$$

Finally, $\vec{b}_i$ and $\vec{b}_j$ are related through Eq.(10)

$$\vec{b}_j = \cos\alpha\vec{b}_i+\sin\alpha\vec{a}\times\vec{b}_i. \tag{20}$$

It is our goal to express $\hat{\vec{b}}_j$ in the form of Eq.(1) in terms of $\hat{\vec{b}}_i$ $\hat{\vec{a}}$, ℓ and α. We begin by an investigation of the dot product $\hat{\vec{b}}_i\cdot\hat{\vec{b}}_j$ and of the cross product $\hat{\vec{b}}_i\times\hat{\vec{b}}_j$ of two line vectors. Keeping in mind that $\varepsilon^2 = 0$ we can rewrite the dot product as

$$\begin{aligned} \hat{\vec{b}}_i\cdot\hat{\vec{b}}_j &= (\vec{b}_i + \varepsilon\vec{r}_i\times\vec{b}_i)\cdot(\vec{b}_j + \varepsilon\vec{r}_j\times\vec{b}_j) \\ &= \vec{b}_i\cdot\vec{b}_j + \varepsilon(\vec{b}_i\cdot\vec{r}_j\times\vec{b}_j + \vec{r}_i\times\vec{b}_i\cdot\vec{b}_j) \\ &= \vec{b}_i\cdot\vec{b}_j + \varepsilon(-\vec{r}_j\cdot\vec{b}_i\times\vec{b}_j + \vec{r}_i\cdot\vec{b}_i\times\vec{b}_j) \\ &= \vec{b}_i\cdot\vec{b}_j - \varepsilon(\vec{r}_j - \vec{r}_i)\cdot\vec{b}_i\times\vec{b}_j \\ &= \cos\alpha - \varepsilon\ell\sin\alpha \end{aligned}$$

In view of Eqs.(16) this is

$$\hat{\vec{b}}_i \cdot \hat{\vec{b}}_j = \cos\hat{\alpha}. \tag{21}$$

This means that the definition $\vec{b}_i \cdot \vec{b}_j = \cos\alpha$ valid for the dot product of ordinary unit vectors can be formally transfered to unit line vectors if line vectors and angles are defined the way they are.

Next, the vector cross product $\hat{\vec{b}}_i \times \hat{\vec{b}}_j$ is considered. Taking into account Eqs.(19) we can write

$$\begin{aligned}
\hat{\vec{b}}_i \times \hat{\vec{b}}_j &= (\vec{b}_i + \varepsilon\vec{r}_i\times\vec{b}_i)\times(\vec{b}_j + \varepsilon\vec{r}_j\times\vec{b}_j)\\
&= \vec{b}_i\times\vec{b}_j + \varepsilon[\vec{b}_i\times(\vec{r}_j\times\vec{b}_j) + (\vec{r}_i\times\vec{b}_i)\times\vec{b}_j]\\
&= \vec{b}_i\times\vec{b}_j + \varepsilon[\vec{b}_i\cdot\vec{b}_j\vec{r}_j - \vec{b}_i\cdot\vec{r}_j\vec{b}_j + \vec{r}_i\cdot\vec{b}_j\vec{b}_i - \vec{b}_i\cdot\vec{b}_j\vec{r}_i]\\
&= \vec{b}_i\times\vec{b}_j + \varepsilon[\vec{b}_i\cdot\vec{b}_j(\vec{r}_j - \vec{r}_i) + \vec{r}_i\times(\vec{b}_i\times\vec{b}_j)]\\
&= \vec{a}\sin\alpha + \varepsilon(\cos\alpha\,\vec{a}\ell + \vec{r}_i\times\vec{a}\sin\alpha)\\
&= (\vec{a} + \varepsilon\vec{r}_i\times\vec{a})(\sin\alpha + \varepsilon\ell\cos\alpha)
\end{aligned}$$

In view of Eqs.(16) and (18) this is

$$\hat{\vec{b}}_i \times \hat{\vec{b}}_j = \hat{\vec{a}}\sin\hat{\alpha}. \tag{22}$$

This means that the definition $\vec{b}_i\times\vec{b}_j = \vec{a}\sin\alpha$ valid for the cross product of ordinary unit vectors can be formally transfered to unit line vectors.

The relationships just established suggest that also an equivalent to Eq.(10) exists in the form

$$\hat{\vec{b}}_j = (\cos\hat{\alpha} + \hat{\vec{a}}\sin\hat{\alpha})\hat{\vec{b}}_i = \cos\hat{\alpha}\,\hat{\vec{b}}_i + \sin\hat{\alpha}\,\hat{\vec{a}}\times\hat{\vec{b}}_i. \tag{23}$$

For checking this we substitute in the last of Eqs.(18) for $\vec{b}_j$ the expression of Eq.(20). Using some of the relationships (19) we obtain

$$\begin{aligned}
\hat{\vec{b}}_j &= \cos\alpha\vec{b}_i + \sin\alpha\vec{a}\times\vec{b}_i + \varepsilon(\vec{r}_i + \vec{a}\ell)\times(\cos\alpha\vec{b}_i + \sin\alpha\vec{a}\times b_i)\\
&= \cos\alpha\vec{b}_i + \varepsilon\cos\alpha(\vec{r}_i\times\vec{b}_i + \ell\vec{a}\times\vec{b}_i) + \sin\alpha\vec{a}\times\vec{b}_i + \varepsilon\sin\alpha(\vec{r}_i\cdot\vec{b}_i\vec{a} - \vec{r}_i\cdot\vec{a}\vec{b}_i - \ell\vec{b}_i).
\end{aligned} \tag{24}$$

On the right hand side of Eq.(23) Eqs.(16) are substituted for the circular functions and Eqs.(18) for $\hat{\vec{a}}$ and $\hat{\vec{b}}_i$:

$$\hat{\vec{b}}_j = (\cos\alpha - \varepsilon\ell\sin\alpha)(\vec{b}_i + \varepsilon\vec{r}_i\times\vec{b}_i) + (\sin\alpha + \varepsilon\ell\cos\alpha)(\vec{a} + \varepsilon\vec{r}_i\times\vec{a})\times(\vec{b}_i + \varepsilon\vec{r}_i\times\vec{b}_i).$$

When this is multiplied out the same expression is obtained as in

Eq.(24). Thus, it is verified that the expression in brackets in Eq.(23) represents the desired screw operator. The operator is a dual quaternion with the norm

$$\cos^2\hat{\alpha}+\vec{\hat{a}}^2\sin^2\hat{\alpha} = (\cos\alpha-\varepsilon\ell\,\sin\alpha)^2+(\vec{a}+\varepsilon\vec{r}_i\times\vec{a})^2(\sin\alpha+\varepsilon\ell\,\cos\alpha)^2$$
$$= (\cos\alpha-\varepsilon\ell\,\sin\alpha)^2+(\sin\alpha+\varepsilon\ell\,\cos\alpha)^2 = 1\ .$$

Dual quaternions consist of eight components. They comprise as special cases ordinary quaternions, dual numbers, line vectors, real numbers and ordinary vectors.

We now return to Eqs.(1) and (2),

$$\vec{\hat{b}}_j = \hat{q}_{ij}\vec{\hat{b}}_i \quad \text{and} \quad \vec{\hat{a}}_{jk} = \hat{q}_j\vec{\hat{a}}_{ij}. \tag{25}$$

The operators have been found to be

$$\hat{q}_{ij} = \cos\hat{\alpha}_{ij}+\vec{\hat{a}}_{ij}\sin\hat{\alpha}_{ij} \quad \text{with} \quad \hat{\alpha}_{ij} = \alpha_{ij}+\varepsilon\ell_{ij}, \tag{26}$$

$$\hat{q}_i = \cos\hat{\varphi}_i + \vec{\hat{b}}_i\sin\hat{\varphi}_i \quad \text{with} \quad \hat{\varphi}_i = \varphi_i + \varepsilon s_i. \tag{27}$$

In analogy to Eq.(13) the multiplication rule for screw operators reads

$$(\hat{u}_2+\vec{\hat{v}}_2)(\hat{u}_1+\vec{\hat{v}}_1) = (\hat{u}_2\hat{u}_1-\vec{\hat{v}}_2\cdot\vec{\hat{v}}_1)+(\hat{u}_2\vec{\hat{v}}_1+\hat{u}_1\vec{\hat{v}}_2+\vec{\hat{v}}_2\times\vec{\hat{v}}_1). \tag{28}$$

This follows from the fact that all products occuring in Eq.(13) are formally transferable to the respective dual quantities. Eqs.(25) through (28) provide the basis for extracting information from the closure condition in one of its forms, for example from Eq.(8).

The fact that the mathematics of finite rotational displacements can be formally transfered to finite screw displacements was discovered by Study [2]. It is referred to as transference principle.

5. Interpretation of Closure Conditions

Since a closure condition is a dual quaternion equation it actually represents eight equations. These are obtained by splitting the equation in a primary and a dual part and each of these parts into one scalar and one vector part the latter representing itself three scalar components. Only six out of the eight equations are independent because there are two constraint equations which state that the primary as well as the dual part of a dual quaternion has the norm one.

Six is precisely the number of dependent variables not only in the mechanism of Fig.1 but in any n-body mechanism forming a single closed loop (provided the constraints in the mechanism are independent). This is seen as follows. Without any constraints the n-1 mobile bodies have 6(n-1) degrees of freedom. If joint j has ν_j variables and, hence, $6-\nu_j$ constraints then the overall number of degrees of freedom, i.e. of independent variables, is

$$N = 6(n-1) - \sum_{j=1}^{n}(6-\nu_j) = \sum_{j=1}^{n}\nu_j - 6 \tag{29}$$

whereas the sum over j represents the total number of variables. This proves the statement.

Before information can be extracted from a closure condition a cartesian reference base of unit line vectors $\hat{\vec{i}}$, $\hat{\vec{j}}$, $\hat{\vec{k}}$ must be chosen and all line vectors of the mechanism must be decomposed in this reference base. The decomposition is done by using repeatedly Eqs.(25). In the case of Fig.3 the reference base is chosen as shown in Fig.7 with

$$\hat{\vec{i}} = \hat{\vec{a}}_{41}, \qquad \hat{\vec{k}} = \hat{\vec{b}}_1, \qquad \hat{\vec{j}} = \hat{\vec{k}}\times\hat{\vec{i}}.$$

This has the advantage that the quaternion $\hat{q}_{41}$ has already the desired form

$$\hat{q}_{41} = \cos\hat{\alpha}_{41} + \hat{\vec{i}}\sin\hat{\alpha}_{41}\,.$$

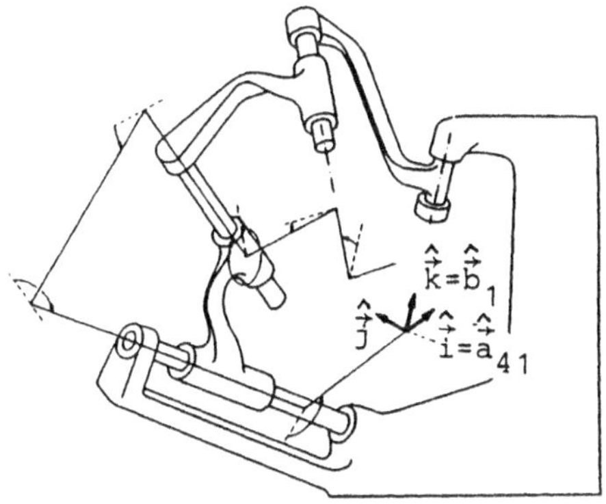

Fig.7: Unit base line vectors $\hat{\vec{i}}$, $\hat{\vec{j}}$, $\hat{\vec{k}}$ for the spatial polygon

The decomposition of the other vectors is achieved as follows. Determine $\hat{\vec{a}}_{12}$ from Eq.(25b), $\hat{q}_{12}$ from Eq.(26), $\hat{\vec{b}}_4$ from Eq.(25a) for j=1 and i=4, $\hat{\vec{a}}_{34}$ from Eq.(25b) for j=4 and i=1, $\hat{q}_{34}$ from Eq.(26), $\hat{\vec{b}}_2$ from Eq.(25a), $\hat{\vec{a}}_{23}$ from Eq.(25b) and $\hat{q}_{23}$ from Eq.(26). In these calculations it is several times necessary to solve an equation of the form

$$\hat{\vec{b}}_j = \hat{q}_{ij}\hat{\vec{b}}_i = (\cos\hat{\alpha}_{ij}+\hat{\vec{a}}_{ij}\sin\hat{\alpha}_{ij})\hat{\vec{b}}_i$$

for $\hat{\vec{b}}_i$. In discussing Fig.4 it was said already that the inverse of a screw operator is obtained by reversing the sign of the unit line vector. Hence, the result reads

$$\hat{\vec{b}}_i = (\cos\hat{\alpha}_{ij}-\hat{\vec{a}}_{ij}\sin\hat{\alpha}_{ij})\hat{\vec{b}}_j .$$

After the transformation of vectors the quaternions $\hat{q}_{41}$, $\hat{q}_{12}$, $\hat{q}_{34}$ and $\hat{q}_{23}$ are expressed in terms of the unit base vectors. Using Eq.(28) the closure condition can now be written in any one of its different forms. In what follows the form of Eq.(8) will be used. The scalar part of the equation results in the relationship (see Yang [6])

$$\hat{A}\sin\hat{\varphi}_4+\hat{B}\cos\hat{\varphi}_4 = \hat{C} \tag{30}$$

with the abbreviations

$$\hat{A} = \sin\hat{\alpha}_{12}\sin\hat{\alpha}_{34}\sin\hat{\varphi}_1$$

$$\hat{B} = -\sin\hat{\alpha}_{34}(\sin\hat{\alpha}_{41}\cos\hat{\alpha}_{12} + \cos\hat{\alpha}_{41}\sin\hat{\alpha}_{12}\cos\hat{\varphi}_1)$$

$$\hat{C} = \cos\hat{\alpha}_{23} - \cos\hat{\alpha}_{34}(\cos\hat{\alpha}_{41}\cos\hat{\alpha}_{12} - \sin\hat{\alpha}_{41}\sin\hat{\alpha}_{12}\cos\hat{\varphi}_1).$$

The equation relates $\hat{\varphi}_4 = \varphi_4+\varepsilon s_4$ to the independent variable $\hat{\varphi}_1 = \varphi_1+\varepsilon s_1$. According to Eq.(14) the primary part of the equation is obtained by omitting everywhere the symbol ^:

$$A(\varphi_1)\sin\varphi_4 + B(\varphi_1)\cos\varphi_4 = C(\varphi_1).$$

This is an equation for φ_4 in terms of φ_1 and of the parameters α_{12}, α_{23}, α_{34} and α_{41}. In accordance with our expectation the parameters s_1, ℓ_{12}, ℓ_{23}, ℓ_{34} and ℓ_{41} do not influence φ_4. With the help of an auxiliary angle β defined by

$$A = \sqrt{A^2+B^2}\cos\beta, \qquad B = \sqrt{A^2+B^2}\sin\beta$$

the equation becomes

$$\sin(\varphi_4+\beta) = C/\sqrt{A^2+B^2} .$$

It has the two solutions

$$\varphi_4 = \begin{cases} -\beta + \sin^{-1}(C/\sqrt{A^2+B^2}) \\ -\beta - \sin^{-1}(C/\sqrt{A^2+B^2}) + \pi . \end{cases} \tag{31}$$

According to Eq.(14) the dual part of Eq.(30) is obtained by differentiation. This results in the explicit expression for s_4 in terms of φ_1 and of $\varphi_4 = \varphi_4(\varphi_1)$

$$s_4 = \frac{D\sin\varphi_4 + E\cos\varphi_4 + F}{A\cos\varphi_4 - B\sin\varphi_4} . \tag{32}$$

Here, A and B are the same quantities as before and D, E and F are defined as follows

$$D = -(\ell_{12}\cos\alpha_{12}\sin\alpha_{34} + \ell_{34}\sin\alpha_{12}\cos\alpha_{34})\sin\varphi_1 - (s_1\sin\alpha_{12}\sin\alpha_{34})\cos\varphi_1 ,$$

$$\begin{aligned} E = {} & \ell_{41}\sin\alpha_{34}(\cos\alpha_{41}\cos\alpha_{12} - \sin\alpha_{41}\sin\alpha_{12}\cos\varphi_1) - \\ & - \ell_{12}\sin\alpha_{34}(\sin\alpha_{41}\sin\alpha_{12} - \cos\alpha_{41}\cos\alpha_{12}\cos\varphi_1) + \\ & + \ell_{34}\cos\alpha_{34}(\sin\alpha_{41}\cos\alpha_{12} + \cos\alpha_{41}\sin\alpha_{12}\cos\varphi_1) - s_1\sin\alpha_{34}\cos\alpha_{41}\sin\alpha_{12}\sin\varphi_1 , \end{aligned}$$

$$\begin{aligned} F = {} & \ell_{41}\cos\alpha_{34}(\sin\alpha_{41}\cos\alpha_{12} + \cos\alpha_{41}\sin\alpha_{12}\cos\varphi_1) + \\ & + \ell_{12}\cos\alpha_{34}(\cos\alpha_{41}\sin\alpha_{12} + \sin\alpha_{41}\cos\alpha_{12}\cos\varphi_1) - \ell_{23}\sin\alpha_{23} + \\ & + \ell_{34}\sin\alpha_{34}(\cos\alpha_{41}\cos\alpha_{12} - \sin\alpha_{41}\sin\alpha_{12}\cos\varphi_1) - s_1\cos\alpha_{34}\sin\alpha_{41}\sin\alpha_{12}\sin\varphi_1 . \end{aligned}$$

In order to arrive at explicit forms also for the remaining dependent variables φ_2, s_2, φ_3 and s_3 the dual vector part of Eq.(8) has to be analyzed. For details the reader is referred to Yang [6].

The mechanism of Fig.1 is a particular case. Not for every spatial mechanism each of the six dependent variables can be expressed in explicit form as a function of the input variable. In order to find either explicit solutions or acceptably simple forms of implicit relationships it is necessary to investigate the closure condition in more than one of its equivalent formulations. The calculations leading to Eqs.(31) and (32) have shown that a very considerable amount of labor is involved in formulating the eight components of any particular form of the equation. With each successive multiplication by a dual quaternion the expressions become substantially more complicated. In order to eliminate not only the labor involved but also the risk of errors an interactive computer program was developed at Karlsruhe by U. Wolz which generates the eight components of any closure condition in analytical form.

6. Overclosure of Mechanisms

In general the dependent variables of a mechanism are functions of the independent input variable and of the system parameters. However, in some cases it is possible to select for the system parameters such a special set of values that one (or more) dependent variable does not depend explicitly on the input variable. Instead, it is constant. Mechanisms exhibiting this property are said to be overclosed. In the case of Fig.3 it is possible to select the parameters such that s_2, s_3 and s_4 are simultaneously identically zero. This was first recognized by Bennett after whom this special family of mechanisms is called. Bennett mechanisms also have the property that φ_4 is identically equal to φ_1. Dimentberg [8] has shown how the conditions to be satisfied by the parameters can be deduced from the conditions $s_4 \equiv 0$, $\varphi_4 \equiv \varphi_1$. The idea is to rewrite the primary and the dual part of Eq.(30) in the form of two coupled polynomial equations by means of the substitution

$$\hat{z}_i = x_i + \varepsilon y_i = \tan\hat{\varphi}_i/2 \qquad i = 1,4.$$

The constraints on s_4 and φ_4 require a certain determinant, the so-called resultant of the two polynomial equations, to be zero. From this condition the constraints on the parameters are deduced.

Other types of overclosed mechanisms can be found in the literature. A six-body mechanism with one independent and five dependent variables was analyzed in ref.[9]. An extreme case of overclosure has been described by Connelly [10]. His mechanism is a polyhedron. Its 14 rigid faces represent the bodies and the 21 edges represent revolute joints. The equivalent of Eq.(29) yields for the total number of degrees of freedom

$$N = 6\cdot 13 - 21\cdot(6-1) = -27.$$

Yet, the actual number is N=1. Connelly's mechanism is a counterexample disproving the conjecture dating back to Euler that all polyhedra are rigid.

Bennett's mechanism was seen to result either from the condition $s_2 = s_3 = s_4 \equiv 0$ or from the condition $s_4 \equiv 0$, $\varphi_4 \equiv \varphi_1$. Each of these conditions is superimposed as an additional constraint upon the six (explicit or implicit) relationships between the dependent and independent variables. It is obvious that not any arbitrarily chosen additional constraint can be satisfied by a suitable choice of system parameters. Admissible additional constraints can be formulated on the basis of a planar mechanism which is known to be a special case of the spatial mechanism under consideration. To give an example, the planar four-bar parallelogram mechanism is a very special case of the mechanism shown in Fig.1 and it has the properties that $s_2 = s_3 = s_4 \equiv 0$ and that $\varphi_1 \equiv \varphi_4$. It follows that these constraints are admissible. In this manner one can choose any special property of any other planar four-bar mechanism. It will always be admissible as additional constraint on the system of Fig.1. The analysis will result in a family of spatial mechanisms which share this particular property with the planar mechanism.

References

1 Clifford, W., "Preliminary Sketch of Biquaternions" Proc. London Math. Soc. Vol. IV, 1873

2 Study, E., "Geometrie der Dynamen" Stuttgart: Teubner 1901-1903

3 Blaschke, W., "Anwendung dualer Quaternionen auf Kinematik" Ann. Acad. Sci. Fenn. Ser.A, 1.Math.250/3, 1958

4 Blaschke, W., "Kinematik und Quaternionen" VEB Deutscher Verlag der Wissenschaften, Berlin 1960

5 Keler, M., "Analyse und Synthese der Raumkurbelgetriebe mittels Raumliniengeometrie und dualer Größen" Diss. München 1958. Auszug: Forsch. Ingenieurwes. 25 (1959) 26-32 u. 55-63

6 Yang, A.T.,"Application of Quaternion Algebra and Dual Numbers to the Analysis of Spatial Mechanisms" Diss.Col. Univ. N.Y., Libr. of Congr. No. Mic.64-2803, Ann Arbor

7 Yang, A.T., Freudenstein, F., "Application of Dual-Number Quaternion Algebra to the Analysis of Spatial Mechanisms" J. Appl. Mech. 86 (1964) 300-308

8 Dimentberg, F.M., "Theory of Screws and its Applications" (in Russian), Moscow NAUKA 1978

9 Wittenburg, J., "Dynamics of Systems of Rigid Bodies" LAMM ser.vol. 33 Teubner 1977

10 Connelly, R., "The Rigidity of Polyhedral Surfaces" Mathematics Magazine 52 (1979) 275-283

QUATERNIONS AND EULER PARAMETERS - A BRIEF EXPOSITION

Roger A. Wehage
US Army Tank-Automotive Command
Warren, Michigan 48090

Abstract. The quaternion concept has found successful applications in many areas of the physical sciences. In the kinematics and dynamics of spatial mechanical systems and synthesis of mechanisms, quaternion theory may be found under the guise of Euler parameters, dual numbers, dual quaternions, rotation tensors, screw axis calculus, etc. Quaternion algebra has been applied to obtain analytical solutions, and to classify single- and multi-degree-of-freedom motions of many closed loop spatial mechanisms. The resulting systems of algebraic equations are generally extremely complex and difficult to interpret or transform to computer programs. The objective of this paper is to look at some of the basic quaternion algebra and identities, and their corresponding matrix representations to aid in the development of mechanism anaysis capabilities and computer algorithms.

1. INTRODUCTION

The quaternion concept has been successfully applied to many areas of the physical sciences. In this regard, quaternions have taken on many interpretations, and consequently many definitions can be found. The various quaternion representations were adapted to fit physical situations, and thus enhance the development of new theories and corresponding governing equations. Quaternions may consist of one, two or three imaginary numbers and correspondingly three, two or one real number, or a scalar such as time and three spatial vectors, etc. The important requirement of quaternions is that they be composed of four

NATO ASI Series, Vol. F9
Computer Aided Analysis and Optimization of Mechanical System Dynamics
Edited by E. J. Haug

orthogonal components. Regardless of the quaternion definition, a number of underlying linear operations involving quaternions and their conjugates exist. Conjugates are obtained by negating the imaginary or vector parts. Quaternion operations ultimately are described by a number of powerful matrix identities involving only scalar amplitudes of the orthogonal components. For example, multiplication and subsequently division are described by two linear 4 by 4 orthogonal matrices, whose individual elements consist of single quaternion components. These matrices are unique, differing only by row and column permutations for the various quaternion definitions.

In light of the above discussion, all spatial vectors are quaternions with zero scalar terms. Euler parameters are normalized quaternions with spatial orientation vectors and scalar normalizing terms. Quaternion theory can thus be applied to develop all the governing equations involving Euler parameters and spatial vectors.

Quaternions have not gained significant popularity in mechanical system dynamic analysis primarily because they are difficult to interpret in a three dimensional space and thus are not well understood. In addition it is generally unrecognized that scalars and spatial vectors are also quaternions and can be included in the established set of quaternion algebra. Thus it has been necessary to transform between four dimensional quaternion space and three dimensional vector space by utilizing left and right inverses. On the other hand quaternion transformations are always orthogonal, their inverses are easy to obtain and only the null quaternion leads to a singular transformation.

Taking quaternions in the context of a linear combination of a scalar and three orthogonal spatial vectors, all the rules of scalar and vector algebra apply. Let $\vec{i}$, $\vec{j}$ and $\vec{k}$ be a set of orthogonal unit vectors and define a quaternion as any quantity of the form

$$\hat{a} = a_0 + a_1 \vec{i} + a_2 \vec{j} + a_3 \vec{k} \tag{1}$$

If $\hat{a} = a_0$, $\hat{a}$ is called a scalar quaternion and if $a_0 = 0$, $\hat{a}$ is called a vector quaternion. In this respect any quaternion used here may be considered as the sum of a scalar and vector, i.e.

$$\hat{a} = a_0 + \vec{a} \tag{2}$$

2. QUATERNION MULTIPLICATION

Let $\hat{b} = b_0 + \vec{b}$ and define the quaternion product

$$\hat{c} = \hat{a}\,\hat{b} \tag{3}$$

Substituting the expressions for $\hat{a}$ and $\hat{b}$ into Eq. 3 yields

$$\hat{c} = a_0\, b_0 + a_0\, \vec{b} + b_0\, \vec{a} + \vec{a}\, \vec{b} \tag{4}$$

The scalar and scalar-vector products are well defined, and the vector product will be defined as

$$\vec{a}\, \vec{b} = \vec{a} \times \vec{b} - \vec{a} \cdot \vec{b} \tag{5}$$

in order to achieve uniqueness, and as a consequence, orthogonality of the quaternion product[1]. Substituting Eq. 5 into Eq. 4 yields the basic definition of quaternion product in terms of vector algebra as

$$\hat{c} = a_0\, b_0 - \vec{a} \cdot \vec{b} + a_0\, \vec{b} + b_0\, \vec{a} + \vec{a} \times \vec{b} \tag{6}$$

Note that the first two terms are scalar quantities and the remaining three terms are vectors.

The conjugate of $\hat{a}$ denoted by $\hat{a}^*$ is obtained by negating its vector component, thus $\hat{a}^* = a_0 - \vec{a}$. If $\hat{d} = \hat{a}^*\, \hat{b}$ then

$$\hat{d} = a_0 b_0 + \vec{a} \cdot \vec{b} + a_0 \vec{b} - b_0 \vec{a} - \vec{a} \times \vec{b} \tag{7}$$

In a similar manner if $\hat{e} = \hat{b}^* \hat{a}^*$ then

$$\hat{e} = a_0 b_0 - \vec{a} \cdot \vec{b} - a_0 \vec{b} - b_0 \vec{a} + \vec{a} \times \vec{b} \tag{8}$$

Comparing Eqs. 6 and 8 reveals that

$$(\hat{a}\,\hat{b})^* = \hat{b}^* \hat{a}^* \tag{9}$$

Likewise

$$(\hat{a}^* \hat{b})^* = \hat{b}^* \hat{a} \tag{10}$$

Substituting $\hat{a}$ for $\hat{b}$ and $\hat{a}^*$ for $\hat{b}^*$ in Eqs. 7 and 10 yields

$$\hat{a}^* \hat{a} = \hat{a}\,\hat{a}^* = a_0^2 + \vec{a} \cdot \vec{a} \tag{11}$$

Observe that if $\hat{a} = \hat{a}^*$, $\hat{a}$ is a scalar quaternion a_0 and if $\hat{a} = -\hat{a}^*$, $\hat{a}$ is a vector quaternion $\vec{a}$. These examples demonstrate that quaternion multiplication generates new quaternions and in some special cases they yield vector quaternions, or as in Eq. 11 scalar quaternions.

It is convenient to represent all quaternion and vector operations in matrix form to simplify equation manipulation and computer programming efforts. Matrices are simply collections of the scalar coefficients multiplying the orthogonal units defining the basis of the quaternion set. According to Eq. 1 this basis set is $(1, \vec{i}, \vec{j}, \vec{k})$ and the coefficients in $\hat{a}$ are (a_0, a_1, a_2, a_3). Define the matrices or column vectors

$$\underset{\sim}{a} = (a_0, a_1, a_2, a_3)^T \tag{12a}$$

$$\underline{a} = (a_1, a_2, a_3)^T \tag{12b}$$

or

$$\underline{a} = (0, a_1, a_2, a_3)^T \tag{12c}$$

where $(\ )^T$ means matrix transpose. Observe that the vector $\vec{a}$ is represented in Eqs. 12b and 12c by 3 x 1 or 4 x 1 vectors and any ambiguity will be removed by the context of its usage.

The product in Eq. 3, defined by Eq. 6 can be represented in matrix form as

$$\underset{\sim}{c} = \begin{bmatrix} c_0 \\ -- \\ \underline{c} \end{bmatrix} = \begin{bmatrix} a_0 \ b_0 - \underline{a}^T \ \underline{b} \\ \text{-----------------} \\ a_0 \ \underline{b} + b_0 \ \underline{a} + \tilde{\underline{a}} \ \underline{b} \end{bmatrix} \tag{13a}$$

$$= \begin{bmatrix} b_0 \ a_0 - \underline{b}^T \ \underline{a} \\ \text{-----------------} \\ b_0 \ \underline{a} + a_0 \ \underline{b} - \tilde{\underline{b}} \ \underline{a} \end{bmatrix} \tag{13b}$$

where the algebraic equivalent of $\vec{a}\cdot$ and $\vec{a}\times$ are defined respectively by $\underline{a}^T$ and the skew-symmetric matrix

$$\tilde{\underline{a}} = \begin{bmatrix} 0 & -a_3 & a_2 \\ a_3 & 0 & -a_1 \\ -a_2 & a_1 & 0 \end{bmatrix} \tag{14a}$$

Since $\vec{a} \times \vec{b} = - \vec{b} \times \vec{a}$, it follows that

$$\underset{\sim}{\underline{a}}\, \underline{b} = - \underset{\sim}{\underline{b}}\, \underline{a} \tag{14b}$$

and these skew-symmetric matrices satisfy the relation $\tilde{\underline{a}}^T = - \tilde{\underline{a}}$. The terms in Eq. 13 can be factored into two equivalent matrix forms

$$\begin{bmatrix} c_0 \\ \hline \underline{c} \end{bmatrix} = \left[\begin{array}{c|c} a_0 & -\underline{a}^T \\ \hline \underline{a} & a_0 I_3 + \tilde{\underline{a}} \end{array}\right] \begin{bmatrix} b_0 \\ \hline \underline{b} \end{bmatrix} \tag{15a}$$

$$= \left[\begin{array}{c|c} b_0 & -\underline{b}^T \\ \hline \underline{b} & b_0 I_3 - \tilde{\underline{b}} \end{array}\right] \begin{bmatrix} a_0 \\ \hline \underline{a} \end{bmatrix} \tag{15b}$$

where I_3 is a 3 x 3 identity matrix. Inspection of the two 4 x 4 matrices in Eq. 15 reveals that they are identical in structure with the exception of a sign change on the 3 x 3 skew-symmetric part (compare Eq. 15 with Eq. 14b). In addition a number of interesting submatrices can be identified. The matrices in Eq. 15 form the basis for almost all of the quaternion operations so it will be convenient to develop compact notation for them. Before doing this, some of their properties are investigated so that meaningful symbols can be assigned. If the two matrices are written out it is easy to verify that every row and column is orthogonal to its neighbor, so they are orthogonal but not necessarily orthonormal. In addition, all rows and columns have lengths a or b. Observe that the two unique skew-symmetric vector quaternion matrices differ only in sign on the 3 x 3 skew-symmetric submatrix. In addition, the quaternion vector and its conjugate can be identified in both matrices. Considering that the two matrices depend only on the elements of a quaternion and differ only by a sign on the 3 x 3 skew-symmetric part, the symbols $\underset{\sim}{\overset{+}{a}}$ and $\underset{\sim}{\overset{-}{b}}$ will be adopted. In addition, when the diagonal matrices are removed, the remaining 4 x 4 skew-symmetric matrices differ only by a sign on the 3 x 3 skew-symmetric submatrix and depend only on the vector quaternions. Therefore the symbols $\overset{+}{\underline{a}}$ and $\overset{-}{\underline{b}}$ will be adopted for these submatrices. It can also be verified that these submatrices are orthogonal but not necessarily orthonormal. Thus

$$\overset{+}{\underline{a}} = \left[\begin{array}{c|c} 0 & -\underline{a}^T \\ \hline \underline{a} & \tilde{\underline{a}} \end{array}\right] \tag{16a}$$

$$\overset{+}{\underset{\sim}{a}} = \left[\begin{array}{c|c} a_0 & -\underline{a}^T \\ \hline \underline{a} & a_0 I_3 + \tilde{\underline{a}} \end{array}\right] \tag{16b}$$

$$= a_0 I_4 + \overset{+}{\underline{a}} \tag{16c}$$

and

$$\overset{-}{\underline{b}} = \left[\begin{array}{c|c} 0 & -\underline{b}^T \\ \hline \underline{b} & -\tilde{\underline{b}} \end{array}\right] \tag{17a}$$

$$\overset{-}{\underset{\sim}{b}} = \left[\begin{array}{c|c} b_0 & -\underline{b}^T \\ \hline \underline{b} & b_0 I_3 - \tilde{\underline{b}} \end{array}\right] \tag{17b}$$

$$= b_0 I_4 + \overset{-}{\underline{b}} \tag{17c}$$

With this notation Eq. 15 can be written more compactly as

$$\underset{\sim}{c} = \overset{+}{\underset{\sim}{a}}\ \underset{\sim}{b} = \overset{-}{\underset{\sim}{b}}\ \underset{\sim}{a} \tag{18}$$

Observe that the quaternion product is not commutative but it is possible to rearrange the quaternion elements in Eq. 18. This property is very important in developing many of the equations for mechanical systems. Again it is interesting to compare Eq. 18 to Eq. 14b.

If $\hat{a}^*$ and $\hat{b}^*$ are substituted into the respective matrices in Eq. 15 it is easy to verify that

$$\overset{+}{\underset{\sim}{a}}{}^{*} = \overset{+}{\underset{\sim}{a}}{}^{T} \tag{19a}$$

$$\overset{-}{\underset{\sim}{b}}{}^{*} = \overset{-}{\underset{\sim}{b}}{}^{T} \tag{19b}$$

and therefore

$$\overset{+}{\underset{\sim}{a}}{}^{T}\ \underset{\sim}{b}^{*} = \overset{-}{\underset{\sim}{b}}{}^{T}\ \underset{\sim}{a}^{*} \tag{19c}$$

Other relations can also be identified such as

$$\overset{+}{\underset{\sim}{a}}\ \underset{\sim}{b}^{*} = \overset{-}{\underset{\sim}{b}}{}^{T}\ \underset{\sim}{a} \tag{20a}$$

and

$$\overset{+}{\underset{\sim}{a}}{}^{T}\ \underset{\sim}{b} = \overset{-}{\underset{\sim}{b}}\ \underset{\sim}{a}^{*} \tag{20b}$$

Using the matrices in Eqs. 16 and 17, it is easily shown that

$$\overset{+}{\underline{a}}\ \overset{+}{\underline{a}}{}^{T} = \overset{-}{\underline{a}}\ \overset{-}{\underline{a}}{}^{T} = (\underline{a}^{T}\ \underline{a})\ I_4 \tag{21a}$$

and

$$\overset{+}{\underset{\sim}{a}}\ \overset{+}{\underset{\sim}{a}}{}^{T} = \overset{-}{\underset{\sim}{a}}\ \overset{-}{\underset{\sim}{a}}{}^{T} = (\underset{\sim}{a}^{T}\ \underset{\sim}{a})\ I_4 \tag{21b}$$

Equations 21 demonstrate orthogonality of the four matrices, and provided $\underline{a}$ or respectively $\underset{\sim}{a}$ are not null,

$$\underline{\overset{+}{a}}^{-1} = \underline{\overset{+}{a}}^{T} / (\underline{a}^{T}\ \underline{a}) \tag{22a}$$

$$\underline{\overset{-}{a}}^{-1} = \underline{\overset{-}{a}}^{T} / (\underline{a}^{T}\ \underline{a}) \tag{22b}$$

or

$$\underset{\sim}{\overset{+}{a}}^{-1} = \underset{\sim}{\overset{+}{a}}^{T} / (\underset{\sim}{a}^{T}\ \underset{\sim}{a}) \tag{22c}$$

$$\underset{\sim}{\overset{-}{a}}^{-1} = \underset{\sim}{\overset{-}{a}}^{T} / (\underset{\sim}{a}^{T}\ \underset{\sim}{a}) \tag{22d}$$

If $\underline{a}$ or respectively $\underset{\sim}{a}$ are normalized to unity then $\underline{a}^{T}\ \underline{a} = 1$ or $\underset{\sim}{a}^{T}\ \underset{\sim}{a} = 1$, and

$$\underline{\overset{+}{a}}^{-1} = \underline{\overset{+}{a}}^{T} \tag{23a}$$

$$\underline{\overset{-}{a}}^{-1} = \underline{\overset{-}{a}}^{T} \tag{23b}$$

or

$$\underset{\sim}{\overset{+}{a}}^{-1} = \underset{\sim}{\overset{+}{a}}^{T} \tag{23c}$$

$$\underset{\sim}{\overset{-}{a}}^{-1} = \underset{\sim}{\overset{-}{a}}^{T} \tag{23d}$$

In this case all of these matrices are orthonormal.

It is interesting to compare Eq. 18 to the cross product operator in Eq. 14. Clearly both equations achieve an interchange of variables. However, the major difference is that Eq. 18 provides unique functional relationships because the matrices $\underset{\sim}{\overset{+}{a}}$ and $\underset{\sim}{\overset{-}{a}}$ are orthogonal, whereas $\underline{\tilde{a}}$ is singular. The dot product operator is the orthogonal complement of the cross product operator and when combined they form unique transformations, which are achieved by the matrices $\overset{+}{\underline{a}}$ and $\underline{\bar{a}}$. Thus quaternion operators obtain their orthogonality by utilizing the complete vector space of dot product and cross product operators. For example, this feature allows the use of Euler parameters to define nonsingular transformations for all relative element orientations, whereas other three-variable representations are subject to numerical singularities.

3. QUATERNION TRIPLE PRODUCTS

Even more interesting results can be obtained when three quaternions are multiplied together. Consider the product of arbitrary quaternions

$$\hat{d} = \hat{a}\ \hat{b}\ \hat{c} \tag{24a}$$

$$= (\hat{b}^*\ \hat{a}^*)^*\ \hat{c} \tag{24b}$$

$$= (\hat{c}^*\ \hat{b}^*\ \hat{a}^*)^* \tag{24c}$$

where Eqs. 9 and 10 are applied. Equation 24 shows that

$$(\hat{a}\ \hat{b}\ \hat{c})^* = \hat{c}^*\ \hat{b}^*\ \hat{a}^* \tag{25}$$

In addition, these operators can be grouped because they are orthogonal,

i.e.

$$(\hat{a}\ \hat{b})\ \hat{c} = \hat{a}\ (\hat{b}\ \hat{c}) \tag{26}$$

Using Eq. 18a, the equivalent matrix form of Eq. 24 can be expressed in a number of different ways such as

$$\underset{\sim}{d} = \overset{+}{\underset{\sim}{a}}\ \overset{+}{\underset{\sim}{b}}\ \underset{\sim}{c} \tag{27a}$$

$$= \overset{+}{\underset{\sim}{a}}\ \overset{-}{\underset{\sim}{c}}\ \underset{\sim}{b} \tag{27b}$$

$$= \overset{-}{(\overset{-}{\underset{\sim}{c}}\ \underset{\sim}{b})}\ \underset{\sim}{a} \tag{27c}$$

$$= \overset{-}{\underset{\sim}{c}}\ \overset{-}{\underset{\sim}{b}}\ \underset{\sim}{a} \tag{27d}$$

$$= \overset{-}{\underset{\sim}{c}}\ \overset{+}{\underset{\sim}{a}}\ \underset{\sim}{b} \tag{27e}$$

Equations 27b and 27e demonstrate commutativity of the matrices $\overset{+}{\underset{\sim}{a}}$ and $\overset{-}{\underset{\sim}{c}}$

$$\overset{+}{\underset{\sim}{a}}\ \overset{-}{\underset{\sim}{c}} = \overset{-}{\underset{\sim}{c}}\ \overset{+}{\underset{\sim}{a}} \tag{28a}$$

and alternately

$$\underset{\sim}{a}^{+T}\ \overset{-}{\underset{\sim}{c}} = \overset{-}{\underset{\sim}{c}}\ \underset{\sim}{a}^{+T} \tag{28b}$$

where Eq. 28b is obtained by substituting $\underset{\sim}{a}^{*}$ for $\underset{\sim}{a}$. The commutativity of these matrices is very useful in the manipulation of equations.

Even more interesting quaternion relations can be obtained by considering the product

$$\hat{a}\ \hat{b} = \hat{c}\ \hat{a} \tag{29}$$

If $\hat{a} \neq 0$ then its inverse can be obtained from Eq. 22 as

$$\hat{a}^{-1} = \hat{a}^{*} \,/\, (\hat{a}^{*}\ \hat{a}) \tag{30}$$

Solving Eq. 29 for $\hat{b}$ or $\hat{c}$ gives

$$\hat{b} = \hat{a}^* \hat{c} \hat{a} / (\hat{a}^* \hat{a}) \tag{31a}$$

or

$$\hat{c} = \hat{a} \hat{b} \hat{a}^* / (\hat{a}^* \hat{a}) \tag{31b}$$

Suppose $\hat{a}$ is normalized to unity. Then from Eq. 11

$$\hat{a}^* \hat{a} = \hat{a} \hat{a}^* = 1 \tag{32}$$

and Eqs. 31 reduce to

$$\hat{b} = \hat{a}^* \hat{c} \hat{a} \tag{33a}$$

or

$$\hat{c} = \hat{a} \hat{b} \hat{a}^* \tag{33b}$$

Equations 31 and 33 are very useful because they represent unique transformations between quaternions. To understand what $\hat{b}$ represents in Eq. 31a, it is useful to look at the square of its magnitude, i.e. $\hat{b}^* \hat{b}$. Substituting Eq. 31a yields

$$\begin{aligned}
\hat{b}^* \hat{b} &= [(\hat{a}^* \hat{c} \hat{a})^* / (\hat{a}^* \hat{a})] \; [(\hat{a}^* \hat{c} \hat{a}) / (\hat{a}^* \hat{a})] \\
&= \hat{a}^* \hat{c}^* (\hat{a} \hat{a}^*) \hat{c} \hat{a} / (\hat{a}^* \hat{a})^2 \\
&= \hat{a}^* (\hat{c}^* \hat{c}) \hat{a} / (\hat{a}^* \hat{a}) \\
&= (\hat{c}^* \hat{c}) (\hat{a}^* \hat{a}) / (\hat{a}^* \hat{a}) \\
&= \hat{c}^* \hat{c}
\end{aligned} \tag{34}$$

Similar results can be obtained for Eqs. 31b and 33. These relationships reveal that quaternion operations of this form transform other quaternions without distortion.

Looking at Eq. 33a in more detail, from Eq. 6

$$\hat{c}\,\hat{a} = (c_0\, a_0 - \vec{c}\cdot\vec{a}) + (c_0\,\vec{a} + a_0\,\vec{c} + \vec{c}\times\vec{a}) \tag{35}$$

and from Eq. 7

$$\hat{b} = \hat{a}^{*}(\hat{c}\,\hat{a})$$

$$= [a_0^2 + \vec{a}\cdot\vec{a}]\, c_0 + [a_0^2\,\vec{c} - (\vec{a}\cdot\vec{a})\,\vec{c} + 2\,\vec{a}\times(\vec{a}\times\vec{c}) - 2\, a_0\,\vec{a}\times\vec{c}] \tag{36}$$

Equation 36 shows that $\hat{c}$ is transformed without mixing its scalar and vector parts. A similar expansion of Eq. 31a is obtained simply by dividing Eq. 36 by $\hat{a}^{*}\,\hat{a}$. Thus

$$\hat{b} = c_0 + [a_0^2\,\vec{c} - (\vec{a}\cdot\vec{a})\,\vec{c} + 2\,\vec{a}\times(\vec{a}\times\vec{c}) - 2\, a_0\,\vec{a}\times\vec{c}] \;/\; (\hat{a}^{*}\,\hat{a}) \tag{37}$$

Considering that the coefficients of c_0 in Eqs. 36 and 37 are unity, and thus $b_0 = c_0$ it is clear that the vector component of $\hat{c}$, namely $\vec{c}$ is also transformed without magnification to $\vec{b}$ since the lengths of $\hat{b}$ and $\hat{c}$ are equal.

Before leaving this subject consider again, Eq. 29, where it is now known that $\hat{b}$ and $\hat{c}$ represent a quaternion pair that are in some way oriented symmetrically about $\hat{a}$ and $\hat{a}^{*}$. To show the latter case, simply pre- and post-multiply Eq. 29 by $\hat{a}^{*}$ giving

$$\hat{b}\,\hat{a}^{*} = \hat{a}^{*}\,\hat{c} \tag{38}$$

Considering that $c_0 = b_0$, let $\vec{c} = \vec{b}'$, correspondingly $\hat{c} = \hat{b}'$ and write Eq. 29 as

$$\hat{a}\,\hat{b} = \hat{b}'\,\hat{a} \tag{39}$$

Now expand using Eq. 6

$$a_0\, b_0 - \vec{a} \cdot \vec{b} + a_0\, \vec{b} + b_0\, \vec{a} + \vec{a} \times \vec{b} = b_0\, a_0 - \vec{b}' \cdot \vec{a} + b_0\, \vec{a} + a_0\, \vec{b}' + \vec{b}' \times \vec{a} \tag{40}$$

Equating scalar components in Eq. 40 yields

$$\vec{a} \cdot \vec{b} = \vec{a} \cdot \vec{b}' \tag{41}$$

Equation 41 is interesting because it shows that $\vec{b}$ and $\vec{b}'$ are always located at equal spatial angles from $\vec{a}$ and the two vectors act as if they are rigidly connected to the axis defined by $\vec{a}$, but are free to rotate about it. Thus the quaternion product in Eq. 39 rotates vectors $\vec{b}$ and $\vec{b}'$ about an axis. To investigate more fully the relationship between $\vec{b}$ and $\vec{b}'$ isolate the vector components of Eq. 40 as

$$a_0\, (\vec{b}' - \vec{b}) = \vec{a} \times (\vec{b}' + \vec{b}) \tag{42}$$

Equations 41 and 42 have an interesting background. The vector $\vec{a}\,/\,a_0$ (more specifically $2\,\vec{a}\,/\,a_0$), called the vector of finite rotations, was first discovered by Rodrigues before 1840[2]. He also established the fundamental identities expressed by Eqs. 41 and 42. This vector is sometimes named after Rodrigues, or Hamilton who developed quaternion theory in the 1830's[3]. It is also called the Gibbsian vector after the applied mathematician Gibbs who made vector calculus popular among many astronomers in this country in the early 1900's[4].

Equation 42 implicitly defines $\vec{b}'$ in terms of $\vec{b}$ and $\vec{a}$. In order to solve explicitly for $\vec{b}'$, cross multiply Eq. 42 by $\vec{a}$ and substitute Eq. 41. This results in

$$a_0\, \vec{a} \times (\vec{b}' - \vec{b}) = 2\,(\vec{a} \cdot \vec{b})\, \vec{a} - a^2\, (\vec{b}' + \vec{b}) \tag{43}$$

which may be solved for $\vec{a} \times \vec{b}'$ and substituted into Eq. 42 yielding

$$\vec{b}' = \vec{b} + 2\, a_0\, \vec{a} \times (\vec{b} + \vec{a} \times \vec{b}) \;/\; (a_0^2 + a^2) \qquad (44)$$

If $\hat{a}$ is normalized to unity then Eq. 44 is simply

$$\vec{b}' = \vec{b} + 2\, a_0\, \vec{a} \times (\vec{b} + \vec{a} \times \vec{b}) \qquad (45)$$

The components of the normalized quaternion vector are taken as

$$a_0 = \cos(\phi \;/\; 2) \qquad (46a)$$

and

$$a = \pm \sin(\phi \;/\; 2) \qquad (46b)$$

because it has been observed that the quaternion multiplication in Eq. 39 is equivalent to rotating a reference vector about $\vec{a}$ into $\vec{b}$ and $\vec{b}'$ by equal and opposite angles ϕ / 2. The net result is a relative rotation of two vectors by $\pm\phi$ / 2. The sign ambiguity on the relative rotation angle is resolved when coordinate systems are assigned to $\vec{b}$ and $\vec{b}'$.

The quaternion multiplications just developed are conveniently represented in matrix form. The equivalent matrix form of Eq. 39 is

$$\overset{+}{\underset{\sim}{a}}\, \underset{\sim}{b} = \overset{+}{\underset{\sim}{b}}{}'\, \underset{\sim}{a} \qquad (47)$$

Using Eq. 18, Eq. 47 can also be written as

$$\overset{+}{\underset{\sim}{a}}\, \underset{\sim}{b} = \overset{-}{\underset{\sim}{a}}\, \underset{\sim}{b}' \qquad (48)$$

Now the inverse of $\overset{+}{\underset{\sim}{a}}$ can be obtained from Eq. 22c giving

$$\underset{\sim}{b} = \overset{+}{\underset{\sim}{a}}{}^{T}\, \overset{-}{\underset{\sim}{a}}\, \underset{\sim}{b}' \;/\; (\underset{\sim}{a}^{T}\, \underset{\sim}{a}) \qquad (49a)$$

and in a similar manner

$$\tilde{b}' = \tilde{\bar{a}}^{T} \tilde{\overset{+}{a}} \tilde{b} \, / \, (\tilde{a}^{T} \tilde{a}) \tag{49b}$$

If $\tilde{a}$ is a normalized quaternion then $\tilde{a}^{T} \tilde{a} = 1$ and Eqs. 49 reduce to

$$\tilde{b} = \tilde{\overset{+}{a}}^{T} \tilde{\bar{a}} \tilde{b}' \tag{50a}$$

and

$$\tilde{b}' = \tilde{\bar{a}}^{T} \tilde{\overset{+}{a}} \tilde{b} \tag{50b}$$

Equations 50 identify the orthonormal quaternion transformations between quaternions $\tilde{b}$ and $\tilde{b}'$ which from Eq. 28b can also be written in the alternate form

$$\tilde{\overset{+}{a}}^{T} \tilde{\bar{a}} = \tilde{\bar{a}} \tilde{\overset{+}{a}}^{T} \tag{51}$$

Expanding Eq. 51 yields

$$\tilde{\overset{+}{a}}^{T} \tilde{\bar{a}} = \left[\begin{array}{c|c} 1 & \underline{0}^{T} \\ \hline \underline{0} & a_0^2 \, I_3 + \underline{a} \, \underline{a}^{T} + \underline{\tilde{a}} \, \underline{\tilde{a}} - 2 \, a_0 \, \underline{\tilde{a}} \end{array} \right] \tag{52}$$

Introducing the additional matrix identity

$$\underline{\tilde{a}} \, \underline{\tilde{a}} = \underline{a} \, \underline{a}^{T} - (\underline{a}^{T} \underline{a}) \, I_3 \tag{53}$$

into Eq. 52 further reduces it to

$$\overset{+T}{\underset{\sim}{a}}\ \overset{-}{\underset{\sim}{a}} = \left[\begin{array}{c|c} 1 & \underline{0}^T \\ \hline \underline{0} & (2\, a_0^2 - 1)\, I_3 + 2\, (\underline{a}\, \underline{a}^T - a_0\, \tilde{\underline{a}}) \end{array}\right] \tag{54}$$

Now Eq. 54 can be used to verify the previous development. Consider Eq. 36 with $\vec{b}'$ substituted for $\vec{c}$. Its matrix equivalent is

$$\left[\begin{array}{c} b_0 \\ \hline \underline{b} \end{array}\right] = \left[\begin{array}{c} b_0 \\ \hline [(2\, a_0^2 - 1)\, I_3 + 2\, (\underline{a}\, \underline{a}^T - a_0\, \tilde{\underline{a}})]\, \underline{b}' \end{array}\right]$$

which according to Eq. 54 factors into

$$\underset{\sim}{b} = \overset{+T}{\underset{\sim}{a}}\ \overset{-}{\underset{\sim}{a}}\ \underset{\sim}{b}'$$

4. EULER PARAMETERS

Since Euler parameters are normalized quaternions the rules of quaternion algebra apply and they yield orthonormal transformations as previously developed. Euler parameters are employed to define the relative angular orientation between coordinate systems. Let two coordinate systems be identified by symbols i and j, and a reference orientation be defined when corresponding axes of the two systems are parallel and have the same orientations. The angular orientation of system j with respect to system i is defined by a single rotation angle ϕ^{ji} about some axis defined by a unit vector $\vec{u}_{ji}$. The direction of this vector defines the positive sense of the relative rotation angle. Define

$$e_0^{ji} = \cos\, (\phi^{ji} / 2) \tag{55a}$$

$$e^{ji} = \sin\, (\phi^{ji} / 2) \tag{55b}$$

and

$$\vec{e}^{ji} = e^{ji}\,\vec{u}_{ji} \tag{55c}$$

The orientation vector $\vec{e}^{ji}$ and the normalizing scalar e_0^{ji} constitute a normalized quaternion $\hat{e}^{ji}$. The ji superscript notation is used as a reminder that Euler parameters $\hat{e}^{ji}$ define the orientation of system j with respect to system i. The double superscript will always be used to denote quantities that relate two systems. Note, however that $\vec{u}_{ji}$ is identified by subscript ji because it may be considered as a vector associated with the two systems i and j.

In order to keep track of the coordinate system that a given vector is projected onto, a superscript will also be appended to it. For example, $\underset{\sim}{a}^j$ and $\underline{a}^j$ represent respectively, projections of $\hat{a}$ and $\vec{a}$ onto system j. If a vector requires other identification, this can be achieved by its symbol designation or by subscripting. In the case above it will be found that $\vec{u}_{ji}$ projects equally onto systems i and j as $\vec{u}_{ji}^{\,i} = \vec{u}_{ji}^{\,j}$ and thus the superscripts i and j are dropped.

If $\underset{\sim}{e}^{ji}$ is substituted into Eq. 54, a spatial transformation matrix designated as A^{ji} relating system j to system i is defined. In general it will be convenient to treat A^{ji} as a 4 x 4 matrix or the 3 x 3 submatrix of Eq. 54. Hence the symbol A^{ji} will be used to represent either matrix and any ambiguity will be eliminated by the context of its usage.

Any quaternion or vector projected onto systems i and j is represented by

$$\underset{\sim}{a}^j = A^{ji}\,\underset{\sim}{a}^i \tag{56a}$$

or correspondingly

$$\underline{a}^j = A^{ji}\,\underline{a}^i \tag{56b}$$

Observe again from Eq. 54 that A^{ji} never changes the scalar component of a quaternion. Since A^{ji} is orthonormal it follows that $A^{ji-1} = A^{jiT} = A^{ij}$ and Eqs. 56 can also be written as

$$\tilde{a}^{i} = A^{ji}\ \tilde{a}^{j} \tag{57a}$$

or

$$\underline{a}^{j} = A^{ji}\ \underline{a}^{i} \tag{57b}$$

The Euler parameter superscripts ji can be reversed giving

$$\phi^{ij} = \phi^{ji} \tag{58a}$$

$$e_0^{ij} = e_0^{ji} \tag{58b}$$

$$e^{ij} = e^{ji} \tag{58c}$$

$$\vec{u}_{ij} = -\vec{u}_{ji} \tag{58d}$$

$$\vec{e}^{ij} = -\vec{e}^{ji} \tag{58e}$$

and

$$\tilde{e}^{ij} = \tilde{e}^{ji*} \tag{58f}$$

which now describes the orientation of system i with respect to system j. Substituting Eq. 58f into Eq. 54 shows immediately that $\tilde{e}^{ij}$ defines $A^{ij} = A^{jiT}$. The same result is also obtained by changing the sign on ϕ^{ji} and keeping the same orientation on the unit vector $\vec{u}_{ji}$.

Before leaving this topic it is interesting to consider the extraction of Euler parameters $\tilde{e}^{ji}$ from the matrix A^{ji}. The problem here is that A^{ji} is quadratic in the elements of $\tilde{e}^{ji}$ and there is no unique quaternion vector unless, for example, e_0^{ji} is accepted as always being positive. With this assumption all elements of $\tilde{e}^{ji}$ can be extracted from the product

$$\underset{\sim}{e}^{ji}\,\underset{\sim}{e}^{jiT} = \begin{bmatrix} e_0e_0 & e_0e_1 & e_0e_2 & e_0e_3 \\ e_1e_0 & e_1e_1 & e_1e_2 & e_1e_3 \\ e_2e_0 & e_2e_1 & e_2e_2 & e_2e_3 \\ e_3e_0 & e_3e_1 & e_3e_2 & e_3e_3 \end{bmatrix}^{ji}$$

The trick is to obtain this matrix form A^{ji}. To see how this might be accomplished, partition the above matrix as

$$\underset{\sim}{e}^{ji}\,\underset{\sim}{e}^{jiT} = \left[\begin{array}{c|c} e_0^2 & e_0\,\underline{e}^T \\ \hline e_0\,\underline{e} & \underline{e}\,\underline{e}^T \end{array}\right]^{ji}$$

and observe that A^{ji} is of the form

$$A^{ji} = \begin{bmatrix} a_{11} & a_{12} & a_{13} \\ a_{21} & a_{22} & a_{23} \\ a_{31} & a_{32} & a_{33} \end{bmatrix}^{ji}$$

$$= [(2\,e_0^2 - 1)\,I_3 + 2\,(\underline{e}\,\underline{e}^T + e_0\,\tilde{\underline{e}})]^{ji}$$

The matrix sum $A^{ji} + A^{ij}$ isolates elements of $\underline{e}\,\underline{e}^T$ plus an additional diagonal term, and the difference $A^{ji} - A^{ij}$ contains elements of $e_0\,\underline{e}$. A simple computer algorithm can now be developed as follows. Let

$$B = \left[\begin{array}{c|ccc} trA & -a_{23} & -a_{31} & -a_{12} \\ \hline a_{32} & a_{11} & a_{12} & a_{13} \\ a_{13} & a_{21} & a_{22} & a_{23} \\ a_{21} & a_{31} & a_{32} & a_{33} \end{array}\right]$$

where

$$\mathrm{trA} = a_{11} + a_{22} + a_{33}$$

$$= 6\, e_0^2 - 3 + 2\, \underline{e}^T\, \underline{e}$$

$$= 4\, e_0^2 - 1$$

Next form the matrix

$$B' = B + B^T + (1 - \mathrm{trA})\, I_4 = 4\, \tilde{e}^{ji}\, \tilde{e}^{jiT}$$

The remaining steps are as follows:

1) Set k to the index of the largest element along the diagonal of B', k= 0, 1, 2, or 3.

2) Evaluate $e_m = b'_{km} / (2 \sqrt{b'_{kk}})$, $m = 0,\ldots,3$

3) $e_m = e'_m \,\mathrm{sgn}\, e'_0$; $\mathrm{sgn}\, e'_0 = \begin{cases} -1 & \text{for } e'_0 < 0 \\ 1 & \text{for } e'_0 \geq 0 \end{cases}$

The last step always insures that $e_0 \geq 0$.

5. SUCCESSIVE COORDINATE SYSTEM TRANSFORMATIONS

Another useful application of Euler parameter quaternions is the development of transformations between other Euler parameter quaternions. Consider a vector $\vec{a}$ projected onto three different coordinate systems i, j and k as $\vec{a}^i$, $\vec{a}^j$ and $\vec{a}^k$ respectively. Let $\hat{e}^{ji}$ define the orientation of system j with respect to system i and $\hat{e}^{kj}$ define the orientation of system k with respect to system j. It can be verified that the transformation in Eq. 57a is equivalent to

$$\vec{a}^i = \hat{e}^{ji}\, \vec{a}^j\, \hat{e}^{ji*} \tag{59}$$

To show this write Eq. 59 in its equivalent matrix form as

$$\underline{a}^{i} = \underset{\sim}{\overset{+}{e}}{}^{ji}\ \underline{\overset{+}{a}}{}^{j}\ \underset{\sim}{e}{}^{ji*}$$

$$= \underset{\sim}{\overset{+}{e}}{}^{ji}\ \underset{\sim}{\overset{-}{e}}{}^{jiT}\ \underline{a}^{j}$$

$$= \underset{\sim}{\overset{-}{e}}{}^{jiT}\ \underset{\sim}{\overset{+}{e}}{}^{ji}\ \underline{a}^{j}$$

$$= A^{jiT}\ \underline{a}^{j}$$

$$= A^{ij}\ \underline{a}^{j}$$

In a similar manner

$$\vec{a}^{j} = \hat{e}^{kj}\ \vec{a}^{k}\ \hat{e}^{kj*} \tag{60}$$

Substituting Eq. 60 into Eq. 59 gives a recursive formula

$$\vec{a}^{i} = \hat{e}^{ji}\ \hat{e}^{kj}\ \vec{a}^{k}\ \hat{e}^{jk*}\ \hat{e}^{ji*} \tag{61a}$$

$$= \hat{e}^{ki}\ \vec{a}^{k}\ \hat{e}^{ki*} \tag{61b}$$

The useful identity from from Eqs. 61 is

$$\hat{e}^{ki} = \hat{e}^{ji}\ \hat{e}^{kj} \tag{62}$$

which has the equivalent matrix form

$$\underset{\sim}{e}{}^{ki} = \underset{\sim}{\overset{+}{e}}{}^{ji}\ \underset{\sim}{e}{}^{kj} \tag{63}$$

An alternate form of Eq. 63 suggested by Eq. 18,

$$\underset{\sim}{e}{}^{ki} = \underset{\sim}{\overset{-}{e}}{}^{kj}\ \underset{\sim}{e}{}^{ji} \tag{64}$$

preserves adjacency of superscripts and may be easier to remember. Equations 63 and 64 are very useful because they provide simple means of determining the relative orientation between any two elements in terms of relative orientations between other elements. Another application of this technique is the reduction of a general orientation or rotation into a sequence of successive rotations. Equations 63 and 64 are also useful in the formulation of revolute joint equations which are developed later. Before leaving this discussion it is noted that Eqs. 63 and 64 can be extended to any number of successive transformations and are very useful in the development of the kinematics and dynamics of robotic manipulators.

The Euler parameter quaternion $\tilde{e}^{ki}$ defines a new spatial transformation matrix A^{ik}. To see this, write the equivalent matrix forms of Eqs. 59 and 60 as

$$\underline{a}^{i} = A^{ij}\,\underline{a}^{j} \qquad (65a)$$

and

$$\underline{a}^{i} = A^{jk}\,\underline{a}^{k} \qquad (65b)$$

Combining these equations then gives

$$\underline{a}^{i} = A^{ij}\,A^{jk}\,\underline{a}^{k}$$

$$= A^{ik}\,\underline{a}^{k}$$

and thus

$$A^{ik} = A^{ij}\,A^{jk} \qquad (66)$$

There are other useful variations of Eq. 66.

6. INTERMEDIATE-AXIS EULER PARAMETERS

Consider two rigid elements interconnected by a revolute or rotational joint defined by a common intermediate axis fixed in both elements. It is desirable to determine the relative angular orientation between the two elements when both element orientations are given, or to specify the orientation of one element when the other element orientation and their relative angular orientation are given. Two cases will be considered; a reference orientation consistent with the constraint exists or does not exist. Recall that a reference orientation exists when corresponding axes of the two systems are parallel and have the same orientations.

The reference orientation situation is the simplest and will be considered first. Because a reference orientation exists, the revolute axis projection onto both coordinate systems is identical. This projection must remain constant for any relative angular orientation because the axis is assumed fixed in both elements. This equal projection onto both systems is the basic requirement of an Euler parameter quaternion. Let systems i and j be embedded in their respective elements, the constant unit vector $\vec{u}_{ji}$ define the common axis and positive orientation, and the angle ϕ^{ji} define the angular orientation of system j with respect to i. The angle $\phi^{ji} = 0$ defines the reference orientation. The following Euler parameters are now defined

$$e_0^{ji} = \cos\,(\phi^{ji}\ /\ 2) \tag{67a}$$

$$e^{ji} = \sin\,(\phi^{ji}\ /\ 2) \tag{67b}$$

$$\vec{e}^{ji} = e^{ji}\ \vec{u}_{ji} \tag{67c}$$

and

$$\hat{e}^{ji} = e_0^{ji} + \vec{e}^{ji} \tag{67d}$$

Clearly Eqs. 67 define the orientation of system j with respect to system i and A^{ij} can be evaluated. Suppose the orientation of system i with respect to another system k is given by $\hat{e}^{ik}$. Then the orientation of system j with respect to k can be determined as $\hat{e}^{jk}$ from Eq. 62. Observing that $\hat{e}^{ki} = \hat{e}^{ik*}$ and $\hat{e}^{kj} = \hat{e}^{jk*}$ then

$$\hat{e}^{jk*} = \hat{e}^{ji}\,\hat{e}^{ik*} \tag{68}$$

The conjugated vectors can be removed by first conjugating Eq. 68

$$\hat{e}^{jk} = \hat{e}^{ik}\,\hat{e}^{ik*} \tag{69}$$

The matrix equivalent of Eq. 69 is

$$\begin{aligned} \underset{\sim}{e}^{jk} &= \underset{\sim}{\overset{+}{e}}{}^{ik}\,\underset{\sim}{e}^{ji} \\ &= \underset{\sim}{\bar{e}}{}^{ji*}\,\underset{\sim}{e}^{ik} \\ &= \underset{\sim}{\bar{e}}{}^{jiT}\,\underset{\sim}{e}^{ik} \end{aligned} \tag{70}$$

where the equivalence of conjugation and matrix transposition is observed. If the orientation of system i with respect to j had been defined then Eq. 70 would become

$$\underset{\sim}{e}^{jk} = \underset{\sim}{\bar{e}}{}^{ij}\,\underset{\sim}{e}^{ik}$$

In the case where $\underset{\sim}{e}^{ik}$ and $\underset{\sim}{e}^{jk}$ are known, and $\tilde{e}^{ji}$ is to be determined, Eq. 70 may be multiplied by $\underset{\sim}{\bar{e}}{}^{ji}$ giving

$$\underset{\sim}{e}^{ik} = \underset{\sim}{\bar{e}}{}^{ji}\,\underset{\sim}{e}^{jk}$$

$$= \overset{+}{\tilde{e}}^{jk} \, \tilde{e}^{ji} \tag{71}$$

Finally multiplying Eq. 71 by $\overset{+}{\tilde{e}}^{jkT}$ yields

$$\tilde{e}^{ji} = \overset{+}{\tilde{e}}^{jkT} \, \tilde{e}^{ik} \tag{72}$$

This example, although appearing somewhat complicated at first, demonstrates but a few of the many potential quaternion matrix manipulations that can be employed to develop simplified equations. To complete the first case, substitute Eqs. 67 into Eq. 72 and the angle ϕ^{ji} can be determined by

$$\begin{bmatrix} \cos\,(\phi^{ji}\,/\,2) \\ \sin\,(\phi^{ji}\,/\,2)\,\underline{u}_{ji} \end{bmatrix} = \overset{+}{\tilde{e}}^{jkT} \, \tilde{e}^{ik}$$

because $\underline{u}_{ji}$ is known. If the form of $\overset{+}{\tilde{e}}^{jk}$ is compared to the matrix in Eq. 15a it can be observed that

$$\cos\,(\phi^{ji}\,/\,2) = \tilde{e}^{jkT} \, \tilde{e}^{ik} \tag{74}$$

and

$$\sin\,(\phi^{ji}\,/\,2) = \underline{u}_{ji}^{T} \, \overset{+}{\tilde{e}}^{jkT} \, \tilde{e}^{ik}$$

$$= \tilde{e}^{jkT} \, \overset{-}{\underline{u}}{}_{ji}^{T} \, \tilde{e}^{ik} \tag{75}$$

Equations 74 and 75 are easy to program and the angle ϕ^{ji} can be determined by using the arc tangent function.

In the second case when no reference orientation exists, consistent with the revolute joint constraint, there can be no intermediate Euler parameter axis that coincides with the joint axis for any relative orientation. If one existed, it would exist for all relative orientations, which is just the first case. One could redefine the orientation of coordinate system i or j within its respective element such that a reference orientation did exist. However this is generally undesirable and often impossible, especially if element i or j is part of a complex system with many constraints, etc. A more practical approach is to consider an intermediate coordinate system, say, i′ embedded in element i that does have a reference orientation with respect to system j. It should be noted that an infinite number of possible orientations exist and the selection can be based on other criteria such as picking the relative orientation between systems i′ and j in the reference configuration or simplifying the constant intermediate axis Euler parameter between systems i and i′. The choice of system i or j for the intermediate frame is also arbitrary in this development. Now the projection of $\vec{u}_{ji}$ onto systems i′ and j is the same and an intermediate axis Euler parameter quaternion $\hat{e}^{ji'}$ can be defined by

$$e_0^{ji'} = \cos\,(\phi^{ji'}\;/\;2) \tag{76a}$$

$$e^{ji'} = \sin\,(\phi^{ji'}\;/\;2) \tag{76b}$$

$$\vec{e}^{ji'} = e^{ji'}\,\vec{u}^{j}_{ji} \tag{76c}$$

and

$$\hat{e}^{ji'} = e_0^{ji'} + \vec{e}^{ji'} \tag{76d}$$

Observe that $\vec{u}^{j}_{ji}$ is used in Eq. 76, not $\vec{u}^{i}_{ji}$. The fixed orientation of system i′ in i is defined by a constant Euler parameter quaternion $\hat{e}^{i'i}$. If system i is oriented relative to k by $\hat{e}^{ik}$ then the following identities can be written

$$\hat{e}^{i'k} = \hat{e}^{ik}\,\hat{e}^{i'i} \tag{77a}$$

and

$$\hat{e}^{jk} = \hat{e}^{i'k}\,\hat{e}^{ji'} \tag{77b}$$

Combining Eqs. 77a and 77b provides the orientation of system j with respect to k in terms of known quantities

$$\hat{e}^{jk} = \hat{e}^{ik}\,\hat{e}^{i'i}\,\hat{e}^{ji'} \tag{78}$$

Equation 78 can be written in a number of equivalent matrix forms, for example,

$$\underset{\sim}{e}^{jk} = \underset{\sim}{\overset{+}{e}}^{ik}\,\underset{\sim}{\overset{+}{e}}^{i'i}\,\underset{\sim}{e}^{ji'} \tag{79}$$

Suppose, on the other hand, that the orientation of systems i and j with respect to k are known. Then Eq. 79 can be manipulated into

$$\underset{\sim}{e}^{ji'} = \underset{\sim}{\overset{+}{e}}^{i'iT}\,\underset{\sim}{\overset{+}{e}}^{ikT}\,\underset{\sim}{e}^{jk} \tag{80}$$

and the angle $\phi^{ji'}$ can be obtained from Eq. 76. Equations 78 and 79 show that the more general case requires only slightly more preprocessing and computational effort. Similar to Eqs. 74 and 75 one finds that

$$\cos\,(\phi^{ji'}\;/\;2) = \underset{\sim}{e}^{i'iT}\,\underset{\sim}{\overset{+}{e}}^{ikT}\,\underset{\sim}{e}^{jk} \tag{81}$$

and

$$\begin{aligned} \sin\,(\phi^{ji'}\;/\;2) &= \underline{u}_{ji}^{jT}\,\underset{\sim}{\overset{+}{e}}^{i'iT}\,\underset{\sim}{\overset{+}{e}}^{ikT}\,\underset{\sim}{e}^{jk} \\ &= \underset{\sim}{e}^{i'iT}\,\overset{-}{\underline{u}}_{ji}^{jT}\,\underset{\sim}{\overset{+}{e}}^{ikT}\,\underset{\sim}{e}^{jk} \end{aligned} \tag{82}$$

The row vectors $\tilde{e}^{i'iT}$ and $\tilde{e}^{i'iT}\,\underline{\bar{u}}^{jT}$ are constant and require evaluation only once.

7. TIME DERIVATIVE OF EULER PARAMETERS

Up to this point many useful quaternion identities have been developed for determining element spatial orientations. Some of these relationships can be differentiated with respect to time in order to investigate the dynamic behavior of systems. Consider the basic property of normalized Euler parameter quaternions given by Eq. 11

$$\hat{e}^{*}\,\hat{e} = \hat{e}\,\hat{e}^{*} = 1 \tag{83}$$

Differentiating Eq. 83 with respect to time yields

$$\hat{e}^{*}\,\dot{\hat{e}} + \dot{\hat{e}}^{*}\,\hat{e} = 0 \tag{84a}$$

and

$$\hat{e}\,\dot{\hat{e}}^{*} + \dot{\hat{e}}\,\hat{e}^{*} = 0 \tag{84b}$$

The quantities in Eq. 84 are vector quaternions because each pair are negative conjugates, i.e. let

$$\vec{c} = \hat{e}^{*}\,\dot{\hat{e}} = -\,\dot{\hat{e}}^{*}\,\hat{e}$$

$$= -\,(\hat{e}^{*}\,\dot{\hat{e}})^{*} \tag{85a}$$

and

$$\vec{d} = \hat{e}\,\dot{\hat{e}}^{*} = -\,\dot{\hat{e}}\,\hat{e}^{*}$$

$$= -\,(\hat{e}\,\dot{\hat{e}}^{*})^{*} \tag{85b}$$

The vectors $\vec{c}$ and $\vec{d}$ clearly depend on time variations of Euler parameters and thus are related to angular velocity vectors $\vec{\omega}^{i}_{ji}$ and $\vec{\omega}^{j}_{ji}$. To investigate these relationships, differentiate the quaternion identity

$$\vec{a}^{i} = \hat{e}^{ji}\,\vec{a}^{j}\,\hat{e}^{ji*}$$

For convenience the superscripts on Euler parameters are dropped. Thus

$$\dot{\vec{a}}^{i} = \hat{e}\,\dot{\vec{a}}^{j}\,\hat{e}^{*} + \dot{\hat{e}}\,\vec{a}^{j}\,\hat{e}^{*} + \hat{e}\,\vec{a}^{j}\,\dot{\hat{e}}^{*} \tag{86}$$

The quaternions $\dot{\hat{e}}$ and $\dot{\hat{e}}^{*}$ can be eliminated from Eq. 86 by rearranging Eq. 85a as $\dot{\hat{e}} = \hat{e}\,\vec{c}$ and $\dot{\hat{e}}^{*} = -\,\vec{c}\,\hat{e}^{*}$ so

$$\dot{\vec{a}}^{i} = \hat{e}\,(\dot{\vec{a}}^{j} + \vec{c}\,\vec{a}^{j} - \vec{a}^{j}\,\vec{c})\,\hat{e}^{*} \tag{87}$$

Now using Eq. 5, one has

$$\vec{c}\,\vec{a}^{j} = \vec{c} \times \vec{a}^{j} - \vec{c} \cdot \vec{a}^{j} \tag{88a}$$

and

$$\vec{a}^{j}\,\vec{c} = \vec{a}^{j} \times \vec{c} - \vec{a}^{j} \cdot \vec{c} \tag{88b}$$

Substituting Eqs. 88 into Eq. 87 and recognizing that

$$\vec{c} \times \vec{a}^{j} = -\,\vec{a}^{j} \times \vec{c}$$

and

$$\vec{c} \cdot \vec{a}^{j} = \vec{a}^{j} \cdot \vec{c}$$

yields

$$\dot{\vec{a}}^{i} = \hat{e}\,(\dot{\vec{a}}^{j} + 2\,\vec{c} \times \vec{a}^{j})\,\hat{e}^{*} \tag{89}$$

Differentiating a vector $\vec{a}^{j}$ in coordinate system j is represented by

$$\dot{\vec{a}}^{j} + \vec{\omega}^{j} \times \vec{a}^{j}$$

so that

$$\dot{\vec{a}}^{i} = \hat{e}\,(\dot{\vec{a}}^{j} + \vec{\omega}^{j} \times \vec{a}^{j})\,\hat{e}^{*} \tag{90}$$

Comparing Eqs. 89 and 90, the identities

$$\vec{c} = \vec{\omega}^{j} / 2$$

or

$$\vec{\omega}^{j} = 2\,\hat{e}^{*}\,\dot{\hat{e}} \tag{91}$$

can be identified.

In a similar manner if one differentiates

$$\vec{a}^{j} = \hat{e}^{ji*}\,\vec{a}^{i}\,\hat{e}^{ji}$$

it follows that

$$\vec{\omega}^{i} = 2\,\dot{\hat{e}}\,\hat{e}^{*} \tag{92}$$

To verify Eq. 92 observe that

$$\vec{\omega}^{i} = \hat{e}\,\vec{\omega}^{j}\,\hat{e}^{*}$$

$$= 2\,\hat{e}\,\hat{e}^{*}\,\dot{\hat{e}}\,\hat{e}^{*}$$

$$= 2\,\dot{\hat{e}}\,\hat{e}^{*}$$

Equation 85 can also be used to show that

$$\vec{\omega}^{j} = -\,2\,\dot{\hat{e}}^{*}\,\hat{e} \tag{93a}$$

and

$$\vec{\omega}^{i} = -\,2\,\hat{e}\,\dot{\hat{e}}^{*} \tag{93b}$$

Some useful matrix identities can be identified from the previous development. Equations 91 and 93a yield

$$\underline{\omega}^j = 2 \overset{+T}{\underset{\sim}{e}} \overset{\bullet}{\underset{\sim}{e}} = -2 \overset{\bullet\;+T}{\underset{\sim}{e}} \underset{\sim}{e} \tag{94}$$

Equations 92 and 93b have the equivalent matrix forms

$$\underline{\omega}^i = 2 \overset{\bullet}{\overset{+}{\underset{\sim}{e}}} \underset{\sim}{e}^* = -2 \overset{+}{\underset{\sim}{e}} \overset{\bullet *}{\underset{\sim}{e}}$$

Equation 18 may be used to obtain a more useful form eliminating the conjugate vectors.

$$\underline{\omega}^i = 2 \overset{-T}{\underset{\sim}{e}} \overset{\bullet}{\underset{\sim}{e}} = -2 \overset{\bullet\;-T}{\underset{\sim}{e}} \underset{\sim}{e} \tag{95}$$

Observe again the equivalence of conjugation and transposition of matrices. Another useful application of Eqs. 94 and 95 is the determination of derivatives of Euler parameters given angular velocity. Equations 94 and 95 yield

$$\overset{\bullet}{\underset{\sim}{e}} = \overset{+}{\underset{\sim}{e}} \underline{\omega}^j / 2 \tag{96a}$$

$$= \underline{\bar{\omega}}^j \underset{\sim}{e} / 2 \tag{96b}$$

and

$$\overset{\bullet}{\underset{\sim}{e}} = \overset{-}{\underset{\sim}{e}} \underline{\omega}^i / 2 \tag{96c}$$

$$= \underline{\overset{+}{\omega}}^i \underset{\sim}{e} / 2 \tag{96d}$$

Some useful applications of the previously derived identities can be developed. Suppose the relative angular velocity of systems i and j with respect to system k are known and it is desired to determine $\dot{\hat{e}}^{ji}$ where $\hat{e}^{ji}$ is known. Equation 91 can be written as

$$\dot{\hat{e}}^{ji} = \hat{e}^{ji} \, \vec{\omega}^{j}_{ji} \; / \; 2 \tag{97}$$

The relative angular velocity vector $\vec{\omega}^{j}_{ji}$ can also be expressed as

$$\vec{\omega}^{j}_{ji} = \vec{\omega}^{j}_{jk} - \vec{\omega}^{j}_{ik} \tag{98}$$

where

$$\vec{\omega}^{j}_{ik} = \hat{e}^{ji*} \, \vec{\omega}^{i}_{ik} \, \hat{e}^{ji} \tag{99}$$

Substituting Eqs. 98 and 99 into Eq. 97 yields

$$\dot{\hat{e}}^{ji} = \hat{e}^{ji} \, (\vec{\omega}^{j}_{jk} - \hat{e}^{ji*} \, \vec{\omega}^{i}_{ik} \, \hat{e}^{ji}) \; / \; 2$$

$$= (\hat{e}^{ji} \, \vec{\omega}^{j}_{jk} - \vec{\omega}^{i}_{ik} \, \hat{e}^{ji}) \; / \; 2 \tag{100}$$

The matrix equivalent of Eq. 100 is

$$\dot{\tilde{e}}^{ji} = (\overset{-}{\underline{\omega}}{}^{j}_{jk} - \overset{+}{\underline{\omega}}{}^{i}_{ik}) \; \tilde{e}^{ji} \; / \; 2 \tag{101}$$

In practical applications the angular velocities in Eq. 101 may come from rate sensors or command signals. Finding the relative angular velocity in either coordinate system i or j is trivial from Eqs. 94 and 95, i.e.

$$\underline{\omega}^{j}_{ji} = 2 \; \overset{+}{\tilde{e}}{}^{jiT} \; \dot{\tilde{e}}^{ji} \tag{102a}$$

and

$$\underline{\omega}^{i}_{ji} = 2 \; \overset{-}{\tilde{e}}{}^{jiT} \; \dot{\tilde{e}}^{ji} \tag{102b}$$

REFERENCES

1. Glaese, J. R. and Kennel, J. F., Torque Equilibrium Attitude Control for Skylab Reentry, NASA Technical Memorandum-78252, George C. Marshal Space Flight Center, Marshal Space Flight Center, Alabama, 1979.

2. Rodrigues, O., Journal de Mathematiques, Vol 5, 1840, pp. 380.

3. Hestenes, D., Space-Time Algebra, Gordon and Breach, 150 Fifth Avenue, New York, NY, 1966.

4. Gibbs, W., Vector Analysis, Yale University Press, New Haven Connecticut, 1913.

Part 2

COMPUTER AIDED FORMULATION OF EQUATIONS OF DYNAMICS

COMPUTER GENERATION OF EQUATIONS OF MOTION

Werner O. Schiehlen
Institute B of Mechanics
University Stuttgart
Stuttgart, F.R.G.

Abstract. The method of multibody systems is discussed with respect to the computerized generation of symbolical equations of motion. The kinematics are presented in an inertial frame and, additionally, in an often very useful moving reference frame. The dynamics include not only spring and damper forces, but also integral forces and contact forces typical for vehicle applications. The equations of motion are found by d'Alembert's principle and Jourdain's principle featuring generalized coordinates and generalized velocities, holonomic and nonholonomic constraints. It is also shown how constraint forces, necessary for the modeling of contact and friction, can be computed using a minimal number of equations. Finally, the symbolical formalism NEWEUL is introduced.

1. INTRODUCTION

Multibody systems are well qualified for dynamical investigations in mechanical engineering and they are applied to machines, manipulators, mechanisms, satellites and all kinds of vehicles. However, the increasing demands on the accuracy of mechanical models result in very complex multibody systems requiring computer generation of equations of motion.

Starting with the space age in the sixties, the derivation of equations appeared as a problem due to the complex structure of spacecrafts. Then, Hooker and Margulies [1] and Roberson and Wittenburg [2] developed formalisms in 1965 and 1966 for the numerical computerized derivation of equations of motion. The restriction to spherical joints was eliminated in the seventies, e.g. by Frisch [3], Wittenburg [4] and Andrews and Kesavan [5]. In addition to space

NATO ASI Series, Vol. F9
Computer Aided Analysis and Optimization of Mechanical System Dynamics
Edited by E. J. Haug

applications, the multibody system approach was also introduced in mechanical and vehicle engineering, e.g. by Paul [6], Orlandea, Chace and Calahan [7] and Haug, Wehage and Barman [8].

Then, difficulties occured with closed kinematical chains and nonholonomic constraints rarely found in space. For their treatment numerical formalisms require Lagrange's multipliers resulting in numerical instabilities. However, the symbolical computerized derivation of equations of motion overcomes all these problems, see Rosenthal and Sherman [9] and Ref. [10]. Due to the limitations of computer memory space, symbolical formalisms have to use the most efficient dynamical principles. These have been discussed recently by Kane and Levinson [11] and in Ref. [12].

The formalisms discussed have to be implemented on the computer. The corresponding computer programs are characterized by acronyms or fantastic names. In Table 1 some of the computer programs are listed.

Program	Institution	Authors
MULTIBODY	DFVLR , F.R.G	Roberson, Wittenburg [2] Schwertassek [13]
DYMAC-G	Paul Associates	Paul [6]
ADAMS	Mechanical Dynamics	Orlandea et.al. [7] Chace [14]
DADS-3D	University of Iowa	Haug et.al. [8] Nikravesh, Chung [15]
NEWEUL	University Stuttgart	Schiehlen, Kreuzer [10]
SD/EXACT	Symbolic Dynamics	Rosenthal, Sherman [9]

Table 1. Computer Programs for Multibody Systems

For computer simulations, the equations of motion have to be generated by numerical formalisms, e.g. by MULTIBODY , thousand and thousand times while the symbolical equations, obtained e.g. by NEWEUL , are generated only once. This means better efficiency of symbolical formalisms for real time simulations.

In this paper the theoretical basis of the program NEWEUL is presented. The kinematics include holonomic and nonholonomic systems using the inertial and a moving reference frame. In particular, the approach of Lagrangian or generalized coordinates is applied. The global system equations are found from the basic laws of motion, Newton's and Euler's law. In detail, the internal, constraint and applied forces are discussed. The dynamical principles of d'Alembert and Jourdain result in the equations of motion of multibody systems, classified as ordinary and general multibody systems. Further, the equations of constraint forces are given even in the case of contact or friction, respectively. The aspects of the computerized derivation of equations of motion are illustrated with respect to symbolical manipulation of formulae as used in the program NEWEUL . The double pendulum serves as a simple example.

2. KINEMATICS OF MULTIBODY SYSTEMS

Multibody systems are characterized by rigid bodies with inertia as well as springs, dashpots and actively controlled servomotors without inertia, Fig. 1. The bodies are interconnected by rigid bearings or any other kind of supports and subject to additional applied forces and torques. Therefore, the method of multibody systems is well qualified for the modeling of mechanical systems with complex geometry and great differences in stiffness usually resulting in motions with frequencies less than 50 Hz.

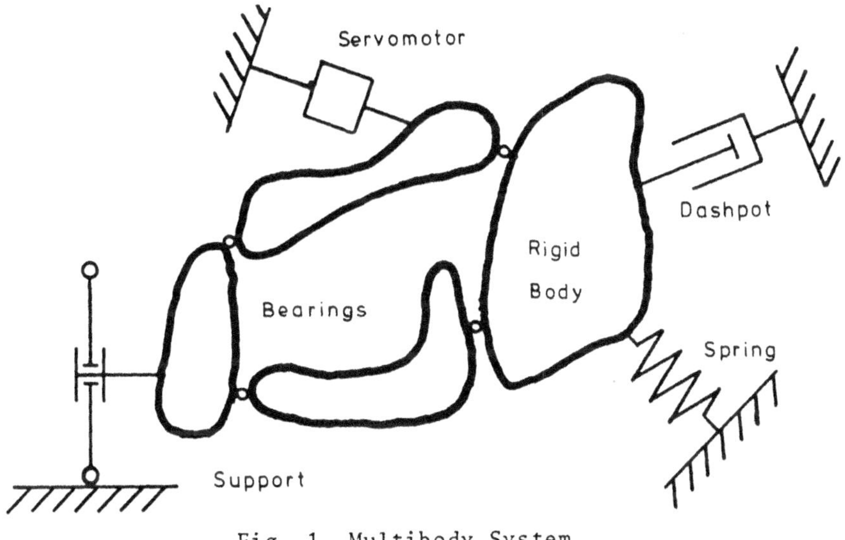

Fig. 1 Multibody System

Kinematics describe the absolute motion of mechanical systems, i.e. position, velocity and acceleration. From a mathematical point of view, the description of motion is convenient in an inertial frame. On the other hand, the technical situation requires often a moving reference frame. Therefore, both frames will be regarded in this chapter.

Free Systems

According to the free body principle, each rigid body or mass point, respectively, of the mechanical system is treated separately and all elements without inertia are replaced by forces, Fig.2 .

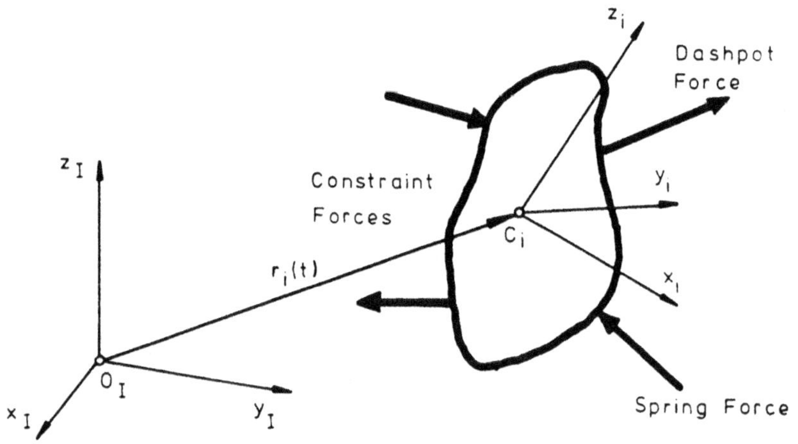

Fig. 2 Position of a Body in the Inertial Frame

The position of a system of p bodies is given relative to the inertial frame I by the 3x1-translation vector

$$r_i(t) = [r_{xi} \quad r_{yi} \quad r_{zi}]^T \qquad , \quad i = 1(1)p \qquad , \quad (1)$$

of the center of mass C_i and the 3x3-rotation tensor

$$S_i(t) = \begin{bmatrix} S_{xxi} & S_{xyi} & S_{xzi} \\ S_{yxi} & S_{yyi} & S_{yzi} \\ S_{zxi} & S_{zyi} & S_{zzi} \end{bmatrix} \quad , \quad i = 1(1)p \ , \qquad (2)$$

written down for each body. The coordinates of the rotation tensor S_i represent the direction cosines relating the inertial frame I and the body-fixed frame i to each other. The nine coordinates of the rotation tensor depend on three angles or generalized coordinates, respectively, due to the constraint conditions $S_i(t)\ S_i^T(t) = E$, $i = 1(1)p$. For more details see the comprehensive presentations in gyrodynamics. A free system of p bodies without any mechanical constraint holds 6p degrees of freedom. Thus, the position of the system can be uniquely described by 6p generalized coordinates summarized in a 6px1-position vector

$$x(t) = [x_1 \ \ x_2 \ \ x_3 \ \dots \ x_{6p}]^T \quad . \qquad (3)$$

Typical generalized coordinates of a free system are translational coordinates, Euler angles or relative distances. If some of the rigid bodies are replaced by mass points, the total number of degrees of freedom has also to be reduced by three for each mass point. The system's position given by (1) and (2) reads now

$$r_i = r_i(x) \quad , \quad S_i = S_i(x) \quad , \quad i = 1(1)p \quad . \qquad (4)$$

It has to be mentioned that the uniqueness of (4) may be lost in singular positions. Then, it is necessary to use a complementary set of generalized coordinates as shown by Rongved and Fletcher [16].

Further, the translational and rotational velocity and acceleration, respectively, of the system are found by differentiation with respect to the inertial frame I as 3x1-vectors:

$$v_i = \frac{\partial r_i}{\partial x} \dot{x} = H_{Ti}(x)\, \dot{x}(t) \quad , \tag{5}$$

$$\omega_i = \frac{\partial s_i}{\partial x} \dot{x} = H_{Ri}(x)\, \dot{x}(t) \quad , \tag{6}$$

$$a_i = H_{Ti}(x)\, \ddot{x}(t) + \frac{\partial v_i}{\partial x} \dot{x}(t) \; , \tag{7}$$

$$\alpha_i = H_{Ri}(x)\, \ddot{x}(t) + \frac{\partial \omega_i}{\partial x} \dot{x}(t) \quad , \tag{8}$$

where the 3x6p-matrices H_{Ti} , H_{Ri} are introduced for abbreviation.

The 3x6p-matrix H_{Ri} requires additional consideration. The infinitesimal 3x1-rotation vector ∂s_i used in (6) follows from the infinitesimal skew-symmetric 3x3-rotation tensor

$$\partial \tilde{s}_i = \partial s_i s_i^T = \begin{bmatrix} 0 & -\partial s_{zi} & \partial s_{yi} \\ \partial s_{zi} & 0 & -\partial s_{xi} \\ -\partial s_{yi} & \partial s_{xi} & 0 \end{bmatrix} \; . \tag{9}$$

However, the matrix H_{Ri} can also be found by a geometrical analysis of the angular velocity vector ω_i with respect to the corresponding generalized coordinates.

Holonomic Systems

A holonomic system of p bodies and q holonomic, rheonomic constraints due to rigid bearings results in f positional degrees of freedom:

$$f = 6p - q \quad .$$

Some holonomic constraints are shown in Fig. 3.

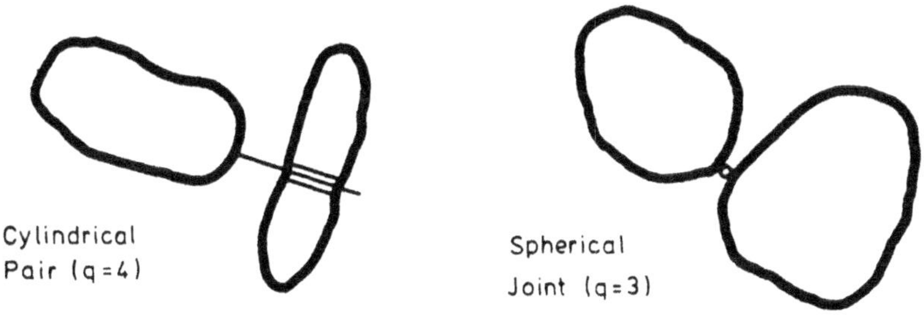

Fig. 3 Holonomic Constraints

The constraints may be given implicity by

$$\phi_j(x,t) = 0 \quad , \qquad j = 1(1)q \quad , \tag{11}$$

or explicitly by

$$x = \underline{f}(y,t) \quad , \tag{12}$$

respectively, where the fx1-position vector

$$y(t) = [y_1 \;\; y_2 \;\; y_3 \;\; \ldots \;\; y_f]^T \tag{13}$$

is used summarizing the f generalized coordinates of the system. Then, from (4) and (12), the system's position reads as

$$r_i = r_i(y,t) \quad , \qquad S_i = S_i(y,t) \quad . \tag{14}$$

Further, by differentiation, the system's absolute velocity and acceleration are obtained as

$$v_i = J_{Ti}(y,t)\,\dot{y}(t) + \frac{\partial r_i}{\partial t} \quad , \tag{15}$$

$$\omega_i = J_{Ri}(y,t)\,\dot{y}(t) + \frac{\partial s_i}{\partial t} \quad , \quad i = 1(1)p \quad , \tag{16}$$

$$a_i = J_{Ti}(y,t)\,\dot{y}(t) + \frac{\partial v_i}{\partial y}\,\dot{y} + \frac{\partial v_i}{\partial t} \, , \tag{17}$$

$$\alpha_i = J_{Ri}(y,t)\,\ddot{y}(t) + \frac{\partial \omega_i}{\partial y}\,\dot{y} + \frac{\partial \omega_i}{\partial t} \quad , \tag{18}$$

where the 3xf-geometrical Jacobian matrices

$$J_{Ti} = \frac{\partial r_i}{\partial y} = \frac{\partial r_i}{\partial x}\frac{\partial x}{\partial y} = H_{Ti}F \quad , \tag{19}$$

$$J_{Ri} = \frac{\partial s_i}{\partial y} = \frac{\partial s_i}{\partial x}\frac{\partial x}{\partial y} = H_{Ri}F \tag{20}$$

and the 6pxf-matrix F are introduced. For scleronomic constraints the partial time-derivatives in (15) to (18) vanish.

Nonholonomic Systems

Additionally to the holonomic constraints, in nonholonomic systems there exist r nonholonomic, rheonomic constraints. The resulting number of motional degrees of freedom is

$$g = f - r \quad . \tag{21}$$

A nonholonomic constraint is shown in Fig. 4.

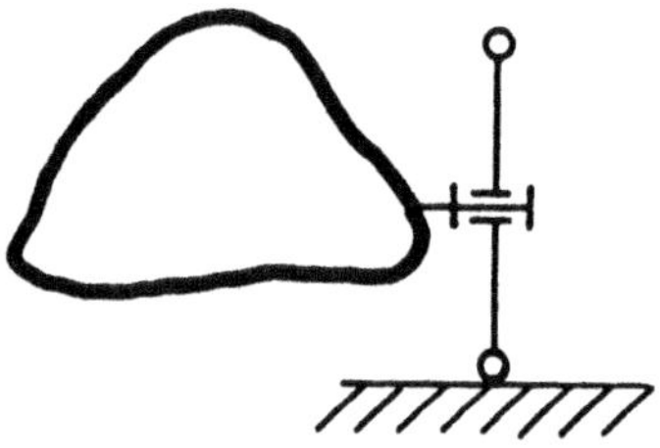

Fig. 4 Nonholonomic Constraint

Implicitly the constraints read as

$$\psi_k(y,\dot{y},t) = 0 \quad , \qquad k = 1(1)r \, , \tag{22}$$

or explicitly as

$$\dot{y}(t) = \underline{g}(y,z,t) \quad , \tag{23}$$

respectively, where the gx1-velocity vector

$$z(t) = [z_1 \; z_2 \; z_3 \; \dots \; z_g]^T \tag{24}$$

characterizes the g generalized velocities of the system. Then, from (15) and (16) it follows for the system's velocity

$$v_i = v_i(y,z,t) \quad , \quad \omega_i = \omega_i(y,z,t) \tag{25}$$

and, by differentiation, for the absolute acceleration

$$a_i = L_{Ti}(y,z,t)\,\dot{z}(t) + \frac{\partial v_i}{\partial y}\,\dot{y} + \frac{\partial v_i}{\partial t} \quad , \tag{26}$$

$$\alpha_i = L_{Ri}(y,z,t)\,\dot{z}(t) + \frac{\partial \omega_i}{\partial y}\,\dot{y} + \frac{\partial \omega_i}{\partial t} \quad , \tag{27}$$

where the 3xg-kinematical Jacobian matrices

$$L_{Ti} = \frac{\partial v_i}{\partial z} = \frac{\partial v_i}{\partial \dot{y}}\,\frac{\partial \dot{y}}{\partial z} = J_{Ti}G = H_{Ti}FG \quad , \tag{28}$$

$$L_{Ri} = \frac{\partial \omega_i}{\partial z} = \frac{\partial \omega_i}{\partial \dot{y}}\,\frac{\partial \dot{y}}{\partial z} = J_{Ri}G = H_{Ri}FG \tag{29}$$

and the fxg-matrix G are used for abbreviation. Further, for scleronomic constraints the partial time-derivatives in (26) and (27) vanish.

Moving Reference Frame

In many applications, a reference frame is given in a natural way. For example, an automobile turning on a circle is naturally described in a moving trackrelated frame. Therefore, the absolute motion will also be presented in a reference frame using the reference motion itself and the bodies' relative motion.

In addition to the inertial frame I and the body fixed frame i there is a moving reference frame R introduced, Fig. 5.

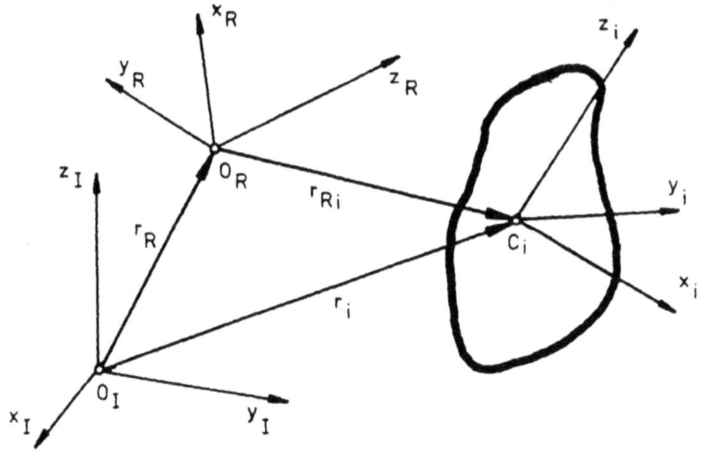

Fig. 5 Position of a Body in a Moving Reference Frame

Then, the absolute position of the system is given by

$$r_i = r_R + r_{Ri} \quad , \qquad S_i = S_R \, S_{Ri} \tag{30}$$

where the reference quantities are characterized by the index R and the relative quantities by the index Ri.

According to the principles of relative motion, the system's absolute velocities and accelerations read now

$$v_i = \overset{x}{r}_i = \overset{x}{r}_R + \tilde{\omega}_R \, r_{Ri} + \dot{r}_{Ri} \quad , \tag{31}$$

$$\omega_i = \overset{x}{s}_i = \omega_R + \omega_{Ri} \quad , \tag{32}$$

$$a_i = \overset{xx}{r}_R + (\dot{\tilde{\omega}}_R + \tilde{\omega}_R^2) \, r_{Ri} + 2\tilde{\omega}_{Ri} \, \dot{r}_{Ri} + \ddot{r}_{Ri} \quad , \tag{33}$$

$$\alpha_i = \dot{\omega}_R + \tilde{\omega}_R \, \omega_{Ri} + \dot{\omega}_{Ri} \quad . \tag{34}$$

The symbol $(^x)$ means differentiation with respect to the inertial frame I and $(\dot{\ })$ denotes the differentiation with respect to the reference frame R . Further, the symbol $(\tilde{\ })$ represents the vector product or the equivalent skew-symmetric matrix, see (9) .

Free and Holonomic Systems: From a mathematical point of view, free systems and holonomic systems without constraints, $(q=0 , x=y)$ are identical. Therefore, free systems will not be treated separately. The generalized coordinates introduced by (13) may be reference or relative coordinates, respectively.

The reference motion is given by

$$r_R = r_R(y,t) \quad , \qquad S_R = S_R(y,t) \tag{35}$$

and the reference velocities read in the reference frame as

$$\dot{r}_R = J_{TR}(y,t)\,\dot{y}(t) + \frac{\partial r_R}{\partial t} \quad , \tag{36}$$

$$\omega_R = J_{RR}(y,t)\,\dot{y}(t) + \frac{\partial s_R}{\partial t} \quad , \tag{37}$$

where J_{TR} and J_{RR} are the 3xf-Jacobian matrices of the reference motion. If only relative coordinates are used, then the Jacobian matrices in (36) and (37) are vanishing. Similar to (17) and (18) the reference accelerations are obtained.

The relative motion of each body is defined as

$$r_{Ri} = r_{Ri}(y,t) \quad , \qquad S_{Ri} = S_{Ri}(y,t) \tag{38}$$

and the relative velocities are

$$\dot{r}_{Ri} = J_{TRi}(y,t)\,\dot{y}(t) + \frac{\partial r_{Ri}}{\partial t} \quad , \tag{39}$$

$$\omega_{Ri} = J_{RRi}(y,t)\,\dot{y}(t) + \frac{\partial s_{Ri}}{\partial t} \quad . \tag{40}$$

Here, J_{TRi} and J_{RRi} are the 3xf-Jacobian matrices of the relative motion. Corresponding to (17) and (18) the relative accelerations are obtained.

The system's absolute velocites and accelerations are now found from (31) to (34) using (35) to (40). For the geometrical Jacobian matrices it yields

$$J_{Ti} = J_{TR} + J_{TRi} - \tilde{r}_{Ri} J_{RR} \quad , \tag{41}$$

$$J_{Ri} = J_{RR} + J_{RRi} \quad . \tag{42}$$

Obviously, an only time-dependent reference frame doesn't affect the Jacobian matrices of the system.

Nonholonomic Systems: The nonholonomic constraints in the explicit form (23) can also be introduced in (36),(37),(39) and (40) affecting the accelerations given by (33) and (34). The resulting equations will be not presented in detail. However, for the kinematical Jacobian matrices, it is found

$$L_{Ti} = L_{TR} + L_{TRi} - \tilde{r}_{Ri} L_{RR} \quad , \tag{43}$$

$$L_{Ri} = L_{RR} + L_{RRi} \quad , \tag{44}$$

corresponding to the results (41) and (42) for holonomic systems.

Comparison of Frames

What are the advantages of the different frames compared to each other? From a mathematical point of view, the expressions of the absolute motion of a multibody system are more complicated in the reference frame than in an inertial frame. However, the choice of the generalized coordinates, the formulation of the applied and constraint forces and the generation of the equations of motion is often simplified by a moving reference frame. It may be even more advantageous to use more than one reference frame as provided in the program NEWEUL.

Formalisms in Literature

The presented definitions allow a simple characterization of formalims known in literature. Some examples will be given. Haug, Wehage and Barman [8] deal with holonomic systems and cartesian coordinates:

$$q \neq 0, \quad r = 0, \qquad f = g = 6p - q \quad ,$$

$$F = [E \,\vdots\, 0] \;, \qquad G = E \quad .$$

Kane and Levinson [11] consider holonomic systems with generalized coordinates and velocities:

$$q \neq 0, \quad r = 0, \qquad f = g = 6p - q \quad ,$$

$$F \neq [E \,\vdots\, 0] \;, \qquad G \neq E .$$

Schwertassek and Roberson [17] investigate nonholonomic systems with cartesian coordinates and generalized velocities:

$$q = 0, \quad r \neq 0, \qquad f = 6p \;, \quad g = 6p - q - r \;,$$

$$F = E \;, \quad G \neq [E \,\vdots\, 0] \quad .$$

In Ref. [10] nonholonomic systems with generalized coordinates and velocities are treated:

$$q \neq 0, \quad r \neq 0, \qquad f = 6p - q, \qquad g = 6p - q - r \,,$$

$$F \neq [E \,\vdots\, 0], \quad G \neq [E \,\vdots\, 0] \quad .$$

3. NEWTON-EULER EQUATIONS

For the application of Newton's and Euler's equations to multibody systems the free body principle has to be used again. Newton's and Euler's equations may be used in an inertial frame or a moving reference frame.

Basic Laws of Motion

Newton's and Euler's equations read for each body in the inertial frame

$$m_i a_i = f_i^e + \sum_{j=1}^{p} f_{ij}^i = f_i^a + f_i^r \quad , \quad i = 1(1)p \quad , \tag{45}$$

$$I_i \alpha_i + \tilde{\omega}_i I_i \omega_i = l_i^e + \sum_{j=1}^{p} l_{ij}^i = l_i^a + l_i^r \quad . \tag{46}$$

The inertia is represented by the scalar mass m_i and the 3x3-inertia tensor I_i with respect to the center of mass C_i of each body. The inertia tensor I_i depending on position and time follows from the constant inertia tensor ${}_iI_i$, given in the body-fixed frame, by the law

$$I_i(y,t) = S_i(y,t) \; {}_iI_i \; S_i^T(y,t) \quad . \tag{47}$$

The forces and torques in (45) and (46) are either composed by external and internal forces and torques or by applied and constraint forces and torques acting on each body. The forces and torques are 3x1-vectors, all torques have to be related to the center of mass C_i of each body. The external forces and torques act from the outside of the system, the internal forces and torques appear always twice within the system. The applied forces and torques, respectively, depend on the motion by different laws and they may be coupled or decoupled to the constraint forces and torques. For the generation of the equations of motion the applied forces and torques are most important.

Newton's and Euler's equations are changed on the left hand side in a moving reference R to

$$m_i \left[\overset{xx}{r}_R + (\dot{\tilde{\omega}}_R + \tilde{\omega}_R^2) \, r_{Ri} + 2\tilde{\omega}_R \, \dot{r}_{Ri} + \ddot{r}_{Ri} \right] = f_i^a = f_i^r \quad , \tag{48}$$

$$I_i \, \dot{\omega}_R + \tilde{\omega}_R \, I_i \, \omega_R + \tilde{\omega}_R \, \omega_{Ri} \, \mathrm{tr} I_i + 2\tilde{\omega}_{Ri} \, I_i \, \omega_R + I_i \, \dot{\omega}_{Ri}$$

$$+ \tilde{\omega}_{Ri} \, I_i \, \omega_{Ri} = l_i^a + l_i^r \quad , \qquad i = 1(1)p \; , \tag{49}$$

where $\mathrm{tr}I_i$ means the trace of the inertia tensor I_i. The force and torque vectors have also to be written in the reference frame, but usually there don't appear new terms on the right hand side of the equations.

Internal Forces and Torques: From the reaction principle it follows

$$f_{ij} + f_{ji} = 0 \quad , \quad f_{ii} = 0 \quad , \tag{50}$$

$$l_{ij}^{Ci} + l_{ji}^{Cj} + \tilde{r}_{ij}\, f_{ji} = 0 \quad , \quad l_{ii}^{Ci} = 0 \quad , \tag{51}$$

where r_{ij} means the 3x1-vector between the center of mass C_i and the center of mass C_j of two bodies i and j, $i = 1(1)p$, $j = 1(1)p$. For the total system, the reaction principle results in

$$\sum_{i=1}^{p} \sum_{j=1}^{p} f_{ij} = 0 \quad , \tag{52}$$

$$\sum_{i=1}^{p} \sum_{j=1}^{p} l_{ij}^{Ci} + \tilde{r}_i\, f_{ij} = 0 \; . \tag{53}$$

Thus, internal forces and torques don't affect the total system's motion.

Constraint Forces and Torques: The constraint forces and torques originate from the reactions in bearings and supports. They can be reduced by distribution matrices to the generalized constraint forces. The number of generalized constraint forces is equal to the total number of constraints $q+r$ in the system. Using the $(q+r)$x1-vector of generalized constraint forces

$$g(t) = [g_1 \;\; g_2 \;\; g_3 \cdots g_{q+r}]^T$$

and the $3x(q+r)$-distribution matrices

$$F_i = F_i(y,z,t) \quad , \quad L_i = L_i(y,z,t) \quad , \tag{54}$$

it turns out

$$f_i^r = F_i g \quad , \quad l_i^r = L_i g \quad , \quad i = 1(1)p \quad , \tag{55}$$

for each body. The generalized constraint forces are characteristic design parameters of bearings and supports, the distribution matrices follow from a geometrical analysis, see Kreuzer and Schmoll [18].

Ideal Applied Forces and Torques: The ideal applied forces are due to the elements of multibody system, e.g. springs, dashpots and further actions on the system.

The proportional forces are characterized by the system's position and timefunctions:

$$f_i^a = f_i^a(x,t) . \tag{56}$$

Conservative spring and weight forces as well as purely time-varying forces are proportional forces, Fig. 6.

The proportional-differential forces depend on the position and the velocity:

$$f_i^a = f_i^a(x,\dot{x},t) . \tag{57}$$

A parallel spring-dashpot configuration is a typical example for this class of forces, Fig. 6.

The proportional-integral forces are a function of the position and the integrals of position:

$$f_i^a = f_i^a(x,w,t) , \quad \dot{w} = \dot{w}(x,w,t) , \tag{58}$$

where the ρx1-vector w describes the position integrals. E.g. serial spring-dashpot configurations and the eigendynamics of servomotors result in proportional-integral forces, Fig. 6.

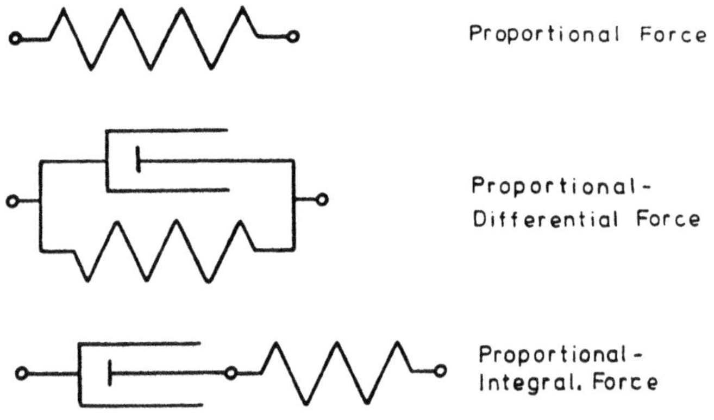

Fig. 6 Ideal Applied Forces

The same laws hold also for the ideal applied torques.

Contact Forces and Torques: In the case of nonideal constraints with sliding friction or contact forces, respectively, the applied forces are coupled with the constraint forces. The contact forces depend on the position and the velocity as well as the constraint forces:

$$f_i^a = f_i^a(y,z,g,t) . \tag{59}$$

For example, the lateral force of an elastic wheel is a typical contact force. Usually, the contact forces are analytically approximated from experimental data. The same law holds also for contact torques.

Global System Equations

The Newton-Euler equations of the global system are summarized in matrix notation by the following vectors and matrices. The inertia properties are written in the 6px6p-diagonal matrix

$$\bar{M} = \text{diag}\{\, m_1E \;\; m_2E \;\ldots\; I_1 \;\ldots\; I_p \} \quad , \tag{60}$$

where the 3x3-identity matrix E is used.

The 6px1-force vectors $\bar{q}^c$ and $\bar{q}^a$ represent the coriolis and centrifugal forces and the applied forces, respectively, in the following scheme

$$\bar{q} = [\, f_1^T \;\; f_2^T \ldots 1_1^T \ldots 1_p^T \,]^T . \tag{61}$$

Further, the 6pxf-matrix $\bar{J}$ and the 6pxg-matrix $\bar{L}$ are introduced as global Jacobian matrices, e.g.

$$\bar{J} = [\, J_{T1}^T \;\; J_{T2}^T \;\ldots\; J_{R1}^T \;\ldots\; J_{Rp}^T]^T , \tag{62}$$

as well as the global 6px(q+r)-distribution matrix of the system

$$\bar{Q} = [\, F_1^T \;\; F_2^T \ldots L_1^T \ldots L_p^T \,]^T . \tag{63}$$

Now, the Newton-Euler equations can be formulated.

Holonomic Systems: The global equations of a holonomic system in the inertial frame follow from (45), (46), (55), (57), (60) to (63) with (17) and (18) as

$$\bar{\bar{M}}\,\bar{J}\,\ddot{y} + \bar{q}^c(y,\dot{y},t) = \bar{q}^a(y,\dot{y},t) + \bar{Q}\,g \quad . \tag{64}$$

In the same way, the global equations in a moving reference frame are obtained from (48), (49) etc. as

$$_R\bar{\bar{M}}\;_R\bar{J}\,\ddot{y} + {}_R\bar{q}^c(y,\dot{y},t) = {}_R\bar{q}^a(y,\dot{y},t) + {}_R\bar{Q}\,g \quad . \tag{65}$$

The lower left index characterizes the reference frame R . Regarding the transformation matrix between frames I and R in the notation (60) ,

$$\bar{S}_R = \text{diag}\,\{\,S_R \;\; S_R \;\ldots\; S_R \;\ldots\; S_R\,\} \quad , \tag{66}$$

and the laws

$$\bar{\bar{M}} = \bar{S}_R\;{}_R\bar{\bar{M}}\;\bar{S}_R^T \quad , \qquad \bar{q} = \bar{S}_R\;{}_R\bar{q} \quad , \tag{67}$$

the equations (64) and (65) are identical.

Nonholonomic Systems: The global equations of a nonholonomic system in the inertial frame, using (26) and (27), read now as

$$\bar{\bar{M}}\,\bar{L}\,\dot{z} + \bar{q}^c(y,z,t) = \bar{q}^a(y,z,t) + \bar{Q}\,g \quad . \tag{68}$$

Eqs. (68) reduce to (64) for $r=0$, $\dot{y}=z$, $G=E$ and $\bar{L}=\bar{J}$, respectively, applying (28) and (29) .

4. DYNAMICAL PRINCIPLES

The Newton-Euler equations are combined algebraical and differential equations. E.g., eqs. (64) represent 6p scalar equations for the q unknown generalizied coordinates. However, by the dynamical principles, the Newton-Euler equations can be separated into purely algebraical and purely differential equations for solution.

D'Alembert's principle states that the virtual work of the constraint forces of the global system is vanishing:

$$\delta W = \delta y^T \bar{J}^T \bar{Q} g = 0 , \tag{69}$$

where δy denotes the virtual displacement. Further, Jourdain's principle says that the virtual power of the constraint forces is zero:

$$\delta P = \delta z^T \bar{L}^T \bar{Q} g = 0 , \tag{70}$$

where δz is the virtual velocity. This means that the equations of motion are obtained from the Newton-Euler equations by premultiplication with the transposed global Jacobian matrices. Three advantages are achieved simultaneously: i) symmetrization of the inertia matrix, ii) reduction to the minimal order of the differential equation system, iii) elimination of the constraint forces and torques.

Equations of Motion of Ordinary Multibody Systems

Multibody systems are called ordinary multibody systems iff they can be characterized by one vector differential equation of the second order with positive definite inertia matrix.

Free or holonomic systems with proportional or proportional-differential applied forces result in ordinary multibody systems. The equations of motion follow from the Newton-Euler equations in the inertial frame or reference frame, applying d'Alembert's or Jourdain's principle.

D'Alembert's Principle in an Inertial Frame: The equations of motion are found from (64) and (69) as

$$M(y,t)\,\ddot{y}(t) + k(y,\dot{y},t) = q(y,\dot{y},t) . \tag{71}$$

Here, the number of equations is reduced from $6p$ to f ,the $f \times f$-inertia matrix

$$M(y,t) = \bar{J}^T \bar{\bar{M}} \bar{J} > 0 \tag{72}$$

is symmetric and positive definite, and the constraint forces and torques are completely eliminated. The remaining fx1-vector k describes the generalized centrifugal and coriolis or gyroscopic forces, respectively, and the fx1-vector q includes the generalized applied forces. Often, the equations of motion can be linearized. Then, one obtains an ordinary rheonomic system of the form

$$M(t)\,\ddot{y}(t) + P(t)\,\dot{y}(t) + Q(t)\,y(t) = h(t) \tag{73}$$

or an ordinary scleronomic system

$$M\,\ddot{y}(t) + (D + G)\,\dot{y}(t) + (K + N)\,y(t) = 0\ . \tag{74}$$

The fxf-matrices $P(t)$ and $Q(t)$ can be partitioned in symmetrical fxf-matrices D, K and skew-symmetrical fxf-matrices G, N in the scleronomic case. The matrix D is due to damping forces, the matrix G characterizes gyroscopic forces, the matrix K describes conservative stiffness forces and the matrix N represents nonconservative forces.

<u>D'Alembert's Principle in a Reference Frame</u>: The equations of motion following from (65) and (69) agree completely with (71). The identity of the equations of motion based on the inertial and the reference frame, respectively, can be seen if (65) and (69) are supplemented by (67) remembering the orthogonality of the matrix S_R :

$$\bar{J}^T \bar{S}_R \left[\bar{S}_R^T \bar{\bar{M}} \bar{S}_R \bar{S}_R^T \bar{J}\,\ddot{y}(t) + \bar{S}_R^T \bar{q}^c \right] = \bar{J}^T \bar{S}_R \left[\bar{S}_R^T \bar{q}^a + \bar{S}_R^T \bar{Q}\,g \right] . \tag{75}$$

Thus, the choice of the reference frame doesn't affect the equations of motion at all. However, kinematics and Newton-Euler equations as well as the application of d'Alembert's principle may be strongly simplified by the choice of a proper reference frame.

<u>Jourdain's Principle in an Inertial Frame</u>: For the dynamical analysis of large ordinary multibody systems, it may be advantageous to introduce in addition to the f generalized coordinates f generalized velocities, see Ref. [12]. Then, it yields $f = g$ and the fxg-matrix G , introduced in (28), (29), can be substituted by the regular fxf-matrix $H = G^{-1}$. From (23) it follows

$$H(y,t)\ \dot{y}(t) = z(t) - \bar{z}(y,t) \quad , \tag{76}$$

where $z(t)$ denotes now the fx1-vector of the generalized velocities. Further, d'Alembert's principle (68) is replaced by Jourdain's principle (69) as well as (64) by (69) resulting in a modified set of equations of motion:

$$M^*(y,t)\ \dot{z}(t) + k^*(y,z,t) = q^*(y,z,t) \quad . \tag{77}$$

It turns out that (77) differs from (71). In particular, the second order differential equations (71) are replaced by two first order differential equations (76) and (77), separating kinematics and dynamics.

However, eqs. (76) and (77) represent still an ordinary multibody system. This can be seen applying the following transformation laws:

$$M = H^T M^* H \quad , \quad q = H^T q^* \ ,$$

$$k = H^T(k + M^*[\dot{H}\dot{y} + \dot{\bar{z}}]) \quad . \tag{78}$$

It is obvious from (78) that a proper choice of H may simplify M^* with respect to M.

Equations of Motion of General Multibody Systems

Multibody systems are called general iff they are not ordinary. Nonholonomic constraints and / or proportional-integral forces produce general multibody systems.

The equations of motion are obtained from the Newton-Euler equations (68), the proportional-integral forces (58) and Jourdain's principle (70). However, the equations of motion are not sufficient, they have to be completed by the nonholonomic constraint equations (23). Thus, the complete equations read as

$$\left.\begin{aligned} &\dot{y}(t) = g(y,z,t) \quad , \\ &M(y,z,t)\ \dot{z}(t) + k(y,z,t) = q(y,z,w,t) \quad , \\ &\dot{w}(t) = \dot{w}(y,z,t) \quad . \end{aligned}\right\} \tag{79}$$

Now, the number of dynamical equations is reduced from 6p to g , characterized by the symmetric positive definite gxg-inertia matrix

$$M(y,z,t) = L^T M L > 0 \tag{80}$$

and the gx1-vectors k and q of the generalized gyroscopic and applied forces.

Equations of Motion of Orbiting Multibody Systems

In satellite dynamics, the orbital motion is separated from the attitude motion. Therefore, special approaches and formalisms for the dynamical analysis of the attitude motion has been developed. It will be shown, see also Ref. [19] , that the equations of motion of orbiting systems can also be obtained by general formalisms.

A body of the orbiting system will be chosen as the basebody, Fig. 7. The basebody number is one and the body-fixed frame 1 coincides with the reference frame R with origin $O_R = C_1$. Then, the position vectors may be decomposed as

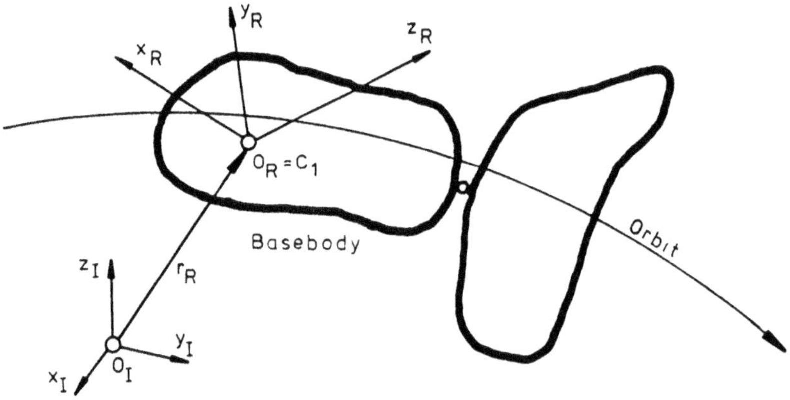

Fig. 7 Orbiting Multibody System

$$x = \begin{bmatrix} r_R \\ \hline x_R \end{bmatrix} , \quad y = \begin{bmatrix} r_R \\ \hline y_R \end{bmatrix} , \tag{81}$$

where r_R follows from (30). Using the abbreviations for the 3px3-matrices

$$\bar{E} = [E \;\; E \; \dots \; E]^T , \quad \bar{0} = [0 \;\; 0 \; \dots \; 0]^T , \tag{82}$$

the global Jacobian matrix (62) can be partitioned as

$$\bar{J} = \left[\begin{array}{c|c} \bar{E} & \bar{J}_{RT} \\ \hline \bar{0} & \bar{J}_{RR} \end{array}\right] . \tag{83}$$

Furthermore, the 6px6p-diagonal matrix (60) is rewritten by four 3px3p-matrices

$$\bar{\bar{M}} = \left[\begin{array}{c|c} \bar{\bar{M}}_T & 0 \\ \hline 0 & \bar{\bar{M}}_R \end{array}\right] . \tag{84}$$

Then, it follows from (64) and (69)

$$m \, \ddot{r}_R + E^T \, \bar{\bar{M}}_R \, \bar{J}_{RT} \, \ddot{y}_R + k_r = q_r , \tag{85}$$

$$(\bar{J}_{RT}^T \, \bar{\bar{M}}_T \, \bar{J}_{RT} + \bar{J}_{RR}^T \, \bar{\bar{M}}_R \, \bar{J}_{RR}) \, \ddot{y}_R + \bar{J}_{RT}^T \, \bar{E} \, \ddot{r}_R + k_y = q_y , \tag{86}$$

where the total mass of the system,

$$m = \sum_{i=1}^{p} m_i = \bar{E}^T \, \bar{\bar{M}}_T \, \bar{E} , \tag{87}$$

has been used. Obviously, the orbital motion r_R in (86) can be eliminated by (85) resulting in (f -3) equations of attitude motion:

$$M_R(y_R,t) \, \ddot{y}_R(t) + k_R(y_R,\dot{y}_R,t) = q_R . \tag{88}$$

It has to be mentioned that the elimination of the orbital motion don't require any matrix inversions since the total mass is a scalar.

Equations of Constraint Forces for Ordinary Multibody Systems

In ordinary multibody systems, the equations of motion and the constraint forces are decoupled. This means that the motion of the system can be found by integration without any knowledge of the constraint forces. However, sometimes the constraint forces are also of technical interest, e.g. for the analysis of the strength of bearings and supports.

The generalized constraint forces are available from Newton-Euler equations (64) as

$$g = \bar{Q}^{+}(\bar{\bar{M}}\,\bar{J}\,\ddot{y} + \bar{q}^{c} - \bar{q}^{a}) \quad , \tag{89}$$

where $(^{+})$ denotes the pseudo inverse of the global distribution matrix $\bar{Q}$. However, eq. (89) is not very convenient since the accelerations $\ddot{y}$ appear and, using (71), the inversion of the fxf-inertia matrix is required. The adequate solution of this problem is due to Schramm [20]. Rewriting d'Alembert's principle (69) as an orthogonality condition

$$\bar{Q}^{T}\,\bar{\bar{M}}^{-1}\,\bar{\bar{M}}\,\bar{J} = 0 \quad , \tag{90}$$

the elimination of the accelerations in (89) means premultiplication with the transposed global distribution matrix and the inverse diagonal matrix of inertia. Three advantages are achieved by Schramm's method: i) symmetrization of the constraint matrix, ii) reduction to the minimal order of the remaining algebraical equations, iii) elimination of the accelerations.

From (89) and (90), one obtains the equations of constraint as

$$N(y,t)\;g + \hat{q}(y,\dot{y},t) = \hat{k}(y,\dot{y},t) \tag{91}$$

where

$$N(y,t) = Q^{T}\,M^{-1}\,Q > 0 \tag{92}$$

is the symmetric and positive definite qxq-constraint matrix. The qx1-vectors $\hat{q}$ and $\hat{k}$ characterize the action of the applied and centrifugal and coriolis forces with respect to the constraints.

The choice of the generalized constraint forces is arbitrary as well as the choice of the generalized coordinates. Kreuzer and Schmoll [18] have shown that there exist natural constraint forces simplifying the generation of the global distribution matrix.

Equations of Constraint Forces in Systems with Contact

In contrary to all proportional forces, the contact forces (59) depend on the constraint forces, too. This means that generally the definition of ordinary multibody systems is violated. The equations of motion (71) and the equations of constraint (91) are coupled

$$
\begin{aligned}
&M(y,t)\,\ddot{y}(t) + k(y,\dot{y},t) = q(y,\dot{y},g,t) \quad , \\
&N(y,t)\,g(t) + \hat{q}(y,\dot{y},g,t) = \hat{k}(y,\dot{y},t) \quad .
\end{aligned}
\tag{93}
$$

However, the differential equations of motion are still of minimal number. A simultaneous solution of the algebraical and differential equations (93) is generally required.

In vehicle dynamics, the contact forces often depend linearly on the constraint or normal forces, Fig. 8 , resulting in the applied forces

$$
\bar{q}^a(y,\dot{y},g,t) = \bar{q}^a(y,\dot{y},t) + \bar{Q}^a(y,\dot{y},t)\,g \tag{94}
$$

with a 6pxq-distribution matrix $\bar{Q}^a$.

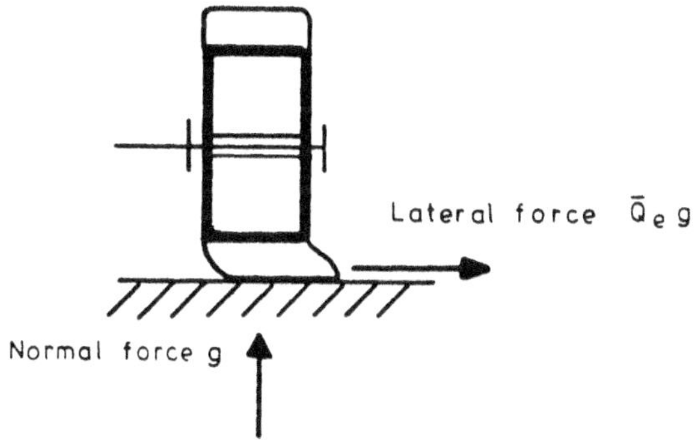

Fig. 8 Lateral Force of an Elastic Wheel

Then, eqs. (93) are also linear with respect to the generalized constraint forces:

$$M(y,t)\,\ddot{y}(t) + k(y,\dot{y},t) = q(y,\dot{y},t) + Q(y,\dot{y},t)\,g(t)$$
$$[N(y,t) + \hat{Q}(y,\dot{y},t)]\,g(t) = \hat{k}(y,\dot{y},t) \qquad (95)$$

resulting in extended equations of motion

$$M\,\ddot{y}(t) + k = q + Q(N + \hat{Q})^{-1}\,\hat{k} \quad , \qquad (96)$$

where Q is a fxq-matrix and $\hat{Q}$ represents an asymmetric qxq-matrix. From a computational point of view, eqs. (96) are not very effective due to the inverse of the matrix sum $(N + \hat{Q})$. But from a theoretical point of view, eqs. (96) show very clearly the behavior and the influence of contact forces.

5. COMPUTERIZED DERIVATION

A main problem in the dynamics of multibody systems is the derivation of equations of motion. Even if the most economic principles like d'Alembert's principle and Jourdain's principle are applied, the equations of motion of large multibody systems can be hardly found by paper and pencil. Computer-aided formalisms represent the adequate solution of the problem.

Mathematical Operations

A formalism executes mathematical operations on the input variables to generate the output. What are the input variables, what are the mathematical operations in multibody dynamics?

For ordinary multibody systems, the following input data have to be prepared.

For the system:

fx1-position vector $y(t)$

and for each body, $i = 1(1)p$:

3x1-translation vector	$r_i(y,t)$	,
3x3-rotation tensor	$S_i(y,t)$	,
1x1-mass	m_i	,
3x3-inertia tensor	${}_iI_i$	,
3x1-applied force vector	$f_i^a(y,\dot{y},t)$	,
3x1-applied torque vector	$l_i^a(y,\dot{y},t)$	.

There are additional input variables, if a moving reference frame is used.

During the generation of equations of motion, the following mathematical operations are executed:

summation of vectors and matrices,
multiplication of vectors and matrices,
differentiation of vectors and matrices,
simplification of trigonometric expressions,
linearization of expressions.

Symbolical Manipulation of Formulae

The mathematical operations may be executed numerically or symbolically. The symbolical manipulation of formulae has the advantage that the equations of motion obtained look like the equations found by paper and pencil. Thus, a physical interpretation is often possible and big savings of computation time during the subsequent numerical integration are at hand.

For the symbolical manipulation of formulae, special languages like MACSYMA or REDUCE can be used. However, due to limited number of mathematical operations, the index coding using FORTRAN may also be applied. The index coding uses for the memorization and summation of terms properly defined arrays of integers. Positive integers represent variables, negative integers are functions. The sign of a term is given by the sign of its numerical factor and all elements of a term (numerical factor, variables and functions) are automatically multiplied. Vectors and matrices are formed using symbolical expressions as matrix elements. Then, the programming of vector and matrix operations completes the routine.

Program NEWEUL

The program NEWEUL applies the index coding for the symbolical generation of equations of motion. The user prepares the required input variables in an interactive dialog with the computer. NEWEUL performs automatically the linearization of expressions and, to some extend, simplifications of trigonometric expressions. The resulting symbolical equations of motion are obtained as a printed listing or as a file ready for numerical solution by any available eigenvalue or integration routine, respectively. For more details see the operating instructions by Kreuzer, Schmoll and Schramm [21].

Example: Due to the limit space for this paper, only the simple example of a double pendulum will be presented, Fig. 9.

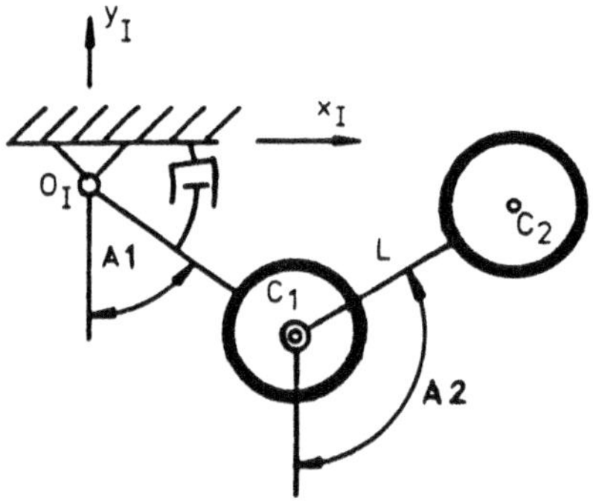

Fig. 9 Double Pendulum

Number of degrees of freedom: $f = 2$.

Position vector: $y = [A1\ A2]^T$, $\dot{y} = [A11\ A21]^T$.

Translation vectors:

$$r_1 = \begin{bmatrix} L * SIN(A1) \\ - L * COS(A1) \\ 0 \end{bmatrix} , \quad r_2 = \begin{bmatrix} L * SIN(A1) + L * SIN(A2) \\ - L * COS(A1) - L * COS(A2) \\ 0 \end{bmatrix} .$$

Rotation tensors:

$$S_i = \begin{bmatrix} COS(Ai) & -SIN(Ai) & 0 \\ SIN(Ai) & COS(Ai) & 0 \\ 0 & 0 & 1 \end{bmatrix} , \quad i = 1,2 .$$

Masses: $m_1 = m_2 = M$.

Inertia tensors:

$${}_i I_i = \begin{bmatrix} IX & 0 & 0 \\ 0 & IX & 0 \\ 0 & 0 & IZ \end{bmatrix} , \quad i = 1,2 .$$

Applied forces and torques due to gravity and viscous damping:

$$f_1 = f_2 = \begin{bmatrix} 0 \\ - M * G \\ 0 \end{bmatrix} , \quad l_1 = \begin{bmatrix} 0 \\ 0 \\ - D*(A11) \end{bmatrix} , \quad l_2 = 0 .$$

From these input data the program NEWEUL generates the following equations of motion, Fig. 10. The elements of the 2x2-inertia matrix are called RM(i,j) . Due to the symmetry, only the upper right elements are printed.

```
INERTIA MATRIX

RM(1,1)=2.*L**2*M+IZ
RM(1,2)=L**2*M*COS(A2)*COS(A1)+L**2*M*SIN(A2)*SIN(A1)

RM(2,2)=L**2*M+IZ

VECTOR OF GENERALIZED GYRO AND CENTRIFUGAL FORCES

K(1)=-L**2*M*A21**2*SIN(A2)*COS(A1)+L**2*M*A21**2*SIN(A1)
     *COS(A2)

K(2)=-L**2*M*A11**2*SIN(A1)*COS(A2)+L**2*M*A11**2*SIN(A2)
     *COS(A1)

VECTOR OF GENERALIZED APPLIED FORCES

Q(1)=-2.*G*L*M*SIN(A1)-D*A11

Q(2)=-G*L*M*SIN(A2)
```

Fig. 10 NEWEUL Output

6. CONCLUSION

An economic generation of equations of motion of multibody systems requires a suitable kinematical description and a proven dynamical principle. The kinematics may be applied in an inertial frame or a moving reference frame. The choice of the frame doesn't affect the equations of motion at all, but a moving frame often reduces the intermediate computational work. As system variables, generalized coordinates and / or generalized velocities may be used. Contrarily to the frames, the system variables affect immediately the equations of motion. However, the different results are related by transformation laws to each other. D'Alembert's and Jourdain's principle, respectively, reduce the generation of equations of motion to a matrix multiplication of the Newton-Euler equations. Therefore, these principles are effective tools for the computerized derivation of equations of motion. Moreover, the equations of motion may be generated completely symbolical as shown by the program NEWEUL .

Additional to the equations of motion, the constraint forces are often needed for a complete dynamical analysis. The generalized constraint forces follow from algebraical equations found by Schramm's method. In the case of contact and friction forces, the equations of motion and the equations of constraint are coupled and they have to be solved simultaneously.

REFERENCES

1. Hooker, W.W., and Margulies, G., "The Dynamical Attitude Equations for an N-Body Satellite", J. Astr. Sci., Vol. 12, 1965, pp. 123-128.

2. Roberson, R.E., and WITTENBURG, J., "A Dynamical Formalism for an Arbitrary Number of Interconnected Rigid Bodies, with Reference to the Problem of Satellite Attitude Control", Proc. 3rd Congress Int. Fed. Auto. Control (IFAC), London, 1966, pp. 46 D. 1-46 D.9.

3. Frisch, H.P., A Vector-Dyadic Development of the Equations of Motion for N-coupled Rigid Bodies and Point Masses, Techn. Note TN D - 7767, Nat. Aeron. Space Adm. (NASA), Washington, D.C. 1974.

4. Wittenburg, J., Dynamics of Systems of Rigid Bodies, Teubner, Stuttgart, 1977.

5. Andrews, G.C., and Kesavan, H.K., "Simulation of Multibody Systems Using the Vector-Network Model", Dynamics of Multibody Systems, Springer-Verlag, Berlin, Heidelberg, New York, 1978, pp. 1-13.

6. Paul, B., Kinematics and Dynamics of Planar Machinery, Prentice-Hall, Englewood Cliffs, 1979 .

7. Orlandea, N., Chace, M.A., and Calahan, D.A., "A Sparsity Oriented Approach to the Dynamic Analysis and Design of Mechanical Systems", ASME J. Eng. Industry, Vol. 99, 1977, pp. 773-784.

8. Haug, E.J., Wehage, R.A., and Barman, N.C., "Dynamic Analysis and Design of Constrained Mechanical Systems", ASME J. Mech. Design, Vol. 103, 1981, pp. 560-570.

9. Rosenthal, D.E., and Sherman, M.A., Symbolic Multibody Equations via Kane's Method, AAS-Paper 83-303, AAS Publ. Office, San Diego, 1983.

10. Schiehlen, W.O., and Kreuzer, E.J., "Symbolic Computerized Derivation of Equations of Motion", Dynamics of Multibody Systems, Springer-Verlag, Berlin-Heidelberg-New York, 1978, pp. 290-305.

11. Kane, T.R., and Levinson, D.A., "Formulation of Equations of Motion for Complex Spacecraft", J. Guidance Control, Vol. 3, 1980, pp. 99-112.

12. Schiehlen, W.O., "Nichtlineare Bewegungsgleichungen großer Mehrkörpersysteme", Z. angew. Math. Mech., Vol. 61, 1981, pp. 413-419.

13. Schwertassek, R., Der Roberson/Wittenburg-Formalismus und das Programmsystem MULTIBODY zur Rechnersimulation von Mehrkörpersystemen, Forschungsbericht DFVLR-FB 78-08, DFVLR, Köln, 1978.

14. Chace, M.A., "Using DRAM and ADAMS Programs to Simulate Machinery, Vehicles", Agricult. Eng., Vol. 59, No. 11, 1981, pp. 18-19 and No. 12, 1981, pp.16-18.

15. Nikravesh, P.E., and Chung, I.S., "Application of Euler Parameters to the Dynamic Analysis of Three-Dimensional Constrained Mechanical Systems", ASME J. Mech. Design, Vol. 104, 1982, pp. 785-791.

16. Rongved, L., and Fletcher, H.J., "Rotational Coordinates", J. Franklin Inst., Vol. 277, 1964, pp. 414-421.

17. Schwertassek, R., and Roberson, R.E., "A State-Space Dynamical Representation for Multibody Mechanical System", Part I and Part II, Acta Mechanica, to appear.

18. Kreuzer, E.J., and Schmoll, K.P., "Zur Berechnung von Reaktionskräften in Mehrkörpersystemen", Z. angew. Math. Mech., to appear.

19. Kreuzer, E.J., and Schiehlen, W.O., Generation of Symbolic Equations of Motion for Complex Spacecraft Using Formalism NEWEUL, AAS-Paper No. 89-302, AAS Publ. Office, San Diego, 1983.

20. Schramm, D., Zur symbolischen Berechnung von Zwangskräften, Zwischenbericht ZB - 10, Institut B für Mechanik, Universität Stuttgart, 1982.

21. Kreuzer, E.J., Schmoll, K.P., and Schramm, D., Programmpaket NEWEUL ' 83, Anleitung AN-7, Institut B für Mechanik, Universität Stuttgart, 1983.

VEHICLE DYNAMICS APPLICATIONS

Werner O. Schiehlen
Institute B of Mechanics
University Stuttgart
Stuttgart, F.R.G.

Abstract. The computerized generation of equations of motion is presented for a simple vehicle with rigid wheels representing a nonholonomic system. Then, the stochastic excitation process of a spatial road, the equations of motion of a complex automobile vehicle, the human sensation of mechanical vibrations and numerical methods for the analysis of random vehicle vibrations are discussed.

1. INTRODUCTION

For the dynamical analyis of handling and ride characteristics of ground vehicles mathematical models of the excitations, the vehicle itself and the rating of the resulting motions are required. The lateral motion is primarily affected by the vehicle's path controlled by the driver. For the rating of the lateral motion, the directional stability of the vehicle is used. The vertical motion is mainly excited by the guideway irregularities representing a Gaussian, ergodic stochastic process. The human sensation of the vertical mechanical vibrations has to be characterized by standard deviations of the resulting random processes. In both cases - handling and ride analysis - the method of multibody systems can be applied for the mathematical modeling of the vehicle itself. The corresponding frequencies of the lateral and vertical motion are less than 50 Hz for ground vehicles.

The dynamical research of ground vehicles requires a broad spectrum of theoretical and experimental methods. They have been presented recently in Ref. [1]. However, due to the long history of vehicle dynamics, there are also many excellent textbooks and proceedings

NATO ASI Series, Vol. F9
Computer Aided Analysis and Optimization of Mechanical System Dynamics
Edited by E. J. Haug

available, e.g. Mitschke [2], Wong [3], Willumeit [4], Wickens [5], Hedrick and Wormley [6].

In this paper, the application of the method of multibody systems to vehicles will be presented. Firstly, the equations of lateral motion of a simple vehicle are derived using the program NEWEUL for nonholonomic systems. Secondly, the vertical motion of a complex automobile is considered resulting in a global system of guideway, vehicle and sensation equations. Thirdly, some aspects of linear vibration analysis are discussed including random processes.

2. HANDLING CHARACTERISTICS OF A SIMPLE VEHICLE

Simple vehicle models don't include all phenomena observed in experiment and/or simulation. But simple models assist the understanding and show the influence of essential parameters more clearly. For the investigation of vehicles with stiff wheels, the model of rigid wheels can be used resulting in nonholonomic constraints.

Vehicle Model

A simple vehicle moving on a rough plane is shown in Fig. 1. The model consists of seven rigid bodies: the car body, the front axle, four wheels and the chassis with the rear axle.

The car body is characterized by the mass MA and the principle moments of inertia IAX, IAY, IAZ with respect to the body-fixed frame 1. The car body is connected to chassis by the joint P with three constraints $(q = 3)$. The generalized coordinates of the car body are the vertical translation Z(t) and the rotation angles AL(t) and BE(t). The applied forces and torques are generated by four spring-dashpot configurations with the parameters KV, DV and KH, DH for the front and rear axle, respectively. Furthermore, the free lengths of the springs are LOV and LOH, respectively. In the equilibrum position the car body may be horizontal, e.g. AL = BE = 0.

The front axle with mass MV and principle moments of inertia IVX, IVY, IVZ = IVX is connected to the chassis by a revolute joint $(q = 5)$. The remaining steering angle is DE(t). The applied force

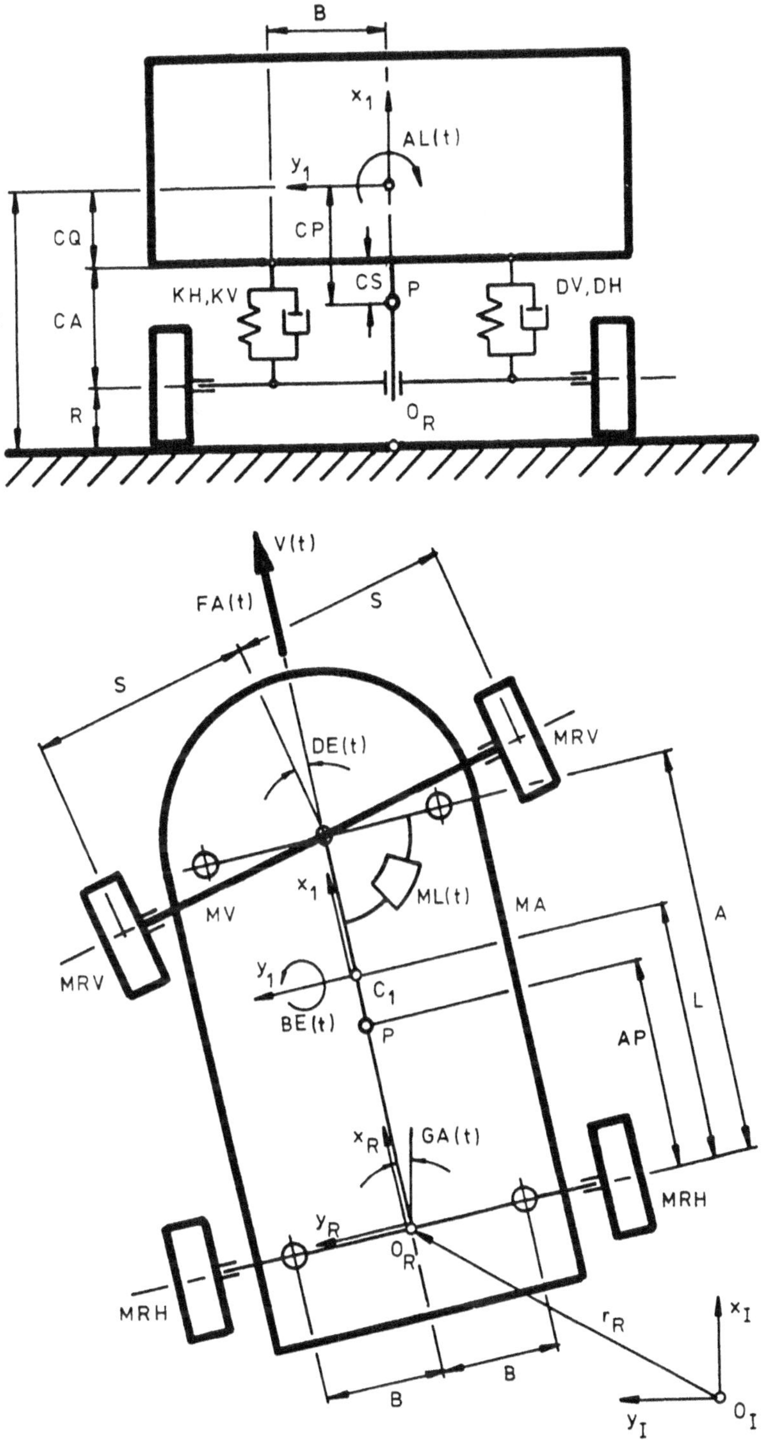

Fig. 1 Simple Vehicle Model

is given by the driving force FA(t) and the applied torque ML(t) is due to a servomotor.

The front wheels have the masses MRV and the principle moments of inertia IRVX = IRVZ, IRVY . The rear wheels are characterized by MRH, IRHX, IRHZ , respectively. Each wheel is mounted with a revolute joint (q = 5) . The generalized coordinates of the wheels are PHl, l = 1(1)4 . There don't exist any applied forces and torques for the wheels.

The chassis is assumed to be massless. Then, only the constraints are of interest. The chassis moves parallel to the rough plan (q = 3) and the generalized coordinates are XR(t), YR(t) and GA(t) . Applied forces and torques are not at hand.

All distances are shown in Fig. 1, representing the equilibrum condition.

Kinematics and Dynamics

From the vehicle description, the following number of positional degrees of freedom is obtained:

$$f = 6 \cdot 7 - (3 + 5 + 4 \cdot 5 + 3) = 11 \quad , \qquad (1)$$

characterized by the vector of the generalized coordinates

$$y(t) = [XR \quad YR \quad GA \quad DE \quad PH1 \quad PH2 \quad PH3 \quad PH4 \quad Z \quad AL \quad BE]^{T}. \qquad (2)$$

The kinematical relations are most easily written in the moving reference frame R , Fig. 1, given by

$$r_R(y) = \begin{bmatrix} XR \\ YR \\ 0 \end{bmatrix} , S_R(y) = \begin{bmatrix} COS(GA) & -SIN(GA) & 0 \\ SIN(GA) & COS(GA) & 0 \\ 0 & 0 & 1 \end{bmatrix} . \qquad (3)$$

Thus, the reference frame is chassis-fixed. With respect to the reference frame, the relative motion of the front axle is given by

$$r_{R2} = \begin{bmatrix} A \\ 0 \\ R \end{bmatrix} , \quad S_{R2}(y) = \begin{bmatrix} COS(DE) & -SIN(DE) & 0 \\ SIN(DE) & COS(DE) & 0 \\ 0 & 0 & 1 \end{bmatrix} . \tag{4}$$

Similar input data are found for the front axle and the wheels. For the chassis no inputs are necessary due to the vanishing forces and masses.

In addition to the holonomic constraints, there exist $r = 6$ nonholonomic constraints due to the front and rear axle and the four wheels. The number of motional degrees of freedom is

$$g = 11 - 6 = 5 \tag{5}$$

and the vector of the generalized velocites reads as

$$z(t) = [\ V \quad DE1 \quad Z1 \quad AL1 \quad BE1\ ]^T \tag{6}$$

where V is the velocity in the x_R-direction and "1" means the first time derivative. The explicit nonholonomic constraint equations follow as

$$y(t) = \begin{bmatrix} V*COS(GA) \\ V*SIN(GA) \\ V*AI*TANDE \\ DE1 \\ RI(V*[\ COS(DE) + TANDE(SIN(DE) - S*AI)\ ] - S*DE1) \\ RI(V*[\ COS(DE) + TANDE(SIN(DE) + S*AI)\ ] + S*DE1) \\ RI(V - S*GA1) \\ RI(V + S*GA1) \\ Z1 \\ AL1 \\ BE1 \end{bmatrix} \tag{7}$$

where "I" means inverse.

The applied force and torques can also be obtained from Fig. 1 where the linearization with respect to the small car body motion $Al(t) << 1$, $BE(t) << 1$, $Z(t) << C$ is helpful. For more details see Ref. [7] .

Computerized Derivation of Equations of Motion

The nonlinear equations of motion for large lateral motions and small vertical motions of the vehicle can be obtained from the above mentioned input data completely automatically by the program NEWEUL . The equations of motion, Fig. 2, are printed in the standard form of general multibody systems:

$$RM(y)\ \dot{z}(t) + K(y,z,t) = Q(y,z,t)) \quad . \qquad (8)$$

Here RM means the symmetrical 5x5-inertia matrix M where only the essential elements of this matrix are shown.

For integration, eqs. (8) have to be completed by eqs. (7). Obviously, eqs. (7) and (8) are strongly coupled.

It is also possible to generate completely linear equations of motion, e.g., with respect to a steady-state turning. Then, the rating of the directional stability is possible using results from stability theory.

3. RIDE CHARACTERISTICS OF A COMPLEX AUTOMOBILE

Vertical vibrations of vehicles are usually characterized by small amplitudes. Then, from the beginning, linearized equations may be generated and the complexity of the model is not such a severe problem as in the nonlinear case. The equations of global vehicle system dynamics follow from the guideway irregularities, the vehicle itself and the passenger sensation to vibration. The dynamical analysis includes free and random vibrations.

Guideway Irregularities

The guideway irregularities for the right and the left trace of a road are given by the random processes $\xi_r(t)$ and $\xi_l(t)$. Thus, a spatial road profile can be represented by a 2x1-vector process

INERTIA MATRIX

```
RM(1,1)=MA+2.*CP*MA*AI*TANDE*AL+L**2*MA*AI**2
        *TANDE**2+2.*CP*L*MA*AI**2*TANDE**2*BE+MV+
        MV*TANDE**2+2.*MRV+2.*MRV*S**2*AI**2
        *TANDE**2+2.*MRV*TANDE**2+2.*MRH+2.*MRH
        *S**2*AI**2*TANDE**2+IAZ*AI**2*TANDE**2+
        IVX*AI**2*TANDE**2+2.*IRVY*RI**2*COS(DE)**2
        +4.*IRVY*RI**2*TANDE*SIN(DE)*COS(DE)+2.
        *IRVY*RI**2*TANDE**2*SIN(DE)**2+2.*IRVY
        *S**2*AI**2*RI**2*TANDE**2+2.*IRVX*AI**2
        *TANDE**2+2.*IRHY*RI**2+2.*IRHX*AI**2
        *TANDE**2
RM(1,2)=2.*MRV*S**2*AI*TANDE+IVX*AI*TANDE+2.*IRVY
        *S**2*AI*RI**2*TANDE+2.*IRVX*AI*TANDE
RM(1,3)=0.
RM(1,4)=-CP*L*MA*AI*TANDE
RM(1,5)=CP*MA


RM(2,2)=2.*MRV*S**2+IVX+2.*IRVY*S**2*RI**2+2.*IRVX
RM(2,3)=0.
RM(2,4)=0.
RM(2,5)=0.


RM(3,3)=MA
RM(3,4)=0.
RM(3,5)=0.


RM(4,4)=CP**2*MA+IAX
RM(4,5)=0.


RM(5,5)=CP**2*MA+IAY
```

VECTOR OF GENERALIZED GYRO AND CENTRIFUGAL FORCES

```
K(1)=CP*MA*AI*TA04*DE1*V*AL+2.*CP*MA*AI*TANDE*AL1
     *V+L**2*MA*AI**2*TA04*TANDE*DE1*V+2.*CP*L*MA
     *AI**2*TA04*TANDE*DE1*V*BE+2.*CP*L*MA*AI**2
     *TANDE**2*BE1*V+MV*TA04*TANDE*DE1*V+MRV*S**2
     *AI**2*TANDE*DE1*V+MRV*S**2*AI**2*TA04*TANDE
     *DE1*V+2.*MRV*TA04*TANDE*DE1*V+MRV*S**2*AI**2
     *TA04*TANDE*DE1*V*SIN(DE)**2+2.*MRH*S**2
     *AI**2*TA04*TANDE*DE1*V+IAZ*AI**2*TA04*TANDE
     *DE1*V+IVX*AI**2*TA04*TANDE*DE1*V-2.*IRVY
     *RI**2*DE1*V*SIN(DE)*COS(DE)+2.*IRVY*RI**2
     *TA04*DE1*V*SIN(DE)*COS(DE)-2.*IRVY*RI**2
     *TANDE*DE1*V*SIN(DE)**4+2.*IRVY*RI**2
     *TANDE**2*DE1*V*SIN(DE)*COS(DE)+2.*IRVY*RI**2
     *TA04*TANDE*DE1*V*SIN(DE)**2+2.*IRVY*S**2
     *AI**2*RI**2*TA04*TANDE*DE1*V+2.*IRVY*RI**2
     *TANDE*DE1*V*COS(DE)**4+2.*IRVX*AI**2*TA04
     *TANDE*DE1*V+2.*IRHX*AI**2*TA04*TANDE*DE1*V
```

Continued

```
K(2)=MRV*S**2*AI*DE1*V+MRV*S**2*AI*TA04*DE1*V+MRV
     *S**2*AI*TA04*DE1*V*SIN(DE)**2+IVX*AI*TA04
     *DE1*V+2.*IRVY*S**2*AI*RI**2*TA04*DE1*V+2.
     *IRVX*AI*TA04*DE1*V

K(4)=L**2*MA*AI*TA04*DE1*V*BE+L*MA*AI*TANDE*V**2
     *BE-L*AP*MA*AI*TA04*DE1*V*BE-AP*MA*AI*TANDE
     *V**2*BE-CP*L*MA*AI*TA04*DE1*V-CP**2*MA*AI
     *TA04*DE1*V*BE-CP**2*MA*AI**2*TANDE**2*V**2
     *AL-2.*CP**2*MA*AI*TANDE*BE1*V-CP*MA*AI*TANDE
     *V**2-IAX*AI*TA04*DE1*V*BE+IAZ*AI*TA04*DE1*V
     *BE-IAY*AI**2*TANDE**2*V**2*AL+IAZ*AI**2
     *TANDE**2*V**2*AL-IAX*AI*TANDE*BE1*V-IAY*AI
     *TANDE*BE1*V+IAZ*AI*TANDE*BE1*V

K(5)=L**2*MA*AI**2*TANDE**2*V**2*BE-L*AP*MA*AI**2
     *TANDE**2*V**2*BE+CP**2*MA*AI*TA04*DE1*V*AL-
     CP*L*MA*AI**2*TANDE**2*V**2-CP**2*MA*AI**2
     *TANDE**2*V**2*BE+2.*CP**2*MA*AI*TANDE*AL1*V+
     L**2*MA*AI*TA04*DE1*V*AL+L*MA*AI*TANDE*V**2
     *AL-L*AP*MA*AI*TA04*DE1*V*AL-AP*MA*AI*TANDE
     *V**2*AL+IAY*AI*TA04*DE1*V*AL-IAX*AI**2
     *TANDE**2*V**2*BE+IAZ*AI**2*TANDE**2*V**2*BE+
     IAX*AI*TANDE*AL1*V+IAY*AI*TANDE*AL1*V-IAZ*AI
     *TANDE*AL1*V
```

VECTOR OF GENERALIZED APPLIED FORCES

```
Q(1)=-2.*CS*EV*KV*BE-2.*EH*KH*CS*BE+2.*CS*EV*KV*L
     *AI*TANDE*AL+2.*EH*KH*CS*L*AI*TANDE*AL+FA

Q(2)=ML

Q(3)=-2.*KV*Z-2.*KH*Z+2.*KV*A*BE-2.*KV*L*BE-2.*KH
     *L*BE-2.*DV*Z1-2.*DH*Z1+2.*DV*A*BE1-2.*DV*L
     *BE1-2.*DH*L*BE1

Q(4)=-2.*CS*EV*KV*CP*AL-2.*EH*KH*CS*CP*AL+2.*CQ*CS
     *EV*KV*AL-2.*CA*CQ*EV*KV*AL+2.*CQ*EH*KH*CS*AL-2.
     *CA*CQ*EH*KH*AL-2.*B**2*KV*AL-2.*B**2*KH*AL-2.
     *B**2*DV*AL1-2.*B**2*DH*AL1

Q(5)=-2.*CS*EV*KV*CP*BE-2.*EH*KH*CS*CP*BE+2.*CQ*CS
     *EV*KV*BE-2.*CA*CQ*EV*KV*BE+2.*CQ*EH*KH*CS*BE-2.
     *CA*CQ*EH*KH*BE+2.*KV*A*Z-2.*KV*L*Z-2.*KV
     *A**2*BE+4.*KV*L*A*BE-2.*KV*L**2*BE-2.*KH*L*Z-2.
     *KH*L**2*BE+2.*DV*A*Z1-2.*DV*L*Z1-2.*DV*A**2
     *BE1+4.*DV*L*A*BE1-2.*DV*L**2*BE1-2.*DH*L*Z1-2.
     *DH*L**2*BE1-ML*AL
```

EV = 1 - LOV/CA

EH = 1 - LOH/CA   LO Undeformed Length of Spring

TAO4 = D(TANDE) / D(DE) = 1/COS(DE)**2

Fig. 2 NEWEUL Output

$$\xi(t) = [\xi_r(t) \quad \xi_l(t)]^T \quad . \tag{9}$$

For the description of the road characteristics, it is more convenient to use the uncorrelated random processes $\xi_M(t)$ and $\xi_D(t)$ of the mean and the difference of the right and left trace, respectively, resulting in another 2x1-vector process

$$v(t) = [\xi_M(t) \quad \xi_D(t)]^T \quad . \tag{10}$$

Then, it yields

$$\xi(t) = H\, v(t) \quad , \quad H = \begin{bmatrix} 1 & -1 \\ 1 & 1 \end{bmatrix} \quad , \tag{11}$$

i.e., the road profile is completely represented by the process (10). On the other hand, the 2x1-vector process of colored noise may be given by a first order shape filter

$$\dot{v}(t) = F\, v(t) + G\, \tilde{w}(t) \quad , \quad \tilde{w}(t) \sim (0,Q) \quad , \tag{12}$$

where F and G are 2x2-coefficient matrices and $\tilde{w}(t)$ is a 2x1-white noise process with zero mean and the 2x2-idensity matrix $Q = qE$. The coefficient matrices are available in literature, see e.g. Rill [8], for various roads.

Vehicle Dynamics

The vehicle model under consideration will include numerous parts as shown in Fig. 3. The model consists of 4 mass points and 7 rigid bodies subject to 35 holonomic constraints resulting in 19 positional degrees of freedom. Altogether, 67 parameters describe this model listed in Ref. [1].

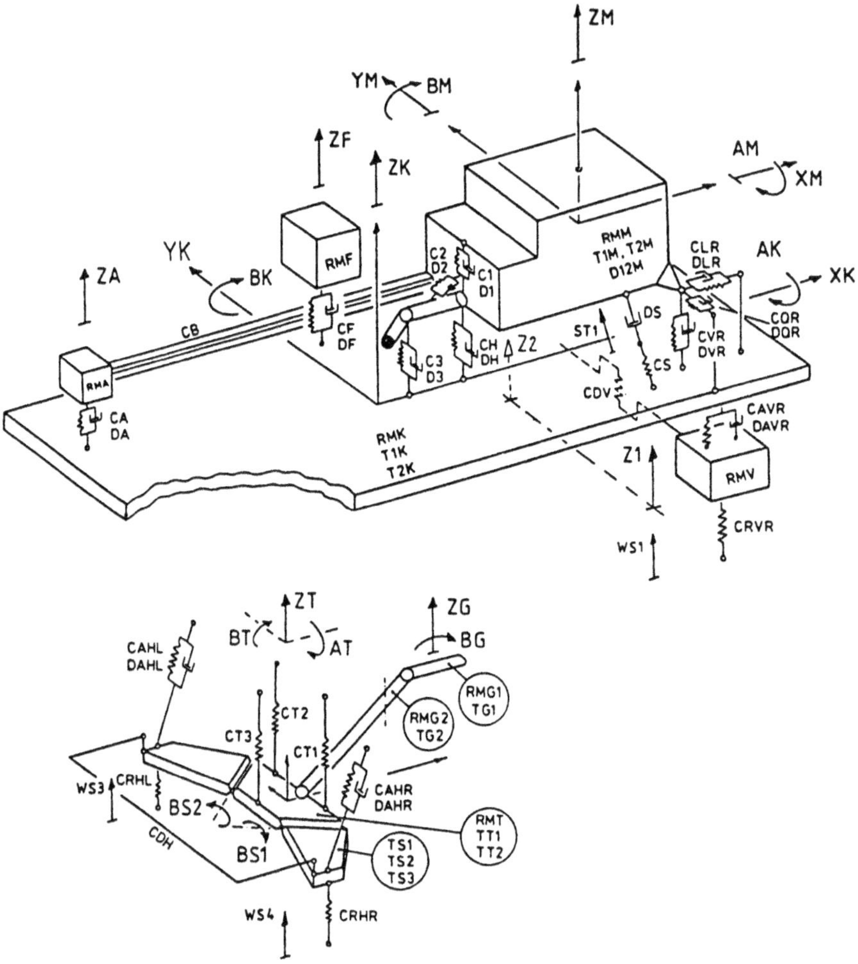

Fig. 3 Vehicle Model

Then, the linear equations of motion generated by the program NEWEUL read as

$$M\,\ddot{y}(t) + P\,\dot{y}(t) + Q\,y(t) = -\,R\,w(t) + S\,\xi(t) \tag{13}$$

where $y(t)$ is the 19x1-position vector and M, P, Q are 19x19-matrices representing inertia, velocity and position forces. In addition to the generalized coordinates, the 2x1-vector

$$w(t) = \left[\, ST1 \quad ST2 \,\right]^T \tag{14}$$

has to be introduced to describe the proportional-integral forces due to the serial spring-dashpot configurations at the engine. The corresponding law reads as

$$\dot{w}(t) + W\,w(t) = -\,\dot{Y}\,\dot{y}(t) \qquad , \tag{15}$$

where the matrices W and $\dot{Y}$ are time-invariant. The excitation of the vehicle is restricted to the guideway irregularities. For each wheel or wishbone, respectively, an independent excitation process is introduced:

$$\xi(t) = [\ WS1 \quad WS2 \quad WS3 \quad WS4\ ]^T \quad . \tag{16}$$

The corresponding input matrix S is also constant.

The state equation of the vehicle reads in the linear case

$$\dot{x}(t) = Ax(t) + B\xi(t) \tag{17}$$

where the 40x1-state vector

$$x(t) = [\ y^T \quad \dot{y}^T \quad w^T\ ] \tag{18}$$

is used. Then, the 40x40-system matrix A and the 40x4-excitation matrix are given by M, P, Q, R, S, W, $\dot{Y}$ completely.

The automobile has two axles. This means a time delay between the excitation at the front and the rear wheel at each trace of the guideway. Regarding (9) and (16) one obtains

$$\xi(t) = [\ \zeta^T(t) \quad \zeta^T(t - \Delta t)\ ]^T \quad , \quad \Delta t = \frac{L}{v} \quad . \tag{19}$$

The time delay Δt depends on the axle distance L and the vehicle speed v . Then, the excitation term in the state equation (17) can be rewritten as

$$B\xi(t) = [\ B_1 \quad B_2\ ]\ [\ \zeta^T(t) \quad \zeta^T(t - \Delta t)\ ]^T \tag{20}$$

where B_1 and B_2 are 40x2-submatrices.

Human Response

The human sensation of mechanical vibrations differs from objective measurable motions. Numerous physiological investigations have shown that the subjective sensation by men is proportional to the acceleration and depends on the dynamics of human organs, see International Standard [9]. However, in this paper, the dynamics of men to be modeled by a low order frequency response or shape filter, respectively, will be neglected. The squared standard deviation

$$P_{\ddot{Z}} = E\{P_{\ddot{Z}}(t)\} \tag{21}$$

at an arbitrary location (U, V) on the car body reads as

$$P_{\ddot{Z}} = P_{\ddot{Z}K} + V^2 P_{\ddot{A}K} + U^2 P_{\ddot{B}K} + 2\ V\ P_{\ddot{Z}K\ddot{A}K} - 2\ U\ P_{\ddot{Z}K\ddot{B}K} - 2\ U\ V\ P_{\ddot{A}K\ddot{B}K} \ , \tag{22}$$

where ZK(t), AK(t), BK(t) are the generalized coordinates of the car body, Fig. 3.

Free Vibrations

Free vibrations of a vehicle exist at vanishing speed. The free vibrations are characterized by the homogeneous state equation derived from (18):

$$\dot{x}(t) = A\ x(t) \quad . \tag{23}$$

The eigenfrequencies ω_k , $k = 1(1)19$, are found by the solution of the eigenvalue problem

$$(A - \lambda E)\ \tilde{x} = 0 \quad , \tag{24}$$

where λ is an eigenvalue and $\tilde{x}$ the corresponding eigenvector, see e.g. Ref. [10]. In addition to the complex eigenvalues $\lambda_j = \delta_k \pm i\omega_k$, $j = 1(1)38$, there exist two real eigenvalues λ_j , $j = 39,40$ due to the integrals of the position caused by the two spring-dashpot configurations. The numerical numbers of the eigenvalues are shown in Ref. [1]. The highest frequency ω_1 = 250 Hz is due to the drive shaft, the medium frequencies ω_7 to ω_{10} with 8 - 10 Hz represent the wheel vibrations and the lowest frequencies ω_{17} to ω_{19} with 0.7 - 1.2 Hz characterize the body vibrations.

Covariance Analysis

Guideway irregularities modeled by colored noise (12) lead to an extended state equation where (11), (12), (17) and (20) are summarized:

$$\dot{\hat{x}}(t) = \hat{A}\,\hat{x}(t) + \hat{B}_1\,\tilde{w}(t) + \hat{B}_2\,\tilde{w}(t-\Delta t) \quad , \tag{25}$$

where the extended state vector

$$\hat{x}(t) = [\, v_1 \;\; v_2 \;\; x \,]^T \tag{26}$$

and extended matrices $\hat{A}$, $\hat{B}_1$, $\hat{B}_2$ are used. Now, the steady-state 44x44-covariance matrix $\hat{P}$ of the vehicle response process $\hat{x}(t)$ can be calculated via the algebraic Ljapunow equation

$$\hat{A}\,\hat{P} + \hat{P}\,\hat{A}^T + \hat{B}_1\,Q\,\hat{B}_1^T + \hat{B}_2\,Q\,\hat{B}_2^T + \hat{\phi}(\Delta t)\,\hat{B}_2\,Q\,\hat{B}_1^T$$
$$\hat{B}_1\,Q\,\hat{B}_2\,\hat{\phi}^T(\Delta t) = 0 \quad , \tag{27}$$

where

$$\hat{\phi}(\Delta t) = e^{\hat{A}\Delta t} \tag{28}$$

is the 44x44-fundamental matrix of the extended system, see Ref. [11].

The rating of the vehicle response process $\hat{x}(t)$ includes the human response $P_{\ddot{z}}$ as well as the wheel load variations P_{Fi} , $i = 1(1)4$. These scalar variances are obtained by 1x44-transformation matrices $\hat{T}$ immediately from the 44x44-covariance matrix $\hat{P}$:

$$P_{\ddot{z}} = \hat{T}_{\ddot{z}}\,\hat{P}\,\hat{T}_{\ddot{z}}^T \quad , \quad P_{Fi} = \hat{T}_{Fi}\,\hat{P}\,\hat{T}_{Fi}^T \quad , \quad i = 1(1)4 \;. \tag{29}$$

A numerical result published by Kreuzer and Rill [12] is shown in Fig. 4. It turns out that the optimal comfort with minimal accelerations appears in the center of the car body and that the rear wheels are subject to larger dynamical loads.

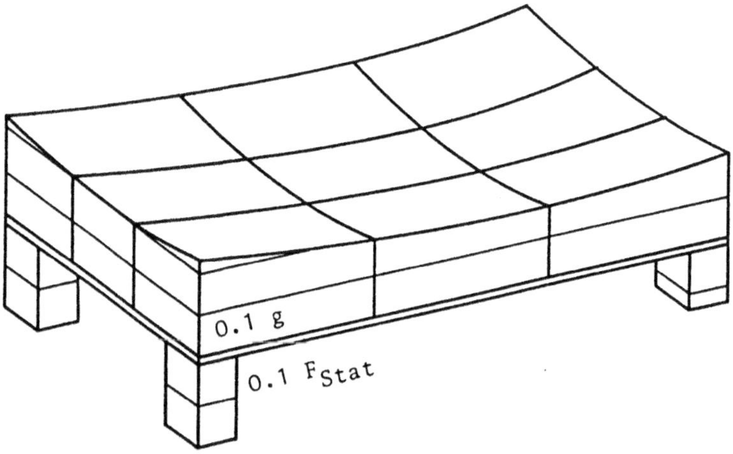

Fig. 4 Acceleration of Car Body and Wheel Loads

Similar results may also be obtained for springs and dashpots with nonlinear behavior, see Ref. [11].

In conclusion, for the analysis of ride characteristics of road vehicles sophisticated methods including spatial guideway models, linear and nonlinear vehicle dynamics and realistic rating criteria are available.

REFERENCES

1. Schiehlen, W.O., Ed., Dynamics of High-Speed Vehicles, CISM Courses and Lectures No. 274, Springer-Verlag, Wien-New York, 1982.

2. Mitschke, M., Dynamik der Kraftfahrzeuge, Springer-Verlag, Berlin-Heidelberg-New Vork, 1972.

3. Wong, J.Y., Theory of Ground Vehicles, Wiley, New York, 1978.

4. Willumeit, H.P., Ed., The Dynamics of Vehicles on Roads and on Rail Tracks, Swets & Zeitlinger, Lisse, 1980.

5. Wickens, A.H., Ed., The Dynamics of Vehicles on Roads and on Railway Tracks, Swets & Zeitlinger, Lisse, 1982.

6. Hedrick, J.K., and Wormley, D.N., Ed., The Dynamics of Roads and Tracks, Swets & Zeitlinger, Lisse, 1984.

7. Schiehlen, W.O., and Schramm, D., Bewegungsgleichungen von nicht-holonomen Fahrzeugmodellen, Institutsbericht IB-7, Institut B für Mechanik, Universität Stutgart, 1983.

8. Rill, G., Instationäre Fahrzeugschwingungen bei stochastischer Erregung, Ph.D. Thesis, Universität Stuttgart 1983.

9. International Standard ISO 2631, Guide for the Evaluation of Human Exposure to Whole-Body Vibrations, Int. Org. Standardization, 1974.

10 Müller, P.C., and Schiehlen, W.O., Lineare Schwingungen, Akad. Verlagsges., Wiesbaden, 1976.

11. Müller, P.C., Popp, K., and Schiehlen, W.O., "Berechnungsverfahren für stochastische Fahrzeugschwingungen", Ing. Arch., Vol. 49, 1980, pp. 235-254.

12. Kreuzer, E., and Rill, G., "Vergleichende Untersuchung von Fahrzeugschwingungen an räumlichen Ersatzmodellen", Ing. Arch., Vol. 52, 1982, pp. 205-219.

METHODS AND EXPERIENCE IN COMPUTER AIDED DESIGN OF LARGE-DISPLACEMENT MECHANICAL SYSTEMS

Milton A. Chace
President
Mechanical Dynamics, Inc.
555 South Forest
Ann Arbor, MI 48103

Abstract. The methods evolved and utilized in the Mechanical Dynamics, Inc., (MDI) DRAM and ADAMS Programs are reviewed, including coordinate choice, analysis modes, integration procedure, and use of sparse matrix methods. A large 3-D vehicle simulation run is included as an example, with problem size, run time, experimental verification and graphic display output cited. System requirements for DRAM and ADAMS are included.

1. INTRODUCTION

This paper reviews the basic methods evolved and utilized in the Mechanical Dynamics, Inc. (MDI) DRAM and ADAMS Programs. Particular attention is devoted to ADAMS because it is most technologically advanced program and because the basic methods in the current version of ADAMS have not been clearly described elsewhere. Topics covered include coordinate choice, analysis modes (kineto static, dynamic, static, quasi-static), predictor-corrector integration procedure and use of sparse matrix methods.

A very important stage of development has been reached in generalized large-displacement simulation, with the successful, fast, accurate simulation of full three-dimensional models of vehicles (e.g., automobiles, trucks, farm and construction vehicles). Application of ADAMS to a light truck handling study is included to illustrate this capability.

The DRAM (Dynamic Response of Articulated Machinery) Program is limited to two-dimensional problems, although it accomodates substantial detail and generality within the 2-d domain. The original version of DRAM was completed in 1969, at The University of Michigan[1,2,3,4] through the efforts of the author and Michael Korybalski. At that time it was named DAMN (Dynamic Analysis

NATO ASI Series, Vol. F9
Computer Aided Analysis and Optimization of Mechanical System Dynamics
Edited by E. J. Haug

of Mechanical Networks). It was historically the first generalized (type-variant) program to provide time response of multifreedom, constrained, mechanical machinery undergoing large displacement behavior. Major improvements and additions were made to the program by D.A. Smith in doctoral thesis work over the period 1968 to 1971.[5] Since then, DRAM (Dynamic Response of Articulated Machinery) has undergone continuous improvement particularly through the efforts of John C. Angell.[8,10,16]

The ADAMS (Automated Dynamic Analysis of Mechanical Systems) Program applies to three-dimensional problems. The original ADAMS was completed in 1973 as doctoral thesis work by Nicolae Orlandea.[7,9] ADAMS utilizes a different coordinate scheme than DRAM and involves sparse matrix methods in the equation solutions.[6,12] Again major improvements and additions have been made to the original ADAMS code by now, most of them by J.C. Angell and Rajiv Rampalli.[11]

DRAM and ADAMS are extensively utilized in industry and several papers describing applications are referenced.[16-20]

Work by MDI has proceeded beyond the rigid body assumption involved in DRAM and ADAMS, to include deformable bodies. Although limits of time and clear expression inhibited review of that work here, some of these methods are described in a recent PhD thesis completed by D.J. Benson.[21]

2. METHODS FOR ANALYSIS AND COMPUTATION

Coordinate Choice

The essential computational problem in large displacement dynamic simulation is the automated construction and numerical evaluation of a simultaneous set of nonlinear ordinary differential equations having known initial conditions. At each integration step all coefficients must be numerically evaluated and a simultaneous set of equations solved for the high order dependent variables. The speed of this process is ultimately very sensitive to coordinate choice.

In the two-dimensional program DRAM, a so-called modified tree branch (mtb) coordinate set was used. This consisted of the relative translational displacements at each single-freedom translational contact, plus the angle of each part measured with respect to ground. For systems having significant closed loop constraint, this coordinate choice results in a small but dense set of equations. Rotational inertia is a negligible computational burden. Translational inertia may be a significant burden, especially for problems involving long chains of parts such as chain drives or crawler tracks. As an example, a single-freedom dynamic four-bar mechanism would be represented by five differential equations (two representing constraint conditions) in mtb coordinates but would require seventeen differential equations with center-of-mass (cm) coordinates. With this small, dense equation set time is saved in determining coefficient values and in solving by Guassian elimination. However application of sparse matrix methods is inappropriate for enough problems that sparse matrix technology is not included in DRAM.

In the three-dimensional program ADAMS, a center-of-mass (cm) coordinate system is used. This consists simply of the three rectangular displacements of each part's center of mass, plus the rotation of each part, with the angles measured with respect to ground. For each part, the corresponding translational and angular velocities and angular momenta are also utilized. This tends to result in a large but sparse system of equations, amenable to use of sparse matrix methods for simultaneous solution. Rotational inertia is now a significant computational burden, but much less than it would have been with a tree branch or relative coordinate set. Constraint expressions are numerous and varied but only involve the coordinates of the two parts immediately paired by a joint, and are therefore sparse. Typical ADAMS problems may generate a set of hundreds of differential equations, but these are solved very quickly by the assembled sparse matrix code.

Analytical Methods in ADAMS

This section describes the general architecture and technical function of the ADAMS Program. A similar description for the DRAM Program is not included because DRAM technology is simpler, earlier, and most of its aspects are a subset of ADAMS technology.

Figure 2.1 shows the general architecture of the ADAMS Program. The discussion of individual technical functions will be sequenced START to STOP through this diagram.

At the outset the user-generated data is input. This is checked element-by-element for obvious errors (e.g., missing element number, inconsistent format, inconsistent spelling of an element name, etc). Next an amount of data storage is assigned according to the numbers of element types present and the storage required for tabulated field and generator characteristics. A second error check is made, this time for situations that are inconsistent from a system perspective (e.g.: a field that connects two markers in the same part, failure to input a generator for a kinematic run).

The system degree of freedom (DOF) is computed here. If DOF = 0, a branch is made to kineto-statics computation. ("Kineto-static" is synonomous with the precise term "Kinematic". Both mean a zero DOF, geometrically determinate system. However, the term "Kinematic" is often applied imprecisely in general usage, connoting any system which moves. Hence, we have used the term "Kineto-static".) If DOF > 0 the program branches to dynamics, statics or quasi-statics. Of course if DOF < 0 the system is a structure and ADAMS does not apply, unless the system constraint is redundant and can be reformulated by the user.

Kineto-static solution: If the system has DOF = 0 it is unnecessary to solve Newton's 2nd Law to determine system motion. However, the system constraint functions must be solved to determine position, velocity, acceleration and reaction force for succeeding moments of time. The position problem is specifically posed as follows:

$$\underline{\phi}\ (\underline{q},\ t_j) = 0 \qquad (2.1)$$

$$\underline{q}_0 \text{ known}$$

Here $\underline{\phi}$ is a vector of non-linear constraint functions of the system coordinates $\underline{q}$. Since DOF = 0, there are as many individual constraint functions ϕ as individual coordinates q. The $\underline{q}$ are to be determined for a succession of values of time t_j corresponding to the required output times of the problem run. At each time t_j this problem is to be solved, a starting estimate of the coordinate values $\underline{q}_0$ is available. When j = 0, $\underline{q}_0$ is available from user input, generator functions and estimates of initial coordinate values. When j > 0, $\underline{q}_0$ is extrapolated from previous solutions of $\underline{q}$.

Equation (2.1) is solved by Newton-Raphson iteration, working from the following form:

$$\underline{\phi}\,(\underline{q}_i, t_j) + \left.\frac{\partial \underline{\phi}}{\partial \underline{q}}\right|_i \underline{\Delta q}_i = 0 \qquad (2.2)$$

Here $\underline{q}_i$ is the ith iterate of $\underline{q}$, i = 0, 1, 2, 3,... . The second term in Equation (2.2) essentially corrects whatever non-zero value $\underline{\phi}(\underline{q}_i, t_j)$ may have due to $\underline{q}_i$ not exactly equaling the solution q. $\left.\frac{\partial \underline{\phi}}{\partial \underline{q}}\right|_i$ is the system Jacobian; $\underline{\Delta q}_i$ is the difference required in $\underline{q}_i$ to provide the next iterate $\underline{q}_{i+1}$:

$$\underline{q}_{i+1} = \underline{q}_i + \underline{\Delta q}_i \qquad (2.3)$$

By rearranging Equation (2.3), the solution for $\underline{\Delta q}_i$ is presented as the solution of a simultaneous set of linear algebraic equations.

$$\left.\frac{\partial \underline{\phi}}{\partial \underline{q}}\right|_i \Delta \underline{q}_i = -\underline{\phi}(q_i, t_j) \qquad (2.4)$$

Before numerical integration begins, a symbolic LU decomposition of the Jacobian is performed, as indicated in the block "Generate Symbolic LU Code for Matrices". The coefficients in the Jacobian and the column vector $-\underline{\phi}$ change significantly from one time step to the next. However, the effect of the symbolic decomposition is to permit explicit evaluation of the $\underline{\Delta q}_i$, once numerical values are assigned to the Jacobian coefficients and to the coefficients of $-\underline{\phi}$. This provides a vital computation time advantage relative to repetitive numerical solution by Guassian elimination. (As Figure 2.1 shows, the symbolic LU decomposition is also a key step in each of the other problem modes: dynamics, statics and quasistatics.)

By looping a few times between Equations (2.4) and (2.3), convergence on a solution $\underline{q}$ for time t_j is obtained.

In the kineto-static mode, solutions for coordinate velocities and accelerations, and for joint reaction forces are also obtained from the constraint functions at time t_j. The basis for the velocity and acceleration solutions is obtained by performing the first and second total derivatives of $\underline{\Phi}$ with respect to t.

$$\frac{d\underline{\Phi}}{dt} = \frac{\partial \underline{\Phi}}{\partial \underline{q}} \dot{\underline{q}} + \frac{\partial \underline{\Phi}}{\partial t} = 0 \tag{2.5}$$

$$\frac{d^2\underline{\Phi}}{dt^2} = \sum_{k=1}^{n}\sum_{l=1}^{n} \frac{\partial^2 \underline{\Phi}}{\partial q_k \partial q_l} \dot{q}_k \dot{q}_l + \sum_{k=1}^{n} \frac{\partial}{\partial t}\left(\frac{\partial \underline{\Phi}}{\partial q_k}\right) \dot{q}_k + \frac{\partial \underline{\Phi}}{\partial \underline{q}} \ddot{\underline{q}} + \sum_{k=1}^{n} \frac{\partial}{\partial q_k}\left(\frac{\partial \underline{\Phi}}{\partial t}\right) \dot{q}_k + \frac{\partial^2 \underline{\Phi}}{\partial t^2} = 0 \tag{2.6}$$

By rearranging, these equations are placed in the form of simultaneous, linear algebraic equations in $\dot{\underline{q}}$ and $\ddot{\underline{q}}$ respectively:

$$\left(\frac{\partial \underline{\Phi}}{\partial \underline{q}}\right) \dot{\underline{q}} = -\frac{\partial \underline{\Phi}}{\partial t} \tag{2.7}$$

$$\left(\frac{\partial \underline{\Phi}}{\partial \underline{q}}\right) \ddot{\underline{q}} = -\left\{ \frac{\partial^2 \underline{\Phi}}{\partial t^2} + \sum_{k=1}^{n}\sum_{l=1}^{n} \frac{\partial^2 \underline{\Phi}}{\partial q_k \partial q_l} \dot{q}_k \dot{q}_l + \frac{\partial}{\partial t}\left(\frac{\partial \underline{\Phi}}{\partial \underline{q}}\right) \dot{\underline{q}} + \frac{\partial}{\partial \underline{q}}\left(\frac{\partial \underline{\Phi}}{\partial t}\right) \dot{\underline{q}} \right\} \tag{2.8}$$

Reaction force is determined from Lagrange's Equation, which applies despite this being the kineto-static mode:

$$\frac{d}{dt}\left(\frac{\partial T}{\partial \dot{\underline{q}}}\right) - \frac{\partial T}{\partial \underline{q}} + \left(\frac{\partial \underline{\Phi}}{\partial \underline{q}}\right)^T \underline{\lambda} = \underline{F} \tag{2.9}$$

Here, the Lagrange multipliers $\underline{\lambda}$ correspond physically to joint reaction forces, for those constraints corresponding to requirements of joint closure or alignment.

Rearranging, another set of simultaneous, linear algebraic equations is obtained, with dependent variables $\underline{\lambda}$.

$$\left(\frac{\partial \underline{\Phi}}{\partial \underline{q}}\right)^T \underline{\lambda} = \left\{ -\frac{d}{dt}\left(\frac{\partial T}{\partial \dot{\underline{q}}}\right) + \frac{\partial T}{\partial \underline{q}} + \underline{F} \right\} \tag{2.10}$$

The only programming challenge in Equations (2.7), (2.8) and (2.10) is to evaluate the coefficient values, given the solution vector $\underline{q}$ already determined from iteration of Equation (2.1). The decomposition of the Jacobian and its transpose is immediately available, providing very rapid solution of $\dot{\underline{q}}$, $\ddot{\underline{q}}$ and $\underline{\lambda}$.

On the diagram of Figure 2.1 the four solutions (2.1), (2.7), (2.8) and (2.10) take place in the "Analyze" block.

Determination of Initial Conditions: If DOF > 0 it is necessary to determine the "initial conditions", that is the values of $\underline{q}$ and $\dot{\underline{q}}$ at $t = t_o$. In ADAMS input data, the user may state initial conditions on some subset of the total coordinate set. The user may further distinguish some of these coordinates as "exact" and others as only approximate. Exact initial coordinates should be consistent with system constraint and generator input; approximate initial coordinates are a means to cause the iteration on the constraint functions to converge to the correct positional mode of possibly many such modes.

Determination of the total set of initial position coordinates $\underline{q}$ is approached as a minimization problem, where the objective function is L_o:

$$L_o = \frac{1}{2}\sum_{i=1}^{n} W_i (q_i - q_{oi})^2 + \sum_{j=1}^{m} \lambda_j^o \phi_j \qquad (2.11)$$

Here n is the total number of system coordinates; m is the number of scalar constraints. The q_{oi} are the coordinates first provided: exact or approximate from the user, or arbitrary by program default.

The W_i are weighting coefficients. Individual coefficients are very large if the corresponding q_{oi} is an "exact" value, small but appreciable if q_{oi} is approximate and zero if q_{oi} is arbitrary. The quantities ϕ_j and λ_j^o are the constraint functions and associated Lagrange multipliers, respectively.

For L_o to be minimum, the partial derivatives $\frac{\partial L_o}{\partial q_i}$ and $\frac{\partial L_o}{\partial \lambda_j^o}$ must be zero. This provides a simultaneous set of n+m non-linear algebraic equations in q_i and λ_j^o:

$$W_i (q_i - q_{oi}) + \sum_{j=1}^{m} \lambda_j^o \frac{\partial \phi_j}{\partial q_i} = 0 \qquad (2.12)$$

$$i = 1, 2, \cdots n$$

$$\Phi_j = 0 \qquad j = 1, 2, \ldots m \tag{2.13}$$

The respective functional forms of these equations are:

$$f_i(q_k, \lambda_\ell^0) = 0 \qquad k = 1, 2, \ldots n \tag{2.14}$$

$$g_j(q_k) = 0 \qquad \ell = 1, 2, \ldots m \tag{2.15}$$

These are to be solved by Newton-Raphson iteration. First establish a form analogous to Equation (2.2):

$$f_i(q_{kp}, \lambda_{\ell p}^0) + \sum_{k=1}^{n} \left.\frac{\partial f_i}{\partial q_k}\right|_p \Delta q_{kp} + \sum_{\ell=1}^{m} \left.\frac{\partial f_i}{\partial \lambda_\ell^0}\right|_p \Delta \lambda_{\ell p}^0 = 0 \tag{2.16}$$

$$g_j(q_{kp}) + \sum_{k=1}^{n} \left.\frac{\partial g_j}{\partial q_k}\right|_p \Delta q_{kp} = 0 \tag{2.17}$$

Here q_{kp} and $\lambda_{\ell p}^0$ are the pth iterates of the variables q_k and λ_ℓ^0
Rearranging in matrix form,

$$\begin{bmatrix} \sum_{k=1}^{n} \frac{\partial f_i}{\partial q_k} & \sum_{\ell=1}^{m} \frac{\partial f_i}{\partial \lambda_\ell^0} \\ \sum_{k=1}^{n} \frac{\partial g_j}{\partial q_k} & 0 \end{bmatrix}_p \begin{Bmatrix} \Delta q_{kp} \\ \Delta \lambda_{\ell p}^0 \end{Bmatrix} = \begin{Bmatrix} -f_i(q_{kp}, \lambda_{\ell p}^0) \\ -g_j(q_{kp}) \end{Bmatrix} \tag{2.18}$$

Recognizing that f_i and g_j are the left sides of Equations (2.12) and (2.13),

$$\left[\begin{array}{c|c} \left(W_i + \sum_{k=1}^{n}\sum_{j=1}^{m} \lambda_j^o \frac{\partial^2 \phi_j}{\partial q_k \partial q_i}\right) & \sum_{j=1}^{m} \frac{\partial \phi_j}{\partial q_i} \\ \hline \sum_{k=1}^{n} \frac{\partial \phi_j}{\partial q_k} & 0 \end{array}\right]_p \left\{\begin{array}{c} \Delta q_{kp} \\ \Delta \lambda_{\ell p}^o \end{array}\right\} = \left\{\begin{array}{c} -W_i(q_{ip} - q_{oi}) - \sum_{j=1}^{m} \lambda_{jp}^o \frac{\partial \phi_j}{\partial q_i}\Big|_p \\ -\phi(q_{kp}) \end{array}\right\} \tag{2.19}$$

Equation (2.19) is iterated to convergence, updating successive q and λ values according to,

$$q_{k,p+1} = q_{kp} + \Delta q_{kp} \tag{2.20}$$

$$\lambda_{\ell,p+1}^o = \lambda_{\ell p}^o + \Delta \lambda_{\ell p}^o \tag{2.21}$$

The coefficient matrix of Equation (2.19) need not be exact for convergence to be obtained. Also, the double summation term in the upper left quadrant never dominates the corresponding terms W_i. Therefore the summation can be neglected with a net improvement in speed of convergence because of faster coefficient evaluation.

Determination of consistent initial velocities, accelerations and reaction forces is a linear problem, once the positions are determined. To determine the initial velocites, minimize L_1:

$$L_1 = \frac{1}{2}\sum_{i=1}^{n} W_i'(\dot{q}_i - \dot{q}_{oi})^2 + \sum_{j=1}^{m} \lambda_j' \frac{d\Phi_j}{dt} \tag{2.22}$$

Here, the notation is the same as, or similar to, that of Equation (2.11). The q_{oi} are the velocities first provided, exact or approximate from the user or zero by program default. The W_i' are a new set of weighting coefficients appropriate to the velocities. As implied by Equation (2.22) the following constraints must be obeyed:

$$\frac{d\Phi_j}{dt} = \sum_{k=1}^{n} \frac{\partial\Phi_j}{\partial q_k}\dot{q}_k + \frac{\partial\Phi_j}{\partial t} = 0 \tag{2.23}$$

The λ_j' are a new set of Lagrange multipliers associated with the velocity constraints.

For L_1 to be minimum,

$$\frac{\partial L_1}{\partial \dot{q}_i} = W_i'(\dot{q}_i - \dot{q}_{io}) + \sum_{j=1}^{m} \lambda_j'\left(\frac{\partial\Phi_j}{\partial q_i}\right) = 0 \qquad i = 1, 2, \ldots n \tag{2.24}$$

$$\frac{\partial L_1}{\partial \lambda_j} = \sum_{k=1}^{n}\left(\frac{\partial\Phi_j}{\partial q_k}\right)\dot{q}_k + \frac{\partial\Phi_j}{\partial t} = 0 \tag{2.25}$$

Rearrange these equations in matrix form:

$$\begin{bmatrix} W_i' & \sum_{j=1}^{m}\frac{\partial\Phi_j}{\partial q_i} \\ \sum_{k=1}^{n}\left(\frac{\partial\Phi_j}{\partial q_k}\right) & 0 \end{bmatrix} \begin{Bmatrix} \dot{q}_k \\ \lambda_j' \end{Bmatrix} = \begin{Bmatrix} W_i'\dot{q}_{oi} \\ -\frac{\partial\Phi}{\partial t} \end{Bmatrix} \tag{2.26}$$

Equation (2.26) is linear in q_k and λ_j. The coefficient matrix contains only position-dependent terms and is therefore essentially constant. Also, a matrix including this pattern of non-zero entries has already been decomposed (ref. Equation 2.19). Hence, the solution for $\dot{q}_k$ and λ_j proceeds quickly and directly.

The initial values of coordinate accelerations and the set of Lagrange multipliers corresponding to reaction forces are of use in the Newton-Raphson iteration associated with the first integration step. These are determined from the constrained form of Newton's second law, plus the second-order expression of system constraint.

$$\sum_{k=1}^{n} \left(m_{ik}(q_k)\right)\ddot{q}_k + \sum_{j=1}^{m} \lambda_j \frac{\partial \Phi_j}{\partial q_i} = g_i(q_k, \dot{q}_k, t) \qquad (2.27)$$

$$\frac{d^2\Phi_j}{dt^2} = \sum_{i=1}^{n}\left(\frac{\partial \Phi_j}{\partial q_i}\right)\ddot{q}_i - h_j(q_k, \dot{q}_k, t) = 0 \qquad (2.28)$$

Here, the functions g_i are the sums of all the terms in Newton's equation dependent only on displacement, velocity and time. The functions h_j are:

$$h_j = -\left\{\frac{\partial^2 \Phi_j}{\partial t^2} + \sum_{i=1}^{n} \frac{\partial}{\partial t}\left(\frac{\partial \Phi_j}{\partial q_i}\right)\dot{q}_i + \sum_{i=1}^{n}\sum_{k=1}^{n}\left(\frac{\partial^2 \Phi_j}{\partial q_k \partial q_i}\right)\dot{q}_k \dot{q}_i\right\} \qquad (2.29)$$

Equations (2.27) and (2.28) can be expressed in a matrix form which has already been symbolically decomposed (ref. Equations (2.19) and (2.26)).

$$\left[\begin{array}{c|c} m_{ik}(q_k) & \sum_{j=1}^{m} \frac{\partial \Phi_j}{\partial q_i} \\ \hline \sum_{i=1}^{n} \frac{\partial \Phi_j}{\partial q_i} & 0 \end{array}\right] \left\{\begin{array}{c} \ddot{q}_k \\ \hline \lambda_j \end{array}\right\} = \left\{\begin{array}{c} g_i \\ \hline h_j \end{array}\right\} \qquad (2.30)$$

Since all position and velocity quantities are already determined the coefficient values are known and the accelerations $\ddot{q}_k$ and reaction forces λ_j can be determined directly.

Statics and Dynamics: Once initial conditions for the system have been determined the user has the option of doing either a static, quasistatic or dynamic analysis of the system. If dynamics is chosen, the user has the choice of using either the implicit Gear algorithm (stiff integration) or an Adams-Moulton algorithm. Only the application of Gear algorithm will be discussed here since it is practically required for problems involving impact and is therefore used most often.

The general expressions for the Gear multistep formula of (k+1)th order are,

$$q^{n+1} = -h\beta_0 \dot{q}^{n+1} + \sum_{j=1}^{k} (\alpha_j q^{n-j+1} + h\beta_j \dot{q}^{n-j+1}) \tag{2.31}$$

where, h ≡ time step from the nth step to the (n+1)th step

$\beta j, \alpha j$ ≡ integration coefficients

K + 1 ≡ order of integration

A set of differential and algebraic equations having the following form[1] is to be solved:

$$\underline{f}(q, \dot{q}, \ddot{q}, \underline{\lambda}, t) = 0 \tag{2.32}$$

$$q(0) \text{ and } \dot{q}(0) \text{ known}$$

State Equation (2.32) in first order form by means of the substitution $\underline{u} = \dot{q}$. Also, apply Equation (2.32) to the n+1 time step. Thus,

$$\underline{f}(q^{n+1}, \underline{u}^{n+1}, \dot{\underline{u}}^{n+1}, \underline{\lambda}^{n+1}, t) = 0 \tag{2.33}$$

[1]The present explanation excludes momentum as a variable although inclusion of at least angular momentum variables may improve sparsity.

Likewise, in Equation (2.31) perform the substitution $\underline{u}^i = \underline{q}^i$ and rearrange.

$$\underline{g}(\underline{q}^{n+1}, \underline{u}^{n+1}, t) = 0 \tag{2.34}$$

$$\underline{g} \equiv \underline{u}^{n+1} + \frac{1}{h\beta_0}\underline{q}^{n+1} + \frac{1}{h\beta_0}\left[\sum_{j=1}^{K}\left(\alpha_j \underline{q}^{n-j+1} + h\beta_j \underline{u}^{n-j+1}\right)\right] \tag{2.35}$$

Constraints may exist between the coordinates, as implied by the existence of the Lagrange multiplier terms $\underline{\lambda}$ in Equation (2.32). For the n+1 integration step, these are stated simply as,

$$\underline{\phi}(\underline{q}^{n+1}) = 0 \tag{2.36}$$

Equations (2.33) and (2.34) can be developed into a form suited for iterative solution, in complete analogy to the development of Equation (2.19):

$$\underline{f}_p + \left.\frac{\partial f}{\partial \underline{q}}\right|_p \Delta \underline{q}_p + \left.\frac{\partial f}{\partial \underline{u}}\right|_p \Delta \underline{u}_p + \left.\frac{\partial f}{\partial \dot{\underline{u}}}\right|_p \Delta \dot{\underline{u}}_p + \left.\frac{\partial f}{\partial \underline{\lambda}}\right|_p \Delta \lambda_p = 0 \tag{2.37}$$

$$\underline{g}_p + \left.\frac{\partial \underline{g}}{\partial \underline{q}}\right|_p \Delta \underline{q}_p + \left.\frac{\partial \underline{g}}{\partial \underline{u}}\right|_p \Delta \underline{u}_p = 0 \tag{2.38}$$

$$\underline{\Phi}_p + \left.\frac{\partial \underline{\Phi}}{\partial \underline{q}}\right|_p \Delta \underline{q}_p = 0 \tag{2.39}$$

Here, the superscript n+1 has been dropped for notational convenience. The index p is an iteration counter.

In Equation (2.37) note that

$$\Delta \dot{\underline{u}}_p = \left.\frac{\partial \dot{\underline{u}}}{\partial \underline{u}}\right|_p \Delta \underline{u}_p \tag{2.40}$$

The term $\frac{\partial \dot{\underline{u}}}{\partial \underline{u}}$ can be replaced by differentiating Equation (2.31) once, rearranging, then performing the respective partial derivative.

$$\frac{\partial \dot{\underline{u}}}{\partial \underline{u}} = -\left[\frac{\diagdown 1}{h\beta_0}\right] \tag{2.41}$$

Also in Equation (2.37) the term $\frac{\partial f}{\partial \lambda}$ can be replaced by $\left(\frac{\partial \Phi}{\partial q}\right)^T$, an mxm diagonal matrix of constraint functions involved as a subset of the functions $\underline{f}$. Thus Equation (2.37) becomes

$$\underline{f}_p + \frac{\partial \underline{f}}{\partial \underline{q}}\bigg|_p \Delta \underline{q}_p + \left(\frac{\partial \underline{f}}{\partial \underline{u}} - \left[\frac{1}{h\beta_0}\right]\frac{\partial \underline{f}}{\partial \dot{\underline{u}}}\right)_p \Delta \underline{u}_p + \left(\frac{\partial \Phi}{\partial \underline{q}}\right)^T_p \Delta \underline{\lambda}_p \tag{2.42}$$

From Equation (2.35),

$$\frac{\partial \underline{g}}{\partial \underline{q}} = \left[\frac{1}{h\beta_0}\right] \tag{2.43}$$

$$\frac{\partial \underline{g}}{\partial \underline{u}} = [1] \tag{2.44}$$

Thus, Equation (2.38) becomes

$$\underline{g}_p + [1]\Delta \underline{q}_p + \left[\frac{1}{h\beta_0}\right]\Delta \underline{u}_p = 0 \tag{2.45}$$

Rearrange Equations (2.41), (2.44) and (2.39) and state them in matrix form.

$$\begin{bmatrix} \frac{\partial \underline{f}}{\partial \underline{q}} & \left(\frac{\partial \underline{f}}{\partial \underline{u}} - \left[\frac{1}{h\beta_0}\right]\frac{\partial \underline{f}}{\partial \dot{\underline{u}}}\right) & \left(\frac{\partial \Phi}{\partial \underline{q}}\right)^T \\ \left[\frac{1}{h\beta_0}\right] & [1] & 0 \\ \frac{\partial \Phi}{\partial \underline{q}} & 0 & 0 \end{bmatrix}_p \begin{Bmatrix} \Delta \underline{q} \\ \Delta \underline{u} \\ \Delta \underline{\lambda} \end{Bmatrix}_p = \begin{Bmatrix} -\underline{f} \\ -\underline{g} \\ -\Phi \end{Bmatrix}_p \tag{2.46}$$

At each integrating step, Equation (2.46) is iterated to convergence, creating successive iterates,

$$\underline{q}_{p+1} = \underline{q}_p + \Delta \underline{q}_p \tag{2.47}$$

$$\underline{u}_{p+1} = \underline{u}_p + \Delta \underline{u}_p \tag{2.48}$$

$$\underline{\lambda}_{p+1} = \underline{\lambda}_p + \Delta \underline{\lambda}_p \tag{2.49}$$

The coefficient matrix in Equation (2.46) is the system Jacobian. Before integration begins, it is symbolically decomposed so that successive solutions of Equation (2.46) can proceed directly.

Equation (2.46) adapts to a static or quasi-static equilibrium solution if only the (1,1), (1,3) and (3,1) partitions are employed. (Quasi-static solution is simply a series of static solutions with corresponding increments in generator input.)

To perform dynamic solution a predictor-corrector process is utilized, incorporating Equation (2.46) as the corrector. The predictor is essentially an explicit polynominal extrapolation of a history of values of $\underline{q}$ and $\underline{u}$. Integration error is measured as the difference between corresponding predicted and corrected values; if the error exceeds user-specified limits the integration step size is reduced and the step is tried again.

If ADAMS encounters numerical singularity in the evaluation of Equation (2.46), the problem may be structural (e.g., a locking or jamming position has been reached) or numerical, involving the pivoting assumed in the symbolic decomposition. For the former case execution is terminated and any output is printed. In the latter case a new symbolic decomposition is performed on the Jacobian, based on an alternate pivoting scheme more suited to current coordinate values.

3. EXAMPLES

Simple Pendulum

Figure 3.1 shows a simple pendulum attached to ground by means of a revolute joint. The pendulum body has mass m and inertia I. At any given time, the body must be in equilibrium with inertial (d'Alembert) forces and forces of constraint.

The problem will be forumulated in three generalized coordinates: x, y, θ. The dynamic equations of motion in these coordinates are:

$$m\ddot{x} + \lambda_1 = 0 \qquad (3.1)$$

$$m\ddot{y} + mg + \lambda_2 = 0 \qquad (3.2)$$

$$I\ddot{\theta} + \lambda_1 r \sin\theta - \lambda_2 r \cos\theta = 0 \qquad (3.3)$$

Here λ_1 and λ_2 are the rectangular components of the reaction force required to maintain joint contact, as exerted on the pendulum from ground.

The system is subject to the vector constraint condition:

$$\vec{R} - \vec{r} - \vec{A} = 0 \tag{3.4}$$

$\vec{A}$ is a constant

$$\vec{r} = r\left[\cos\theta\,\hat{i} + \sin\theta\,\hat{j}\right] \tag{3.5}$$

$$\vec{R} = x\hat{i} + y\hat{j} \tag{3.6}$$

The system constraints are the two rectangular components of Equations (3.4).

$$x - r\cos\theta - \vec{A}\cdot\hat{i} = 0 \tag{3.7}$$

$$y - r\sin\theta - \vec{A}\cdot\hat{j} = 0 \tag{3.8}$$

Equations (3.1) - (3.3) can be recast in first-order form, using the following defining relations:

$$\dot{x} - u = 0 \tag{3.9}$$

$$\dot{y} - v = 0 \tag{3.10}$$

$$\dot{\theta} - \omega = 0 \tag{3.11}$$

In matrix form, the complete set of eight equations is:

$$\begin{bmatrix} m & & & & & & 1 & \\ & m & & & & & & 1 \\ & & I & & & & C_{37} & C_{38} \\ & & & 1 & & & & \\ & & & & 1 & & & \\ & & & & & 1 & & \\ & & & & & & & \\ & & & & & & & \end{bmatrix} \begin{Bmatrix} \dot{u} \\ \dot{v} \\ \dot{\omega} \\ \dot{x} \\ \dot{y} \\ \dot{\theta} \\ \lambda_1 \\ \lambda_2 \end{Bmatrix} = \begin{Bmatrix} 0 \\ -mg \\ 0 \\ 1 \\ 1 \\ 1 \\ b_7 \\ b_8 \end{Bmatrix}. \tag{3.12}$$

Here,

$$c_{37} = r\sin\theta \tag{3.13a}$$

$$c_{38} = -r\cos\theta \tag{3.13b}$$

$$b_7 = x - r\cos\theta - (\vec{A}\cdot\hat{i}) = 0 \tag{3.13c}$$

$$b_8 = y - r\sin\theta - (\vec{A}\cdot\hat{j}) = 0 \tag{3.13d}$$

The Jacobian for this example can be constructed following the form of Equation (2.44):

$$J = \begin{bmatrix} \frac{\partial f}{\partial q} & \left(\frac{\partial f}{\partial u} - \frac{1}{h\beta_0}\frac{\partial f}{\partial \dot{u}}\right) & \left(\frac{\partial \Phi}{\partial q}\right)^T \\ \left[\diagdown \frac{1}{h\beta_0} \diagdown\right] & \left[\diagdown 1 \diagdown\right] & 0 \\ \left(\frac{\partial \Phi}{\partial q}\right) & 0 & 0 \end{bmatrix} \tag{3.14}$$

The expansions for three partitions in the top row are:

$$\frac{\partial f}{\partial q} = \begin{bmatrix} 0 & 0 & 0 \\ 0 & 0 & 0 \\ 0 & 0 & (\lambda_1 r\cos\theta + \lambda_2 r\sin\theta) \end{bmatrix} \tag{3.15}$$

$$\left(\frac{\partial f}{\partial u} - \frac{1}{h\beta_0}\frac{\partial f}{\partial \dot{u}}\right) = -\frac{1}{h\beta_0}\begin{bmatrix} m & 0 & 0 \\ 0 & m & 0 \\ 0 & 0 & I \end{bmatrix} \tag{3.16}$$

$$\frac{\partial \Phi}{\partial q} = \begin{bmatrix} 1 & 0 & r\sin\theta \\ 0 & 1 & -r\cos\theta \end{bmatrix} \tag{3.17}$$

The intention with this example has been to provide a sense of the analytical foundation for an ADAMS dynamic run. Numerical results are not included.

Vehicle Simulation

Large-displacement dynamic analysis has evolved to the point that three-dimensional vehicle models can be simulated with all inertial, steering, suspension, shock, bushing, and tire effects included. Vehicles can be shown in crash and roll-over, although such drastic examples tend to be restricted as proprietary information by client companies. The example chosen here is a Ford Bronco II involved in severe handling.

A total of 580 equations were formulated by ADAMS for this vehicle, representing a 42 degree-of-freedom model. The vehicle model was subjected to 210° steering wheel ramp input in 0.4 seconds while traveling at a speed of 45 mph. Figures 3.3 through 3.5 compare simulation results with experimentally measured results for vehicle lateral acceleration, roll angle and yaw rate. The model was run from zero to five seconds of real time, requiring approximately 175 cpu seconds of simulation time on a CDC Cyber 176 computer. This run time is representative of many other vehicle examples run under varying circumstances.

One of the most sensitive elements of a vehicle simulation is the tire model. Present state of the art requires experimental data on the particular tires involved. In this instance data was obtained from the Calspan Corporation.

Computer graphics is a very helpful interface between the voluminous, detailed output of a simulation run and the capability of a human analyst to appreciate an overall situation. Figure 3.6 shows a succession of three frames of the Bronco as simulated on an Evans and Sutherland PS300 graphic display terminal. In actual graphic display the motion appears continuous, involving rapid generation of a succession of several hundred frames.

4. SYSTEMS REQUIREMENTS

The ADAMS program is currently running on a variety of mainframe and minicomputers as can be seen in Table 3.1. ADAMS is written entirely in ANSI standard FORTRAN IV; subsequent versions will be written in FORTRAN 77.

Random access memory requirements (32-bit words) may be summarized as follows:

ADAMS: 232K

IGL (Interactive Graphics Library): 36K

Postprocessor: 50K

ADAMS uses one large integer array to store all data. Arrays with problem-dependent size are allocated from this array. A double precision dummy array is equivalenced to this integer array to facilitate the storage and retrieval of double precision arrays. Permanent space is allocated from the left end of the array, and temporary space from the right end of the array.

By means of this storage arrangement, ADAMS is variably dimensioned, and larger problems may be accommodated by simply changing the size of the large integer array, which is defined in the block data.

Graphics display using the ADAMS postprocessor is supported on the Tektronix 40XX series and 4100 series terminals. The graphic routines used are Tektronix PLOT10 or IGL, if available at the user's site, or CKL1B which is supplied by MDI.

Deere and Company have been utilizing Megatek and Ramtek terminals as graphics output devices from the postprocessor. The Evans and Sutherland PS300 is another ADAMS postprocessor graphics device with the special capability of real-time coordinate transformation. Once simulation has been completed and down loaded, the PS300 operates stand-alone and the user is provided with a wide variety of viewing options using the control dials and the function buttons.

MDI is presently completing a preprocessor for both DRAM and ADAMS using the PS300. The preprocessor will allow the user to add/alter/delete component types such as parts, joints, forces, requests and generators using the attached digitizing tablet.

TABLE 3.1
INSTALLATIONS OF DRAM AND ADAMS

HARDWARE	O/S	MAIN MEMORY DRAM	MAIN MEMORY ADAMS
VAX	VMS	Virtual	Virtual
Prime 50 Series	PRIMOS	Virtual	Virtual
Amdahl 5860	MTS	Virtual	Virtual
IBM 43XX	VM/CMS	4 MByte	6 MByte
IBM 370 Series	MVS	4 MByte	6 MByte
Cyber 176 Series	NOS/BE	192K 60-bit words*	192K 60 bit words*
Cray-1	COS	300K 64-bit words	400K 64-bit words

*Programs are overlayed to fit in main memory

ACKNOWLEDGEMENT

In this paper, methods and experience in the computer-aided design of large displacement mechanical systems, have been reviewed. The structuring and broad distribution of this capability has required an interplay of talent of many individuals. I wish particularly to acknowledge the technical contributions of John Angell and Rajiv Rampalli in program development. As respects the present paper, the Figure 2.1 flow chart and basic scheme of explanation were contributed by Rampalli. The accomplishments of John Angell and Jim Vincke in achieving a generalized three-dimensional vehicle dynamic simulation capability are also acknowledged. Editorial assistance by Hal Burchfield, Sandy Reich and Vic Sohoni is appreciated. The original versions of DRAM and ADAMS were primarily the work of Don Smith and Nicolae Orlandea, respectively.

REFERENCES

1. M.A. Chace, "A Network-Variational Basis for Generalized Computer Representation of Multifreedom, Constrained, Mechanical Systems." Design Automation Conference, Miami, Florida, 1969.

2. M.A. Chace, "DAMN-A Prototype Program for the Dynamic Analysis of Mechanical Networks.: 7th Annual Share Design Automation Workshop, San Francisco, California, June, 1970.

3. M.A. Chace and M.E. Korybalski, "Computer Graphics in the Schematic Representation of Nonlinear, Constrained, Multifreedom Mechanical Systems." Computer Graphics 70 Conference, Brunel University, April, 1970.

5. D.A. Smith, "Reaction Forces and Impact in Generalized Two-Dimension Mechanical Dynamic Systems." Ph.D. Thesis, University of Michigan, Ann Arbor, Michigan, September, 1971.

6. G. Hachtel, R. Brayton and P. Gustavson, "The Sparse Tableaux Approach to Network Analysis and Design" IEEE Transactions on Circuit Theory, Vol. 18, No. 1, 1971.

7. N. Orlandea, "Development and Application of Node-Analogous Sparsity-Oriented Methods for Simulation of Mechanical Dynamic Systems." Ph.D. Thesis, University of Michigan, 1973.

8. M.A. Chace and J.C. Angell, "Interactive Simulation of Machinery With Friction and Impact Using DRAM," SAE Paper No. 770050, February, 1977.

9. N. Orlandea, M.A. Chace and D.A. Calahan, "A Sparsity-Oriented Approach to the Dynamic Analysis and Design of Mechanical Systems - Parts I and II." Paper Nos. 76-DET-19 and 76-DET-20, presented at ASME Mechanisms Conference, Montreal, Quebec, Canada, October, 1976.

10. Mechanical Dynamics, Inc., "DRAM User's Guide", Mechanical Dynamics, Inc., 555 South Forest, Ann Arbor, Michigan, December 1979.

11. Mechanical Dynamics, Inc., "ADAMS User's Guide", Mechanical Dynamics, Inc., 555 South Forest, Ann Arbor, Michigan, March, 1981.

12. D.A. Calahan, "A Vectorized General Sparsity Solver." Systems Engineering Laboratory Report No. 168., The University of Michigan, October 1, 1982.

13. C.W. Gear, "Simultaneous Numerical Solution of Differential-Algebraic Equations." IEEE Transactions on Circuit Theory CT-18, No. 1, January 1971:89-95.

14. L.F. Shampine and C.W. Gear, "A User's View of Solving Stiff Ordinary Differential Equations, SIAM Review, Vol. 21, No. 1, January 1979.

15. L.R. Petzold, and C.W. Gear, "ODE Methods for the Solution of Differential/Algebraic Equations." Sandia Report SAND82-8051, October 1982.

16. P.A. Erickson, G.M. Ferguson, and E.W. Kenderline, "Design and Simulation of a Unique Signal Mechanism", ASME Design Engineering Division, Paper No. 82-DET-15, September 1982.

17. R.E. Kaufman, "Mechanism Design by Computer", Machine Design Magazine, October 26, 1978.

18. E.W. Smith, et. al., "Automated Simulation and Display of Mechanism and Vehicle Dynamics" American Society of Agricultural Engineering, Paper No. 82-5019, 1982.

19. N.S. Rai, A.R. Solomon, and J.C. Angell, "Computer Simulation of Suspension Abuse Tests Using ADAMS", SAE Paper No. 820079, 1982.

20. J.B. McConville and J.C. Angell, "The Dynamic Simulation of a Moving Vehicle Subject to Transient Steering Input Using the ADAMS Computer Program", paper submitted to ASME for publication, February 1983.

21. D.J. Benson, "The Simulation of Deformable Mechanical Systems Using Vector Processors", PhD Thesis, University of Michigan, 1983.

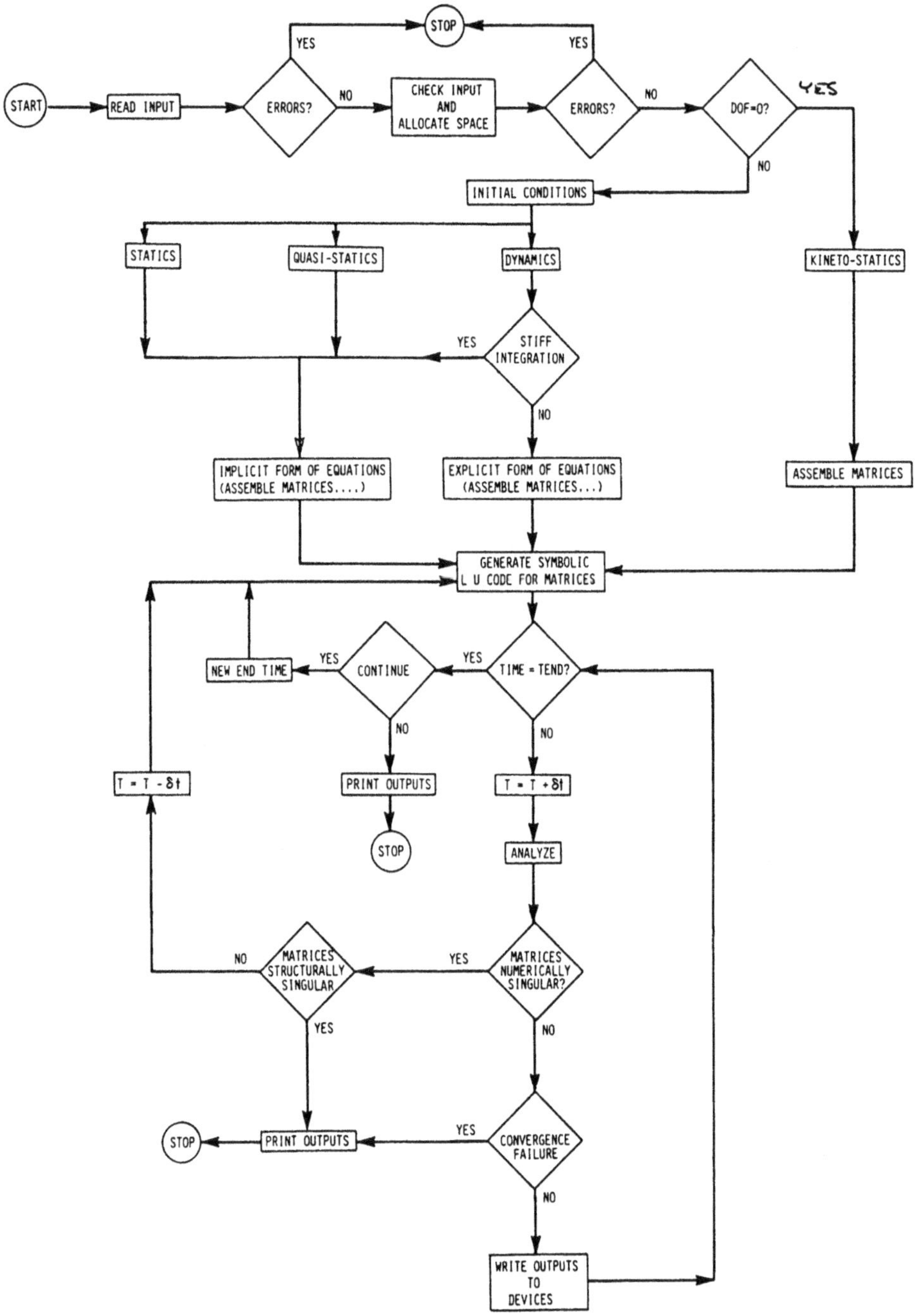

Figure 2.1. Flow chart of the operation of the ADAMS Program. Four different solution modes are possible. Kineto-static, static, quasi-static and dynamic. All modes use a symbolic decomposition prior to numerical evaluation.

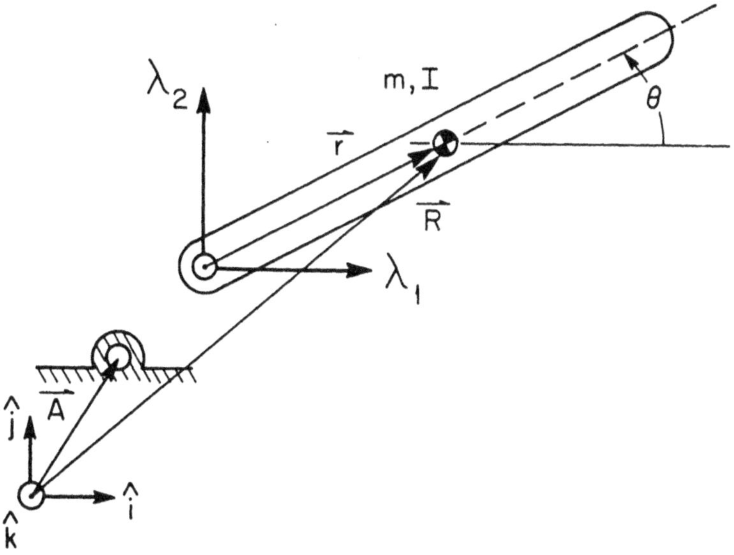

Figure 3.1. Example planar pendulum formulated in center-of-mass coordinates x, y, θ, where x and y are the rectangular components of vector $\vec{R}$.

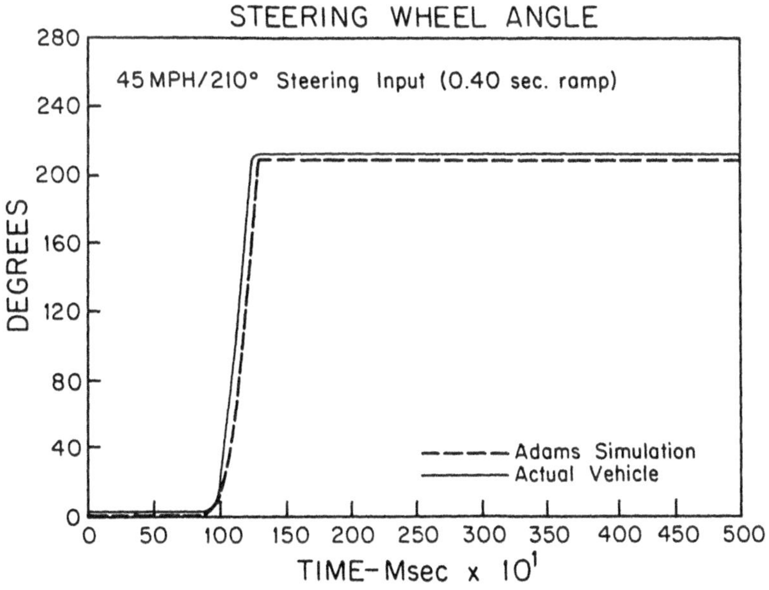

Figure 3.2. Steering wheel angle versus time; ADAMS input and test input, to a Ford Bronco II.

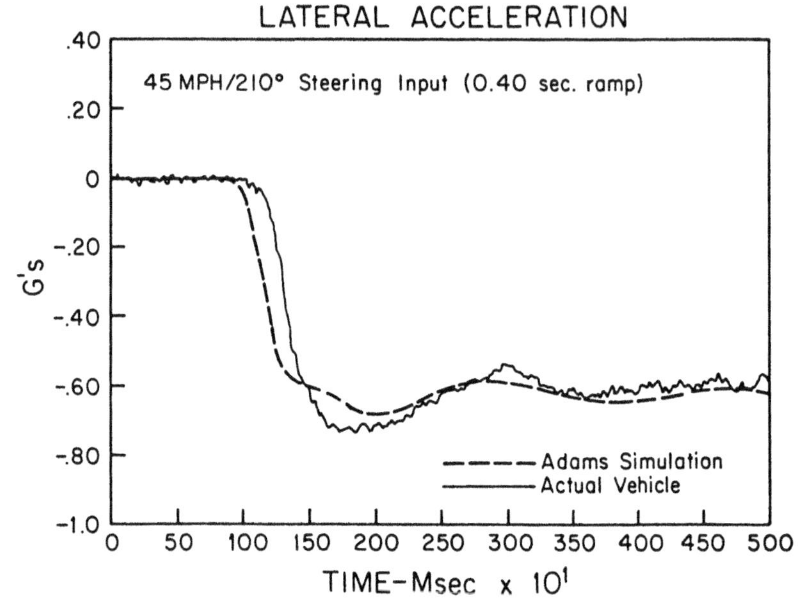

Figure 3.3. Lateral acceleration versus time at 45 mph with steering input of Figure 3.2; ADAMS and test results.

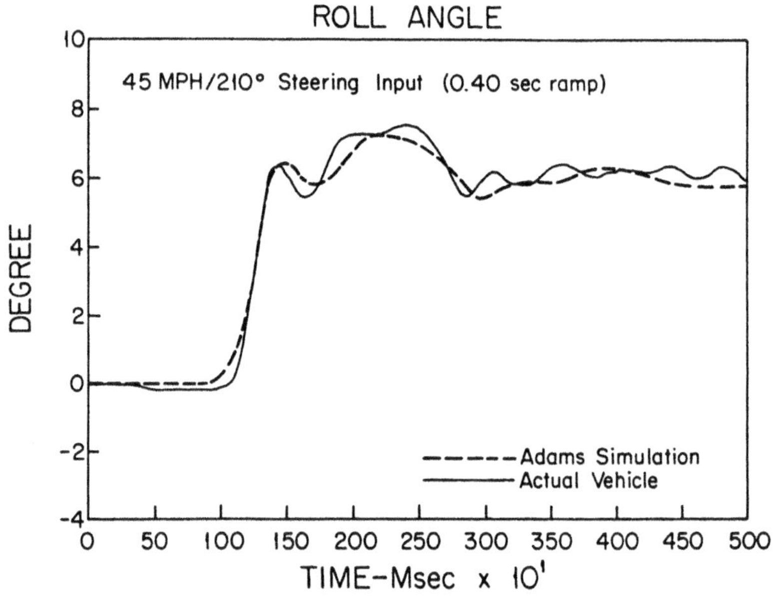

Figure 3.4. Roll angle versus time at 45 mph with steering input of Figure 3.2; ADAMS and test results.

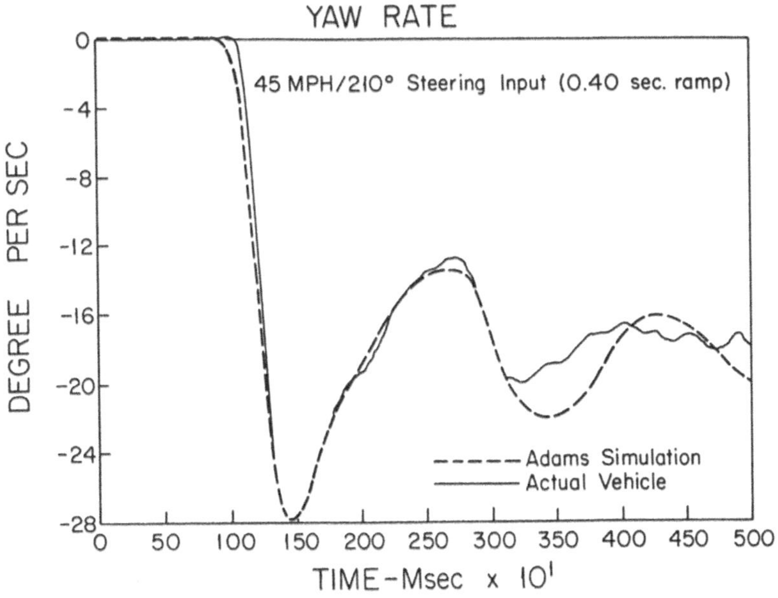

Figure 3.5. Yaw rate versus time at 45 mph with steering input of Figure 3.2; ADAMS and test results.

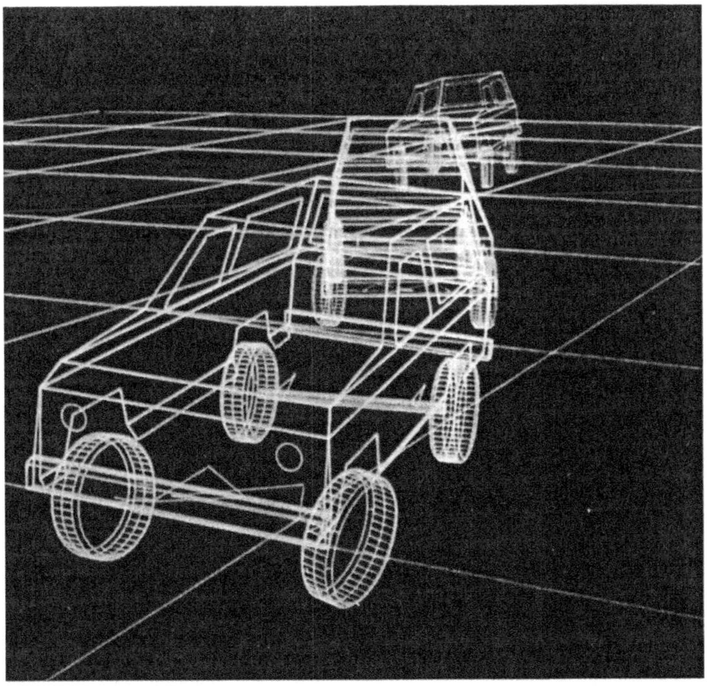

Figure 3.6. Succession of three frames of a Bronco II as output on an Evans and Sutherland PS300 computer graphic display.

SPATIAL KINEMATIC AND DYNAMIC ANALYSIS WITH EULER PARAMETERS

Parviz E. Nikravesh
Center for Computer Aided Design
College of Engineering
The University of Iowa
Iowa City, Iowa 52242

Abstract. This paper is devoted to developing the mathematical tools involved in describing angular orientation and equations of motion of rigid bodies, which are of considerable interest in themselves. Euler parameters are employed to derive interesting and useful identities. Physical interpretation of Euler parameters and their corresponding transformation matrices are discussed. Finally, Lagrange's equations of motion in terms of Euler parameters for a rigid body is presented.

1. INTRODUCTION

Although the method of analysis for spatial kinematics is not any different from the planar case, spatial kinematics requires much more powerful tools for analysis than planar kinematics. One of the major differences between the two analyses is the mathematical techniques used to describe the angular orientation of a rigid body in a global coordinate system. As the title suggests, this paper concentrates on a set of parameters known as Euler parameters, which eliminate drawbacks of other commonly used angular coordinates such as Euler angles. At the beginning, it may appear that Euler parameters have no physical significance and that they are just mathematical tools. However, when the subject is thoroughly understood, their physical interpretation also becomes evident. Furthermore, for large-scale computer programs considering angular orientation of bodies, rigid or deformable, the use of Euler parameters may, in many cases, drastically simplify the mathematical formulation.

In the following sections, useful identities between Euler parameters and various transformation matrices are derived. Some of

NATO ASI Series, Vol. F9
Computer Aided Analysis and Optimization of Mechanical System Dynamics
Edited by E. J. Haug

the identities are used in the derivation of Lagrange's equation of motion for an unconstrained rigid body in terms of Euler parameters.

2. EULER PARAMETERS

Euler parameters are generalized coordinates for angular orientation of a local Cartesian (moving or body-fixed) coordinate system ξ-η-ζ with respect to a global (non-moving or inertial) coordinate system x-y-z. If the two coordinate systems coincide at the origin, then based on Euler's theorem the transformation between the two coordinate systems can be accomplished by a single rotation about a unique axis that is referred to as orientational axis of rotation. If the transformation matrix from local to global coordinates is denoted by $\mathbf{A}$, then

$$\mathbf{s} = \mathbf{A}\ \mathbf{s}' \tag{2.1}$$

where $\mathbf{s}$ and $\mathbf{s}'$ are the global and local components of a vector $\vec{s}$ that is fixed in the local ξ-η-ζ coordinate system.

If the direction of the orientational axis of rotation is specified by unit vector $\vec{u}$ and the angle of rotation about this axis is ϕ, as shown in Fig. 2.1, then a vector $\vec{e}$ on this axis is defined as

$$\mathbf{e} = \mathbf{u} \sin \frac{\phi}{2} \tag{2.2}$$

where the components of $\vec{e}$ are

$$\mathbf{e} = [e_1, e_2, e_3]^T \tag{2.3}$$

These components are the same in the x-y-z and ξ-η-ζ axes. A fourth parameter is defined as

$$e_0 = \cos \frac{\phi}{2} \tag{2.4}$$

These four parameters

$$\mathbf{p} = [e_0, e_1, e_2, e_3]^T \tag{2.5}$$

are known as Euler parameters. These parameters are not independent. If Eqs. 2.2 and 2.4 are combined, it is found that

$$p^T p = 1 \tag{2.6}$$

The transformation matrix **A** in terms of Euler parameters is of the form [1]

$$\mathbf{A} = 2\begin{bmatrix} e_0^2 + e_1^2 - \frac{1}{2} & e_1e_2 - e_0e_3 & e_1e_3 + e_0e_2 \\ e_1e_2 + e_0e_3 & e_0^2 + e_2^2 - \frac{1}{2} & e_2e_3 - e_0e_1 \\ e_1e_3 - e_0e_2 & e_2e_3 + e_0e_1 & e_0^2 + e_3^2 - \frac{1}{2} \end{bmatrix} \tag{2.7}$$

or, in compact form,

$$\mathbf{A} = (2e_0^2 - 1)\mathbf{I} + 2(\mathbf{e}\mathbf{e}^T + e_0\tilde{\mathbf{e}}) \tag{2.8}$$

where **I** is a 3×3 identity matrix and the skew-symmetric matrix $\tilde{\mathbf{e}}$ is defined as

$$\tilde{\mathbf{e}} = \begin{bmatrix} 0 & -e_3 & e_2 \\ e_3 & 0 & -e_1 \\ -e_2 & e_1 & 0 \end{bmatrix} \tag{2.9}$$

It may be verified that the matrix **A** is orthonormal; i.e.,

$$\mathbf{A}^T \mathbf{A} = \mathbf{I} \tag{2.10}$$

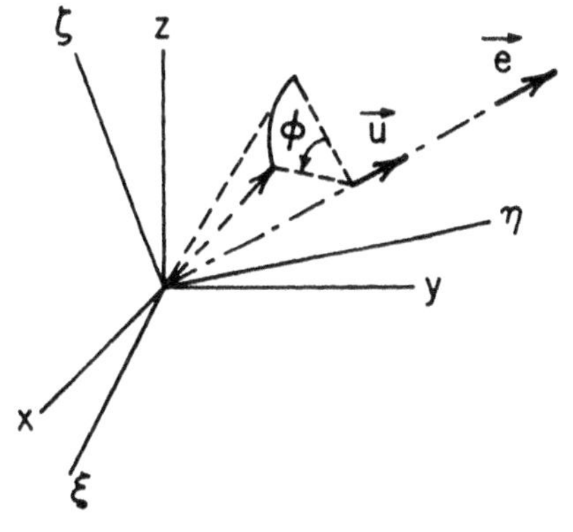

Figure 2.1 Angular rotation of ξ-η-ζ coordinate system about $\vec{u}$ axis

3. IDENTITIES WITH EULER PARAMETERS

In this section useful and interesting formulas and identities between Euler parameters, their time derivatives, and transformation matrices are presented. These identities become useful in derivation of kinematic constraints and the equations of motion for rigid bodies in a mechanism.

It can be shown, simply by inspection, that the following relations involving the skew-symmetric matrix $\tilde{\mathbf{e}}$ hold:

$$\tilde{\mathbf{e}}\,\mathbf{e} = \mathbf{o} \tag{3.1}$$

and

$$\begin{aligned}\tilde{\mathbf{e}}\,\tilde{\mathbf{e}} &= \mathbf{e}\,\mathbf{e}^T - (\mathbf{e}^T\,\mathbf{e})\,\mathbf{I} \\ &= \mathbf{e}\,\mathbf{e}^T - (1 - e_0^2)\,\mathbf{I}\end{aligned} \tag{3.2}$$

Matrices $\mathbf{E}$ and $\mathbf{G}$ are defined as

$$\begin{aligned}\mathbf{E} &= \begin{bmatrix} -e_1 & e_0 & -e_3 & e_2 \\ -e_2 & e_3 & e_0 & -e_1 \\ -e_3 & -e_2 & e_1 & e_0 \end{bmatrix} \\ &= [-\mathbf{e},\ \tilde{\mathbf{e}} + e_0\,\mathbf{I}]\end{aligned} \tag{3.3}$$

and

$$\begin{aligned}\mathbf{G} &= \begin{bmatrix} -e_1 & e_0 & e_3 & -e_2 \\ -e_2 & -e_3 & e_0 & e_1 \\ -e_3 & e_2 & -e_1 & e_0 \end{bmatrix} \\ &= [-\mathbf{e},\ -\tilde{\mathbf{e}} + e_0\,\mathbf{I}]\end{aligned} \tag{3.4}$$

It is observed that every row of $\mathbf{E}$ and $\mathbf{G}$ is orthogonal to $\mathbf{p}$, using Eq. 3.1, so

$$\mathbf{E}\mathbf{p} = \mathbf{G}\mathbf{p} = \mathbf{o} \tag{3.5}$$

In addition, the rows of $\mathbf{E}$ are orthogonal and the rows of $\mathbf{G}$ are orthogonal, so

$$\mathbf{E}\mathbf{E}^T = \mathbf{G}\mathbf{G}^T = \mathbf{I} \tag{3.6}$$

However, $\mathbf{E}^T\mathbf{E}$ and $\mathbf{G}^T\mathbf{G}$ are of the form

$$\mathbf{E}^T\mathbf{E} = \mathbf{G}^T\mathbf{G} = -\mathbf{p}\ \mathbf{p}^T + \mathbf{I}^* \tag{3.7}$$

where $\mathbf{I}^*$ is the 4 × 4 identity matrix. It can be verified that

$$\mathbf{E}\mathbf{G}^T = (2e_0^2 - 1)\ \mathbf{I} + 2(\mathbf{e}\ \mathbf{e}^T + e_0\ \tilde{\mathbf{e}}) \tag{3.8}$$

Comparing Eq. 3.8 with the transformation matrix **A** of Eq. 2.8 reveals that

$$\mathbf{A} = \mathbf{E}\mathbf{G}^T \tag{3.9}$$

Equation 3.9 demonstrates that the transformation matrix A can be considered as the result of two successive linear transformations. This is one of the most useful relationships between the **E** and **G** matrices and is a powerful property of Euler parameters.

The first time derivative of Eq. 2.6 yields

$$\mathbf{p}^T\ \dot{\mathbf{p}} = \dot{\mathbf{p}}^T\ \mathbf{p} = 0 \tag{3.10}$$

Similarly, the first time derivative of Eq. 3.5 results in

$$\mathbf{E}\ \dot{\mathbf{p}} = -\dot{\mathbf{E}}\ \mathbf{p} \tag{3.11}$$

and

$$\mathbf{G}\ \dot{\mathbf{p}} = -\dot{\mathbf{G}}\ \mathbf{p} \tag{3.12}$$

The products $\dot{\mathbf{E}}\ \dot{\mathbf{p}}$ and $\dot{\mathbf{G}}\ \dot{\mathbf{p}}$ are found to be

$$\dot{\mathbf{E}}\ \dot{\mathbf{p}} = \dot{\mathbf{G}}\ \dot{\mathbf{p}} = \mathbf{o} \tag{3.13}$$

It can be shown that

$$\mathbf{E}\ \dot{\mathbf{G}}^T = \dot{\mathbf{E}}\ \mathbf{G}^T \tag{3.14}$$

which can be used to demonstrate that the time derivative of the transformation matrix $\mathbf{A} = \mathbf{E}\mathbf{G}^T$ is

$$\begin{aligned}\dot{\mathbf{A}} &= 2\mathrm{E}\,\dot{\mathrm{G}}^T \\ &= 2\dot{\mathrm{E}}\,\mathrm{G}^T\end{aligned} \tag{3.15}$$

Furthermore, the time derivative of Eq. 3.15 results in

$$\begin{aligned}\ddot{\mathbf{A}} &= 2\dot{\mathrm{E}}\,\dot{\mathrm{G}}^T + 2\ddot{\mathrm{E}}\,\mathrm{G}^T \\ &= 2\dot{\mathrm{E}}\,\dot{\mathrm{G}}^T + 2\ddot{\mathrm{E}}\,\mathrm{G}^T\end{aligned} \tag{3.16}$$

from which

$$\ddot{\mathrm{E}}\,\mathrm{G}^T = \mathrm{E}\,\ddot{\mathrm{G}}^T \tag{3.17}$$

4. GENERAL IDENTITIES

A number of important identities are valid for a transformation matrix **A**, that may be presented in terms of Euler parameters or any other sets of rotational generalized coordinates. If the vector product of $\vec{s}$ with an arbitrary vector $\vec{a}$ results in vector $\vec{b}$, then, in terms of the global and local components, this vector product is expressed as

$$\mathbf{b} = \tilde{\mathbf{s}}\,\mathbf{a} \tag{4.1}$$

and

$$\mathbf{b}' = \tilde{\mathbf{s}}'\,\mathbf{a}' \tag{4.2}$$

Since $\mathbf{b} = \mathbf{A}\,\mathbf{b}'$ and $\mathbf{a} = \mathbf{A}\,\mathbf{a}'$, Eq. 4.1 becomes

$$\mathbf{A}\,\mathbf{b}' = \tilde{\mathbf{s}}\,\mathbf{A}\,\mathbf{a}' \tag{4.3}$$

Substituting Eq. 4.2 into Eq. 4.3, and noting that the result holds for vectors $\mathbf{a}'$, it follows that

$$\tilde{\mathbf{s}} = \mathbf{A}\,\tilde{\mathbf{s}}'\,\mathbf{A}^T \tag{4.4}$$

Assume that the ξ-η-ζ coordinate system is rotating relative to the x-y-z system with an angular velocity $\vec{\omega}(t)$. The components of $\vec{\omega}$ along the x-y-z and ξ-η-ζ coordinate systems are $\boldsymbol{\omega} = [\omega_{(x)}, \omega_{(y)}, \omega_{(z)}]^T$ and $\boldsymbol{\omega}' = [\omega_{(\xi)}, \omega_{(\eta)}, \omega_{(\zeta)}]^T$ respectively. The time derivative of Eq. 2.1 yields

$$\dot{\mathbf{s}} = \dot{\mathbf{A}}\,\mathbf{s}' + \mathbf{A}\,\dot{\mathbf{s}}' \tag{4.5}$$

If $\vec{s}$ is fixed in the body-fixed ξ-η-ζ axes, then $\dot{\mathbf{s}}' = \mathbf{o}$ and

$$\dot{\mathbf{s}} = \dot{\mathbf{A}}\,\mathbf{s}' \tag{4.6}$$

The vector $\dot{\vec{s}}$ may also be expressed [1] in terms of angular velocity $\vec{\omega}$ as $\dot{\vec{s}} = \vec{\omega} \times \vec{s}$; i.e.,

$$\dot{\mathbf{s}} = \tilde{\boldsymbol{\omega}}\,\mathbf{s} \tag{4.7}$$

or by using the transformation of Eq. 2.1, Eq. 4.7 becomes

$$\dot{\mathbf{s}} = \tilde{\boldsymbol{\omega}}\,\mathbf{A}\,\mathbf{s}' \tag{4.8}$$

If Eqs. 4.6 and 4.8 are combined, yielding an identity that must hold for all vectors $\mathbf{s}'$, it follows that

$$\dot{\mathbf{A}} = \tilde{\boldsymbol{\omega}}\,\mathbf{A} \tag{4.9}$$

Substitution of $\tilde{\boldsymbol{\omega}} = \mathbf{A}\,\tilde{\boldsymbol{\omega}}'\,\mathbf{A}^T$ from Eq. 4.4 into Eq. 4.9 yields

$$\dot{\mathbf{A}} = \mathbf{A}\,\tilde{\boldsymbol{\omega}}' \tag{4.10}$$

5. ANGULAR VELOCITY

In order to find a simple relationship between the components of angular velocity $\boldsymbol{\omega}$ and the time derivative of Euler parameters $\dot{\mathbf{p}}$, Eq. 4.9 can be used. Post-multiplying this equation by $\mathbf{A}^T$ yields

$$\dot{\mathbf{A}}\,\mathbf{A}^T = \tilde{\boldsymbol{\omega}} \tag{5.1}$$

Substituting from Eqs. 3.9 and 3.15 for $\mathbf{A}$ and $\dot{\mathbf{A}}$, Eq. 5.1 becomes

$$2\,\dot{\mathbf{E}}\,\mathbf{G}^T\,\mathbf{G}\,\mathbf{E}^T = \tilde{\boldsymbol{\omega}} \tag{5.2}$$

Using Eq. 3.7 and 3.6, the above equation becomes

$$2\,\dot{\mathbf{E}}\,\mathbf{E}^T = \tilde{\boldsymbol{\omega}} \tag{5.3}$$

Post-multiplying by an arbitrary vector **a** yields,

$$2\,\dot{\mathbf{E}}\,\mathbf{E}^T\,\mathbf{a} = \tilde{\boldsymbol{\omega}}\,\mathbf{a} = -\tilde{\mathbf{a}}\,\boldsymbol{\omega} \tag{5.4}$$

After some manipulation Eq. 5.4 results in

$$-2\,\tilde{\mathbf{a}}\,\dot{\mathbf{E}}\,\mathbf{p} - 2\,\mathbf{a}\,\dot{\mathbf{p}}^T\,\mathbf{p} = \tilde{\mathbf{a}}\,\boldsymbol{\omega}$$

which, upon application of Eqs. 3.10 and 3.11, gives

$$2\,\tilde{\mathbf{a}}\,\mathbf{E}\,\dot{\mathbf{p}} = \tilde{\mathbf{a}}\,\boldsymbol{\omega} \tag{5.5}$$

Since **a** is arbitrary, the coefficients of $\tilde{\mathbf{a}}$ must be equal, yielding

$$\boldsymbol{\omega} = 2\,\mathbf{E}\,\dot{\mathbf{p}} \tag{5.6}$$

Equation 5.6 can be combined with Eq. 3.10 to obtain

$$\begin{bmatrix} 0 \\ \omega_{(x)} \\ \omega_{(y)} \\ \omega_{(z)} \end{bmatrix} = 2 \begin{bmatrix} e_0 & e_1 & e_2 & e_3 \\ -e_1 & e_0 & -e_3 & e_2 \\ -e_2 & e_3 & e_0 & -e_1 \\ -e_3 & -e_2 & e_1 & e_0 \end{bmatrix} \begin{bmatrix} \dot{e}_0 \\ \dot{e}_1 \\ \dot{e}_2 \\ \dot{e}_3 \end{bmatrix} \tag{5.7}$$

or

$$\boldsymbol{\omega}^* = 2\mathbf{E}^*\,\dot{\mathbf{p}} \tag{5.8}$$

where $\boldsymbol{\omega}^* \equiv [0,\ \omega_{(x)},\ \omega_{(y)},\ \omega_{(z)}]^T$. A direct calculation shows that the matrix $\mathbf{E}^*$ is orthonormal, so the inverse transformation of Eq. 5.8 is simply

$$\dot{\mathbf{p}} = \frac{1}{2}\,\mathbf{E}^{*T}\,\boldsymbol{\omega}^* \tag{5.9}$$

An approach similar to the above steps yields a relationship between $\boldsymbol{\omega}'$ and $\dot{\mathbf{p}}$; i.e.,

$$\boldsymbol{\omega}' = 2\,\mathbf{G}\,\dot{\mathbf{p}} \tag{5.10}$$

Equation 5.10 can also be combined with Eq. 3.10 to obtain

$$\begin{bmatrix} 0 \\ \omega_{(\xi)} \\ \omega_{(\eta)} \\ \omega_{(\zeta)} \end{bmatrix} = 2 \begin{bmatrix} e_0 & e_1 & e_2 & e_3 \\ -e_1 & e_0 & e_3 & -e_2 \\ -e_2 & -e_3 & e_0 & e_1 \\ -e_3 & e_2 & -e_1 & e_0 \end{bmatrix} \begin{bmatrix} \dot{e}_0 \\ \dot{e}_1 \\ \dot{e}_2 \\ \dot{e}_3 \end{bmatrix} \tag{5.11}$$

or

$$\boldsymbol{\omega}'^{*} = 2\,\mathbf{G}^{*}\,\dot{\mathbf{p}} \tag{5.12}$$

where $\boldsymbol{\omega}'^{*} \equiv [0,\ \omega_{(\xi)},\ \omega_{(\eta)},\ \omega_{(\zeta)}]^T$. Since the matrix $\mathbf{G}^{*}$ is also orthonormal, the inverse transformation is simply

$$\dot{\mathbf{p}} = \frac{1}{2}\,\mathbf{G}^{*T}\,\boldsymbol{\omega}'^{*} \tag{5.13}$$

A simple relationship between the magnitudes of vectors $\boldsymbol{\omega}$ and $\dot{\mathbf{p}}$ can be found by using Eq. 5.6 as

$$\boldsymbol{\omega}^T\,\boldsymbol{\omega} = 4\,\dot{\mathbf{p}}^T\,\mathbf{E}^T\mathbf{E}\,\dot{\mathbf{p}} \tag{5.14}$$

Substitution of Eq. 3.7 into Eq. 5.14 and employing Eq. 3.10 yields

$$\boldsymbol{\omega}^T\,\boldsymbol{\omega} = 4\,\dot{\mathbf{p}}^T\,\dot{\mathbf{p}} \tag{5.15}$$

6. PHYSICAL INTERPRETATION OF EULER PARAMETERS

The concept of Euler parameters as rotational generalized coordinates may appear, to the unfamiliar reader, as a mathematical tool without any physical meaning. However, careful study of these parameters will prove the contrary. Physical interpretation of Euler parameters is simple and more natural to implement than any other set of rotational generalized coordinates, such as Euler or Bryant angles.

The angular orientation of one coordinate system relative to another coordinate system, such as global, can be considered by Eulers theorem as the result of a single rotation about the so-called orientational axis of rotation by an angle ϕ. Now, as a first example shown in Fig. 6.1(a), consider an observer standing along the axis of rotation in the global x-y-z system. If the x-y-z and ξ-η-ζ axes are initially coincident, then to move the ξ-η-ζ system to its final it will have rotated by angle ϕ as seen by the observer. A positive rotation, to the observer, will be a clockwise rotation of ξ-η-ζ about $\vec{u}$. If the observer is in the ξ-η-ζ coordinate system as shown in Fig. 6.1(b), the above rotation will be viewed as a counter-clockwise rotation of the x-y-z system by an angle ϕ about $\vec{u}$. A third case can exist, in which the observer is in a semi-rotating coordinate system designated by b_1-b_2-b_3, as shown in Fig. 6.1(c). For the observer who is standing along the axis of rotation in the b_1-b_2-b_3 coordinate system, the above rotation will be viewed as a clockwise rotation of ξ-η-ζ system by an angle $\phi/2$ and a counterclockwise rotation of x-y-z system by an angle $\phi/2$.

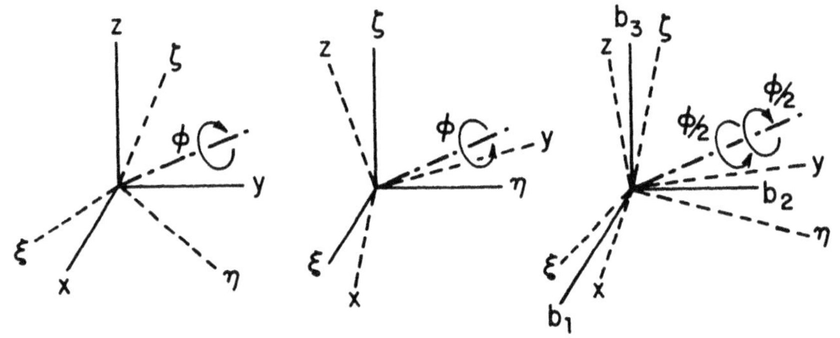

Figure 6.1 Interpretation of the coordinates rotation for an observer standing along the orientational axis of rotation; (a) attached to x-y-z axes, (b) attached to ξ-η-ζ axes, and (c) attached to b_1-b_2-b_3 axes.

If the orientation of ξ-η-ζ with respect to b_1-b_2-b_3 is described by a set of Euler parameters $[b_0, b_1, b_2, b_3]^T \equiv [b_0, \mathbf{b}^T]^T$, then a transformation matrix $\mathbf{B}$ can be defined as

$$\mathbf{B} = (2b_0^2 - 1)\mathbf{I} + 2(\mathbf{b}\,\mathbf{b}^T + b_0\,\tilde{\mathbf{b}}) \tag{6.1}$$

This matrix transforms the components of an arbitrary vector $\vec{s}$ from ξ-η-ζ to b_1-b_2-b_3 coordinates as

$$\mathbf{s}^{(b)} = \mathbf{B}\,\mathbf{s}' \tag{6.2}$$

The Euler parameters b_0, b_1, b_2, and b_3, in terms of the angle of rotation $\phi/2$ and unit vector $\vec{u}$ along the axis of rotation, are defined as

$$\begin{aligned} b_0 &= \cos\frac{\phi}{4} \\ \mathbf{b} &= \mathbf{u}\sin\frac{\phi}{4} \end{aligned} \tag{6.3}$$

Since $e_0 = \cos\phi/2$ and $\mathbf{e} = \mathbf{u}\sin\phi/2$, Eq. 6.2 yields

$$\begin{aligned} e_0 &= 2b_0^2 - 1 \\ \mathbf{e} &= 2b_0\,\mathbf{b} \end{aligned} \tag{6.4}$$

Using Eq. 6.4 in Eq. 6.1 gives

$$\mathbf{B} = e_0\,\mathbf{I} + \tilde{\mathbf{e}} + \frac{\mathbf{e}\,\mathbf{e}^T}{1 + e_0} \tag{6.5}$$

Similarly, it can be shown that the transformation matrix from b_1-b_2-b_3 to x-y-z is also $\mathbf{B}$; i.e.,

$$\mathbf{s} = \mathbf{B}\,\mathbf{s}^{(b)} \tag{6.6}$$

Substituting Eq. 6.2 into Eq. 6.6 yields

$$\mathbf{s} = \mathbf{B}\,\mathbf{B}\,\mathbf{s}' \tag{6.7}$$

The transformation matrix $\mathbf{A}$ from ξ-η-ζ to x-y-z; i.e.,

$$\mathbf{s} = \mathbf{A}\,\mathbf{s}'$$

may thus be written as

$$\mathbf{A} = \mathbf{B}\,\mathbf{B} \tag{6.8}$$

Equation 6.8 illustrates the almost obvious fact that any rotation is the result of two successive semi-rotations.

Equation 3.9 states that the transformation matrix **A** is the result of two successive transformations; i.e., **A** can be expressed as the product of two 3×4 matrices **E** and **G** as $\mathbf{A} = \mathbf{E}\mathbf{G}^T$. Comparison of Eqs. 3.9 and 6.8 shows that the product **BB** is analogous to the product $\mathbf{E}\mathbf{G}^T$. Equation 6.7 is interpreted as a transformation of **s**' from ξ-η-ζ by a semi-rotation to the intermediate 3-space semi-rotating system b_1-b_2-b_3 to obtain $\mathbf{s}^{(b)}$. Then, $\mathbf{s}^{(b)}$ is transformed from the semi-rotating system by another semi-rotation to x-y-z. With this interpretation of Eq. 6.7, Eq. 3.9 or the relation

$$\mathbf{s} = \mathbf{E}\,\mathbf{G}^T\,\mathbf{s}' \tag{6.9}$$

may be interpreted as transforming **s**' from ξ-η-ζ to an intermediate 4-space semi-rotating system to obtain $\mathbf{s}^{(4)}$; i.e.,

$$\mathbf{s}^{(4)} = \mathbf{G}^T\,\mathbf{s}' \tag{6.10}$$

Then $\mathbf{s}^{(4)}$ is transformed from the 4-space semi-rotating system to x-y-z by a second semi-rotation; i.e.,

$$\mathbf{s} = \mathbf{E}\mathbf{s}^{(4)} \tag{6.11}$$

7. GENERALIZED COORDINATES OF A RIGID BODY

An unconstrained rigid body in space requires six independent generalized coordinates to determine its orientation. Three generalized coordinates specify its translation and three specify its rotation. The six generalized coordinates define the location of a ξ-η-ζ Cartesian coordinate system fixed in the body, relative to the global coordinate axes. The coordinates of the origin of the body-fixed axes are the translational generalized coordinates. The rotational generalized coordinates define the orientation of the local ξ-η-ζ axes, relative to the global x-y-z coordinate axes.

When the generalized coordinates of a body are known, the global coordinates of a point P on the body, as shown in Fig. 7.1, can be written as

$$\mathbf{r}^P = \mathbf{r} + \mathbf{s} = \mathbf{r} + \mathbf{A}\,\mathbf{s}' \tag{7.1}$$

where $\mathbf{r}^P = [x^P, y^P, z^P]^T$ are the global coordinates of P, $\mathbf{r} = [x, y, z]^T$ are the translational generalized coordinates of the body, $\mathbf{s}' = [\xi^P, \eta^P, \zeta^P]^T$ are the local coordinates of P, and $\mathbf{A}$ is a transformation matrix that depends on the angular generalized coordinates of the body. If the transformation matrix $\mathbf{A}$ is described in terms of four Euler parameters $\mathbf{p}$, then there are seven generalized coordinates, three translational and four rotational, that describe the orientation of the body in space. The vector of generalized coordinates for the body may be denoted as

$$\mathbf{q} = [\mathbf{r}^T, \mathbf{p}^T]^T = [x, y, z, e_0, e_1, e_2, e_3]^T \tag{7.2}$$

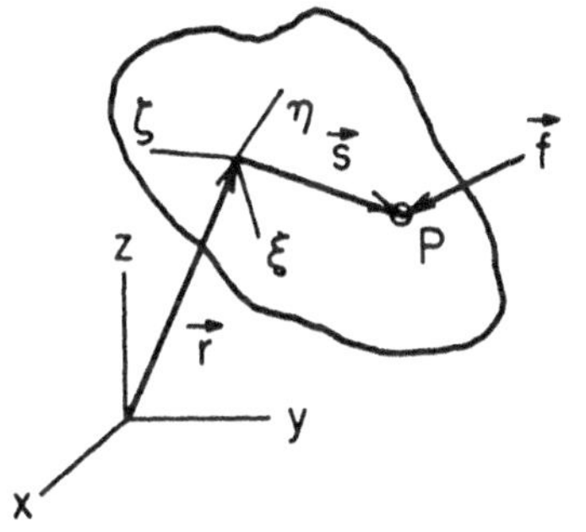

Figure 7.1 Translation and rotation of a body in three-dimensional space.

The velocity of a point P on the body can be found from Eq. 7.1 as

$$\dot{\mathbf{r}}^P = \dot{\mathbf{r}} + \dot{\mathbf{s}} = \dot{\mathbf{r}} + \tilde{\boldsymbol{\omega}}\,\mathbf{s} = \dot{\mathbf{r}} - \tilde{\mathbf{s}}\,\boldsymbol{\omega} \tag{7.3}$$

where Eq. 4.7 is employed. The Euler parameter transformation of Eq. 5.6 may be applied to obtain

$$\dot{r}^P = \dot{r} - 2\,\tilde{s}\,E\,\dot{p} \tag{7.4}$$

Substituting the identity of Eq. 4.4 into Eq. 7.4 yields

$$\dot{r}^P = \dot{r} - 2A\,\tilde{s}'\,G\,\dot{p} \tag{7.5}$$

Equations 7.4 and 7.5 provide explicit relationships for the velocity of point P in terms of time derivatives of Euler parameters.

8. GENERALIZED FORCES

Since Euler parameters are employed as rotational coordinates, the three moment components of the generalized force vector must be converted to four moment components associated with Euler parameters. In order to obtain explicit expressions for generalized forces, it is first necessary to write virtual displacements of points of application of forces on a body in terms of variations in generalized coordinates. Since generalized coordinates are functions of time, a virtual displacement may be interpreted as the total differential of a displacement vector; i.e.,

$$\begin{aligned} \delta r^P &= \dot{r}^P\,\delta t \\ \delta r &= \dot{r}\,\delta t \\ \delta p &= \dot{p}\,\delta t \\ \delta s &= \dot{s}\,\delta t \end{aligned} \tag{8.1}$$

Multiplying through Eq. 7.4 by an infinitesimal δt, the following relationship between the virtual displacement of point P is obtained, in terms of variations in Euler parameters:

$$\delta r^P = \delta r - 2\tilde{s}\,E\,\delta p \tag{8.2}$$

$$\delta r^P = \delta r - 2A\,\tilde{s}'\,G\,\delta p \tag{8.3}$$

For a force $\vec{f}$ acting on point P of the body, as shown in Fig. 7.1, the virtual work is

$$\delta w = f^T\,\delta r^P \tag{8.4}$$

Substituting from Eq. 8.1 in terms of Euler parameters, the virtual work is

$$\delta w = \mathbf{f}^T(\delta \mathbf{r} - 2\tilde{\mathbf{s}}\ \mathbf{E}\ \delta \mathbf{p}) \tag{8.5}$$

The seven components of generalized force corresponding to $\vec{f}$ are obtained as coefficients in Eq. 8.5 of the variations in the seven generalized coordinates **r** and **p**; i.e.,

$$\mathbf{g} = \begin{bmatrix} \mathbf{g}_{(r)} \\ \mathbf{g}_{(p)} \end{bmatrix} = \begin{bmatrix} \mathbf{f} \\ 2\mathbf{E}^T\ \tilde{\mathbf{s}}\ \mathbf{f} \end{bmatrix} \tag{8.6}$$

Since $\mathbf{f}' = \mathbf{A}^T\ \mathbf{f}$, Eq. 8.6 can be written as

$$\mathbf{g} = \begin{bmatrix} \mathbf{g}_{(r)} \\ \mathbf{g}_{(p)} \end{bmatrix} = \begin{bmatrix} \mathbf{f} \\ 2\mathbf{G}^T\ \tilde{\mathbf{s}}'\ \mathbf{f}' \end{bmatrix} \tag{8.7}$$

The quantities $\tilde{\mathbf{s}}'\ \mathbf{f}'$ and $\tilde{\mathbf{s}}\ \mathbf{f}$ are simply moments $\mathbf{n}'$ and $\mathbf{n}$, with respect to the origin of the ξ-η-ζ coordinates, expressed in local and global coordinate systems, respectively. Thus,

$$\mathbf{g} \equiv \begin{bmatrix} \mathbf{g}_{(r)} \\ \mathbf{g}_{(p)} \end{bmatrix} = \begin{bmatrix} \mathbf{f} \\ 2\mathbf{G}^T\ \mathbf{n}' \end{bmatrix} = \begin{bmatrix} \mathbf{f} \\ 2\mathbf{E}^T\ \mathbf{n} \end{bmatrix} \tag{8.8}$$

In the case of a pure moment, Eq. 8.8 becomes

$$\mathbf{g} \equiv \begin{bmatrix} \mathbf{g}_{(r)} \\ \mathbf{g}_{(p)} \end{bmatrix} = \begin{bmatrix} \mathbf{0} \\ 2\mathbf{G}^T\ \mathbf{n}' \end{bmatrix} = \begin{bmatrix} \mathbf{0} \\ 2\mathbf{E}^T\ \mathbf{n} \end{bmatrix} \tag{8.9}$$

Equations 8.8 and 8.9 indicate the simplicity of obtaining generalized forces in terms of Euler parameters, from arbitrary forces or moments.

9. THE INERTIA TENSOR

Let the origin of the body-fixed ξ-η-ζ axes be located at the center of mass of the body (centroidal coordinates), as shown in Fig. 9.1. Vector $\vec{s}$ locates an infinitesimal mass dm in the body. It is assumed that the body has volume v. The inertia tensor is defined as the integral

$$\mathbf{J}' = -\int_{(v)} \tilde{\mathbf{s}}' \, \tilde{\mathbf{s}}' \, dm \tag{9.1}$$

which is a 3×3 matrix. Another form of the inertia tensor may be defined in terms of the global coordinate axes; i.e.,

$$\mathbf{J} = -\int_{(v)} \tilde{\mathbf{s}} \, \tilde{\mathbf{s}} \, dm \tag{9.2}$$

By substituting from Eq. 4.4 into Eq. 9.2, it is found that

$$\mathbf{J} = \mathbf{A} \, \mathbf{J}' \, \mathbf{A}^T \tag{9.3}$$

Note that in contrast to $\mathbf{J}'$, which is a constant matrix, $\mathbf{J}$ is a function of angular orientation of the body.

The time derivatives of $\mathbf{J}'$ and $\mathbf{J}$ are

$$\dot{\mathbf{J}}' = 0 \tag{9.4}$$

and

$$\dot{\mathbf{J}} = \tilde{\boldsymbol{\omega}} \, \mathbf{J} - \mathbf{J} \, \tilde{\boldsymbol{\omega}} \tag{9.5}$$

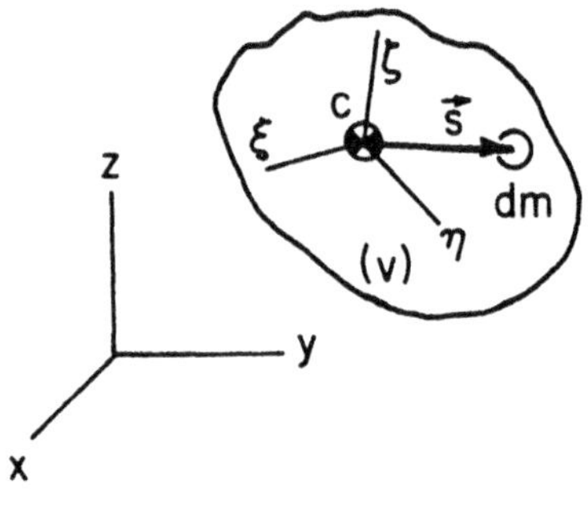

Figure 9.1 A rigid body as a collection of differential masses dm

10. KINETIC ENERGY

Consider a rigid body with a centroidal body fixed coordinate system. The kinetic energy for this body is written as

$$T = \frac{1}{2} m \dot{\mathbf{r}}^T \dot{\mathbf{r}} + \frac{1}{2} \boldsymbol{\omega}'^T \mathbf{J}' \boldsymbol{\omega}' \tag{10.1}$$

where m is the mass of the body. Equation 10.1 may also be written as

$$T = \frac{1}{2} m \dot{\mathbf{r}}^T \dot{\mathbf{r}} + \frac{1}{2} \boldsymbol{\omega}^T \mathbf{J} \boldsymbol{\omega} \tag{10.2}$$

The kinetic energy expression, using the identity $\boldsymbol{\omega} = 2\mathbf{E}\,\dot{\mathbf{p}}$, can be written as

$$T = \frac{1}{2} m \dot{\mathbf{r}}^T \dot{\mathbf{r}} + 2\, \dot{\mathbf{p}}^T \mathbf{E}^T \mathbf{J} \mathbf{E} \dot{\mathbf{p}} \tag{10.3}$$

or

$$T = \frac{1}{2} m \dot{\mathbf{r}}^T \dot{\mathbf{r}} + 2\, \dot{\mathbf{p}}^T \mathbf{G}^T \mathbf{J}' \mathbf{G} \dot{\mathbf{p}} \tag{10.4}$$

In compact form Eq. 10.3 may be expressed as

$$T = \frac{1}{2} \dot{\mathbf{q}}^T \mathbf{M} \dot{\mathbf{q}} \tag{10.5}$$

where $\dot{\mathbf{q}} = [\dot{\mathbf{r}}^T, \dot{\mathbf{p}}^T]^T$,

$$\mathbf{M} = \begin{bmatrix} \mathbf{N} & \mathbf{0} \\ \mathbf{0} & 4\mathbf{E}^T\mathbf{J}\mathbf{E} \end{bmatrix} \tag{10.6}$$

and $\mathbf{N} = \mathrm{diag}[m,m,m]$. The term $\mathbf{E}^T\mathbf{J}\mathbf{E}$ in the mass matrix is equal to

$$\mathbf{E}^T\mathbf{J}\mathbf{E} = \mathbf{G}^T\mathbf{J}'\mathbf{G} \tag{10.7}$$

The mass matrix of Eq. 10.6 may thus be written as

$$\mathbf{M} = \begin{bmatrix} \mathbf{N} & \mathbf{0} \\ \mathbf{0} & 4\mathbf{G}^T\mathbf{J}'\mathbf{G} \end{bmatrix} \tag{10.8}$$

Note that $\mathbf{M}$ is a 7×7 matrix.

11. EQUATIONS OF MOTION

The familiar form of Lagrange's equations of motion for an unconstrained rigid body is expressed as

$$\frac{d}{dt}\left(\frac{\partial T}{\partial \dot{q}}\right)^T - \left(\frac{\partial T}{\partial q}\right)^T = g \tag{11.1}$$

In order to derive Lagrange's equations of motion in terms of Euler parameters, the required partial derivatives of Eq. 10.4 are written as

$$\left(\frac{\partial T}{\partial \dot{r}}\right)^T = m\,\dot{r} \tag{11.2}$$

$$\left(\frac{\partial T}{\partial \dot{p}}\right)^T = 4G^T\,J'\,G\,\dot{p} \tag{11.3}$$

and

$$\left(\frac{\partial T}{\partial r}\right)^T = 0 \tag{11.4}$$

where it is observed that **G** depends only on elements of **p**. Before carrying out the remaining partial derivatives, it is helpful to employ Eq. 3.12 in Eq. 10.4 to obtain

$$T = \tfrac{1}{2}\,m\,\dot{r}^T\,\dot{r} + 2p^T\,\dot{G}^T\,J'\,\dot{G}\,p \tag{11.5}$$

Observing that $\dot{G}$ depends only on elements of $\dot{p}$ and not on **p** explicitly, it follows that

$$\left(\frac{\partial T}{\partial p}\right)^T = 4\,\dot{G}^T\,J'\,\dot{G}\,p \tag{11.6}$$

Differentiating Eqs. 11.2 and 11.3 with respect to time yields

$$\frac{d}{dt}\left(\frac{\partial T}{\partial \dot{r}}\right)^T = m\,\ddot{r} \tag{11.7}$$

and

$$\frac{d}{dt}\left(\frac{\partial T}{\partial \dot{p}}\right)^T = 4\,G^T\,J'\,G\,\ddot{p} - 4\,\dot{G}^T\,J'\,\dot{G}\,p \tag{11.8}$$

where Eq. 3.13 is used.

Equations 11.4, 11.6, 11.7, and 11.8 may be combined to form Lagrange's equations of motion. However, Lagrange's equations of motion of Eq. 11.1 is based on the assumption that the generalized coordinates of the body are independent. Since the four Euler parameters of the body are dependent, by the constraint equation of Eq. 2.6; i.e.,

$$\Phi \equiv \mathbf{p}^T \mathbf{p} - 1 = 0 \tag{11.9}$$

the effect of Eq. 11.9 must be included in Eq. 11.1. This can be done by using the Lagrange multiplier technique, justified by Farkas Lemma [2]. The partial derivative of Eq. 11.9 with respect $\mathbf{p}$ is $\Phi_{\mathbf{p}} = 2\mathbf{p}^T$. Using this result, Eq. 11.1 for translational and rotational equations of motion of the body is modified as

$$\begin{aligned} &\frac{d}{dt}\left(\frac{\partial T}{\partial \dot{\mathbf{r}}}\right)^T - \left(\frac{\partial T}{\partial \mathbf{r}}\right)^T = \mathbf{g}_{(r)} \\ &\frac{d}{dt}\left(\frac{\partial T}{\partial \dot{\mathbf{p}}}\right)^T - \left(\frac{\partial T}{\partial \mathbf{p}}\right)^T + 2\mathbf{p}\,\sigma = \mathbf{g}_{(p)} \end{aligned} \tag{11.10}$$

where σ is the Lagrange multiplier associated with the constraint of Eq. 11.9 and $\mathbf{g}_{(r)}$ and $\mathbf{g}_{(p)}$ are the components of $\mathbf{g}$, corresponding to $\mathbf{r}$ and $\mathbf{p}$, respectively. Substitution of Eqs. 11.4, 11.6, 11.7, and 11.8 into Eq. 11.10 yields

$$\begin{bmatrix} N & 0 & o \\ \mathbf{0} & 4G^T J' G & 2\mathbf{p} \end{bmatrix} \begin{bmatrix} \ddot{\mathbf{r}} \\ \ddot{\mathbf{p}} \\ \sigma \end{bmatrix} = \begin{bmatrix} g_{(r)} \\ g_{(p)} + 8\dot{G}^T J' \dot{G} p \end{bmatrix} \tag{11.11}$$

Equation 11.11 may be written in matrix form as

$$\mathbf{M}\,\ddot{\mathbf{q}} + [o^T,\ 2\mathbf{p}^T]^T\,\sigma = \mathbf{g} + [o^T,\ 8\mathbf{p}^T\dot{G}^T\ J'\dot{G}^T]^T \tag{11.12}$$

where $\ddot{\mathbf{q}} = [\ddot{\mathbf{r}}^T,\ \ddot{\mathbf{p}}^T]^T$.

Equation 11.11 cannot be solved for $\ddot{\mathbf{r}}$, $\ddot{\mathbf{p}}$, and σ, since the matrix on the left of Eq. 11.11 is a 7 × 8 matrix; i.e., it is singular. The second time derivative of Eq. 11.9; i.e.,

$$\ddot{\Phi} = 2\mathbf{p}^T\,\ddot{\mathbf{p}} + 2\dot{\mathbf{p}}^T\,\dot{\mathbf{p}} = 0 \tag{11.13}$$

can be appended to Eq. 11.12 to obtain

$$\begin{bmatrix} \mathbf{N} & 0 & o \\ 0 & 4\mathbf{G}^T\mathbf{J}'\mathbf{G} & 2\mathbf{p} \\ o^T & 2\mathbf{p}^T & 0 \end{bmatrix} \begin{bmatrix} \ddot{\mathbf{r}} \\ \ddot{\mathbf{p}} \\ \sigma \end{bmatrix} = \begin{bmatrix} g_{(r)} \\ g_{(p)} + 8\dot{\mathbf{G}}^T\mathbf{J}'\dot{\mathbf{G}}\mathbf{p} \\ -2\dot{\mathbf{p}}^T\dot{\mathbf{p}} \end{bmatrix} \quad (11.14)$$

The matrix on the left of Eq. 11.14 is nonsingular, provided that $\mathbf{N}^{-1}$ and $\mathbf{J}'^{-1}$ exist, which is the case for well posed problems. The solution of Eq. 11.14 for $\ddot{\mathbf{r}}$, $\ddot{\mathbf{p}}$, and σ is then

$$\begin{bmatrix} \ddot{\mathbf{r}} \\ \ddot{\mathbf{p}} \\ \sigma \end{bmatrix} = \begin{bmatrix} \mathbf{N}^{-1} & 0 & o \\ 0 & (\frac{1}{4})\mathbf{G}^T\mathbf{J}'^{-1}\mathbf{G} & (\frac{1}{2})\mathbf{p} \\ o^T & (\frac{1}{2})\mathbf{p}^T & 0 \end{bmatrix} \begin{bmatrix} g_{(r)} \\ g_{(p)} + 8\dot{\mathbf{G}}^T\mathbf{J}'\dot{\mathbf{G}}\mathbf{p} \\ -2\dot{\mathbf{p}}^T\dot{\mathbf{p}} \end{bmatrix} \quad (11.15)$$

12. MECHANICAL SYSTEMS

The identities and equations of motion derived in terms of Euler parameters can be used to formulate constraints and equations of motion for a mechanical system. Kinematic joints that connect the rigid bodies in a system yield, in general, nonlinear algebraic equations. These algebraic equations, in addition to the equations of motion, result in a mixed system of algebraic and differential equations. When these equations are derived in terms of Euler parameters, the following attractive features are observed: (1) the singularity problem and trigonometric function evaluations associated with other sets of three rotational coordinates; e.g., Euler angles, is circumvented, (2) a general-purpose computer program can be developed to generate these equations automatically in a simple systematic order, (3) computational efficiency is obtained when this program is compared with similar programs that are written in terms of Euler angles, and (4) computational overhead associated with the extra constraint equation on each set of Euler parameters is negligible, compared with the computational efficiency gained elsewhere.

REFERENCES

1. Wittenburg, J., Dynamics of Systems of Rigid Bodies, B.G. Teubner, Stuttgart, 1977.

2. Fiacco, A.V. and McCormick, G.P., Nonlinear Programming: Sequential Unconstrained Minimization Technique, Wiley, New York, 1968.

APPLICATION OF SYMBOLIC COMPUTATION TO THE ANALYSIS OF MECHANICAL SYSTEMS, INCLUDING ROBOT ARMS

M.A. Hussain
General Electric Company
Corporate Research and Development
Schenectady, New York

AND

B. Noble
Mathematics Research Center
University of Wisconsin

Abstract. This paper illustrates the application of symbolic computation in connection with three aspects of mechanical systems:

1. The derivation of dynamical equations by Lagrangian methods.

2. The analysis and synthesis of kinematic mechanisms.

3. A robot manipulator arm.

NATO ASI Series, Vol. F9
Computer Aided Analysis and Optimization of Mechanical System Dynamics
Edited by E. J. Haug

INTRODUCTION

This paper illustrates the *potential* of symbolic computation in connection with the formulation and analysis of equations for dynamical systems, sensitivity analysis, linkages and mechanisms, and robot manipulator arms.

We use MACSYMA (project MAC's SYmbolic MAnipulation system), a large-scale computer program for symbolic mathematical computation. MACSYMA can handle polynomial manipulation, simplification and substitution with symbolic expressions, symbolic solution of algebraic and differential equations, and matrix manipulation. Although we have found MACSYMA particularly convenient to use, other symbolic programs such as REDUCE and SMP could be used.

This report deals with three topics:

1. The derivation of dynamical equations by Lagrangian methods, including sensitivity analysis (Sections 1, 2, 3, 4, and 5).
2. The analysis and synthesis of kinematic mechanisms, including dual-number quaternions (Sections 7 and 8).
3. The direct and inverse problem involving robot manipulator arms (Sections 9 and 10).

In order to make the presentation clearer to the general reader who lacks specialized knowledge of symbolic manipulation, we explain the mathematical aspects in the main text (namely the kind of problem for which we feel symbolic computation is useful), and give the detailed MACSYMA programs in appendices.

In a certain sense the real "meat" of this paper is the detailed programs which appear in the appendices. The reader interested in symbolic manipulation should solve the problems outlined in the text using MACSYMA, or any other suitable program, with the appendices as a guide.

The objective of this paper is to encourage the use of symbolic manipulation in the analysis of mechanical systems. It is clear that the complexity of the problems being tackled is increasing to the point where symbolic manipulation must play an important role in their formulation and solution. In this paper we have simply picked out the tedious parts of well known methods and examples, and illustrated the ease of performing the manipulation using MACSYMA.

1. DESIGN OF A 5-DEGREES-OF-FREEDOM VEHICLE SUSPENSION

The objective of this example is to illustrate how MACSYMA handles Lagrange's equation of motion in the form:

$$\frac{d}{dt}\left(\frac{\partial T}{\partial \dot{q}_i}\right) - \frac{\partial T}{\partial q_i} + \frac{\partial V}{\partial q_i} - Q_i = 0, \qquad i = 1, \cdots, n \tag{1}$$

where T and V are quadratic forms representing kinetic energy and potential energy, respectively, expressed in terms of generalized coordinates q_i. Q_i are nonconservative generalized forces. Consider the 5-degrees-of-freedom model of a vehicle suspension system shown in Figure 1 and dealt with in Haug and Arora [6] (pp. 25, 200):

$$T = \frac{1}{2} m_1 \dot{Z}_1^2 + \frac{1}{2} m_2 \dot{Z}_2^2 + \frac{1}{2} I \dot{Z}_3^2 + \frac{1}{2} m_4 \dot{Z}_4^2 + \frac{1}{2} m_5 \dot{Z}_5^2$$

$$V = \frac{1}{2} K_1 d_1^2 + \frac{1}{2} K_2 d_2^2 + \frac{1}{2} K_3 d_3^2 + \frac{1}{2} K_4 d_4^2 + \frac{1}{2} K_5 d_5^2 \tag{2}$$

where $d_1 = Z_2 + \frac{L}{12} Z_3 - Z_1 \qquad d_2 = Z_4 - Z_2 - \frac{L}{3} Z_3$

$d_3 = Z_5 - Z_2 + \frac{2L}{3} Z_3 \qquad d_4 = Z_4 - f_2(t)$

$d_5 = Z_5 - f_1(t)$

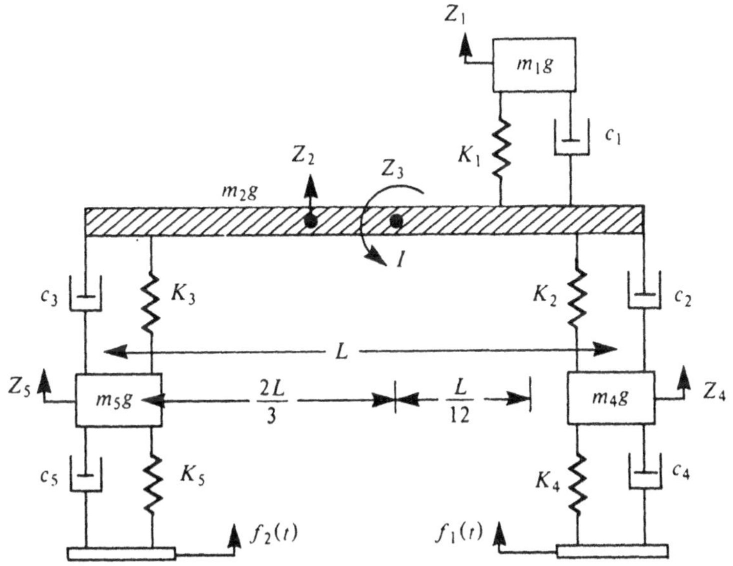

Figure 1.

The Q_i are found from

$$\delta W = \sum_{i=1}^{5} Q_i \delta Z_i = - c_1 \dot{d}_1 \delta d_1 - c_2 \dot{d}_2 \delta d_2 \cdots - c_5 \dot{d}_5 \delta d_5 \tag{3}$$

The MACSYMA program and output for the above problem is given in its entirety in Appendix A-I. We comment on the key commands. (C2) establishes the vector Q of generalized coordinates Z1, · · · Z5. (C3) establishes the dependence of the elements of Q on time. (C4)-(C7) establish mass, spring, damping and displacement vectors (for DISP equations following Eq. 2 above). (C11),(C12) derive the generalized forces Qn (=QQ(n) in the program) by picking out the coefficients of δZ_n (=DEL(Q(N)) in the program) in δW in Eq. 3 above (=DW in the program). (C9),(C10) establish kinetic and potential energies defined in Eq. 2 above. (C13) forms and displays the equations of motion by evaluating Eq. 1 above. The key command here is DIFF(EXPR,T) where EXPR is some function of T, which takes the derivative of EXPR with respect to T. Thus

$$\text{DIFF(DIFF(TT,DIFF(Q(N),T)),T)} \equiv \frac{d}{dt}\left(\frac{\partial T}{\partial \dot{q}_n}\right)$$

As requested, the computer then displays the equation, of which we have shown only the first, namely

$$m_1 \ddot{Z}_1 + c_1 \dot{Z}_1 - c_1 \dot{Z}_2 - \frac{L}{12} c_1 \dot{Z}_3 + K_1 Z_1 - K_2 Z_2 - \frac{L}{12} K_1 Z_3 = 0$$

2. SLIDER CRANK PROBLEM

This example illustrates the derivation of equations of motion when constraints are present. The appropriate Lagrange equations are:

$$\frac{\partial}{\partial t}\left(\frac{\partial T}{\partial \dot{q}_i}\right) - \frac{\partial T}{\partial q_i} + \frac{\partial V}{\partial q_i} + \left(\frac{\partial \Phi}{\partial q_i}\right)^T \lambda = Q_i \qquad i = 1, \cdots, n \tag{4}$$

where T, V, and Q_i are as previously defined and the constraints are represented by k algebraic equations:

$$\Phi(q) = 0$$

and λ are k values of Lagrange multipliers or undetermined coefficients.

For the slider crank mechanism shown in Figure 2 we have

$$q = [\phi_1, x_2, y_2, \phi_2, x_3]^T$$

$$T = \frac{1}{2} J_1 \dot{\phi}_1^2 + \frac{1}{2} m_2 (\dot{x}_2^2 + \dot{y}_2^2) + \frac{1}{2} J_2 \dot{\phi}_2^2 + \frac{1}{2} m_3 \dot{x}_3^2$$

$$\delta W = f(t)\, \delta x_3$$

$$\Phi_1 \equiv r \sin\phi_1 - y_2 - l \sin\phi_2 = 0 \tag{5}$$

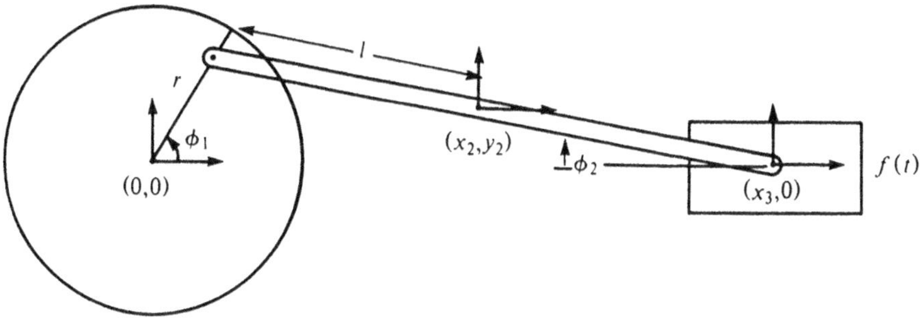

Figure 2.

$$\Phi_2 \equiv r\cos\phi_1 - (x_2 - l\cos\phi_2) = 0$$

$$\Phi_3 \equiv x_2 + l\cos\phi_2 - x_3 = 0$$

$$\Phi_4 \equiv y_2 - l\sin\phi_2 = 0$$

Again the procedure outlined is easily handled by MACSYMA (see the symbolic program given in Appendix A-II). From the output of this program we have the following equations of motion for the system:

$$J_1\ddot{\phi}_1 + [\lambda_1 r\cos\phi_1 - \lambda_2 r\sin\phi_1] = 0$$

$$m_2\ddot{x}_2 + [-\lambda_2 + \lambda_3] = 0$$

$$m_2\ddot{y}_2 + [-\lambda_1 + \lambda_4] = 0 \tag{6}$$

$$J_2\ddot{\phi}_2 + [\lambda_1 l\cos\phi_2 - \lambda_2 l\sin\phi_2 - \lambda_3 l\sin\phi_2 - \lambda_4 l\cos\phi_2] = 0$$

$$m_3\ddot{x}_3 - \lambda_3 = f(t)$$

Note that Eqs. 5,6 are a differential-algebraic system consisting of nine equations in nine unknowns.

This problem has only a single degree of freedom. If this is chosen as ϕ_1, we find from Eq. 5 that x_2, y_2, ϕ_2, and x_3 can be expressed in terms of ϕ_1. If these are substituted in the kinetic energy ($=TT$) expression we find $TT = TT(\phi_1,\dot{\phi}_1)$. The equation of motion is now given by

$$\frac{d}{dt}\left(\frac{\partial TT}{\partial\dot{\phi}_1}\right) - \frac{\partial TT}{\partial\phi_1} = Q_1 \tag{7}$$

i.e., one single differential equation in one unknown, with no constraints. A MACSYMA program for deriving this equation is given in Appendix A-III.

Equation 7 must be equivalent to nine equations in Eqs. 5,6, though derived independently. To deduce Eq. 7 directly from Eq. 6 we can proceed as follows:

Write Eqs. 6 and 5 in matrix notation as

$$M\ddot{q} + A^T(q)\lambda = f \tag{8}$$

$$\Phi(q) = 0 \tag{9}$$

$A(q)$ is a 4×5 matrix, so that the equation $A(q)x = 0$ has a solution of the form $x = Cx_0(q)$, where C is an arbitrary constant. Multiplying Eq. 8 by x_0^T obtaining

$$x_0^T M\ddot{q} = x_0^T f$$

expresses $x_0(q)$ in terms of ϕ_1 only. This leads to the single differential equation given by Eq. 7.

In this problem, another approach would be to choose two generalized coordinates with one side constraint. This would lead to two ordinary differential equations involving ϕ_1, ϕ_2 and one Lagrange's multiplier and one side constraint. These can be obtained either directly (as Eq. 7) or by eliminating three of the λ's in Eq. 6.

3. JACOBIANS

If, instead of looking at specific examples as in the last two sections, we consider general formulations, then the following type of situation arises. Suppose that cartesian x-components x_i $(i=1, \cdots, n)$ depend on generalized coordinates q_j $(j=1, \cdots, p)$. Then

$$\dot{x}_i = \sum_{j=1}^{p} \frac{\partial x_i}{\partial q_j} \dot{q}_j$$

$$\ddot{x}_i = \sum_{j=1}^{p} \frac{\partial x_i}{\partial q_j} \ddot{q}_j + \sum_{j=1}^{p} \sum_{k=1}^{p} \frac{\partial^2 x_i}{\partial q_j \partial q_k} \dot{q}_j \dot{q}_k$$

The quantity $d\Phi/dq$ in Eq. 4 has the same form as the Jacobian $[\partial x_i/\partial q_j]$ occurring above, and the MACSYMA command for $\partial\phi/\partial q$ is given in the last line of Appendix A-II.

It is convenient to use the MACSYMA subroutine, or BLOCK to obtain $\partial^2 x_i/\partial q_j \partial q_k$, and this is done in Appendix A-IV.

4. SENSITIVITY ANALYSIS

The objective of this section is to illustrate how MACSYMA deals with some aspects of sensitivity analysis, with particular reference to the paper by Haug and Ehle [7]. Consider a dynamical system described by design variables $b = [b_1, \cdots, b_k]^T$ and a state variable $z(t) = [z_1(t), \cdots, z_k(t)]^T$ which is the solution of an initial value problem of the form

$$\dot{z} = f(z,b), \qquad 0 < t < T$$

$$z(0) = h(b)$$

where $\dot{z} = dz/dt$ and T is determined by the condition

$$\Lambda(T, z(T)) = 0$$

A typical function that may arise in a design formulation is

$$\psi = g(z(T), b) + \int_0^T F(t, z, b)\, dt \tag{10}$$

It is required to find $d\psi/db$, which is a k-vector. This is done by considering an adjoint variable λ satisfying

$$\dot{\lambda} + f_z^T \lambda = F_z^T$$

and then

$$\frac{d\psi}{db} = g_b - \lambda^T(0) h_b + \int_0^T (F_b - \lambda^T f_b)\, dt \tag{11}$$

The procedure outlined above is carried out by the MACSYMA procedure given in Appendix A-V. This also illustrates the MACSYMA solution of linear equations by Laplace transform. Consider an example given in Ref. 7, namely a simple oscillator governed by the equation

$$\begin{array}{ll} \ddot{x} + kx = 0 & 0 < t < \pi/2 \\ x(0) = 0 & \dot{x}(0) = v \end{array} \tag{12}$$

with $\psi = x(\pi/2)$, $b = [k, v]^T = [b_1, b_2]^T$.

The results of the MACSYMA procedure give the first derivative of the functional ψ as

$$\frac{d\psi}{db} = \begin{bmatrix} -\dfrac{b_2}{2b_1^{3/2}} \sin \dfrac{\pi\sqrt{b_1}}{2} + \dfrac{\pi b_2}{4b_1} \cos \dfrac{\pi\sqrt{b_1}}{2} \\ \dfrac{1}{\sqrt{b_1}} \sin \dfrac{\pi\sqrt{b_1}}{2} \end{bmatrix}^T \tag{13}$$

Higher-order sensitivity analysis requires the Jacobian for which a BLOCK MACSYMA command is given in Appendix A-IV as discussed in the last section.

Note: MACSYMA is awkward for differentiating functions having a definite integral; e.g., from Eq. 10 we have

$$\frac{d\psi}{db} = g_z[z_b(T) + \dot{z}(T)(T)_b] + g_b + \int_0^T [F_z z_b + F_b]\, dt + F(T)(T)_b \tag{14}$$

However, MACSYMA does not perform the derivative under the integral sign (a possible dialogue with MACSYMA is given in Appendix A-VI).

5. A SPACECRAFT PROBLEM

Levinson [10] has described in detail an application of the symbolic language FORMAC to formulate the spacecraft problem shown in Figure 3, consisting of two rigid bodies with a common axis of rotation *b*.

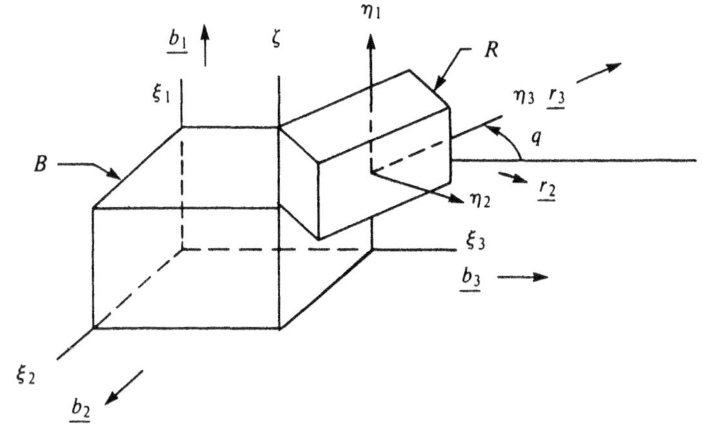

Figure 3.

The equations are given in Ref. 10 in complete detail, and are translated into MACSYMA in Appendix A-VII. To illustrate the point we give typical equations with MACSYMA equivalents:

Equations from Ref. 10		MACSYMA
$\underline{r_2} = \cos q\, \underline{b_2} + \sin q\, \underline{b_3}$	(1)	R[2]:COS(Q)*B[2]+SIN(Q)*B[3];
$\omega^B = u_1 \underline{b_1} + u_2 \underline{b_2} + u_3 \underline{b_3}$	(3)	WB:U[1]*B[1]+U[2]*B[2]+U[3]*B[3];
$u_4 = \dot{q}$		U[4]:DIFF(Q,T);
$\alpha^R = \frac{d}{dt}(\omega^R) + \omega^B \times \omega^R$	(7)	

We could implement this last mathematical expression, (7) [10], by converting the vector product into matrix form but it was simpler to write a BLOCK function to do this, as in Appendix A-VII. The MACSYMA equivalent of (7) is now

```
ALPR:DIFF(WR,T)+CROSS(WB,WR);
```

We discuss only one other correspondence. Equation (27) in Ref. 10 is

$$F_r = \frac{\partial v^{B^*}}{\partial u_r} \cdot (F)_B + \frac{\partial \omega^B}{\partial u_r} \cdot (T)_B \qquad (r = 1, \cdots, 7)$$

becomes in MACSYMA

```
F(R):=DOT(DIFF(VBS,U[R]),FB)

        +DOT(DIFF(WB,U[R]),TB);
```

Dot in the above is defined by another block in Appendix A-VII. The distinction between := and :, as used in this command, is discussed in Appendix B.

The complete set of equations given in Ref. 10 is generated by Appendix A-VII. The reader should compare the corresponding FORMAC program given in Levinson [10].

6. AN EXAMPLE OF MANIPULATION AND SIMPLIFICATION USING MACSYMA

In previous sections we have not found it necessary to use powerful commands in MACSYMA concerned with the simplification of complicated expressions. As an introduction to the manipulation needed in the later sections, we present the following simple dialog:

(C1) F:(X+Y+Z)^2/Y;

(D1) $$\frac{(Z+Y+X)^2}{Y}$$

(C2) EXPAND(%);

(D2) $$\frac{Z^2}{Y} + \frac{2XZ}{Y} + 2Z + Y + \frac{X^2}{Y} + 2X$$

(C3) COMBINE(%);

(D3) $$\frac{Z^2+2XZ+X^2}{Y} + 2Z + Y + 2X$$

(C4) XTHRU(%);

(D4) $$\frac{Z^2+Y(2Z+Y+2X)+2XZ+X^2}{Y}$$

(C5) RATSIMP(%);

(D5) $$\frac{Z^2+(2Y+2X)Z+Y^2+2XY+X^2}{Y}$$

(C6) EV(%,Z=0);

(D6) $$\frac{Y^2+2XY+X^2}{Y}$$

(C7) SUBST(SIN(2*TH),X,%);

(D7) $$\frac{Y^2+2\,SIN(2TH)\,Y+SIN^2(2TH)}{Y}$$

(C8) TRIGEXPAND(%);

(D8) $$\frac{Y^2+4\,COS(TH)\,SIN(TH)\,Y+4\,COS^2(TH)\,SIN^2(TH)}{Y}$$

```
(C9) TRIGREDUCE(%);
```

(D9) $$Y - \frac{COS(4TH)}{2Y} + \frac{1}{2Y} + 2\,SIN(2TH)$$

7. THE FOUR-BAR LINKAGE COUPLER CURVE

The objective of this section is to illustrate how MACSYMA performs algebraic and trigonometric manipulations encountered in the analysis and synthesis of linkages.

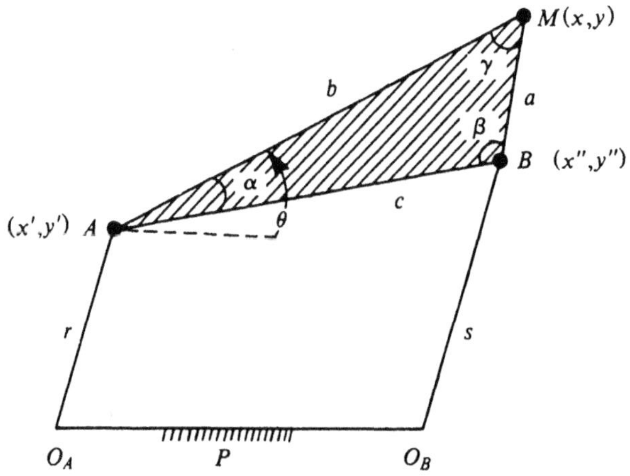

Figure 4.

Consider first, following Hartenberg and Denavit [5] (p. 150), the four-bar linkage AO_AO_BB shown in Figure 4, where the bars lie in a plane and are pin-jointed at A, O_A, O_B, and B, and the positions of O_A and O_B are fixed. MAB is a lamina, so that M is fixed relative to A and B. If BO_B is rotated about O_B, the point M will trace a planar curve, the equation of which we wish to determine.

Using the linkage parameter shown in Figure 4 we have:

$$\begin{aligned} x' &= x - b\cos\theta & x'' &= x - a\cos(\theta+\gamma) \\ y' &= y - b\sin\theta & y'' &= y - a\sin(\theta+\gamma) \\ r^2 - x'^2 - y'^2 &= 0 & s^2 - (x''-p)^2 - y''^2 &= 0 \end{aligned} \tag{15}$$

The required equation for the motion of $M(x,y)$ is obtained by eliminating (x',y'), (x'',y''), and θ from Eq. 15. This is done in Appendix A-VIII. Elimination of (x',y') and (x'',y'') leads to the equation of the form

$$\begin{aligned} N\cos\theta - L\sin\theta &= \phi \\ -P\cos\theta + M\sin\theta &= \psi \end{aligned} \tag{16}$$

where $L = (x-p)\sin\gamma - y\cos\gamma\,, \quad N = (x-p)\cos\gamma + y\sin\gamma$

$$M = y\,, \quad P = -x$$

$$\phi = \frac{1}{B}\,(y^2+x^2-r^2-b^2)\,, \quad \psi = \frac{1}{2A}\left[(x-p)^2+y^2+a^2-s^2\right]$$

Eliminating θ from Eq. 16 gives

$$(P\psi+N\phi)^2 + (M\psi+L\phi)^2 = (LP-MN)^2 \tag{17}$$

This sixth-degree polynomial in (x,y) is called the tricircular sextic. The determinant of Eq. 16 vanishes when $LP-NM = 0$, i.e.,

$$x(x-p) + y^2 - py\cot\gamma = 0 \tag{18}$$

The above equation is called a circle of singular foci.

We next derive a basic relation used to synthesize four-bar linkages, namely the so-called displacement equation which gives the output angle ψ for a given input angle ϕ in Figure 5.

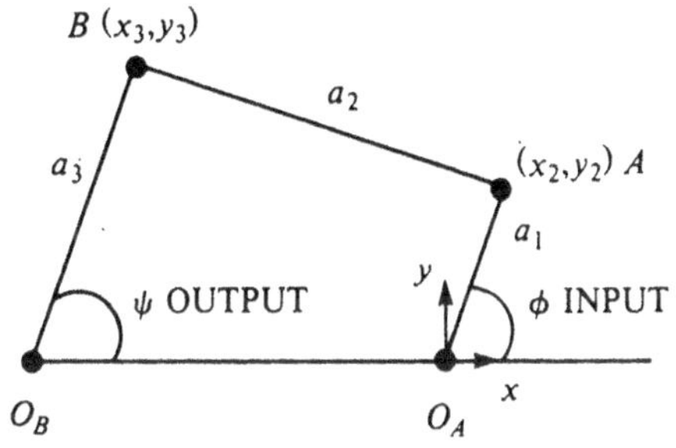

Figure 5.

$$\begin{aligned} &x_2 = a_1\cos\phi\,, \quad y_2 = a_1\sin\phi \\ &x_3 = -a_4 + a_3\cos\psi\,, \quad y_3 = a_3\sin\psi \\ &a_2^2 - (x_2-x_3)^2 - (y_2-y_3)^2 = 0 \end{aligned} \tag{19}$$

Eliminating (x_2,y_2) and (x_3,y_3) from Eq. 19 leads to

$$A\sin\psi + B\cos\psi = C \tag{20}$$

where $A = 2a_1a_3\sin\phi\,, \quad B = 2a_1a_3\cos\phi + 2a_3a_4$

$$C = 2a_1a_4\cos\phi + (a_4^2 + a_3^2 - a_2^2 + a_1^2)$$

To solve Eq. 20 for ψ set

$$\sin\psi = \frac{2\tan\psi/2}{1+\tan^2\psi/2} \qquad \cos\psi = \frac{1-\tan^2\psi/2}{1+\tan^2\psi/2} \tag{21}$$

Substitution into Eq. 20 leads to a quadratic in $\tan\psi$, the solution of which is

$$\tan\frac{\psi}{2}=\frac{1}{(B+C)}\left\{A\pm\sqrt{A^2+B^2-C^2}\right\}$$

The two solutions correspond to the two ways to close the four-bar linkage shown in Figure 6.

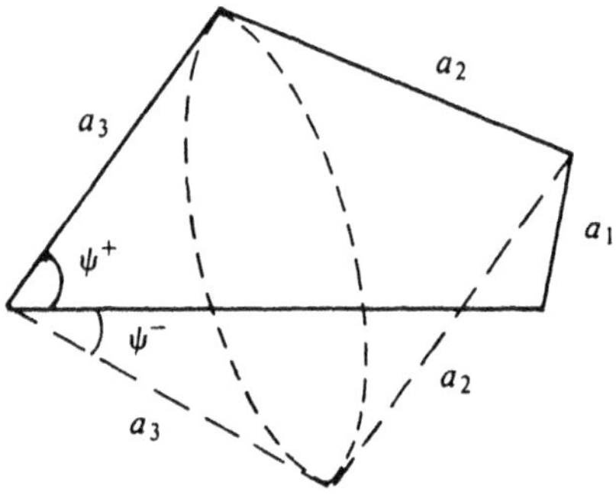

Figure 6.

For the purpose of synthesis, Eq. 20 can be rewritten as:

$$K_1\cos\phi - K_2\cos\psi + K_3 = \cos(\phi-\psi) \tag{22}$$

with $$K_1=\frac{a_4}{a_3},\qquad K_2=\frac{a_4}{a_1},\qquad K_3=a_2^2=(a_1^2+a_3^2+a_4^2-2a_1a_3K_3)/a_1a_3 \tag{23}$$

Hartenberg and Denavit [5] (p. 297) discuss the problem of designing a planar four-bar linkage such that, to three given positions ϕ_1, ϕ_2, and ϕ_3 of the crank O_AA, there correspond three prescribed positions ψ_1, ψ_2, and ψ_3 of the follower O_BB. The form of Eq. 22 is well-suited for this purpose. The solution in this case is obtained by solving the set of three simultaneous equations for K_1, K_2, and K_3 obtained by substituting $\phi=\phi_i$, $\psi=\psi_i$, $i=1,2,3$ in Eq. 22, and then obtaining a_3, a_1, and a_2 from Eq. 23 (a_4 can be selected equal to one). Appendices A-VIII and A-IX give the MACSYMA program to carry out the above procedures.

8. DUAL-NUMBER QUATERNIONS

Next, consider a laborious calculation contained in the appendix to the Yang and Freudenstein paper [19] in connection with the analysis of a spatial four-bar mechanism. We are given

$$A(\hat\theta_1)\sin\hat\theta_4 + B(\hat\theta_1)\cos\hat\theta_4 = C(\hat\theta_1) \tag{24}$$

where $A(\hat{\theta}_1) = \sin\hat{\alpha}_{12}\sin\hat{\alpha}_{34}\sin\hat{\theta}_1$

$$B(\hat{\theta}_1) = -\sin\hat{\alpha}_{34}(\sin\hat{\alpha}_{41}\cos\hat{\alpha}_{12} + \cos\hat{\alpha}_{41}\sin\hat{\alpha}_{12}\cos\hat{\theta}_1) \tag{25}$$

$$C(\hat{\theta}_1) = \cos\hat{\alpha}_{23} - \cos\hat{\alpha}_{34}(\cos\hat{\alpha}_{41}\cos\hat{\alpha}_{12} - \sin\hat{\alpha}_{41}\sin\hat{\alpha}_{12}\cos\hat{\theta}_1)$$

Here

$$\begin{aligned} \hat{\alpha}_{12} &= \alpha_{12} + \epsilon a_{12}, & \hat{\theta}_1 &= \theta_1 + \epsilon s_{11} \\ \hat{\alpha}_{23} &= \alpha_{23} + \epsilon a_{23}, & \hat{\theta}_2 &= \theta_2 + \epsilon s_2 \\ \hat{\alpha}_{34} &= \alpha_{34} + \epsilon a_{34}, & \hat{\theta}_3 &= \theta_3 + \epsilon s_3 \\ \hat{\alpha}_{41} &= \alpha_{41} + \epsilon a_{41}, & \hat{\theta}_4 &= \theta_4 + \epsilon s_4 \end{aligned} \tag{26}$$

where ϵ is a symbol with the property $\epsilon^2 = 0$. This implies that, if $\hat{\theta} = \theta + \epsilon s$, then

$$\sin\hat{\theta} = \sin\theta + \epsilon s\cos\theta, \qquad \cos\hat{\theta} = \cos\theta - \epsilon s\sin\theta.$$

It is then clear that Eq. 24 can be reduced to the form

$$P + \epsilon Q = R + \epsilon S \tag{27}$$

where P, Q, R, and S are independent of ϵ. It is required to find the explicit form of P, Q, R, and S. To calculate this by hand is extremely laborious, but straightforward in MACSYMA. The program is given in Appendix A-X.

Three-dimensional problems in kinematics and dynamics involve laborious calculations involving Euler angles and Euler parameters. (See, for instance, Nikravesh, et al. [13], and Wittenburg [18].) These calculations are easily handled in MACSYMA in a routine fashion. The techniques involved are illustrated in connection with other examples in this paper, so we do not elaborate further.

9. ROBOT ARMS — THE DIRECT PROBLEM

Robot arm manipulators can be considered to consist of a series of links connected together by joints. It is convenient to use cartesian coordinates by assigning a separate coordinate frame to each link. Without going into detail (which can be found in Paul [15], for instance), the relation between the coordinate frames assigned to one link and the next, consisting of translations and rotations, can be derived by a 4×4 matrix of the form

$$A = \begin{bmatrix} n_x & o_x & a_x & p_x \\ n_y & o_y & a_y & p_y \\ n_z & o_z & a_z & p_z \\ 0 & 0 & 0 & 1 \end{bmatrix} \tag{28}$$

where the elements of the top left 3×3 submatrix are the direction cosines representing the rotations and (p_x, p_y, p_z) is the translation.

The position and orientation of the coordinate frame of the end of the manipulator is specified by six parameters (3 translations, 3 rotations). A general manipulator can be designed using six links, each having one degree of freedom. If T_6 denotes the A-matrix corresponding to the end of the manipulator, and A_i $(i=1, \cdots, 6)$ are the A-matrices for the individual links, T_6 is given terms of the A_i by

$$T_6 = A_1 A_2 A_3 A_4 A_5 A_6$$

A typical A-matrix for a link is

$$A_2 = \begin{bmatrix} \cos\theta_2 & 0 & \sin\theta_2 & 0 \\ \sin\theta_2 & 0 & -\cos\theta_2 & 0 \\ 0 & 1 & 0 & d_2 \\ 0 & 0 & 0 & 1 \end{bmatrix}$$

If d_2 is fixed and θ_2 is a variable representing rotation of the second link, this is called a *revolute* joint. If θ_2 is fixed and the translation d_2 is varying, this is called a *prismatic* joint.

The so-called "direct" problem is: given the A_i, find T_6—this is obviously straightforward, although algebraically laborious (see Paul [15], p. 59 and Appendix A-X).

The main computational problem connected with the direct problem is the question of differential motion discussed in Paul [15] (Chapter 4). These are important in connection with dynamic analysis of manipulators, sensitivity analyses, and small adjustments of the end manipulator.

Without going into detail (which can be found in Ref. 15), the computational problem involved is the following. Suppose that the six parameters representing degrees of freedom are denoted by a 6-vector x, small changes in these parameters are denoted by Δx, and the corresponding small changes in the three displacements and three rotational parameters of the end point frame of the manipulator are denoted by the 6-vector A, then we have a relation of the form

$$\Delta A = J \Delta x$$

where the ith column of J is $\begin{bmatrix} d_i \\ \delta_i \end{bmatrix}$; where, for $i=1, \cdots, 6$, and for a *revolute* joint,

$$d_i = \begin{bmatrix} (-n_{ix}p_{iy} + n_{iy}p_{ix}) \\ (-o_{ix}p_{iy} + o_{iy}p_{ix}) \\ (-a_{ix}p_{iy} + a_{iy}p_{ix}) \end{bmatrix} \qquad \delta_i = \begin{bmatrix} n_z \\ o_z \\ a_z \end{bmatrix}$$

and for a *prismatic* joint

$$d_i = \begin{bmatrix} n_z \\ o_z \\ a_z \end{bmatrix} \qquad \delta_i = \begin{bmatrix} 0 \\ 0 \\ 0 \end{bmatrix}$$

The MACSYMA program for the symbolic computation of A and the numerical example in Ref. 15 are given in Appendix A-XI (see Ref. 15, pp. 104-107).

10. ROBOT ARMS — THE INVERSE PROBLEM

The "inverse" problem consists of obtaining the A_i, $i=1, \cdots, 6$, given numerical values of T_6. In theory this can be done from

$$A_1 A_2 A_3 A_4 A_5 A_6 = T_6 \tag{29}$$

which gives 12 equations in 6 unknowns (the 6 degrees of freedom of the links). These equations are redundant, and there are only six independent equations in the six unknowns. However, the equations are highly complicated. The method used in practice is to consider also the following equations which are completely equivalent to Eq. 29:

$$\left.\begin{aligned} A_2 A_3 A_4 A_5 A_6 &= A_1^{-1} T_6 \\ A_3 A_4 A_5 A_6 &= A_2^{-1} A_1^{-1} T_6 \\ A_4 A_5 A_6 &= A_3^{-1} A_2^{-1} A_1^{-1} T_6 \\ A_5 A_6 &= A_4^{-1} A_3^{-1} A_2^{-1} A_1^{-1} T_6 \\ A_6 &= A_5^{-1} A_4^{-1} A_3^{-1} A_2^{-1} A_1^{-1} T_6 \end{aligned}\right\} \tag{30}$$

[In practice the A_i^{-1} are usually easily obtained from the A_i.] Equations 29 and 30 give 72 equations for the 6 unknowns. The procedure is now to pick out the simplest 6 independent equations from the set of 72. The simplest solution occurs when one of the equations involves only one unknown, say x_1, another equation involves x_1 and a second unknown x_2, a third equation involves only x_1, x_2, x_3, and so on. The system can then be solved sequentially. This is the solution with Stanford and the elbow manipulators described by Paul [15].

A more complicated situation occurs in the robot arm discussed by Lumelsky [11], where such a simple sequence of equations cannot be found. Instead, the simplest set is of the form

$$\begin{aligned} x_1 &= f_1(x_3, x_4) \qquad & x_2 &= f_2(x_3, x_4) \\ x_3 &= f_3(x_1, x_2) \qquad & x_4 &= f_4(x_1, x_2) \end{aligned} \tag{31}$$

These can be solved by straightforward iteration.

In Appendix A-XII we give a MACSYMA program for selecting the basic 6 equations from the 72 available. This can be done automatically by using the command FREEOF to print out a dependency table showing which of the variables occur in each of the 72 equations. Equation 31 can be deduced directly from this dependency table.

ACKNOWLEDGMENTS

It is a pleasure to acknowledge the help of Maria Barnum in preparing this difficult manuscript for publication.

MACSYMA work of the Mathlab group is currently supported, in part, by the National Aeronautics and Space Administration under grant NSG 1323, by the Office of Naval Research under grant N00014-77-C-0641, by the U.S. Department of Energy under grant ET-78-C-02-4687, and by the U.S. Air Force under grant F49620-79-C-020. Funding for one of the authors [B.N.] was provided by the Army Research Office.

REFERENCES

1. Bordoni, L. and Golarossi, A., "An Application of Reduce to Industrial Machinery," *ACM Sigsam Bull. No. 58,* 115, 1981, pp. 8-12.
2. Bottema, O. and Roth, B., *Theoretical Kinematics,* North-Holland, 1979.
3. Drinkard, R.D. and Sulinski, N.K., *MACSYMA: A Program for Computer Algebraic Manipulations,* Naval Underwater Systems Center, Newport, Rhode Island, NUSC Tech. Doc. 6401, 10 March 1981.
4. Goldstein, H., *Classical Mechanics,* Addison-Wesley, 1959.
5. Hartenberg, R.S. and Denavit, J., *Kinematic Synthesis of Linkages,* McGraw-Hill, 1964.
6. Haug, E.J. and Arora, J.S., *Applied Optimal Design,* Wiley, 1979.
7. Haug, E.J. and Ehle, P.E., "Second-Order Design Sensitivity Analysis of Mechanical System Dynamics," *Int. J. Num. Meth. Eng.,* Vol. 18, 1982, pp. 1699-1717.
8. Hussain, M.A. and Noble, B., "Application of MACSYMA to Calculations in Applied Mathematics," General Electric Company Report No. 83CRD054, March 1983.
9. Kreuzer, E.J., "Dynamical Analysis of Mechanisms Using Symbolical Equation Manipulation," *Proc. Fifth World Congress on Theory of Machines and Mechanisms,* ASME, 1979.
10. Levinson, D.A., "Equations of Motion for Multiple-Rigid-Body Systems via Symbolic Manipulation," *J. Spacecraft and Rockets,* Vol. 14, 1977, pp. 479-487.
11. Lumelsky, V.J., "Iterative Procedure for Inverse Coordinate Transformation for One Class of Robots," General Electric Company Report No. 82CRD332, February 1983.
12. *MACSYMA: The Reference Manual,* Version 10, 1983, Math Lab Group, Laboratory for Computer Science, MIT. *See also MACSYMA Primer.*

    Information, plus the MACSYMA tape (available to colleges and universities at special rates), is available from: Symbolics Inc., 257 Vassar St., Cambridge, MA 02139.
13. Nikravesh, P.E., Wehage, R.A., and Haug, E.J., *Computer Aided Analysis of Mechanical Systems* (1982), to be published.
14. Paul, B., *Kinematics and Dynamics of Planar Machinery,* Prentice-Hall, 1979.
15. Paul, R.P., *Robot Manipulators,* MIT Press, 1981.
16. Schiehlen, W.O. and Kreuzer, E.J,. "Symbolic Computerized Derivation of Equations of Motion," in *Dynamics of Multibody Systems 1,* IUTAM Conf. (K. Magnus, ed.), Springer-Verlag, 1978, pp. 290-305.

17. Vukobratovic, M. and Potkonjak, V., *Dynamics of Manipulation Robots,* Springer-Verlag, 1982.

18. Wittenberg, J., *Dynamics of Systems of Rigid Bodies,* Teubner, 1977.

19. Yang, A.T. and Freudenstein, F., "Application of Dual-Number Quaternion Algebra to the Analysis of Spatial Mechanisms," *Trans. ASME, J. Appl. Mech.,* 1964, pp. 300-308.

APPENDIX A

APPENDIX A-I

(C2) Q:[Z1,Z2,Z3,Z4,Z5];

(D2) [Z1, Z2, Z3, Z4, Z5]

(C3) DEPENDS(Q,T);

(D3) [Z1(T), Z2(T), Z3(T), Z4(T), Z5(T)]

(C4) MASS:[M1,M2,I,M4,M5];

(D4) [M1, M2, I, M4, M5]

(C5) SPRING:[K1,K2,K3,K4,K5];

(D5) [K1, K2, K3, K4, K5]

(C6) DASH:[CC1,CC2,CC3,CC4,CC5];

(D6) [CC1, CC2, CC3, CC4, CC5]

(C7) DISP:[Z2+L/12*Z3-Z1,Z4-Z2-L/3*Z3,Z5-Z2+2*L/3*Z3,Z4-F1(T),Z5-F2(T)];

(D7) $$\left[\frac{L\,Z3}{12}+Z2-Z1,\ Z4-\frac{L\,Z3}{3}-Z2,\ Z5+\frac{2\,L\,Z3}{3}-Z2,\ Z4-F1(T),\ Z5-F2(T)\right]$$

(C8) DERIVABBREV:TRUE;

(D8) TRUE

(C9) TT:1/2*(MASS.DIFF(Q,T)^2);

(D9) $$\frac{M5(Z5_T)^2+M4(Z4_T)^2+I(Z3_T)^2+M2(Z2_T)^2+M1(Z1_T)^2}{2}$$

(C10) POT:1/2*(SPRING.DISP^2);

(D10) $$\left(K3\left(Z5+\frac{2\,L\,Z3}{3}-Z2\right)^2+K5\left(Z5-F2(T)\right)^2+K2\left(Z4-\frac{L\,Z3}{3}-Z2\right)^2+K4\left(Z4-F1(T)\right)^2+K1\left(\frac{L\,Z3}{12}+Z2-Z1\right)^2\right)/2$$

(C11) DW:-DIFF(DISP,T)*DIFF(DISP);

(D11) $$\Bigg[-\left(\frac{L\,Z3_T}{12}+Z2_T-Z1_T\right)\left(DEL(Z2)+\frac{L\,DEL(Z3)}{12}-DEL(Z1)+\left(\frac{L\,Z3_T}{12}+Z2_T-Z1_T\right)DEL(T)+\frac{Z3\,DEL(L)}{12}\right),$$
$$-\left(Z4_T-\frac{L\,Z3_T}{3}-Z2_T\right)\left(DEL(Z4)-DEL(Z2)-\frac{L\,DEL(Z3)}{3}+\left(Z4_T-\frac{L\,Z3_T}{3}-Z2_T\right)DEL(T)-\frac{Z3\,DEL(L)}{3}\right),$$
$$-\left(Z5_T+\frac{2\,L\,Z3_T}{3}-Z2_T\right)\left(DEL(Z5)-DEL(Z2)+\frac{2\,L\,DEL(Z3)}{3}+\left(Z5_T+\frac{2\,L\,Z3_T}{3}-Z2_T\right)DEL(T)+\frac{2\,Z3\,DEL(L)}{3}\right),$$
$$-\left(Z4_T-F1(T)_T\right)\left(DEL(Z4)+\left(Z4_T-F1(T)_T\right)DEL(T)\right),$$
$$-\left(Z5_T-F2(T)_T\right)\left(DEL(Z5)+\left(Z5_T-F2(T)_T\right)DEL(T)\right)\Bigg]$$

(C12) FOR N THRU 5 DO QQ[N]:RATCOEF(DW,DEL(Q[N]));

(D12) DONE

(C13) FOR N THRU 5 DO LDISPLAY(EQUATION[N]
= DIFF(DIFF(TT,DIFF(Q[N],T)),T)-DIFF(TT,Q[N])+DIFF(POT,Q[N])-DASH . QQ[N]);

(E13) $$EQUATION_1=-K1\left(\frac{L\,Z3}{12}+Z2-Z1\right)+M1\,Z1_{TT}-\frac{CC1\,(L\,Z3_T+12\,Z2_T-12\,Z1_T)}{12}$$

(E14) $$EQUATION_2=\frac{-K3\left(Z5+\frac{2\,L\,Z3}{3}-Z2\right)-2\,K2\left(Z4-\frac{L\,Z3}{3}-Z2\right)+2\,K1\left(\frac{L\,Z3}{12}+Z2-Z1\right)}{2}+M2\,Z2_{TT}-\frac{CC3\,(3\,Z5_T+2\,L\,Z3_T-3\,Z2_T)}{3}-\frac{CC2\,(3\,Z4_T-L\,Z3_T-3\,Z2_T)}{3}+\frac{CC1\,(L\,Z3_T+12\,Z2_T-12\,Z1_T)}{12}$$

(E15) $$EQUATION_3=\frac{\frac{4K3L\left(Z5+\frac{2LZ3}{3}-Z2\right)}{3}-\frac{2K2L\left(Z4-\frac{LZ3}{3}-Z2\right)}{3}+\frac{K1L\left(\frac{LZ3}{12}+Z2-Z1\right)}{6}}{2}+I\,Z3_{TT}-\frac{CC2\,(3\,L\,Z4_T-L^2\,Z3_T-3\,L\,Z2_T)}{9}+\frac{CC3\,(6\,L\,Z5_T+4\,L^2\,Z3_T-6\,L\,Z2_T)}{9}+\frac{CC1\,(L^2\,Z3_T+12\,L\,Z2_T-12\,L\,Z1_T)}{144}$$

(E16) $$EQUATION_4=M4\,Z4_{TT}+\frac{2\,K2\left(Z4-\frac{L\,Z3}{3}-Z2\right)+2\,K4\left(Z4-F1(T)\right)}{2}+\frac{CC2\,(3\,Z4_T-L\,Z3_T-3\,Z2_T)}{3}-CC4\left(-Z4_T+F1(T)_T\right)$$

(E17) $$EQUATION_5=M5\,Z5_{TT}+\frac{2\,K3\left(Z5+\frac{2\,L\,Z3}{3}-Z2\right)+2\,K5\left(Z5-F2(T)\right)}{2}\quad\frac{CC3\,(3\,Z5_T-2\,L\,Z3_T-3\,Z2_T)}{3}-CC5\left(-Z5_T+F2(T)_T\right)$$

```
/* APPENDIX A-I (CONT.)  MACSYMA PROGRAM */
/*.....GENERALIZED COORDINATES...*/
Q:[Z1,Z2,Z3,Z4,Z5];
DEPENDS(Q,T);
/*.....GENERALIZED MASS.........*/
MASS:[M1,M2,I,M4,M5];
/*.....SPRING CONSTANT..........*/
SPRING:[K1,K2,K3,K4,K5];
/*.....DAMPING CONSTANT.........*/
DASH:[CC1,CC2,CC3,CC4,CC5];
/*.....GENERILIZED DISPLACEMENT..*/
DISP:[Z2+L/12*Z3-Z1,Z4-Z2-L/3*Z3,Z5-Z2+2*L/3*Z3,Z4-F1(T),Z5-F2(T)];
DERIVABBREV:TRUE;
/*.....KINETIC AND POTENTIAL ENERGIES..*/
TT:1/2*MASS.DIFF(Q,T)^2;
POT:1/2*SPRING.DISP^2;
/*.......DW.........*/
DW:-DIFF(DISP,T)*DIFF(DISP);
FOR N THRU 5 DO QQ[N]:RATCOEF(DW,DEL(Q[N]));
/*.....EQUATIONS OF MOTION............*/
FOR N:1 THRU 5 DO LDISPLAY (EQUATION[N]=
DIFF(DIFF(TT,DIFF(Q[N],T)),T)
-DIFF(TT,Q[N])
+DIFF(POT,Q[N])
-DASH.QQ[N]);
```

APPENDIX A-II

```
/* CO-ORDINATES */
  Q:[PH1,X2,Y2,PH2,X3];
  DEPENDS(Q,T);
  MASS:[J1,M2,M2,J2,M3];
  CONSTRAINT:[R*SIN(PH1)-Y2+L*SIN(PH2),
     R*COS(PH1)-X2-L*COS(PH2),
     X2+L*COS(PH2)-X3,
     Y2-L*SIN(PH2)];
/* LAGRANGE MULTIPLIERS      */
  LAM:[LAM1,LAM2,LAM3,LAM4];
/* KINETIC ENERGY            */
  TT:(1/2*DIFF(Q,T)^2).MASS;
/* EQUATIONS OF MOTION       */
  FOR I:1 THRU 5 DO LDISPLAY (QQ[I]=DIFF(DIFF(TT,DIFF(Q[I],T)),T)
     -DIFF(TT,Q[I]) +
     LAM.(DIFF(CONSTRAINT,Q[I])));
```

APPENDIX A-III

```
/* ALTERNATE WAY TO DO SLIDER CRANK PROBLEM */
  Q:[PH1,X2,Y2,PH2,X3];
  DEPENDS(Q,T);
  Y2:1/2*R*SIN(PH1);
  X2:R*COS(PH1)+L*SQRT(1-R**2/(4*L**2)*SIN(PH1));
  X3:X2+L*COS(PH2);
  PH2:ASIN(R/(2*L)*SIN(PH1));
  MASS:[J1,M2,M2,J2,M3];
  DERIVABBREV:TRUE;
  TT:(1/2*DIFF(Q,T)^2).MASS;
  TT:EV(TT,DIFF);
  EQUATION:DIFF(DIFF(TT,DIFF(PH1,T)),T)-DIFF(TT,PH1)
```

APPENDIX A-IV

```
/*..............TEST FOR JACOBIAN......HIGHER DERIVATIVES..
  DIMENSION OF A X AND Y ARE P K AND M RESPECTIVELY
    AND SECOND ORDER DERIVATIES ARE FORMED   */
 JAC2(A,P,X,K,Y,M):=BLOCK(
 FOR L:1 THRU P DO (
 DEPENDENT:DETERMINANT(A[L]),
 FOR I:1 THRU K DO (
 VARIABLE1:DETERMINANT(X[I]),
 FOR J:1 THRU M DO (
 VARIABLE2:DETERMINANT(Y[J]),
 PART[L,I,J]:DIFF(DIFF(DEPENDENT,VARIABLE1),VARIABLE2)))));
/*  FOLLOWING IS A SIMPLE EXAMPLE......   */
 A:MATRIX([Z1*B1**2-2*B1*Z3],[B2**2*Z2],[3*B1**2*Z3-B1**3*Z1]);
 X:MATRIX([B1],[B2],[B3]);
 Y:MATRIX([Z1],[Z2],[Z3]);
 JAC2(A,3,X,3,Y,3);
 FOR L:1 THRU 3 DO ( FOR I:1 THRU 3 DO ( FOR J:1 THRU 3 DO
 (LDISPLAY (PART[L,I,J]))));
/*  IN THE FOLLOWING CASE A HAS DIMENSION N BY M
   AND B HAS P BY 1 AND FIRST DERIVATIVES ARE FORMED.  */
 JAC3(A,N,M,B,P):=BLOCK(FOR L:1 THRU P DO (
 VARIABLE:DETERMINANT(B[L]),
 FOR I:1 THRU N DO (
 FOR J:1 THRU M DO (
 DEPENDENT:(A[I,J]),
 PART[L,I,J]:DIFF(DEPENDENT,VARIABLE)))));
 AA:MATRIX([Z1**2*B3,Z2**2*B1,Z3*B1],[Z2**3*B3,Z1*Z2,Z1*B3],
 [Z3**2+Z2,Z2*B2*Z3,Z1*B3**2]);
 JAC3(AA,3,3,X,3);
 FOR L:1 THRU 3 DO (
 FOR I:1 THRU 3 DO (
 FOR J:1 THRU 3 DO LDISPLAY(
 PART[L,I,J])));
```

APPENDIX A-V

```
/*..THIS IS THE FIRST PROBLEM OF HAUG, SEE REF. [7]
  B IS THE DESIGN VARIABLE AND Z IS STATE VARIABLE
  F IS THE RIGHT HAND SIDE OF DIFFERENTIAL EQUATION
  AND H IS B.C  */
H:MATRIX([0],[B2]);
B:MATRIX([B1],[B2]);
F:MATRIX([Z2(T)],[-B1*Z1(T)]);
Z:MATRIX([Z1(T)],[Z2(T)]);
/*..SET UP DIFF EQ AND SOLVE BY LAPLACE TRANSFORM..*/
/*.............................................*/
EQ1:DIFF(Z,T)-F;
/*  INITIAL VALUES ARE GIVEN HERE  */
ATVALUE(Z1(T),T=0,0);
ATVALUE(Z2(T),T=0,B2);
P1:DETERMINANT(EQ1[1]);
P2:DETERMINANT(EQ1[2]);
EQ2:LAPLACE(P1,T,S);
EQ3:LAPLACE(P2,T,S);
LINSOLVE([EQ2,EQ3],[LAPLACE(Z1(T),T,S),
LAPLACE(Z2(T),T,S)]);
PP1:RHS(FIRST(%));
PP2:RHS(LAST(%TH(2)));
/* INVERSE LAPLACE TRANSFORM  */
Z1(T):=ILT(PP1,S,T);
Z2(T):=ILT(PP2,S,T);
Z1(T);
Z1(T):=''%;
Z2(T);
Z2(T):=''%;
JAC(A,N,B,K):=BLOCK([PART],
FOR I:1 THRU N DO (
DEPENDENT:DETERMINANT(A[I]),
FOR J:1 THRU K DO (
VARIABLE:DETERMINANT(B[J]),
PART[I,J]:DIFF(DEPENDENT,VARIABLE))),
GENMATRIX(PART,N,K,1,1));
/* NOW WE SOLVE FOR ADJOINT VARIABLES */
DEPENDS([LAM1,LAM2],[T]);
LAM:MATRIX([LAM1(T)],[LAM2(T)]);

EQ1:DIFF(LAM,T)+TRANSPOSE(JAC(F,2,Z,2)).LAM;
/* SINCE LAPLACE TRANSFORM SOLVES WITH INITIAL VALUES
  ONLY WE ASUUME ALPHA AND BETA AS THE INITIAL VALUES
  AND SOLVE FOR ALPHA AND BETA FROM THE FINAL VALUES
  OF THE SOLUTON */
ATVALUE(LAM1(T),T=0,ALPHA);
ATVALUE(LAM2(T),T=0,BETA);
P1:DETERMINANT(EQ1[1]);
P2:DETERMINANT(EQ1[2]);
EQ2:LAPLACE(P1,T,S);
EQ3:LAPLACE(P2,T,S);
LINSOLVE([EQ2,EQ3],[LAPLACE(LAM1(T),T,S),
LAPLACE(LAM2(T),T,S)]);
PP1:RHS(FIRST(%));
PP2:RHS(LAST(%TH(2)));
LAM1(T):=ILT(PP1,S,T);
LAM2(T):=ILT(PP2,S,T);
/*  THIS THE SOLUTION OF THE ADJOINT VARIABLES  */
LAM1(T);
LAM1(T):=''%;
LAM2(T);
LAM2(T):=''%;
LINSOLVE([LAM1(%PI/2)+1,LAM2(%PI/2)],[ALPHA,BETA]),GLOBALSOLVE:TRUE;
/* ALPHA BETA HAVE TO BE SUBSTITUTED IN THE ABOVE SOLUTION */
```

APPENDIX A-VI

```
/*  DESIGN SENSITIVITY ANALYSIS  */
DEPENDS([T,Z],[B]);
DEPENDS([ZZ],[T,B]);
DEPENDS([G],[ZZ,B]);
PSI:G+INTEGRATE(FF(Y,Z,B),Y,0,T);
DIFF(PSI,B);
EV(%,DIFF);
/*  NOTE: THE LAST COMMAND GIVES UNDESIRABLE RESULTS.
   (SEE REF. [7]) THE PROBLEM CAN POSSIBLY BE HANDLED
   BY GRADEF COMMAND  */
```

APPENDIX A-VII

```
/*........UNIT VECTORS ARE B1 B2 B3......SEE LEVINSON [10] */
/*........DEFINE DOT AND CROSS PRODUCTS....*/
DOT(V1,V2):=BLOCK([P,PP],
FOR I:1 THRU 3 DO P[I]:RATCOEFF(V1,B[I]),
FOR I:1 THRU 3 DO PP[I]:RATCOEFF(V2,B[I]),
P[4]:SUM(P[I]*PP[I],I,1,3),
RETURN(P[4]))$
CROSS(V1,V2):=BLOCK([P,PP,PPP],
FOR I:1 THRU 3 DO P[I]:RATCOEFF(V1,B[I]),
FOR I:1 THRU 3 DO PP[I]:RATCOEFF(V2,B[I]),
PPP[1]:(P[2]*PP[3]-P[3]*PP[2]),
PPP[2]:(-P[1]*PP[3]+P[3]*PP[1]),
PPP[3]:(P[1]*PP[2]-P[2]*PP[1]),
PPP[4]:B[1]*PPP[1]+B[2]*PPP[2]+B[3]*PPP[3],
RETURN(PPP[4]))$
/*....NOW WE INPUT EQUATIONS FROM LEVINSON'S PAPER  */
DEPENDS(U,T);
DEPENDS(Q,T);
R[2]:COS(Q)*B[2]+SIN(Q)*B[3];
R[3]:-SIN(Q)*B[2]+COS(Q)*B[3];
WB:U[1]*B[1]+U[2]*B[2]+U[3]*B[3];
DERIVABBREV:TRUE;
U[4]:DIFF(Q,T);
WR:(U[1]+U[4])*B[1]+U[2]*B[2]+U[3]*B[3];
ALPB:DIFF(U[1],T)*B[1]+DIFF(U[2],T)*B[2]+DIFF(U[3],T)*B[3];
ALPR:DIFF(WR,T)+CROSS(WB,WR);
PPBS:B1*B[1]+B2*B[2]+B3*B[3];
PPRS:R1*B[1]+R2*R[2]+R3*R[3];
PRSBS:PPBS-PPRS;
VBS:U[5]*B[1]+U[6]*B[2]+U[7]*B[3];
VRS:VBS +DIFF(PRSBS,T)+CROSS(WB,PRSBS);
ABS:DIFF(VBS,T)+CROSS(WB,VBS);
ARS:DIFF(VRS,T)+CROSS(WB,VRS);
IBBSWB:BET1*B[1]*DOT(B[1],WB)+BET2*B[2]*DOT(B[2],WB)+BET3*B[3]*DOT(B[3],WB);
IRRSWR:RHO1*B[1]*DOT(B[1],WR)+RHO2*B[2]*DOT(B[2],WR)+RHO3*B[3]*DOT(B[3],WR);
IBBSALPB:BET1*B[1]*DOT(B[1],ALPB)+BET2*B[2]*DOT(B[2],ALPB)
       +BET3*B[3]*DOT(B[3],ALPB);
IRRSALPR:RHO1*B[1]*DOT(B[1],ALPR)+RHO2*B[2]*DOT(B[2],ALPR)
       +RHO3*B[3]*DOT(B[3],ALPR);
FSB:-MB*ABS;
FSR:-MR*ARS;
TSB:CROSS(IBBSWB,WB)-IBBSALPB;
TSR:CROSS(IRRSWR,WR)-IRRSALPR;
FB:F1*B[1]+F2*B[2]+F3*B[3];
TB:T1*B[1]+T2*B[2]+T3*B[3];
F[R]:=DOT(DIFF(VBS,U[R]),FB)+DOT(DIFF(WB,U[R]),TB);
FS[R]:=DOT(DIFF(VBS,U[R]),FSB)+DOT(DIFF(VRS,U[R]),FSR)
 +DOT(DIFF(WB,U[R]),TSB)+DOT(DIFF(WR,U[R]),TSR);
EQ[R]:=F[R]+FS[R];
EQ[1];
FOR I:1 THRU 7 DO LDISPLAY ( X[1,I]:RATCOEFF(EQ[1],DIFF(U[I],T)));
```

APPENDIX A-VIII

```
XP:X-B*COS(TH);
YP:Y-B*SIN(TH);
XPP:X-A*COS(TH+GAM);
YPP:Y-A*SIN(TH+GAM);
EQ1:R**2-XP**2-YP**2;
EQ2:S**2-(XPP-P)**2-YPP**2;
EQ1:EXPAND(EQ1);
EQ1:RATSUBST(1-SIN(TH)^2,COS(TH)^2,%);
EQ1:EQ1/(2*B);
EQ2:EXPAND(EQ2);
EQ2:TRIGEXPAND(EQ2);
EQ2:RATSUBST(1-COS(TH)^2,SIN(TH)^2,EQ2);
EQ2:RATSUBST(1-COS(GAM)^2,SIN(GAM)^2,%);
EQ2:EQ2/(2*A);
LL:RATCOEF(EQ2,-SIN(TH));
NN:RATCOEF(EQ2,COS(TH));
MM:RATCOEF(EQ1,SIN(TH));
PP:RATCOEF(EQ1,-COS(TH));
PHPH:EQ1-MM*SIN(TH)+PP*COS(TH);
PSIPSI:EQ2+LL*SIN(TH)-NN*COS(TH);
PSIPSI:RATSIMP(%);

/*  APPENDIX A-VIII (CONT.)  */

EQ11:-P*COS(TH)+M*SIN(TH)-PH;
EQ22:N*COS(TH)-L*SIN(TH)-PSI;
SET:LINSOLVE([EQ11,EQ22],[COS(TH),SIN(TH)]);
CTH:RHS(PART(EV(SET),1));
STH:RHS(PART(EV(SET),2));
STH^2+CTH^2;
%-1;
XTHRU(%);
NUM(%);
```

APPENDIX A-IX

```
/*..........SNTHESIS BY ANALYTIC METHODS........*/
X2:A1*COS(PH);
Y2:A1*SIN(PH);
X3:-A4+A3*COS(PSI);
Y3:A3*SIN(PSI);
F1:A2^2-(X2-X3)^2-(Y2-Y3)^2;
EXPAND(%);
F11:%;
AA:RATCOEF(F11,SIN(PSI));
BB:RATCOEF(F11,COS(PSI));
F11-AA*SIN(PSI)-BB*COS(PSI);
RATSUBST(1-SIN(PH)^2,COS(PH)^2,%);
RATSUBST(1-SIN(PSI)^2,COS(PSI)^2,%);
T11:%;
T11:T11/(2*A1*A3);
/*.........SOLUTION BY DENAVIT METHOD...........*/
F1:A*SIN(PSI)+B*COS(PSI)-C;
RATSUBST(2*TAN(X)/(1+TAN(X)**2),SIN(PSI),%);
RATSUBST((1-TAN(X)**2)/(1+TAN(X)**2),COS(PSI),%);
RATSIMP(%);
SOLVE(%,TAN(X));

/*........SOLUTION BY HARTENBERG AND DENAVIT....*/
FF(PH,PSI):=K1*COS(PH)-K2*COS(PSI)+K3-COS(PH-PSI);
/*.............NOW A NUMERICAL EXAMPLE..........*/
/*........CHEBYCHEV SPACING IS GIVEN BY
XK=A+H*COS(2K-1)*PI/2N
    WHERE A=MEAN H=HALF THE INTERVAL OF X......*/
KEEPFLOAT:TRUE;
X(K):=3/2+1/2*COS((2*K-1)*%PI/6);
X1:X(3),NUMER;
X2:X(2);
X3:X(1),NUMER;
Y1:LOG(X1)/LOG(10),NUMER;
Y2:LOG(X2)/LOG(10),NUMER;
Y3:LOG(X3)/LOG(10),NUMER;
DELPH:60/180*%PI;
DELPSI:60/180*%PI;
PH1:0;
PSI1:0;
YF:LOG(2)/LOG(10),NUMER;

PH2:(X2-X1)*DELPH,NUMER;
PH3:(X3-X1)*DELPH,NUMER;
PSI2:(Y2-Y1)/YF*DELPSI,NUMER;
PSI3:(Y3-Y1)/YF*DELPSI,NUMER;
A4:1;
EQ1:FF(PH1,PSI1);
EQ2:FF(PH2,PSI2);
EQ3:FF(PH3,PSI3);
LINSOLVE([EQ1,EQ2,EQ3],[K1,K2,K3]),GLOBALSOLVE:TRUE;
A3:A4/K1;
A1:A4/K2;
A2:SQRT(A1**2+A3**2+A4**2-2*A1*A3*K3);
```

APPENDIX A-X

```
/*..ALGEBRA FOR QUATERNIONS FROM YANG'S PAPER..*/
NNPRED(N):=IS(N>=2);
NNPRED(2);
NNPRED(3);
MATCHDECLARE(NN,NNPRED);
TELLSIMPAFTER(EP^NN,0);
AL12H:AL12+EP*A12;
AL23H:AL23+EP*A23;
AL34H:AL34+EP*A34;
AL41H:AL41+EP*A41;
 TH1H:TH1+EP*S11;
 TH2H:TH2+EP*S2;
 TH3H:TH3+EP*S3;
 TH4H:TH4+EP*S4;
SAL12H: EXPAND(TAYLOR(SIN(AL12H),EP,0,1));
SAL23H: EXPAND(TAYLOR(SIN(AL23H),EP,0,1));
SAL34H: EXPAND(TAYLOR(SIN(AL34H),EP,0,1));
SAL41H: EXPAND(TAYLOR(SIN(AL41H),EP,0,1));
STH1H:  EXPAND(TAYLOR(SIN(TH1H),EP,0,1));
STH2H:  EXPAND(TAYLOR(SIN(TH2H),EP,0,1));
STH3H:  EXPAND(TAYLOR(SIN(TH3H),EP,0,1));
STH4H:  EXPAND(TAYLOR(SIN(TH4H),EP,0,1));
CAL12H:EXPAND(TAYLOR(COS(AL12H),EP,0,1));
CAL23H:EXPAND(TAYLOR(COS(AL23H),EP,0,1));
CAL34H:EXPAND(TAYLOR(COS(AL34H),EP,0,1));
CAL41H:EXPAND(TAYLOR(COS(AL41H),EP,0,1));
CTH1H: EXPAND(TAYLOR(COS(TH1H),EP,0,1));
CTH2H: EXPAND(TAYLOR(COS(TH2H),EP,0,1));
CTH3H: EXPAND(TAYLOR(COS(TH3H),EP,0,1));
CTH4H: EXPAND(TAYLOR(COS(TH4H),EP,0,1));
AATH1H:SAL12H*SAL34H*STH1H;
BBTH1H:-SAL34H*(SAL41H*CAL12H+CAL41H*SAL12H*CTH1H);
CCTH1H:CAL23H-CAL34H*(CAL41H*CAL12H-SAL41H*SAL12H*CTH1H);
EQ1:AATH1H*STH4H+BBTH1H*CTH4H-CCTH1H;
PRIMARY:EV(EQ1,EP=0);
DUAL:RATCOEFF(EQ1,EP);
A:RATCOEFF(PRIMARY,SIN(TH4));
B:RATCOEFF(PRIMARY,COS(TH4));
C:EXPAND(PRIMARY-A*SIN(TH4)-B*COS(TH4));
DUAL1:DUAL-S4*(A*COS(TH4)-B*SIN(TH4));
A0:RATCOEFF(DUAL1,SIN(TH4));
B0:RATCOEFF(DUAL1,COS(TH4));
CC0:EXPAND(DUAL1-A0*SIN(TH4)-B0*COS(TH4));
CC0:RATSIMP(CC0);
```

APPENDIX A-XI

```
/*....THIS PROGRAM SETS UP THE COMPLETE MATRIX
    EQATION FOR ROBOT....FOR DIFFERENTIAL MOTION....*/
/*....modified for Stanford manipulator....*/
/*CTH[I]:=COS(TH[I]);
STH[I]:=SIN(TH[I]); */
GCPRINT:FALSE;
CAL[I]:=COS(AL[I]);
SAL[I]:=SIN(AL[I]);
AL[1]:AL[4]:-%PI/2;
AL[2]:AL[5]:%PI/2;
AL[3]:AL[6]:0;
AA[1]:AA[3]:AA[4]:AA[5]:AA[6]:0;
AA[2]:0;
DD[1]:DD[4]:DD[5]:DD[6]:0;
CTH[3]:1;
STH[3]:0;
A[I]:=MATRIX([CTH[I],-STH[I]*CAL[I],STH[I]*SAL[I],AA[I]*CTH[I]],
[STH[I],CTH[I]*CAL[I],-CTH[I]*SAL[I],AA[I]*STH[I]],
[0,SAL[I],CAL[I],DD[I]],
[0,0,0,1]);
A[1];
RATSUBST(1-(STH[1])**2,(CTH[1])**2,%);
%^^-1;
IA[1]:%$
RATSUBST(1-(STH[1])**2,(CTH[1])**2,%);
IA[1]:%$
A[2];
%^^-1;
RATSUBST(1-(STH[2])**2,(CTH[2])**2,%);
IA[2]:%$
A[3];
%^^-1;
RATSUBST(1-(STH[3])**2,(CTH[3])**2,%);
IA[3]:%$
A[4];
%^^-1;
RATSUBST(1-(STH[4])**2,(CTH[4])**2,%);
IA[4]:%$
A[5];
%^^-1;
RATSUBST(1-(STH[5])**2,(CTH[5])**2,%);
```

```
IA[5]:%$
A[6];
%^^-1;
RATSUBST(1-(STH[6])**2,(CTH[6])**2,%);
IA[6]:%;
T56:A[6];
T46:A[5].A[6];
T36:A[4].A[5].A[6];
T26:A[3].A[4].A[5].A[6];
T16:A[2].A[3].A[4].A[5].A[6];
T6:A[1].A[2].A[3].A[4].A[5].A[6]$
NX:T6[1,1]$
NY:T6[2,1]$
NZ:T6[3,1]$
OX:T6[1,2]$
OY:T6[2,2]$
OZ:T6[2,3]$
AX:T6[1,3]$
AY:T6[2,3]$
AZ:T6[3,3]$
PX:T6[1,4]$
PY:T6[2,4]$
PZ:T6[3,4]$
TT6:MATRIX([NNX,OOX,AAX,PPX],
[NNY,OOY,AAY,PPY],
[NNZ,OOZ,AAZ,PPZ],
[0,0,0,1]);
IA1T6:IA[1].TT6;
IA2T6:IA[2].%;
IA3T6:IA[3].%;
IA4T6:IA[4].%$
IA5T6:IA[5].%$
IA6T6:IA[6].%$

/*....differential relations...this may be used
  for obtaining the sensitivity analysis...we
  follow the agorithm provided by R.P. Paul
  PAGE no.103....REVOLUTE=TDR PRISMATIC=TDP..*/

TDR(MAT):=BLOCK([NX,NY,NZ,PX,PY,PZ,OX,OY,OZ,AX,AY,AZ],
NX:MAT[1,1],NY:MAT[2,1],NZ:MAT[3,1],
OX:MAT[1,2],OY:MAT[2,2],OZ:MAT[3,2],
AX:MAT[1,3],AY:MAT[2,3],AZ:MAT[3,3],
PX:MAT[1,4],PY:MAT[2,4],PZ:MAT[3,4],
TRANSPOSE(MATRIX([-NX*PY+NY*PX,-OX*PY+OY*PX,-AX*PY+AY*PX,NZ,OZ,AZ])));
TDP(MAT):=BLOCK([NX,NY,NZ,PX,PY,PZ,OX,OY,OZ,AX,AY,AZ],
NX:MAT[1,1],NY:MAT[2,1],NZ:MAT[3,1],
OX:MAT[1,2],OY:MAT[2,2],OZ:MAT[3,2],
AX:MAT[1,3],AY:MAT[2,3],AZ:MAT[3,3],
PX:MAT[1,4],PY:MAT[2,4],PZ:MAT[3,4],
TRANSPOSE(MATRIX([NZ,OZ,AZ,0,0,0])));
/*....NOW WE SET UP COMPLETE DIFFERENTIAL MATRIX....*/
COL[1]:TDR(T6)
COL[2]:TDR(T16);
COL[3]:TDP(T26);
COL[4]:TDR(T36);
COL[5]:TDR(T46);
COL[6]:TDR(T56);
FOR J:1 THRU 6 DO (FOR I:1 THRU 6 DO (DIFFARRAY[I,J]:COL[J][I]));
JACOBIAN:GENMATRIX(DIFFARRAY,6,6);
/*....NUMERICAL EXAMPLE PAGE 107 ROBERT PAUL....*/
TH[1]:TH[4]:0;
TH[2]:TH[5]:TH[6]:%PI/2;
TH[3]:0;
DD[3]:20;

FOR I:1 THRU 6 DO (CTH[I]:COS(TH[I]),STH[I]:SIN(TH[I]));
JAC:EV(JACOBIAN,NUMER);
JAC:EV(%,NUMER);
DQ:TRANSPOSE(MATRIX([0.1,-0.1,2.0,0.1,0.1,0.1]));
JAC.DQ;
```

APPENDIX A-XII

```
/*   EXAMPLE OF ROBOT CONSIDERD BY LUMELSKY   */

CTH[I]:=COS(TH[I]);
STH[I]:=SIN(TH[I]);
GCPRINT:FALSE;
CAL[I]:=COS(AL[I]);
SAL[I]:=SIN(AL[I]);
AL[1]:AL[2]:AL[4]:%PI/2;
AL[3]:AL[6]:0;
AL[5]:-%PI/2;
AA[1]:AA[3]:AA[4]:AA[5]:AA[6]:0;
AA[2]:A2;
DD[1]:DD[2]:DD[4]:0;
CTH[3]:1;
STH[3]:0;
A[I]:=MATRIX([CTH[I],-STH[I]*CAL[I],STH[I]*SAL[I],AA[I]*CTH[I]],
[STH[I],CTH[I]*CAL[I],-CTH[I]*SAL[I],AA[I]*STH[I]],
[0,SAL[I],CAL[I],DD[I]],[0,0,0,1]);
A[1];
RATSUBST(1-STH[1]^2,CTH[1]^2,%);
%^^(-1);
A[1]:%$
RATSUBST(1-STH[1]^2,CTH[1]^2,%);
A[1]:%$
A[2];
%^^(-1),
RATSUBST(1-STH[2]^2,CTH[2]^2,%);
A[2]:%$
A[3];
%^^(-1);
RATSUBST(1-STH[3]^2,CTH[3]^2,%);
A[3]:%$
A[4];
%^^(-1);
RATSUBST(1-STH[4]^2,CTH[4]^2,%);
A[4]:%$
A[5];
%^^(-1);
RATSUBST(1-STH[5]^2,CTH[5]^2,%);
A[5]:%$
A[6];
%^^(-1);
RATSUBST(1-STH[6]^2,CTH[6]^2,%);
A[6]:%;
T56:A[6];
T46:A[5] . A[6];
T36:A[4] . A[5] . A[6];
T26:A[3] . A[4] . A[5] . A[6];
T16:A[2] . A[3] . A[4] . A[5] . A[6];
T6:A[1] . A[2] . A[3] . A[4] . A[5] . A[6]$
NX:T6[1,1];
NY:T6[2,1];
NZ:T6[3,1];
OX:T6[1,2];
OY:T6[2,2];
OZ:T6[2,3];
AX:T6[1,3];
AY:T6[2,3];
AZ:T6[3,3];
PX:T6[1,4];
PY:T6[2,4];
PZ:T6[3,4];
TT6:MATRIX([NNX,OOX,AAX,PPX],[NNY,OOY,AAY,PPY],[NNZ,OOZ,AAZ,PPZ],[0,0,0,1])
A1T6:IA[1] . TT6;
A2T6:IA[2] . %;
A3T6:IA[3] . %;
A4T6:IA[4] . %;
A5T6:IA[5] . %;
A6T6:IA[6] . %;
EQ1:TT6-T6$
EQ2:IA1T6-T16$
EQ3:IA2T6-T26$
EQ4:IA3T6-T36$
EQ5:IA4T6-T46$
EQ6:IA5T6-T56$
TABLE(MAT,VAR):=BLOCK([EQ],FOR I THRU 4 DO (FOR J THRU 4 DO EQ[I,J]:0),
FOR I THRU 4 DO (FOR J THRU 4 DO (FOR L THRU 6 DO
(IF FREEOF(VAR[L],MAT[I,J]) = FALSE THEN EQ[I,J]:EQ[I,J]+T[L]
 ELSE FALSE))),GENMATRIX(EQ,4,4));
VAR[1]:TH[1];
VAR[2]:TH[2];
VAR[3]:DD[3];
VAR[4]:TH[4];
VAR[5]:TH[5];
VAR[6]:TH[6];
TABLE(EQ1,VAR);
TABLE(EQ2,VAR);
TABLE(EQ3,VAR);
TABLE(EQ4,VAR);
TABLE(EQ5,VAR);
TABLE(EQ6,VAR);
```

APPENDIX A-XII (CONT.)

FOLLOWING IS A PARTIAL OUTPUT FROM ABOVE PROGRAM:
T1 T2 ETC REPRESENTS PRESENCE OF VARIABLE 1 2 ETC.

TABLE(EQ2,VAR);

$$\text{COL 1} = \begin{bmatrix} T_6+T_5+T_4+T_2+T_1 \\ T_6+T_5+T_4+T_2 \\ T_6+T_5+T_4+T_1 \\ 0 \end{bmatrix} \qquad \text{COL 2} = \begin{bmatrix} T_6+T_5+T_4+T_2+T_1 \\ T_6+T_5+T_4+T_2 \\ T_6+T_5+T_4+T_1 \\ 0 \end{bmatrix}$$

$$\text{COL 3} = \begin{bmatrix} T_5-T_4+T_2+T_1 \\ T_5+T_4+T_2 \\ T_5+T_4+T_1 \\ 0 \end{bmatrix} \qquad \text{COL 4} = \begin{bmatrix} T_5+T_4+T_3+T_2+T_1 \\ T_5+T_4+T_3+T_2 \\ T_5+T_4+T_1 \\ 0 \end{bmatrix}$$

TABLE(EQ3,VAR);

$$COL\,1 = \begin{bmatrix} T_6+T_5+T_4+T_2+T_1 \\ T_6+T_5+T_4+T_1 \\ T_6+T_5+T_2+T_1 \\ 0 \end{bmatrix} \quad COL\,2 = \begin{bmatrix} T_6+T_5+T_4+T_2+T_1 \\ T_6+T_5+T_4+T_1 \\ T_6+T_5+T_2+T_1 \\ 0 \end{bmatrix}$$

$$COL\,3 = \begin{bmatrix} T_5+T_4+T_2+T_1 \\ T_5+T_4+T_1 \\ T_5+T_2+T_1 \\ 0 \end{bmatrix} \quad COL\,4 = \begin{bmatrix} T_5+T_4+T_2+T_1 \\ T_5+T_4+T_1 \\ T_5+T_3+T_2+T_1 \\ 0 \end{bmatrix}$$

TABLE(EQ4,VAR);

$$COL\,1 = \begin{bmatrix} T_6+T_5+T_4+T_2+T_1 \\ T_6+T_5+T_4+T_1 \\ T_6+T_5+T_2+T_1 \\ 0 \end{bmatrix} \quad COL\,2 = \begin{bmatrix} T_6+T_5+T_4+T_2+T_1 \\ T_6+T_5+T_4+T_1 \\ T_6+T_5+T_2+T_1 \\ 0 \end{bmatrix}$$

$$COL\,3 = \begin{bmatrix} T_5+T_4+T_2+T_1 \\ T_5+T_4+T_1 \\ T_5+T_2+T_1 \\ 0 \end{bmatrix} \quad COL\,4 = \begin{bmatrix} T_5+T_4+T_2+T_1 \\ T_5+T_4+T_1 \\ T_5+T_3+T_2+T_1 \\ 0 \end{bmatrix}$$

TABLE(EQ5,VAR);

$$COL\,1 = \begin{bmatrix} T_6+T_5+T_4+T_2+T_1 \\ T_6+T_5+T_2+T_1 \\ T_6+T_4+T_2+T_1 \\ 0 \end{bmatrix} \quad COL\,2 = \begin{bmatrix} T_6+T_5+T_4+T_2+T_1 \\ T_6+T_5+T_2+T_1 \\ T_6+T_4+T_2+T_1 \\ 0 \end{bmatrix}$$

$$COL\,3 = \begin{bmatrix} T_5+T_4+T_2+T_1 \\ T_5+T_2+T_1 \\ T_4+T_2+T_1 \\ 0 \end{bmatrix} \quad COL\,4 = \begin{bmatrix} T_5+T_4+T_2+T_1 \\ T_5+T_3+T_2+T_1 \\ T_4+T_2+T_1 \\ 0 \end{bmatrix}$$

TABLE(EQ6,VAR);

$$COL\,1 = \begin{bmatrix} T_6+T_5+T_4+T_2+T_1 \\ T_6+T_4+T_2+T_1 \\ T_5+T_4+T_2+T_1 \\ 0 \end{bmatrix} \quad COL\,2 = \begin{bmatrix} T_6+T_5+T_4+T_2+T_1 \\ T_6+T_4+T_2+T_1 \\ T_5+T_4+T_2+T_1 \\ 0 \end{bmatrix}$$

$$COL\,3 = \begin{bmatrix} T_5+T_4+T_2+T_1 \\ T_4+T_2+T_1 \\ T_5+T_4+T_2+T_1 \\ 0 \end{bmatrix} \quad COL\,4 = \begin{bmatrix} T_5+T_4+T_3+T_2+T_1 \\ T_4+T_2+T_1 \\ T_5+T_4+T_3+T_2+T_1 \\ 0 \end{bmatrix}$$

APPENDIX B

SOME REMARKS ON MACSYMA COMMANDS

We assume that the reader is familiar with the introduction given in the MACSYMA Primer, which is an introduction for beginners (see Refs. 8 and 12).

We will use *EXPR* to denote any symbolic expression such as $X+Y$, SIN(X), etc.

```
F: EXPR;
```

assigns *EXPR* to the variable F. (In the reference manual, F would be called an atomic variable.) Note that $F(X)$ has no meaning in this context. However, if we write

```
F(X):= EXPR
```

we are now defining a function $F(X)$. For example:

```
F(X):= SIN(X);
F(Z);                  (Machine prints SIN(Z))
```

This can also be achieved by the LAMBDA notation

```
F. LAMBDA([X],SIN(X));
```

We *can* now use $F(Z)$. For example:

```
F(Z);                  (Machine prints SIN(Z))
```

One advantage of this procedure is that F can be an argument to another function, for example as in SIMPSON (see Ref. 8).

An abbreviated, partial list of the commands that have been used in this paper follows. For details, see the MACSYMA Manual.

DEPENDS([R,P],[RHO]); $\equiv$ R and P are functions of ρ.

DEPENDENCIES(R(RHO)); $\equiv$ $R = R(\rho)$, as in the last command.

DIFF(SIN(X),X); $\equiv$ $d\sin(x)/dx$

EV(EXPR,X=0); $\equiv$ Evaluate *EXPR* with $x=0$. EV is a powerful command in MACSYMA and takes multiple arguments. See Manual.

GRADEF(R,RHO,P/R); $\equiv$ set $\dfrac{\partial R}{\partial \rho} = P/R$

INTEGRATE(SIN(X),X); $\equiv$ $\int \sin(x)\,dx$

LDISPLAY $\equiv$ Display with equation numbers.

LINSOLVE([EQ1,EQ2],[X,Y]);	≡	Solve set of linear equations $Eq_1 = 0$, $Eq_2 = 0$, for x and y.
GLOBALSOLVE:TRUE;	≡	Assigns the values to the variables obtained by LINSOLVE command.
MAP(FACTOR,EXPR1);	≡	Factors each part of *EXPR*1 separately.
RATCOEFF(EXPR,X^I)	≡	Obtain the coefficient of x^i in *EXPR*.
RATCOEFF(EXPR,X,I)	≡	Same as above. If the third argument is not specified, as in the last command, it is taken as 1 by default.
RATSIMP(EXPR);	≡	Obtain rational simplification of expression *EXPR*.
SUBST(0,X,EXPR);	≡	Substitute 0 for x in the expression *EXPR*. (See also RATSUBST in manual.)
SUM(P[I]X^I,I,0,M);	≡	$\sum_{i=0}^{M} P_i x^i$
TAYLOR(SIN(X),X,0,5);	≡	Obtain Taylor expansion of $\sin x$ around $x = 0$ up to the fifth power. Can also be used for multivariate functions.
TRIGEXPAND(EXPR1);	≡	Expand the trigonometric functions in the expression *EXPR*1.

FOR I:1 STEP 1 THRU N DO (ANY MACSYMA COMMAND); (Involving set of I's)

≡ This is a DO loop for 1 to n in steps of 1. Note that the default for starting is 1 and that the default step is 1; FOR I THRU N DO(...) will accomplish the same task.

Part 3

NUMERICAL METHODS IN DYNAMICS

NUMERICAL METHODS FOR SYSTEMS OF INITIAL VALUE PROBLEMS THE STATE OF THE ART

W.H. Enright
Dept. of Computer Science
University of Toronto
Toronto, M5S 1A4

Abstract. State of the art software for initial value problems will be surveyed and recent developments and the current and future implications of these developments will be identified. Software libraries such as IMSL, NAG and SLATEC are now available at most major computing sites and, with a little guidance, the initial value software that is provided can prove invaluable to practitioners. Using routines from these libraries as examples we will identify the important problem characteristics and details of the programming environment that will determine the most appropriate method. In particular characteristics such as the form of the differential equations, whether it is stiff, the size of the system and the accuracy desired will be shown to be important. The extent to which numerical integration methods can be considered as interchangeable modules or 'black boxes' will also be discussed and future developments which should make this approach more feasible will be identified. On the other hand, the importance of special purpose methods which exploit the special structure of particular classes of problems will be acknowledged and examples of systems where this can be critical will be presented.

1. INTRODUCTION

In this paper we will attempt to survey the current state of the art regarding the numerical solution of systems of initial value problems. We will begin in the next section by investigating exactly what it is that can now be expected of a numerical method. In doing this we will distinguish a method from a formula and a working code from general purpose numerical software. Several general purpose soft-

NATO ASI Series, Vol. F9
Computer Aided Analysis and Optimization of Mechanical System Dynamics
Edited by E. J. Haug

ware packages are now in use and we will identify a selection of the most widely available and point out types of problems that are not yet covered by such software. The extent to which software can be conveniently interchanged will be discussed and it will be argued that this will likely become easier in the future. We will then discuss how one should go about choosing the most appropriate method for his particular problem. We will describe the relevant problem-dependent features and how they affect this choice. A revised testing package which can aid in this process will be described. We will then discuss under what circumstances it would be worth modifying methods to exploit special features or structure which might be inherent in the problem. Examples will be given where such exploitation can result is significant improvements in efficiency. Finally we will consider where future developments are most likely and the implications these developments will have on the user community.

2. NUMERICAL METHODS: FORMULAS, WORKING CODES, SOFTWARE

One of the major difficulties associated with the use of numerical methods for initial value problems is that there is no general agreement as to precisely what a reliable numerical method should attempt to accomplish. The mathematical problem is completely specified by:

$$y' = f(x,y),\ y(x_o) = y_o \text{ over } [x_o,x_F]. \qquad (1)$$

Restrictions on $f(x,y)$ which lead to a unique solution are well understood and the mathematical conditioning of the problem (sensitivity of the solution to perturbations in y_o or $f(x,y)$) can be quantified and is a well developed area of research. It is important to appreciate that any underlying mathematical ill-conditioning of (1) must be inherited (as numerical instability) by a numerical method since round-off error alone will result in the computed representation of y_o and $f(x,y)$ being perturbations of the corresponding exact quantities. Until the mid-sixties the situation was relatively straightforward with a numerical method for initial value problems consisting of a fixed stepsize implementation of a specific formula. The question of reliability was then a question of whether the formula was coded correctly and possibly whether a check was made to ensure that round-off error did not dominate truncation error. Asymptotic results guaranteed that as $h \to 0$ the numerical solution converged to the exact solution and the numerical stability of the formula was essentially equivalent to the mathematical conditioning of the problem. Adaptive (variable-stepsize) methods began to appear in the early sixties and their ability to dynamically adapt the stepsize to match the behaviour of the differential equation lead to methods which were more efficient and more convenient to use. For such methods a user is asked to specify an accuracy parameter, TOL, rather than the stepsize, h. Asymptotic results ($h\to 0$) are then not directly relevant. The interesting question for these methods is to understand what happens as $TOL\to 0$, but results of this type are rare.

Today virtually all effective methods for (1) are of the adaptive type. In this case the method will produce a discrete set of approximations $(x_i,y_i)_{i=0}^{N}$ with $x_0<x_1<x_2\ldots<x_N=x_F$ and y_i an approximation to $y(x_i)$. It would be ideal if a method could attempt to ensure that the maximum global error, $\max_{i=1,2,\ldots N}\{\|y_i-y(x_i)\|\}$, is bounded by TOL. While some methods do attempt to satisfy this requirement, it is inherently very difficult as well as prohibitively expensive since the maximum global error will depend critically on the mathematical conditioning of (1) as well as on various properties of the method itself. Typically these methods will attempt to ensure that the maximum global error is proportional to TOL, although the way this is accomplished various considerably. We will return to this point later.

A numerical method for (1) must then include a step-choosing strategy. This strategy is usually based on monitoring and controlling the size of the local error introduced on each step of the integration. If we let $y(x;x_i,y_i)$ denote the solution of the mathematical problem:

$$y' = f(x,y),\ y(x_i) = y_i, \tag{2}$$

then the local error on step i is defined to be $y(x_{i+1};x_i,y_i) - y_{i+1}$.

We can now distinguish a basic formula, a rule that specifies how the approximation y_{i+1} is determined given the partitioning $x_0\ x_1\ldots x_{i+1}$, from a method which will include, in addition to the basic formula, error estimators and stepsize choosing strategies. Methods themselves fall into two broad classes, working codes and general purpose software. Working codes are often produced in an academic environment to illustrate the potential of a new approach or technique. Such a method will not usually be well-documented and will generally require a detailed knowledge of the underlying approach before it can be used effectively. On the other hand, general purpose software will be well documented, robust, convenient to use as a black-box, and flexible enough so that a variety of options are available without a user having to access the program source. It should not be too surprising that, while there have been well over one hundred potentially useful formulas for initial value problems suggested in the past decade, fewer than half of them have been implemented in working codes and only about ten to twenty can be considered implemented as general purpose software.

A method can also be classified according to the type of its underlying basic formula or according to the problem domain over which it is suitable. In the former case the traditional classifications are multistep, Runge-Kutta and extrapolation while in the latter case the most common classification is based on whether or not a method is suitable for solving stiff equations. We will discuss the concept of stiffness and what makes a method suitable for stiff problems in the next section.

Multistep methods are based on formulas which, when the stepsize is constant, have the general form:

$$y_{i+1} = \sum_{j=1}^{k} \alpha_j y_{i+1-j} + h \sum_{j=0}^{k} \beta_j y'_{i+1-j} . \tag{3}$$

The two most widely used families of multistep formulas are:

i) the Adam's formula of order (k+1):

$$y_{i+1} = y_i + h \sum_{j=0}^{k} \beta_j y'_{i+1-j},$$

which is extended to a variable stepsize formula by:

$$y_{i+1} = y_i + \int_{x_i}^{x_{i+1}} P_k(s)\, ds ,$$

where, on step i, $P_k(x)$ is the polynomial of degree $\le k$ interpolating y'_{i+1-j} at x_{i+1-j} for $j=0,1\ldots k$ and

ii) the backwards differentiation formulas (BDF) of order k:

$$y_{i+1} = \sum_{j=1}^{k} \alpha_j y_{i+1-j} + h\beta_0 y'_{i+1} ,$$

which is extended to a variable stepsize formula by

$$y_{i+1} = \frac{d}{ds}\{Q_{k-1}(s)\}\Big|_{s=x_{i+1}}$$

where $Q_{k-1}(s)$ is the polynomial of degree $\le k$ interpolating y_{i+1-j} at x_{i+1-j}, for $j=0,1\ldots k$.

To solve the implicit set of equations (3) on each time step multistep methods usually employ functional iteration (or predictor corrector iteration) which will converge if $h\|\frac{\partial f}{\partial y}\| < |1/\beta_0|$. Such an iteration scheme works well for non-stiff problems where convergence is usually observed in one or two iterations. For methods designed for stiff problems one cannot expect $h\|\frac{\partial f}{\partial y}\|$ to be small and a modified Newton iteration is usually adopted.

Variable stepsize Runge Kutta methods are generally based on a formula pair which can be written as:

$$y_{i+1} = y_i + h \sum_{r=1}^{s} \omega_r k_r,$$

$$\hat{y}_{i+1} = y_i + h\sum_{r=1}^{s}\hat{\omega}_r k_r ,$$

where

$$k_r = f(x_i+\alpha_r h, y_i+h\sum_{j=1}^{r}\beta_{rj}k_j).$$

Generally these approximations will satisfy

$$y_{i+1} = y(x_{i+1};x_i,y_i) + O(h^{p+1}),$$

$$\hat{y}_{i+1} = y(x_{i+1};x_i,y_i) + O(h^{p+2}),$$

and $\hat{y}_{i+1}-y_{i+1}=h\sum_{r=1}^{s}(\hat{\omega}_r-\omega_r)k_r$ will provide a convenient estimate of the local error. Note that the formula pair will be explicit if $\beta_{rr}=\hat{\beta}_{rr}=0$ for r=1,2...s. If the formula pair is not explicit, the solution of at least one system of nonlinear equations will be required. A method based on an explicit formula pair cannot be suitable for stiff problems and those that are suitable will generally use a modified Newton iteration to solve the resulting system of nonlinear equations.

Extrapolation methods are based on a formula which, when applied with a fixed stepsize has an error expansion satisfying:

$$y_r = y(x_r) + \sum_{j=1}^{k}\gamma_j h^j + O(h^{k+1}), \tag{4}$$

where the γ_js are independent of h. On the ith step, for a sequence of integers $N_1<N_2...<N_s$ (with s<k), we define $Y_{i+1}(h_j)$ to be the approximation to $y(x_{i+1};x_i,y_i)$ generated using N_j steps of size $h_j=(\frac{x_{i+1}-x_i}{N_j})$ of the basic formula, starting from (x_i,y_i). One then can consider the approximation $Y_{i+1}(h_j)$ to be a discrete sampling of a function $Y_{i+1}(h)$. If we let $R_{s-1}(h)$ be the polynomial (in h) interpolating $Y_{i+1}(h_j)$ for j=1,2...s, it is possible to show that $R_{s-1}(0)=y(x_{i+1};x_i,y_i)+O(h^{s+1})$. This observation is the basis for variable-order, variable-step extrapolation methods. If the basic formula is such that the error expansion (4) contains only even powers of h,

$$y_r = y(x_r) + \sum_{j=1}^{k}\gamma_j h^{2j} + O(h^{2k+2}),$$

then one can interpolate $Y_{i+1}(h_j)$ by an even polynomial of degree $\leq 2s-2$, $Q_s(h)$, and it can be shown that

$$Q_s(0) = y(x_{i+1};x_i,y_i) + O(h^{2s+2}).$$

Note that each 'step' of an extrapolation method requires $(\sum_{j=1}^{s}N_j)$ steps of the basic formula.

In the next section we will consider in detail what factors contribute to the choice of the most appropriate method for a particular problem. Nevertheless there are some general observations that can be made at this time with respect to the relative advantages of each of the above classes of methods. Multistep methods generally use the fewest number of function evaluations to obtain a given accuracy but the overhead per step and the storage requirements will be larger than for other types of method. Multistep methods are easily implemented in a variable-order format and this results in methods that are efficient over a wide range of accuracy requests. Another distinct advantage of these methods is that they provide a piecewise polynomial approximation to the solution. Runge-Kutta methods have low overhead and are easily understood and modified. They are usually fixed order and hence a given method will only be efficient over a limited range of accuracy requests. Extrapolation methods have modest overhead and a high cost per step. They are generally most competitive at stringent accuracy requests.

Standard libraries now include good general purpose software for initial value problems. A summary of some of the widely available software is presented in table 1.

Table 1. Summary of available initial value software.

A) IMSL:
- DGEAR - BDF/Adams combination
- DVERK - 8 stage order 6/5 Runge Kutta formula pair

B) NAG (Gladwell (1980)
- D02__ family of codes - includes BDF, Adams and Rung-Kutta methods

C) SLATEC (Shampine and Watts (1980))
- DEABM - Adams
- DEBDF - BDF
- DERKF - 6 stage order 5/4 Runge-Kutta formula pair

D) ODEPACK (Hindmarsh (1983))
- LSODE - BDF/Adams combination
- LSODES - BDF/Adams combination for problem with sparse Jacobian
- LSODI - BDF/Adams combination for implicit ODEs.

E) Other packages

EPISODE (Byrne & Hindmarsh (1975))	STRIDE (Butcher et al. (1981))
STIFEQ (Klopfenstein (1971))	MTAN1 (Denflhardt (1982))
SECDER (Addison (1981))	DODES (Schryer (1975))
STINT (Tendler et al. (1978))	

Given the availability of good general purpose software the question of choosing the most appropriate method for a particular problem and whether or not one can conveniently switch from one to another must be addressed. We will first consider the latter of these questions. Although the calling sequences differ considerably, most software provides a compatable set of options. These options include the ability to monitor the solution after each step, the ability to specify how the error is to be measured (whether in the absolute or relative sense or a mixture) and the ability to follow the solution until a specified condition arises (eg. the solution satisfies $g(y)=0$ where g is specified by the user). Because methods impose different requirements on the routine supplied by the user to specify the differential equation, an interface routine may be necessary when switching methods.

It is generally possible to write a special purpose driver routine which accepts as arguments a set of parameters appropriate to the application and sets the appropriate options, workspace etc. before invoking the method. Usually this method-dependent driver can be written without understanding the details of the underlying code.

A major impediment to the convenient switching of software is the fact that the vague requirement of keeping global error proportional to the specified TOL is interpreted differently by different methods. While there is some agreement as to the alternatives that should be available to measure the 'size' of the global error there is little agreement as to how this 'size' is to be related to TOL. One promising approach which could lead to a standard interpretation is to consider local errors in terms of a defect or residual. That is, for most existing methods it is possible to show that the numerical solution $(x_i,y_i)_{i=1}^{N}$ lies on the exact solution $z(x)$ to a perturbed initial value problem:

$$z' = f(x,z) + \delta(x), \quad z(x_o) = y_o,$$

where $\delta(x)$ can be estimated and controlled in an effective way. That is on step i it is reasonable to expect the method to attempt to ensure that $\sup_{x_i<x\leq x_{i+1}} \{\|\delta(x)\|\} \leq TOL$ (see Hanson and Enright (1983)). This approach has the definite advantage that the size of the global error can be related to the size of $\delta(x)$ in a method-independent way using only properties of the mathematical conditioning of (1). In this way we can separate the effects of the mathematical conditioning of the problem (the relationship between the size of $\delta(x)$ and the size of the global error) from the numerical stability of the method (the relationship between the size of $\delta(t)$ and TOL).

3. CHOOSING THE RIGHT METHOD

There are several factors that affect the choice of method for a particular class of problems. Such site-dependent features as local expertise and experience, hardware configuration and level of available documentation are often the most important considerations. We will assume that a collection of tested and documented routines are available and we will identify the relevant problem-dependent features that will generally determine the most suitable method. We will consider four features of a problem that should be considered in detail and we will also consider other characteristics of a problem that can cause difficulties for some methods.

The first feature one should consider is the form of the differential equation. Standard software will usually handle only explicit first order systems of the form (1). If the problem arises as a higher order equation or arises implicitly as:

$$A(x,y)y' = f(x,y),\ y(x_o) = y_o, \qquad (5)$$

then it may be better to solve the problem directly, using a special purpose method, rather than applying a preliminary transformation to an equivalent first order system. Experience (eg. Krogh (1975), Addison (1980)) indicates that reducing higher order equations to an equivalent first order system leads to little loss of efficiency unless the right hand side of the equation depends only on x and y (as in $y''=f(x,y)$). In this latter case direct methods can often be better by a factor of two. Similarly if one is solving a large implicit system (5) and $A(t,y)$ is weakly dependent on y and sparse or banded then one should consider direct methods such as LSODI for this class of problems. If $A(t,y)$ is nearly singular then one should consider treating the problem as a system of algebraic differential equations.

An important feature that is widely discussed in the literature but very difficult to define precisely is that of stiffness. Stiffness is a characteristic often present in differential equations arising from models of real systems where there are interactions taking place on more than one time scale. One example would be in a chemical kinetics model where some transient reactions are occurring on a time scale of a few microseconds or less while a slower steady-state reaction takes place on a time scale of a second or larger. These problems are mathematically very well conditioned. Unfortunately standard numerical techniques are inappropriate for this type of problem since, in order to assure numerical stability, the stepsize of these methods must be severely constrained.

One should suspect stiffness when a problem is very expensive to integrate with standard methods and when the cost of the integration (or average stepsize used) seems to be relatively insensitive to the requested accuracy. Stiffness is inevitable if the eigenvalues of largest magnitude of the Jacobian matrix have negative real parts and the magnitude of these negative real parts are much larger than the reciprocal of the length of integration. Methods designed for stiff systems allow much larger stepsizes, but require more work per step. The cost of these methods is often dominated by the cost of setting up and solving the linear equations on each step associated with the modified Newton iteration.

A third feature then, which is particularly important in the solution of stiff systems, is the size or dimension of the system. For standard non-stiff methods the operation count per step is usually $O(n)$, where n is the size of the system, but for stiff methods the operation count per step is at least $O(n^2)$ and can also be $O(n^3)$ on some steps. When solving large stiff systems (say more than about forty equations) it is essential that the linear algebra cost be minimized and any structure in the linear equations be exploited. For such problems methods which require the solution of more than one linear equation or which cannot exploit a sparse or banded Jacobian-structure are inappropriate. Methods based on the backwards differentiation formulas, such as LSODE, DGEAR or D02E_F, are generally used for these large systems but other methods can also be appropriate. For smaller systems one need not be as concerned about extra linear algebra operations.

The remaining general problem-dependent feature which can have a significant affect on the choice of method is the level of accuracy required. Again this can be particularly crucial for stiff systems. If stringent accuracy is required (say more than six significant figures) then one should choose a method that can employ formulas of order greater than four. For non-stiff methods this includes virtually all standard general purpose software but for stiff methods this reduces the choice to only a few (the BDF family, SECDER and STRIDE). While this is of interest to some users there are many more interested in relaxed accuracy solutions (say one or two significant figures). This latter class of users are not well served by existing software. Methods seem to be tuned to be reliable and efficient at moderate to stringent tolerances and they often tend to be unreliable, inefficient or both at relaxed tolerances. Low order single step methods show the most promise here but they tend to exist more as working codes (often developed by practitioners). One stiff method we have found effective for this class of problems is STIFEQ.

In addition to the four general features discussed above a problem can exhibit some special characteristics which cause severe difficulties for some methods. For example it is well known that most of the backward differentiation family experience difficulties when some of the large magnitude eigenvalues of the Jacobian are near the imaginary axis. In this case, if the corresponding high frequency components are not significant, other methods (such as STRIDE or SECDER) can perform much better. In a recent study (Gaffney (1982)) a class of problems exhibiting this difficulty is identified and the performance of three stiff methods is investigated with the general conclusion that none is entirely satisfactory. If, on the other hand, the high frequency components are significant then the problem is mathematically ill-conditioned and other interpretations of what an acceptable solution is may be called for. (This difficulty is discussed in detail in Gear (1983).) Problems that exhibit discontinuities or near discontinuities in the differential equation can impair the performance of a method. For example the step choosing strategy of multistep methods is severely tested by such problems and the inevitable drop in order of such methods will usually require several extra steps. Note that in the case of discontinuities the difficulty can often be reduced if one uses a special purpose driver (more about this later), but the 'near discontinuity' case is much more difficult to handle.

Often, after considering the relevant site-dependent and problem-dependent features one must still choose between two or more candidate methods. At this point one should choose a representative problem or class of problems and investigate the performance of each of the candidate methods on this representative class of problems. We have recently revised a testing package of subroutines which can aid in this process and make this final task relatively painless and straightforward. For several years we have been interested in the assessment and evaluation of codes for initial value problems. During this time testing packages have been developed to aid in the assessment of such methods. In the last year we have completely rewritten and documented these packages so they can easily be used by others (Enright and Pryce (1983)). The packages STDTST for assessing the performance of stiff methods and NSDTST for assessing the performance of non-stiff methods are designed to easily permit the introduction of new problems and hence would be useful in assessing a method on representative problems.

4. WHEN SHOULD METHODS BE MODIFIED

There are situations where one should at least consider modifying an existing method to take advantage of special features of the problem and improve the overall effectiveness of the approach. This is a task that requires detailed knowledge of the various components that make up the method and how these components interact. It should not be attempted unless there is a reasonable prospect of significant improvement. We will consider three situations where modifications can increase the effectiveness of a method and we will attempt to quantify the improvement that is possible using numerical examples.

The first example is that of explicitly exploiting the fact that the differential equation is discontinuous when a specified function of the dependent variable, say g(y), is equal to zero. For methods which provide flexible options one can write a special purpose driver to exploit this knowledge and need not modify the source itself. We will present an example of how this can be done using the method DVERK. Note that this approach is not new, but it is a simple example of what gains are possible if one exploits knowledge about the problem, and it can be applied to any method. The special purpose driver uses the option of DVERK to return after every step and a bisection technique is used to locate the discontinuity (the point at which $g(y)=0$). At that point the method is restarted. The problem considered was problem 1:

$$y' = \begin{cases} y \text{ if } y<10 \\ y/2 \text{ otherwise} \end{cases}, \quad y(0) = 1 \quad \text{over } [0,10]$$

and the results illustrating the improvement possible is presented in figure 1. where the number of function evaluations (FCN), the number

Figure 1. Use of a special purpose driver to handle discontinuous problems.

	DVERK - standard			DVERK - special driver		
TOL	FCN	STEPS	MAX ERR	FCN	STEPS	MAX ERR
10^{-2}	48	6	6.2×10^{-2}	40	5	$.26\times10^{-2}$
10^{-3}	108	10	5.7×10^{-3}	40	5	$.09\times10^{-3}$
10^{-4}	177	16	$54.\times10^{-4}$	56	7	$.22\times10^{-4}$
10^{-5}	254	23	$16.\times10^{-5}$	72	9	$.33\times10^{-5}$
10^{-6}	392	35	55×10^{-6}	104	13	$.31\times10^{-6}$
10^{-7}	494	46	18×10^{-7}	152	19	$.14\times10^{-7}$
10^{-8}	673	64	69×10^{-8}	216	27	$.29\times10^{-8}$

of steps (STEPS) and maximum observed global error (MAX ERR) is given for both the unmodified DVERK and the version of DVERK with a special driver over a range of accuracy requests (TOL).

Another feature that can be exploited is linearity. We have analyzed this question in detail for different classes of linear problems in Enright (1980). The first observation one can make is that unlike other areas of numerical analysis, one must work very hard to exploit linearity and most of the time the improvement in efficiency would not be significant. For example if we consider the nonhomogeneous constant coefficient equation:

$$y' = Ay + h(x)$$

then, unless the system is large and stiff, no significant improvement is possible. If the problem is non-stiff then standard multistep or extrapolation methods cannot be improved at all and while it is possible to derive special purpose Runge Kutta formulas using fewer function evaluations the gain is unlikely to be significant. For stiff systems significant gains are possible and solution time can be reduced by more than 50%.

As a final example we observe that a stiff method can be modified to exploit natural partitioning that may exist in the differential equation. This can be accomplished using an automatic partitioning or using a user-supplied partitioning. Such techniques are discussed in detail in Kamel and Enright (1982) and are still under active investigation.

5. FUTURE DEVELOPMENTS

Over the next few years we should see more experimental codes transformed into robust software. In particular methods which are effective at low accuracy requests and methods which can effectively solve stiff problems with eigenvalues near the imaginary axis will be improved and become widely available.

Another development that could change our view of numerical methods is that some methods are now producing a piecewise continuous approximation to the solution y(x) over the range of integration. While it is possible to associate such an interpolant with any method, they are generally very expensive to evaluate. Investigations are now underway to develop inexpensive and effective interpolants for all types of methods. This approach may well lead to a more uniform interpretation of the accuracy of a numerical solution based on the defect (or residual) of such an interpolant.

Finally the software packages for stiff systems are becoming more modular. In particular standard linear algebra modules are being used and improved interface routines are being adopted. Sparse or banded solvers can replace standard solvers in such an organization and as new techniques, such as those based on partitioning are introduced, it should be much easier to replace the appropriate modules and investigate the potential improvements.

REFERENCES

Addison, C.A. (1979), 'Implementing a stiff method based upon the second derivative formulas', Dept. of Computer Science Tech. Rep. No. 130/79, University of Toronto.

Addison, C.A. (1980), 'Numerical methods for a class of second order ODEs arising in structural dynamics', Ph.D. thesis, Dept. of Computer Science Tech. Rep. No. 147/80, University of Toronto.

Burrage, K., Butcher, J.C. and Chipman, F.H. (1979), 'STRIDE: a stable Runge Kutta integration for differential equations, private communication.

Byrne, G.D. and Hindmarsh, A.C. (1975), 'A polyalgorithm for the numerical solution of ordinary differential equations', ACM Trans. on Math. Software, 1, pp. 71-96.

Deuflhard, P. (1982), 'Recent progress in extrapolation methods for ordinary differential equations', Invited talk presented at SIAM 30th anniversary meeting, Stanford, July 1982.

Enright, W.H. (1979), 'On the efficient and reliable numerical solution of large linear systems of ODEs', IEEE Trans. on Automatic Control, AC-24, pp. 905-910.

Enright, W.H. and Pryce, J.D. (1983), 'A collection of programs for assessing initial value methods', Dept. of Computer Science Tech. Rep. No. 167/83, University of Toronto.

Gafney, P. (1982), 'A survey of Fortran subroutines suitable for solving stiff oscillating ordinary differential equations', Tech. Memo. 134, Oak Ridge.

Gear, C.W. (1983), 'Recent results on the numerical solution of differential-algebraic equations', this volume.

Gladwell, I. (1980). 'Initial value routine in the NAG library, ACM Trans. on Math. Software, 5, pp. 386-394.

Hanson, P. and Enright, W.H. (1983), 'Relating the defect to the error tolerance in existing variable order Adams methods, ACM Trans. on Math. Software, 9, pp. 71-97.

Hindmarsh, A.C. (1983), 'ODEPACK, a systematized collection of ODE solvers', in Numerical Methods for Scientific Computation, R.S. Stepleman (ed.), North Holland, to appear.

Klopfenstein, R.W. (1971), 'Numerical differentiation formulas for stiff systems of ordinary differential equations', RCA Review, Vol. 32, No. 3, pp. 447-462.

Krogh, F.T. (1975), 'Summary of test results with variants of a variable order Adams method', Computing Memo. No. 376, Section 914, Jet Propulsion Laboratory, CIT, Pasadena.

Schryer, N.L. (1975), 'A user's guide to DODES, a double precision ordinary differential equation solver', Bell Labs. Comp. Science Tech. Rep. No. 33.

Shampine, L.F. and Watts, H.A. (1980), 'DEPAC - Design of a user oriented package of ODE solvers', Rep. SAND79-2374, Sandia National Laboratories, Albuquerque, New Mexico.

Tendler, J.M., Bickart, T.A. and Picel, Z. (1978), 'A stiffly stable integration process using cyclic composite methods, ACM Trans. on Math. Software 1, pp. 399-403.

DIFFERENTIAL-ALGEBRAIC EQUATIONS

C. W. Gear
Department of Computer Science
University of Illinois at Urbana-Champaign
Urbana, Illinois 61801

Abstract. In this paper we study the numerical solution of the differential/algebraic systems $F(t, y, y') = 0$. Many of these systems can be solved conveniently and economically using a range of ODE methods. Others can be solved only by a small subset of ODE methods, and still others present insurmountable difficulty for all current ODE methods. We examine the first two groups of problems and indicate which methods we believe to be best for them. Then we explore the properties of the third group which cause the methods to fail. A reduction technique is described which allows systems to be reduced to ones which can be solved. It also provides a tool for the analytical study of the structure of systems.

1. INTRODUCTION

We are interested in initial value problems for the differential/algebraic equation (DAE)

$$F(t, y, y') = 0 , \tag{1}$$

where F, y, and y' are N-dimensional vectors. F will be assumed to be suitably differentiable. Many of these problems can be solved conveniently and economically using numerical ODE methods. Other problems cause serious difficulties for these methods. Our purpose in this paper is first to examine those classes of problems that are solvable by ODE methods, and to indicate which methods are most advantageous for this purpose. Secondly, we want to describe the problems which are not solvable directly by ODE methods, and the properties of these problems which cause the methods to fail. Finally, we want to discuss some analytical techniques for rewriting systems in a form which can be solved by numerical methods.

The idea of using ODE methods for solving DAE systems directly was introduced in [3], and is best illustrated by considering the simplest possible algorithm, based on the backward Euler method. In this method the derivative $y'(t_{n+1})$ at time t_{n+1} is approximated by the first backward divided difference of y(t), and the resulting system of nonlinear equations is solved for y_{n+1},

$$F(t_{n+1}, y_{n+1}, (y_{n+1} - y_n)/(t_{n+1} - t_n)) = 0 . \tag{2}$$

In this way the solution is advanced from time t_n to time t_{n+1}. Higher order techniques such as backward differentiation formulas (BDF), Runge-Kutta methods, and extrapolation methods are generalizations of this simple idea. For example, the k-step BDF method is used by substituting

Supported in part by the U.S. Department of Energy, Grant DOE DEAC0276ERO2383.

NATO ASI Series, Vol. F9
Computer Aided Analysis and Optimization of Mechanical System Dynamics
Edited by E. J. Haug

$$y'_{n+1} = \frac{1}{h} \sum_{i=0}^{k} \alpha_i \; y_{n+1-i}$$

into (1).

The advantages of using ODE methods directly for solving DAE systems are that these methods preserve the sparsity of the system and require no prior manipulation of the equations. For example, one set of DAE systems which is particularly simple to solve consists of systems which are really ODEs in disguise. If, in (1), $\partial F/\partial y'$ is nonsingular, then the system can, in principle, be inverted to obtain an explicit system of ODEs

$$y' = f(t, y) \, . \tag{3}$$

However, if $\partial F/\partial y'$ is a sparse matrix, its inverse may not be sparse, and hence $\partial f/\partial y = [\partial F/\partial y]^{-1} \, \partial F/\partial y$ may not be sparse. Since the solution of (2) or (3) often involves the solution of linear equations involving those Jacobians, considerable computer time can be saved by solving (2) instead of (3) when the system is large. Thus it is preferable to solve the system directly in its original form.

The most challenging difficulties for solving DAE systems occur when $\partial F/\partial y'$ is singular. These are the systems with which we are concerned here. In some sense the simplest, or at least the best understood, class of DAE systems is that which is linear with constant coefficients. These systems,

$$Ay'(t) + By(t) = g(t) \, , \tag{4}$$

can be completely understood via the Kronecker canonical form of the matrix pencil (A,B). The important characteristic of equation (4) that determines the behavior of the system and numerical methods is the *index of nilpotency* of the matrix pencil (A,B). Numerical methods such as (2) can be used to solve linear and nonlinear systems of index no greater than one with no great difficulty. Algorithms based on these methods may experience problems when the index is greater than one. We will introduce a scheme for determining if a system has index greater than one. This scheme can be used in a code to warn the user of probable difficulty.

One might hope that the study of (4) could be used as a guide for understanding more complicated DAE systems. In general this fails to be true. The structure of the local constant-coefficient system may not describe the behavior of solutions to the DAE for nonlinear or even linear, nonconstant-coefficient systems whose index is greater than one. Numerical methods which work for (4) break down when the matrices are time-dependent and the index is greater than one. In fact, we are not aware of any numerical methods (based on ODE techniques or otherwise) for solving the most general linear DAE systems, let alone nonlinear systems. In Section 4 we examine the structure of time-dependent problems and show where the difficulties with conventional methods arise. In the Section 5 we describe some analytical techniques for rewriting systems in a form which can be solved by numerical methods. These techniques are useful not only for simplifying systems in practice, but also as theoretical tools for exposing the underlying structure of high index systems. In the last section we apply the ideas to Euler-Lagrange equations with constraints and see that these always lead to index 3 problems.

2. INDEX ONE PROBLEMS

In this section we describe the index of a problem and discuss problems whose index does not exceed one. These problems are solvable by ODE methods.

The existence and solution of linear constant-coefficient systems (4) is easily understood by transforming the system to Kronecker canonical form (KCF). For details see [14]. We give only an overview. The main idea is that there exist nonsingular matrices P and Q which reduce (A,B) to canonical form. When P and Q are applied to the constant-coefficient problem (4), we obtain

$$PAQQ^{-1}y' + PBQQ^{-1}y = Pg(t)\,. \tag{5}$$

where (PAQ,PBQ) is the canonical form. When $A + \lambda B$ is singular for all values of λ, no solutions exist, or infinitely many solutions exist. It is not even reasonable to try to solve these systems numerically. Fortunately, numerical ODE methods reject these problems almost automatically because they have to solve a linear system involving the matrix $A + h\beta B$ (where h is the stepsize and β is a scalar which depends on the method and recent stepsize history) which is singular for all values of h. When $det(A + \lambda B)$ is not identically zero, the system is "solvable" by the following definition, which was introduced in [14]. Here we give it for the time varying linear problem.

Definition. A linear system $A(t)y' + B(t)y = g(t)$ is *solvable* iff for any sufficiently smooth input function g(t), solutions to the differential/algebraic equation exist, and solutions which have the same initial values are identical.

In the following we will deal only with solvable systems.

For solvable systems the KCF form (5) of a constant-coefficient problem can be written as

$$z_1'(t) + Cz_1(t) = g_1(t)\,, \tag{6a}$$

$$Ez_2'(t) + y_2(t) = g_2(t)\,, \tag{6b}$$

where

$$z = Q^{-1}y(t) = \begin{vmatrix} z_1(t) \\ z_2(t) \end{vmatrix}, Pg(t) = \begin{vmatrix} g_1(t) \\ g_2(t) \end{vmatrix},$$

and E has the property that either there exists an integer m such that $E^m = 0$, $E^{m-1} \neq 0$ or E is the "empty" (or zero by zero) matrix. In the latter case, m is defined as 0. The value of m is defined to be the index of nilpotency of the system. The matrix E is composed of Jordan blocks of the form

$$\begin{vmatrix} 0 & & & \\ 1 & \ddots & & \\ & \ddots & \ddots & \\ & & 1 & 0 \end{vmatrix}, \tag{7}$$

and m is the size of the largest of these blocks. Because (4) is linear, the application of a linear method to compute y_n, $n = 1, ,...$ will yield exactly the equivalent of the application of a linear method to (6); that is, if the initial values are transforms ($z_0 = Q^{-1}y_0$) then $z_n = Q^{-1}y_n$.

The behavior of numerical methods for solving standard ODE systems (6a) is well understood and will not be discussed here. Since the systems (6a) and (6b) are completely uncoupled and the methods we are interested in are linear, it suffices for understanding (4) to study the action of numerical methods on subsystems of the form (6b), where E is a single block of form (7). When E is a matrix of the form (7) and size m, the system is referred to as a canonical ($index = m$) subsystem.

Systems whose index does not exceed one are the most easily understood, and they seem to occur far more frequently in solving practical problems than the other ($index > 1$) subsystems. When the index is one, the matrix E in (6b) is identically zero. Thus the system reduces to a system of standard form ODEs plus some variables which are completely determined by simple linear relations.

It is clear that the application of a BDF method to index ≤ 1 problems is equivalent to solving for z_1 using BDF and solving for z_2 using (6b) directly because z_2' does not appear in (6b). In fact, index ≤ 1 problems cause no difficulties even in the nonlinear case. A proof is given in [6] of a result which essentially says:

Theorem 2.1. If the linearized form of $F(y, y', t) =$ is of index 1 in a neighborhood of the solution and certain technical smoothness properties hold, then the k-step BDF method converges with order k for this problem if $k \leq 6$.

It should be noted that while the ODE methods behave basically as expected for the $index = 1$ problems, there are still some practical difficulties involved in implementing these methods for this class of problems. Some of these problems are discussed in [10], [11]; we will not discuss most of these difficulties here.

Most automatic codes for solving DAE systems [11] are designed to handle nonlinear systems of $index \leq 1$. These codes cannot handle systems of higher index, and it would be desirable in such codes to detect higher index problems and stop. The next algorithm, which is actually an application of a more general scheme (Algorithm 5.1) which will be introduced in Section 5, is a tool for detecting high index problems. We will describe it in terms of linear problems, but the technique extends easily to nonlinear problems.

Theorem 2.2. Let $A(t)y' + B(t)y = g(t)$. Choose a nonsingular R(t) such that

$$RA = \begin{vmatrix} A_1 \\ 0 \end{vmatrix},$$

where the $q \times s$ matrix A_1 has rank q. Then, if $q = s$, the index is zero. If $q < s$, define

$$RB = \begin{vmatrix} B_1 \\ B_2 \end{vmatrix},$$

where B_1 is $q \times s$, and examine the matrix

$$\begin{vmatrix} A_1 \\ B_2 \end{vmatrix}$$

If it is nonsingular, the index is one.

Proof. If $q = s$, the result is trivial. If $q < s$ and

$$\begin{vmatrix} A_1 \\ B_2 \end{vmatrix}$$

is nonsingular, it has an inverse which we denote by $[Q_1^T \; Q_2^T]$ where Q_1^T is $q \times s$. We have

$$\begin{vmatrix} A_1 \\ B_2 \end{vmatrix} [Q_1^T \; Q_2^T] = \begin{vmatrix} A_1Q_1^T & A_1Q_2^T \\ B_2Q_1^T & B_2Q_2^T \end{vmatrix} = I.$$

Hence, $A_1Q_1^T = I$, $B_2Q_1^T = 0 = A$, $B_2Q_2^T = I_2$. Let $P = R$ and $Q = [Q_1^T \; Q_2^T]$. We have

$$PAQ = \begin{vmatrix} I_1 & 0 \\ 0 & 0 \end{vmatrix}$$

$$PBQ = \begin{vmatrix} B_1Q_1^T & B_1Q_2^T \\ 0 & I_2 \end{vmatrix},$$

and this implies that the index is one as Lemma 2.3 below shows.

Q.E.D.

Lemma 2.3. If E has nilpotency m then

$$index \begin{vmatrix} I_1 & 0 \\ 0 & E \end{vmatrix}, \begin{vmatrix} C & D \\ 0 & I_2 \end{vmatrix} = index\ (E,\ I_2) = m$$

Proof. The result follows by simple reductions to nullify D. Premultiply the pencil by

$$\begin{vmatrix} I_1 & -D \\ 0 & I_2 \end{vmatrix}$$

and postmultiply by

$$\begin{vmatrix} I_1 & DE \\ 0 & I_2 \end{vmatrix}$$

to obtain

$$\begin{vmatrix} I_1 & 0 \\ 0 & E \end{vmatrix}, \begin{vmatrix} C & CDE \\ 0 & I_2 \end{vmatrix}.$$

A similar transformation can be applied to reduce the upper right corner to $C(CDE)E = C^2DE^2$. This can be repeated m times to obtain $C^m DE^m = 0$.

Q.E.D.

3. LINEAR CONSTANT-COEFFICIENT PROBLEMS OF HIGH INDEX

Systems of index greater than one have several properties which are not shared by the lower index systems. The properties of these high index constant-coefficient systems which cause codes to fail are discussed in much greater detail in [10]; we give only a brief outline here. We can understand many of the properties of (4) and of numerical methods by studying the simplest index 3 problem,

$$\begin{aligned} z_1 &= g(t) \\ z_1' - z_2 &= 0 \\ z_2' - z_3 &= 0\,. \end{aligned} \tag{8}$$

The solution to this problem is $z_1 = g(t)$, $z_2 = g'(t)$, $z_3 = g''(t)$. If initial values are specified for the z_i, the solution has a jump discontinuity unless these initial values are compatible with the solution. If the driving term g(t) is not twice differentiable everywhere, the solution will not exist everywhere. For example, if g(t) has a simple jump discontinuity at some point, z_2 is a multiple of the dirac delta function, and z_3 is its derivative.

What happens when a numerical method is applied to one of these problems? It is surprising that some of the numerical ODE methods work so well on these problems which are so unlike ODEs. We can best explain how the methods work by example. When the backward Euler method is used to solve the index = 3 problem (8), we find that the solution at time t_n is given in terms of the solution at time t_{n-1} by

$$z_{1,n} = g_n \tag{9}$$

$$z_{2,n} = (z_{1,n} - z_{1,n-1})/h$$

$$z_{3,n} = (z_{2,n} - z_{2,n-1})/h$$

The values of z_1 will be correct at all steps (if roundoff error is ignored), although the initial value $z_{1,0}$ may be incorrect. If the initial values (which need not be specified for the original problem but must be specified for the numerical procedure) are inconsistent, the values of $z_{2,1}$ and $z_{3,1}$ are incorrect. In fact, as $h \to 0$ they diverge. However, after two steps we obtain an O(h) correct value of $z_{2,2}$ because it is obtained by the divided difference of g(t). Finally, after the third step we obtain a good approximation to z_3 which is given by the second divided difference of g(t). After the third step all the components will be O(h) accurate.

The behavior of a general BDF method is very similar to that of backward Euler for fixed stepsize as shown in the following theorem, proved in [14].

Theorem 3.1. If the k-step, constant-stepsize BDF method is applied to the constant-coefficient linear problem (4) with $k < 7$, the solution is $O(h^k)$ accurate globally after a maximum of $(m-1)k + 1$ steps.

Unfortunately, these results for BDF break down when the stepsize is not constant, as shown in the next theorem, proved in [4].

Theorem 3.2. If the k-step BDF method is applied to (4) with $k < 7$ and the ratio of adjacent steps is bounded, then the global error is $O(h_{\max}^q)$, where $q = min(k, k-m+2)$.

The result follows from the argument that the k-step BDF, considered as a numerical differentiation technique, gives an $O(h^k)$ error. When the result of such a computation is itself numerically differentiated, the $O(h^k)$ terms are multiplied by the differentiation coefficients which are $O(h^{-1})$, so one order is lost for each additional differentiation. In an index m problem an algebraic variable (which can be found exactly) is differentiated m-1 times. The first of these has error $O(h^k)$ and the last will be $O(h^{k-m+2})$.

The difficulty can be seen by considering (9) for variable stepsizes. In that case we get

$$z_{1,n} = g_n$$

$$z_{2,n} = (z_{1,n} - z_{1,n-1})/h_n$$

$$z_{3,n} = (z_{2,n} - z_{2,n-1})/h_n$$

Even after the initial errors have disappeared we find that

$$z_{3,n} = \frac{\dfrac{g_n - g_{n-1}}{h_n} - \dfrac{g_{n-1} - g_{n-2}}{h_{n-1}}}{h_n}$$

If this were to be an O(h) correct approximation to g_n'' is the denominator should be $(h_n + h_{n-1})/2$. Hence the error is

$$\frac{1}{2}\left(\frac{h_{n-1}}{h_n} - 1\right)g_n''$$

which is O(1) if $h_n = 0(h_{n-1})$ but $h \neq h_{n-1}$.

Although, in principle, a problem of index no greater than seven could be solved by the six-step BDF method with variable stepsize, the hypothesis in Theorem 3.2, that the ratio of adjacent steps is bounded, is not a reasonable model in practice. When a code is attempting to take the next step, all previous stepsizes are now fixed, and the next step must be chosen to achieve the desired error. In this model the error of a BDF formula used for numerical differentiation is O(h), where h is the current stepsize. Consequently, if the index exceeds 2, the error of one step does not converge as that stepsize goes to zero, and diverges if the index exceeds 3. This can be seen in the above example in which the error in $z_{3,n}$, namely $(h_{n-1}/h_n - 1)g_n''$, behaves like $O(h_n^{-1})$.

The above results suggest that variable-stepsize BDF is not a suitable method for solving constant-coefficient DAEs with arbitrary index.

4. TIME DEPENDENT PROBLEMS

In this section we study nonconstant-coefficient linear problems,

$$A(t)y'(t) + B(t)y(t) = g(t)\,, \tag{10}$$

We explore the underlying structure of these systems, and examine the reasons why they have proven to be so difficult to solve.

When the coefficients are not constant, as in (10), there are several possible ways to define the index of the system. We can clearly define the *local index*, $l(t) = index(A(t), B(t))$, whenever the pencil $(A(t), B(t))$ is nonsingular. We can also define the *global index*, when it exists, in terms of possible reductions of the DAE to a semi-canonical form. By making a change of variables $y = H(t)z$ and scaling the system by G(t), where G(t) and H(t) are nonsingular, we obtain from (10)

$$G(t)A(t)H(t)z' + (G(t)B(t)H(t) + G(t)A(t)H'(t))z = G(t)g(t) \tag{11}$$

Now, if there exist G(t) and H(t) so that

$$G(t)A(t)H(t) = \begin{vmatrix} I_1 & 0 \\ 0 & E \end{vmatrix} \tag{12}$$

$$G(t)B(t)H(t) + G(t)A(t)H'(t) = \begin{vmatrix} C(t) & 0 \\ 0 & I_2 \end{vmatrix},$$

and the nilpotency of E is m, we will say that the system has global index of m. Note that the global index is the local index of this semi-canonical form.

Clearly, it is the global structure that determines the behavior of the solution. If the global structure is a constant, we know that n_1 independent initial values can be chosen, where n_1 is the dimension of the "differential" part of the system, and that the driving term can be subject to differentiation m-1 times. (Changes in the index or the structure of the system are called turning points. Problems with turning points are of importance in electrical network analysis. See Sastry, et al. [13] for a discussion in that context, and Campbell [1] for a discussion of types of turning points.)

The local index in some sense governs the behavior of the numerical method. For example, if the matrix pencil is singular, then numerical ODE methods cannot solve the problem because they will be faced with the solution of singular linear equations. In understanding why numerical ODE methods break down, it is natural to ask how the local index and global index are related. The next theorem answers this question.

Theorem 4.1. If the local index is not greater than one, then it is not changed by a smooth transformation. If the local index is greater than one, then almost all smooth nonconstant transformations of variables in (10) will yield a system whose local index is two. A restricted set of transformations will cause the index to be greater than two, or the pencil to be singular.

When the transformation to semi-canonical form (12) is used, this shows the relationship between the local and global indices, namely that they are the same only if they do not exceed one. The proof of this result and some examples can be found in [5].

Whenever the global index exists, we have a good understanding of the behavior of the solutions to the system. Thus it is important to know when this index exists. That is, when does there exist a nonsingular scaling and change of variables transforming (10) to the semi-canonical form (12)? The following theorem, whose proof can be found in [6], answers this question.

Theorem 4.2. Suppose A(t) and B(t) are sufficiently smooth. Then, except possibly at a finite number of points, there exist nonsingular matrices G(t) and H(t) transforming (10) to the semi-canonical form (12), if and only if the system is solvable.

Since solvable systems are so closely related to systems of the form (12) (where the singular part of the system has constant coefficients), we might hope that some of the same techniques which work for solving constant-coefficient problems numerically might also be effective for general linear problems. Unfortunately, this turns out not to be the case.

We have seen that the constant stepsize BDF method can be used for constant-coefficient problems. What happens when it is applied to nonconstant-coefficient problems? If the local index is two we may have a stability problem depending on the rate of change of the coefficients. If the local index is greater than two, we almost always have a stability problem. We want to stress that this is a stability problem and not an accuracy question, so it does not appear that higher order methods will help. Also note that it depends on the local index while the behavior of the underlying equation depends on the global index. The details are given in [6].

5. REDUCTION TECHNIQUES

What can the person with a high index system do? It is sometimes possible to use analytical techniques to rewrite the system in a form with lower index which can be solved numerically. In this section we discuss a reduction technique. It is useful for reducing the index of systems (and also determining their index).

The reduction technique is described below for linear systems (10), but it applies directly to nonlinear problems (1) when F is linear in y'.

Algorithm 5.1.

(1) If A in (10) is nonsingular, then we are done.

(2) Otherwise premultiply (10) by a nonsingular matrix P(t) to zero out a maximal number of rows of A and permute the zero rows to the bottom to obtain:

$$\begin{vmatrix} A_{11} \\ 0 \end{vmatrix} y' + \begin{vmatrix} B_{11} \\ B_{12} \end{vmatrix} y = g(t)$$

(3) Differentiate the bottom half of the system to obtain the new system

$$\begin{vmatrix} A_{11} \\ B_{12} \end{vmatrix} y' + \begin{vmatrix} B_{11} \\ B'_{12} \end{vmatrix} y = \hat{g}(t)$$

Now apply the process to this new system.

Intuitively, the idea behind this algorithm is that by differentiating the "algebraic constraints" of the system we can reduce its index without changing the solution to the system. If this is repeated, as in Algorithm 5.1, eventually we should produce a system of ODEs which can be solved by numerical methods. That this intuition is correct is shown in Theorem 5.2 below.

Of course, by differentiating we have introduced a number of constants of integration, which means that we must determine the correct initial conditions. This can be done by satisfying the initial system and each of its differentiated forms at the initial point.

Theorem 5.2. For solvable linear systems (10) with no turning points, Algorithm 5.1 terminates in m iterations iff the global index is m. Algorithm 5.1 does not terminate for systems which are not solvable.

The proof of this can be found in [6].

An alternative approach is to append the additional equations found by differentiation and to use these to eliminate the "algebraic" variables. This, in effect, is the technique used in [15]. They compute the matrix $P(t)$ numerically at each step using the special structure of Euler-Lagrange equations with constraints to be discussed in the next section.

6. EULER-LAGRANGE EQUATIONS

A mechanical system subject to constraints using the Lagrangian equations of motion: If the kinetic energy is $T(y, y')$, the external forces are $f(y, y', t)$ and the constraints are $c(y, t) = 0$ we have

$$\frac{d}{dt}\left(\frac{\partial T}{\partial y'}\right)^* - \left(\frac{\partial T}{\partial y}\right)^* + f + \left(\frac{\partial c}{\partial y}\right)^* \lambda = 0 \tag{13}$$

where $*$ is the transpose operator and λ is a Lagrange multiplier. By the usual substitution of $y' = z$, equation (13) can be reduced to first order for analysis and then written in the form (1). Typically $T(y, y')$ is quadratic positive definite in y', so that $\partial T/\partial y'$ is linear in y'. In this case we can replace y' by z in equation (3) and solve for z' to get

$$\begin{vmatrix} I & 0 & 0 \\ 0 & H & 0 \\ 0 & 0 & 0 \end{vmatrix} \begin{vmatrix} y' \\ z' \\ \lambda' \end{vmatrix} + \begin{vmatrix} -z \\ q(y,z,t) + \left(\frac{\partial c}{\partial y}\right)^* \lambda \\ c(y,t) \end{vmatrix} = 0 \tag{14}$$

where $H = \partial^2 T/\partial y'^2$ and is positive definite.

A particular example of this is furnished by a simple pendulum. Suppose it swings in a plane with rectangular coordinates (x, y) and velocities (u, v). The kinetic energy is $(u^2 + v^2)/2$ for a unit mass, while the external force is $[0, -1]^T$ for a unit gravitational force in the negative y direction. If the mass is unit distance from the pivot, the constraint is

$x^2 + y^2 - 1 = 0$ so equation (5) takes the form

$$\begin{vmatrix} 1 & & & & \\ & 1 & & & \\ & & 1 & & \\ & & & 1 & \\ & & & & 0 \end{vmatrix} \begin{vmatrix} x' \\ y' \\ u' \\ v' \\ \lambda' \end{vmatrix} + \begin{vmatrix} -u \\ -v \\ 2\lambda x \\ 2\lambda y - 1 \\ x^2 + y^2 - 1 \end{vmatrix} = 0 \tag{15}$$

In fact, -2λ is the tension in the pendulum rod.

If we apply Algorithm 5.1 to (14) we will see that the index is either 3 or the system is singular as follows:

In the first step the matrix A is singular (hence *index* ≥ 1) and A is already in the desired form. Hence we differentiate the constraints to get the equations

$$\frac{\partial C}{\partial y} y' + \frac{\partial c}{\partial t} = 0 \tag{16}$$

The new A matrix is

$$\begin{vmatrix} A_{11} \\ B_{12} \end{vmatrix} = \begin{vmatrix} I & 0 & 0 \\ 0 & H & 0 \\ \frac{\partial c}{\partial y} & 0 & 0 \end{vmatrix}$$

and is still singular. Hence *index* ≥ 2. We can zero out the last group of rows corresponding to the constraints. Because of the structure of the problem this is the same as replacing y' in (16) by z.

Now the last group of equations is

$$\frac{\partial c}{\partial y} z + \frac{\partial c}{\partial t} = 0$$

and must be differentiated again to get

$$\frac{\partial^2 c}{\partial y^2} z^2 + \frac{2\partial^2 c}{\partial t \partial y} z + \frac{\partial c}{\partial y} z' = 0 \tag{17}$$

The new A matrix is

$$\begin{vmatrix} I & 0 & 0 \\ 0 & H & 0 \\ 0 & \frac{\partial c}{\partial y} & 0 \end{vmatrix}$$

and still singular so *index* ≥ 3. This time the reduction is equivalent to substituting

$$z' = -H^{-1}(\frac{\partial c}{\partial y})^{*} \lambda + \cdots$$

from equation (14) into (17) to get the equation

$$(\frac{\partial c}{\partial y}) H^{-1} (\frac{\partial c}{\partial y})^{*} \lambda = \hat{q}(y, z, t) \tag{18}$$

Differentiating, we get a new A matrix of

$$\begin{vmatrix} I & 0 & 0 \\ 0 & H & 0 \\ X & X & M \end{vmatrix}$$

where $M = \frac{\partial c}{\partial y} H^{-1} (\frac{\partial c}{\partial y})^*$. Hence the index of this system is at least 3, and is exactly 3 if and only if

$$\frac{\partial c}{\partial y} H (\frac{\partial c}{\partial y})^*$$

is nonsingular. If the constraints c are independent (so that $\partial c / \partial y$ has full rank), this matrix is positive definite, so in most cases the index is exactly 3. If the constraints are not independent, the Lagrange multipliers λ are nonunique so the problem is not solvable in our definition, although the variables of interest, y and z, can be determined. See Lotstedt [8].

7. CONCLUSIONS

This paper has described a number of theoretical results which depend on the index of a system. In general, the index of a system, like the rank of a matrix, is not something one should attempt to compute numerically, so what does the ordinary user with the DAE (1) do?

If the index does not exceed 1, automatic codes such as [11] can solve the problem with no trouble. If the problem has index greater than one, an automatic code will usually fail—the stepsize is reduced repeatedly but it cannot satisfy its error tolerance criterion. In that case it would be desirable to apply the technique of Theorem 2.2 to determine if the failure was due to a high index. An integrator for (1) will have computed approximations to $A = \partial F / \partial y'$ and $B = \partial F / \partial y$. Theorem 2.2 can be applied to these approximations. It requires a rank determination which we know is not reasonable. However, if the problem is "near" to a high index problem, it will cause numerical difficulties. Hence, in determining the "rank" we should treat values below appropriately scaled error tolerances as zero. (We have not investigated ways to scale appropriately since we do not yet fully understand how to scale the differential equations.) If Theorem 2.2 suggests that the index is greater than one, the user should be encouraged to reduce it.

The reduction described in Algorithm 5.1 can be applied in many cases because the index is determined by the nonzero structure of the matrices rather than the actual values of their entries as in the Euler-Lagrange equations discussed in Section 6.

Acknowledgment

Much of the work summarized was done jointly with L. Petzold of Sandia National Laboratories, Livermore and is reported in [6].

REFERENCES

[1] Campbell, S. L., Linear time varying singular systems of differential equations, Dept. Mathematics, North Carolina State Univ., Raleigh, 1981.

[2] Deuflhard, P., Order and stepsize control in extrapolation methods, Preprint No. 93, Univ. Heidelberg, 1980.

[3] Gear, C. W., The simultaneous numerical solution of differential-algebraic equations, *IEEE Trans. Circuit Theory TC-18,* (1), 1971, 89-95.

[4] Gear, C. W., Hsu, H. H. and L. Petzold, Differential-algebraic equations revisited, Proc. Numerical Methods for Solving Stiff Initial Value Problems, Oberwolfach, W. Germany, June 28-July 4, 1981.

[5] Gear, C. W. and L. R. Petzold, Differential/algebraic systems and matrix pencils, Proc. Conference on Matrix Pencils, Pitea, Sweden, March 1982; also, Dept. Rpt. UIUCDCS-R-82-1086, 1982.

[6] Gear, C. W. and L. R. Petzold, ODE methods for the solution of differential/algebraic systems, to appear *SIAM J. Num. Anal.*

[7] Gear, C. W. and K. Tu, The effect of variable mesh size on the stability of multistep methods, *SIAM J. Numer. Anal. 11,* (5), 1974, 1025-1043.

[8] Lotstedt, Per, A numerical method for the solution of mechanical systems with unilateral constraints, Report TRITA-NA-7920, Royal Inst. Technology, Stockholm, 1979.

[9] Painter, J. F., Solving the Navier-Stokes equations with LSODI and the method of lines, Lawrence Livermore Laboratory Rpt. UCID-19262, 1981.

[10] Petzold, L. R., Differential/algebraic equations are not ODEs, Rpt. SAND81-8668, Sandia National Laboratories, Livermore, CA, April 1981.

[11] Petzold, L. R., A description of DASSL: A differential/algebraic system solver, to appear, Proceedings of IMACS World Congress, Montreal, Canada, August 1982.

[12] Starner, J. W., A numerical algorithm for the solution of implicit algebraic-differential systems of equations, Tech. Rpt. 318, Dept. Mathematics and Statistics, Univ. New Mexico, May 1976.

[13] Sastry, S. S., Desoer, C. A. and P. P. Varaiya, Jump behavior of circuits and systems, Memorandum No. UCB/ERL M80/44, Electronics Research Laboratory, University of California-Berkeley, CA, October 1980.

[14] Sincovec, R. F., Dembart, B., Epton, M. A. Erisman, A. M., Manke, J. W. and E. L. Yip, Solvability of large-scale descriptor systems, Final Report DOE Contract ET-78-C-01-2876, Boeing Computer Services Co., Seattle, WA.

[15] Wehage, R. A. and E. J. Haug, Generalized coordinate partitioning for dimension reduction in analysis of constrained dynamic systems, *J. Mech. Design 104,* January 1982, 247-255.

THE NUMERICAL SOLUTION OF PROBLEMS WHICH MAY HAVE HIGH FREQUENCY COMPONENTS

C. W. Gear
Department of Computer Science
University of Illinois at Urbana-Champaign

Abstract. This talk surveys the state of the art of methods for solving problems which have the potential for high frequency oscillations. In some cases the oscillatory components can be damped, but in other cases they must be followed. In the latter situation it is sometimes possible to separate the system into fast and slow components. If not, direct methods for nearly periodic solutions can be used.

1. INTRODUCTION

When we say that a problem "may have high frequency components" we can mean a number of different things, each of which requires a different numerical approach, so it is important that we understand these differences. First, by "high-frequency" some people mean just "rapidly changing" while others mean "exhibiting nearly periodic behavior," in other words, "oscillating." By "components" we sometimes really mean "variables," that is, some variables change more rapidly than others, while other times we are thinking of "eigencomponents," that is, the system could, in principle, be transformed into one in which some variables are fast and some are not, but that is not the way it is given to us. Even the word "may" in the title has a number of interpretations. In some cases the differential equations are such that the solution could have high frequency components but the initial conditions are such that the solution we want does not. In other cases we do not know if the solution has these components--but we are not interested in them, and in still further cases we want to know about the fast components. ("Fast," by the way, can only be defined with respect to our interval of integration. If the behavior of a component is such that the integration time step by conventional methods is extremely small compared to the interval of integration, that component is fast.)

Using a title with so many interpretations is not to endorse sloppy language use; rather it is to stress the fact that there are a number of different but related problems whose solutions require different techniques. Because these terms are used with their different meanings by different people, there is sometimes misunderstanding concerning the applicability of various methods and techniques. We will start by looking at some classifications of these problems in an attempt to distinguish subclasses that can be treated by the same or related methods. Then we will examine some of the methods that are available and point out some of the unsolved problems. The first classification we can make concerns the presence, or otherwise, of the first components. In the three cases of importance, the components are:

Supported in part by the U.S. Department of Energy under grant DOE DEAC0276ERO2383.

NATO ASI Series, Vol. F9
Computer Aided Analysis and Optimization of Mechanical System Dynamics
Edited by E. J. Haug

(1) not present in desired solution;

(2) present, but we are not interested in their detail, only in the behavior of the slow components;

(3) present and we are interested in some detail.

The first case could properly be called the *stiff* problem: the fast component is not present in the desired solution, but it is present in nearby solutions. In the easiest form of stiff equations studied, the nearby fast solutions decayed rapidly to the slow solution. This type of problem is *not* stiff in the region in which the fast solution is active (the *transient* region), but becomes stiff after the transient has decayed. The simplest problem with this behavior is the linear equation

$$y' = \lambda(y - F(t)) + F'(t) \tag{1}$$

whose solution is

$$y = ce^{\lambda t} + F(t) \tag{2}$$

If λ is large and negative, the first component decays in the transient region, leaving the second, slow component in the stiff region. If λ is purely imaginary (or, for a real problem, λ is a matrix with purely imaginary eigenvalues), the first component oscillates forever unless the initial values are such that c is zero. Strictly speaking, this problem should only be called stiff if c is zero; otherwise, a small step method that follows the oscillations or special methods that take advantage of their linearity must be used. (There has been an unfortunate tendency to call almost any problem with fast components "stiff," destroying the utility of the classification.)

In the second case we may be able to ignore the fast components in several ways, and thus avoid taking small steps to follow them. We will call this the *quasi-stiff* case. One way is to numerically damp them so they are no longer present in the computed solution. This changes the model, so we must be sure that the components of interest are not changed. For example, in equation (1), λ is large and imaginary, the solutions oscillate around $F(t)$. Damping the oscillation numerically will yield $F(t)$. A second way is to use a method that recognizes the special nature of the fast solution ($e^{\lambda t}$); there are, for example, method which are exact for such components.

In the third case we have to follow some details of the fast components, but want to take some short cuts. This is the situation in which other classifications play an important role. If the fast behavior appears in only a few variables, it is possible to handle those variables with a different methods and/or smaller stepsize to speed integration. If the fast behavior is oscillatory--that is, the solution is nearly periodic, special methods which make use of this are possible.

Each of these three major cases will be discussed below.

2. THE STIFF CASE

Recall we are using "stiff" to mean that the desired solution is slow but that there are nearby solutions that are fast. This means that along the solution $y(t)$, $f(y(t), t)$ is of reasonable size, but that it is large for the nearby value of y. Hence, at least some of the eigenvalues of $\partial f / \partial y$ are large near the solution. (This is true whether or not we have a linear problem.) The approach that must be used with stiff equations is to find methods that do not perturb the solution from the slow one desired to nearby fast ones. For an introduction to standard methods for stiff problems, see [15].

The problem that stiff methods have to deal with is that any numerical integration method that is accurate for some class of equations must have integration errors for other equations. These errors will move the numerical solution away from true solution and introduce the fast components. A stiff method must damp these. The methods for stiff equations make use of the fact that λh is large in the stiff region and use this fact in an implicit method to damp unwanted components. The same approaches can be used if λ is large but imaginary, although this tends to restrict the methods more than if λ is real.

The details of stiff methods are reported in many places in the literature, and we will not repeat them here. The key points are that they must be implicit and that their order is severely restricted as eigenvalues approach the imaginary axis (corresponding to oscillatory behavior). If the eigenvalues are well away from the imaginary axis, BDF methods are usually the most efficient (see [7], but if they are close to the imaginary axis, we may be better off with various forms of implicit Runge-Kutta methods (see [2]).

3. THE QUASI-STIFF CASE

If the high frequency components are present but we are not interested in them, we would like to get rid of them in one way or another. High frequency components can arise in a number of ways, including

(1) a fast *driving term:* the t-component in $y' = f(y, t)$

(2) a (nearly) linear oscillator: a purely imaginary eigenvalue of the Jacobian

(3) a non-linear oscillator: these are usually characterized by eigenvalues that alternate rapidly between very negative and very positive

(4) a fast initial transient which damps out: a very negative eigenvalue

If the fast components have only a *linear effect* on the rest of the system, then we can attempt to compute the time average value of the solution over a period of time which is long compaied to the fast components but short compared to the slow components. Since, in the linear case, the averaging operator commutes with the equation, we can then find the averaged slow response by solving the slow equations using the averaged value of the fast components. If the fast components are oscillatory, their average value is their steady-state value, so in this case, it suffices to simply damp the oscillations. Therefore, if we have a fast, linear oscillator, we can use a method that is stable and damping along the imaginary axis (such as backward Euler, second order BDF, or the q-stage implicit Runge-Kutta methods of order $2q-1$ based on the subdiagonal Pade approximate to e^{λ}). If we have a strongly nonlinear oscillator, it may only be necessary to use a method that is stable for very positive eigenvalues as this will prevent the oscillation from starting. (A typical nonlinear oscillator, such as the Van der Pol equation, has positive eigenvalues while the solution has small oscillations. As the solution becomes large, the eigenvalues become negative, limiting the oscillation.) Most stiffly stable methods (BDFs, for example) are also stable in the positive half plane.

It must be emphasized that we can only damp the oscillation (that is, use the average) without ill-effect if the effect of the oscillation is linear. The example,

$$u' = \omega v$$

$$v' = -\omega u$$

$$y' = (u^2 + v^2)^{1/2}$$

illustrates this. The solution is $u = A\sin(\omega t + b)$, $v = A\cos(\omega t + b)$, $y = At + c$, so that the solution of the slow component, y, depends on the amplitude of the oscillation. In this situation we must use techniques to be discussed in the next section.

If there is a fast initial transient, we can use the conventional methods to get through the transient, then switch to stiff methods. Usually the transient is sufficiently short that there is not much point in looking for faster methods. If the transient is reasonably short, it is possible to damp it even more rapidly without introducing significant error. In this case it may be possible to use stiff methods with large stepsizes, even in the transient.

4. THE FAST CASE

This is the case of greatest difficulty (and interest). The fast components are present, and either of direct interest or cannot be ignored because of their effect on the slow components. We will look at two types of methods that can handle some of these problems more effectively than conventional methods: *multirate methods,* which can sometimes be used when the fast terms are confined to a few variables, and *multirevolutionary methods,* which can be used when the fast components are nearly periodic.

Multirate Methods

One area in which multirate methods have been considered is simulation [1] [12], although there is little documentation on the subject. Some work has been done on partitioning systems into subsystems for treatment by different methods, as in [13] where the nonlinear part is integrated by a conventional Runge Kutta method and the linear part is handled by methods that use precomputed matrix exponentials. In our case, we are interested in partitioning based on the relative speeds of subsystems. In such applications, the engineer is often able to use a large amount of "engineering intuition" in the design of an integrator. For example, consider the simulation of an aircraft. It consists of a number of subsystems. For simplicity we will consider two shown in Figure 1.

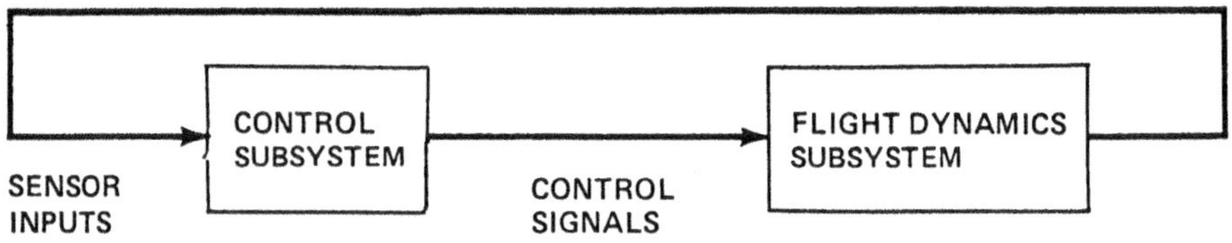

Figure 1. Simplified simulation model

The control subsystem reacts very rapidly (being electronic), whereas the flight dynamic subsystem reacts relatively slowly, being a mechanical change. Consequently, a simulation system tailored to this model might choose to integrate the control with small stepsizes, and the dynamics with large stepsizes. The rationale sometimes heard for this is that because the dynamics are slow, we can break the feedback loop from the dynamics to the control and handle each subsystem separately. "Breaking the feedback loop" refers to assuming that there is no change in the output of the slow system over one of its steps while the fast system is being integrated over several smaller steps. What is actually happening? Suppose the outputs of the fast and slow systems are designated as y and z respectively, and they satisfy the differential equations

$$y' = f(y, z, t) \tag{3}$$

$$z' = g(y, z, t)$$

Let us suppose that the stepsizes used for the y and z integrations are h and Mh respectively, where M is an integer, and that the values of y and z are known for time value t_{Mn}, where $t_k = t_0 + kh$. The process consists of integrating y over M small steps of size h while z is held constant at its value of $z(t_{Mn})$. Then z is integrated over one step of size Mh so that we have information at time $t_{M(n+1)}$. Figure 2 shows a possible response of the numerical system where $f(y, z, t) = -100(y - z)$. The y values change rapidly after z has been updated, then settle to a new "stable" state based on the value of z at the beginning of the group of M small steps. In fact, we would expect that the true value of y at the end of the set of M small steps should be determined by the value of z at the end, and this suggests that it might be better to perform the integration of z over its big step before doing the M steps of y. This would correspond to "breaking the feedback loop" between the control and dynamic subsystems; intuitively less appealing but possibly more accurate. At first glance it might appear that this simply transfers the problem to z which now will be integrated over one step Mh based on the value of y at the start of that step. This is not a problem if the integration method used for z is explicit because the value of $g(y, z, t)$ is needed only from the beginning of the step, but there is a more important reason why it is not a problem. If the behavior of y is fast but of z is slow, *there can be very little coupling from y to z.* That is to say, the dependence of $g(y, z, t)$ on y must be very small. Hence the link from y to z is a natural place to "break the feedback loop." This point will be important in later discussions.

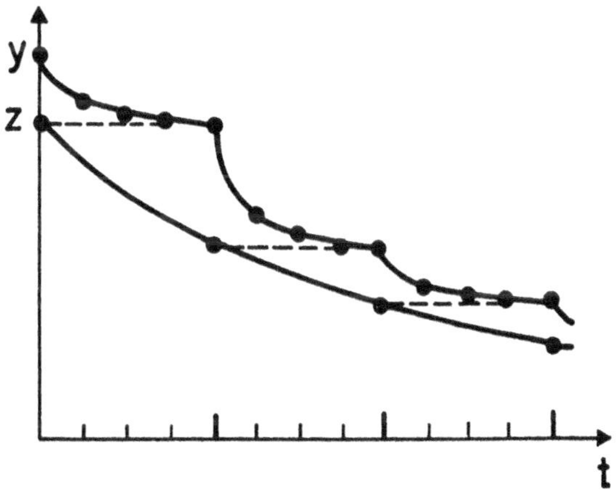

Figure 2. Integration after breaking the feedback loop

Wherever the feedback loop is broken, a constant value of one variable is being used in the integration of the other variable over an interval of length Mh . This "approximation" by a constant function will introduce an error of size $O(Mh)$, and its integration over an interval of size Mh contributes a local error of $O(Mh)^2$ which causes a global error of size $O(Mh)$. The natural thing to do is to interpolate or extrapolate the value of z (or y) while integrating y (or z) thus reducing the approximation error. If the extrapolation is done by a polynomial of degree p-1 it introduces an extrapolation error of $O(Mh)^p$. In that case the global error introduced will also be $O(Mh)^p$, which would be a natural ally of an order p integration method. It follows that if methods are used for extrapolation (or interpolation) and integration, the method is order p convergent.

Speed Considerations: Multirate methods appear attractive because they should use less computer time. The reasoning is that the total time is roughly proportional to the number of integration steps taken for each equation. Since a conventional method must use an amount of work proportional to the number of steps for the fastest component times the number of equations, a multirate method can reduce the amount of work for the slow components. However, multirate programs can be more complex than straightforward methods, so we must be certain that increased program overhead does not override any savings. The computational cost of a numerical integration program is due to several factors. They are:

Evaluations of derivatives
Application of integration formulas
Interpolations/extrapolations if a multirate method is used
Estimating errors
Logic for automatic step control
Repeated steps when a failure occurs and nearly completed steps must be discarded
Solution of implicit equations if they are used

Let us assume that the same method is used for each equation over stepsizes of h and Mh, and that it requires q function evaluations per step. We will consider the cost of one compound step of size Mh. Suppose that c_f, c_g, c_z, and c_i are the costs of an evaluation of f, an evaluation of g, an extrapolation for of z, and an integration step, respectively. Then the cost of one compound step is

$$C_M = (qc_f + c_z)M + qc_g + (M + 1)c_i \tag{4}$$

This assumes that only one interpolation/extrapolation is needed for a z value in each integration step for y, as would be the case in a multistep method. If intermediate values are used, as in a Runge Kutta method, the cost must be increased by $M(q-1)c_z$. This must be compared with the cost of using stepsize h for both components over M steps, which is

$$C = (c_f + c_g)Mq + 2Mc_i \tag{5}$$

because no extrapolations will be necessary. The difference is

$$C - C_M = (qc_g + c_i)(M - 1) - M[q]c_z \tag{6}$$

where the [q] term appears if a Runge Kutta-like method is used. In Runge-Kutta methods, c_i is small and q is large (4 to 13, depending on the order of the method). In this case we can approximate the difference by

$$C - C_M \approx (c_g - c_z)(M - 1)q - qc_z \tag{7}$$

Therefore, there is nothing to be gained from multirate Runge Kutta methods unless $c_g > c_z$ in which case there is a saving if

$$M > 1 + \frac{c_z}{c_g - c_z} \tag{8}$$

If c_g is only a little larger than c_z, the slight savings for large M will be lost to the undoubtedly higher overhead of a more complex program. On the other hand, for multistep methods, q is small, typically 2. Also, the number of prior points used for extrapolation is naturally the number of points used in the multistep method for two reasons. The first is that we are saving values at those points anyway. The second is that the predictor step of a multistep method is an extrapolation; we have chosen the number of values and the stepsize in the multistep method such that the error of the prediction is not too large, and this implies that the error in the extrapolation will also be appropriately small if the same number of past values are used. Consequently, the cost of a multistep integration step is only slightly higher than the cost of an extrapolation because the former involves a prediction step followed by a correction. Therefore,

$$c_z < c_i$$

Using this in eq. (6), we get

$$C - C_M > q(M-1)c_g - c_z \tag{9}$$

Therefore, we may see a saving as long as

$$M > 1 + \frac{c_z}{qc_g} \tag{10}$$

Eqs. (8) and (10) indicate that multistep methods are more likely to yield a savings in multirate schemes than Runge Kutta methods, but that unless c_g is large compared to c_z, a large M will be needed, which means that there must be a very large disparity in behavior of the two components.

Similar comments apply to larger systems, but sparsity can have a positive effect. Suppose that y and z above each represent systems of r equations, but that the evaluation of f requires the value of only s of the z variables. Then the extrapolation cost is sc_z whereas the cost of evaluation of all of the components of g is rc_g if c_z and c_g are the costs for each component. Thus the sparsity ratio s/r multiplies c_z in equations (8) and (10), reducing the value M for which savings can be achieved. Clearly this analysis can be extended to cases in which more than two differents stepsizes are employed and the same general result will be obtained, namely that the ratio of the cost of extrapolation to the cost of function evaluation is critical. Also note that it is very important that where possible different components are evaluated at the same point. If, for example, the two components in eq. (1) were integrated with the same stepsize on different meshes, say $t_0 + nh$ and $t_0 + (n+1/2)h$, two unnecessary extrapolations would be done in each step, and the method would take longer than if the equations were "kept in synchronization."

Variable-step Methods: Very few problems are such that the user can predetermine the stepsize to be used throughout an integration, so most modern codes vary the order and stepsize during the integration based on local error estimates. In this section we look at the problem of varying the stepsize in a multirate method.

There are problems for which it is known that some components are always faster than others, and for these problems it may be possible to permanently set the ratio of the stepsizes. Then it is a question of choosing one step size and letting the other adjust correspondingly. A typical automatic code performs a single step integration, estimates the error, and then decides whether to accept the step because the estimated error is within tolerance, or to reject it and repeat it with a smaller stepsize. A multirate method with a fixed ratio between stepsizes in different components could be viewed as a type of cyclic method with a cycle equal to the largest stepsize used in any component. Error estimations could be made over that cycle and either the cycle accepted or rejected and repeated. Unfortunately this introduces two inefficiencies: the loss of a relatively large amount of work when a cycle is rejected, and the need to back-up a number of steps in some components. This means that a considerable number of additional past values must be saved.

If large-scale back-up and loss of work is to be avoided, errors must be estimated as each step is made. In that case, there does not seem to be any great reason for fixing the stepsize ratio; indeed in general we cannot decide a priori on a ratio of stepsizes, some components may be the fastest at some times and the slowest at others. Three approaches to the organization of automatic multirate methods have been tried. In all approaches the local error is estimated for each component separately, and then the step in that component is either accepted or rejected. If it is rejected, it will be tried with a smaller stepsize later. The approaches are:

(1) An "event-driven" model in which each mesh point is viewed as an event, and the next event in time is chosen for execution.

(2) A "recursive" view in which all equations are attempted with the largest possible stepsize, and those that are rejected are integrated with a pair of recursive calls at half the stepsize.

(3) A "largest-step-first" approach which retains some of the advantages of the recursive approach but reduces function evaluations.

The first approach was tried in an experimental code described in [10]. Each component of the system was integrated with a possibly different stepsize and order. They were selected by the integrator for each component independently of the behavior of the other components. Consequently, at any given time each component i had been integrated to a different t value t_i, and the integrator had suggested a different stepsize h_i for the next step for each. The basic strategy was to select the component with the smallest value of $t_i + h_i$ as the component to be integrated next. It would be integrated one step, its error estimated, and a value for its next stepsize recommended. The idea behind this is that the extrapolation performed in the other components is to points within the range of their next recommended steps and within this range the extrapolation should be reasonably accurate. The snag to this arrangement is that the recommended step may be too large. If the recommended step is far too large, the extrapolation can be badly in error causing large errors to be propogated to other components. The difficulties do not stop there. If a recommended step turns out to be too large and must be greatly reduced, we will find that we need to approximate a fast component at a t value that is too far back to safely extrapolate backwards. Consequently, an arbitrary amount of back-up may be necessary. Clearly, it is not feasible to save information for backing up any component to arbitrary earlier values, so we tried various techniques to prevent step failures. Consider the integration of z for a moment. The recommended step may be too large because of a change of behavior in g that is not predicted by earlier values of z, or by a change in the behavior of y that couples into z through g. The first can be predicted by doing a trial integration of z with a simple extrapolation for y. We used this to see if the recommended step seemed reasonable. It is clearly expensive because it almost doubles the number of integration steps. The second source of problems was examined by trying to estimate the effect of changes in the behavior of y on z. We assumed that the stepsize chosen for z had been based on the current knowledge of y. This is equivalent to saying that it is based on the extrapolated value of y. When y was integrated, the predictor-corrector difference is the difference between the extrapolated value of y and the new value. The effect of this difference on the z integration was estimated. This required knowledge of the Jacobian g_y. The method consisted of integrating one component one step using the selection algorithm given above, and then determining if the recommended steps of any other components had to be reduced because of the error in the step just performed. The amount of work was made manageable for a system of several equations by keeping the sparsity structure of the Jacobian. Only components which had nonzero entries in the row of the Jacobian corresponding to the equation just integrated had to be considered. The results of numerical tests of this technique indicated that it is feasible if the evaluation costs of g are sufficiently high and the ratio of stepsizes that could be used in different components was high enough. For a number of test problems of 6 to 20 equations, the number of individual evaluations of derivatives for the multirate method was about half that of the standard method. However, the overhead was very much higher.

The principal difficulty in the event-driven method is due to the integration of fast components before one is certain that the selected stepsize for the slow components is not too large. Recall from Section 2 that we said the coupling between fast and slow components is necessarily small and it might make sense to integrate the slower component first. In the recursive method we do just that. All equations are integrated using the largest recommended stepsize. Then the error in each is estimated and the results for those with small errors are accepted. The remainder are reintegrated using two steps of the half size by the same technique. The advantages of the recursive approach are that it is conceptually simple and not too difficult to program even if recursion is not provided in the language. Also, more accurate interpolation rather than extrapolation is used to get the intermediate values of the slow variables when integrating the faster ones. Its disadvantage is that there are many unnecessary function evaluations, integrations, and interpolations for those components whose stepsizes have to be reduced several times. In fact, there are almost twice as many evaluations, integrations, and interpolations because if a step of $H/2^N$ is used when the largest step is H, the number of unsuccessful steps is $2^N - 1$ to the number of successful ones 2^N.

The third technique uses a similar principle to the recursive technique, that is, it integrates the slowest components with the largest stepsizes first. A maximum stepsize H is chosen, and all stepsizes are $H/2^N$ for some N. Initially, all components are known at a given time value t_0. The components whose stepsizes are H are integrated first. A step is halved if its estimated error is too large. Next the steps of size H/2 are integrated, and so on. Finally, the step with the smallest size, $H/2^N$, are integrated. This process is repeated for all components whose current time value is least. The immediate effect of this is to integrate the components with the smallest step to $t_0 + H/2^{(N-1)}$. Repeating the process again causes the components with stepsizes $H/2^{(N-1)}$ and $H/2^N$ to be integrated, in that order. If the local error estimate in a component is very small, the stepsize can be doubled. However, a step of size $H/2^M$ for $M > 0$ can only be doubled when the time status of its differential equation is on a mesh point corresponding to steps of size $H/2^{(M-1)}$. The effect of this is that components using stepsize $H/2^M$ are evaluated only on the mesh points $T_0 + mH/2^M$ for m = 0, 1, ..., 2^M. When time $T_0 + H$ is reached, a new maximum stepsize H is chosen on the basis of the error estimates for the slowest components.

The last method appears to be the most efficient on the basis of preliminary tests. Note that it has one hidden additional cost. It is necessary to extrapolate the values of the fast components in order to evaluate the derivatives of the slower components which are integrated first. Because of the assumed small coupling between the fast and slow components, we have been using very low order extrapolation such as linear methods, even when higher-order integration schemes are used. Additional studies and tests on this are reported in [17].

Stability, Stiffness and Implicit Methods: It must be stressed that the existence of slow and fast components has nothing to do with stiffness per se. A system may be neutrally stable and still profit from multirate methods, but the existence of stiffness can be a complicating factor because of the need to use implicit integration methods and to solve nonlinear equations at each step. When a stiff problem is to be solved, the Jacobian of all equations being integrated to the same mesh point must be formed, and this Jacobian must be used in a quasi-Newton iteration for the solution of the implicit integration formulas. In the event approach above, the values of the other variables will be extrapolated and fixed. In the recursive method, all components with larger stepsizes can be interpolated, and the system is solved for all components with the current and smaller stepsizes. In the largest-step-first method, values of variables with smaller stepsizes are not yet known. Consequently they must either be extrapolated--a risky business in stiff equations--or can be held constant. This corresponds to a zero-th order extrapolation and seems to be the best choice.

The use of multirate methods is attractive when function evaluation They are particularly attractive when the stepsizes can be fixed *a priori*, as in real-time simulation. However, the organization of an automatic code is not simple, and it is very easy to allow the overhead to become a major part of the cost. A particular problem is that of avoiding back-up over several steps at a step failure, and the counter-intuitive approach of integrating the slowest components first seems to be most effective in this.

Multirevolutionary Methods

These are methods designed to handle problems whose solutions have high frequency, nearly periodic components which cannot be ignored.

The problem of highly oscillatory ODEs has some parallels with that of stiff ODEs: often the solution is not nearly periodic initially, and maybe not even oscillatory, so conventional methods are best in this transient phase, but after awhile the solution exhibits a nearly periodic behavior and the objective may be to determine the average behavior, the waveform, or its envelope over many millions of cycles. There are some methods that are applicable in the latter nearly periodic phase, for example, [8], [11], and [14]. However, these methods cannot be used in the transient phase, so we must detect the onset of nearly periodic behavior. Conversely, a nearly periodic system may cease to be so. This also must be detected so that a switch back to a conventional integrator can be made, just as detection of the termination of stiffness is also desirable, although there it is for the sake of efficiency, not necessity.

It is important to realize that the difficulty of highly oscillatory ODEs is, unlike stiff equations, *not* due to the presence of large eigenvalues. Large eigenvalues may be present and be responsible for the oscillatory behavior, but in the more interesting cases the system is nonlinear and we must track the amplitude and waveshape of the oscillation. (Note that tracking the phase over billions of cycles is an inherently ill-conditioned problem unless the phase is locked to an oscillatory input.)

The methods we consider for nearly periodic problems are generally known as *multirevolutionary* from their celestial orbit background. The idea of such methods is to calculate, by some conventional integrator, the change in the solution over one orbit. If the period of an orbit is T (for a moment assumed fixed), then a conventional integrator is used to compute the value of

$$D(t, y) = d(t) = y(t + T) - y(t)$$

by integrating the initial value problem $y' = f(t, y)$ over one period T. If we consider the sequence of times $t = mT$, m integral, we have a sequence of values $y(mT)$ which are slowly changing if y is nearly periodic. The conventional integrator allows us to compute the first differences $d(mT)$ of this sequence at any time mT. Under appropriate "smoothness" conditions (whatever that means for a sequence) we can interpolate or extrapolate for values of $d(mT)$ from a subset of all values of d, for example from $d(kqT)$, $k = 1, 2, 3, \ldots$, where q is an integer > 1, and thus estimate $y(mT)$ by integrating only over occasional orbits.

In a satellite orbit problem it is fairly easy to define the meaning of "one period." For example, one could use a zero crossing of a particular coordinate, or even a fixed period based on a first order theory. In her thesis, Petzold [14] considered problems for which it is difficult to find physical definitions of the period and examined a method for determining the approximate period by minimizing a function of the form

$$I(t, T) = \int_{t}^{t+T} ||y(\tau + T) - y(\tau)||d\tau .$$

The value of T which minimizes $I(t, T)$ is a function of g, and $T(t)$ was said to be the period of the solution. This enabled $d(t) = y(t + T(t)) - y(t)$ to be calculated and multirevolutionary methods to be used. The variable period was handled easily by a change of independent variables to s in which the period is constant, say 1. The equation

$$t(s + 1) - t(s) = T(t(s))$$

was appended to the system

$$z(s + 1) - z(s) = g(s, z)$$

where $z(s) = y(t(s))$ and $g(s, z) = D(t(s), z)$ for integer values of s. (When T is constant, this is the analog of the old device for converting a non-autonomous system to an autonomous system by appending the differential equation $t' = 1$.)

The scheme for period calculation used by Petzold suffers from three drawbacks. The first drawback is that it is fairly expensive, requiring a numerical approximation to the first two derivatives of $I(t,T)$ by quadrature which itself requires the values of $y(\tau)$, $y'(\tau)$, and $y''(\tau)$ over two periods. The second drawback is that a reasonably accurate period estimate is needed for the iteration to converge. Outside the region of convergence a search scheme for a minimum could be used but this would be very expensive because of the computation involved in each quadrature even if all previously computed values could be saved. This makes the approach very unattractive for initial period detection when there is no starting estimate. The third drawback is that minimizing a function subject to several sources of error (including truncation errors in the integration and quadrature, and roundoff errors revealed by considerable cancellation in $||y(\tau + T) - y(\tau)||$) is likely to yield a fairly inaccurate answer. Since the value of $d(t) = g(s, z)$ is quite sensitive to small absolute changes in the period T, the function $g(s, z)$ may not appear to be very smooth.

An alternate approach to determination of the period was described in Gear [6]. It also allows for the onset of nearly periodic behavior to be detected and a decision to be made when to switch to multirevolutionary methods. This method can also be used to decide when the solution is no longer nearly periodic. It should be noted that in this case, $T(t)$ and hence $D(t, y)$ and $g(t, y)$ are no longer defined. This method has been implemented in [3]. As Gallivan points out, it is important to use the same technique to decide when to invoke the multirevolutinary methods as used in these methods to control their continued use, or the program may repeatedly switch back and forth. The multirevolutionary and periodic detection/determination techniques will be summarized below.

The Quasi-envelope and Multirevolutionary Methods: Suppose, for a moment, that the period $T(t)$ is known. To simplify the discussion we will also take it to be a constant, although neither of these suppositions is necessary. A *period T quasi-envelope*, $z(t)$, of a function $y(t)$ is *any* function that agrees with y at the *periodic points* $t = mT$. We are interested in the case in which the function y(t) is the solution of the initial value problem $y' = f(t, y)$, $y(0) = y_0$, which is nearly periodic with period T, and in a smooth quasi-envelope. For example, if $y(t)$ is periodic, then the best quasi-envelope for our purposes is a constant. The importance of the quasi-envelope is that when we know it we have a low-cost way of computing the solution of the original problem at any point: to find the value of $y(t^*)$ choose the largest integer m such that $mT \leq t^*$ and integrate $y' = f(y, t)$ from $t = mT$, $y(mt) = z(mT)$ to $t = t^*$. If m is very large, this is much less expensive $z(t + T) - z(t) = d(t)$. Hence, from the quasi-envelope and the differential equation we can compute information such as the waveform, amplitude, energy, etc., at any point at a low cost. Note that if the original ODE is autonomous, we can integrate it from any starting point $(t, z(t))$ to determine a waveshape which evolves continuously (and differentiably) in time. The same is approximately true if the ODE is nearly autonomous, that is, $\partial f / \partial t$ is small compared to 1/T. In these cases it is not necessary to start the integration at a periodic point. We call this the ***unsynchronized*** mode. Otherwise we can either determine the phase from the driving term in what we call the ***synchronized mode,*** or the phase is unimportant.

A multirevolutionary method is a technique for computing a quasi-envelope given a way to compute $z(t = T) - z(t) = d(t)$. For small T this says $z'(t) = d(t)/T$, hence, it is not surprising that the numerical interpolation for $z(t)$ given a technique for computing $d(t)/T$ is very similar to a numerical integration technique. In the new coordinate system, the basic structure of the program is an ***outer integrator*** which solves the equations

$$z(s + 1) - z(s) = g(t(s), z(s))$$

$$t(s + 1) - t(s) = T(t(s))$$

using an outer stepsize H. The method varies the order and stepsize just as an ordinary integrator does. See [3] for details. It calls a subroutine to evaluate g and T given z and t. This is done by integrating the underlying ordinary differential equation $y' = f(y)$ starting from $y(t) = z$, determining when a period has elapsed and computing $g(t, z) = y(t + T(t)) - y(t)$.

Periodic Behavior Detection: We have been deliberately imprecise about the meaning of "nearly periodic," and will continue that way with the working definition in our minds of the type of problem that can be handled efficiently by multirevolutionary methods. We have been equally imprecise about the definition of the "period" of a nearly periodic function. We could use some intuitively reasonable mathematical description, in which case we would have to seek computational algorithms for its approximation. However, the period is most easily defined in terms of the algorithm used to calculate it. It should, of course, yield the exact period for periodic functions and be close for small perturbations of periodic functions. This replaces an analysis of the accuracy of period calculation with an analysis of the efficiency of the multirevolutionary method with respect to different period definitions. This latter may be an easier task.

Petzold's period definition, based on minimizing a norm, is very expensive to apply and cannot be considered as a technique for determining if an arbitrary output of an integrator is nearly periodic. Therefore, we look for alternate definitions of the period. First, note that is the oscillation is due to a periodic driving function, we probably know its period or can examine the system which generates the driving function directly. Hence, we can restrict ourselves to autonomous systems or nearly autonomous systems. A nearly autonomous system can be made autonomous by the substitution $t = v/\epsilon$ and the additional equation $v' = \epsilon$. Since v is slowly changing, the enlarged autonomous system may also be nearly periodic.

The solution of an autonomous system is completely determined by the specification of the value of the solution vector y at one time. That is to say, if we identify two times on the solution such that $y(t_1) = y(t_2)$, we know that the solution is periodic with period $t_2 - t_1$. This first suggests determining the period by looking for minimum of $||y(t_1) - y(t_2)||$. The cost of this is not particularly low and it requires a clever adaptive program with a lot of heuristics to determine the onset of nearly periodic behavior because we know neither t_1, the value when the behavior first occurs, not $t_2 - t_1$, the period.

A more reliable way of defining the period is to identify certain points on the solution at which a simple characterization is repeated, such as zero crossing. The solution itself may not have zero crossings and, if it consists of a periodic function superimposed on a slowly growing function, there may be difficulty in choosing any value which is crossed periodically. However, its derivative will have periodic sign changes, so we have experimented with a definition of period based on the zero crossings of $c^T y'$ where c is a vector of constants. The program examines the integrator output for positive-going zero crossings of $c^T y'$. (Currently, c is a vector of the weights provided by the user for error norm calculations.) Anything but a simple periodic solution may lead to more than one zero crossing in a single period, so the norm $||y'(t_1) - y'(t_2)||$ is also examined, where t_1 and t_2 are a pair of zero crossings. If the norm is small, the possibility of a period is considered. The procedure used is as follows:

(1) Identify a positive going sign change in $c^T y'$.

(2) Interpolate to find the t value, $t_{current}$, of the zero crossing. Also compute interpolated values of y and y' at the zero crossing.

(3) Save these values. (Up to ten prior values are saved in the experimental program.)

(4) Compare the current values of y' with each prior value in turn until a small $||y'_{old} - y'_{current}||$ is found.

(5) Save $T = t_{current} - t_{old}$.

(6) Continue to calculate additional periods, T, starting from the latest $t_{current}$ each time. Examine the backward differences of T over several periods. When they are small, indicating a smoothly varying period, consider switching to multirevolutionary methods. Details are given in [3].

The decision on when to switch to multirevolutionary methods is based on estimates of the stepsize H that can be used in the outer integrator. Because the ODE has been integrated over several periods, we have backward differences of $g(t(s_n), z(s_n))$ based on a stepsize in s of $H = 1$. These are used to estimate the order and stepsize that can be used by multirevolutionary methods. Next, the work factor, H/W, is calculated where H is the stepsize and W is an estimate of the cost of the multirevolutionary method compared to a non-multirevolutionary method when H is 1. If H/W exceeds one, a switch is made to multirevolutionary methods.

Period Calculation and Stiffness Detection: When using the multirevolutionary method, we need to compute $g(t(s), z)$ and $T(t(s))$ given $z = z_n$ and $t(s) = t_n$. This uses the same technique as the periodic detection except that the vector c must be chosen so that $c^T y_n' \equiv c^T f(t_n, z_n) = 0$. The first step of the inner integrator is executed so that y_n' and y_n'' can be calculated and estimated, respectively. Then c is chosen to maximize $c^T y_n''$ subject to $||c|| = 1$ and $c^T y_n' = 0$. (A single equation requires special treatment here, but it can only be oscillatory if there is an oscillatory driving term.) The inner integrator continues and positive going zero crossings of $c^T y'$ are checked to find one such that $||y' - y_n'||$ is small. If a period is not found within 30 of the previously calculated period, it is decreed that the function is no longer nearly periodic from the assigned starting values. This will cause a stepsize reduction in the outer integrator until the periodic detection is successful or the outer stepsize H is so small that the work factor is less than one. This causes a switch back to a conventional method, as would be appropriate if the solution were no longer nearly periodic.

The outer integrator initially uses a generalized Adams method because there is no knowledge of the Lipschitz constant. Two corrector iterations are used, enabling a Lipschitz estimate to be obtained. The step/order selection algorithm is basically that described in the previous section. Whenever the stepsize is estimated, the decision between stiff and nonstiff methods is made based on the current Lipschitz estimate.

5. CONCLUSION

Results reported in [6] and [3] indicate that some highly oscillatory problems can be integrated very efficiently by these methods. The types of problems that are amenable to these techniques are those with a single oscillation, either due to a driving term or a nonlinear oscillator whose behavior is "stable," that is, whose amplitude and waveform are not sensitive to small perturbations. Essentially this means that the problem is reasonably well posed. The important problem of two or more oscillations at different frequencies cannot be currently handled by these techniques.

When the fast behavior is confined to a few variables, multirate methods can be considered. The use of these methods is attractive when function evaluation costs are very high, or if there is a high degree of sparsity. They are particularly attractive when the stepsizes can be fixed *a priori*, as in real-time simulation. However, the organization of an automatic code is not simple, and it is very easy to allow the overhead to become a major part of the cost. A particular problem is that of avoiding back-up over several steps at a step failure, and the counter-intuitive approach of integrating the slowest components first seems to be most effective in this.

REFERENCES

[1] Andrus, J. F., Numerical solution of systems of ordinary differential equations separated into subsystems, *SIAM J. Numerical Analysis 16* (4), August 1979, 605-611.

[2] Gaffney, P. W., A survey of FORTRAN subroutines for solving stiff oscillatory ordinary differential equations, ORNL/CSD/TM-134, Oak Ridge National Laboratory, 1981.

[3] Gallivan, K. A., An algorithm for the detection and integration of highly oscillatory ordinary differential equations using a generalized unified modified divided difference representation, Dept. Rept. R-83-1121; also, Ph.D. thesis, May 1983.

[4] Gear, C. W., Tu, K-W., The Effect of Variable Mesh Size on the Stability of Multistep Methods, *SIAM J. Numerical Analysis 11,* (4), October 1974, 1025-1043.

[5] Gear, C. W., Runge-Kutta starters for multistep methods, *TOMS 6* (3), September 1980, 263-279.

[6] Gear, C. W., Automatic treatment of stiff and/or oscillatory equations, Dept. Rept. R-80-1019; *Proc. Bielefeld Conference on Numerical Methods in Computational Chemistry,* Bielefeld, W. Germany, 1980, in *Lecture Notes in Mathematics 968,* 1982, 190-206.

[7] Gear, C. W., Stiff software: what do we have and what do we need?, *Proc. Intl. Conf. Stiff Computation,* Salt Lake City, Utah, April 12-14, 1982.

[8] Graff, O. F. and D. G. Bettis, Modified multirevolution integration methods for satellite orbit computation, *Celestial Mechanics 11,* 1975, 443-448.

[9] Graff, O. F., Methods of orbit computation with multirevolution steps, Applied Mechanics Research Laboratory Report 1063, University of Texas at Austin, 1973.

[10] Orailoglu, A., A Multirate Ordinary Differential Equation Integrator, Dept. Rept. R-79-959, March 1979.

[11] Mace, D. and L. H. Thomas, An extrapolation method for stepping the calculations of the orbit of an artificial satellite several revolutions ahead at a time, *Astronomical J. 65* (5), June 1960.

[12] Palusinski, O. A., Simulation of dynamic systems using multirate techniques, CSRL Memo #333, Engineering Experiment Station, University of Arizona, Tuscon, Nov. 1979.

[13] Palusinski, O. A. and J. V. Wait, Simulation methods for combined linear and nonlinear systems, *Simulation 30* (3), March 1978, 85-94.

[14] Petzold, L. R., An efficient numerical method for highly oscillatory ordinary differential equations, Dept. Rept. R-78-933; also, Ph.D. Thesis 1978.

[15] Shampine, L. F. and C. W. Gear, A user's view of solving stiff ordinary differential equations, *SIAM Review 21* (1), January 1979, 1-17.

[16] Skelboe, S., Computation of the periodic steady state response of nonlinear networks by extrapolation methods, *IEEE Trans. Circuits and Systems CAS-27*, (3), 1980, 161-175.

[17] Wells, D. R., Multirate linear multistep methods for the solution of systems of ordinary differential equations, Dept. Rept. R-82-1093; also, Ph.D. Thesis, July 1982.

SOME METHODS FOR DYNAMIC ANALYSIS OF CONSTRAINED MECHANICAL SYSTEMS: A SURVEY

Parviz E. Nikravesh
Center for Computer Aided Design
College of Engineering
The University of Iowa
Iowa City, Iowa 52242

Abstract. Three algorithms are presented for dynamic analysis of constrained mechanical systems. The first algorithm integrates the differential equations of motion without any consideration for constraint violation. The other two algorithms consider the violation of the kinematic constraints and correct the violation in two different ways. A brief comparison between these algorithms is also provided.

1. INTRODUCTION

Transient dynamic solution of equations of motion for constrained mechanical systems requires solution of a mixed set of algebraic and differential equations. The algebraic equations are the constraint equations that describe the kinematic joints in the system, which are generally nonlinear in terms of the generalized coordinates, and the differential equations are of second order. Except for simple problems, exact closed form solution for these equations cannot be found. Therefore, numerical methods must be employed.

The subject of numerical methods for solving mixed systems of algebraic and differential equations has not yet been fully understood. Several methods for solving mixed algebraic and differential equations, with reference to the transient dynamic analysis of mechanical systems, have been suggested and tested in the past decade. One method converts the algebraic equations to second order differential equations, then solves these equations with the differential equations of motion, without considering the integration numerical error that results in constraint violation. This method is discussed in Section 3. A second method, which is discussed in

NATO ASI Series, Vol. F9
Computer Aided Analysis and Optimization of Mechanical System Dynamics
Edited by E. J. Haug

Section 4, introduces constraint violations as feed-back terms to correct the violations in the next integration step. Finally, a third method is presented in Section 5. In this method, the generalized coordinates are partitioned into independent and dependent sets. Numerical integration is carried out for independent generalized coordinates. Then, constraint equations are solved for dependent generalized coordinates. In this paper, prior to the presentation of these three methods and algorithms, the general form of algebraic and differential equations of motion is formulated. Finally a brief comparison of the three algorithms is presented in the last section.

2. SYSTEM EQUATIONS OF MOTION

Consider a mechanical system that is modeled by n generalized coordinates

$$\mathbf{q} = [q_1, q_2, \ldots, q_n]^T \tag{2.1}$$

The vectors of generalized velocities and accelerations are denoted by

$$\dot{\mathbf{q}} = [\dot{q}_1, \dot{q}_2, \ldots, \dot{q}_n]^T \tag{2.2}$$

and

$$\ddot{\mathbf{q}} = [\ddot{q}_1, \ddot{q}_2, \ldots, \ddot{q}_n]^T \tag{2.3}$$

Presume there are m holonomic constraint equations in the system, expressed as

$$\boldsymbol{\Phi}(\mathbf{q}, t) = \mathbf{0} \tag{2.4}$$

where t is the time. These equations are, in general, nonlinear in terms of $\mathbf{q}$. The m constraint equations cause the n generalized coordinates to be dependent. Similarly, the velocities and accelerations are dependent, according to the kinematic velocity and acceleration equations,

$$\boldsymbol{\Phi}_{\mathbf{q}}\, \dot{\mathbf{q}} = \boldsymbol{\nu} \tag{2.5}$$

and

$$\Phi_q \ddot{q} = \gamma \tag{2.6}$$

where

$$\nu = -\Phi_t \tag{2.7}$$

and

$$\gamma = -(\Phi_q \dot{q})_q \dot{q} - 2\Phi_{qt} \dot{q} - \Phi_{tt} \tag{2.8}$$

Note that Eqs. 2.5 and 2.6 consist of m equations each, which are linear in $\dot{q}$ and $\ddot{q}$ respectively.

In addition to the kinematic constraint, velocity, and acceleration equations, there are n second order differential equations of motion. These equations are expressed as

$$\mathbf{M} \ddot{\mathbf{q}} + \Phi_q^T \lambda = \mathbf{g} \tag{2.9}$$

where the vector $\boldsymbol{\lambda}$ contains m Lagrange multipliers associated with m holonomic constraints. In Eq. 2.9, $\mathbf{M} = \mathbf{M}(\mathbf{q})$ is the generalized mass matrix that can be a function of $\mathbf{q}$, $\Phi_q \equiv \frac{\partial \Phi}{\partial q}$ is the Jacobian matrix of the kinematic constraint equations, and $\mathbf{g} = \mathbf{g}(\mathbf{q}, \dot{\mathbf{q}}, t)$ is the vector of <u>modified</u> generalized forces. The term 'modified' is used since; e.g., when Euler parameters are used, some additional terms are included in the vector of generalized forces.

To complete the equations of motion, an appropriate set of initial conditions must be defined. The proper set of initial conditions is

$$\mathbf{q}(0) = \mathbf{q}^0 \tag{2.10}$$

and

$$\dot{\mathbf{q}}(0) = \dot{\mathbf{q}}^0 \tag{2.11}$$

These initial conditions cannot be given arbitrary values, since the initial conditions $\mathbf{q}^0$ and $\dot{\mathbf{q}}^0$ must satisfy the constraint and velocity equations of Eqs. 2.4 and 2.5, respectively.

In the next sections, three methods for solving the above set of constraint and differential equations are presented. The advantages and disadvantages of each method are discussed to some extent. These algorithms are stated to determine the transient response for $\mathbf{q}$, $\dot{\mathbf{q}}$, $\ddot{\mathbf{q}}$, and $\boldsymbol{\lambda}$ from an initial time t^0, normally $t^0 = 0$, to a final time t^e. The time parameter t is incremented from t^0 to t^e by constant or variable increments Δt(or h).

3. DIRECT INTEGRATION METHOD

At a given instant in time, if the generalized coordinates and velocities, $\mathbf{q}$ and $\dot{\mathbf{q}}$, are known, then the accelerations $\ddot{\mathbf{q}}$ and Lagrange multipliers $\boldsymbol{\lambda}$ can be calculated. The equations of motion of Eq. 2.9 provide n equations for n + m unknowns $\ddot{\mathbf{q}}$ and $\boldsymbol{\lambda}$. However, Eq. 2.6 provides m equations for n unknowns $\ddot{\mathbf{q}}$ that can be appended to Eq. 2.9 to give

$$\begin{bmatrix} \mathbf{M} & \boldsymbol{\Phi}_{\mathbf{q}}^T \\ \boldsymbol{\Phi}_{\mathbf{q}} & 0 \end{bmatrix} \begin{bmatrix} \ddot{\mathbf{q}} \\ \boldsymbol{\lambda} \end{bmatrix} = \begin{bmatrix} \mathbf{g} \\ \boldsymbol{\gamma} \end{bmatrix} \tag{3.1}$$

Equation 3.1 is a set of n + m linear equations in n + m unknowns $\ddot{\mathbf{q}}$ and $\boldsymbol{\lambda}$. The matrix on the left of Eq. 3.1 is function of $\mathbf{q}$ and the vector on the right is function of $\mathbf{q}$ and $\dot{\mathbf{q}}$ that can be evaluated, since $\mathbf{q}$ and $\dot{\mathbf{q}}$ are known.

To show that the coefficient matrix on the left of Eq. 3.1 is nonsingular, it is sufficient to prove that

$$\begin{bmatrix} \mathbf{M} & \boldsymbol{\Phi}_{\mathbf{q}}^T \\ \boldsymbol{\Phi}_{\mathbf{q}} & 0 \end{bmatrix} \begin{bmatrix} \boldsymbol{\alpha} \\ \boldsymbol{\beta} \end{bmatrix} = \mathbf{o} \tag{3.2}$$

implies $\boldsymbol{\alpha} = 0$ and $\boldsymbol{\beta} = 0$. In Eq. 3.2, $\boldsymbol{\alpha}$ and $\boldsymbol{\beta}$ are arbitrary n and m vectors respectively. Premultiplying Eq. 3.2 by $[\boldsymbol{\alpha}^T, \boldsymbol{\beta}^T]$, it is found that

$$\boldsymbol{\alpha}^T \mathbf{M} \boldsymbol{\alpha} + \boldsymbol{\alpha}^T \boldsymbol{\Phi}_{\mathbf{q}}^T \boldsymbol{\beta} + \boldsymbol{\beta}^T \boldsymbol{\Phi}_{\mathbf{q}} \boldsymbol{\alpha} = 0 \tag{3.3}$$

The second row of Eq. 3.2 shows that $\boldsymbol{\Phi}_{\mathbf{q}} \boldsymbol{\alpha} = \mathbf{o}$, then the last two terms of Eq. 3.3 are zeros. Hence, Eq. 3.3 becomes

$$\boldsymbol{\alpha}^T \mathbf{M} \boldsymbol{\alpha} = 0 \tag{3.4}$$

If $\boldsymbol{\alpha}$ is interpreted as a nonzero velocity that is consistent with kinematic constraints, then there must be an associated positive kinetic energy. Therefore, if $\boldsymbol{\alpha}$ is nonzero, then $\boldsymbol{\alpha}^T \mathbf{M} \boldsymbol{\alpha} > 0$. Hence, Eq. 3.4 can be true only for $\boldsymbol{\alpha} = \mathbf{0}$. Now, substitution of $\boldsymbol{\alpha} = \mathbf{0}$ into the first row of Eq. 3.2 yields

$$\boldsymbol{\Phi}_{\mathbf{q}}^T \boldsymbol{\beta} = \mathbf{o} \tag{3.5}$$

Premultiplying Eq. 3.5 by $\boldsymbol{\Phi}_{\mathbf{q}}$ results in

$$\boldsymbol{\Phi}_{\mathbf{q}} \boldsymbol{\Phi}_{\mathbf{q}}^T \boldsymbol{\beta} = \mathbf{o} \tag{3.6}$$

The matrix $\boldsymbol{\Phi}_{\mathbf{q}} \boldsymbol{\Phi}_{\mathbf{q}}^T$ is positive definite, therefore Eq. 3.6 implies that $\boldsymbol{\beta} = \mathbf{o}$. By the Lax-Milgram theorem [1], there exists a unique solution of Eq. 3.2. Since $\boldsymbol{\alpha} = \mathbf{o}$, $\boldsymbol{\beta} = \mathbf{o}$ is a solution, it is the only solution and the coefficient matrix in Eq. 3.2 is nonsingular.

A variety of numerical methods, such as Gaussian elimination or L-U factorization, may be employed to solve Eq. 3.1. Following the solution of Eq. 3.1 for $\ddot{\mathbf{q}}$, the numerical integration process may begin. If the velocity vector $\dot{\mathbf{q}}$ is renamed as

$$\mathbf{s} = \dot{\mathbf{q}} \tag{3.7}$$

then,

$$\dot{\mathbf{s}} = \ddot{\mathbf{q}} \tag{3.8}$$

Now both vectors $\dot{\mathbf{q}}$ and $\dot{\mathbf{s}}$ at time t are integrated to obtain $\mathbf{q}$ and $\mathbf{s}$ at the next time step; i.e.,

$$\begin{aligned} \dot{\mathbf{q}}^t &\xrightarrow{\text{integrate}} \mathbf{q}^{t+\Delta t} \\ \dot{\mathbf{s}}^t &\xrightarrow{\text{integrate}} \mathbf{s}^{t+\Delta t} \end{aligned} \tag{3.9}$$

Equation 3.9 simply indicates that 2n variables, $\dot{\mathbf{q}}$ and $\ddot{\mathbf{q}}$, at time t are integrated to obtain $\mathbf{q}$ and $\dot{\mathbf{q}}$ at time $t + \Delta t$. The reason for introducing the new vector $\mathbf{s}$ is that most numerical integration algorithms deal with first order differential equations. When $\mathbf{q}$ and $\dot{\mathbf{q}}$ are determined at the next time step, the process of solving Eq. 3.1 and the numerical integration step can be repeated. This method is summarized in the following algorithm:

Algorithm I:

(a) Main Routine

(a.1) set a time step counter i to i = 0 and initialize $t^i = t^0$

(a.2) use initial conditions $\mathbf{q}^i = \mathbf{q}^0$ and $\dot{\mathbf{q}}^i = \dot{\mathbf{q}}^0$

(a.3) define vector $\mathbf{y}$ as $\mathbf{y}^i = [\mathbf{q}^{iT}, \dot{\mathbf{q}}^{iT}]^T$

(a.4) enter a predictor/corrector (or some other method) numerical integration routine

(b) Numerical Integration Routine

This routine integrates (solves) initial-value problems of the form $\dot{\mathbf{y}} = \mathbf{f}(\mathbf{y}, t)$, from $t^i = t^0$ to $t^i = t^e$

.
.
.

(b.1) during the prediction and the correction steps $\mathbf{f}(\mathbf{y}^i, t^i)$ must be evaluated;
call FUNCTION routine with known $\mathbf{y}^i$ and t^i to obtain $\dot{\mathbf{y}}^i$

.
.
.

(c) FUNCTION Routine

(c.1) transfer $\mathbf{y}^i$ to $\mathbf{q}^i$ and $\dot{\mathbf{q}}^i$

(c.2) determine Jacobian matrix $\mathbf{\Phi}^i_{\mathbf{q}}$, generalized mass matrix M^i, vectors $\boldsymbol{\gamma}^i$ and $\mathbf{g}^i$, then solve Eq. 3.1 to obtain $\ddot{\mathbf{y}}^i$ and $\boldsymbol{\lambda}^i$

(c.3) transfer $\dot{\mathbf{q}}^i$ and $\ddot{\mathbf{q}}^i$ to vector $\dot{\mathbf{y}}^i$

(c.4) return

It can be seen that this algorithm does not use the kinematic constraint and velocity equations of Eqs. 2.4 and 2.5. Numerical integration algorithms provide only an approximate solution, instead of an exact solution to the differential and algebraic equations under consideration. Therefore, when $\dot{\mathbf{q}}^0$ and $\ddot{\mathbf{q}}^0$ at $t = t^0$ are integrated to obtain $\mathbf{q}^1$ and $\dot{\mathbf{q}}^1$ at $t^1 = t^0 + \Delta t$, $\mathbf{q}^1$ and $\dot{\mathbf{q}}^1$ will contain some error. Hence, $\mathbf{q}^1$ and $\dot{\mathbf{q}}^1$ may not satisfy Eqs. 2.4 and 2.5 precisely; i.e.,

$$\mathbf{\Phi}(\mathbf{q}^1, t^1) = \boldsymbol{\varepsilon} \tag{3.10}$$

and

$$\mathbf{\Phi}_{\mathbf{q}}\,\dot{\mathbf{q}}^1 + \mathbf{\Phi}_t = \boldsymbol{\sigma} \tag{3.11}$$

where $\boldsymbol{\varepsilon}$ and $\boldsymbol{\sigma}$ are referred to as constraint violations. When $\boldsymbol{\varepsilon} = \mathbf{0}$ and $\boldsymbol{\sigma} = \mathbf{0}$, the constraints are satisfied. Since Eq. 3.1 is a function of $\mathbf{q}$ and $\dot{\mathbf{q}}$, $\ddot{\mathbf{q}}^1$ is an approximation to the exact accelerations at $t = t^1$. When $\dot{\mathbf{q}}^1$ and $\ddot{\mathbf{q}}^1$ are integrated to find $\mathbf{q}^2$ and $\dot{\mathbf{q}}^2$, in many cases $\mathbf{q}^2$ and $\dot{\mathbf{q}}^2$ contain more error than in the previous step. In the first few time steps, the constraint violations $\boldsymbol{\varepsilon}$ and $\boldsymbol{\sigma}$ are usually small and negligible. However, as time progresses, the error in computed values for $\mathbf{q}$, $\dot{\mathbf{q}}$, and $\ddot{\mathbf{q}}$, accumulates and constraint violations increase.

4. CONSTRAINT VIOLATION STABILIZATION METHOD

It was shown in Section 3 that the direct integration method may yield large numerical error in the solution vectors $\mathbf{q}$, $\dot{\mathbf{q}}$, and $\ddot{\mathbf{q}}$. Since the correct values of these vectors are not known, the amount of numerical error in computed values cannot be determined directly. However, since these vectors must satisfy the kinematic constraint equations, their correctness can be tested by monitoring violation of the constraints. The constraint violation correcting method discussed in this section allows constraints to be violated slightly before corrective action can take place, in order to force the violation to vanish [2].

The generalized coordinate and velocity vectors $\mathbf{q}$ and $\dot{\mathbf{q}}$ must satisfy the constraint and velocity equations

$$\boldsymbol{\Phi} \equiv \boldsymbol{\Phi}(q, t) = \mathbf{o} \tag{4.1}$$

and

$$\dot{\boldsymbol{\Phi}} \equiv \boldsymbol{\Phi}_{\mathbf{q}}\, \dot{\mathbf{q}} - \boldsymbol{\nu} = \mathbf{o} \tag{4.2}$$

respectively. The acceleration vector $\ddot{\mathbf{q}}$ always satisfies the acceleration equations

$$\ddot{\boldsymbol{\Phi}} \equiv \boldsymbol{\Phi}_{\mathbf{q}}\, \ddot{\mathbf{q}} - \boldsymbol{\gamma} = \mathbf{o} \tag{4.3}$$

since Eq. 4.3 combined with the equations of motion; i.e.,

$$\begin{bmatrix} \mathbf{M} & \boldsymbol{\Phi}_{\mathbf{q}}^T \\ \boldsymbol{\Phi}_{\mathbf{q}} & \mathbf{0} \end{bmatrix} \begin{bmatrix} \ddot{\mathbf{q}} \\ \boldsymbol{\lambda} \end{bmatrix} = \begin{bmatrix} \mathbf{b} \\ \boldsymbol{\gamma} \end{bmatrix} \tag{4.4}$$

are solved for $\ddot{q}$ and λ. At the ith time step, $\dot{q}^i$ and $\ddot{q}^i$ are integrated to obtain q^{i+1} and $\dot{q}^{i+1}$, then Eq. 4.4 is solved for $\ddot{q}^{i+1}$; i.e.,

$$\begin{array}{l} \dot{q}^i \xrightarrow{\text{integrate}} q^{i+1} \\ \ddot{q}^i \xrightarrow{\text{integrate}} \dot{q}^{i+1} \end{array} \xrightarrow{\text{Eq. 4.4}} \ddot{q}^{i+1} \tag{4.5}$$

At the new time step, the process of Eq. 4.5 is repeated as

$$\begin{array}{l} \dot{q}^{i+1} \xrightarrow{\text{integrate}} q^{i+2} \\ \ddot{q}^{i+1} \xrightarrow{\text{integrate}} \dot{q}^{i+2} \end{array} \xrightarrow{\text{Eq. 4.4}} \ddot{q}^{i+2} \tag{4.6}$$

From Eqs. 4.5 and 4.6, it can be deduced that

$$\ddot{q}^i \xrightarrow{\text{integrate}} \dot{q}^{i+1} \xrightarrow{\text{integrate}} q^{i+2} \tag{4.7}$$

The coordinate, velocity, and acceleration vectors of Eq. 4.7 must satisfy Eqs. 4.1, 4.2, and 4.3 respectively; i.e.,

$$\Phi^{i+2} \equiv \Phi(q^{i+2}, t^{i+2}) = 0 \tag{4.8}$$

$$\dot{\Phi}^{i+1} \equiv \Phi_q^{i+1} \dot{q}^{i+1} - \nu = 0 \tag{4.9}$$

and

$$\ddot{\Phi}^i \equiv \Phi_q^i \ddot{q}^i - \gamma = 0 \tag{4.10}$$

However, because of numerical errors, Eqs. 4.8 and 4.9 may not be satisfied. Equations 4.8, 4.9, and 4.10 may be interpreted as indirectly integrating $\ddot{\Phi}^i$ to obtain $\dot{\Phi}^{i+1}$ and, similarly, indirectly integrating $\dot{\Phi}^{i+1}$ to obtain Φ^{i+2}; i.e.,

$$\ddot{\Phi}^i \xrightarrow[\text{integrated}]{\text{indirectly}} \dot{\Phi}^{i+1} \xrightarrow[\text{integrated}]{\text{indirectly}} \Phi^{i+2} \tag{4.11}$$

Then, the violations of Eqs. 4.8 and 4.9 may be interpreted as indirect integration error caused by the process of Eq. 4.11.

The process of integration, direct or indirect, and the numerical error accumulation between Eqs. 4.7 and 4.11 are analogous. Therefore, the constraint violation correcting method attempts to consider and correct the numerical error in terms of the process shown in Eq. 4.11, instead of the process stated in Eq. 4.7. Correcting the error in the process of Eq. 4.11, in turn, will correct the numerical error in the process of Eq. 4.7.

In control systems and circuit theory, it is well known that circuits described by second order differential equations such as

$$\ddot{y} = 0 \tag{4.12}$$

are unstable, since outside disturbances such as noise can be amplified. In contrast to Eq. 4.12, known as an open loop system, a closed loop system defined as

$$\ddot{y} + 2\alpha\dot{y} + \beta^2 y = 0 \tag{4.13}$$

where α and β are positive nonzero constants, is stable. The terms $2\alpha\dot{y}$ and $\beta^2 y$ in Eq. 4.13 play the role of control terms that achieve stability for the differential equation of Eq. 4.13.

Using the above idea and considering the numerical integration error as an outside disturbance, the differential equations $\ddot{\Phi} = \mathbf{0}$ of Eq. 4.3 may be replaced by

$$\ddot{\Phi} + 2\alpha\,\dot{\Phi} + \beta^2\,\Phi = 0 \tag{4.14}$$

or

$$\Phi_q\,\ddot{q} = \gamma - 2\alpha\,\dot{\Phi} - \beta^2\,\Phi \tag{4.15}$$

Hence, the modified acceleration equations and the equations of motion are combined and written as

$$\begin{bmatrix} M & \Phi_q^T \\ \Phi_q & 0 \end{bmatrix} \begin{bmatrix} \ddot{q} \\ \lambda \end{bmatrix} = \begin{bmatrix} b \\ \gamma - 2\alpha\,\dot{\Phi} - \beta^2\,\Phi \end{bmatrix} \tag{4.16}$$

At time t^i, after $\mathbf{q}^i$ and $\dot{\mathbf{q}}^i$ are obtained, Eqs. 4.1 and 4.2 may yield some violation; i.e., nonzero Φ and $\dot{\Phi}$. By specifying constant values to α and β, Eq. 4.16 is solved for $\ddot{\mathbf{q}}^i$ and λ^i. Note that when constraint and velocity equations are satisfied, Φ and $\dot{\Phi}$ are zero and Eq. 16 is identical to Eq. 4.4. However, when the constraints are violated, these correcting terms provide adjustments in $\ddot{\mathbf{q}}^i$. These adjustments are such that, when $\dot{\mathbf{q}}^i$ and $\ddot{\mathbf{q}}^i$ are integrated, $\mathbf{q}^{i+1}$

and $\dot{\mathbf{q}}^{i+1}$ move toward response values that are consistent with constraints.

When both α and β are considered to be zero, which is exactly the method of Algorithm I, the numerical result may diverge from the exact solution. In contrast, for nonzero values of α and β, the solution vector oscillates about the exact solution. Magnitude and frequency of the oscillation, due to the correcting effect, depend on the values of α and β. Experience has shown that, for most practical problems, a range of values between 5 to 50 for α and β is adequate. When $\alpha = \beta$, critical damping is achieved, which usually stabilizes the response more quickly. This method is summarized in the following algorithm:

Algorithm II:

(a) Main Routine

- (a.1) set a time-step counter i to $i = 0$ and initialize $t^i = t^0$
- (a.2) use initial conditions $\mathbf{q}^i = \mathbf{q}^0$ and $\dot{\mathbf{q}}^i$ to $\dot{\mathbf{q}}^0$
- (a.3) define vector $\mathbf{y}$ as $\mathbf{y}^i = [\mathbf{q}^{iT}, \dot{\mathbf{q}}^{iT}]^T$
- (a.4) specify α and β parameters
- (a.5) enter the predictor/corrector numerical integration routine

(b) Numerical Integration Routine

this routine integrates (solves) initial-value problems of the form $\dot{\mathbf{y}} = \mathbf{f}(\mathbf{y}, t)$, from $t^i = t^0$ to $t^i = t^e$

•
•
•

- (b.1) during the prediction and the correction steps $\mathbf{f}(\mathbf{y}^i, t^i)$ must be evaluated;
  call FUNCTION routine with known $\mathbf{y}^i$ and t^i to obtain $\dot{\mathbf{y}}^i$

•
•

(c) FUNCTION Routine

- (c.1) transfer $\mathbf{y}^i$ to $\mathbf{q}^i$ and $\dot{\mathbf{q}}^i$
- (c.2) determine constraint violations $\boldsymbol{\Phi}^i$ and $\dot{\boldsymbol{\Phi}}^i$ from Eqs. 4.1 and 4.2
- (c.3) determine generalized mass matrix $\mathbf{M}^i$, Jacobian matrix $\boldsymbol{\Phi}_{\mathbf{q}}^i$, vectors $\boldsymbol{\nu}^i$ and $\mathbf{g}^i$, then solve Eq. 4.16 for $\ddot{\mathbf{q}}^i$ and $\boldsymbol{\lambda}^i$
- (c.4) transfer $\dot{\mathbf{q}}^i$ and $\ddot{\mathbf{q}}^i$ to vector $\dot{\mathbf{y}}^i$
- (c.5) return

5. GENERALIZED COORDINATE PARTITIONING METHOD

The generalized coordinate partitioning method [3] controls the accumulated response error quite differently from the method of Section 4. In this method, at every time step, the constraint, velocity, and acceleration equations are satisfied within a specified error tolerance.

This method makes use of the fact that the n generalized coordinates $\mathbf{q}$ are not independent. The generalized coordinates are dependent, through the m independent constraint equations

$$\mathbf{\Phi}(\mathbf{q}, t) = \mathbf{0} \tag{5.1}$$

where $n > m$. Since the m constraint equations are independent, the Jacobian matrix for Eq. 5.1; i.e.,

$$\mathbf{\Phi}_{\mathbf{q}} \equiv \frac{\partial \mathbf{\Phi}}{\partial \mathbf{q}} \tag{5.2}$$

has full row rank. The implicit function theorem [4] guarantees that if $k = n - m$ of the generalized coordinates are specified, then Eq. 5.1 may be solved for the remaining m generalized coordinates. The k generalized coordinates that are given specified values are denoted by $\mathbf{v}$ and are called <u>independent generalized coordinates</u>. The remaining m quantities are denoted by $\mathbf{u}$ and are called <u>dependent generalized coordinates</u>. With this notation, the vector $\mathbf{q}$ is partitioned as

$$\mathbf{q} = [\mathbf{u}^T, \mathbf{v}^T]^T \tag{5.3}$$

A method for partitioning $\mathbf{q}$ into $\mathbf{u}$ and $\mathbf{v}$ is discussed in the Appendix. Now, Eq. 5.1 can be expressed as

$$\mathbf{\Phi}(\mathbf{u}, \mathbf{v}, t) = \mathbf{0} \tag{5.4}$$

From the implicit function theorem, it is deduced that if there is a point $\mathbf{q}$ that satisfies Eq. 5.1, then in a neighborhood of $\mathbf{q}$ there exist continuously differentiable functions $\boldsymbol{\psi}(\mathbf{v}, t)$ such that

$$\mathbf{u} = \boldsymbol{\psi}(\mathbf{v}, t) \tag{5.5}$$

satisfies Eq. 5.1 or Eq. 5.4; i.e., $\mathbf{\Phi}(\boldsymbol{\psi}(\mathbf{v}, t), \mathbf{v}, t) = \mathbf{0}$.

The vector of generalized velocities $\dot{\mathbf{q}}$ is partitioned, according to the partitioning of $\mathbf{q}$, into

$$\dot{\mathbf{q}} = [\dot{\mathbf{u}}^T, \dot{\mathbf{v}}^T]^T \tag{5.6}$$

where $\dot{\mathbf{u}}$ and $\dot{\mathbf{v}}$ are called dependent and independent generalized velocities, respectively. Now the velocity relation of Eq. 2.5 is written as

$$\Phi_{\mathbf{u}} \dot{\mathbf{u}} + \Phi_{\mathbf{v}} \dot{\mathbf{v}} = \nu \tag{5.7}$$

where $\Phi_{\mathbf{u}}$ and $\Phi_{\mathbf{v}}$ are obtained by partitioning the Jacobian matrix $\Phi_{\mathbf{q}}$ as

$$\Phi_{\mathbf{q}} \equiv [\Phi_{\mathbf{u}}, \Phi_{\mathbf{v}}] \tag{5.8}$$

Note that $\Phi_{\mathbf{u}}$ is an $m \times m$ matrix and $\Phi_{\mathbf{v}}$ is an $m \times k$ matrix. Further note that the columns of $\Phi_{\mathbf{q}}$ may be permuted, according to the partitioning of $\mathbf{q}$ into $\mathbf{u}$ and $\mathbf{v}$, in order to obtain matrices $\Phi_{\mathbf{u}}$ and $\Phi_{\mathbf{v}}$. It is shown in the Appendix that $\Phi_{\mathbf{u}}$ is a nonsingular matrix. Therefore, Eq. 5.7 can be written as

$$\dot{\mathbf{u}} = \Phi_{\mathbf{u}}^{-1}(\nu - \Phi_{\mathbf{v}} \dot{\mathbf{v}}) \tag{5.9}$$

Equation 5.9 shows that if $\dot{\mathbf{v}}$ is known, then $\dot{\mathbf{u}}$ can be found. Since matrix inversion is not computationally efficient, methods such as Gaussian elimination or L-U factorization may be performed to solve Eq. 5.9 for $\dot{\mathbf{u}}$.

The acceleration vector $\ddot{\mathbf{q}}$ is partitioned as

$$\ddot{\mathbf{q}} = [\ddot{\mathbf{u}}^T, \ddot{\mathbf{v}}^T]^T \tag{5.10}$$

The combined acceleration equations and equations of motion are kept in the form

$$\begin{bmatrix} M & \Phi_{\mathbf{q}}^T \\ \Phi_{\mathbf{q}} & 0 \end{bmatrix} \begin{bmatrix} \ddot{\mathbf{q}} \\ \lambda \end{bmatrix} = \begin{bmatrix} b \\ \gamma \end{bmatrix} \tag{5.11}$$

which can be solved for $\ddot{\mathbf{q}}$ and λ.

If the independent velocity vector is renamed as

$$\mathbf{s} = \dot{\mathbf{v}} \tag{5.12}$$

then

$$\dot{\mathbf{s}} = \ddot{\mathbf{v}} \tag{5.13}$$

Now, the vectors $\dot{\mathbf{v}}$ and $\dot{\mathbf{s}}$ at time t^i are integrated to obtain $\mathbf{v}$ and $\mathbf{s}$ at the next time step t^{i+1}; i.e.,

$$\begin{aligned} \dot{\mathbf{v}}^i &\xrightarrow{\text{integrate}} \mathbf{v}^{i+1} \\ \dot{\mathbf{s}}^i &\xrightarrow{\text{integrate}} \mathbf{s}^{i+1} \end{aligned} \tag{5.14}$$

Equation 5.14 indicates that 2k variables $\dot{\mathbf{v}}^i$ and $\ddot{\mathbf{v}}^i$ at time t^i are integrated to obtain $\mathbf{v}^{i+1}$ and $\dot{\mathbf{v}}^{i+1}$ at time t^{i+1}. Following the determination of $\mathbf{v}^{i+1}$ and $\dot{\mathbf{v}}^{i+1}$, the process of solving Eqs. 5.5, 5.9, and 5.11 is repeated at time t^{i+1}. This method is summarized in the following algorithm:

Algorithm III:

(a) Main Routine
- (a.1) set a time step counter i to i = 0 and initialize $t^i = t^0$
- (a.2) use initial conditions $\mathbf{q}^i = \mathbf{q}^0$ and $\dot{\mathbf{q}}^i = \dot{\mathbf{q}}^0$ that are consistent with constraints
- (a.3) partition $\mathbf{q}$ into $\mathbf{u}$ and $\mathbf{v}$
- (a.4) define vector $\mathbf{y}$ as $\mathbf{y}^i = [\mathbf{v}^{iT}, \dot{\mathbf{v}}^{iT}]^T$
- (a.5) enter the predictor/corrector numerical integration routine

(b) Numerical Integration Routine
this routine integrates (solves) initial-value problems of the form $\dot{\mathbf{y}} = \mathbf{f}(\mathbf{y}, t)$, from $t^i = t^0$ to $t^i = t^e$

.
.
.

- (b.1) during the prediction and the correction steps $\mathbf{f}(\mathbf{y}^i, t^i)$ must be evaluated;
  call FUNCTION routine with known $\mathbf{y}^i$ and t^i to obtain $\dot{\mathbf{y}}^i$

.
.
.

(c) FUNCTION Routine
- (c.1) transfer $\mathbf{y}^i$ to $\mathbf{v}^i$ and $\dot{\mathbf{v}}^i$

(c.2) having $\mathbf{v}^i$ solve Eq. 5.4 for $\mathbf{u}^i$

(c.3) having $\mathbf{v}^i$, $\mathbf{u}^i$, and $\dot{\mathbf{v}}^i$ solve Eq. 5.7 for $\dot{\mathbf{u}}^i$

(c.4) determine generalized mass matrix $\mathbf{M}^i$, Jacobian matrix Φ_q, vectors γ^i and $\mathbf{g}^i$, then solve Eq. 5.11 for $\ddot{\mathbf{q}}^i$ and λ^i, and split $\ddot{\mathbf{q}}^i$ into $\ddot{\mathbf{u}}^i$ and $\ddot{\mathbf{v}}^i$

(c.5) transfer $\dot{\mathbf{v}}^i$ and $\ddot{\mathbf{v}}^i$ to vector $\dot{\mathbf{y}}^i$

(c.6) return

Numerical experiments have shown that one of the most troublesome parts of this algorithm is step (c.2). In this step the independent coordinates $\mathbf{v}^i$ are known and the constraint equations are solved for the dependent coordinates $\mathbf{u}^i$. Since the algebraic constraint equations are, in general, highly nonlinear, an iterative numerical method such as Newton-Raphson iteration must be employed. Such iterative methods require an estimate to the solution vector $\mathbf{u}^i$. This estimate cannot be too far away from the exact solution, since the iterative process may not converge. An estimate for $\mathbf{u}^i$ at step (i), at the present form of Algorithm III, is the values of $\mathbf{u}$ from the previous time step; i.e., $\mathbf{u}^{i-1}$. If Δt is large enough, then $\mathbf{u}^{i-1}$ may not be a close estimate for $\mathbf{u}^i$, therefore it may cause divergence. To overcome this difficulty, a much better estimate than $\mathbf{u}^{i-1}$ is needed. This objective, in most cases, can be accomplished by the use of a polynomial interpolation (extrapolation) technique. This technique requires limited history of the dependent variables $\mathbf{u}$; i.e., $\mathbf{u}^{i-1}$, $\mathbf{u}^{i-2}$, ... These values are used and extrapolated ahead in time to determine a better estimate of $\mathbf{u}^i$. Note that there is no need to extrapolate for $\dot{\mathbf{u}}^i$, since Eq. 5.7 is a set of linear equations in $\dot{\mathbf{u}}$.

Proper partitioning of the generalized coordinates $\mathbf{q}$ into $\mathbf{u}$ and $\mathbf{v}$ is critical in controlling the amount of the numerical error. If at $t = t^0$ the generalized coordinates are partitioned into $\mathbf{u}^0$ and $\mathbf{v}^0$, at a later time, since the system is in motion, this set of independent and dependent coordinates may not be adequate; i.e., calculating $\mathbf{u}^i$ from $\mathbf{v}^i$ may include unacceptable numerical error. Two methods for testing the amount of error in calculating $\mathbf{u}$ is suggested here:

(1) Check the norm of each row of the influence matrix

$$\mathbf{H} = -\Phi_u^{-1}\,\Phi_v \tag{5.15}$$

The numerical error in **u** and **v**, namely $\Delta\mathbf{u}$ and $\Delta\mathbf{v}$ respectively, are related by

$$\Delta\mathbf{u} = \mathbf{H}\ \Delta\mathbf{v} \tag{5.16}$$

If the elements of **H** are less than or equal unity, then the numerical error in **v** will not be magnified into **u** and the partitioning of **q** into **u** and **v** is still accepted. Otherwise, **q** must be partitioned into a new set of **u** and **v**.

(2) If the numerical integration method used is a variable time step algorithm with error control, then the time step taken by the algorithm can be used as a criterion. In successive time steps as long as Δt is increasing or remains unchanged, the partitioning of **q** into **u** and **v** is still accepted. As soon as Δt decreases, **q** must be partitioned into a new set of **u** and **v**, since accumulation of error has been detected by the numerical integration algorithm.

One of the above tests can be included in Algorithm III, after each successful time step taken by the integration routine. If repartitioning is needed, the algorithm should return to step (a.3) and continue the process from that instant of time.

6. COMPARISON

The Dynamic Analysis and Design System (DADS-2D) computer program [5] was used to simulate several small and large-scale mechanical systems for transient response. The DADS program can employ any of the three algorithms discussed in previous sections, upon the user's request. Each problem was simulated by all three methods and the findings are presented in following paragraphs.

The error accumulation and possible divergence of Algorithm I (A-I) is somewhat proportional to the size (number of bodies and elements) of the problem. If highly stiff force elements; e.g. stiff springs, or sudden large variation in external forces are exhibited, then the divergence rate is faster. Correct initial conditions on positions and velocities; i.e., no initial violation of constraints and velocity equations, is very crucial. It is suggested here that use of this method should be avoided.

Algorithm I can be converted to Algorithm II with minor additional programming. The high frequency response, contributed to the system response, from the stabilization terms is almost undetectable in the position curves, but noticeable on the accelation curves. The most important factor on stability and error accumulation in this method is the correct initial condition. Depending on the problem at hand, the initial constraint violation may increase, remain unchanged, or vanish during dynamic analysis. When correct initial conditions are given, the algorithm is almost insensitive to the values of the parameters α and β within the range of 1 to 20. This algorithm remained stable for about 75% of the simulations.

The third algorithm proved to be the most stable among the three algorithms. The choice of correct initial condition is not critical, as long as the initial condition on the independent variables at time $t = 0$ is correct. A comparison of approximate CPU time used by each algorithm, by averaging over all of the problems, is given in Table 1.

Table 1
CPU Comparison with
Reference to Algorithm I

Algorithm	I	II	III
CPU	1.0	1.5	2.0

REFERENCES

1. Aubin, J.P., Applied Functional Analysis, Wiley, New York, 1979.

2. Baumgarte, J., "Stabilization of Constraints and Integrals of Motion," Comput. Meth. in Appl. Mech. Eng., 1, 1972.

3. Wehage, R.A. and Haug, E.J., "Generalized Coordinate Partitioning for Dimension Reduction in Analysis of Constrained Dynamic Systems," ASME J. of Mech. Design, vol. 104, no. 1, 1982.

4. Goffman, C., Calculus of General Variables, Harper and Row, New York, 1965.

5. Nikravesh, P.E. and Park, T., "Dynamic Analysis and Design System Computer Program for Planar Motion (Modular-DADS-2D)," Center for Computer Aided Design, The University of Iowa, 1983.

APPENDIX

Consider the m×n matrix **A** with full row rank. A process of L-U factorization with full pivoting converts matrix **A** into

$$[\bar{\mathbf{A}}] \rightarrow [\mathbf{L}/\mathbf{U}|\mathbf{R}]$$

where $\bar{\mathbf{A}}$ is a permutation of **A** by row and column interchanging, **L** is a lower triangular matrix with unit diagonal elements, **U** is an upper triangular matrix with nonzero diagonal terms, and **R** is an m×(n-m) matrix.

If the constraint Jacobian $\Phi_{\mathbf{q}}$ is assumed to be the matrix **A**, then the products **LU** and **LR** are

$$\Phi_{\mathbf{u}} \equiv \mathbf{LU}$$

and

$$\Phi_{\mathbf{v}} \equiv \mathbf{LR}$$

where **u** and **v** are called the dependent and independent variables and can be used as partitioning variables of vector **q**. Since matrices **L** and **U** are found by factorization with full pivoting of the matrix $\Phi_{\mathbf{q}}$, $\Phi_{\mathbf{u}}$ is nonsingular.

An influence matrix $\mathbf{H} = -\Phi_{\mathbf{u}}^{-1}\Phi_{\mathbf{v}}$ can be determined from

$$\mathbf{H} = -\mathbf{U}^{-1}\mathbf{L}^{-1}\mathbf{LR} = -\mathbf{U}^{-1}\mathbf{R}$$

In order to calculate the elements of H by the above equation, $\mathbf{U}^{-1}$ is not needed. Instead,

$$\mathbf{UH} = -\mathbf{R}$$

can be used.

APPLICATION OF ANIMATED GRAPHICS IN LARGE SCALE MECHANICAL SYSTEM DYNAMICS

Parviz E. Nikravesh
Center for Computer Aided Design
College of Engineering
The University of Iowa
Iowa City, Iowa 52242

Abstract. A computerized method for response post-processing in dynamic analysis of mechanical systems is presented. The transient response from planar and spatial dynamic analysis programs are processed. Output in print, plot, single-frame and animated graphics can be obtained. Several methods for generating animated graphic displays are reviewed.

1. INTRODUCTION

In the past decade, several large-scale computer programs for dynamic analysis of mechanical systems have been developed. These programs accept input data for a model describing a mechanical system and carry out kinematic and dynamic simulations. As output, position, velocity, and acceleration of links and bodies in the system are generated. The output, in general, is in numerical form printed at every integration time step. For planar systems with few moving bodies, interpretation and understanding of dynamic response from printed output is possible, although it is to some extent time consuming. This task becomes even more time consuming when the number of moving bodies in the system increases, or when spatial motion is considered instead of planar motion.

The difficulty of interpreting transient dynamic output can be resolved by developing a post-processing computer program. This program must be able to take the transient dynamic output from the analysis package and convert it into other forms of output; e.g., a table, plot, graph, etc. A general-purpose post-processor program can be developed to accept data from more than one dynamic analysis

NATO ASI Series, Vol. F9
Computer Aided Analysis and Optimization of Mechanical System Dynamics
Edited by E. J. Haug

package. This would require proper interfacing between the post-processor and other packages.

The following is the description and methodology used in development of a post-processor for the Dynamic Analysis and Design System program (DADS) [1,2]. The procedure is general and can be used for any other dynamic analysis package.

2. DYNAMIC ANALYSIS

A dynamic analysis package accepts input data describing a mechanical system. The input data consists of inertial properties of the bodies in the system, connectivity and interaction between the bodies, and forces that act on the bodies. The program generates algebraic and differential equations of motion that are solved numerically to predict the transient dynamic response. The analysis package may employ a numerical integration algorithm with variable time step. The program may output the response in unequal time steps, as they are computed, or they may be reported in equal time steps by using interpolation techniques.

The process of input and output to and from a dynamic analysis package, such as DADS, is shown in Fig. 1. The output is saved on a disk file in binary (unformatted) form. The binary output is in a condensed form and saves disk space.

The binary output file contains large amounts of information. At the beginning of the file, information regarding the model such as the number of bodies, number and types of kinematic joints and connectivity, and spring-damper force elements are stored. Then, at

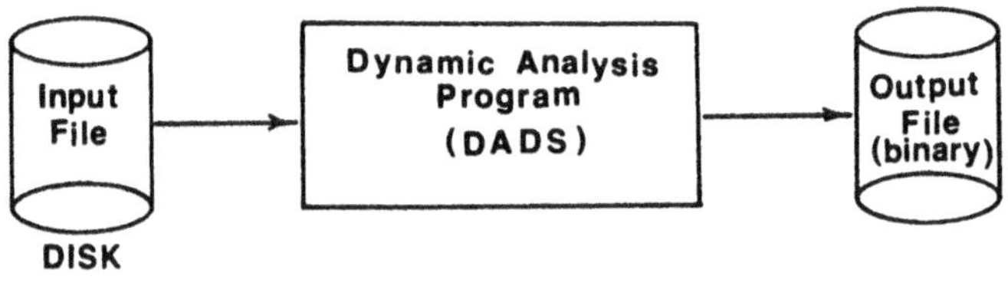

Figure 1. Input/output to and from a dynamic analysis program.

every time step, the integration time, generalized coordinates, velocities, and accelerations for each body, reaction forces at the joints (Lagrange multipliers in the equations of motion), and the spring-damper information are saved.

3. POST-PROCESSOR

The post-processor that is considered in this paper contains several modules. The main processor communicates with sub-processors; print, plot, graphic, or other processors based on the application. The sub-processors are able to read data from the binary file via a stripper program. The flow of data from the binary file to sub-processors is shown in Fig. 2, as solid lines. In this figure, the flow of commands from one processor to another processor is shown by dotted lines. The modules in the post-processor are explained in the following.

Main: This program interactively communicates with the user. The user specifies the name of the binary file of interest and the types of output that is needed. The main routine then transfers to the proper module for data processing.

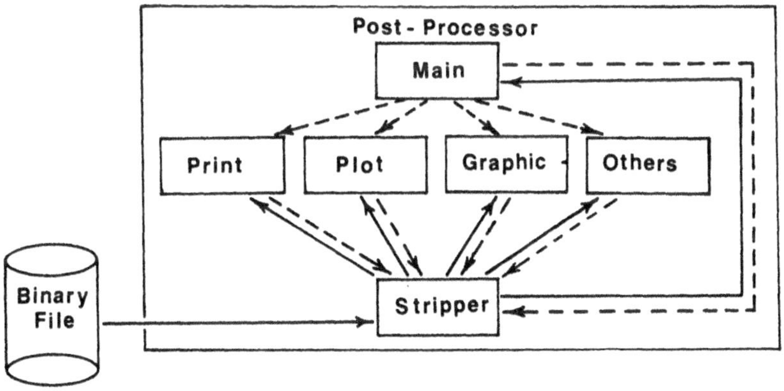

Figure 2. Flow of data from binary file to post-processor

Stripper: The stripper program can receive commands from any of the processors. Based on these commands, the stripper locates any particular section of data in the binary file and reads in the data. This data is sent directly to the processor that submitted the command. For specific application, the stripper may be required to interpolate the response for a particular time of interest. Linear or higher order interpolation functions, such as cubic spline functions, can be employed.

Print: This program can provide printed output in different formats. Based on a request from the user, this module sends commands to the stripper module to obtain particular pieces of data from the binary file. This information is then printed in tabular form.

Plot: This program can provide a plot of the response of any variable versus any other variable or time. Multiple curves can be plotted on one chart for easy comparison. This program makes use of the available plot software routines in the library of the computing machine.

Graphic: This program is discussed in detail in the next sections.

Others: This option allows the user to write and append additional programs to the post-processor. Examples of such programs are Fast-Fourier-Transform analysis, Power-Spectral-Density, and computation of system energy.

4. GRAPHICS

One of the most expressive types of output to aid in analyzing the motion of a system can be obtained by computer graphics. The process involved in displaying a graphic representation of a system, at time t^i of the dynamic simulation, is shown in Fig. 3. The user

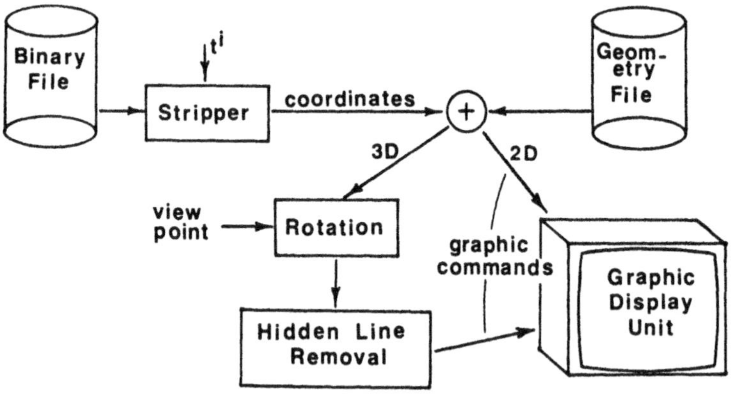

Figure 3. Process of single frame graphic display

must specify the time of interest t^i and also a file containing the outline geometry of the bodies in the system. The geometry file describes the outline of each body by a set of points that are defined with respect to the body-fixed coordinate system and connectivity between the points. The lines produced from the connectivity information, when displayed, represent surfaces and edges of the shape of the body.

On the command of the graphic module, the stripper locates and strips the generalized coordinates for all of the bodies in the system at time t^i. The generalized coordinates are combined with the geometry file data. This process locates and orients each body in the system with respect to a global reference frame. If the simulation is the result of a planar analysis, no further processing is required at this point. However, if the simulation is the result of a spatial analysis, the combined geometry data and generalized coordinates are rotated to the requested view point. This information can then be sent to another package to remove the hidden lines. This process makes the final display appear in a realistic form. The final output of the graphics module is a file that contains graphic commands that are ready to be displayed on a graphic display terminal. The graphic commands are in general machine (graphic display unit) dependent.

The three dimensional rotation is performed by using the angular generalized coordinates of the bodies at time t^i. This process is mostly done by a simple software package. However, in recent years, several graphic systems have become available with hardware rotation capability. Rotation by means of hardware is instantaneous and more efficient than rotation through software.

The three dimensional hidden line removal by means of software is computationally the most time consuming part of the graphic process. Several well written packages for hidden line removal have been developed in recent years. Few of these packages have been developed for arbitrary body shapes. The majority of these packages have been developed for finite element mesh display. Hidden line removal by means of hardware is the state-of-the-art in computer graphics on some high quality graphic units.

5. ANIMATED GRAPHICS

When a sequence of graphic displays of the orientation of the system, at small and successive increments of time, is generated and displayed at a rate of at least 30 frames-per-second (fps), an animation of the motion is created. In this section, two methods for animated graphics are presented.

The first method makes use of a high-speed graphic terminal. A high-speed graphic terminal is considered as one that can display several thousands of lines, flicker free, in one second. The graphic commands for every frame of simulation are generated and stored on the disk, as shown in Fig. 4. For example, for 10 seconds of simulation with 30 fps, 300 frames of graphic commands are generated and stored. After completion of this step, the computer reads and sends the graphic commands, one frame at a time, to the high-speed graphic unit to be displayed. The computer reads and transfers the frames normally at the rate of 30 fps. By varying the frame rate of generating and displaying the graphic commands, slow or fast motion display can be achieved.

The second method for animated graphics uses a standard graphic terminal. The graphics are generated and displayed one frame at the time in a slow rate. After completion of each frame, the graphic

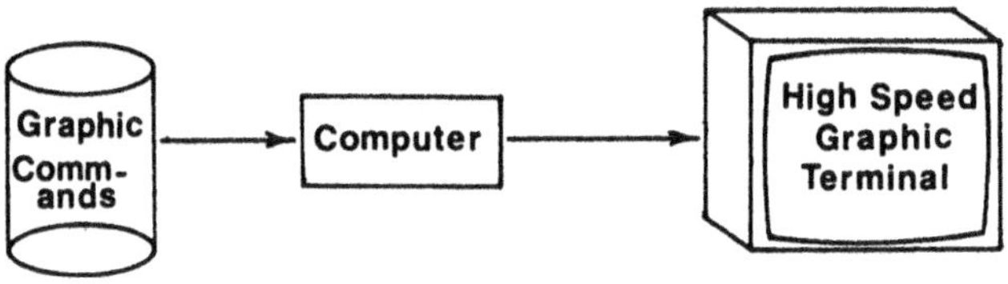

Figure 4. Animated graphics with a high-speed graphic terminal

display can be recorded on film or on video tape (cassette), as shown in Fig. 5. In the case of recording on film, a movie camera facing the graphic display is triggered by a signal received from the terminal at the completion of the display. After the display is shot, the camera is advanced automatically and waits for the next signal. This method is the least expensive technique for animated graphics at present time. In the case of using a video tape, after completion of each frame, the display is sent to a video cassette recorder (VCR). The display is recorded for 1/30 of a second. In this process, a control unit is needed for synchronization of the recorded frames. The video signal received by the controller is sent either directly by the graphic terminal or by a video camera monitoring the screen of the graphic display.

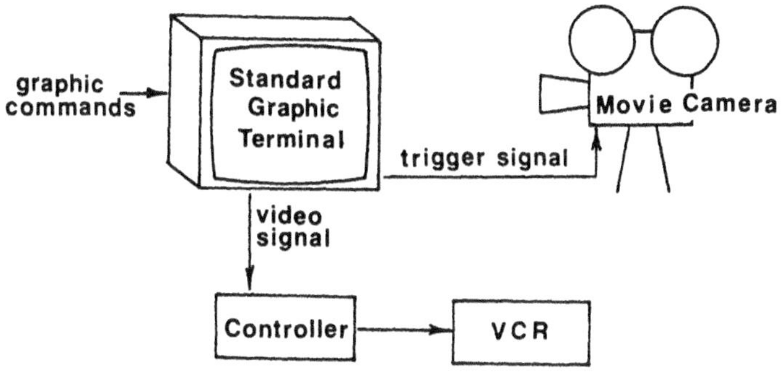

Figure 5. Animated graphics with a standard graphic terminal

Figure 6 shows two examples of dynamic simulation and animated graphics of the response. In Fig. 6(a), a complete gait cycle of a walking robot is illustrated. The simulation is performed in planar motion and the graphics is done without removing hidden lines. The second example, shown in Fig. 6(b), is a model of a truck in spatial motion driven over rough terrain. In the post-processing step, the hidden lines are removed.

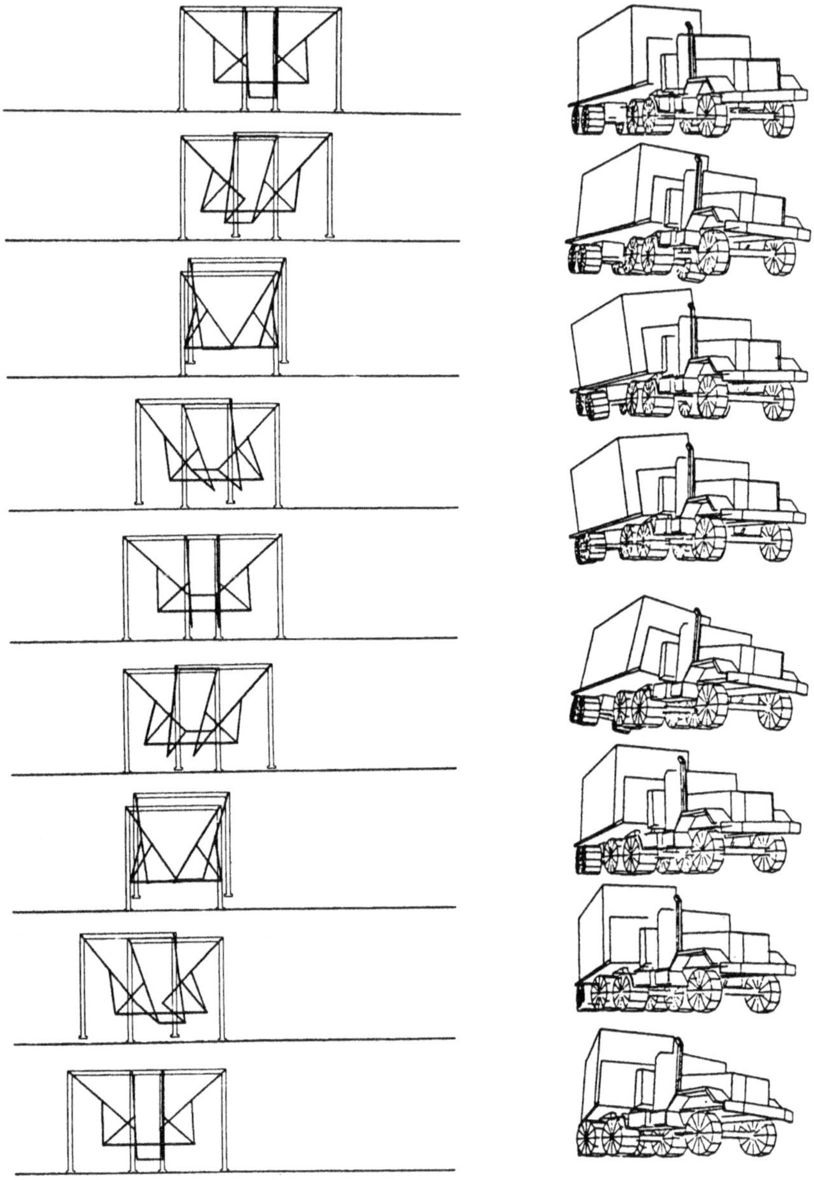

(a) (b)

Figure 6. Examples of animated graphics, (a) a walking robot, (b) a truck driven over rough terrain.

REFERENCES

1. Nikravesh, P.E. and Park, T., "Dynamic Analysis and Design System Computer Program for Planar Motion (Modular-DADS-2D)," Center for Computer Aided Design, The University of Iowa, 1983.

2. Nikravesh, P.E. and Kwon, O.K., "Dynamic Analysis and Design System Computer Program for Spatial Motion (Modular-DADS-3D)," Center for Computer Aided Design, The University of Iowa, 1983.

Part 4

INTERDISCIPLINARY PROBLEMS

DYNAMICS OF FLEXIBLE MECHANISMS

K. van der Werff and J.B. Jonker
Delft University of Technology
The Netherlands

Abstract. The description of the kinematics of structures as given in the finite element method is a good starting point for the numerical treatment of mechanism analysis. In this analysis the relations between deformation and displacements play a central role. Methods are presented for the calculation of the transfer functions of multi-degree of freedom mechanisms. The theory is completed with the formulation of dynamics. Due to the approach with finite element notions, the method presented includes the description of the mechanical behaviour of flexible link mechanisms. The equations derived are therefore not only applicable to rigid link mechanism dynamics but to flexible link mechanisms as well.

NATO ASI Series, Vol. F9
Computer Aided Analysis and Optimization of Mechanical System Dynamics
Edited by E. J. Haug

1. INTRODUCTION

Characteristic for a finite element method is the decomposition of the mechanism into finite elements of a simple geometry by means of suitable sections. In many cases the elements coincide with the links of a mechanism, but this need not be the case. This is illustrated in fig. 1, where a fourbar linkage is decomposed in 7 elements.

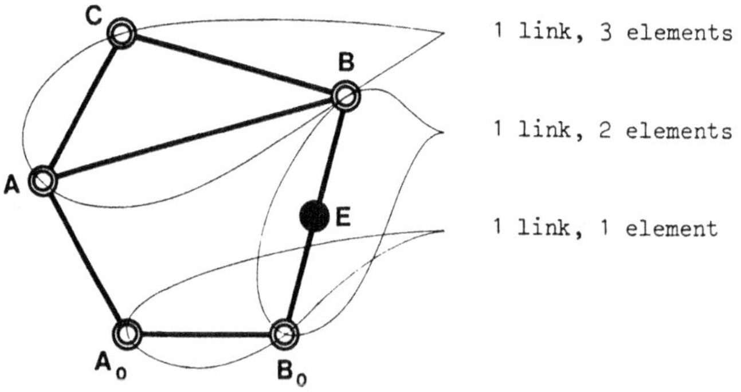

Fig. 1. Links and elements.

A number of different elements can be used for the description of mechanisms, provided they can adequately describe finite displacements and rotations. A number of possible elements is presented in section 2.

The position of each mechanism element, representing a link or part of it, is described by position parameters $x^k \in X^k$. X^k is the space of element position parameters (coordinates) of the k-th element. The position parameter spaces of the individual elements are subspaces of the mechanism position parameter space X, that is

$$X = \bigcup_k X^k. \tag{1}$$

In the choice of the element position parameters possible deformation of the element is taken into account. Rigid link mechanisms are considered as a special case by imposing the undeformability condition a posteriori for the assembled mechanism. The deformation of the elements is described by a vector of deformation parameters $\varepsilon^k \in E^k$. The element deformation spaces form together the space of the mechanism deformation parameters E. As elements don't share deformation parameters, E is the direct sum of the E^k's, that is

$$E = \bigoplus_k E^k. \tag{2}$$

The spaces X and E can both separated into subspaces according to the function of their vector components:

$$X = X^o \oplus X^m \oplus X^c,$$
$$E = E^o \oplus E^m \oplus E^c. \tag{3}$$

The mechanism in fig. 2 contains all six types of components as listed in table 1.

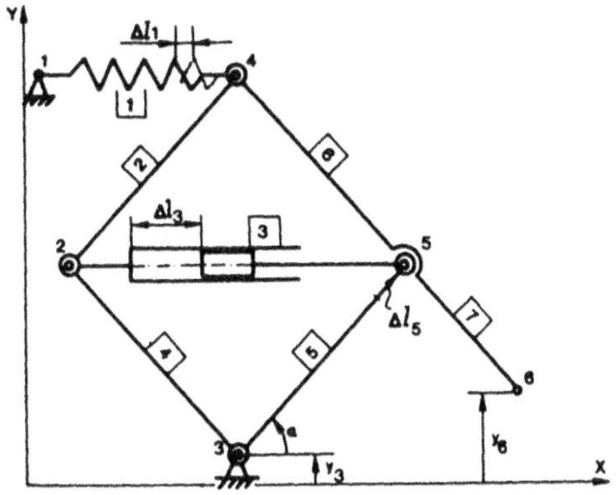

Fig. 2. Subspaces of X and E.

	index	x components	ε components
invariant	o	$(\ldots, y_3, \ldots) \in X^o$	$(\ldots, \Delta\ell_5, \ldots) \in E^o$
input	m	$(\ldots, \alpha_1, \ldots) \in X^m$	$(\ldots, \Delta\ell_3, \ldots) \in E^m$
calculable	c	$(\ldots, y_6, \ldots) \in X^c$	$(\ldots, \Delta\ell_1, \ldots) \in E^c$

Table 1. Subspaces of X and E.

The most general problem formulation for the kinematical analysis is the determination of the calculable position and deformation parameters for given values of the input position and deformation parameters.
Hence determine the map:

$$F : X^m \times E^m \rightarrow X \times E, \tag{4}$$

which maps the imposed position parameters and imposed deformation parameters to the spaces X and E describing the position and deformation state of the mechanism completely.
The map F is called the transfer function of the mechanism. F will be determined for a discrete number of input parameter values. For these positions of the mechanism also the first and second order derivatives of F, DF and D^2F, will be calculated. These derivatives give the possibility to express the mechanism velocities $\dot{x}$ and accelerations $\ddot{x}$ in terms of the input variables according:

$$(\dot{x},\ \dot{\varepsilon}) = DF(x^m,\ \varepsilon^m)\cdot(\dot{x}^m,\ \dot{\varepsilon}^m), \tag{5a}$$

$$(\ddot{x},\ \ddot{\varepsilon}) = (D^2F(x^m,\ \varepsilon^m)\cdot(\dot{x}^m,\ \dot{\varepsilon}^m))\cdot(\dot{x}^m,\ \dot{\varepsilon}^m) + DF(\dot{x}^m,\ \dot{\varepsilon}^m)\cdot(\ddot{x}^m,\ \ddot{\varepsilon}^m). \tag{5b}$$

In the following section a number of element types will be described. For each element type the vector spaces X^k and E^k are defined and the relation between them is described in terms of a map:

$$\mathcal{D}^k : X^k \rightarrow E^k. \tag{6}$$

This map, which is known as the continuity map plays a major role in the derivation of the theory. The derivatives of $\mathcal{D}^k$, $D\mathcal{D}^k$ and $D^2\mathcal{D}^k$, are also important, however no explicit expressions will be given for all of them. Their determination is lengthy but straightforward.

2. FINITE ELEMENTS FOR KINEMATICAL ANALYSIS

In this part a number of finite elements, appropriate to describe kinematics of spatial mechanisms are presented. As will be shown in the next section, the description of deformation of the elements is essential in our method, although the theory, as it is presented here, only deals with mechanisms containing rigid links.
The element descriptions shall always define the position parameters and the deformation parameters of the element type considered. The choice of the position parameters must allow large displacements and rotations of the element. The deformation parameters must describe at least small deformation of an element. In some cases however also large deformations must be described, especially when a deformation is used to describe the input motion.

The planar truss element

The position of the planar truss element is described by the cartesian coordinates $\vec{x}^p = (x^p,\ y^p)$ and $\vec{x}^q = (x^q,\ y^q)$ of the end nodes p and q. These coordinates are the only position parameters for this element, hence

$$x^k_{\text{planar truss}} = (\vec{x}^p,\ \vec{y}^p). \tag{7}$$

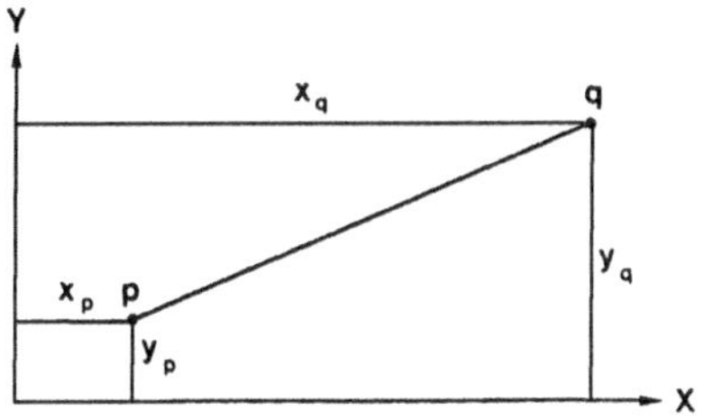

Fig. 3. Planar truss element.

The superscript k is added to show that a specific element k is considered. The number of rigid body degrees of freedom is three, which gives rise to a single deformation parameter, the elongation.
This elongation can be expressed in the momentaneous values of the element position parameters contained in x^k and the original truss length ℓ_o. This expression is called the continuity equation for this element type and is written as a map

$$\mathcal{D}^k : x^k \to \varepsilon^k = ||\vec{x}^p - \vec{x}^q|| - \ell_o . \tag{8}$$

The general form (8) is

$$\mathcal{D}^k : x^k \to \varepsilon^k = \mathcal{D}^k(x^k), \tag{9}$$

which expresses in abstracto that the deformation parameters are determined from the position parameters of an element.

The planar beam element

In the planar beam element we consider in addition to the elongation the in-plane bending deformation. The position parameters of this element are given in fig. 4.

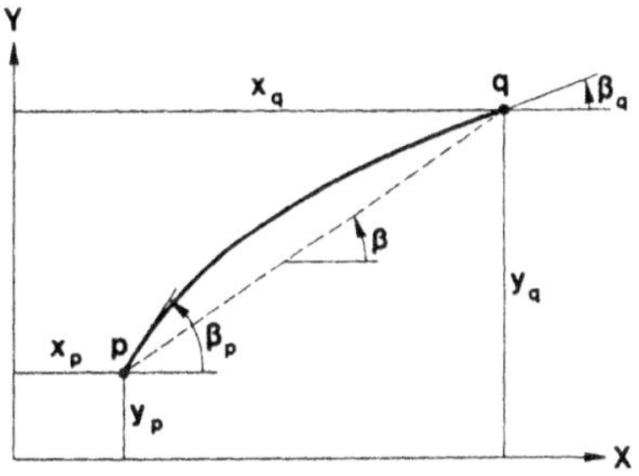

Fig. 4. The planar beam element.

Hence

$$x^k_{\text{planar beam}} = (x^p,\ y^p,\ \beta^p,\ x^q,\ y^q,\ \beta^q). \tag{10}$$

The planar beam element deformation parameters describe the elongation and two bending modes.

The continuity map for this element is given here in full detail:

$$\mathcal{D}^k(x) : \mathcal{X}^k \to \mathcal{E}^k \qquad x^k \mapsto \varepsilon^k$$

$$\begin{cases} \mathcal{D}^k_1(x) & : \quad x^k \mapsto \quad \varepsilon^k_1 = [(x_q-x_p)^2+(y_q-y_p)^2]^{\frac{1}{2}} - [(x^r_q-x^r_p)^2+(y^r_q-y^r_p)^2]^{\frac{1}{2}} \\ \mathcal{D}^k_2(x) & : \quad x^k \mapsto \quad \varepsilon^k_2 = [\ \beta_p - \arccos\{(x_q-x_p)\ [(x_q-x_p)^2+(y_q-y_p)^2]^{-\frac{1}{2}}\}]\cdot\ell^r \\ \mathcal{D}^k_3(x) & : \quad x^k \mapsto \quad \varepsilon^k_3 = [-\beta_q + \arccos\{(x_q-x_p)\ [(x_q-x_p)^2+(y_q-y_p)^2]^{-\frac{1}{2}}\}]\cdot\ell^r \end{cases}$$

$$D\mathcal{D}^k(x) : \mathcal{X}^k \to \mathcal{E}^k \qquad \Delta x^k \mapsto \Delta\varepsilon^k$$

$$\begin{cases} D\mathcal{D}^k_1(x) & : \ \Delta x^k \mapsto \Delta\varepsilon^k_1 = [\ -\cos\beta \quad -\sin\beta \quad 0 \quad +\cos\beta \quad +\sin\beta \quad 0\ ]\ \Delta x^k \\ D\mathcal{D}^k_2(x) & : \ \Delta x^k \mapsto \Delta\varepsilon^k_2 = [\ -\sin\beta \quad +\cos\beta \quad \ell \quad +\sin\beta \quad -\cos\beta \quad 0\ ]\ \Delta x^k \\ D\mathcal{D}^k_3(x) & : \ \Delta x^k \mapsto \Delta\varepsilon^k_3 = [\ +\sin\beta \quad -\cos\beta \quad 0 \quad -\sin\beta \quad +\cos\beta \quad -\ell\ ]\ \Delta x^k \end{cases}$$

$$D^2\mathcal{D}^k(x) \ : \ \mathcal{X}^k \times \mathcal{X}^k \to \mathcal{E}^k \qquad (\Delta x^k, \Delta x^k) \mapsto \Delta^2\varepsilon^k$$

$$\begin{cases} D^2\mathcal{D}^k_1(x) : (\Delta x^k,\Delta x^k) \mapsto \Delta^2\varepsilon^k_2 = \dfrac{\Delta x^{kT}}{\ell^r} \begin{pmatrix} +\sin^2\beta & -\sin\beta\cos\beta & 0 & -\sin^2\beta & +\sin\beta\cos\beta & 0 \\ -\sin\beta\cos\beta & +\cos^2\beta & 0 & +\sin\beta\cos\beta & -\cos^2\beta & 0 \\ 0 & 0 & 0 & 0 & 0 & 0 \\ -\sin^2\beta & +\sin\beta\cos\beta & 0 & +\sin^2\beta & -\sin\beta\cos\beta & 0 \\ +\sin\beta\cos\beta & -\cos\beta & 0 & -\sin\beta\cos\beta & +\cos\beta & 0 \\ 0 & 0 & 0 & 0 & 0 & 0 \end{pmatrix} \Delta x^k \\ D^2\mathcal{D}^k_2(x) : (\Delta x^k,\Delta x^k) \mapsto \Delta^2\varepsilon^k_2 = \dfrac{\Delta x^{kT}}{\ell^r} \begin{pmatrix} -\sin2\beta & +\cos2\beta & 0 & +\sin2\beta & -\cos2\beta & 0 \\ +\cos2\beta & +\sin2\beta & 0 & -\cos2\beta & -\sin2\beta & 0 \\ 0 & 0 & 0 & 0 & 0 & 0 \\ +\sin2\beta & -\cos2\beta & 0 & -\sin2\beta & +\cos2\beta & 0 \\ -\cos2\beta & -\sin2\beta & 0 & +\cos2\beta & +\sin2\beta & 0 \\ 0 & 0 & 0 & 0 & 0 & 0 \end{pmatrix} \Delta x^k \\ D^3\mathcal{D}^k_3(x) = -D^2\mathcal{D}^k_2(x) \end{cases} \tag{11}$$

<u>Gear pair element</u>

Experiences with the description of slider pair, have led to a gear pair element which is very similar to the planar beam element discussed above, when we take the same beam element but with degree of freedom equal four and only two independant deformation parameters instead of three. The first deformation parameter of the beam element, describing the change of distance of the nodes is also for the gear pair useful because it describes the change of distance of the gear's axes. The second deformation parameter is expressed in terms of the planar beam element deformation parameters ε_2 and ε_3:

$$\varepsilon_2^g = - \varepsilon_2\, r_1/\ell_o + \varepsilon_3 r_2/\ell_o . \tag{12}$$

The geometric meaning of this deformation parameter is illustrated in fig. 5.

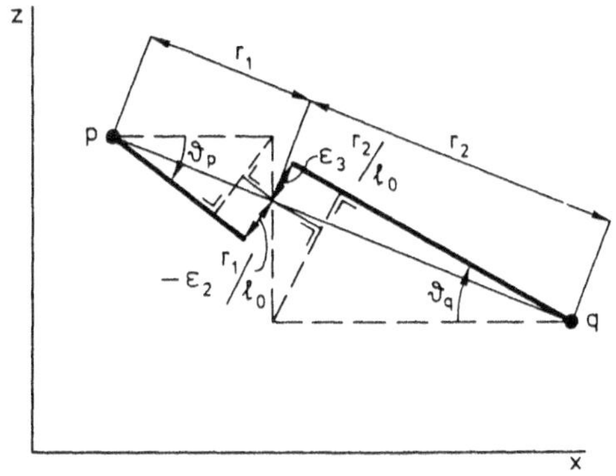

Fig. 5. Deformation parameter of gear pair element.

The space truss element

The position of the spatial truss element is described by the carthesian coordinates $\vec{x}^p$ and $\vec{x}^q$ of the end nodes p and q. These coordinates are the only position parameters for this element, hence

$$x^k_{truss} = (\vec{x}^p, \vec{x}^q). \tag{13}$$

A possible rotation of the truss around pq is not considered. The number of degrees of freedom is thus five, which gives rise to a single deformation parameter, the elongation. This elongation can be expressed in the momentaneous values of the element position parameters contained in x^k and the original truss length ℓ_o. The continuity map for this element type

$$\mathcal{D}^k : x^k \rightarrow \varepsilon^k = \| \vec{x}^q - \vec{x}^p \| - \ell_o . \tag{14}$$

Intermezzo: rotations

Before presenting the elements for spatial motions it is necessary to give some attention to the description of rotations.
The description of the rotation of mechanism components is a keypoint in the analysis of spatial mechanisms. We shall describe the rotational part of a motion by means of a rotation matrix R, which describes how a triad $(\vec{e}_x, \vec{e}_y, \vec{e}_z)$ attached to a rigid body is transformed to a new position $(\vec{e}_{x'}, \vec{e}_{y'}, \vec{e}_{z'})$:

$$R: (\vec{e}_x, \vec{e}_y, \vec{e}_z) \mapsto (\vec{e}_{x'}, \vec{e}_{y'}, \vec{e}_{z'}) = (R\vec{e}_x, R\vec{e}_y, R\vec{e}_z). \tag{15}$$

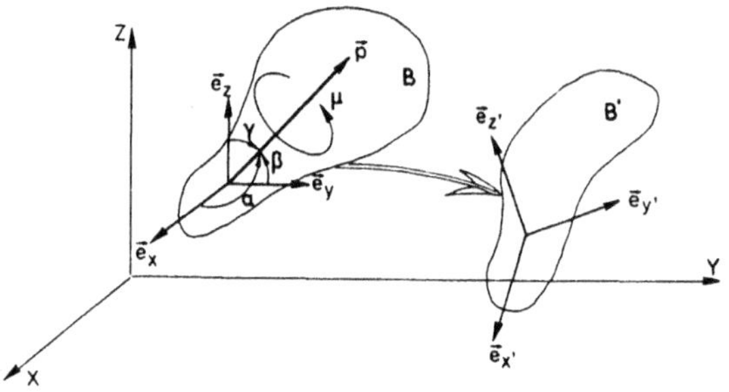

Fig. 6. Triad in initial and new position.

Initially the triad is oriented according to the fixed coordinate system. With the transformation denoted by R, it should be possible to describe all finite rotations in space. In order to describe a proper rotation, the components of R are determined in principle by three independant parameters. Several possibilities are known, such as a description with (modified) Euler angles, Euler parameters, Rodriguez parameters and others [1,2]. In our approach the description with Euler parameters was chosen because this method suited our requirements best, while they don't show singular behaviour for certain values of the parameters.
In the Euler parameter description the rotation is described by stating the direction of the rotation axis $\vec{p}$ and the angle of rotation μ, given rise to four describing parameters λ_i and a normalizing condition:

$$\vec{\lambda} = \begin{pmatrix} \lambda_0 \\ \lambda_1 \\ \lambda_2 \\ \lambda_3 \end{pmatrix} = \begin{pmatrix} \cos\frac{1}{2}\mu \\ \cos\alpha\sin\frac{1}{2}\mu \\ \cos\beta\sin\frac{1}{2}\mu \\ \cos\gamma\sin\frac{1}{2}\mu \end{pmatrix} \quad \text{see fig. 6.}$$

$$\text{sub } \lambda_0^2 + \lambda_1^2 + \lambda_2^2 + \lambda_3^2 = 1. \tag{16}$$

The relation between $\vec{\lambda}$ and R is given by the map

$$R : \vec{\lambda} \to R(\vec{\lambda}),\ R(\vec{\lambda}) \in L(\mathbb{R}^3; \mathbb{R}^3) \tag{17}$$

with

$$R(\vec{\lambda}) = \begin{bmatrix} \lambda_0^2 + \lambda_1^2 - \lambda_2^2 - \lambda_3^2 & 2(\lambda_1\lambda_2 - \lambda_3\lambda_0) & 2(\lambda_1\lambda_3 + \lambda_2\lambda_0) \\ 2(\lambda_1\lambda_2 + \lambda_3\lambda_0) & \lambda_0^2 - \lambda_1^2 + \lambda_2^2 - \lambda_3^2 & 2(\lambda_2\lambda_3 - \lambda_1\lambda_0) \\ 2(\lambda_1\lambda_3 - \lambda_2\lambda_0) & 2(\lambda_2\lambda_3 + \lambda_1\lambda_0) & \lambda_0^2 - \lambda_1^2 - \lambda_2^2 + \lambda_3^2 \end{bmatrix} \tag{18}$$

For detailed information about the Euler parameters we refer to [1,2].
Derivatives of R can easily be determined because R contains only quadratic terms in the λ_i.

The space beam element

In the space beam element also bending and torsion deformation is considered. As bending and torsion deformation is expressed in terms of endpoint rotations, the beam element position parameters contain, apart from the nodal coordinates $\vec{x}^p$ and $\vec{x}^q$, also two sets of Euler parameters describing the rotation of the nodes. In the initial position to each and node a triad is attached which is oriented according the fixed coordinate system. The rotation part of the motion of the deformable beam is then defined by the rotation of these triads which in turn are determined by rotation matrices $R(\vec{\lambda}^p)$ and $R(\vec{\lambda}^q)$ respectively. If the beam is rigid then the rotation matrices are the same. The position parameters of the beam element are thus:

$$x^k_{beam} = (\vec{x}^p, \vec{\lambda}^p, \vec{x}^q, \vec{\lambda}^q). \tag{19}$$

The deformation parameters of an element are chosen such that they have a clear physical meaning. This allows their use lateron for the description of strength and stiffness. The bending deformation in the principal directions of the cross section can serve this purpose. For this reason the principal axes of the beam cross section must be given and it will be assumed that they are given in the initial position as a triad $(\vec{x}, \vec{y}, \vec{z})$ where $\vec{x}$ is directed along pq and $\vec{y}$ and $\vec{z}$ according the principal axes of the beam's cross section. In the deformed state the endnodes are affected by rotations $R(\vec{\lambda}^p)$ and $R(\vec{\lambda}^q)$ and there will exist two rotated principal axes triads

$$P = R(\vec{\lambda}^p) \cdot (\vec{x}, \vec{y}, \vec{z}),$$

$$Q = R(\vec{\lambda}^q) \cdot (\vec{x}, \vec{y}, \vec{z}).$$

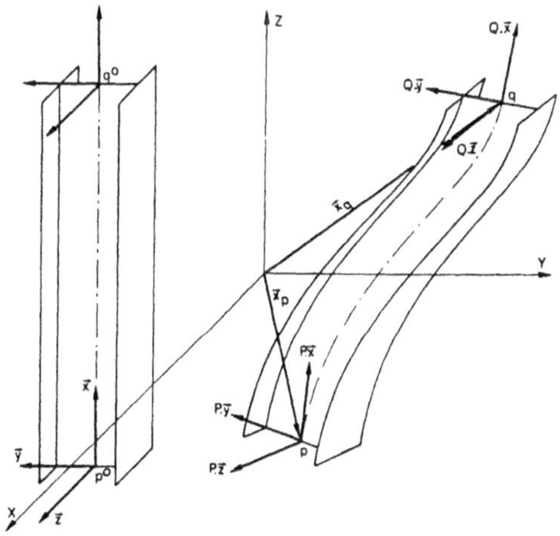

Fig. 7. Beam element: Initial and deformed position.

The six deformation parameters of the beam element can now be expressed in terms of $P\vec{y}$, $P\vec{z}$, $Q\vec{y}$, $Q\vec{z}$ and $\vec{\Delta}x = \vec{x}^q - \vec{x}^p$. As the $\vec{y}$ and $\vec{z}$ are the principal axis in the given initial position they are constant. The deformation parameters are thus only functions of the position parameters given in (19).
The expressions for the deformations are given in table 2.

elongation: $\mathcal{D}_1^k = \|\vec{x}^q - \vec{x}^p\| - \ell_o.$

torsion : $\mathcal{D}_2^k = \{(P\vec{z}, Q\vec{y}) - (P\vec{y}, Q\vec{z})\}\ell_o/2.$

bending : $\mathcal{D}_3^k = -(P\vec{z}, \Delta\vec{x}),$

$\mathcal{D}_4^k = (Q\vec{z}, \Delta\vec{x}),$

$\mathcal{D}_5^k = (P\vec{y}, \Delta\vec{x}),$

$\mathcal{D}_6^k = -(Q\vec{y}, \Delta\vec{x}).$ (20)

Table 2. Deformation parameters of beam element.

In these expressions ( ) stands for the inner product of two vectors. Similar expressions are used by BESSELING for the nonlinear analysis of structures [3]. The bending deformation component $\mathcal{D}_3$ is proportional to the projection of unit vector along the z-principal axis on the connecting line pq. It may be understood that $\mathcal{D}_3$ is an adequate deformation parameter for small deformations. The torsion deformation is also depicted in table 1, where the situation is sketched when looking along pq. $\mathcal{D}_2$ also gives a good description only for small torsion deformation.

Hinge element

The hinge element is necessary in order to be able to connect beam elements with a cilindrical hinge. With properly chosen deformation parameters it will be possible to prescribe the relative rotation of the hinge nodes p and q.

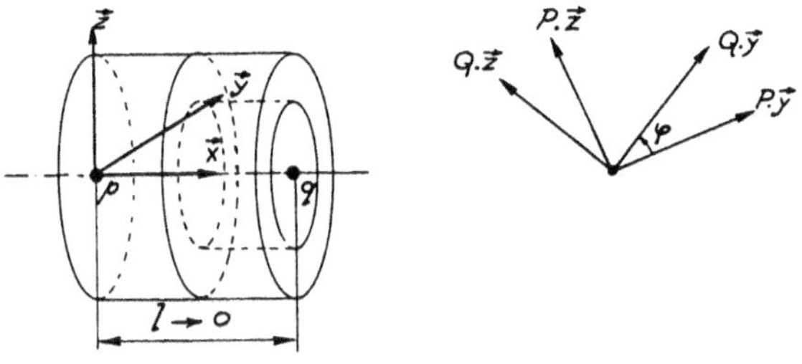

Fig. 8. Hinge element.

As shown in fig. 8 the hinge has two nodes p and q. Their carthesian coordinates are immaterial since this element only deals with rotations. Similar to the beam element the rotations in p and q are described by Euler parameters $\vec{\lambda}^p$ and $\vec{\lambda}^q$, hence

$$x^k_{hinge} = (\vec{\lambda}^p, \vec{\lambda}^q). \tag{21}$$

The initial direction of the hinge axis is stored as the $\vec{x}$ component of a triad $(\vec{x}, \vec{y}, \vec{z})$ attached to the hinge in the same way as it was done with the beam element. The directions of $\vec{y}$ and $\vec{z}$ are chosen arbitrarely.

In the deformed state the nodes p and q may have a different rotation so that the hinge is deformed. The rotations of p and q are given by rotation matrices

$$\begin{aligned} P &= R(\vec{\lambda}^p) \cdot (\vec{x}, \vec{y}, \vec{z}), \\ Q &= R(\vec{\lambda}^q) \cdot (\vec{x}, \vec{y}, \vec{z}). \end{aligned} \tag{22}$$

The deformation parameters of the hinge element are the relative rotation around the hinge axis and two orthogonal bending deformations. The latter two deformations describe the bending of the hinge pin. For the deformation parameters the following expressions are used:

$$\begin{aligned} \mathcal{D}^k_1 &= \phi = \arctan\{-(P\vec{y}, Q\vec{z})/(Py, Qy)\} \\ \mathcal{D}^k_2 &= (P\vec{y}, Q\vec{x}) \\ \mathcal{D}^k_3 &= (P\vec{z}, Q\vec{x}). \end{aligned} \tag{23}$$

For the relative rotation $\mathcal{D}^k_1$ an expression is chosen which allows a complete revolution of the hinge. The arctan function uses both values of $\sin\phi$ and $\cos\phi$, so that at least a 2π domain for ϕ is obtained.

General theory of kinematics

Consider a mechanism composed of a number of finite elements. It is assumed that the deformation parameters of all elements can be varied independently. The motion of the mechanism is then determined by a map

$$\begin{aligned} F &: X^m \times E^m \to X \times E \\ &(x^m, \varepsilon^m) \mapsto (x, \varepsilon) = F(x^m, \varepsilon^m), \end{aligned} \tag{24}$$

which maps the imposed input position and deformation parameters to the position parameter space X and the deformation space E.

The map F (24) is called the transfer function of the mechanism because it relates the mechanism position to the input. The kinematical analysis is directed to the determination of F and its derivatives DF and D^2F.

X and E are composed from the element position parameter spaces X^k and the element deformation parameter spaces E^k as follows:

$$X = \bigcup_k X^k,$$
$$E = \bigoplus_k E^k. \tag{25}$$

Apparantly E^k and X^k are both subspaces of E and X respectively, however E is the direct sum of the E^k's, while X is just the union of the X^k's because some position parameters may be shared by more than one element.
Having defined the mechanism bound spaces X and E, the continuity conditions as given in the previous section of the elements constituting the mechanism, can be taken together in a continuity condition for the whole mechanism:

$$\begin{aligned} &\mathcal{D} : X \to E \\ &\quad x \mapsto \varepsilon = \mathcal{D}(x),\ x \in X,\ \varepsilon \in E. \end{aligned} \tag{26}$$

For each mechanism position also the differential maps $D\mathcal{D}(x)$ and $D^2\mathcal{D}(x)$ can be determined. These maps are (bi) linear maps $X^{(2)} \to E$ and they can be composed from the element contributions. $D\mathcal{D}(x)$ and $D^2\mathcal{D}(x)$ are defined as:

$$\begin{aligned} D\mathcal{D}(x) \ &: X \to E,\ D\mathcal{D}(x) \in \mathcal{L}in(X;\ E) \\ &\quad \Delta x \mapsto \Delta\varepsilon = D\mathcal{D}(x) \cdot \Delta x,\ \Delta x \in X,\ \Delta\varepsilon \in E, \\ D^2\mathcal{D}(x) \ &: X \times X \to E,\ D^2\mathcal{D}(x) \in \mathcal{L}in(X^2;\ E) \\ &\quad (\Delta x_1, \Delta x_2) \mapsto \Delta\varepsilon = (D^2\mathcal{D}(x) \cdot \Delta x_1) \cdot \Delta x_2, \\ &\qquad\qquad (\Delta x_1, \Delta x_2) \in X \times X. \end{aligned} \tag{27}$$

Consider the following maps derived from (24) and (26)

$$\begin{aligned} F^{\varepsilon} \ &: (x^m, \varepsilon^m) \to \varepsilon = F^{\varepsilon}(x^m, \varepsilon^m), \\ \mathcal{D}\circ F^x &: (x^m \cdot \varepsilon^m) \to \varepsilon = \mathcal{D}\circ F^x(x^m, \varepsilon^m), \end{aligned} \tag{28}$$

in which F^x denotes the partial map

$$F^x : (x^m, \varepsilon^m) \to x = F^x(x^m, \varepsilon^m). \tag{29}$$

Clearly F^{ε} and $\mathcal{D}\circ F^x$ are idential maps and must therefore satisfy the equation

$$F^{\varepsilon}(x^m, \varepsilon^m) = \mathcal{D}\circ F^x(x^m, \varepsilon^m). \tag{30}$$

When F^x is known this equation may be used for the determination of the calculable deformation components ε^c:

$$F^{\varepsilon^c}(x^m, \varepsilon^m) = \mathcal{D}^{\varepsilon^c} \circ F^x(x^m, \varepsilon^m). \tag{31}$$

The remaining equations, related to the deformation components ε^o and ε^m, will be used for the determination of the calculable position parameters x^c. We have left:

$$\begin{Bmatrix} F^{\varepsilon^o} \\ F^{\varepsilon^m} \end{Bmatrix} = \begin{Bmatrix} \mathcal{D}^{\varepsilon^o} \\ \mathcal{D}^{\varepsilon^m} \end{Bmatrix} \circ F^x. \tag{32}$$

Due to the nonlinear character of the equations the unknown part of F, i.e. F^{x^c} cannot be calculated directly from this equation.
It will be shown however that expressions for the derivatives DF^{x^c} and $D^2F^{x^c}$ can be obtained. F^{x^c} itself is then found by integration and subsequent iteration. Consider the first derivative of (32).
The derivative of (32) yields

$$\begin{Bmatrix} DF^{\varepsilon^o} \\ DF^{\varepsilon^m} \end{Bmatrix} = \begin{Bmatrix} D\mathcal{D}^{\varepsilon^o} \\ D\mathcal{D}^{\varepsilon^m} \end{Bmatrix} \circ DF^x. \tag{33}$$

When (33) is separated into its subspaces according table I we have

$$\begin{Bmatrix} DF^{\varepsilon^o} \\ DF^{\varepsilon^m} \end{Bmatrix} = \begin{Bmatrix} D^{x^c}\mathcal{D}^{\varepsilon^o} \\ D^{x^c}\mathcal{D}^{\varepsilon^m} \end{Bmatrix} \cdot DF^{x^c} + \begin{Bmatrix} D^{x^m}\mathcal{D}^{\varepsilon^o} \\ D^{x^m}\mathcal{D}^{\varepsilon^m} \end{Bmatrix} \cdot DF^{x^m} + \begin{Bmatrix} D^{x^o}\mathcal{D}^{\varepsilon^o} \\ D^{x^o}\mathcal{D}^{\varepsilon^m} \end{Bmatrix} \cdot DF^{x^o}. \tag{34}$$

The only unknown in this equation is the map DF^{x^c}. For the other partial maps we have:

$$DF^{\varepsilon^o} = \left\{ \frac{\partial \varepsilon^o}{\partial x^m}, \frac{\partial \varepsilon^o}{\partial \varepsilon^m} \right\} = \left\{ 0, 0 \right\},$$

$$DF^{\varepsilon^m} = \left\{ \frac{\partial \varepsilon^m}{\partial x^m}, \frac{\partial \varepsilon^m}{\partial \varepsilon^m} \right\} = \left\{ 0, I \right\},$$

$$DF^{x^o} = \left\{ \frac{\partial x^o}{\partial x^m}, \frac{\partial x^o}{\partial \varepsilon^m} \right\} = \left\{ 0, 0 \right\},$$

$$DF^{x^m} = \left\{ \frac{\partial x^m}{\partial x^m}, \frac{\partial x^m}{\partial \varepsilon^m} \right\} = \left\{ I, 0 \right\}. \tag{35}$$

Due to the choice of the subspaces according table I the map $\begin{Bmatrix} D^{x^c}\mathcal{D}^{\varepsilon^o} \\ D^{x^c}\mathcal{D}^{\varepsilon^m} \end{Bmatrix}$ is nonsingular and DF^{x^c} can be calculated by:

$$DF^{x^c} = \begin{Bmatrix} D^{x^c}\mathcal{D}^{\varepsilon^o} \\ D^{x^c}\mathcal{D}^{\varepsilon^m} \end{Bmatrix}^{-1} \cdot \left[\begin{Bmatrix} DF^{\varepsilon^o} \\ DF^{\varepsilon^m} \end{Bmatrix} - \begin{Bmatrix} D^{x^m}\mathcal{D}^{\varepsilon^o} \\ D^{x^m}\mathcal{D}^{\varepsilon^m} \end{Bmatrix} \cdot DF^{x^m} \right]. \tag{36}$$

Expressions for the second order transfer function are derived in a similar way. Differentiation of (30) leads to:

$$DF^{\varepsilon} = D\mathcal{D} \cdot DF^x,$$

$$D^2F^{\varepsilon} = (D^2\mathcal{D} \cdot DF^x) \cdot DF^x + D\mathcal{D} \cdot D^2F^x. \tag{37}$$

Making the same separation in subspaces the expressions for the nonzero transfer-function parts become:

$$D^2F^{x^c} = - \begin{Bmatrix} D^{x^c}\mathcal{D}^{\varepsilon^o} \\ D^{x^c}\mathcal{D}^{\varepsilon^m} \end{Bmatrix}^{-1} \cdot \begin{bmatrix} (D^2\mathcal{D}^{\varepsilon^o} \cdot DF^x) \cdot DF^x \\ (D^2\mathcal{D}^{\varepsilon^m} \cdot DF^x) \cdot DF^x \end{bmatrix}, \tag{38a}$$

$$D^2F^{\varepsilon^c} = (D^2F^{\varepsilon^c} \cdot DF^x) \cdot DF^x + D\mathcal{D}^{\varepsilon^c} \; D^2F^x. \tag{38b}$$

When a starting position $x_1 = F(x_1^m, \varepsilon_1^m)$ is given, new positions $x_2 = F(x_2^m, \varepsilon_2^m)$ can be approximated by integration

$$x_2 = F(x_1^m, \varepsilon_1^m) + DF(x_1^m, \varepsilon_1^m) \cdot (x_2^m - x_1^m, \varepsilon_2^m - \varepsilon_1^m) + \\ + (D^2F(x_1^m, \varepsilon_1^m) \cdot (x_2^m - x_1^m, \varepsilon_2^m - \varepsilon_1^m)) \cdot (x_2^m - x_1^m, \varepsilon_2^m - \varepsilon_1^m). \tag{39}$$

An iteration process is then applied in order to quarantee that ultimately $\mathcal{D}(x_2) = \varepsilon_2$.

General theory of dynamics

The primary objective of the dynamical analysis is the evaluation of the velocities and accelerations of the mechanism subjected to a given time-varying load. In order to describe the dynamic characteristics of the mechanism we must introduce the inertia properties of the elements.

The simplest form of the mathematical model for the inertia properties is the lumped-mass representation. In this idealization concentrated mass and rotational inertias are attached to the end nodes of the elements. They are calculated by assuming that the elements behave like a rigid body. This assumption therefore excludes dynamic coupling between the translational and rotational motions. The distribution of the element mass to the node mass and rotational inertias is determined according to the following three conditions which are necessary and sufficient for dynamical equivalence.

1. the mass of the element should be equal to the total mass of the concentrated masses.
2. the centre of gravity of the element and of the discrete mass model should be coincide, and
3. the inertial ellipsoid of the element and that of the discrete model should be equal.

Since the rotations are described in terms of Euler parameters, it follows that the rotational equations of motion should be expressed in terms of these parameters.

Rotational equations of motion

Consider a rigid body with mass-centre c subjected to an external torque $\vec{k}$. See figure (9).

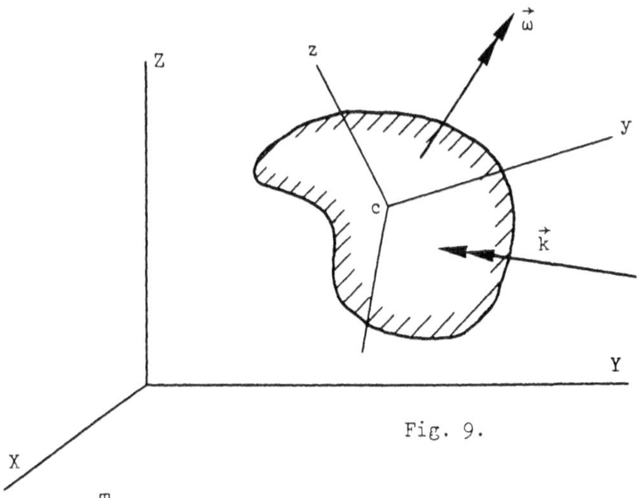

Fig. 9.

Let $\omega = (\omega_x, \omega_y, \omega_z)^T$ be the projections of the angular velocity vector on a body-fixed triad $(x, y, z)^T$ which initially coincided with the fixed axes $(X, Y, Z)^T$. Then the kinetic energy of the body can be written as

$$T = \frac{1}{2}\, \omega^T J\, \omega, \tag{40}$$

where J is the rotational inertia about the x, y, z-axes. The components $(\omega_x\ \omega_y\ \omega_z)^T$ in terms of Euler parameters can be expressed as [1, 2],

$$\omega = 2\,\Lambda\,\dot{\lambda}\,, \tag{41}$$

where $\dot{\lambda} = (\dot{\lambda}_0, \dot{\lambda}_1, \dot{\lambda}_2, \dot{\lambda}_3)^T$, satisfy the constraint equation

$$\lambda^T\,\dot{\lambda} = 0, \tag{42}$$

and $\Lambda = \begin{bmatrix} -\lambda_1 & \lambda_0 & \lambda_3 & -\lambda_2 \\ -\lambda_2 & -\lambda_3 & \lambda_0 & \lambda_1 \\ -\lambda_3 & \lambda_2 & -\lambda_1 & \lambda_0 \end{bmatrix}$.

According to the principle of virtual power we then have

$$< \left\{ f^\lambda - \frac{d}{dt}\left(\frac{\partial T}{\partial \dot{\lambda}}\right) + \frac{\partial T}{\partial \lambda} \right\},\ \dot{\lambda} > - < \sigma^*,\ \lambda^T\,\dot{\lambda} > = 0,\ \forall\ \dot{\lambda},$$

where σ^* is a Lagrange multiplier representing the constraint force corresponding to the constraint condition of the Euler parameters. The bracket <,> operation on the left-hand side is the scalar product. Hence with (40) and (41) four equations of motion for the body are written as

$$f^\lambda - 4\,\Lambda^T J\,\Lambda\,\ddot{\lambda} - 8\,\dot{\Lambda}^T J\,\Lambda\,\dot{\lambda} = \sigma^*\,\lambda. \tag{43}$$

Using eqs. (42) and (43) a solution for the λ's as well as σ^* can be obtained. The dual quantities f^λ can be derived from the virtual power equation

$$< k, \omega > = < f^{\lambda}, \dot{\lambda} >, \tag{44}$$

which must hold for all ω, satisfying the equation

$$\omega = 2\Lambda' \dot{\lambda}, \tag{45}$$

where $\Lambda' = \begin{bmatrix} -\lambda_1 & \lambda_0 & -\lambda_3 & \lambda_2 \\ -\lambda_2 & \lambda_3 & \lambda_0 & -\lambda_1 \\ -\lambda_3 & -\lambda_2 & \lambda_1 & \lambda_0 \end{bmatrix}$.

Here $\omega = (\Omega_x, \Omega_y, \Omega_z)^T$ are the components of the angular velocity vector on the fixed axes $(X, Y, Z)^T$. By substituting of (45) into (44) we obtain with the dual of the transformation Λ', the expression

$$2\Lambda'^T k = f^{\lambda},$$

where $k = (K_x, K_y, K_z)^T$ are components of the external torque vector on the fixed axes $(X, Y, Z)^T$.

Equations of motion of the mechanism

In view of the different treatment of the linear and angular velocities in the derivation of the equations of motion, it is useful to separate the space X of mechanism position parameters into subspaces X^x and X^{λ}, where X^x is the space of position parameters x and X^{λ} is the space of Euler parameters λ. In such a way the map F is expressible as

$$\begin{aligned} F^x &: X^m \times E^m \to X^x, \\ F^{\lambda} &: X^m \times E^m \to X^{\lambda}, \\ F^{\varepsilon} &: X^m \times E^m \to E. \end{aligned} \tag{46}$$

From (46) we can calculate the velocities $\dot{x}$, $\dot{\lambda}$ and $\dot{\varepsilon}$ by

$$\dot{x} = DF^x \,.\, (\dot{x}^m, \dot{\varepsilon}^m), \tag{47a}$$
$$\dot{\lambda} = DF^{\lambda} \,.\, (\dot{x}^m, \dot{\varepsilon}^m), \tag{47b}$$
$$\dot{\varepsilon} = DF^{\varepsilon} \,.\, (\dot{x}^m, \dot{\varepsilon}^m). \tag{47c}$$

Let F^x be the space of externally applied node forces and let F^{λ} be the space of quantities f^{λ}. Both of the spaces are dual with the spaces of $\dot{x}$ and $\dot{\lambda}$. In order to describe the loading state of the mechanism elements we introduce the vector of generalized stresses $\sigma \in \Sigma$. The space Σ is dual with the space of deformation velocities $\dot{\varepsilon}$. Note that in the vector σ also σ^* is included.
If M and J are the lumped mass matrix and the matrix of rotational inertias of the mechanism, then the principle of virtual power for the external loads f^x and f^{λ} states that

$$< (f^x - M\ddot{x}),\ \dot{x} > + < (f^\lambda - 4\Lambda^T J \Lambda \ddot{\lambda} - 8\dot{\Lambda}^T J \Lambda \dot{\lambda}),\ \dot{\lambda} > = < \sigma,\ \dot{\varepsilon} >,$$

holds for all kinematically admissible velocities. By substituting of (47) we may rewrite this condition as

$$< (f^x - M\ddot{x}),\ DF^x.(\dot{x}^m, \dot{\varepsilon}^m) > + < (f^\lambda - 4\Lambda^T J \Lambda \ddot{\lambda} - 8\dot{\Lambda}^T J \Lambda \dot{\lambda}),\ DF^\lambda \cdot (\dot{x}^m, \dot{\varepsilon}^m) > =$$
$$= < \sigma,\ DF^\varepsilon \cdot (\dot{x}^m, \dot{\varepsilon}^m) >,$$

for all arbitrary velocities $\dot{x}^m$, $\dot{\varepsilon}^m$. Hence with the duals of DF we obtain

$$\left| DF^{xT} DF^{\lambda T} \right| \left| \begin{matrix} f^x - M\ddot{x} \\ f^\lambda - 4\Lambda^T J \Lambda \ddot{\lambda} - 8\dot{\Lambda}^T J \Lambda \dot{\lambda} \end{matrix} \right| = DF^{\varepsilon T} \sigma . \tag{48}$$

From (47^a) and (47^b) we can calculate the accelerations $\ddot{x}$ and $\ddot{\lambda}$ by

$$\ddot{x} = (D^2F^x.(\dot{x}^m, \dot{\varepsilon}^m)).(\dot{x}^m, \dot{\varepsilon}^m) + DF^x.(\ddot{x}^m, \ddot{\varepsilon}^m),$$
$$\ddot{\lambda} = (D^2F^\lambda.(\dot{x}^m, \dot{\varepsilon}^m)).(\dot{x}^m, \dot{\varepsilon}^m) + DF^\lambda.(\ddot{x}^m, \ddot{\varepsilon}^m).$$

Substituting into (48) we have the equations of motion of the mechanism:

$$\left| DF^{xT} DF^{\lambda T} \right| \left| \begin{matrix} M & 0 \\ 0 & 4\Lambda^T J \Lambda \end{matrix} \right| \left| \begin{matrix} DF^x \cdot (\ddot{x}^m, \ddot{\varepsilon}^m) \\ DF^\lambda \cdot (\ddot{x}^m, \ddot{\varepsilon}^m) \end{matrix} \right| = \left| DF^{xT} DF^{\lambda T} DF^{\varepsilon T} \right| \left| \begin{matrix} f^x \\ f^\lambda - 8\dot{\Lambda}^T J \Lambda \dot{\lambda} \\ \sigma \end{matrix} \right| +$$

$$- \left| DF^{xT} DF^{\lambda T} \right| \left| \begin{matrix} M & 0 \\ 0 & 4\Lambda^T J \Lambda \end{matrix} \right| \left| \begin{matrix} (D^2F^x \cdot (\dot{x}^m, \dot{\varepsilon}^m)) \cdot (\dot{x}^m, \dot{\varepsilon}^m) \\ (D^2F^\lambda \cdot (\dot{x}^m, \dot{\varepsilon}^m)) \cdot (\dot{x}^m, \dot{\varepsilon}^m) \end{matrix} \right| .$$

There are as many equations involved as there are given degree's of freedom x^m, ε^m. In the equations all forces have been reduced to the degrees of freedom by means of the momentaneous kinematical relations as reflected in the transfer functions.

The degree's of freedom can chosen such that the equations of motion describe the rigid link mechanism. Flexible mechanisms are introduced in the following way. The force vector contains externally applied forces applied to the nodes, but also the generalized stresses σ corresponding to the ε. Flexible links are characterized by a linear elastic law

$$\sigma = S\,\varepsilon \tag{49}$$

in which S is the appropriate finite element stiffness matrix. Other relations between σ and ε are also possible for example describing damper characteristics. The approach followed leads us to a description of flexible multidegree of freedom spatial mechanisms and opens the possibility to treat the problems of vibration and internal stress calculation in the same way as it is done with kinematically determinate structures.

The formalisms used have led to equations which can be generated automatically by a computer. The computer programs can be structured according to the structure present in the theory. The addition of new element types in the computer programme requires in fact only the programming of the continuity equations for that element.

Example

As an example we shall derive the equations of motion for the plane system of fig. 10.

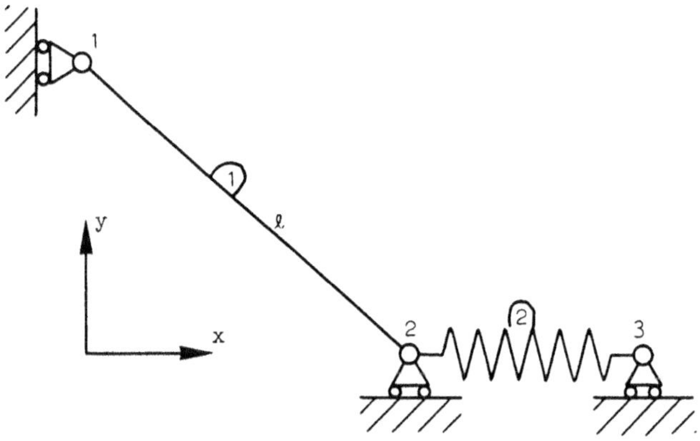

Fig. 10. Example.

The input motion is y_1 and ε_2. Let $\ell = 5$, $y_1 = y = 3$ and $x_2 = x = 4$ then $\frac{\partial x}{\partial y} = -\frac{3}{4}$ and $\frac{\partial^2 x}{\partial y^2} = -\frac{25}{64}$.

The vectors (x, ε), (x^m, ε^m), the first order transfer function are then as given:

$$(x^m, \varepsilon^m) = |y_1, \varepsilon_2|$$

$$(x, \varepsilon) = \begin{vmatrix} x_1 \\ y_1 \\ x_2 \\ y_2 \\ x_3 \\ y_3 \\ \varepsilon_1 \\ \varepsilon_2 \end{vmatrix}, \quad DF = \begin{bmatrix} 0 & 0 \\ 1 & 0 \\ -\frac{3}{4} & 0 \\ 0 & 0 \\ -\frac{3}{4} & 1 \\ 0 & 0 \\ 0 & 0 \\ 0 & 1 \end{bmatrix}, \quad D^2F.(\dot{x}^m, \dot{\varepsilon}^m)^2 = \left| (\dot{y}_1, \dot{\varepsilon}_2) \left\{ \begin{pmatrix} \begin{bmatrix} 0 & 0 \\ 0 & 0 \end{bmatrix} \\ \begin{bmatrix} 0 & 0 \\ 0 & 0 \end{bmatrix} \\ \begin{bmatrix} a & 0 \\ 0 & 0 \end{bmatrix} \\ \begin{bmatrix} 0 & 0 \\ 0 & 0 \end{bmatrix} \\ \begin{bmatrix} a & 0 \\ 0 & 0 \end{bmatrix} \\ \begin{bmatrix} 0 & 0 \\ 0 & 0 \end{bmatrix} \\ \begin{bmatrix} 0 & 0 \\ 0 & 0 \end{bmatrix} \\ \begin{bmatrix} 0 & 0 \\ 0 & 0 \end{bmatrix} \end{pmatrix} \right\} \begin{vmatrix} \dot{y}_1 \\ \dot{\varepsilon}_2 \end{vmatrix} \right| .$$

$$a = \frac{-25}{64}\,\dot{y}_1^2;$$

Concentrated masses are assumed in the nodes so that the mass matrix M becomes:

$$M = \text{diag}\,\{M_1, M_1, M_2, M_2, M_3, M_3\}.$$

The external forces are denoted with Fx_i, Fy_i for the forces acting in node i in x resp. y direction.

For the terms in the equations of motion we then find:

$$DF^{x^T} M\, DF\cdot(\ddot{x}^m, \ddot{\varepsilon}^m) = \begin{bmatrix} M_1 + \frac{9}{16} M_2 + \frac{9}{16} M_3 & -\frac{3}{4} M_3 \\ -\frac{3}{4} M_3 & M_3 \end{bmatrix} \begin{vmatrix} \ddot{y}_1 \\ \ddot{\varepsilon}_2 \end{vmatrix},$$

$$DF^{x^T}\cdot f = \begin{vmatrix} F_{y_1} - \frac{3}{4} F_{x_2} - \frac{3}{4} F_{x_3} \\ F_{x_3} \end{vmatrix},$$

$$DF^{\varepsilon}_{\cdot}\sigma = \begin{vmatrix} 0 \\ \sigma^2 \end{vmatrix}$$

$$DF^{x^T} M\, D^2F^x_{\cdot}(\dot{x}^m, \dot{\varepsilon}^m)^2 = \begin{vmatrix} \frac{75}{256} (M_2 + M_3)\, \dot{y}_1^2 \\ -\frac{25}{64} M_3\, \dot{y}_1^2 \end{vmatrix}.$$

The equations can be formulated completely in y_1 and ε_2 when the constitutive equation $\sigma_2 = S\varepsilon_2$ is introduced.

References

1. Bottema, O. and B. Roth, Theoretical Kinematics NHPC, Amsterdam, 1979.
2. Whittaker, E.T., A treatise on the analytical dynamics of particles and rigid bodies.
   Dover, New York, 1936.
3. Besseling, J.F., Stijfheid en Sterkte 2, Toepassingen.
   Oosthoek, Scheltema & Holkema, Utrecht, 1975.

Nomenclature

D	Differentiation operator
E	Deformation parameter space
J	Rotational inertia tensor
M	Mass matrix
R	Rotation matrix
S	Stiffness matrix
T	Kinetic energy
X	Position parameter space
$\vec{e}$	unit vector
f	generalized force vector
k	torque vector
x	vector of position parameters
$\mathcal{D}$	continuity map
$\mathcal{F}$	transfer function
$\mathcal{L}$	linear transformation
ε	vector of deformation parameters
λ	vector of Euler parameters
μ	rotation angle
σ	vector of generalized stresses
ω	vector of angular velocities
Λ	see (42)

CONTROLLED DYNAMIC SYSTEMS MODELING[1]

M. Vanderploeg and G.M. Lance
Center for Computer Aided Design
College of Engineering
The University of Iowa
Iowa City, IA 52242

Abstract. In recent years, many researchers have studied simulation of combined control and mechanical systems. Interest in this area is growing as more mechanical systems utilize some type of control system. This is especially evident in the field of robotics. To date, the disciplines of control theory and mechanical system dynamics have been brought together in only an adhoc manner. This paper describes a general purpose simulation program which models combined control and mechanical systems.

Several general purpose computer programs for dynamic analysis of mechanisms and mechanical systems have become available in recent years. These programs formulate and automatically integrate the equations of motion, requiring that the user input only the geometry and governing parameters of the mechanical system. Several of the more developed programs include: IMP, DRAM, ADAMS, and DADS.

To date, simulation of a control system coupled with a mechanical system, using the above mentioned programs, requires at least the preparation of a user supplied subroutine. The subroutine typically contains the dynamics of the control system and defines the connectivity between the control system and the mechanical system. This requires the user to write a FORTRAN program and have substantial understanding of the analysis program, thus greatly reducing the number of potential users.

[1]Research partially sponsored by the U.S. Army Tank and Automotive Research and Development Command.

NATO ASI Series, Vol. F9
Computer Aided Analysis and Optimization of Mechanical System Dynamics
Edited by E.J. Haug

This paper presents a control package designed to be an integral part of the DADS program. The package includes a library of the more common control blocks, a standardized connectivity scheme, and a convenient format for the addition of user-supplied functions and differential equations. Control behavior can then be incorporated into the mechanical system simulation with a minimum of user input.

Several examples are presented to demonstrate the effectiveness of the package. The examples indicate that for applications involving control, equations of motion of the mechanical system need to be derived using relative Lagrangian coordinates.

INTRODUCTION

Recently, several general purpose computer programs for dynamic analysis of mechanical systems have become available. Because these programs automatically formulate and integrate the equations of motion, dynamic analysis of mechanical systems has become available to a wide range of engineers and designers. For this reason, these programs have become valuable design tools, allowing the engineer or designer to analyze a mechanical system without building costly prototypes.

In mechanical system design there is an increasing need to control some elements of the mechanical system. This is especially evident in the field of robotics. The currently available programs do not contain provisions for formulating equations for control elements. Therefore, a user must prepare a user supplied subroutine containing the control equations and the connectivity between the control system and the mechanical system. This requires the user to formulate complex control equations and to have substantial understanding of the mechanical dynamics analysis program. These requirements greatly reduce the number of potential users.

On the other hand, several computer programs, such as EASY5, CSMP, and TOTAL are currently available for control system analysis and optimization. A review of computer aided control system design software packages is presented in Reference 1. These programs are excellent for control system analysis, but do not contain provisions for mechanical systems modeling.

This paper presents a control package designed to be an integral part of the Dynamic Analysis and Design System [2,3], or DADS mechanical system analysis program. The package contains a library of the more common control blocks, a standardized connectivity scheme, and a convenient format for the addition of user-supplied functions and differential equations. The DADS program then formulates and integrates the combined mechanical and control system equations. This capability will allow a wide range of users to model coupled control and mechanical systems.

Several examples are presented to demonstrate the effectiveness of the package. The examples indicate that for applications involving control, equations of motion of the mechanical system need to be derived using relative Lagrangian coordinates.

REVIEW OF DYNAMIC ANALYSIS CODES

The first general purpose computer programs for dynamic analysis of large scale mechanical systems appeared around 1970. Several of the earliest and more developed programs in this area are DYMAC (1970) [4], DAMN-DRAM (1971) [5], IMP (1972) [6] ADAMS-3D (1973) [7], and DADS (1979) [2,3]. Although similar in some respects, they are quite different in others. Each program requires similar input consisting of mechanism geometry and physical properties such as inertia and spring stiffness. The programs then formulate and integrate the equations of motion. Major differences include 2D vs. 3D, Lagrangian vs. Cartesian coordinates, and methods of equation formulation and solution. A good review and comparison of these programs is presented by Nikravesh et.al. [8]. At this point it is again noted that none of the above mentioned codes have provisions for including control systems. The next section will describe a controls package developed to be an integral part of the DADS program.

THE DADS CONTROL PACKAGE

The DADS control package was developed to serve as either a stand alone program for simulation of linear and nonlinear control systems or in conjunction with the original DADS program as an integrated simulation program. The package contains a library of common control functions called control blocks. Presently available control functions are presented in Table 1 and discussed in subsequent sections. The user assembles a control system simulation working directly from a system block diagram. After defining the necessary control blocks, the program input requires only the specification of parameter values and definition of the connectivity between blocks. DADS then automatically formulates and integrates the system equations, yielding a time history of the system states. While the usual control systems involve loops containing some elements described by differential equations, the package has been designed to identify and solve algebraic loops as well. Also available are several common control input functions such as a step, ramp, etc. In addition, a convenient method for inputing user supplied control input functions is provided. Output is easily displayed using the DADS post-processor. Internal control variables as well as control output can be plotted interactively.

Several of the common nonlinear functions encountered such as saturation, dead band, hysteresis and time delay are available in the library of control elements. Also, user supplied functions and differential equations are easily input. In addition to individual functions, it is feasible to create a module that represents a complete subsystem. An example of this is the module representing an electrohydraulic servoactuator. This "super module" is used in a system simulation in the same way as the less complex elements, requiring only a definition of the servoactuator parameters and the system connectivity.

Although the controls package has a stand alone capability, the primary objective was to develop a user friendly connectivity scheme to merge the controls simulation with the mechanical simulation. Mechanical system state variables, accelerations, or forces can be used as inputs to the control system, and control system outputs can be used as force or position inputs to the mechanical system.

Definition of the control system and the connectivity to the DADS mechanical model is broken into three divisions: 1) definition of the block diagram in terms of the DADS library of control elements and user supplied functions; 2) definition of inputs to the control system. These consist of the controller input and feedback of DADS state variables to the control system; and 3) definition of control system outputs. These normally take the form of actuator forces or positions when used with mechanical systems.

Table 1

Type	Block Diagram	Type	Block Diagram
Input (INPT)	→ Controller ← DADS	Switch (SWIT)	X → [Y, 1.0, a_1, X] → Y
Summer (SUMM)	X_1 (+), X_2, X_3 (−) → ⊗ → Y	Dead-Zone (DZON)	X → [a_2, Y, 1, 1, a_1, X] → Y
Linear Amplifier (AMPL)	X → [Y, K, X] → Y	Hysteresis (HYST)	X → [X, Y, 1, 1, a_2, a_1] → Y
First Order Integrator (FRST)	X → $\frac{b_1 s + b_0}{a_1 s + a_0}$ → Y	User Supplied Amplifier (UAMP)	X → [Y, X] → Y
Second Order Integrator (SCND)	X → $\frac{b_2 s^2 + b_1 s + b_0}{a_2 s^2 + a_1 s + a_0}$ → Y	Delay (DELY)	X → e^{-as} → Y
Simple Integrator (INTG)	X → $\frac{1}{s}$ → Y	Hydraulic Servo-Actuator (HYSV)	X → [ ] → Y, L'
User Supplied Differential Equations (UDIF)	X → $\dot{Z} = f(Z,V,t)$, $Y = g(Z,X)$ → Y	Output (OUTP)	Controller → DADS
Limiter (LIMT)	X → [Y, a_2, 1, 1, a_1, X] → Y	Notations: a, b, K = constants; L' = load velocity; X = input; Y = output; Z = internal state variable	

EXAMPLES

The DADS control system structure is best illustrated through the use of example problems. The first example chosen is a stabilized gun mounted on a tracked vehicle. The gun is driven by an electro-hydraulic actuator. Feedback of gun tube angular position and velocity and actuator pressure differential are used. The use of the control package is illustrated by simulating the gun tube and control system with the vehicle chassis treated as fixed (i.e. ground). Figure 1 shows the elevation axis for the simplified two body system. Figure 2 is a block diagram of the elevation axis control system.

Table 2 presents the input data for the combined mechanical and control system. The control system definition starts with the data input statement "CNTR". Statements above this define the mechanical system and a discussion of the input format is presented in Reference [9]. For further discussion of the control system it should be noted that the gun tube is defined as body 2 in the mechanical system.

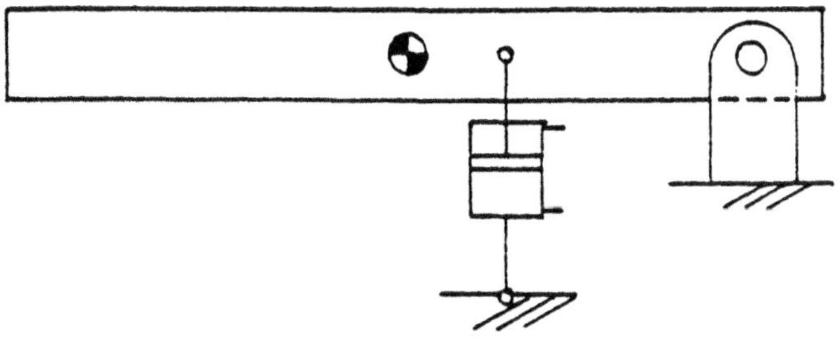

Figure 1. Schematic of simplified gun tube model

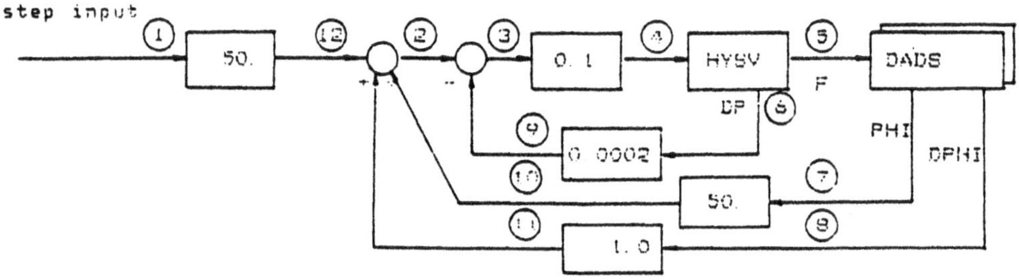

Figure 2. Block diagram of control system

The control system definition starts by assigning a number to each node of the block diagram as illustrated in Figure 2. Node numbers are used for all three phases of control system definition. The first input statement defines the controller input at node 1 to be a step of magnitude .05 applied to time t = 0. Input statements two and three define feedback variables from the DADS model. Node 7 is defined to be ϕ of body 2, the gun tube, and node 8 is defined to be $\dot{\phi}$ of the gun tube. Next the control elements making up the control system are defined. For example summer #1 is defined as having nodes 10, 11, and 12 as inputs and node 2 as output. A negative node number implies subtraction. Therefore summer #2 defines node 3 as equal to node 2 minus node 9. Linear amplifiers 1 through 5 are defined by the input node, output node, and the gain in that sequence. Finally, the hydraulic servoactuator is defined with node 4 being an input and nodes 5 and 6 being outputs of force and pressure differential. The parameters required are system supply pressure, a combined flow coefficient, fluid bulk modulus, maximum spool valve displacement, actuator cross sectional area and initial volumes of fluid under compression on the two sides of the actuator. The third line of data defines where the force output is applied to the mechanical system.

Table 2

```
LIST
HEAD
      3BODIES MODEL
SYST
A  1    3
B  1       1               0.0         2.0
D  1       1               0.02
BODY
    1   1                102.29   1832.55
    1   2                 12.5       5.33
    2   1                234.57   1413.08
    2   2                 17.22      5.371
CSTR
    1   2          G
RVLT
    1         1    2        4.72       0.041
TSDA
    1   1     2    1       -3.72      -3.041      1.0         0.0
    1   2                 100.0                   44.757
CNTR
  INPT
    1   1   STEP            0.0        0.05
    2   7   PH     1
    3   8   PHD    1
  SUMM
    1  12   10    11    2
    2   2   -9          3
  AMPL
    1   6    9             0.0002
    2   7   10            50.0
    3   8   11             1.0
    4   3    4             0.1
    5   1   12            50.0
  HYSV
    1   4    5     6    2000.0      78.0     200000.0     0.05
    1                     12.0     150.0        150.0
    1   2    1            -3.72     -3.041        1.0      0.0
END
```

In the case of the hydraulic servoactuator super module, output definitions which actuate the mechanical system are contained within the super element. In general, an output statement is required which indicates which control signal is output, and where it acts on the mechanical system.

A complete system model was assembled by appending the above control loop to a DADS simulation of a full scale tracked vehicle. A similar control is applied to the azimuth axis when using a 3D simulation. Figure 3 shows the complete vehicle model. The system response was illustrated by simulating the vehicle traversing a specified terrain profile with the command input to the control calling for the gun to be maintained in an inertially fixed orientation.

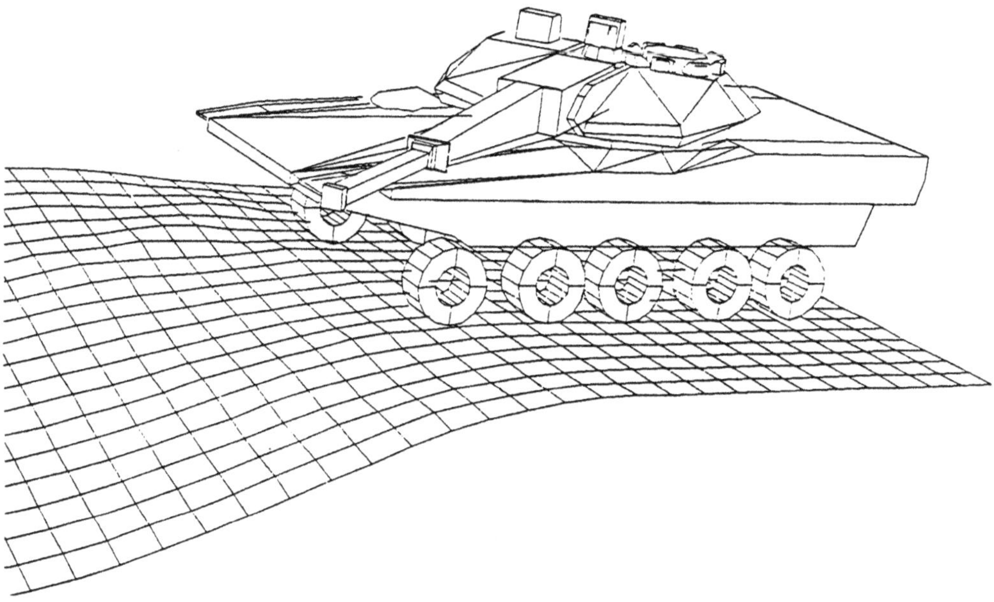

Figure 3. Gun tube model appended to full vehicle model

A transient response was used to illustrate the behavior of the simplified system. The simplified model elevation angle, ϕ, response to a step input of -0.05 radians is shown in Figure 4. The loop gains were adjusted to give a rapid rise time with approximately critical damping. These parameters were then retained for the 3D simulation.

The over terrain response of the 3D model in the pitch plane is shown in Figure 5. It is clear at this point that, while plots of

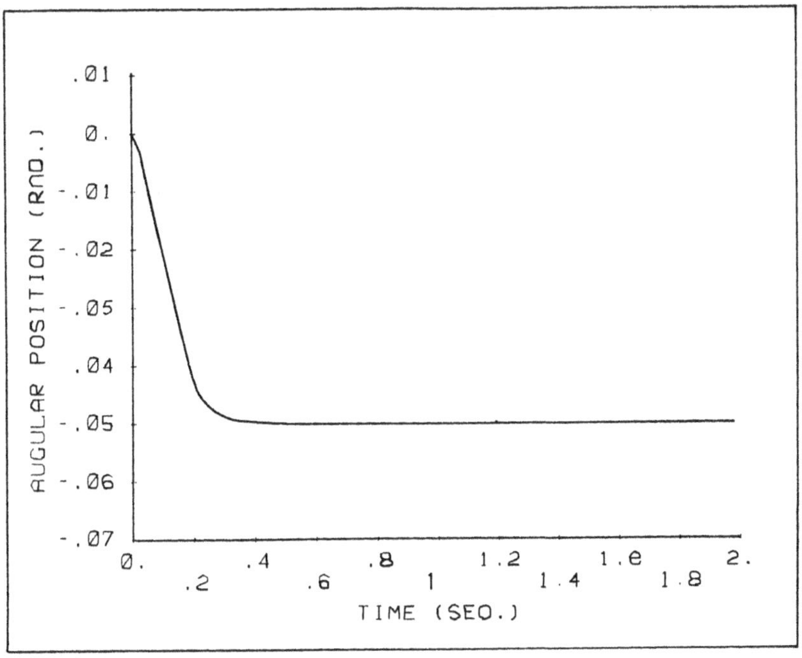

Figure 4. Step response of simplified gun tube model

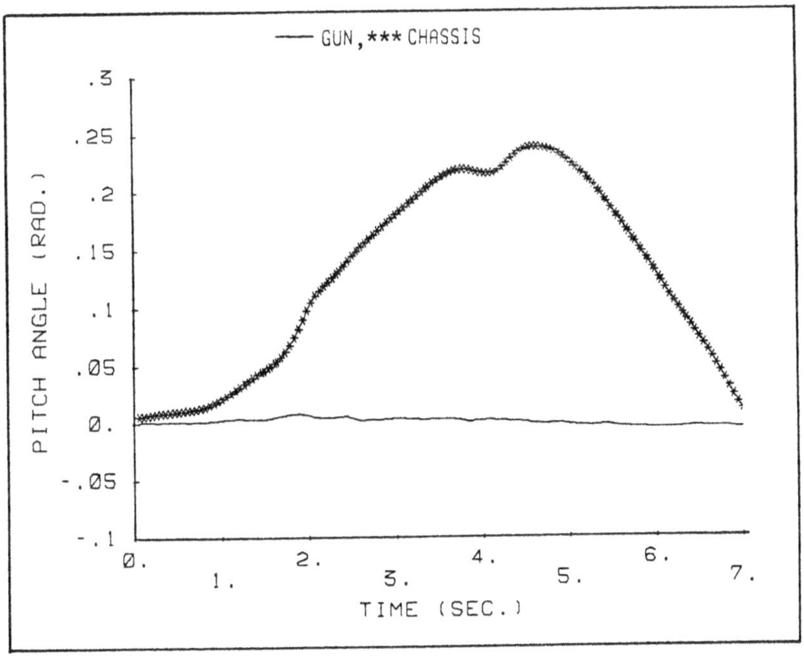

Figure 5. Time history of chassis pitch angle and gun tube pitch angle

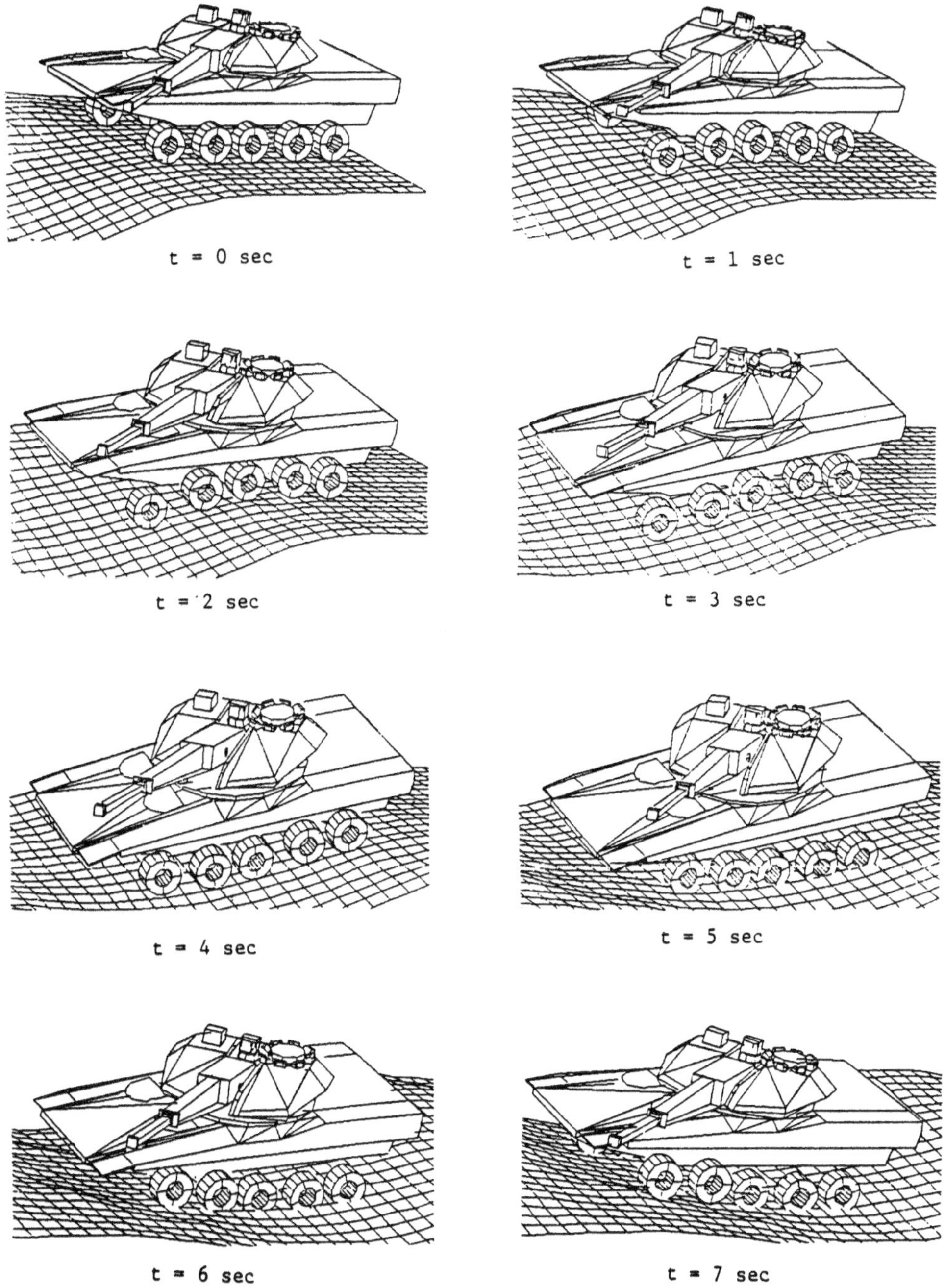

Figure 6. Single frames from vehicle animation

system behavior in any desired form can be made, a much better interpretation of system behavior can be obtained through an animation of the full system in three dimensions. Figure 6 presents a series of frames from a video animation of the vehicle with the gun stabilized.

A second example is presented in Figure 7. A magnet pulls a steel block in the x direction with the force between the two being inversely proportional to the distance between them, σ. Viscous damping is assumed between the block and ground. The system has an equilibrium velocity V_0, and relative displacement σ_0, as shown in the figure. Without control, the equilibrium point is unstable and the block falls behind, or catches up with the magnet. The block diagram of a feedback control system which stabilizes the system is shown in Figure 8. Note that the relative displacement and velocity between the magnet and the block are used as feedback to stabilized the system through a force on the magnet. Since the DADS formulation uses a set of generalized cartesian coordinates for each body in the system, the relative displacement and velocity of the two bodies is not directly available from DADS, and must be computed in the control portion of the simulation. Because of the simple nature of this example, this is not a difficulty, but for more complicated problems, it is a concern. The next example will better illustrate the problem.

A final example, a two degree of freedom robot arm, is presented in Figure 9. The arm rotates about the origin of the coordinate system and extends in the radial direction. The block diagram of a control system to control the arm is shown in Figure 10. Inputs for controlling the system are functions representing the desired time history of the rotation and the extension of the arm. Actuators provide a torque at the axis of rotation, and a force along the axis of extension. Feedback variables required for control are angular position and velocity, and the position and velocity of the extension

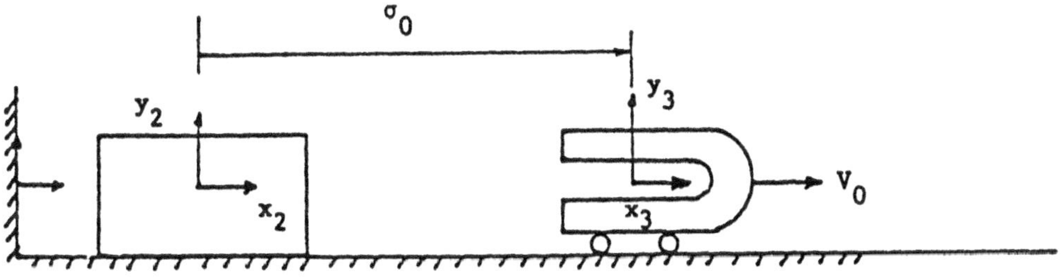

Figure 7. Magnet block example

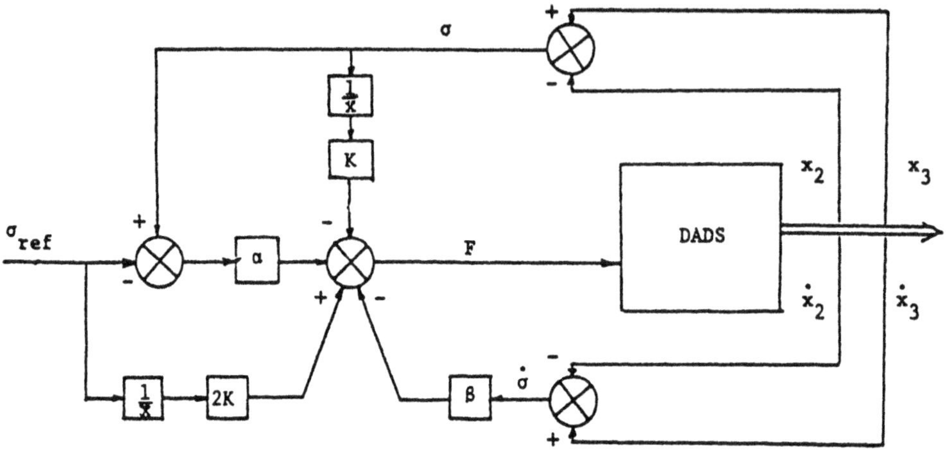

Figure 8. Control system for magnet block

coordinate. From DADS, the angular position and velocity is directly obtained. However, the relative extension variables are not internal variables in the DADS formulation. Computing the relative extension and velocity are complicated functions of the cartesian coordinates and velocities of the body which extends. These functions are shown as f_1 and f_2 in Figure 10. In this case, the functions are complicated, but can be computed. For more complicated cases, this is not always true. For example, if another arm which rotates and

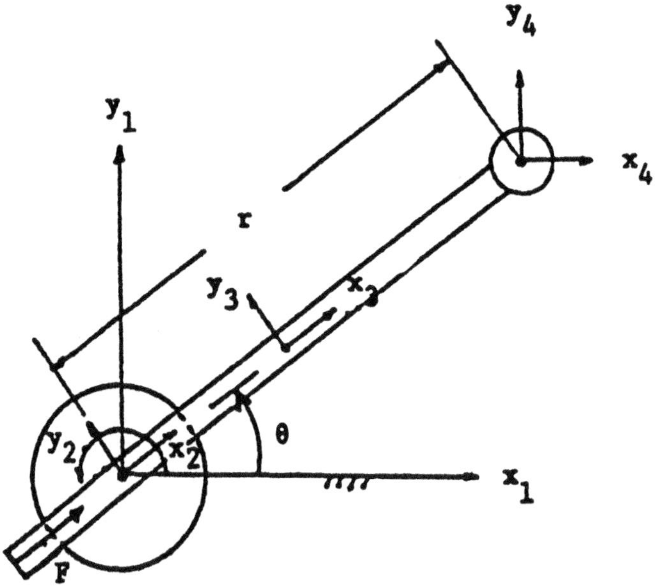

Figure 9. Robot arm example

extends was attached to the end of the existing arm, the functions required to compute relative coordinates from body fixed local coordinates are very complicated. From these preliminary cases, it has become clear that for the general purpose addition of control systems to mechanical systems, a formulation for the mechanical equations of motion using relative coordinates is desirable. Thus, relative coordinates would be directly available for feedback to the control system.

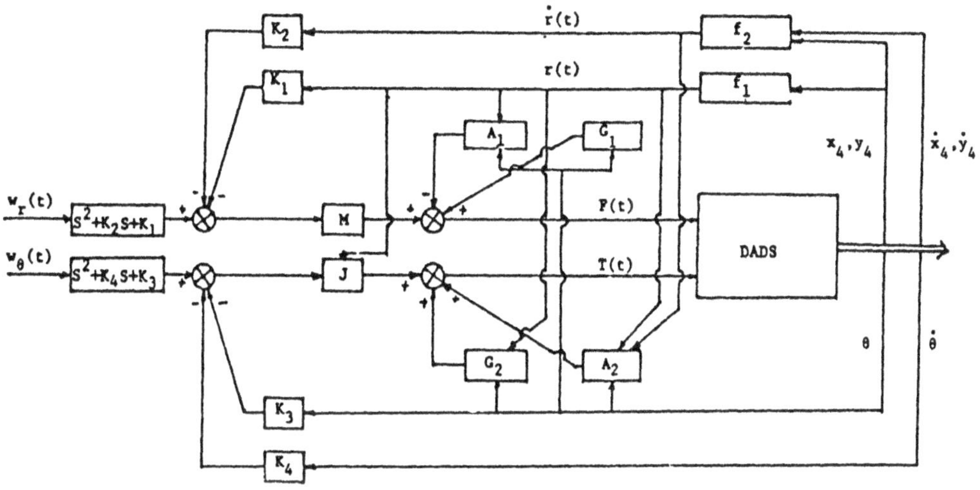

Figure 10. Control system for robot arm

CONCLUSIONS

The user friendly control package designed as an integral part of the DADS program has proven to be a robust tool for the study of controlled mechanical systems. The required user input is straightforward and an extensive number of control elements are currently available.

Preliminary experience with the package indicates that a formulation for the mechanical equations of motion should be derived using relative coordinates. Feedback variables for the control system will then be directly available from the mechanical system simulation. This problem is currently being studied by the authors.

Work is also underway on the development of a preprocessor for control system definition. This would include an interactive input mode and a graphical echo of the block diagram for control system verification.

REFERENCES

1. Frederick, D.K., "Software Summaries", Control Systems Magazine, Vol. 2, No. 4, Dec. 1982.

2. Wehage, R.A. and Haug, E.J., "Generalized Coordinate Partitioning for Dimension Reduction in Analysis of Constrained Dynamic Systems", ASME, J. Mech. Design, Vol. 104, 1981, pp. 247-255.

3. Nikravesh, P.E. and Chung, I.S., "Application of Euler Parameters to the Dynamic Analysis of Three Dimensional Constrained Mechanical Systems", ASME, J. Mech. Design, July 1982.

4. Paul, B. and Krajcinovic, D., "Computer Analysis of Machines with Planar Motion - Part I: Kinematic; Part II: Dynamics", Journal of Applied Mechanics, Transactions of ASME, Ser. E, Vol. 37, pp. 697-712, 1970.

5. Chace, M.A. and Smith, D.A., "DAMN-A Digital Computer Program for the Dynamic Analysis of Generalized Mechanical Systems", SAE paper 710244, January 1971.

6. Sheth, P.N. and Uicker, J.J., Jr., "IMP (Integrated Mechanisms Program), A Computer Aided Design Analysis System for Mechanisms and Linkages", Journal of Engineering for Industry, Transactions of ASME, Ser. B, Vol. 94, pp. 454-464, 1972.

7. Orlandea, N., Chace, M.A., and Calahan, D.A., "A Sparsity-Oriented Approach to the Dynamic Analysis and Design of Mechanical Systems, Parts I and II", Journal of Engineering for Industry, Transactions of ASME, Ser. B, Vol. 99, pp. 773-784, 1977.

8. Nikravesh, P.E., Haug, E.J., and Wehage, R.A., Computer Aided Analysis of Mechanical Systems, Center for Computer Aided Design, The University of Iowa, Iowa City, Iowa.

9. DADS User Manual, Center for Computer Aided Design, The University of Iowa, Iowa City, Iowa.

NUMERICAL INTEGRATION OF SYSTEMS WITH UNILATERAL CONSTRAINTS

G.P. Ostermeyer
Institut für Mechanik
Technische Universität
D-3300 Braunschweig, FRG

INTRODUCTION

If a mechanical system with coordinates $\underline{q} = (q_1 \dots q_n)^t$ is subject to the unilateral constraint

$$f(\underline{q},t) \geq 0, \quad f \in C^2[U \subset \mathbb{R}^{n+1}], \quad t \text{ - physical time} \tag{1}$$

then discontinuities can occur in the velocities (in contrast to systems with the classical bilateral constraint $f(\underline{q},t) = 0$ which does not cause such discontinuities). For the velocity jumps algebraic equations are known (see refs.[1, 2]). When a system of differential equations of motion subject to the constraint (1) is solved numerically the usual procedure is as follows. First the impact time t_i is determined iteratively. The numerical integration is then interrupted at $t=t_i$ and the velocity jumps are computed. With new initial conditions the numerical integration is then continued. The present paper describes a technique for computing highly accurate solutions without having to interrupt the numerical integration and without having to solve algebraic equations for velocity jumps. Any standard integration routine with stepsize control can be used.

1. APPROXIMATION OF UNILATERAL CONSTRAINTS BY POTENTIALS

The constraint force resulting from a classical holonomic constraint $f(\underline{q},t) = 0$ has the potential $V = \lambda f(\underline{q},t)$ where λ is a Lagrangian multiplier. A well known approximation of this potential is $V_1 = \frac{1}{\alpha} k f^\alpha$ where the positive parameter k can be interpreted as a spring stiffness. The "bilateral" potential V_1 can be represented as a sum of two "unilateral" potentials V_1^- and V_1^+ as follows

$$V_1 \equiv V_1^- + V_1^+ = \frac{1}{\alpha} k \left[\frac{1}{2}(f-|f|)\right]^\alpha + \frac{1}{\alpha} k \left[\frac{1}{2}(f+|f|)\right]. \tag{2}$$

The term V_1^- alone represents an approximation for the potential V_2

NATO ASI Series, Vol. F9
Computer Aided Analysis and Optimization of Mechanical System Dynamics
Edited by E. J. Haug

associated with the unilateral constraint (1), $V_2 = V_1^-$. In fact, this potential implies the action of a force in the equations of motion which has the form

$$-V_{2\underline{q}} = -k\left[\frac{1}{2}(f-|f|)\right]^{\alpha-1} \cdot f_{\underline{q}}, \qquad (\alpha-1)\in\mathbb{R}^+ . \tag{3}$$

This force is nonzero only if the unilateral constraint is violated. The direction of the force is that of the inner normal to the limit surface $f(\underline{q},t) = 0$ in the configuration space. When k tends to infinity the time integral of the force yields the impact impulse which is used in the classical theory of impulsive motion (coefficient of restitution $e = 1$). A highly accurate approximation is obtained by choosing for the stiffness k a value of 10^{20} or 10^{30} or even more. In order to get a solution of the "numerical" Dirac deltafunction in (3) by a standard integration routine we need a transformation $t=t(s)$ from physical time t to a fictitious time s which is identical with t when the system is far away from an impact constellation, but which is equal to εt with $\varepsilon \ll 1$ when an impact occurs. This is achieved by putting $dt/ds = t' = \phi(\underline{q},t)$ with some appropriately chosen function ϕ. As a result of this transformation the short time interval associated with large impact forces is spread out so as to render integration routines applicable.

The mechanical system is specified by its Lagrangian

$$L(\underline{q},\dot{\underline{q}},t) = \frac{1}{2}\,\dot{\underline{q}}^T \underline{\underline{A}}\dot{\underline{q}} + B^T\dot{\underline{q}} - V(\underline{q},t),$$

by forces $\underline{Q}$ not contained in the potential and by the constraint (1). With the new Lagrangian

$$L^* = L - \frac{1}{\alpha}k\,[\tfrac{1}{\alpha}(f-|f|)]^{\alpha}$$

the equations of motion written in terms of s take the form

$$\begin{aligned} \underline{q}' &= \underline{\underline{M}}^{-1}(\underline{p}-\underline{B})\cdot\phi \\ \underline{p}' &= (L^*_q+\underline{Q}) \quad \cdot\phi \\ t' &= \phi . \end{aligned} \tag{4}$$

A possible form for ϕ is

$$\phi = \frac{2}{\pi}\tan^{-1}\left((f+|f|)^{\beta}\right)+\varepsilon . \tag{5}$$

Important is a suitable choice for β. If β is too large then the procedure requires too much computation time. If it is too small then the integration routine stops. An "optimal" value for β is obtainable by applying the regularization technique of impulsive motion described in [4].

2. REGULARIZATION OF IMPACT; A PHYSICAL INTERPRETATION

First, we need a reformulation of the constraint (1). In Sec.1 we have transformed the inequality (1) into a (nonholonomic) equality constraint $g(q,t) = 0$, $g := f-|f|$. This is equivalent to the two constraints $\hat{g} = \bar{f}-w = 0$ and $h=w-|w| = 0$. By choosing a new "generalized" variable z such that $w := z^2$ the constraint h is identically satisfied and we get the well known equivalent of (1)

$$\hat{f} = f - z^2 = 0. \tag{6}$$

This constraint serves as starting point for a very general theory of impulsive motion (see[2]). In the form (6) the unilateral constraint can be physically interpreted as follows (see Fig.1a). Let the masspoint m which is moving along a straight line be constrained by $f = \ell^2(t) - x^2 > 0$. The equivalent constraint in terms of the parameter z is

$$\hat{f}(x,t,z) = \ell^2(t)-x^2-z^2 = 0. \tag{7}$$

Equation (7) describes the pendulum shown in Fig.1b with a massless "suspension point" P. When m approaches the position $x = -\ell(t)$, P moves toward $z = 0$. At z=0 the velocity of P is unbounded. At this

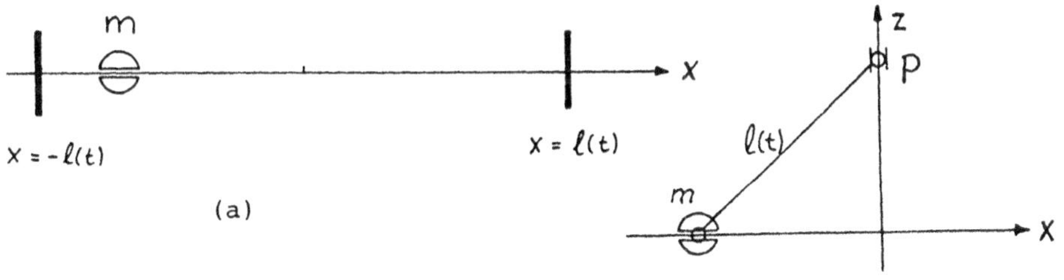

Fig.1

point the velocity of m jumps. In order to avoid the singularities at $z = 0$, a new independent variable must be chosen such that the velocity of P continues to be bounded. A suitable new independent variable is z itself, $s \equiv z$ ($z' = 1$ = velocity of P). The transformation technique is similar to the one described in [3]. The introduction of z as an independent variable leads to a singular time transformation $t = t(s)$:

$$t' \sim |z| = + \sqrt{|\ell^2(t) - x^2| - \ell^2(t) + x^2} \ . \qquad (8)$$

It can be shown in full generality (c.f.[4]) that the resulting equations of motion are regular. They can be integrated without difficulty by any standard integration routine. The transformation back to physical time t is simple. It yields the motion of the system with impact.

3. NUMERICAL INTEGRATION OF SYSTEMS WITH IMPACT

An optimal value for the parameter β results from (8): $\beta = 1/2$. The parameter ε in (5) depends on α in the potential (3) and on the integration routine itself. If there is more than one unilateral constraint in the system then a potential must be formulated for each constraint separately. If two or more impacts occur in the same time interval then the impact forces are superimposed (see [5]).

References

1 Wittenburg, J., "Dynamics of Systems of Rigid Bodies", BG Teubner (1977).

2 Baumgarte, J., "Analytische Mechanik der beschränkten Konfigurationsräume", to appear.

3 Baumgarte, J., Ostermeyer, G.P., "Transformation der unabhängigen Variablen in einer verallgemeinerten Hamiltonschen Formulierung", ZAMM 61, T16-T18 (1981).

4 Ostermeyer, G.P., "Regularisation of Impulsive Motion, to appear.

5 Ostermeyer, G.P., "Mechanische Systeme mit beschränktem Konfigurationsraum", Diss. TU Braunschweig (1983).

Part 5

SYNTHESIS AND OPTIMIZATION

SYNTHESIS OF MECHANISMS

Ing. H. Rankers
Delft University of Technology
Delft, The Netherlands

SUMMARY

NATO ASI Series, Vol. F9
Computer Aided Analysis and Optimization of Mechanical System Dynamics
Edited by E. J. Haug

CHAPTER 1: DESIGN PHILOSOPHY

1.1. The task of a designer in a CAD/CAM-environment

The effort of designers is directed to creation of new or to improvement of existing products including production machines with respect to failure, efficient use or manufacturing costs. The high level quality of the product ensures continuous operation without any risk for the operator or the surroundings [1.1].

To achieve this goal, the development of so called design methods combined with appropriate education and training, to work systematically, has found to be very helpful. Different approaches of design methods are published by Rodenacker [1.2], Koller [1.3], Steinwachs [1.4], Beitz/Pahl [1.5] and the VDI-recommendation nr.2222 [4.6] (VDI = Verein Deutscher Ingenieure, German Society of Engineers).

The entire content of all these methods is the statement, that a design is growing in seven main steps, as shown in figure 1.01a, accordingly [1.6].
Figure 1.01b shows that every design step is associated with a different number of possibilities and modes. Obviously every step requires decisions, which possibilities have to be maintained to fulfill the goal function and which possibilities have to be rejected. The rejection can be done by means of different techniques, for example by error analysis and other decision making investigations, finally the definite design has been accomplished.

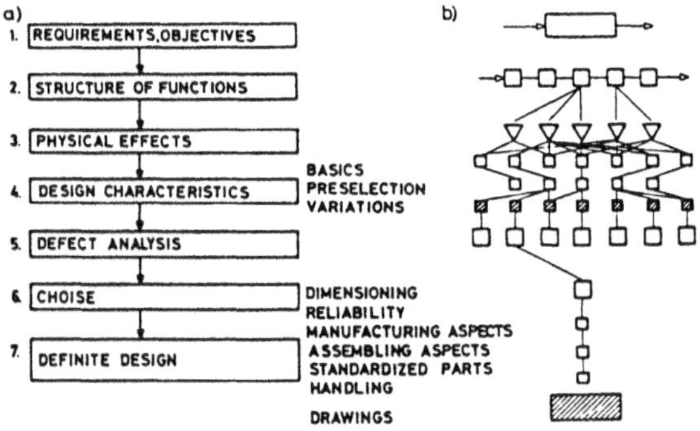

Figure 1.01. Design method.
a) Simplified view on VDI-recommendation nr. 2222, the seven main steps of the design method.
b) Different possibilities in every step.

Figure 1.01 gives the impression of a straight line design process in sequential steps. However, an experienced designer knows that every design process consists of compromises, and that the evaluation of achieved results leads to an interactive process and the necessity of compromises respectively re-interpretation of requirements. The loop back in the real design process is contrarely to the straight line idea of figure 1.01.

In this way figure 1.02 shows the principles of a systematically structured design process with several loops, which make it possible to return with new instructions or recommendations to decision taken earlier with the objective to make a better design [1.7]. This figure expresses the results of the authors investigations in the design structure of machinery with moving parts like packaging machinery.

Figure 1.02 makes two things obvious: firstly the entire job of designing a new product with a complex task is divided into several sub-tasks directed to unit operations or assembly units. This is equivalent to step 1 of figure 1.01. Each of the sub-tasks is well defined by a goal function, which might be expressed identically by step 2 in figure 1.01. Secondly the design process for every sub-task consists of certain sequences of activities, mixed or alternated by creative as well as routine work. Unfortunately, during the educational stay at a University, students learn all about the creative part of their future work, but hear almost nothing about the tremendous amount of routine work, which normally has to be done by them after having finished the education.

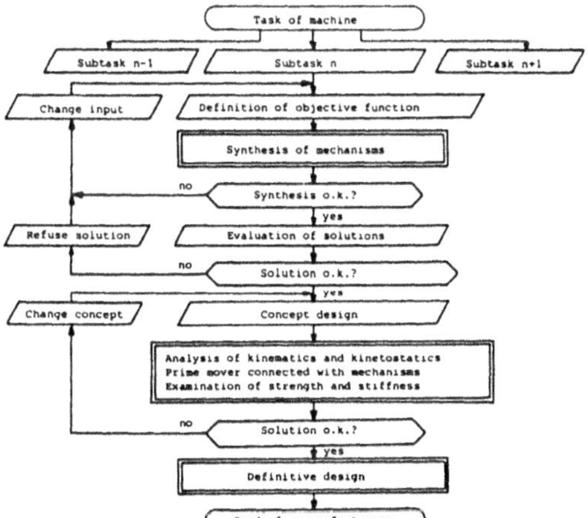

Figure 1.02. Structure of the design process of a mechanism. ▭ = routine work, ▱ = creative work, input generation or output evaluation, ⬭ = decisions.

Generally, a sub-task can be realized by different physical effects as mentioned in step 3. In the design of a machine with moving parts this step is identical with both the type synthesis and the dimension synthesis of mechanisms, which are able to fulfill the goal function (desired motion). Type synthesis means choice or determination of the mechanism type [3.11]. Dimension synthesis means calculation of kinematic and assembly parameters. It might happen that the kinematic synthesis step does not result in sufficient approximation of the goal function and that therefore the obtained solution can not be accepted. In this case it is necessary to check wether or not a certain part of the goal function must be changed to achieve an acceptable result, which allows to leave the loop. The changed goal function has to be entered again into the procedure and the loop has to be restarted at its beginning. If the synthesis procedure results in more than one solution, it becomes necessary to evaluate and to compare the different solutions to choose the best one for further use, as explained in step 4.

During the evaluation step it might happen that all solutions found have to be rejected. Also in this case it is necessary to change the goal function and to go back to the top of the loop, because else a properly defined synthesis procedure would generate the same results. Finally the solution with the best goal function approximation will be chosen as explained in step 5.

This solution forms the basis of the concept design (step 6). The concept design includes

- global mass distribution,
- stiffness of the materials,
- diameter of shafts, and
- other dimensions of the elements.

Together with the applied forces and the mathematical model of the prime mover they specify the (routine) problem to analyze the internal forces and deformations of all designed elements and machine parts. The results of this dynamic examination make clear which changes in the concept design are to be made. The designer is only allowed to finish this loop after having achieved satisfactory results. The definite design drawings and the generation of part lists in accordance with step 7 may be started finally.

Conclusion

* The design process of any sub-task is divided into seven steps and consists at least of two loops.
* The loops are directed to kinematic synthesis and dynamic analysis.
* In both loops the sequence of activities consists of both creative and routine work.
* The routine work should be done by computer, supposed that appropriate software is available.
* The designer's task will change dramatically when he decides to use the computer to solve routine work. The saving in time makes it possible to pay more attention to all creative work, like the definition of goal function, the evaluations and decisions, which will eventually increase the quality of his design results.

As a professor in Mechanical Engineering I would like to point out that teaching the above mentioned design philosophy made it necessary to change the traditional written examination of students. You can not tell them during the course that all routine work should be done by a computer, because a computer will do it faster and more reliable than the designer itself, but during the written examination every student should do his routine work by hand and without any hard- or software assistance. So the written examination had to be cancelled. Instead of this the course has to be finished with an individually formulated exercise which involves the use of the computer, but leaves many creative aspects to the student [1.08]. In order to concentrate the examination on mechanism problems, it was decided to use an interactive input program version. As a result of this decision, the student does not come into contact with any program language. He does not need any knowledge or experience in computing or programming. He has only to do with the different aspects of design philosophy and with the theory of mechanisms and mechanical systems. The examination must be completed with a brief report. Most students needs 10-20 hours of time to do the whole exercise. The software package used at the Delft University of Technology and all Dutch Technical Schools is called CADOM (= Computer Aided Design Of Mechanisms), a development of the CADOM-task-group of the Delft University of Technology [1.09, 1.10, 1.11, 1.12, 1.13, 1.14, 1.15, 1.16]. This software package supports not only synthesis but also kinematic, kinetostatic and dynamic analysis of both linkages and cam mechanisms.

1.2. Specification of software needs

To involve the computer in the designer's work, the appropriate hardware and software has to be available. Synthesis, analysis and design of mechanical systems needs specific software. Knowledge about the task of a designer in such a CAD/CAM-environment forms the basis for specification of software needs. If the mechanical system consists of mechanisms, actuators, prime mover and moving tools, as for example applied in packaging machinery, then an appropriate software package has to handle the following activities:

* All kinds of motion definition:
  - coordination of input and output link motion,
  - coordination of point positions,
  - coordination of plane positions.
* All types of mechanisms:
  - linkages, cams, gears,
  - cam controlled linkages,
  - geared linkages.
* All kinds of mechanisms:
  - plane, spherical or spatial mechanisms.

As is generally known, there are a great number of one-sided software-packages directed to analysis purposes only. Of course the analysis task is very important, but it would be better, if a program package should help both the analysis and synthesis tasks in systems design. The CADOM-package is such a package, see figure 1.3, which is especially designed to serve both synthesis and analysis of the mentioned problems.

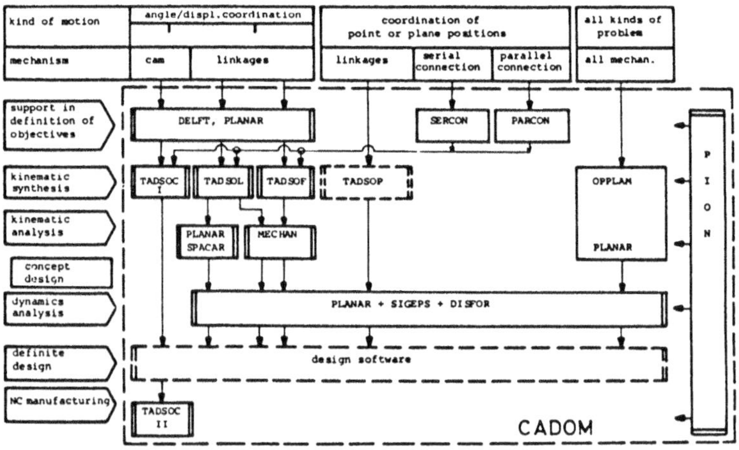

Figure 1.03. CADOM-modules and design activities

Besides this, the CADOM-package is extremely helpful in all routine work in the context of defining the objective function (goal function), if periodic or non-periodic motions are desired or if serial or parallel connection of mechanisms is to be recommended. Therefore, in the CADOM software package four specific programmes are available:

DELFT = Standard functions for definition of intervals,
PLANAR = Calculation of non-linear transformations,
SERCON = Serial connection of two mechanisms,
PARCON = Parallel/serial connection of two mechanisms.

* All kinds of kinematic synthesis procedures:
  TADSOC = Type And Dimension Synthesis Of Cam mechanisms,
  TADSOL = Type And Dimension Synthesis Of Linkages,
  TADSOF = Type And Dimension Synthesis Of Function generators
  TADSOP = Type And Dimension Synthesis Of Path generators,
  OPPLAM = Optimization of Planar Mechanisms.

* All kinds of kinematic analysis:
  PLANAR = Plane mechanisms analysis program,
  SPACAR = Spatial mechanisms analysis program,
  MECHAN = Mechanisms Analysis programmes.

* Kinetostatic and dynamic analysis:
  PLANAR = Plane mechanisms analysis program,
  SIGEPS = Sigma and Epsilon (strength and stiffness),
  DISFOR = Displacement and Forces program.

* Papertape to be used in NC-controlled machine tools:
  TADSOC(II) knows postprocessors for ISO-8 coded papertapes to control cam milling machines with either x-y or R-ϕ coordinates.

* All mentioned program-modules are programmed in FORTRAN-IV (ANS), well structured and portable. They are running on IBM, Amdahl, DEC, PRIME under different operating systems, but there exists an interactive version under CP/M 2.2 too, which is called MICRO-CADOM.

* All program-modules use the general purpose program library PION (= Periodic functions, Input, Output and Numerical treatment).

* The whole CADOM-program-package is an open end system and will be enriched continuously.

CHAPTER 2: DESIGN OBJECTIVES AND GOAL FUNCTIONS IN SYNTHESIS OF MECHANISMS

Design objectives of mechanisms are related to

- kinematics or
- kinetostatics or
- dynamics.

Kinematic synthesis solves problems of motion. In most cases this motion is concerned with coordination of output versus input link, with point position coordination or with plane position coordination. It is remarkable, that all these tasks can be handled with the same type of mechanism. Simplified, often is spoken about "motion generation" and "path generation". Figure 2.1 shows some definitions with respect to the above mentioned kinds of motion.

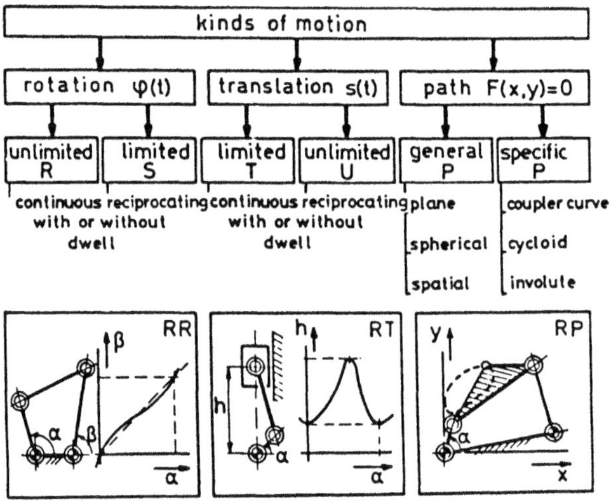

Figure 2.1. Different kinds of motion in mechanisms.

Figure 2.2 shows the 4x5-matrix of all input-output motion combinations that exist. The abbreviations R, S, T, U and P are explained in figure 2.1. All mechanisms with R-input give periodic output motions. Mechanisms located in the middle of the matrix generate closed curves, there is no complete relative rotation in these mechanisms. Usually, they are called "all-rocker mechanisms" [2.1].

According to the IFToMM-terminology, kinematic synthesis is defined by: "Design of a mechanism which satisfies various prescribed combinations of positions, velocity ratios, acceleration ratios, etc., assuming all members as rigid and massless" [2.2]. Because this ratio

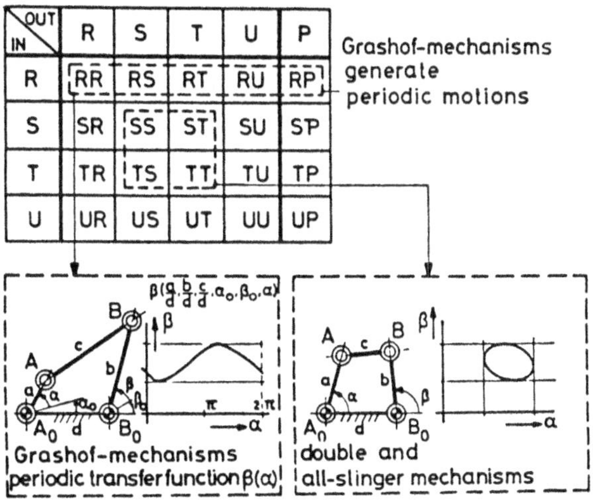

Figure 2.2. All the input-output motion combinations in link mechanisms.

is time-independent, it is called "objective transfer function" or "goal function" [2.2]. Such geometrically determined motion quantities serve as (discrete) conditions for the kinematic synthesis process. "Uniform conditions" means that one type of conditions, for example positions only (= zero order transfer function), is prescribed. To the contrary "mixed conditions" means that for example positions and velocities are prescribed (= zero and first order transfer functions).

Kinetostatic synthesis solves design problems of forces and torques, especially of force and torque transmission. A typical example is the compensation of torque caused by the potential energy of link masses or by a spring loaded cam mechanism [2.3]. Kineto-elastostatic synthesis means "design of mechanisms which satisfies various combinations of prescribed positions, velocity ratios, acceleration ratios, force and torque transmissions, etc., while mechanism members are assumed to be elastic" [2.2]. The kineto-elastodynamic synthesis is directed to design of a mechanism which satisfies various combinations of positions, velocities, accelerations, force and torque transmissions, stresses, strains, etc., at a predetermined running speed. The mechanism members are assumed to be elastic and have either concentrated or distributed masses [2.2].

Dynamic synthesis solves design problems of "mechanisms which satisfie various prescribed combinations of position, velocities and accelerations, etc., considering the members as rigid and as having either concentrated or distributed masses" [2.2]. Typical examples are "balancing a mechanism" with respect to a specified goal: constant angular velocity of the input crank, no shaking forces, etc. [2.4].

The objective transfer function (goal function) is defined either by

- discrete conditions or by
- continuous conditions.

The prescribed conditions might be

- periodic conditions, like $f(\alpha) = f(\alpha + 2\pi)$, or
- non-periodic conditions.

The strategy for kinematic synthesis is directed to receive two highly important informations [2.5]:

- the types of appropriate mechanisms (= type synthesis), and
- the kinematic parameters and the assembly parameters of those types (= dimension synthesis).

A catalogue of mechanisms might be very helpful to interprete the computer output [2.6].

All synthesis activities have to begin with kinematic synthesis as mentioned before in chapter 1: "design philosophy". Therefore a presentation of different synthesis methods will be started now.

Figure 2.3 shows an arbitrarily chosen crank and rocker mechanism with crank A_0A, the coupler AB and two arbitrarily chosen points C and D in the coupler plane, the rocker B_0B and the frame A_0B_0. The position of the input link (= crank A_0A) is marked by angle α . The output link position is defined by angle β . The entire mechanism is located in a YOX-plane.

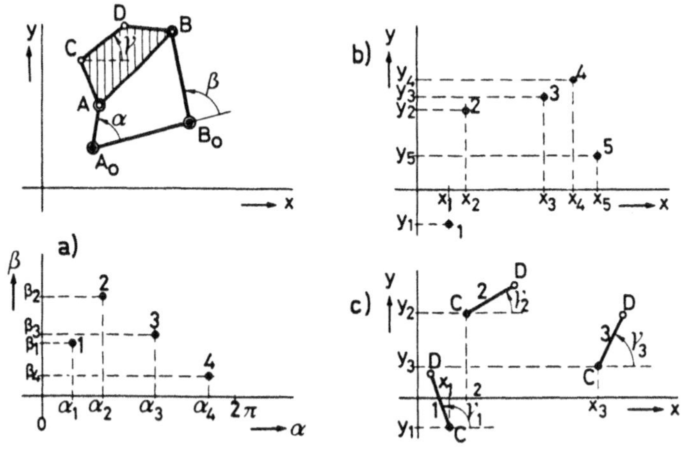

Figure 2.3. Discrete conditions:
a) coordination of angles and/or displacements,
b) coordination of point positions,
c) coordination of plane positions.

2.1. Coordination of angles and/or displacements

The coordination of angles and/or displacements of output versus input link positions is a daily problem of a designer of mechanisms.

The goal function might be given either by discrete or by continuous conditions, and the goal functions might be a periodic or non-periodic function.

If discrete conditions are given, the maximum number of them may not exceed the number of mechanism parameters p. The total number p of mechanism parameters is the sum of the k kinematic parameters and the p = k + m. Looking at a mechanism with one degree of freedom, normally there are m = 2 assembly parameters: to define the assembly situation at the input link respectively at the output link, while the number k of kinematic parameters depends on the topology of mechanism with n links: k = 3(n-3). The relation between the number n of links and the number g of joints is given by the well known formula [2.7]: g = (3n-2)/2. Tabel 2.1 shows the total number p of parameters in accordance with the topological definition of the mechanism using n and g:

number of links n	number of joints g	number of assembly parameters m	maximum number of kinematic parameters k	maximum total number of all parameters p=k+m
4	4	2	3	5
6	7	2	9	11
8	10	2	15	17
10	13	2	21	23

Tabel 2.1. Topology and number of parameters of mechanisms with one degree of freedom.

The kinematic synthesis procedure, that can handle the above mentioned type of goal function, is outlined in chapter 3.1. The number p of prescribed discrete conditions (goal function given in precision points) will be exactly fulfilled.

If the goal function is given by continuous conditions, another procedure has to be used to synthesize a mechanism. There is no limitation in the number of mechanism parameters, because of the "overshooting" information that has to be handled. In the TADSOL-synthesis procedure, the information abundance is used to realize the type of mechanism selection with the aid of presence of some significant Fourier coefficients. Continuous conditions are specified for example at the problem to compensate the resulting torque of the input shaft. This torque is caused by the potential energy of all link masses of a mechanism and has to be compensated by another mechanism, that is spring loaded. To obtain a solution, when continuous conditions are prescribed, a certain optimization technique will be applied, as explained later on in chapter 3.2 respectively in chapter 3.4.

Between the above mentioned two kinds of prescribing the goal function, either discrete or continuous, exists the possibility to connect the precision points (= discrete conditions) by a certain spline function to obtain the goal function in continuous conditions. A second possibility is the use of so called standard functions to prescribe with their aid several intervals, the sequence of them define the total goal function. This approach is commonly known in connection with cam mechanism synthesis, but there are some reasons to extend this approach and to use them as a part of the general synthesis technique.

2.2. Coordination of point positions

The coordination of point positions is a more pretentious case in the field of mechanism synthesis.
Point positions are used to prescribe the path of a certain point of a certain link in a XoY-plane. Point positions are defined by prescribed values x and y. In so far, point position is a problem with two degrees of freedom. Consequently mechanisms with two degrees of freedom are able to solve this problem in general [2.8]. As worked out in figure 2.4, there are 17 different five link topologies with closed chains, when all variations of turning and sliding joints are applied. Considering open chains the same joint variation leads to 4 different topologies, as shown in figure 2.5 [2.9].

Theoretically point position can be specified either with or without crank angle. If the crank angle α is specified, the coordination of point positions is defined by the well known parameter representation $x(\alpha)$, $y(\alpha)$ of a curve. This case will be regarded also as "uniform conditions". This means, that the prescribed conditions are all of the same kind, in this case, points of the zero order path function. In a more complex situation the point positions and their derivatives with respect to the crank angle α are prescribed. In that case is spoken about "mixed conditions".

Under certain conditions, the coordination of point positions including crank angles prescribed to the coupler point of a four-link mechanism leads to succesful synthesis of the six kinematic parameters of the linkage. I suppose, that George N. Sandor and F. Freudenstein wrote in 1958 the first computer program to determine the dimensions of pivoted four-link mechanisms in the plane to generate a path given by five precision points. By means of a quality index, the computer program calculates all possible solutions with a maximum of 12 different mechanisms [2.10].

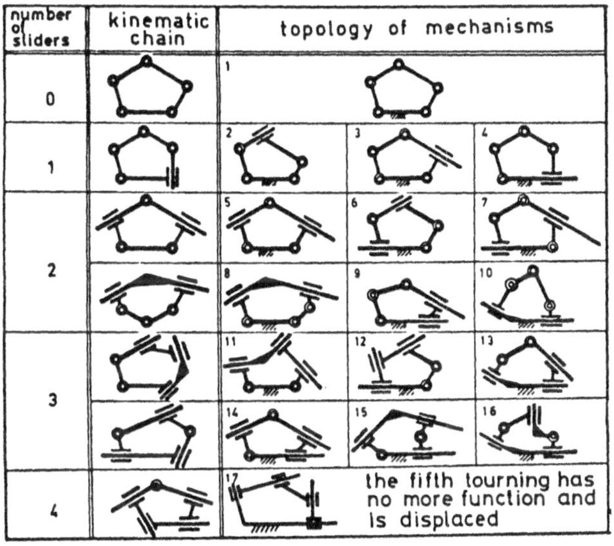

Figure 2.4. 17 closed chain mechanisms with two degrees of freedom.

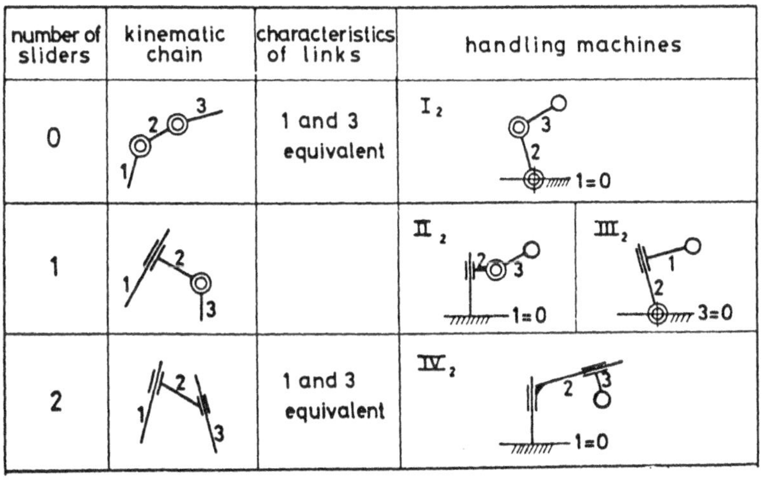

Figure 2.5. 4 open chain mechanisms with two degrees of freedom.

In all other cases, for instance when coupler curves are not able to fulfill the uniform demands, or when mixed conditions are prescribed, the above mentioned synthesis technique will fail. A general solution of this class of problems is not yet known, but there are two different special methods to come to a solution: by means of parallel connection (PARCON-subset of CADOM) or by means of series connection (SERCON-subset of CADOM) of two appropriate mechanisms.

Concerning parallel connection, a special solution with twin-cam controlled linkages was given by the author in 1969 [2.11]. The basic idea is to describe the path as the zero order path function and to consider the first order path function to be dependent of the zero order function. Now it becomes possible to resolve this first order path function into two new, separated (periodic) goal functions, using one of the 17 possible two degrees of freedom mechanisms as a superposition mechanism. The two input motions must be generated by two separated mechanisms, which are parallel connected by their crankshaft. In [2.11] this was realized by two cam-and-follower mechanisms. In this way the problem of the position point coordination is reduced to twice the normal cam synthesis procedure. In the CADOM-software package the subset PARCON is available, see figure 1.3, to generate the necessary (new cam follower) goal functions automatically. Also in plane position coordinations this PARCON plays an important role, see chapter 2.3.

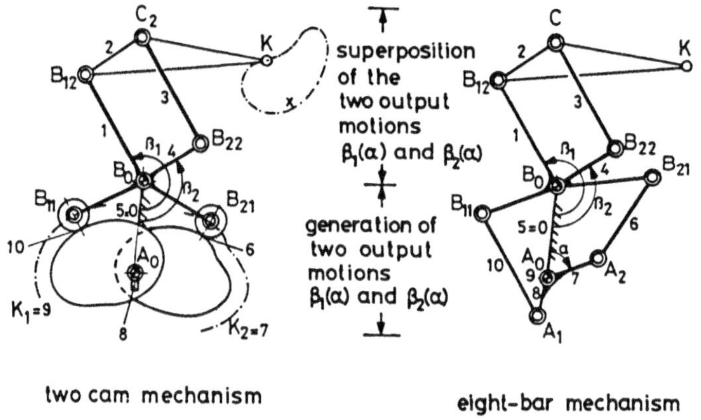

Figure 2.6. Two different realisations of mixed conditions prescribed by point positions and point velocities:
a) twin-cam controlled five-bar linkages with two coincident pivots in the frame,
b) a special 8-link mechanism.

Figure 2.6 gives an impression of two possible solutions of point position coordination. The solution at the left part consists of a twin-cam on the cam shaft and a five-bar superposition mechanism, see nr.1 in figure 2.4. Each degree of freedom is controlled by a cam. The cams itself are the results of twice the TADSOC-synthesis procedure. The solution at the right part consists for example of two different four-bar linkages, each of them is the result of the TADSOL-synthesis procedure.

The basic idea of the above mentioned series connection of two different linkages is, that the zero order path function of a mechanism $f(x,y) = 0$ depends only on the kinematic parameters k_2 of the second mechanism. The first order path function however can be influenced by placing a certain mechanism with kinematic parameters k_1 in front of the path generating mechanism, to realize such an input-output angle coordination $\beta(\alpha)$, that the demands of the mixed conditions will be fulfilled. Obviously, the product rule describes the series connection:

$$x(\alpha) = x\{\beta(\alpha)\} \,, \qquad y(\alpha) = y\,\{\beta(\alpha)\} \,,$$

$$dx/d\alpha = dx/d\beta \,.\, d\beta/d\alpha \,, \qquad dy/d\alpha = dy/d\beta \,.\, d\beta/d\alpha \,,$$

$$d^2x/d\alpha^2 = d^2x/d\beta^2 \,.\, (d\beta/d\alpha)^2 \,, \qquad d^2y/d\alpha^2 = d^2y/d\beta^2 \,.\, (d\beta/d\alpha)^2 \,,$$

$$+ \, dx/d\beta \,.\, d^2\beta/d\alpha^2 \,, \qquad + \, dy/d\beta \,.\, d^2\beta/d\alpha^2 \,.$$

An example of a series connection, applicated at the egg container closing station of an egg packaging machine is reported in [2.12]. A straight line generating mechanism A_0ABB_0 is chosen to guide the "pressing element", see figure 2.7, along the desired path. But the velocity along the straight line interval has no constant value. To influence the first order motion without changing the path itself (= zero order motion), series connection of two mechanisms is worked out. The synthesis procedure has pointed out, that a double-crank mechanism with certain kinematic and assembly parameters could help to correct the first order path function. Figure 2.7 shows the resulting 6-link mechanism of type Watt-2 to guide the "pressing element" along the desired path with constant velocity on the straight line part of that path.

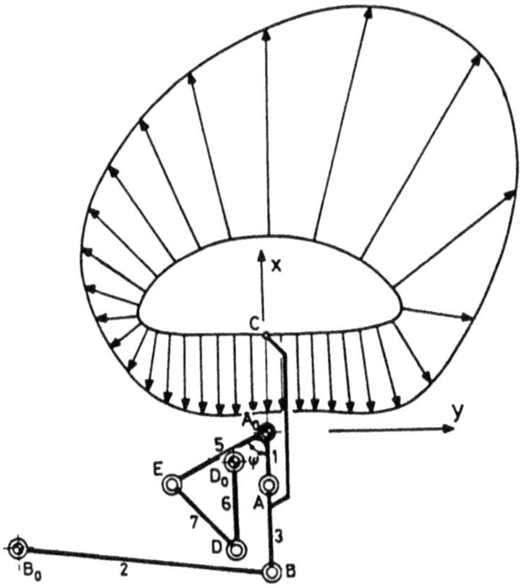

Figure 2.7. Egg container closing station of an egg packaging machine [2.12].

2.3. Coordination of plane positions

The coordination of plane positions is the most unpopular problem in kinematics. Plane position is prescribed by three coordinates $x(\alpha)$, $y(\alpha)$ and $\phi(\alpha)$. To move a plane, a mechanism needs normally three degrees of freedom. As reported in [2.7], for those mechanisms exists only one structure with six links (n=6) and six tourning pairs (g=6) , but two different structures with eight links (n=8) and nine joints (g=9) exist, as shown in figure 2.8. These structures are defined by

a) $n4 = 0, \quad n3 = 2, \quad n2 = 6$ respectively

b) $n4 = 1, \quad n3 = 0, \quad n2 = 7$.

Based on specification a) a design of a three degrees of freedom NC-controlled sawing machine is worked out by one of my former students [2.13]. Figure 2.9 shows to the left part a schematic drawing of the mechanism with three co-incident pivots $A_0=B_0=C_0$ and the plane DEF, connected by three couplers AD, BE and CF to the three cranks A_0A, B_0B, C_0C with input angles α, β and γ respectively. To the right part of figure 2.9 a raw idea of the concept realization is scetched with three center-free rings instead of real cranks, driven by three motors M1, M2 and M3.

Based on specification b) à year ago a special low cost welding machine was designed and constructed to be used in heat exchanger element manufacturing. It was found, that application of CAD/CAM-techniques to synthesize, design and construct a set of three cams to control the three degrees of freedom of a certain, product related plane position coordination gives the same flexibility as first generation industrial robots, the large advantage of preprogramming outside of the welding machine, and the availability for a tenth of the investment in comparison with a robot installation including NC-controlled parts handler. Besides this advantages, it is obvious, that welding of a big class of products is possible only by changing the triplet-cam, and that the cam synthesis procedure and the cam manufacturing takes place independently of the welding machinery. The constant welding velocity along the weld is guaranteed by the synthesis procedure [2.14].

In engineering practice it is of great importance to try reduction of the degree of freedom from F=3 to F=2 by specifying and/or using certain relations between point coordinations in the plane. Of course, this approach requires the re-definition of the motion problem. About three different applications of plane position coordination synthesis with a reduced degree of freedom is briefly reported in [2.15]. The cases are worked out by postgraduate students during a three-month-stay in industries to qualify themselves in design of mechanisms and mechanization of production.

2.4. Relationship between different kinds of coordination

The relationship will be explained by using four-bar linkages. But the relationship between coordinations of either angle or displacements, point or plane positions is a common relationship between linkages in general.

How many point position coordinations do exist? In figure 2.3 moves the coupler-point C along N points (X_i, Y_i) of the prescribed path. The four-bar linkages has, as previously explained, 9 independent parameters, namely four parameters (XA_0, YA_0, XB_0, YB_0) to define the frame position and further on three link lengths (a, b, c) and two parameters (e, f) to define the coupler-point position. For all N positions of point C, it is to be written:

$$X_i = X(XA_0,\ YA_0,\ XB_0,\ YB_0,\ a,\ b,\ c,\ e,\ f,\ \alpha_i)\ ,$$
$$Y_i = Y(XA_0,\ YA_0,\ XB_0,\ YB_0,\ a,\ b,\ c,\ e,\ f,\ \alpha_i)\ , \qquad i=1(1)N.$$

If the parameter α_i is the unknown crank angle, for which the coupler point C has to lie in position (X_i, Y_i), then the 2N equations have 9+N unknown parameters. There might be a single solution if 2N = 9+N respectively if N = 9. Thus, a coupler curve is able to pass through nine prescribed points without crank angle definition.

If the crank angle belongs to the prescribed conditions and if it is expected, that every point position will be reached at a certain crank position $\alpha_i+\beta$, then the problem has 10 parameters, all above mentioned parameters plus the assembly parameter β . Now, there are 2N equations with 10 unknowns. Thus, N=5 and a maximum of five point positions with crank angle is to be prescribed.

Figure 2.3 is helpfull to determine the number of possible plane positions. The plane, marked by the line CD, may lie in N positions (X_i, Y_i, γ_i) . Thus, we have to write three equations for each position:

$$X_i = X(XA_0,\ YA_0,\ XB_0,\ YB_0,\ a,\ b,\ c,\ e,\ f,\ \gamma_0,\ \alpha_i)\ ,$$
$$Y_i = Y(XA_0,\ YA_0,\ XB_0,\ YB_0,\ a,\ b,\ c,\ e,\ f,\ \gamma_0,\ \alpha_i)\ ,$$
$$\gamma_i = \gamma(XA_0,\ YA_0,\ XB_0,\ YB_0,\ a,\ b,\ c,\ e,\ f,\ \gamma_0,\ \alpha_i)\ , \qquad i=1(1)N.$$

These are 3N equations with 10+N unknowns. There might be a single solution for five plane positions without crank angle, because 3N = 10+N or N = 5.

Finally, the question is, how many angle or displacement coordinations between the crank plane and the rocker plane are possible, if the mechanism has five parameters, as earlier explained? For each coordination is prescribed

$$\beta_i = \beta(\lambda,\ \sigma,\ \kappa,\ \delta,\ \psi,\ \alpha_i), \qquad i=1(1)N.$$

Now, the crank angle is prescribed too, he does not belong to the unknowns. There are N equations with 5 unknowns. Thus, it is possible to prescribe a maximum of five coordinations, N=5.

It is not astonishing, when each of the coordination cases

- plane position,
- point positions with crank angle and
- angle or displacement

require a maximum of five conditions, because the three cases belong to different formulations of the same problem, as explained with the aid of figure 2.8. A point position coordination is transformed into a plane position coordination, if a ROBERTS-cognate is examined. The position of the coupler-point K of the four-bar linkage A_0ABB_0 is identical with the position of plane A'K, that lies parallel to the crank A_0A under the crank angle α , as sketched in figure 2.8a. And angle or displacement coordinations change into plane position coordinations, if the so called "frame exchange" is applied and the crank A_0A is exchanged to frame A_0B_0 respectively the rocker B_0B becomes coupler AB, as sketched in figure 2.8b.

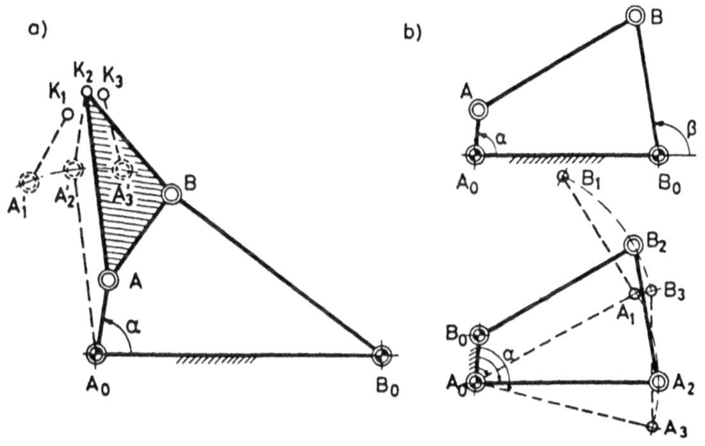

Figure 2.8. Relationship between plane position coordination and a) point positions coordination with crank angle or b) angle/displacement coordination.

CHAPTER 3: DESIGN TECHNIQUES IN SYNTHESIS OF MECHANISMS

With respect to the topics of the 1983 Iowa City Advanced Study Institute and in continuation of previously given design objectives and goal functions, now it is time to switch over to four major design techniques in synthesis of mechanisms. The four techniques are:

3.1. Linear algebra,

3.2. Fourier series,

3.3. Non-linear transformations,

3.4. Optimization techniques.

3.1. Linear algebra, theory, examples

A large number of publications (books and papers) treating the precision point synthesis exist. The traditional approach deals with plane position problems, because, as it is explained in chapter 2.4, the two other problems of point position respectively of angle/displacement coordination are included due to simple transformations. The traditional approach is best known as Burmester-Theory [3.1]. Burmester himself and many successors have worked out the high level of sophistication that has been reached. But never-the-less, in the curriculum of most of the Universities of Technology this subject is missing and teaching this theory has been stopped, because it is so time consuming. Even in industrial applications the Burmester-Theory was found to be too expensive and its use demands highly skilled people. The entrance of computers in general, but especially the entrance of micro-computers in drawing offices begins to change the situation dramatically. If a designer is familiar with the design philosophy and the synthesis procedure, and if an interactive computer program package is available, the synthesis itself is done in a few seconds or minutes. This tremenduous saving of time opens the door for outlining of design variations, that has never been done in the past. Evaluation of design alternatives is the way to find better solutions and to minimize design risks. The evaluation and interpretation of synthesis results form the new and very important creative task of the designer, while the routine work, for example outlining of several loci of points or centres, to find suitable shaft positions, link lengths and so on, that has formed the major task of a designer of mechanisms till now, is displaced.

But investigations done by several scientists have shown, that it is very convenient to start with angle/displacement coordination and translate the results to the two other problems, because angle/displacement coordination is easily formulated with the aid if linear algebra. In this context I will name the contributions of F. Freudenstein, W. Meyer zur Capellen, R.S. Hartenberg and J. Denavit, J.N. Nieto. K. Luck and K.H. Modler [3.2, 3.3, 3.4, 3.5, 3.6], which have influenced the authors own work [3.7, 3.8, 3.9, 3.10].

3.1.1. The crank-and-slider mechanism

The well known crank-and-slider mechanism will be used to explaine the linear algebra synthesis approach [3.8]. The topology of that mechanism is sketched twice in figure 3.1. The difference between the two drawings is lain in the position of the assembly parameter x_0. The same manipulation should have been possible with the other assembly parameter α_0 , too.

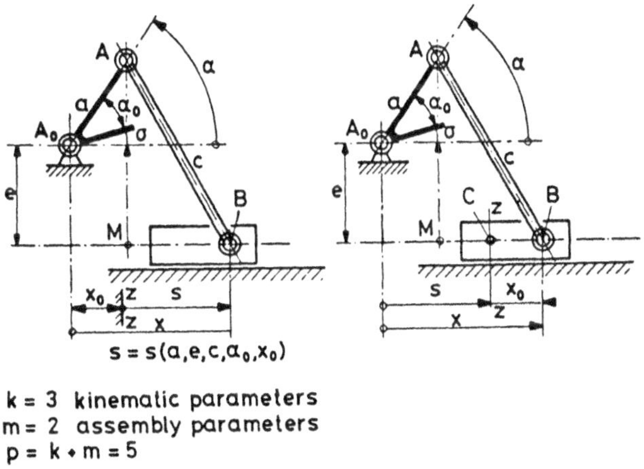

Figure 3.1. Kinematic and assembly parameters of the crank-and-slider mechanism.

Let $AM = a\sin(\alpha_0 + \sigma) + e$ $\qquad \alpha_i = \alpha_0 + \sigma_i$

$MB = s + x_0 - a\cos(\alpha_0 + \sigma)$ $\qquad x_i = x_0 + s_i$

$AB = c$

Than $$c^2 - \{e + a\sin(\alpha_0 + \sigma)\}^2 - \{s + x_0 - a\cos(\alpha_0 + \sigma)\}^2 = 0$$

Combining and regrouping leads to

$$2a_0\{x_0 \cos x_0 - e \sin\alpha_0\}\cos\sigma - 2a\{x_0 \sin\alpha_0 + e \cos\alpha_0\}\sin\sigma$$

$$+2a \cos\alpha_0\ s \cos\sigma - 2a \sin\alpha_0\ s \sin\sigma + c^2 - a^2 - c^2 - x_0^2 = 2x_0 s + s^2$$

or rewritten

$$p_1 \cos\sigma - p_2 \sin\sigma + p_3 s \cos\sigma - p_4 s \sin\sigma + p_5 = p_6 s + s^2 \qquad \text{(I)}$$

with p_i, $i=1(1)6$ pseudoparameters
and a, c, e, α_0, x_0, the 5 mechanism parameters.

To calculate the 5 mechanism parameters 5 linear equations with the values of 5 prescribed goal function precision points $s_i(\sigma_i)$, $i=1(1)5$, have to be given.

The pseudoparameters can easily be defined by the 5 mechanism parameters:

$$p_1 = 2a\{x_0 \cos\alpha_0 - e \sin\alpha_0\}$$

$$p_2 = 2a\{x_0 \sin\alpha_0 + e \cos\alpha_0\}$$

$$p_3 = 2a \cos\alpha_0$$

$$p_4 = 2a \sin\alpha_0$$

$$p_5 = c^2 - a^2 - e^2 - x_0^2$$

$$p_6 = 2x_0$$

Now it is important to investigate the inverse relationship between the five mechanism parameters and the six pseudoparameters and to try to define the mechanism parameters by the pseudoparameters. It becomes obvious that the missing sixth relationship has to be established by formulating a "compatibility equation". It is easy to show that

$$a = \frac{1}{2}\sqrt{p_3^2 + p_4^2}\ ,$$

$$c = \sqrt{p_5 + a^2 + e^2 + x_0^2}\ ,$$

$$e = \frac{p_2 - p_4 x_0}{p_3}$$

$$\tan\alpha_0 = \frac{p_4}{p_3}\ ,$$

but for determination of the value x_0 two formulas are available:

$$x_0 = \frac{1}{2} p_6 \qquad \text{and} \qquad x_0 = \frac{p_1 p_3 + p_2 p_4}{p_3^2 + p_4^2} \; .$$

Elimination of x_0 gives the desired compatibility equation

$$p_6 \, (p_3^2 + p_4^2) = 2(p_1 p_3 + p_2 p_4) \; .$$

In mathematics it is usual to reformulate the equation (I) to be, under certain conditions, the sum of two other equations:

$$p_1 \cos\sigma - p_2 \sin\sigma + p_3 s \cos\sigma - p_4 s \sin\sigma + p_5 = p_6 s + s^2 \qquad \text{(II)}$$

as

$$q_1 \cos\sigma - q_2 \sin\sigma + q_3 s \cos\sigma - q_4 s \sin\sigma + q_5 = s \qquad \text{(IIIa)}$$

$$r_1 \cos\sigma - r_2 \sin\sigma + r_3 s \cos\sigma - r_4 s \sin\sigma + r_5 = s^2 \qquad \text{(IIIb)}$$

or $\mathbb{A}\,\underline{q} = \underline{s}$, $\qquad \underline{q}^T = |q_1 \; q_2 \; q_3 \; q_4 \; q_5|$,

$\mathbb{A}\,\underline{r} = \underline{t}$, $\quad t = s^2$, $\qquad \underline{r}^T = |r_1 \; r_2 \; r_3 \; r_4 \; r_5|$

and

$$\mathbb{A} = \begin{bmatrix} \cos\sigma_1 & -\sin\sigma_1 & s_1 \cos\sigma_1 & -s_1 \sin\sigma_1 & 1 \\ \cos\sigma_2 & -\sin\sigma_2 & s_2 \cos\sigma_2 & -s_2 \sin\sigma_2 & 1 \\ \cos\sigma_3 & -\sin\sigma_3 & s_3 \cos\sigma_3 & -s_3 \sin\sigma_3 & 1 \\ \cos\sigma_4 & -\sin\sigma_4 & s_4 \cos\sigma_4 & -s_4 \sin\sigma_4 & 1 \\ \cos\sigma_5 & -\sin\alpha_5 & s_5 \cos\sigma_5 & -s_5 \sin\sigma_5 & 1 \end{bmatrix}$$

so that

$$\begin{aligned} p_1 &= p_6 q_1 + r_1 \, , \\ p_2 &= p_6 q_2 + r_2 \, , \\ p_3 &= p_6 q_3 + r_3 \, , \\ p_4 &= p_6 q_4 + r_4 \, , \\ p_5 &= p_6 q_5 + r_5 \, . \end{aligned}$$

After some rearrangements a third order equation in x_0 is obtained

$$x_0^3 + u x_0^2 + v x_0 + w = 0$$

with the three coefficients

$$u = \frac{q_3r_3 + q_4r_4 - q_1q_3 - q_2q_4}{q_3^2 + q_4^2}$$

$$v = \frac{r_3^2 + r_4^2 - 2(q_1r_3 + q_3r_1 + q_2r_4 + q_4r_2)}{4(q_3^2 + q_4^2)}$$

$$w = \frac{-(r_1r_2 + r_2r_4)}{4(q_3^2 + q_4^2)}$$

Reduction with $x_0 = x - \frac{1}{3}u$ gives a new, normalized third order equation

$$x^3 + ax + b = 0$$

with

$$a = v - \frac{1}{3}u^3$$

$$b = w + \frac{2}{27}u^3 - \frac{1}{3}uv$$

whereby the value of the discriminant is $\Delta = (\frac{b}{2})^2 + (\frac{a}{3})^3$.

Depending on the value of that discriminant, wether greater, equal or less than zero, solution of the third order equation gives three results as shown in tabel 3.A. With respect to the mechanism synthesis, only the real roots are important. In so far, there are one, two or even three real solutions. With other words, the five position synthesis leads always to at least one solution.

Δ	conj. compl. roots	real number	roots formula	number	formula
>0	2	1	$x_1 = \alpha + \beta$ $\alpha = \sqrt[3]{-\frac{b}{2} + \sqrt{\Delta}}$ $\beta = \sqrt[3]{-\frac{b}{2} - \sqrt{\Delta}}$	1	x_1
=0	-	3	$x_1 = 2\alpha$ $x_2 = x_3 = -\alpha$	2	x_1 x_2
0	-	3	$\cos 3\varphi = \frac{-b}{2\sqrt{(-\frac{a}{3})^3}}$ $x_1 = 2\sqrt{-\frac{a}{3}}\cos\varphi$ $x_2 = 2\sqrt{-\frac{a}{3}}\cos(\varphi + 120)$ $x_3 = 2\sqrt{-\frac{a}{3}}\cos(\varphi + 240)$	3	x_1 x_2 x_3

Tabel 3.A. Roots of the third order equation in x_0 of the crank-and-slider mechanism.

A check of the angle/displacement coordination synthesis procedure is carried on with the following five pairs of arbitrarily chosen precision points, see tabel 3.B.

	argument degree	goal function value
1	0.0	6.19739
2	30.0	4.64516
3	60.0	2.92816
4	90.0	1.64517
5	120.0	1.00126

Tabel 3.B. Five arbitrarily chosen goal function precision points.

The calculation gives in a few seconds the following three results, see tabel 3.C.

parameter number	parameter value	result nr. 1	result nr. 2	result nr. 3
1	a	3.0	1.14836	1.49786
2	e	1.0	12.12568	0.52778
3	c	8.0	13.55134	2.02912
4	α_0	0.52361	1.75816	1.00322
5	x_0	4.0	-9.23472	-4.43800
figure #		.1	.2	.3

Tabel 3.C. Three results of the synthesis procedure.

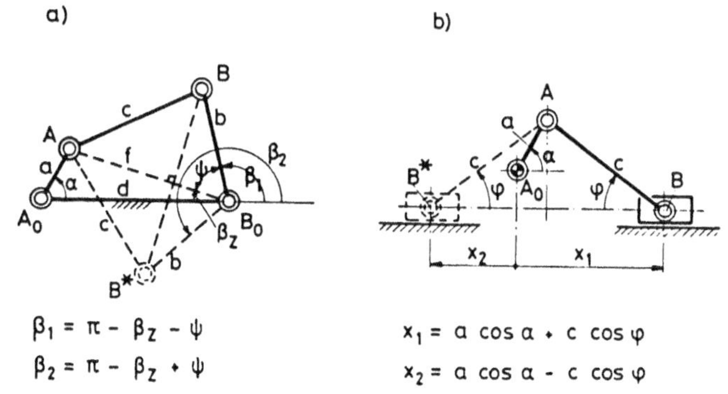

Figure 3.2. Dual mechanisms generate identical transfer functions

Evaluation of the results obtained makes it clear, that it is absolutely necessary to introduce the so-called dual mechanisms, see figure 3.2. The crank-and-slider mechanism A_0AB with crank A_0A of the length a moves the coupler AB of the length c, so that the projection of crank and coupler on the sliding direction is given by the value

$$x_1 = a \cos\alpha + c \cos\phi.$$

Because $\cos(+\phi) = \cos(-\phi)$, it exists a second form of the same mechanism with the same link length, but in this second form the position of the slider B is reflected to position B*. The projection of crank and coupler on the sliding direction now is given by the value

$$x_2 = a \cos\alpha - c \cos\phi.$$

Each of the solutions calculated has to be presented and has to be interpreted for both the dual mechanisms. It became evident, that only one of the dual mechanisms generates the prescribed goal function precision points, while the other dual mechanism moves in a different way. Thus, the synthesis procedure gives now twice three results in correspondence with the six solutions of the classical Burmester theory. Figures 3.3 till 3.5 show the synthesis procedure results. Number 1 and 2 indicate not only the plotted curves, but also the associated parameters, mentioned above. Number 3 marks the precision point positions and gives the list of parameters used for the calculation of these points.

Figure 3.3 shows the generation of the goal function precision points by the first form of a mechanism with the same parameters as used in the calculation of the goal function.

Figure 3.4 shows the case that line nr.2 is passing through the five prescribed precision points. This indicates that the second form of the dual mechanism is able to generate the goal function.

The third result of the synthesis procedure, see figure 3.5, is almost surprising, because two of the prescribed precision points lie on curve nr.1, while the other three points are connected by curve nr.2. This result is still subject of further investigation.

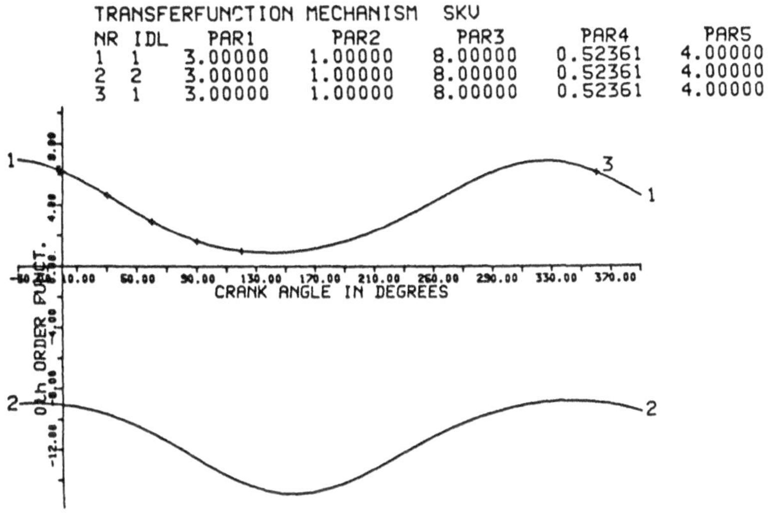

Figure 3.3. First crank-and-slider synthesis result.

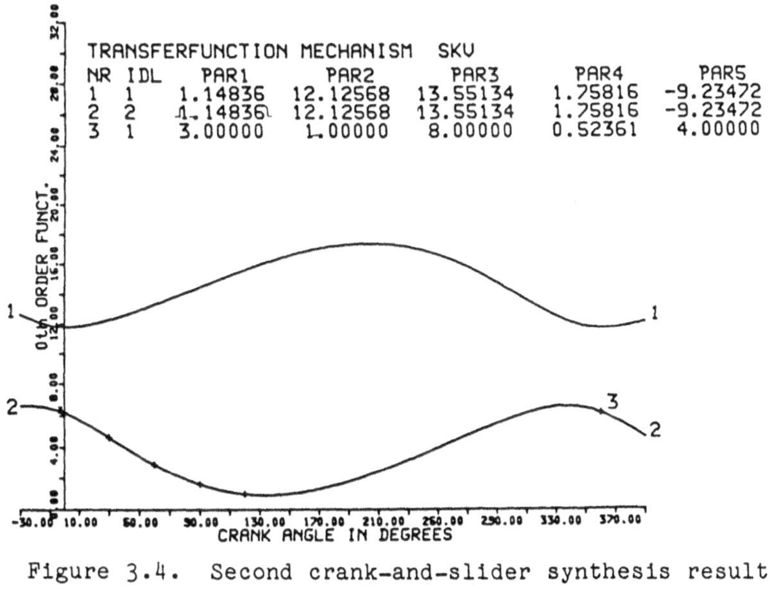

Figure 3.4. Second crank-and-slider synthesis result.

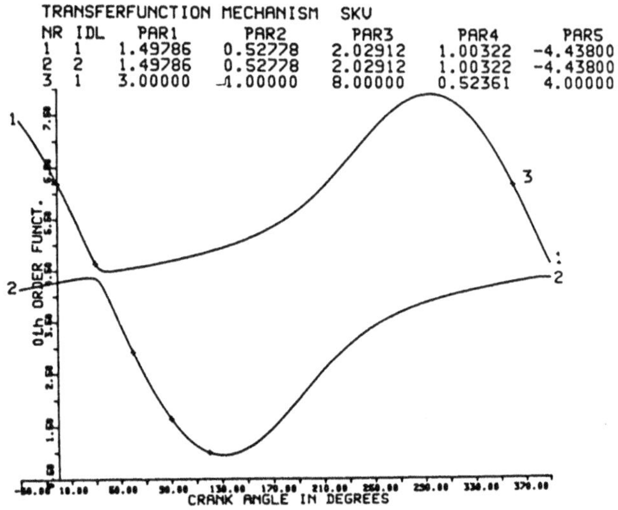

Figure 3.5. Third crank-and-slider synthesis result.

3.1.2. Structural error

The behaviour of zero order transfer function of the mechanism in the intervals between the prescribed precision points is to be established by the calculation of the transfer function. Application of the so called Tschebyschev-spacing leads towards minimization of deviations with respect to an appropriate description of the desired function [3.4]. In practical applications of the precision point synthesis procedure it becomes obvious, that definition respectively prescribing of process dependent precision points is a very difficult and responsible task, and that application of the Tschebyschev-spacing makes things more complex than necessary. In the contect of the above mentioned design philosophy another, unconventional and new approach seems to be more attractive, as explained in the next chapter.

3.1.3. Multiple results synthesis method [3.9].

In addition to the already mentioned definition the TADSOF-approach allows the specification of the goal function as to be periodic or non-periodic. The periodic goal function asks a mechanism, that obeys the so-called Grashof-condition: Due to appropriate link lengths the crank may completely rotate. In the synthesis procedure of non-periodic goal function generators the Grashof condition is not checked, and synthesis requirements are satisfied even if so-called Non-Grashof mechanisms could be found.

If in the synthesis procedure successively different mechanisms are involved, it might happen, that some or all of them seem to be able to move in accordance with the goal function precision points. In [3.9] will be reported about the results of synthesis if three precision points of an input angle and output sliding coordination are prescribed.

The involved mechanisms are:

nr.1: Scotch-Yoke mechanism, see figure 3.6,
a four-bar linkage with 3 parameters [3.7].

nr.2: Crank-and-rocker mechanism, see figure 3.1,
a four-bar linkage with 5 parameters, as described in chapter 3.1.1. and in [3.8].

nr.3: Inverted crank-and-rocker mechanism coupled with and followed by a rocker-and-slider mechanism,
a six-bar Watt-2 mechanism with 4 parameters [3.8],
see figure 3.7.

1) <u>The Scoth-Yoke mechanism</u> generates the zero order transfer function, see figure 3.6,

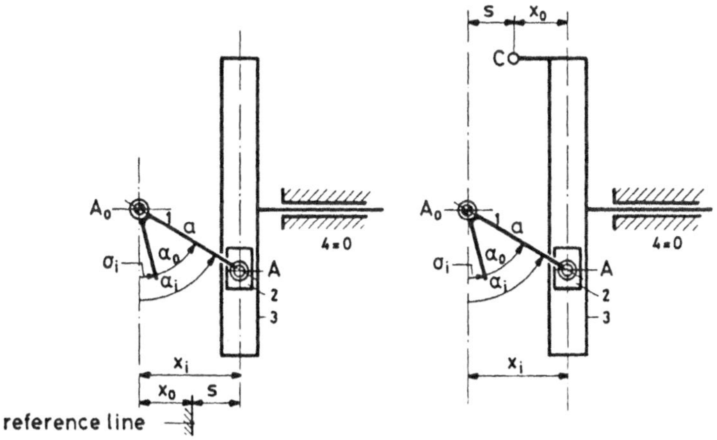

Figure 3.6. Scotch Yoke mechanism.

$$s = a\sin(\alpha_0 + \sigma) - x_0 , \qquad (1.1)$$

or rewritten

$$p_1 \cos\sigma_i + p_2 \sin\sigma_i - p_3 = s_i , \qquad i=1(1)3 \qquad (1.2)$$

with

$$a = \sqrt{p_1^2 + p_2^2} , \qquad (1.3)$$

$$\alpha_0 = \arctan\left(\frac{p_1}{p_2}\right) , \qquad (1.4)$$

$$x_0 = p_3 , \qquad (1.5)$$

or written in matrix-notation $\mathbb{A}\,\underline{p} = \underline{s}$ (1.6)

with

$$\underline{p}^T = [p_1, p_2, p_3] , \tag{1.7}$$

$$\underline{s}^T = [s_1, s_2, s_3] . \tag{1.8}$$

and

$$\mathbb{A} = \begin{bmatrix} \cos\sigma_1 & \sin\sigma_1 & -1 \\ \cos\sigma_2 & \sin\sigma_2 & -1 \\ \cos\sigma_3 & \sin\sigma_3 & -1 \end{bmatrix} , \det [\ \mathbb{A}\ ] \neq 0 \tag{1.9}$$

With the aid of three linear equations it is possible to define the three pseudoparameters p and finally the mechanism parameters a, α_0 and x_0.

2) The crank-and-slider mechanism has been explained for five prescribed precision points, see figure 3.1. Each missing point decreases the number of parameters that can be determined, as long as kinematic parameters remain determinable.
If $\alpha_0 = 0$ and $x_0 = 0$ one needs to determine with the aid of three linear equations the remaining three kinematic parameters. In this case the transfer function equation becomes very simple:

$$2a\ s_i\ \cos\sigma_i\ -\ 2ae\ \sin\sigma_i\ +\ c^2 - a^2 - e^2\ =\ s^2 \tag{2.1}$$

or rewritten as

$$p_1 s_i \cos\ s_i - p_2 \sin\sigma_i + p_3 = t_i ; \qquad t_i = s_i^2 ; \qquad i=1(1)3 \tag{2.2}$$

with

$$a = \frac{p_1}{2} \tag{2.3}$$

$$e = \frac{p_2}{p_1} \tag{2.4}$$

$$c = \sqrt{p_3 + \left(\frac{p_2}{p_1}\right)^2 + \left(\frac{p_1}{2}\right)^2} \tag{2.5}$$

or rewritten in matrix-notation $\mathbb{A}\,\underline{p} = \underline{t}$ (2.6)

$$\text{with}\quad \underline{p}^T = [p_1, p_2, p_3] , \tag{2.7}$$

$$\underline{t}^T = [s_1^2, s_2^2, s_3^2] , \tag{2.8}$$

and

$$\mathbb{A} = \begin{bmatrix} s_1\ \cos\sigma_1 & \sin\sigma_1 & 1 \\ s_2\ \cos\sigma_2 & \sin\sigma_2 & 1 \\ s_3\ \cos\sigma_3 & \sin\sigma_3 & 1 \end{bmatrix} ; \quad \det[\ \mathbb{A}\ ] \neq 0 . \tag{2.9}$$

With the aid of three linear equations it is possible to determine the three kinematic parameters while the two assembly parameters of the crank-and-slider mechanism are chosen to be both zero. There is always one solution.

3) The six-bar mechanism might be seen as a combination of two parts: A four-bar inverted crank-and-slider mechanism and a four-bar rocker-and-slider mechanism, see figure 3.7. The assembling is done in such a way that the direction of the output slider motion is to be perpendicular to the frame of the first mechanism part. The entire mechanism has four parameters, two kinematic parameters h and $\lambda = a/d$ and two assembly parameters α_0 and x_0 . The zero order transfer function is given by the equation

$$s = h \frac{\lambda \sin(\alpha_0 + \sigma)}{1 - \lambda \cos(\alpha_0 + \sigma)} - x_0 \quad . \tag{3.1}$$

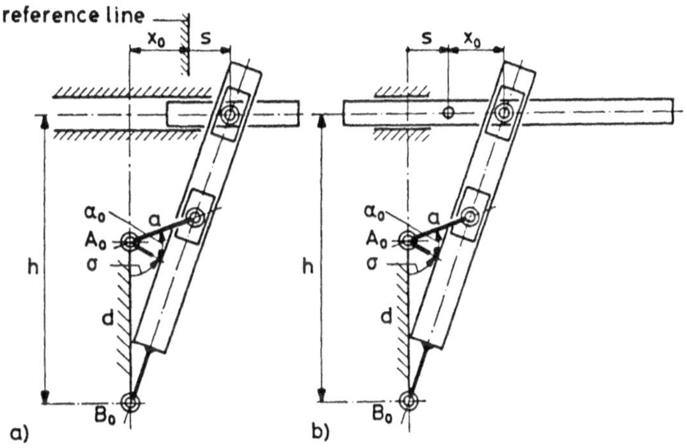

Figure 3.7. Six-bar Watt-2 mechanism.

Because the goal function is defined by three precision points, two different solutions will be possible, one with $\alpha_0 = 0$ and the other with $x_0 = 0$. Both of these cases require special compatibility equations which are briefly explained:

Case $\alpha_0 = 0$:

$$p_1 s_1 \cos\sigma_1 + p_2 \sin\sigma_1 + p_3 \cos\sigma_1 = p_4 + s_1 \tag{3.1.1}$$

$$p_1 = \lambda \quad ; \qquad \lambda = p_1 = p_3/p_4$$
$$p_2 = \lambda h \quad ; \qquad h = p_2/p_1$$
$$p_3 = \lambda x_0 \quad ; \qquad \alpha_0 = 0$$
$$p_4 = x_0 \quad ; \qquad x_0 = p_4$$

and the required compatibility equation $p_4 = p_3/p_1$.
Finally one obtaines a quadratic equation in x_0 of the form

$$x_0^2 + ax_0 + b = 0$$

with $a = (r_1 - q_3)/q_1$

$$b = -r_3/q_1$$

$$\underline{q} = B^{-1}\ \underline{1};\ \underline{q}^T = |q_1,\ q_2,\ q_3|$$

$$\underline{r} = B^{-1}\ \underline{s};\ \underline{r}^T = |r_1,\ r_2,\ r_3|$$

$$B = \begin{bmatrix} s_1 \cos\sigma_1 & \sin\sigma_1 & \cos\sigma_1 \\ s_2 \cos\sigma_2 & \sin\sigma_2 & \cos\sigma_2 \\ s_3 \cos\sigma_3 & \sin\sigma_3 & \cos\sigma_3 \end{bmatrix}$$

The number of real roots to be found depends on the value of the discriminant $D = (a/2)^2 - b$. If $D > 0$, there are two real roots, $D = 0$ results in one real root (two identical solutions). If $D < 0$, no real roots. Only two conjugate complex roots occur.

Case $x_0 = 0$:

In a way similar as explained in the case of $\alpha_0 = 0$, a quadratic equation in $\lambda \sin\alpha_0$ is obtained. A compatibility equation helps to generate the fourth equation (non-linear) and enables us to calculate the parameters. It depends on the value of the discriminant, how many real roots will be obtained. For more details see [3.8].

The goal function, which is given in three precision points, can be generated as well by a mechanism with $\alpha_0 = 0$ as by a mechanism with $x_0 = 0$. In both the cases a quadratic equation has to be solved. In each case there can be a maximum of two real roots, one real root or no real root. So, it might happen, that in total four real roots can be found.

Conclusion:

The example with three prescribed precision points of a non-periodic motion has six different solutions as shown in figure 3.8 [3.9]. If the goal function would have been declared to be periodic, only two solution would have been found as shown in figure 3.9 [3.9].

RESULTS T A D S O F

FUNCTION VALUES OF TEST RA77

INDEX	ARGUMENT	FUNCTION VALUE
1	0.00000	1.00000
2	.52360	1.15470
3	1.04720	.50000

INPUT CHARACTERISTIC R
OUTPUT CHARACTERISTIC T

SOLUTIONS AFTER TYPE- AND DIMENSION SYNTHESIS

NO	MECH	PAR1	PAR2	PAR3	PAR4	PAR5
1	F031	3.06181	1.21123	1.86601		
2	F034	.71891	-1.12125	-1.23001	0.00000	

Figure 3.8. Six solutions for a non-periodic goal function, defined by 3 precision points.

RESULTS T A D S O F

FUNCTION VALUES OF TEST RA77

INDEX	ARGUMENT	FUNCTION VALUE
1	0.00000	1.00000
2	.52360	1.15470
3	1.04720	.50000

INPUT CHARACTERISTIC S
OUTPUT CHARACTERISTIC T

SOLUTIONS AFTER TYPE- AND DIMENSION SYNTHESIS

NO	MECH	PAR1	PAR2	PAR3	PAR4	PAR5
1	F071	3.06181	1.21123	1.86601		
2	F073	.88490	-.37669	.39388	0.00000	0.00000
3	F074	1.00000	.32735	0.00000	.06699	
4	F074	1.59309	-.07373	0.00000	-1.00000	
5	F074	1.00740	.27595	-.02546	0.00000	
6	F074	.71891	-1.12125	-1.23001	0.00000	

Figure 3.9. Two solutions for a periodic goal function, defined by three precision points.

All the shown solutions fulfill prescribed goal function precision points exactly, but the chapes of the transfer functions are quite different. This spreading makes it possible to realize other respectively additional requirements than defined by the precision points itselves. The additional requirements may concern first or second order transfer function behaviour or number of elements and complexity of the type of mechanism or space demands or so on. Thus, the multiple results synthesis method, introduced here, can help to optimize the mechanical system.

3.2. Fourier series, theory, examples

The precision point synthesis, as valual as it is, fail when the entire period of a goal function is prescribed. Because of the overshoot of information with respect to the limited number of mechanism parameters, no exact solution exists. But it is possible to search the best approximation of the entire goal function by the transfer function of an appropriate mechanism. Accordingly the basic idea, published by the author in 1958 [3.11], the available amount of informations may be used to solve two problems subsequently: first of all to determine or to match respectively the type of mechanism or a group of mechanisms suitable for the goal function approximation, and secondly to determine all mechanism parameters to reach the best approximation. The basic idea has inspired the acronym TADSOL (= Type And Dimension Synthesis Of Linkages) [3.12, 3.13, 3.14, 3.15, 3.16]. TADSOL markes the periodic goal function synthesis procedure within the CADOM-software package [3.15, 3.16].

The used method is to develop as well the goal function as all mechanisms transfer functions into Fourier series. The application area of Fourier series is the field of periodic functions $f(\alpha) = f(2\pi+\alpha)$. Instead of different representations, for example formulas of mechanisms transfer functions or lists with measured goal functions, a uniform representation in terms of Fourier coefficients is chosen. The coefficients of the goal function can be compared with coefficients of mechanisms transfer function. If both the Fourier coefficients are similar, the mechanism makes a chance to approximate the goal function. If the Fourier coefficients of goal function and transfer function are quite different, there is no chance to approximate the goal function by a mechanisms transfer function, and every attempt, to try it nevertheless, is accompanied by a systematic error and fails [3.11].

3.2.1. Fourier representation of a periodic function

All the periodical functions $f(x)$, that fulfill the Dirichlet conditions, can be represented by a Fourier series. This series has two forms of representation: the so called A-B-representation (=cos-sin-representation) or the so called amplitude-phase(angle)-representation (= R-ϕ-representation).

The A-B-representation is formulated as

$$f(x) = a_0 + a_1\cos x + a_2\cos 2x + a_3\cos 3x + \\ + b_1\sin x + b_2\sin 2x + b_3\sin 3x + \ldots. \qquad (1.1)$$

$$f(x) = a_0 + \sum_{n=1}^{\infty} (a_n\cos nx + b_n\sin nx) \qquad (1.2)$$

with the interpretation that

a_n is the coefficient of cos of the n-th harmonics,

b_n is the coefficient of sin of the n-th harmonics.

They are found by multiplying left and right hand terms with cospx or sinpx respectively and after that by integrating over $[0,2\pi]$.

Hereby the following two equations are obtained:

$$\int_0^{2\pi} f(x)\cos px\,dx = \sum_{n=0}^{\infty}\int_0^{2\pi} a_n\cos nx\cos px\,dx + \sum_{n=0}^{\infty}\int_0^{2\pi} b_n\sin nx\cos px\,dx \qquad (2.1)$$

$$\int_0^{2\pi} f(x)\sin px\,dx = \sum_{n=0}^{\infty}\int_0^{2\pi} a_n\cos nx\sin px\,dx + \sum_{n=0}^{\infty}\int_0^{2\pi} b_n\sin nx\sin px\,dx \qquad (2.2)$$

Using the well known solutions of integrals

$$\int_0^{2\pi} \cos nx\cos px\,dx = 0 \quad \text{if } n \neq p$$
$$= \pi \quad \text{if } n = p \neq 0$$
$$= 2\pi \quad \text{if } n = p = 0$$

$$\int_0^{2\pi} \sin nx\cos px\,dx = 0$$

$$\int_0^{2\pi} \sin nx\sin px\,dx = 0 \quad \text{if } n \neq p$$
$$= 0 \quad \text{if } n = p = 0$$
$$= \pi \quad \text{if } n = p \neq 0$$

and the choice of $p = 0$ results in the coefficient

$$a_n = \frac{1}{2\pi}\int_0^{2\pi} f(x)\,dx \quad \text{and} \quad b_n = 0, \qquad (3.1,\ 3.2)$$

but for p = n is the result

$$a_n = \frac{1}{\pi}\int_0^{2\pi} f(x)\ \cos nx\ dx \qquad (3.3)$$

$$b_n = \frac{1}{\pi}\int_0^{2\pi} f(x)\ \sin nx\ dx\ . \qquad (3.4)$$

The <u>C-ϕ-representation</u> is given by

$$f(x) = c_0 + c_1\sin(x+\phi_1) + c_2\sin(2x+\phi_2) + c_3\sin(3x+\phi_3) + \ldots \qquad (4.1)$$

$$f(x) = c_0 + \sum_{n=1}^{\infty} c_n\sin(nx+\phi_n) \qquad (4.2)$$

Between the two representations exists a simple relationship

1) $c_n = \sqrt{a_n^2 + b_n^2}$; $\tan\phi_n = a_n/b_n$; (5)

2) $a_n = c_n\sin\phi_n$; $b_n = c_n\cos\phi_n$. (6)

3.2.2. <u>Fourier coefficients in different coordinate systems</u>

If a function is given in a $\{x,f(x)\}$-coordinate system it is possible to shift the coordinate system in x- and/or f(x)-direction. Then $x^*,f(x^*)$ is the new coordinate system.

<u>Shifting in f(x)-direction over $-u_0$:</u>

Shifting of the coordinate system in $\{-f(x)\}$-direction results in

$$f(x^*) = f(x) + u_0\ . \qquad (7.1)$$

The values of c_0- respectively a_0-Fourier coefficients change only:

$$c_0^* = a_0^* = a_0 + u_0\ . \qquad (7.2)$$

All other Fourier coefficients remain unchanged.

Shifting in x-direction over $-\tau$ (figure 3.10):

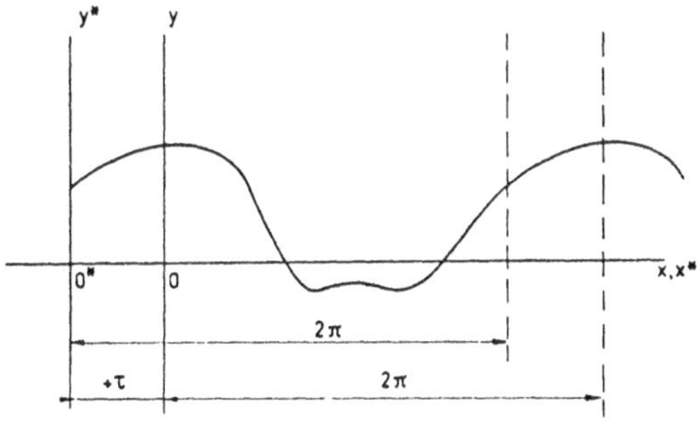

Figure 3.10. Shifting of the coordinate system (y,x) to (y*,x*).

With $x = x^* - \tau$ and $a_{n,0} = a_n$ respectively $b_{n,0} = b_n$ the above mentioned Fourier series mutate into

$$f(x^*) = a_0^* + \sum_{n=1}^{\infty} (a_{n,i} \cos nx^* + b_{n,i} \sin nx^*) \tag{8.1}$$

$$= a_0^* + \sum_{n=1}^{\infty} c_{n,i} \sin (nx^* + \phi_{n,i}) \tag{8.2}$$

whereby is

$$a_{n,\tau} = a_{n,0} \cos n\tau - b_{n,0} \sin n\tau$$
$$b_{n,\tau} = a_{n,0} \sin n\tau + b_{n,0} \cos n\tau$$

respectively

$$c_{n,\tau} = c_{n,0}$$
$$\phi_{n,\tau} = \phi_{n,0} - n\tau$$

It is obvious that the C-Fourier coefficients has always a positive value and they do not change, when the coordinate system is shifted. But shifting can be used to change all other coefficient values. Of importance is that, due to shifting, certain strategies can be used to look at special coefficient value combinations: for example that as many as possible zero values occure and all non-zero coefficients become visible. These coefficients are "significant" and their information content is used in solving the synthesis procedure. The TADSOL-procedure uses nine different shift strategies.

3.2.3. Reflection of function f(x) on coordinate axis

Another kind of coordinate transformation is the reflection of a function on coordinate axis, see figure 3.11. Due to the simply geometric relationship the following results are obtained:

1) Reflection on the x-axis, i.e. f(x) becomes $\{-f(x)\}$, changes all signs of f(x)-values, a_n and b_n.

2) Reflection on the f(x)-axis, i.e. f(x) becomes $\{f(-x)\}$, changes the signs of all b-coefficients only. The signs of all a-coefficients remain unchanged.

3) Reflection on both the axis, i.e. f(x) becomes $\{-f(-x)\}$, changes all signs of funtion value f(x) and of all a-coefficients, but has no effect on b-coefficient signs.

	a1	a2	a3	a4	a5	a6	b1	b2	b3	b4	b5
original	+	+	+	+	+	+	+	+	+	+	+
S1 (x-axis)	-	-	-	-	-	-	-	-	-	-	-
S2 (y-axis)	+	+	+	+	+	+	-	-	-	-	-
S3=S1 & S2	-	-	-	-	-	-	+	+	+	+	+

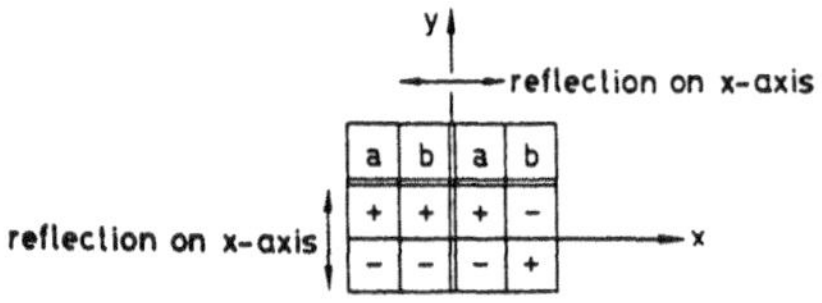

Figure 3.11. Reflection of function and signs of Fourier coefficients.

3.2.3. Real and symbolic coefficient value

If two boundary values e_{max} and e_{min} are used to check the Fourier coefficient value, four combinations may occure, see figure 3.12.

1) Real zero coefficient values are associated with a symbolic coefficient value "0".

2) Coefficient values $c < e_{min}$ are associated with a symbolic value "0" too.

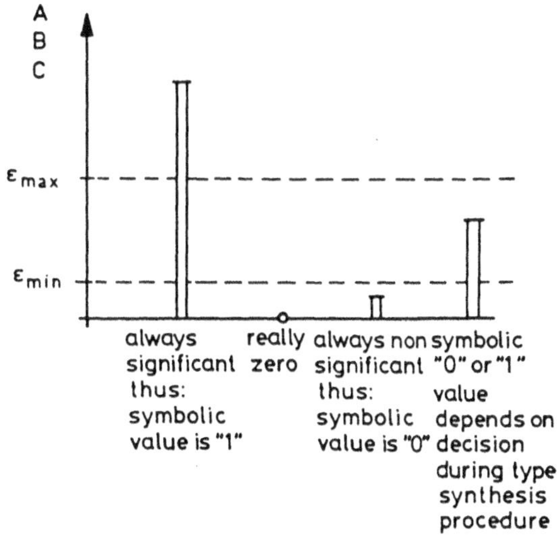

Figure 3.12. Boundary values and values of Fourier coefficients.

3) If a coefficient value exceeds e_{max}, it is always a significant value and always will be associated with a symbolic value "1".

4) Remains the definition of symbolic coefficient values if the real coefficient value is lain within the area marked by e_{max} and e_{min}. Hereby the strategy is followed that the demands of synthesis procedure control the decision to associate such coefficients either with a symbolic value "1" or with "0" as later shown in the example.

3.2.4. Fourier coefficients of mechanisms transfer functions

In mathematics algebraical and numerical methods are available to calculate Fourier coefficients. Accordingly the algebraical approach the integrals of the equations (3.2.3) are solved analytically. In general it is impossible to follow this way. But numerical methods make it always possible to determine the Fourier coefficients. In the CADOM-software package the subroutine FOURAN does this job.

For a few mechanisms the Fourier coefficients of the zero order transfer function will be determined. The coordinate system which is used is called reference position.

A cycloidal controlled Scotch-Yoke mechanism:

Figure 3.13 shows the scheme of a cycloidal controlled Scotch-Yoke mechanism with five kinematic parameters.

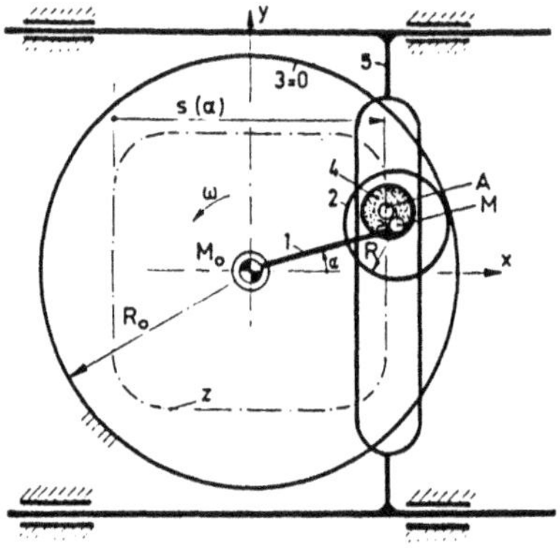

Figure 3.13. Cycloidal controlled Scotch-Yoke mechanism.

The algebraic analysis of the transfer function of the yoke results in the very simple Fourier series

$$s(\delta,\alpha) = A_0 + A_1 \cos\alpha + A_m \cos m\alpha + B_1 \sin\alpha + B_m \sin m\alpha \;. \qquad (10)$$

with the Fourier coefficients as functions of the mechanism parameters

$A_0 = (a - p)\cos\delta$
$A_1 = -a\cos\delta$
$B_1 = a\sin\delta$
$A_m = \xi\, p\cos\delta$
$B_m = p\sin\delta$
ξ = sign of hypo- or epi-cycloid: $\xi = -1$ or $\xi = +1$.

But contrary to that the synthesis procedure is asking an equation in which the mechanisms parameter are functions of Fourier coefficients. Because of simple relations between Fourier coefficients and parameters of the Scoth-Yoke mechanism the inverse functions are, see figure 3.14.

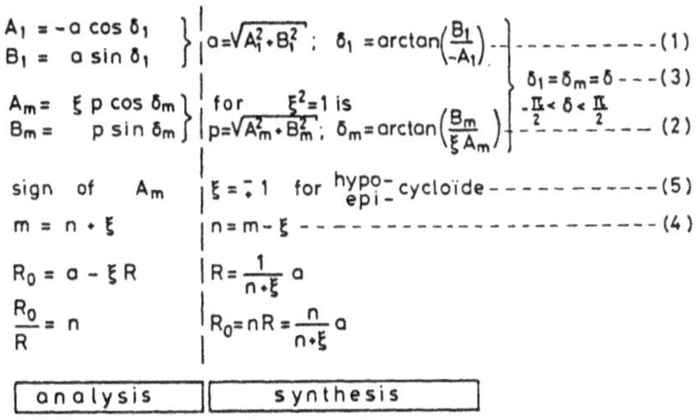

Figure 3.14. Relations between Fourier coefficients and parameters of the Scotch-Yoke mechanism.

The inverse crank-and-slider mechanism:

Figure 3.15 shows the scheme of the inverted crank-and-slider mechanism with two kinematic parameters λ and i lain in the complex plaine. As in [3.18] is developed the transfer function consists of sin-coefficients only:

$$\delta(\alpha) = \sum_{n=1} i\ B_n \sin n\alpha, \qquad B_n = \frac{1}{n}\lambda^n .$$

The inverse relation is easily be found because

$$\lambda = 2B_2/B_1 \qquad \text{and} \qquad i = B_1^2/(2B_2) .$$

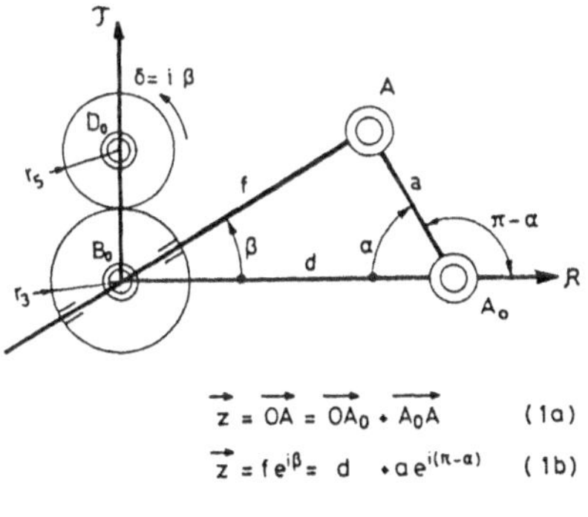

Figure 3.15. Inverted crank-and-slider mechanism with gears. Transfer function $\delta(\alpha) = i\beta(\alpha)$.

This example makes clear that the calculation of the two parameters of the inverted crank-and-slider mechanism needs two Fourier coefficients only. To obtain a good synthesis result the approximation of the other, non used coefficients has to be checked as explained in chapter 3.28.

3.2.5. Ranking of a set of mechanisms

The Fourier coefficients of a large number of mechanisms are determined. It was found that ranking of a set of mechanisms should be possible using six C-coefficients only if they are available. Because of the fact that all transfer functions have to be periodically with the period 2π, the C1-coefficient has to be present always! Therefore the most simple subset of mechanisms is the subset which contains the C1-coefficient generators only: The sin-motion generating Scotch-Yoke mechanism and its plaine, spherical of spatial cognates. The other five Fourier coefficients are absent. Thus, there is one combination z only.

If there are n Fourier coefficients found in the transfer function $z = \binom{5}{n-1}$ combinations of ranking exist for every n as shown in figure 3.16. The total number of combinations is 32. Every combination of n Fourier coefficients can be coded by a number m with n numerals in such a way that the highest coefficient gets the place value 10exp0, and subsequently increased values for every lower coefficient. Thus, if C1, C3 and C5 are significant Fourier coefficients of a transfer function, the mechanism is coded by n = 3 and by m = 135.

This approach makes a simple arrangement of sets and subsets of mechanisms possible using the Fourier coefficients of their transfer functions.

Spectrum of Fourier coefficients		Name of selection	Number of significant harmonics	Number z of comb.
C_1	—	SINGLE	—	1
C_1	C_K	DOUBLE	$2 \leq K \leq 6$	5
C_1	C_K, C_L	TRIPLE	$2 \leq K < L \leq 6$	10
C_1	C_K, C_L, C_M	QDRPLE	$2 \leq K < L < M \leq 6$	10
C_1	C_K, C_L, C_M, C_N	QNTPLE	$2 \leq K < L < M < N \leq 6$	5
C_1	C_2 C_3 C_4 C_5 C_6	SXTPLE	—	1

Figure 3.16. Ranking of mechanisms using the Fourier coefficients of their transfer functions

Sometimes the total system is called a mosaik system because completion is always possible [3.11].

3.2.6. The catalogue of mechanisms

In the catalogue of mechanisms all those mechanisms are collected from which the synthesis procedure is already worked out [3.19]. It is provided that the zero, first and second order transfer functions and the Fourier coefficients of the transfer functions are known. In case of S- and T-mechanisms the Fourier coefficients are taken from the zero order transter function, but in case of R-mechanisms the Fourier coefficients belong to the first order transfer function.

The catalogue contains separate pages for each mechanism. Each page shows all important informations, statements and declarations:

- a schematic drawing of the mechanism with the definition of the reference position,
- formulae of the zero, first and second order transfer functions,
- Fourier coefficients of the periodic transfer function,
  in case of S- and T-mechanisms: of the zero order transfer function,
  in case of the R-mechanisms: of the first order transfer function,
- the characteristic configurations of the C- and the A-B-Fourier coefficients,
- definifion of the successive parameters and formulae to calculate them,
- formula to calculate the shift angle value.

The use of the catalogue of mechanisms will be explained in chapter 3.2.10.

3.2.7. The goal function specification

The periodic goal function may be given on several modes as mentioned hereafter:

- a large number of discrete points, see figure 3.17 with a schedule for discrete points and tolerances. For practical reasons only the number of points is limited with 120. With respect to the necessary number of six sin- and cos-coefficients the minimum number is put on 12 points. The argument values may be equidistant or non-equidistant values.
- a number of intervals with "standard functions" as normally used in cam follower motion definition,
- a number of Fourier coefficients,
- an explicite or implicite formula,
- other not yet mentioned specification modes.

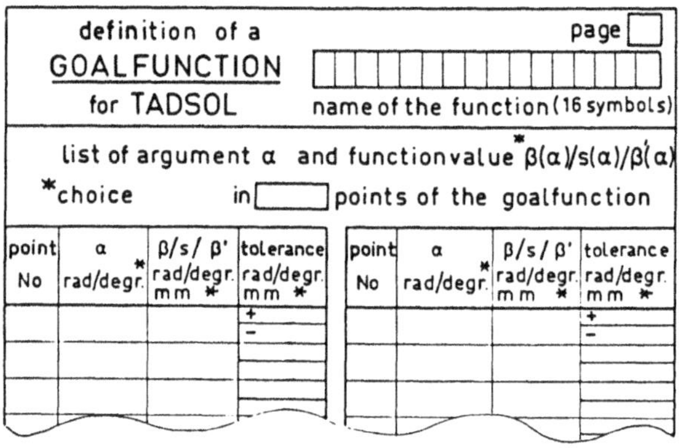

Figure 3.17. Schedule for the definition of a goal function by discrete points including tolerances.

Because of the required periodicity of the goal function all input motion characteristics has to be R, see chapter 2, figures 2.1 and 2.2. It is enough to code the output motion characteristic by one of the three letters R, S and T. Figure 3.18 gives examples of mechanisms and transfer functions for each code. The code letter U is not yet in use.

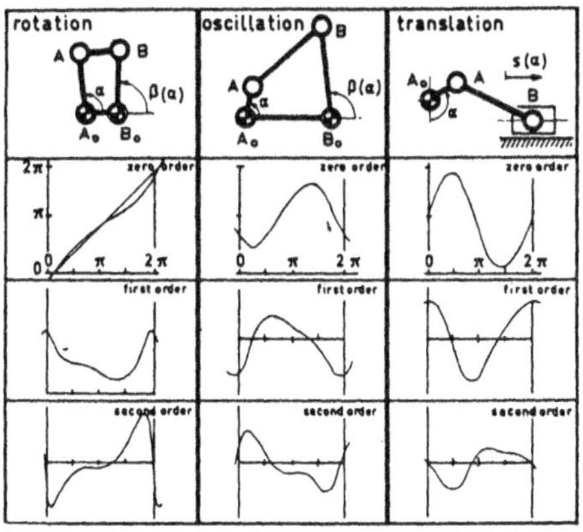

Figure. 3.18. Definitions of three kinds of output motion including examples of mechanisms with their zero, first and second order transfer function.

It is very important to recognize that the zero order R-coded output rotation is a non-periodic motion because of $f(x) \neq f(x + 2\pi)$. If the R-code is used to define the character of the goal function, automatically the first order goal function will be generated and used in the synthesis procedure.

With respect to the synthesis procedure requirements it is enough to determine the fixed number of 24 Fourier coefficients. They are printed for information purposes. A few of them will be used to select appropriate mechanisms and some of them enter the calculation of the parameters of the mechanisms.

3.2.8. Synthesis procedure

As already mentioned the synthesis procedure is divided into two steps [3.11]:

- the first step is directed to the selection of appropriate types of mechanism, the type synthesis (TYPSYN), and
- the second step which is directed to the calculation of the dimensions, link length or gear ratios, the dimension synthesis (DIMSYN).

<u>The type synthesis (TYPSYN):</u>

The type of a mechanism capable to approximate the goal function is to be found or selected by comparison the informations and characteristics of the goal function and the appropriate set of transfer functions. The characteristics are expressed by the Fourier coefficients. The other informations concern the number of significant goal function Fourier coefficients and their amplitude numbers. Figure 3.19 shows the mentioned comparison as the intersection of the vertical line "1" that represents all informations about the goal function, with the horizontal line "2" that represents all stored information about transfer functions of the reference position. Both the functions are available in Fourier series form.

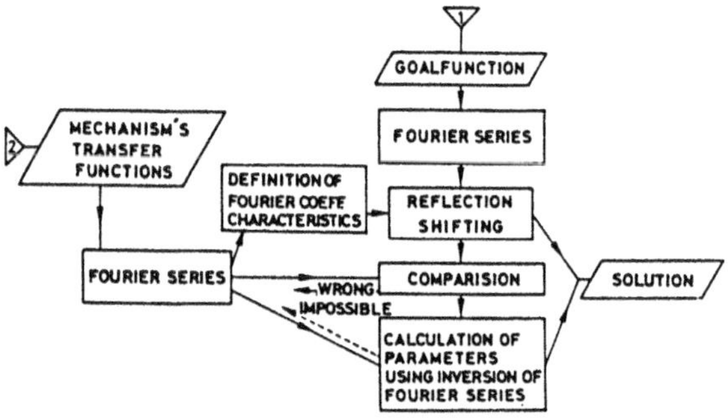

Figure 3.19. Flow chart of the TADSOL synthesis procedure.

In order to generate better chances during the type synthesis procedure the goal function is automatically three times reflected. Therefore the synthesis procedure has to do with four different goal functions. During the comparison procedure the selection of appropriate types takes place with the aid of certain criteria which will be explained step by step:

- <u>First</u> criterion of selection:
  With respect to the fact that the Fourier coefficients of higher order decrease very rapidly and converge upon zero, if the dynamic behaviour of the goal function is acceptable and therefore a chance exists to find an approximation mechanism, it is permitted to look at the first six C-Fourier coefficients only as explained above. Thus, the first criterion of selection is the number n of significant C-coefficients found in the goal function. The range of n is n=1(1)6.

- The second criterion of selection:
  Now the number k of possible combinations of the n significant C-Fourier coefficients found in the goal function will be determined. The real value will be displaced by the symbolic values "0" of "1". The symbolic goal function characteristic leads to one of the 32 subsets of mechanisms with the same characteristic. The mechanisms of such a subset fulfill the suppositions of a good approximation of the goal function without systematic error only.

  In general the numbers k or n say nothing about the quantity of mechanisms which are brought together in a certain subset. It is possible to meet an empty subset.

- The third criterion of selection:
  In this stage it becomes necessary to switch over from the -C-ϕ-representation to the A-B-representation and to refine the symbolic value of both the transfer functions of mechanisms and the goal function. The A-B-representation gives more information about the functions to be compared. The refinement of the symbolic value is necessary to watch over the numerical calculation of the parameters and to avoid division by zero. Figure 3.20 shows the strategy of recognition of the type of mechanism using the Fourier coefficients of the goal function "d" and the transfer functions "m" of a certain mechanism of an already selected subset.

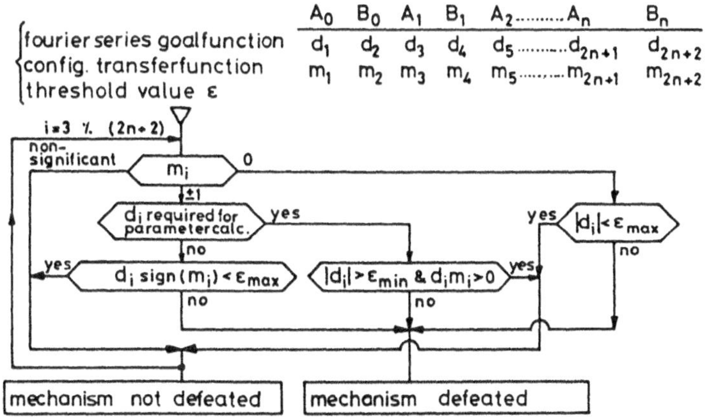

Figure 3.20. Strategy of recognition of the type of mechanism for a good approximation of the goal function.

Shifting of the above mentioned four goal function towards a reference position of the transfer function plays an important role in this stage too. Even here the different shift strategies to bring the goal function Fourier coefficients into a certain form, for example all B-coefficients into a geometric series, are used.

- The fourth criterion of selection:
  Within the above mentioned subset which has been collected because the transfer functions of all mechanisms have shown the same significant C-Fourier coefficients, differentiation occure among the mechanisms if the A-B-Fourier coefficients will be inspected. Now mechanisms with the same significant A-B-Fourier coefficients belongs to one family, they are approximately equivalent each other and they are candidates in the competition to approximate the goal function. This criterion is very important in all cases when in a subset more than one mechanism is "lodged".

It is not necessary that all these four criteria are used:
- For n=1 the number k of possible combinations is k=1. The use of the first criterion is enough.
- For n=2, 3, 4 or 5 is the number of combinations k=5, 10, 10 and 5. All the mentioned criteria are used.
- For n=6 the number k is again k=1 and all criteria are important, especially the fourth criterion.

- The fifth criterion of selection:
  A fifth criterion can be used after having calculated the kinematic parameters or the link lenght respectively. For each mechanism it might be useful to have boundaries with respect to the values of kinematic parameters, for example in connection with applications in practice, space limitations or with dynamic behaviour. In this context the pressure angle can be used as a fifth criterion. Further more searching equivalent mechanisms the fifth criterion is of great importance. There are mechanisms of exact equivalence, the so called cognates and alternative mechanisms. Cognates have the same topology, alternative mechanisms differ in topology. Exact equivalence has to do with: The twofold generation of cycloids [3.19], the threefold generation of a coupler curve of four-bar mechanisms [3.20] and the manifold generation of pathes of six-bar mechanisms [3.21]. About alternative mechanisms see for example [3.21, 3.22, 3.23, 3.24].

If the equivalence is not exact but approximate, the mechanisms are called nearly equivalent mechanisms or substitute mechanisms with respect to the whole goal function period [3.25, 3.11]. Examples with even small differences in the second order transfer function are known [3.11, 3.15].

The dimension synthesis (DIMSYN):

The calculation of the values of the several parameters of the chosen mechanism is called dimension synthesis. The calculation makes use of algorithms of the parameters as functions of the large Fourier coefficients, as explained above. In some cases very difficult calculations of explicite equations have to be carried out. In other cases exist implicite equations only which have to be solved by numerically procedures like regula falsi for one or two variables, iterations or expensive computer optimization procedures [3.26].

The scheme of the total synthesis procedure TADSOL:

Figure 3.21 shows a simplified scheme of the selection procedure for a real example:

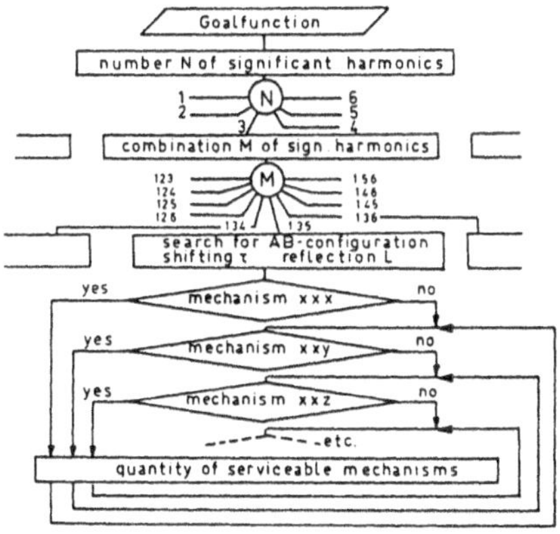

Figure 3.21. Simplified scheme of the selection procedure of suitable mechanisms for a real example.

- A goal function is offered. Fourier analysis makes clear that there are three significant Fourier coefficients. Thus, n=3. Having passed the first criterion the selection of those mechanisms is possible which are able to generate transfer functions with three significant Fourier coefficients.

- After that the combination of three Fourier coefficients has to be found out. In this example the combination may show the first, third and fifth C-Fourier coefficient, thus C1, C3 and C5. This combination is one of the above mentioned 10 combinations in figure 3.16 and will be marked by the code m=135. Having passed the second criterion selection of those mechanisms which are able to generate transfer functions with the wanted Fourier coefficients C1, C3, C5 etc., is possible only.
- In the following step the A-B-representation is used and it is tried to calculate a shift angle τ accordingly the strategy of searching a conformity between the symbolic Fourier coefficient values of transfer function of the selected group of mechanisms and goal function. If a sufficient conformity is reached first of all a value for shifting the goal function is known, the third criterion is fulfilled and the selected mechanism is a real candidate for the goal function approximation.
- Now the parameters of the mechnism are to be calculated using the real Fourier coefficient values of the shifted and/or reflected goal function. After that it has to be checked if the transfer function of the mechanism with the mentioned parameters containes exactly or with an acceptable approximation all those Fourier coefficients which not have been used for the calculation of the parameter values. If the fourth criterion is not fulfilled the solution has to be refused after the dimension synthesis.
- The fifth criterion is not yet programmed and installed. It is to the designer to make his choice.

3.2.9. Presentation of results

The results of the type and dimension synthesis procedure will be given by the periphery of the computer uniformally for each selected mechanism. The type of mechanism is indicated by means of a code number which corresponds with the code of mechanisms in the already mentioned mechanism catalogue [3.27].

On top of the output list first of all the code number of mechanism is printed. Besides this the list contains the values of shift angle, reflection modes and assembly value. Further on the values of all kinematic parameters are printed together with informations about the maximum deviation between the transfer function and the goal function. If desired the printer gives a first view on the transfer functions which are generated by the chosen mechanism with the just calculated parameters.

Using all the printed information makes clear to the designer:

- the starting position of the mechanism to fulfil the goal function in its coordinate system,
- the direction of motion of the crank input motion,
- the assembly situation at the output link that generates the goal function output motion.

3.2.10. Synthesis example

Figure 3.22 shows an arbitrary, in 12 equidistant points given periodic goal function s(α*). As result of the Fourier analysis the most important harmonics are C1, C2 and C4. In figure 3.23 they are drawn in scale.

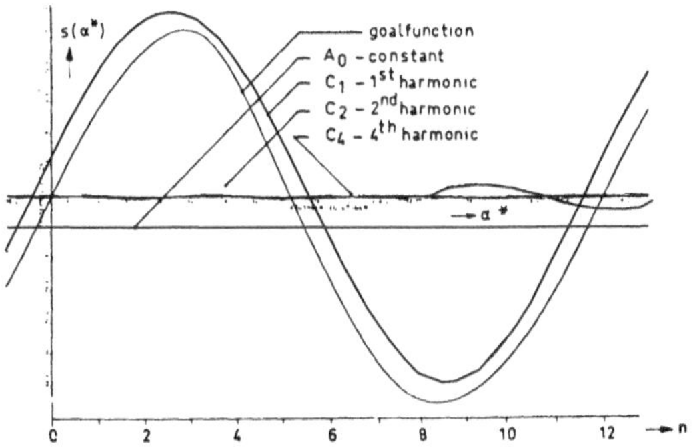

Figure 3.22. Arbitrary given periodic goal function and its harmonics.

It may be concluded that the two Fourier coefficients C1 and C2 are the most important coefficients. If the C4-coefficient will be neglected, then there are n=2 Fourier coefficients and the code becomes m=12. The synthesis procedure selects the mechnism T002 and determines the five parameters of that mechanism. Figure 3.24 shows under nr. 3 the transfer function of mechanism T002.

If the three (of the six allowed) Fourier coefficients C1, C2 and C4 would taken into account, then the synthesis procedure is marked by n=3 and m=124. The type synthesis selects the mechanism T004 and the dimension synthesis calculates the eight kinematic parameter values. In figure 3.24 markes nr.4 the transfer function of mechanism T004.

If all six Fourier coefficients are declared to be significant, the problem is code by n=6 and m=123456. In the subset of the 1982 state of the catalogue of mechanisms, the synthesis procedure has found four mechanisms with the ability to approximate the goal function with small deviations: The mechanism code numbers are: T006, T008 and twice T014.

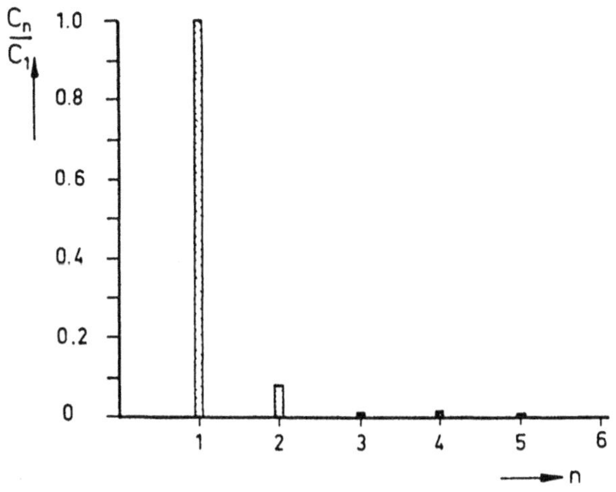

Figure 3.23. Amplitude spectrum of Fourier coefficient ratios Cn/C1, belongs to figure 3.22.

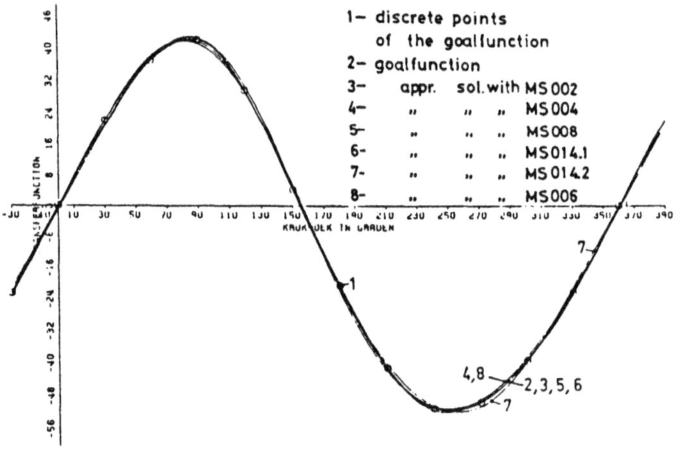

Figure 3.24. Six approximations of the goal function (nr.2).

Conclusion:

The example with an arbitrary prescribed periodic goal function shows that the problem has accidentally six different solutions, see figure 3.24. All the shown solutions approximate very well the goal function. The high quality of the approximation is obviously. Besides this the approximate equivalence of different mechanisms becomes clear because their second order transfer functions are still close together.

Now it is the task of the designer to evaluate the results and to make his finale choice using additional requirements. This requirements may concern the first or second order transfer function behaviour or number of elements and simplicity of mechanism or space demands respectively. In so far the multiple result synthesis method TADSOL can help to optimize the mechanical system.

After having choosen the designer has to work out the assembly position of the mechanism in the machine. This task is more or less new because of the new design philosophy. Three sources of information are available:

- The schematic drawing of the choosen mechanism in the catalogue,
- the results of the computer synthesis,
- a schematic drawing of the machine in which the mechanism has to work.

Details are discussed in [3.16].

3.3 Non-linear transformation

In general the goal function describes the motion of a tool or a machine part that has to fulfill a certain task, for example to fold a sheet of paper or to pick up an egg from the weighting station and bring it to a gripper on the conveyor chain. But normally the tool is mounted on a certain machine element and this element is located so far from crankshaft that a certain transition mechanism is neseccary to span the distance. Very often this transition mechanism has to transform a large displacement into a smaller displacement which is more suitable to the transfer function possibilities of a linkage. Then the transition mechanism transforms non-linearly the "tool oriented goal function" into a "mechanism oriented goal function". The mathematical procedure is characterized by the multiplication of two functions (chain rule) which interprete the series connection of two mechanisms. If the "tool oriented goal function" is prescribed and type as well as dimensions of the transition mechanism are known, it is possible to calculate the "mechanism oriented goal function" which will be used in the synthesis procedure.

3.3.1. Theory

In series connection of two mechanisms nr.1 and nr.2 with transfer functions $\beta(s)$ and $s(\alpha)$, the resulting mechanism oriented goal transfer function $\beta(\alpha)$ is defined by the formula

$$\beta(\alpha) = \beta\,\{s\,(\alpha)\} \tag{1}$$

$$\frac{d\beta}{d\alpha} = \frac{d\beta}{ds}\,\frac{ds}{d\alpha} \tag{2}$$

$$\frac{d^2\beta}{d\alpha^2} = \frac{d^2\beta}{ds^2}\left(\frac{d\sigma^2}{d\alpha}\right) + \frac{d\beta}{ds}\,\frac{d^2 s}{d\alpha^2} \tag{3}$$

where s, $ds/d\alpha$ and $d^2s/d\alpha^2$ represent the prescribed tool oriented goal function but $\beta(s)$, $d\beta/ds$ and $d^2\beta/ds^2$ define the transfer function of the transition mechanism.

3.3.2. Example [3.11]

In figure 3.25 the principle of a cardboard printing machine is scetched. The sheets of cardboard are stored in a flat magazine and pushed subsequently to the printing cylinder. The task of the pusher is to accelerate the sheet from dwell position towards a stationary velocity which is the same as the velocity of the circumference of the printing cylinder. The velocity of the pusher has to be stationary in an interval that the front of the sheet is located 10 mm before and behind the printing position. The transportation of the sheet will be taken over by the printing cilinder and other driven rollers. The pusher may return in its starting position. The pusher stroke is limited because of practical size demands. The "tool oriented goal function" consists of a straight line only which markes the absolutely necessary stationary velocity, as shown in figure 3.26.

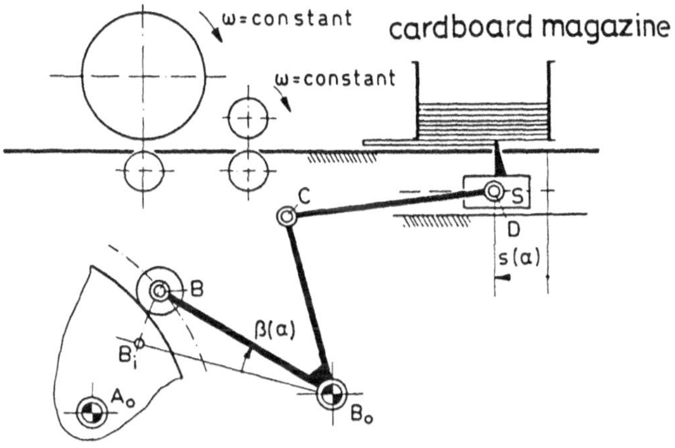

Figure 3.25. Scheme of a cardboard printing machine.

With respect to the cylinder diameter D = 300 mm the printing cylinder circumference velocity is $\dot{s} = s'\dot{\alpha}$ and the known transfer function interval represents a first order transfer function with the stationary value $s' = ds/d\alpha = 150$ mm/rad , see figure 3.26. The short interval forms a part of a not yet known periodic transfer function. To accomplish the period an attempt has been made with both a polynome of the fifth degree and the cycloidal function. The best result was achieved with the polynome. Now the "tool oriented goal function" is ready, see figure 3.27 [3.28].

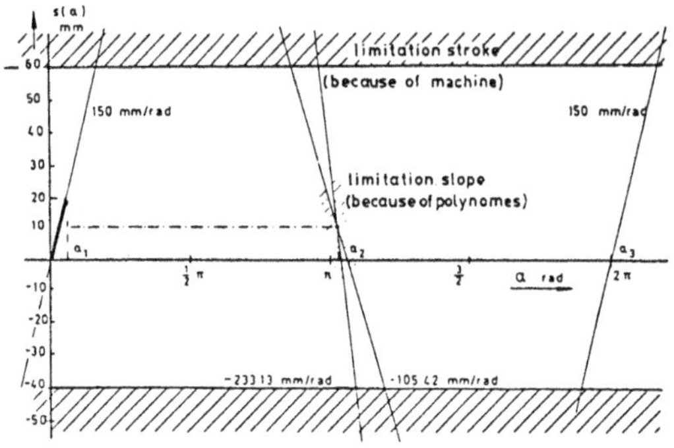

Figure 3.26. Strictly prescribed tool oriented goal function interval.

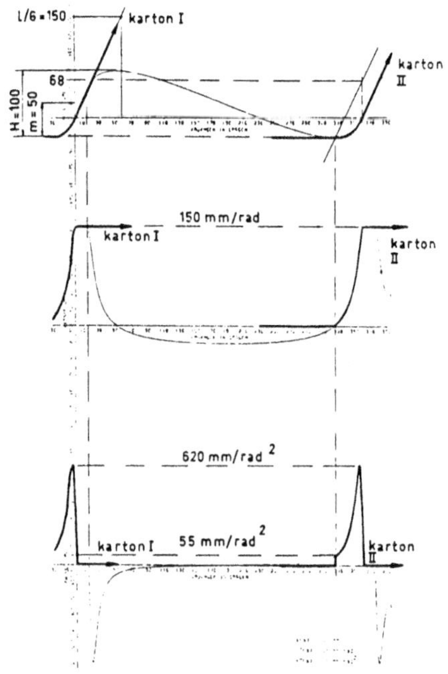

Figure 3.27. Final definition on the total period of the tool oriented goal function.

The transition mechanism will be a rocker-and-slider mechanism. To be sure that optimal pressure angle are present the kinematic parameters are selected using the VDI-standard nr.2115 [3.29]. With respect to the real dimensions of the printing machine and the planned location of the camshift the following link lengths are choosen, see figure 3.28: xo = 0.264 m, b = 0.110 m, c = 0.200 m and e = 0.112 m.

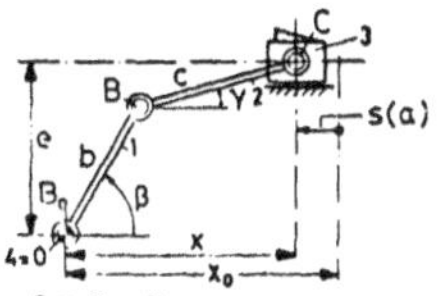

pseudo- parameters	mechanism- parameters
$p_1 = 2b$	$b = \frac{1}{2} p_1$
$p_2 = 2be$	$e = \frac{p_2}{p_1}$
$p_3 = c^2 - b^2 - e^2$	$c = \sqrt{p_3 + b^2 + e^2}$

$s = x_0 - x$

$x = b \cos\beta + c \cos\gamma$

$e = b \sin\beta + c \sin\gamma$

squaring, adding and arranging result in

$2bx \cos\beta + 2be \sin\beta + c^2 - b^2 - e^2 = x^2$

$p_1 x \cos\beta + p_2 \sin\beta + p_3 = x^2$

slider mechanism
$x = x(p_1, p_2, p_3, \beta)$ solution explicit

rocker mechanism
$\beta = \beta(p_1, p_2, p_3, x)$ solution implicit e.g. with Regula Falsi

Figure 3.28. Rocker-and slider transition mechanism.

If the crank of the crank-and-slider mechanism acts as the input link and angle α is the independent variable, the transfer function of the transition mechanism was already written in chapter 3.1.1. It was found x = x(p1, p2, p3, β) with $x = x_0 - s$. But in the application as a transition mechanism in the printing machine the slider displacement s has to be the independent variable because the slider will be moved accordingly the already known "tool oriented goal function". The angle β of the rocker has to be calculated, but there is an implicite function β = β(p1, p2, p3, s) only. Regula falsi or another procedure will solve this equation. But it is to prefer to calculate the zero, first and second order transfer function. To do this the most simpliest way is to use a tested numerical method [3.30], that is available in the CADOM-software package the program PLANAR. The non-linear transformation leads towards the "mechanism oriented goal function" β(α) as shown in scale in figure 3.29. The intervals with proportional motion and fifth degrees polynominal respectively are numerically transformed into other, no longer definable functions. But this fact is not important because the "mechanism oriented goal function" will be used as input in the synthesis procedure of linkage (TADSOL) or cam mechanisms (TADSOC) [3.31].

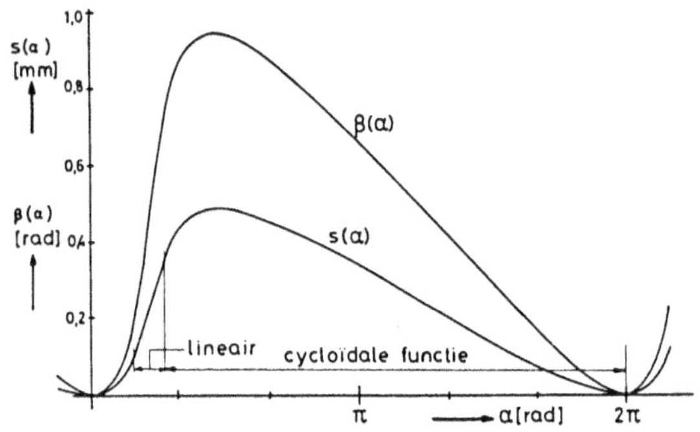

Figure 3.29. Results of non-linear transformation. The prescribed displacement s(α) is transformed into the angle β(α).

Figure 3.30 shows the cam mechanism A_0B_0B as the result of the cam mechanism synthesis procedure TADSOC. The relative path of the roller centre is marked by "1", the so called kinematic profile. The cam profile "2" itself is an equidistant profile and its definition is available as soon as the diameter of the roller follower is choosen. The points of the evolute "3" are very helpful indicators because the optimum roller diameter is lain at approximately half the value of the smallest difference between curve "3" and curve "2" respectively.

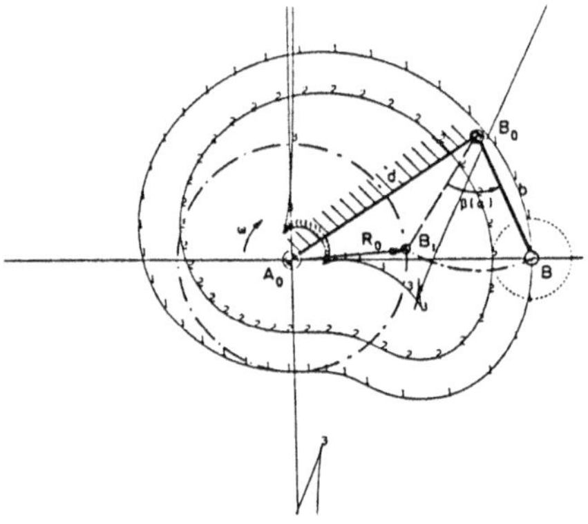

Figure 3.30. The resulting cam kinematic profile (1), cam profile (2) and its evolute (3).

Conclusion

The non-linear transformation of a tool oriented goal function into a mechanism oriented goal function is an important procedure if the goal function is prescribed for another element than is used in the sythesis procedure. The technique is worked out for applications with one and more degrees of freedom [3.31].

3.4. Optimization

In the context of the design philosophy, optimization is the universal dimension synthesis method, which will be used, of the methods, as mentioned in the chapters 3.1 and 3.2, fails.

As input is asked not only the type of mechanism and the goal function, but also a start solution.
Unfortunately, because of the non-linearity of the parameter space of the mechanism, the optimization process may miss the desired result.
More details of our philosophy are published in [3.32].

3.5. Other approaches

There are a few other design approaches well known as for example:

* the series connection of two mechanisms and
* the parallel connection of two mechanisms with superposition of their functions.

In some cases of synthesis of link mechanisms a single mechanism is unable to give a close approximation of the goal function.
The deviation between the mechanism's transfer function and the goal function yields as a new goal function of RR-type for setting up the synthesis procedure of a preset mechanism in front of the already known single mechanism with unsatisfactory approximation. Both the mechanisms acting in series connection will approximate the goal function much more better.
The program SERCON is created to help in series connection synthesis.

In other cases of linkage synthesis a good approximation will be achieved by superposition of two functions generated each by a certain mechanism. The two mechanisms are driven with the same degree of freedom (independent variable) and act in parallel connection. The

superposition may be a linear (gears of chain) or a non-linear superposition (linkages with two input links).
The program PARCON is created to help in parallel synthesis.

The special technique to split the problem in two or more subproblems, which will be solved by steps, sometimes is called "Partial Synthesis Procedure" [3.33].

CHAPTER 4: EVALUATION AND INTERPRETATION OF SYNTHESIS RESULTS*)

Accordingly to the CADOM design philosophy the synthesis procedure which has been the major contents of the designer's task in the past will be displaced to the computer because this routine work is better and much more faster done by a computer. But the evaluation and interpretation of the computer results forms the new and highly important task content of the designer because it seems to be impossible to formulate algorithm for all the decisions which have to be taken during this stage.

4.1. The type of mechanism and its parameters

As a result of the computer calculations the types of appropriate mechanisms become known. In general one goal function leads to more than one solution. Therefore it has been spoken about "multiple result synthesis procedure". A catalogue of mechanisms has to be available to indentify the type of mechanism, to know the definition of the kinematic as well as the assembly parameters and to interpret the results of the computer calculations. In figure 4.1 a protocol of the result of the TADSOL-synthesis procedure is presented. The protocol contains always

- the indentification number of the mechanism,
- the assembly parameters τ, L and U_0,
- a list with the values of the kinematic parameters.

*) Chapter 6.7 of [4.1] has been written by ir. A. van Dijk, member of scientific staf, Production Automation Lab., Mech. Eng. Dept., Delft University of Technology.

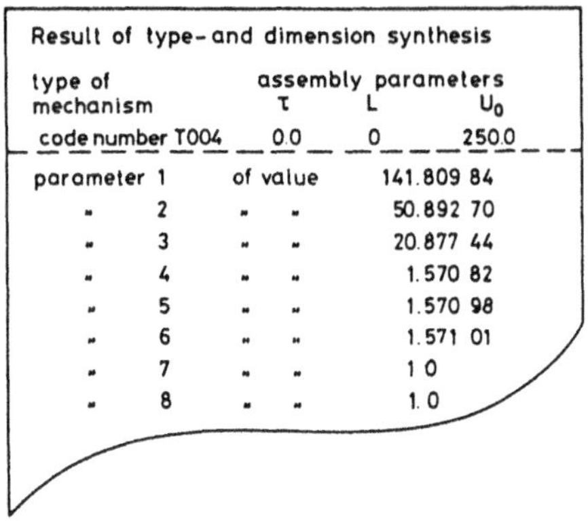

Result of type- and dimension synthesis

type of mechanism	assembly parameters τ	L	U_0
code number T004	0.0	0	250.0

parameter 1	of value	141.809 84
" 2	" "	50.892 70
" 3	" "	20.877 44
" 4	" "	1.570 82
" 5	" "	1.570 98
" 6	" "	1.571 01
" 7	" "	1 0
" 8	" "	1. 0

Figure 4.1. Protocol of the results of a TADSOL-synthesis procedure.

In figure 4.2 the top of the page with mechanism T004 is reproduced. The top contains a schematic drawing of the mechanism and a list with the parameter identification. The mechanism T004 is defined by eight kinematic parameters. The missing part of the page contains the mathematical formulation of the zero, first and second order transfer functions, the synthesis approach and in the case of the MECAT-L catalogue of linkages the Fourier series of the transfer function. In many cases the mechanism identification page of the catalogue contains additional information with respect to cognates and alternative mechanisms as well as references.

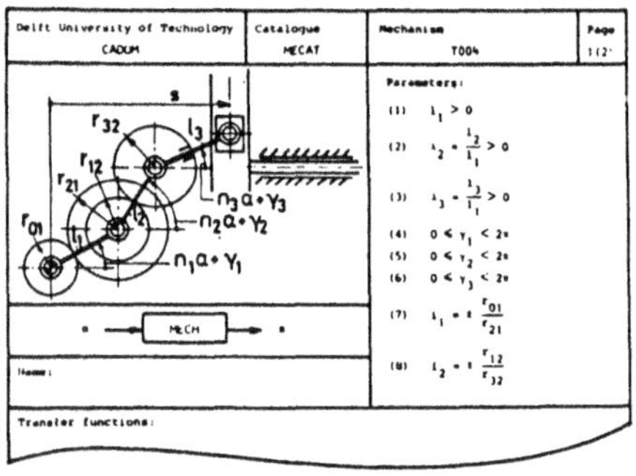

Figure 4.2. Top of page of catalogue of mechanisms MECAT-L with mechanism T004.

The pages of the catalogues of function generators and of cam mechanisms are similar to that shown pages.

4.2. Interpretation of parameter value i of gear ratio

In mechanisms with gear trains the parameter value i occurs. The value may be positive or negative. The parameter value determines for a pair of gears or a gear-and-rack combination the type and the gear ratio. A negative sign means that a pair of gears consists of both outside toothed gears and the direction of rotation is contrary. A positive sign means that the direction of rotation is the same and that one of the two wheels is an inside thoothed gear. With respect to the gear-and-rack combination a negative sign of the parameter i indicates that the rack is mounted on top of the gear, but a positive sign is related with a rack mounted below the gear. Figure 4.3 shows the definition of gear ratios as they are used to interpret the synthesis results.

It is evident that parameter i stands for the quotient of the pitch circle radius of a pair of gears or for the radius of the gear itself if a rack-and-gear combination is used.

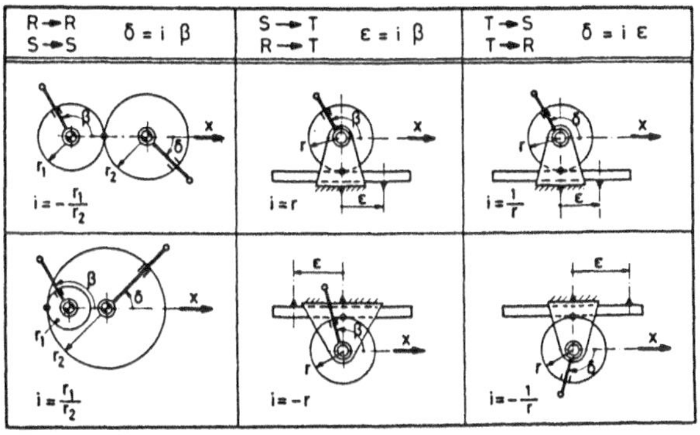

Figure 4.3. Definition of gear ratios for pairs of gears and for gear-and-rack combinations.

4.3. Determination of the assembly position of the mechanism

The "assembly position of a mechanism in a machine" is solved if three informations are available:

- The type of mechanism and its reference position,
- the kinematic as well as the assembly parameters τ, 1 and U_0,
- a scheme of the machine in which the mechanism has to work.

In figure 4.4 an example with a geared six-bar linkage is presented. The left part reproduces the scheme of mechanism S011 with the definition of the reference position of crank and output gear. In the middle of the figure the assembly parameter values are printed. The left part shows the scheme of the machine with the positions of input and output shafts as well as the machine dependent coordinate system in which the goal function is prescribed.

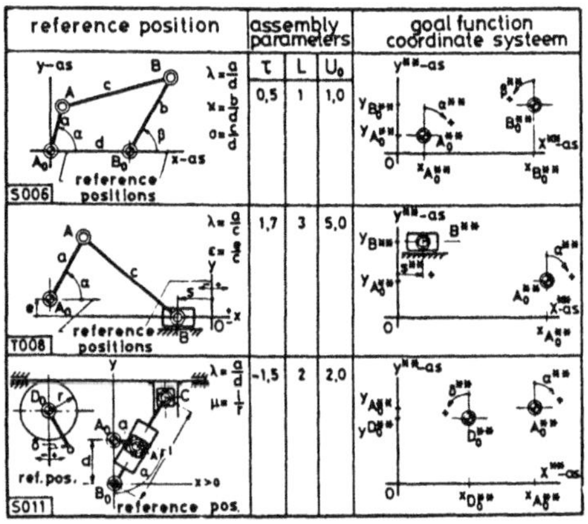

Figure 4.4. Three information sources to be used to determine the mechanism assembly position.

Now the assembly position will be reached in two steps: The first step has to do with the transformation of the coordinate system used in the catalogue of mechanisms. The second step involves the transformation of the mechanism in a situation which takes into account the definition of the coordinate system of the machine.

First step:

The results of the synthesis procedure are used to transform the coordinate sytem of the catalogues of mechanisms in such a manner that the zero order transfer function in the revised coordinate system coincides with the goal function in its prescribed coordinate system. The necessary transformation follows the route which is set out in figure 4.5.

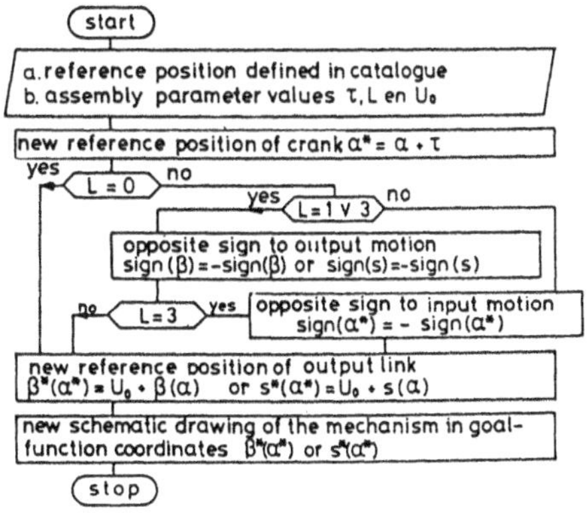

Figure 4.5. Flow chart of the coordinate transformation procedure.

The sequence of use of the values τ, L and U_0 influenced the result. Therefore it is recommended to follow the outlined routing.

- With the value τ the new reference line of the crank will be determined. The new crank angle is $\alpha^* = \alpha + \tau$. The reference line of the crankshaft itself remains unchanged. The new reference line is found if the angle $\alpha = -\tau$ is set out in the scheme of the catalogue of mechanisms. In this position the new crank angle α^* has the value $\alpha^* = 0$.
- If L = 1 or L = 3 the direction of motion β or s of the output link as defined in the catalogue of mechanisms has to be changed into the opposite direction.
- If L = 2 or L = 3 the direction of rotation α^* of the input link as defined in the catalogue of mechanisms has to be changed into the opposite direction.
- U_0 indicates the new reference line of motion of the output link so that $\beta^* = \beta(\alpha)+U_0$ or $s^* = s(\alpha)+U_0$ respectively. The new reference line will be found if $\beta = -U_0$ or $s = -U_0$ respectively is set out in the scheme of the catalogues of mechanisms. Attention is asked because in the case of L = 1 or L = 3 the direction of motion is already revised.

Figure 4.6 presents the results of the first step of transformation of the coordinate system.

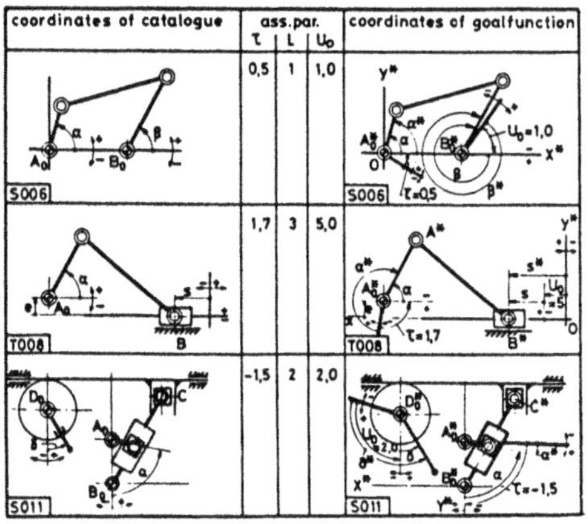

Figure 4.6. First step of transformation of coordinate system. All goal function coordinates are marked with *.

Second step:

The second step deals with the transformation of the goal function coordinates in accordance with the first taken step into the assembly position in the machine coordinate system. In the scheme of the machine is difined how to connect the input as wel as the output shaft with the appropriate links of the mechanism. Unfortunately till now it is impossible to present a simple procedure which described the necessary steps, the problem seems to be too complex.

Therefore it is choosen to explaine what can be done and what the consequences are with respect to the position of the mechanism in the coordinate system and what the consequences with respect to the transfer function are.

Transformation is possible with the aid of:

- Reflection on a line, see figure 4.7,
- rotation around a point,
- translation along a line,
- increase or decrease the scale factor of the length of all elements.

REFLECTION of a R- or S-mechanism on a line 1 will alter the positive direction. Reflection of a T-mechanism will change the positive direction of the output motion in the case only that the reflection line stands perpendicular on the direction of motion, but in any case the input motion will alter. Figure 4.7 shows examples for reflection of R- or S-mechanisms and of a T-mechanism on a line. The reflected mechanisms are drawn with stippled lines and marked by an index s (s = Spiegelbild). The reference lines r_i and r_u of respectively the input and the output motion are reflected too.

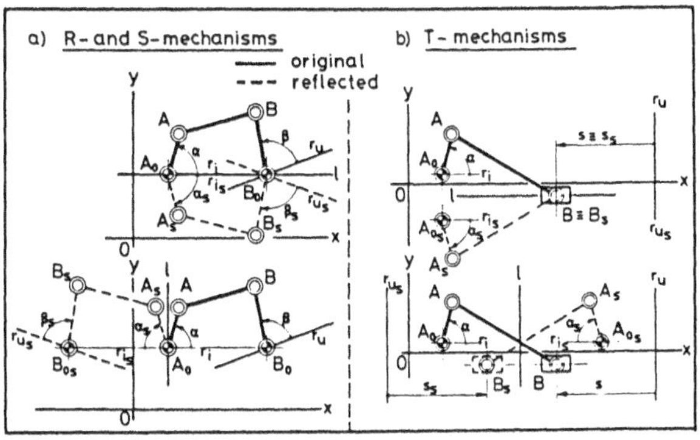

Figure 4.7. Examples for reflection of mechanisms on a line.

ROTATION and TRANSLATION may be used to obtain the desired positions of the frame pivots. T-mechanisms are changing their positive direction of output motion by rotation with 180 degrees.

Change of SCALE is used to fit R- and S-mechanisms so that frame related pivots coincide with the fixed points of the mechanism.

It is the designer's decision to make a choice of the mentioned possibilities and to evaluate which choice gives an optimal result of his problem.

Figure 4.8 shows the results of the second step of transformation of the coordination, based on the situation as recieved after the first step, see figure 4.6.

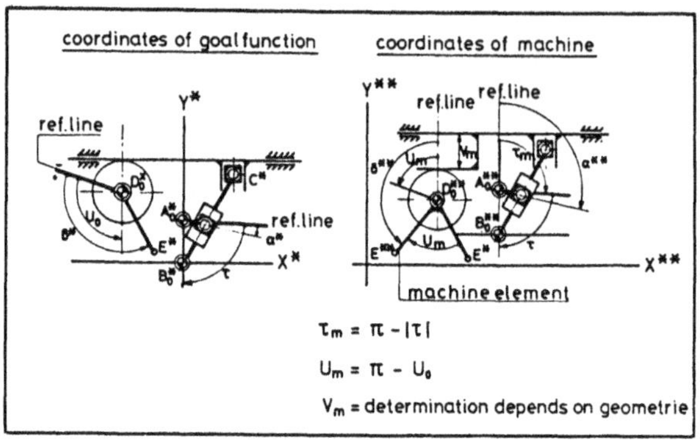

Figure 4.8. Second step of tranformation of coordinate system.

As the result of the transformation the angle in which defines the assembly position of the crank relative to the input shaft reference line becomes available. The same result has to achieve for the assembly position of the output link of the mechanism relative to the element of the machine. The distance between these two members is U_m. The crank has to be fixed with respect to the reference line on the shaft under the angle τ_m. The values of τ_m and U_m are derived from the drawing of the mechanism in the machine coordinate system. What concerns mechanism S011 the second step of the assembly position is taken by translation of crank point A_0* to the prescribed pivot A_0**. Thereby the goal function coordinate system (x*, y*) is changed into the machine coordinate system (x**, y**). All directions of rotation remain unchanged. To receive the proper position of point D_0* an extra translation of the left part of the mechanism over the distance V_m seems to be necessary. This distance has to be found from given positions of the shafts in relation to the real link dimensions of the mechanism.

The driven machine element Do**E** is lain under the angle U_m with respect to the gear fixed rocker $D_0E = D_0$**E*. The final result is:

$$\tau_m = \pi - \tau \ ,$$
$$U_m = \pi - U_0 \ ,$$
$$v_m = \text{designers decision} \ .$$

CHAPTER 5: MECHANISM'S CONCEPT DESIGN

The synthesis phase has to be completed with a concept design, worked out by an experienced designer. This work will never been done automatically.

Here the designer can influence the details of the solution, for example:

- If a pivot was guided on a straight line, he may try to use the approximately straight part of a coupler curve of a four-bar linkage.
- If input and output links of a mechanism do not hit each other, they may be placed in the same plane.
- If a pair of gears has been found in the schematic drawing of the mechanism's catalogue, equivalent solutions using chain and chain weels may be examined.

The concept design makes available:

- dimension and cross sections of all links, pivots and axis,
- mass and center of gravity of every link,
- inertia,
- necessary space.

Based on this concept drawing the examination may be started with checks of material strength and stiffness as planed in the second loop of the design process.

CHAPTER 6: DEMONSTRATION OF CADOM-SOFTWARE PACKAGE

The CADOM-software is implemented on several computers as AMDAHL, DEC, IBM and PRIME, but there exists a "micro-CADOM" software package, that runs under the CP/M operating system respectively the CP/M-compatible operating system CDOS. This package was demonstrated during 8 hours of evenings sessions on a Cromemco System Three 8-bit micro-computer under CDOS.

Figure 6.1 shows the configuration that Cromemco Inc., Mountain View, California (USA), made available. This configuration consists of the computer with one 8" floppy disk drive and a harddisk (to the right), the terminal (in the center) and the desy-wheel-printer (to the left).

All modules of CADOM were demonstrated and the participants get the chance to input their own mechanism problem.

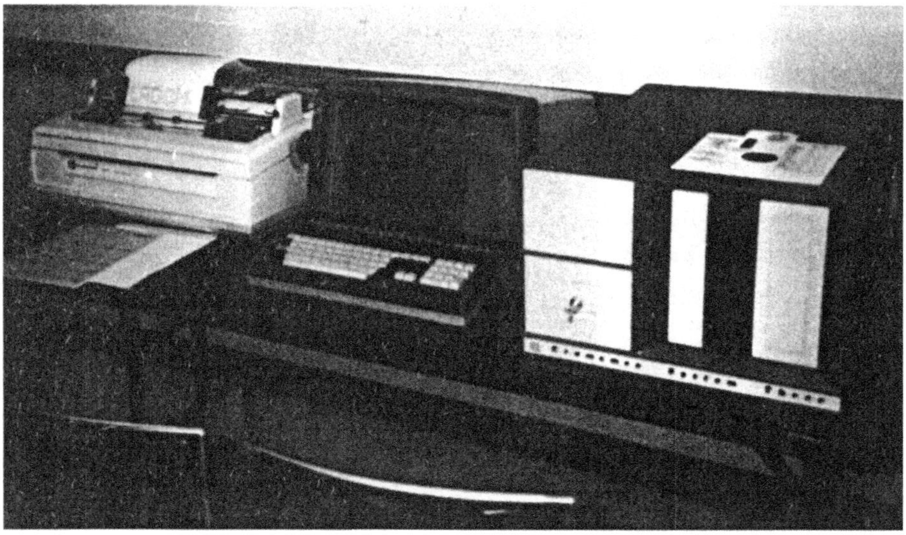

Figure 6.1. Micro-CADOM demonstration on CS3.

CHAPTER 7: FINAL REMARKS

The design philosophy and the synthesis approach is the result of very effective teamwork done in the CADOM-task group of the Delft University of Technology. The task group has been established in 1972 based on a written contract between two laboratories of the Department of Mechanical Engineering:

- the laboratory of Engineering Mechanics, and
- the laboratory of Production Automation and Mechanisms (PAM).

The ordinary members are:

- Chr.B. van den Berg,
- K.H. Drent,
- A. van Dijk,
- A.J. Klein Breteler,
- H. Rankers.
- B. Tanuwidjaja,
- K. van der Werff.

The members are coaching students during their post-graduate projects, if the project is involved in mechanism's synthesis and/or analysis. Most of the projects are real industrial problems, found in the field of Production Mechanisation/Automation.

Finally, I would like to thank whole my scientific and technical staff, who has helped me to finish this presentation.

Whole the content of this paper is part of the author's textbook used in the course "Design of Mechanisms" at the Department of Mechanical Engineering at the Delft University of Technology, Delft/The Netherlands.

REFERENCES

[1.01] HANSEN, F.: Konstruktionssystematik. Berlin, Technik, 1968.

[1.02] RODENACKER, W.: Methodisches Konstruieren. Berlin, Springer, 1970.

[1.03a] KOLLER, R.: Konstruktionsmethode für den Maschinen-, Geräte- und Apparatebau. Berlin, Springer, 1976.

[1.03b] KOLLER, R.: Eine algorithmisch-physikalisch orientierte Konstruktionsmethode. VDI-Zeitschrift 115(1973), S.147-152, S.309-317, S.843-847.

[1.04] STEINWACHS, H.: Praktische Konstruktionsmethode. Würzburg, Vogel, 1976.

[1.05] PAHL, G. and W. BEITZ: Konstruktionslehre - Handbuch für Studium und Praxis. Berlin, Springer, 1977.

[1.06] VDI-Richtlinie Nr.2222, Blatt 1: Konstruktionsmethodik. Berlin/Köln, Beuth, 1973.

[1.07] RANKERS, H.: Ontwerpen van Mechanismen. Collegedictaat W76 (Design of Mechanisms), Dept. Mech. Eng., Delft University of Technology, 1981.

[1.08] RANKERS, H.: Education in Computer Aided Design, Philosophy, Consequences, Experiences. Speach on SEMEMATRO, Bulgaria, Sept.1982.

[1.09] RANKERS, H. et al: Computer Aided Design Of Mechanisms, the CADOM-Project of the Delft University of Technology. Proceedings of the 5th World Congress on Theory of Machines and Mechanisms, Montreal 1979, P.667-672.

[1.10] RANKERS, H.: Anwendung numerischer Methoden in der Getriebetechnik, dargestellt an Beispielen. VDI-Berichte Nr.321, S.1-8. Düsseldorf, VDI, 1979.

[1.11] RANKERS, H. and K. van der WERFF: Getriebetyp-unabhängige Methode der Analyse der Kinematik und Dynamik der Räderkurbelgetriebe. VDI-Berichte Nr.321, S.9-16.

[1.12] RANKERS, H. and A. van DIJK: Entwurf eines Manipulators, Kinematik und Dynamik. VDI-Berichte Nr.321, S.17-26. Düsseldorf, VDI, 1979.

[1.13] RANKERS, H. and A.J. KLEIN BRETELER: Über das Programmieren von Getriebeproblemen - Vorschläge zur Strukturierung und Vereinheitlichung. VDI-Berichte Nr.321, S.27-32. Düsseldorf, VDI, 1979.

[1.14] KLEIN BRETELER, A.J.: Partial derivatives in kinematic optimization. Proceedings of the 5th World Congress on Theory of Machines and Mechanisms. Montreal 1979, p.883-888.

[1.15] WERFF, K. van der : Dynamics of flexible mechanism. Proceedings NATO Advanced Study Institute, Iowa 1983: Computer-Aided Analysis and Optimization of Mechanical Systems Dynamics. New York/Berlin, Springer, 1983.

[1.16] WERFF, K. van der : Kinematic and dynamic analysis of mechanisms. A finite element approach. Thesis, Delft University of Technology. Delft, University Press, 1977.

[2.01] HAIN, K.: Übersicht über sämtliche Umlauf- und Schwingbewe=gungen an Gelenkvierecken. Grundlagen der Landtechnik 15(1965)Nr.4, S.97-106.

[2.02] BÖGELSACK, G. et altera: Terminology for the theory of Machines and mechanisms. IFToMM-Commission A: Standards of Terminology. Fourth Draft 1982.

[2.03] BERG, Chr.B. van den: Torque compensating mechanisms. (Dutch: Ontwerpen van een nokmechanisme voor het compenseren van een aandrijfmoment). Internal reports nr. 6/80 and 8/80, Section Production Automation and Mechanisms, Department of Mechanical Engineering, Delft University of Technology.

[2.04] KEIJZER, H.A. de: Dynamic balancing of mechanisms (Dutch: Dynamisch balanceren van mechanismen). Internal report, Section Production Automation and Mechanisms, Department of Mechanical Engineering, Delft University of Technology, 1982.

[2.05] RANKERS, H.: Angenäherte Getriebe-Synthese durch harmonische Analyse der vorgegebenen periodischen Bewegungsverhältnisse. Doktor-Thesis T.H. Aachen 1958.

[2.06] KLEIN BRETELER, A.J.: Mechanism Catalogue Of Linkages, MECAT-L. Department of Mechanical Engineering, Delft University of Technology, 1983.

[2.07] KIPER, G. and D. SCHIAN: Sammlung der Grübler'schen kinema= tischen Ketten mit bis zu zwölf Gliedern. Fortschrittsbericht VDI-Z. Reihe 1, Nr.44. Düsseldorf: VDI-Verlag 1976.

[2.08] KIRCHHOF, M.: Das Gelenkfünfeck und sein Bewegungsbereich. Maschinenbautechnik (Getriebetechnik) 12(1963)H.2, S.99-106.

[2.09] DIJK, A. van and H. RANKERS: Entwurf eines Manipulators, Kinematik und Dynamik. VDI-Berichte Nr.321. Düsseldorf: VDI-Verlag, 1979.

[2.10] SANDOR, G.N. and F. FREUDENSTEIN: Kinematik Synthesis of Path Generating Mechanisms by means of the IBM/650 Computer. Columbia University: IBM/650 Profram Library, February 1958.

[2.11] RANKERS, H.: Getriebe für Bahnkurven mit vorgeschriebenem Geschwindigkeitsverlauf. VDI-Berichte Nr.140. Düsseldorf: VDI-Verlag 1970.

[2.12] KLEIN BRETELER, A.J.: Partial derivatives in kinematic optimization. Proceedings of the 5th World Congress on Theory of Machines and Mechanisms. New York: ASME 1979.

[2.13] TOL, C.J.M.: Design of a teach-in and work unit for sawing wooden puzzle toys (Dutch: Ontwerp voor een teach-in voor het zagen van kinderpuzzels). Internal report, Section Production Automation and Mechanisms, Department of Mechanical Engineering, Delft University of Technology, 1978.

[2.14] BOENDER, C.C.: Design of a cam-controlled welding machine (Dutch: Verbeteren lasautomaat). Internal report, Section Production Automation and Mechanisms, Department of Mechanical Engineering, Delft University of Technology, 1983.

[2.15] RANKERS, H.: Spezielle Ebenen-Führungen mit Zwei-Kurven-Mechanismen, Voraussetzungen und Synthese. Proceedings of the 6th IFToMM World Congress on Theory of Machines and Mechanisms, New Delhi 1983. New York: ASME 1983.

[3.1] BURMESTER, L.: Lehrbuch der Kinematik. Leipzig: Felix 1888.

[3.2] FREUDENSTEIN, F.: Approximate synthesis of four-bar linkages. Trans. ASME, vol.77, p.853-861, August 1955.

[3.3] MEYER ZUR CAPELLEN, W. and K.A. RISCHEN: Lagenzuordnungen an ebenen Viergelenkgetrieben in analytischer Darstellung. Eine Mass-Synthese. Forschungsbericht Nr.923 des Landes Nordrhein-Westfalen. Köln und Opladen, Westdeutscher Verlag, 1961.

[3.4] HARTENBERG, R.S. and J. DENAVIT: Kinematic synthesis of linkages. New York, McGraw-Hill, 1964.

[3.5] NIETO, J.N.: Sintesis de Mecanismos. Madrid, Editorial 1978.

[3.6] LUCK, K. and K.H. MODLER: Computersynthese von Viergelenk-Getrieben bei vorgegebenen Lagenzuordnungen. Wiss.Z. der Techn. Uni. Dresden 22.(1973)H.3, p.509-513.

[3.7] RANKERS, H.: Weg-Winkel-Zuordnung für den Kreuzschubkurbel Mechanismus. Eine exakte kinematische Mass-Synthese. Department of Mechanical Engineering, Delft University of Technology. WTHD-Rapport Nr. 155/1983.

[3.8] RANKERS, H.: Precision point synthesis of plane mechanisms. Department of Mechanical Engineering, Delft University of Technology. WTHD-Rapport (under preparation).

[3.9] RANKERS, H.: Mass-Synthese einfacher Mechanismen mit Dreh-Schub-Umwandlung bei vorgegebenen Präzisionspunkten der Ziel-Ubertragungsfunktion. Proceedings Sixth IFToMM World Congress on Theory of Machines and Mechanisms. New York, ASME, 1983.

[3.10] RANKERS, H.: Design of mechanisms (Dutch: Ontwerpen van Mechanismen). Textbook W76, Department of Mechanical Engineering, Delft University of Technology, 1981.

[3.11] RANKERS, H.: Angenäherte Getriebe-Synthese durch harmonische Analyse der vorgegebenen periodischen Bewegungsverhältnisse. Dr.-Dissertation, Technische Hochschule Aachen, 1958.

[3.12] RANKERS, H. et altera: TADSOL - Type And Dimension Synthesis Of Link mechanisms. A user oriented description of the computer program. Proceedings of the Symposium on Computer Aided Design In Mechanical Engineering. Milan: Clup 1976.

[3.13] RANKERS, H.: Ziel-Uebertragungsfunktion und Getriebetyp. Rechnerunterstützte Typen- und Mass-Synthese einfacher und zusammengesetzter Mechanismen. VDI-Berichte Nr. 28, pp. 119-131. Düsseldorf, VDI-Verlag 1977.

[3.14] RANKERS, H.: Computer Aided Design Of Mechanisms. Programmapakketten voor analyse en synthese van stangen- en nokmechnismen beschikbaar. Aandrijftechniek, 13 January 1978.

[3.15] RANKERS, H. and coll.: Computer Aided Design Of Mechanisms. The CADOM-project of the Delft University of Technology. Proceedings of the Fifth IFToMM World Congress on the Theory of Machines and Mechanisms, Montreal, 1979. New York, ASME 1979.

[3.16] RANKERS, H.: Design Of Mechanisms (Dutch: Ontwerpen van mechanismen), Textbook W76, chapter 6. Department of Mechanical Engineering, Delft University of Technology, 1981.

[3.17] FLES, H.S.M.: User Manual "STAM", Graphics version of TADSOL. Internal report 1/80, Lab Production Automation and Mechanisms, Department of Mechanical Engineering, Delft University of Technology.

[3.18] Van DIJK, A. and K van der WERFF: TADSOL-Introduction (Dutch: TADSOL-Introductie). Internal report nr.5/76, Section Production Automation and Mechanisms, Department of Mechanical Engineering, Delft University of Technology.

[3.19] De La HIRE: Traite' des roulettes. 1706, pp.348.17.

[3.20] ROBERTS, S.: Three-bar Motion In Plaine Space. London Math. Soc. Proc., Vol. 7 (1875), pp. 14-23.

[3.21] DIJKSMAN, E.A.: Motion Geometry Of Mechanisms. London: Cambridge University Press 1976.

[3.22] RÖSSNER, W.: Sechsgliedrige Gelenkgetriebe zur Erzeugung einer bestimmten Koppelkurve. Maschinenbautechnik 8 (1959) 2, pp. 105-107. Abwandlung ebener Gelenkgetriebe zur Anpassung an praktische Bedingungen. TZ. f. prakt. Metallbearbeitung 55. (1961) 7, pp. 332-340.

[3.23] LUCK, K.: Zur Erzeugung von Koppelkurven viergliedriger Getriebe. Maschinenbautechnik 8(1959)2, pp. 97-104.

[3.24] RANKERS, H.: Ausgleich der ungleichförmigen Bewegung langgliedriger Ketten. Ind.-Anz. 89. (1967) nr. 34, pp. 723.

[3.25] MEYER ZUR CAPELLEN, W.: Ueber gleichwertige periodische Getriebe. Fette, Seifen, Anstrichmittel. Die Ernährungsindustrie 59. (1957) nr. 4, pp. 257-266.

[3.26] KLEIN BRETELER, A.J.: Partial Derivatives In Kinematic Optimization. Proc. fifth IFToMM world congress on Theory of machines and mechanisms, 1979 Montreal, pp. 883-888.

[3.27] KLEIN BRETELER, A.J.: MECAT-L, Mechanism Catalogue Of Linkages. Section Automation of Production and Mechanisms, Dept. of Mech. Engineering, Delft University of Technology, 1983.

[3.28] BERG, Chr.B. van den: Application of polynomes and cycloidal functions in transfer functions of cam mechanisms (Dutch: Toepassingsmogelijkheden van polynomen en cycloïdale functies in overdrachtsfuncties van nokmechanismen). Internal report nr. 5/80, Section Production Automation and Mechanisms, Department of Mechanical Engineering, Delft University of Technology.

[3.29] VDI-RICHTLINIEN: Bewegungsgesetze für Kurvengetriebe. Theoretische Grundlagen. VDI-Richtlinie Nr. 2115, Blatt 1. Düsseldorf: VDI, 1980.

[3.30] WERFF, K. van der: Kinematic and Dynamic Analysis of Mechanisms - A Finite Element Approach. Thesis, Delft University of Technology, Delft: University Press, 1977.

[3.31] RANKERS, H.: Special Twin-Cam Controlled Plaine Position Coordination Mechanisms (German: Spezielle Ebenen-Führungen mit Zwei-Kurven-Mechanismen, Voraussetzungen und Synthese).
Proceedings of the Sixth IFToMM World Congress on Theory of Machine and Mechanisms, New Delhi 1983. New York: ASME 1983.

[3.32] KLEIN BRETELER, A.J.: Partial Derivatives in Kinematic Optimization. Proceedings of the Fifth IFToMM World Congress on Theory of Machines and Mechanisms, Montreal, 1979, p.883-888.

[3.33] LICHTENHELD, W. und K. LUCK: Konstruktionslehre der Getriebe. Chapter 4.5, p.147-153. Berlin, Akademie-Verlag, 1979.

[4.1] Rankers, H.: Design of Mechanisms (Dutch: Ontwerpen van mechanismen). Textbook w76, Section Production Automation and Mechanisms, Department of Mechanical Engineering, Delft University of Technology, 1981.

DESIGN SENSITIVITY ANALYSIS AND OPTIMIZATION OF KINEMATICALLY DRIVEN SYSTEMS

Edward J. Haug and Vikram N. Sohoni
Center for Computer Aided Design
College of Engineering
The University of Iowa

Abstract. A state space design sensitivity analysis and optimization method is presented in which problems of optimal design of machines are formulated in a setting that allows treatment of general design objectives and constraints. Synthesis of machines to perform both kinematic and kinetic functions is considered. A Cartesian coordinate formulation is employed for position, velocity, acceleration, and force analysis. An adjoint variable technique is employed to compute derivatives with respect to design of general cost and constraint functions that involve kinematic, force, and design variables. Linearization (or sequential quadratic programming) and gradient projection optimization algorithms are employed, using the design sensitivity analysis method developed, for design optimization. Four optimal design problems are solved to demonstrate use of the method.

1. INTRODUCTION

Optimization of mechanisms and machines has generally been pursued using techniques that are oriented toward design of specific systems. The increasing complexity of machines, particularly as required for programable action, requires general purpose techniques for design of large scale, multidegree of freedom systems. The purpose of this paper is to present a general purpose theory for design sensitivity analysis and optimization of planar mechanisms and machines. A computer code that was developed to implement the theory

NATO ASI Series, Vol. F9
Computer Aided Analysis and Optimization of Mechanical System Dynamics
Edited by E. J. Haug

is used to solve a number of example problems, to demonstrate the wide range of applicability of the technique.

Existing Methods of Mechanism Optimization

Optimization techniques for mechanisms and machines have generally been aimed at designing systems in which some member is required to describe a desired path or generate a function of the input. The objective in such design situations is to determine member lengths and other geometrical parameters to minimize the difference between the desired and actual path generated by the mechanism. Precision point approaches [1,2] have been applied for solution of such design problems. The basic idea underlying such approaches is to design a mechanism to minimize deviation between actual and desired paths at specified points on the path.

Balancing of machines is another area that has received considerable attention [3,4,5]. Machines being considered in these investigations are generally high-speed, inertia variant, rotating devices. Due to the inertia variant nature of such machines, the support frame experiences large shaking forces and moments. The design objective for such systems is to redistribute mass of links or to add counterweights to minimize shaking forces or moments.

Synthesis of one or two degree-of-freedom mechanisms has been the subject of many papers, e.g.; Refs. 6 and 7. Synthesis of a mechanism so that it will occupy less than a prescribed amount of space has been studied in Ref. 8.

In the synthesis methods noted above, constraints have generally been imposed on design variables, such as link lengths, or on generalized coordinates, such as angles between links. Constraints on force transmission angle have also been extensively used. Methods for stress and deformation constrained design of mechanisms have recently appeared in the literature [9]. Minimum weight design has generally been the objective of such design schemes.

As is evident from this brief survey, most available optimal synthesis schemes have been oriented toward design of a specific type of mechanism, to perform a specific task. Some efforts have been made in developing synthesis methods that are more general than those described above [10,11,12]. The generality of these methods, however, is limited to a particular class of problem.

Modeling Techniques for Large Scale Mechanisms and Machines

Modeling techniques that are general enough to predict performance of large scale dynamic mechanical systems have been developed only in the 1970's [13]. Dynamic modeling techniques for large scale electronic and structural systems, on the other hand, have been available for some time. Two modeling methods for dynamic mechanical systems are considered appropriate for analysis of kinematic mechanical systems. One of them, the loop closure method [14], is embodied in the computer codes such as IMP [15]. This modeling method has been used extensively for analysis of kinematic systems [16]. A cartesian formulation is the basis for computer codes such as ADAMS [17] and DADS [18,19]. This modeling method involves writing equations of motion for individual members and then adjoining equations of constraint through Lagrange multipliers. This method, though not yet used for optimization of kinematic systems, has attractive features for doing so.

Techniques for Design Optimization of Large Scale Systems

Most methods that have been employed for optimization of structural and mechanical systems belong to the field of nonlinear programming, in which the design problem is formulated directly in terms of design variables that are to be selected. Optimization methods such as the Sequential Unconstrained Minimization Technique (SUMT) [8] and the Optimality criteria method [9] have been used for Kinematic synthesis. Performance constraints, however, are most naturally stated in terms of system state or response variables. Ad-hoc techniques have been used to reduce small scale design problems to standard nonlinear programming form, but this approach is not feasible for large scale systems.

Numerical methods used in optimal control and optimal design theory [20] employ a state space formulation that explicitly treats design and state variables. The state variable is generally the solution of an algebraic or differential equation, for which an adjoint variable is defined as the solution of a related problem. The adjoint variable method provides design sensitivity information that is required for virtually all iterative methods of design optimization.

Scope of the Paper

In this paper, a cartesian coordinate formulation is used for analysis and optimization of large scale, planar kinematic systems. Velocities and accelerations of members and reaction forces in joints are computed and constrained. The basic assumption is that kinematics of the system are independent of externally applied forces; i.e., systems treated are kinematically driven. A general cost function may be minimized, subject to constraints on position, velocity, force (hence stress and shaking forces), and design variable magnitudes.

2. KINEMATIC ANALYSIS

Before mechanism optimization schemes are considered, a technique for kinematic analysis of mechanisms must be adopted. A cartesian coordinate technique that has been successful in modeling large scale open and closed loop mechanical systems [17,18] is employed here.

Position Analysis

The cartesian coordinate approach embeds a local coordinate system in each link of the mechanism or machine, at the center of mass, as shown in Fig. 2.1. Since only planar systems are being considered in this paper, the position and orientation of any body in the system can be described by three generalized coordinates x_i, y_i, and θ_i. These quantities can be represented by the vector

$$q^{(i)} = \begin{bmatrix} x_i \\ y_i \\ \theta_i \end{bmatrix} \qquad (2.1)$$

As shown in Fig. 2.1, any point p on body i (which is associated with body i) of the system can be represented by coordinates ξ_{ij} and η_{ij} that are measured in the body-fixed coordinate system (considered as a "drafting board coordinate system").

A mechanical system generally consists of many members that are connected by mechanical joints. These joints could be looked upon as constraints on the relative motion of connected pairs of bodies. The Cartesian coordinate formulation represents joints as algebraic constraints between bodies that make up the system.

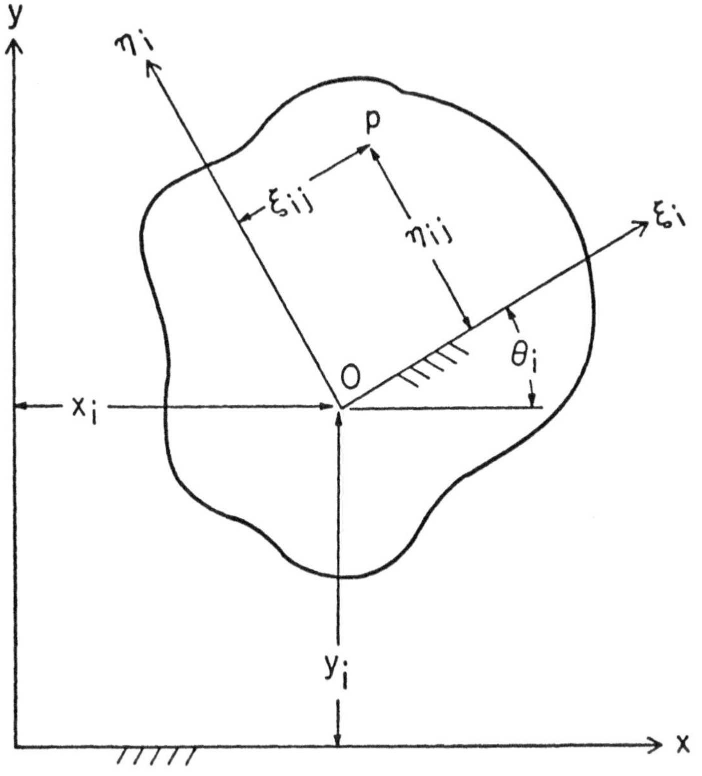

Figure 2.1 Definition of Generalized Coordinates for Body i

It is necessary to determine the explicit form of the equations of constraint. Since kinematic equations of constraint occur in a general form; i.e., the equations of constraint for all joints of the same type have the same general form, it is sufficient to consider a typical joint of each type. This paper considers only revolute and translation joints. A typical joint of each type is treated in each of the following subsections.

Constraint Equations for a Revolute Joint: Figure 2.2 shows adjacent bodies i and j, with body-fixed coordinate systems $0_i x_i y_i$ and $0_j x_j y_j$, respectively. The origins of these reference frames are located in the global reference frame by vectors $\bar{R}^i$ and $\bar{R}^j$, respectively. Let point p_{ij} on body i be located by a body-fixed

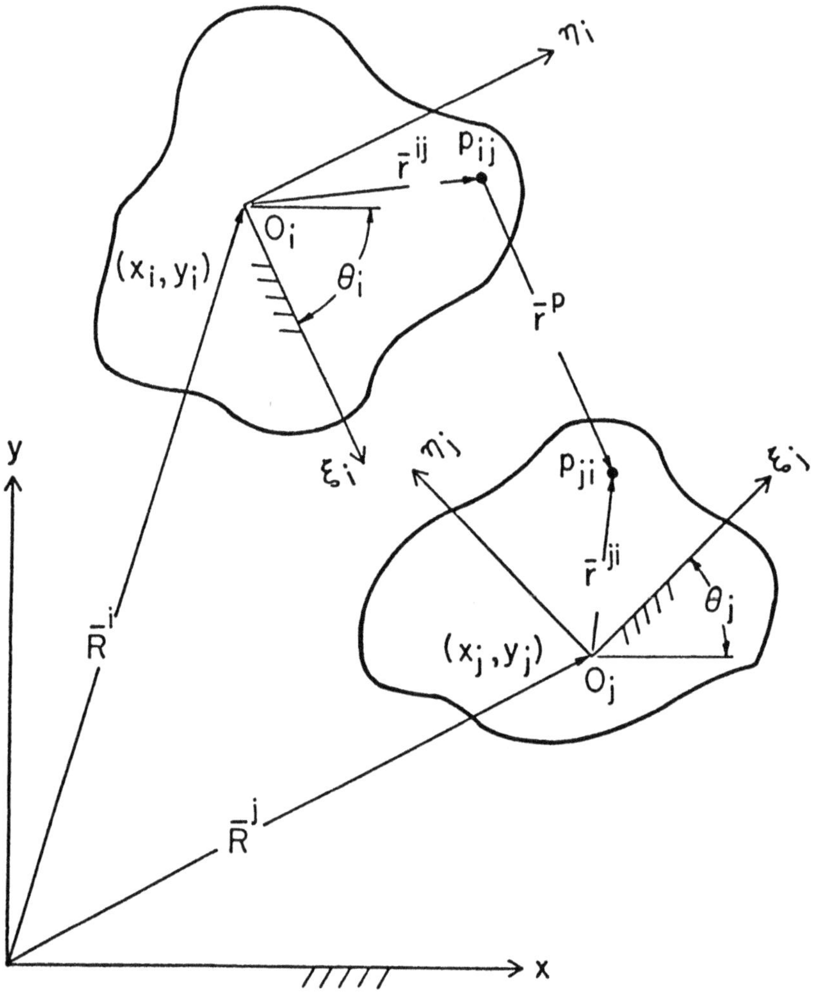

Figure 2.2 Revolute Joint

vector $\bar{r}^{ij}$ and point p_{ji} on body j be located by a body-fixed vector $\bar{r}^{ji}$. Points p_{ij} and p_{ji} are, in turn, connected by a vector $\bar{r}^{p}$,

$$\bar{r}^{p} = \bar{R}^{j} + \bar{r}^{ji} - \bar{r}^{ij} - \bar{R}^{i} \tag{2.2}$$

Demanding that points p_{ij} and p_{ji} are coincident prescribes a rotational joint between bodies i and j at this common point. This is equivalent to the vector equation $\bar{r}^{p} = \bar{0}$. Thus, Eq. 2.2 yields

$$\bar{R}^i + \bar{r}^{ij} - \bar{r}^{ji} - \bar{R}^j = 0 \tag{2.3}$$

In component form, Eq. 2.3 can be written as the pair of scalear equations

$$\Phi_x = x_i + \xi_{ij} \cos \theta_i - \eta_{ij} \sin \theta_i - x_j - \xi_{ji} \cos \theta_j + \eta_{ji} \sin \theta_j = 0 \tag{2.4}$$

$$\Phi_y = y_i + \xi_{ij} \sin \theta_i + \eta_{ij} \cos \theta_i - y_j - \xi_{ji} \sin \theta_j - \eta_{ji} \cos \theta_j = 0 \tag{2.5}$$

In Eqs. 2.4 and 2.5, x_i, y_i, θ_i, x_j, y_j, and θ_j are state variables. The parameters ξ_{ij}, η_{ij}, ξ_{ji}, and η_{ji} are related to the length of the members and hence to the design variables.

Constraint Equations for a Translational Joint: Figure 2.3 shows two bodies that are connected by a translational joint. For this type of joint, points p_{ij} and p_{ji} lie on a line that is parallel to the path of relative motion between the two bodies. These points are located by nonzero body-fixed vectors $\bar{r}^{ij}$ and $\bar{r}^{ji}$ that are perpendicular to the line of relative motion. A scalar equation of constraint can be written by taking the scalar product of $\bar{r}^{ij}$ with $\bar{r}^p$. Since these two vectors are perpendicular, their scalar product must vanish; i.e.,

$$\bar{r}^p \cdot \bar{r}^{ij} = 0 \tag{2.6}$$

Using Eq. 2.2 for $\bar{r}^p$, Eq. 2.6 becomes

$$[(\bar{R}^j + \bar{r}^{ji}) - (\bar{R}^i + \bar{r}^{ij})] \cdot \bar{r}^{ij} = 0 \tag{2.7}$$

Writing $\bar{r}_{ji}$ and $\bar{r}_{ij}$ in the global reference frame, Eq. 2.7 can be written as

$$\Phi_n = (U_i - x_i)(U_i - U_j) + (V_i - y_i)(V_i - V_j) = 0 \tag{2.8}$$

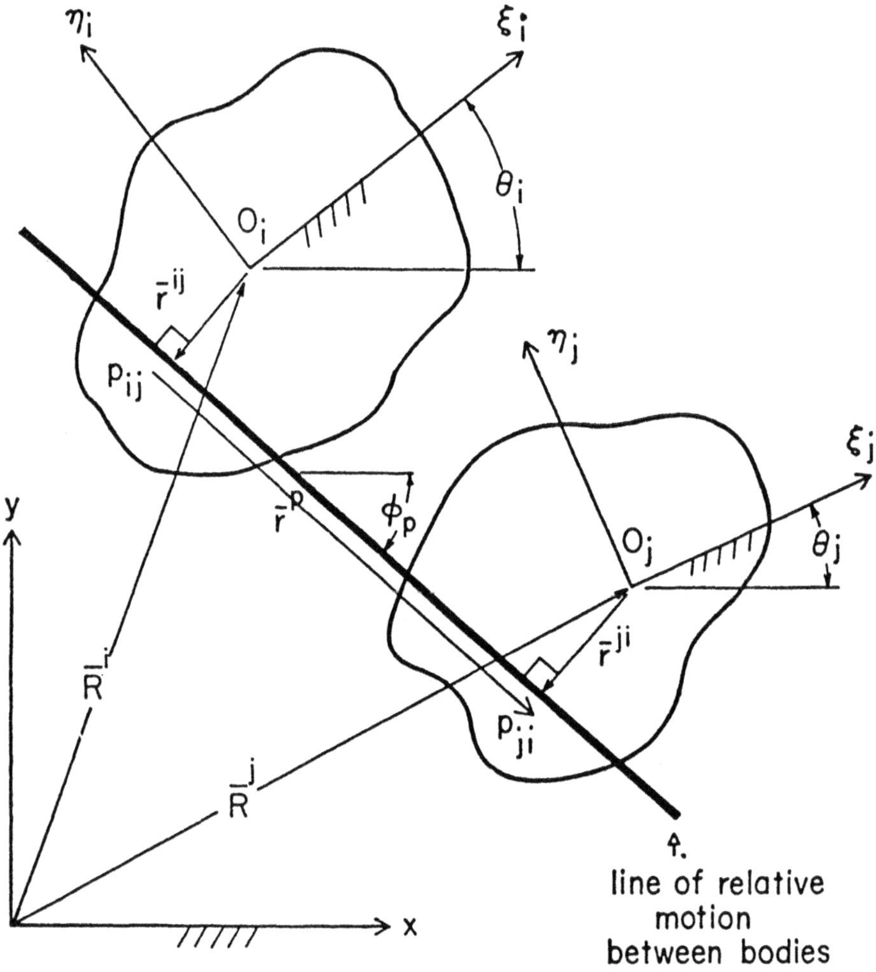

Figure 2.3 Translation Joint

where

$$\left.\begin{aligned}
U_i &= x_i + \xi_{ij} \cos \theta_i - \eta_{ij} \sin \theta_i \\
U_j &= x_j + \xi_{ji} \cos \theta_j - \eta_{ji} \sin \theta_j \\
V_i &= y_i + \xi_{ij} \sin \theta_i + \eta_{ij} \cos \theta_i \\
V_j &= y_j + \xi_{ji} \sin \theta_j + \eta_{ji} \cos \theta_j
\end{aligned}\right\} \tag{2.9}$$

The second scalar equation of constraint can be obtained by noting that $\bar{r}^{ij}$ and $\bar{r}^{ji}$ must be parallel, since both of these vectors are perpendicular to $\bar{r}^{p}$. In three dimensions, this condition can be expressed as

$$\bar{r}^{ij} \times \bar{r}^{ji} = \bar{0} \tag{2.10}$$

Expanding this equation, using the notation of Eq. 2.9, the component perpendicular to the x-y plane is

$$\Phi_\theta = (U_i - x_i)(V_j - y_j) - (V_i - y_i)(U_j - x_j) = 0 \tag{2.11}$$

Equations 2.8 and 2.11 prescribe a translational constraint between bodies i and j. Note that coefficients in these equations may depend on design variables.

<u>System Kinematic Equations of Constraint:</u> Consider a general system of n bodies that are connected by a total of ℓ independent (revolute and translational) joints, each giving rise to two equations of constraint. A system of n bodies in the plane has a total of 3n generalized coordinates. If this system has ℓ independent joints, there are 2ℓ equations of constraint between the 3n generalized coordinates. Thus, the number of free-degrees-of-freedom can be written as

$$m = 3n - 2\ell \tag{2.12}$$

n = number of bodies in the system

ℓ = number of independent joints in the system

The condition $m > 0$ must be satisfied by kinematic systems. The generalized coordinates for the entire system can be denoted by the position state vector $q \in R^{3n}$, defined as

$$q = \begin{bmatrix} q^{(1)} \\ q^{(2)} \\ \vdots \\ q^{(n)} \end{bmatrix} \tag{2.13}$$

Assuming that the system has a set of design variables $b \in R^{s}$, the kinematic equations of constraint can be written as

$$\Phi^{k}(q,b) = 0 \tag{2.14}$$

where

$$\Phi^k(q,b) = \begin{bmatrix} \Phi_1^k(q,b) \\ \Phi_2^k(q,b) \\ \vdots \\ \Phi_{2\ell}^k(q,b) \end{bmatrix}$$

where $\Phi_{2i-1}^k(q,b)$ is the first kinematic constraint and $\Phi_{2i}^k(q,b)$ is the second kinematic constraint due to joint i.

Kinematic Driving Equations: Equation 2.14 comprise a system of 2ℓ equations in 3n variables. For a kinematic system with $3n > 2\ell$, Eq. 2.14 is a system of fewer equations than unknowns. To solve for q from Eq. 2.14, m additional equations are required. These equations can be developed by observing that the purpose of a kinematic system is to transmit motion from input links to output links. The mechanism or machine can only be given input motion through a set of free degrees-of-freedom. These free degrees-of-freedom can be specified as functions of some free parameter, or they may be specified by some relationship between the 3n state variables. Since all the free degrees-of-freedom must be specified to drive the mechanism in a unique way, m additional driving equations of constraint arise, in the form

$$\Phi^d(q,b,\alpha) \equiv g(q,b) - h(b,\alpha) = 0 \tag{2.15}$$

where

$$\Phi^d(q,b,\alpha) = \begin{bmatrix} \Phi_1^d(q,b,\alpha) \\ \vdots \\ \Phi_m^d(q,b,\alpha) \end{bmatrix}$$

$\alpha \in R^p$ is a vector of input parameters, and $\Phi_i^d(q,b,\alpha)$ represents the ith driving constraint equation. The notation of Eq. 2.15 is selected to emphasize that no products of q and α appear.

State Equation for Position: Combining Eqs. 2.14 and 2.15, one has 3n independent equations, which may be written as

$$\Phi\,(q,b,\alpha) = \begin{bmatrix} \Phi^k(q,b) \\ \Phi^d(q,b,\alpha) \end{bmatrix} = 0 \tag{2.16}$$

Equation 2.16 is the state equation for position of the mechanism. Specifying the design variable vector b and the input parameter vector α makes Eq. 2.16 a system of 3n independent equations in 3n unknowns, $q \in R^{3n}$. Since these equations are highly nonlinear, more than one solution for q is possible. Conversely, since the equations are nonlinear, for some designs and inputs no solution may exist.

Solution Technique for State Equations: Constraint equations for the two typical joints considered here, Eqs. 2.4, 2.5, 2.8, and 2.11, are nonlinear. The position state equation is thus nonlinear and a solution technique that is applicable to nonlinear equations must be employed. One of the commonly used techniques for solution of nonlinear equations is the Newton-Raphson method [21].

Consider the position state equation of Eq. 2.16 for the entire system. Before any attempt is made to solve this nonlinear system, variables b and α must be specified. This is reasonable, since in most iterative design algorithms the design variable b is estimated before the synthesis procedure is initiated. The vector of input parameters α is also a part of the problem specification. The only unknowns in Eq. 2.16 for kinematic analysis are the state variables q. Equation 2.16 is thus a system of 3n nonlinear equations in 3n variables and a unique solution of this system will exist locally, if the hypotheses of the Implicit Function Theorem are satisfied; i.e., if the constraint Jacobian matrix

$$\frac{\partial \Phi(q,b,\alpha)}{\partial q} = \Phi_q \equiv \left[\frac{\partial \Phi_i}{\partial q_j}\right]_{3n\times 3n} \tag{2.17}$$

is nonsingular. This condition is satisfied if the system of constraints has no redundant joints.

The Newton-Raphson method [21] requires that the state variable q be initially estimated. The method then computes updates Δq to this state, to obtain an improved value for the state. The improved approximation is given by [21]

$$q^{(i+1)} = q^{(i)} + \Delta q^{(i)} \tag{2.18}$$

where $i \geq 0$ is an iteration counter, and $\Delta q^{(i)}$ is the solution of

$$\Phi_q(q^i,b,\alpha)\ \Delta q^i = -\ \Phi(q^i,b,\alpha) \tag{2.19}$$

For large scale mechanisms, the system of Eq. 2.16 can be quite large, giving rise to a large system of linear equations in Eq. 2.19. Examining the kinematic constraint equations for the two types of joints, (Eqs. 2.4, 2.5, 2.8, and 2.11), it is noted that these pairs of constraint equations involve only the state variables of the two bodies that they connect. These equations are thus weakly coupled and the Jacobian matrix on the left side of Eq. 2.19 is sparse. Efficient sparse matrix codes [22] can thus be used for solution of Eq. 2.19.

Repeated solution of systems of equations of the form of Eq. 2.19 are often required in the following sections. To perform these computations efficiently, the sparse matrix code initially does a symbolic LU factorization of the coefficient matrix. Subsequent solutions of linear systems with the same coefficient matrix, but with different right sides, can be carried out very efficiently.

Since mechanisms are to be synthesized to perform over a range of input parameters α, solution of the state equations would be required at many values of α. The process of obtaining the solution of the position state equation for a specified value of α can be repeated to obtain the solution for a desired sequence of input variables α^j. The numerical efficiency of such a sequence of calculations is good if α^{j+1} is close to α^j, since $q(\alpha^j)$ serves as a good starting estimate for computation of $q(\alpha^{j+1})$. If, however, α^{j+1} and α^j are not close, then an update $\delta q(\alpha^j)$ to $q(\alpha^j)$ is required to produce a reasonable estimate for this computation. One such update can be obtained by linearizing the position state equation of Eq. 2.16, keeping b fixed; i.e.,

$$\Phi_q\,\delta q(\alpha^j) = -\ \Phi_\alpha\,\delta\alpha \tag{2.20}$$

where $\delta\alpha = \alpha^{j+1} - \alpha^j$.

An improved estimate for $q(\alpha^{j+1})$ can thus written as

$$q(\alpha^{j+1}) \approx q(\alpha^j) + \delta q(\alpha^j) \tag{2.21}$$

where $\delta q(\alpha^j)$ is the solution of Eq. 2.20. Note the Eq. 2.20 has the same coefficient matrix as Eq. 2.19, so its numerical solution is quite efficient.

Velocity Analysis

Most mechanisms are driven by input sources that give input links of the mechanism finite velocity. It is then necessary to determine velocities of the remaining links in the mechanism. Since the state equation for the mechanism, of Eq. 2.16 is required to hold for all time, it can be differentiated with respect to time to obtain

$$\frac{d}{dt}\Phi(q,b,\alpha) = \Phi_q\dot{q} + \Phi_\alpha\dot{\alpha} = 0 \tag{2.22}$$

The above equation can be rewritten as

$$\Phi_q\dot{q} = -\Phi_\alpha\dot{\alpha} = \begin{bmatrix} \bar{0} \\ \Phi^d_\alpha \end{bmatrix}\dot{\alpha} \tag{2.23}$$

where $\dot{q} \in R^{3n}$ is the vector of generalized velocities of members of the system and the form of Φ_α follows since Φ^k does not depend on α.

Equation 2.23 is the velocity state equation. This equation is linear in velocities and has the same coefficient matrix as Eq. 2.19. As noted above, this matrix is sparse and its symbolically factorized LU form has been determined and stored. The solution of Eq. 2.23 is thus the same as solving Eq. 2.19, with a different right side, which is very efficient.

Acceleration Analysis

Whenever a velocity input is supplied to a mechanism, some links experience acceleration. Computation of accelerations is important, since forces acting on links in the mechanism depend on acceleration.

Since the velocity state equation of Eq. 2.23 is required to hold over the entire range of inputs, Eq. 2.23 can be differentiated once again with respect to time, using Eq. 2.15, and defining

$$\Omega = \begin{bmatrix} \bar{0} \\ (\Phi^d_\alpha\dot{\alpha})_\alpha\,\dot{\alpha} + \Phi^d_\alpha\ddot{\alpha} \end{bmatrix} \tag{2.24}$$

to obtain

$$\Phi_q \ddot{q} = - [\Phi_q \dot{q}]_q \dot{q} - \Omega \tag{2.25}$$

where $\ddot{\alpha} \in R^p$ is the vector of second time derivatives of input parameters and $\ddot{q} \in R^{3n}$ is the vector of generalized accelerations.

Equation 2.25 is the acceleration state equation. This is a system of linear equations with the same coefficient matrix as Eq. 2.19. All the desirable properties of this coefficient matrix still hold, so the solution of Eq. 2.25 is efficient. Since the right side of Eq. 2.25 involves $\dot{q}$, the velocity state equation of Eq. 2.23 must be solved before Eq. 2.25 can be solved.

3. FORCE ANALYSIS

For mechanical design of mechanisms, it is necessary to impose stress constraints in links and force constraints in joint bearings. This requires that a force equation be derived to determine internal forces on links, in terms of the externally applied forces and system velocities and accelerations. The applied forces could be forces due to gravity, spring-damper-actuator forces, or forces from other external sources.

Generalized Force

Figure 3.1 shows body i, with a body-fixed coordinate system $0_i x_i y_i$. Externally applied forces $\overline{F}^{ik}$ and external moments $\overline{T}^{i\ell}$ act on this body. The point of application of force $\overline{F}^{ik}$ is located by the vector $\overline{S}^{ik}$. The virtual work of all external forces that act on body i can be written as [23]

$$\delta W_i = \sum_{k=1}^{N_i} \overline{F}^{ik} \cdot \delta(\overline{R}^i + \overline{S}^{ik}) + \sum_{\ell=1}^{M_i} \overline{T}^{i\ell} \cdot \delta\overline{\theta}^i \tag{3.1}$$

where

N_i = total number of forces acting on body i

M_i = total number of moments acting on body i

Since $\overline{R}^i$, $\overline{S}^{ik}$, and $\overline{\theta}^i$ are functions of the position state variables z, Eq. 3.1 can be written as

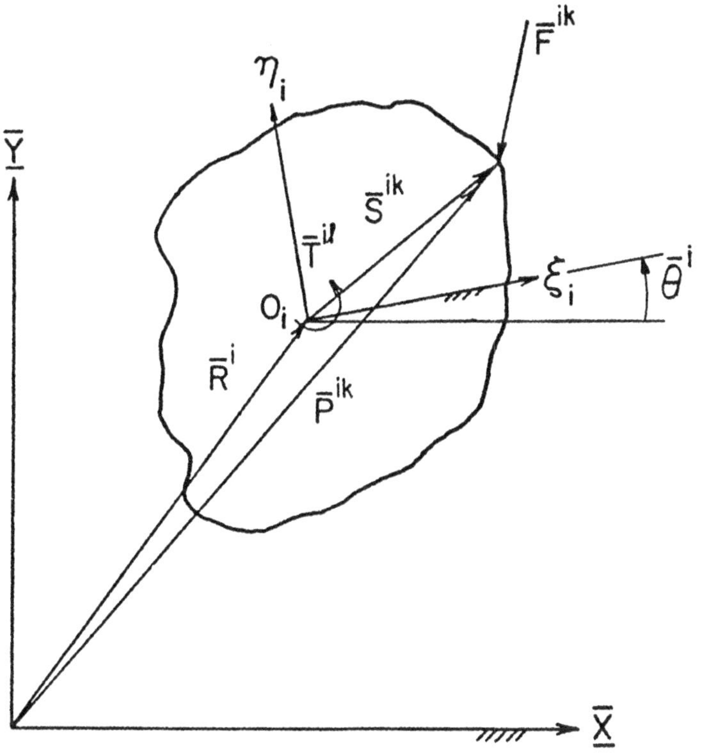

Figure 3.1 Force Acting on Body i

$$\delta W_i = \sum_{k=1}^{N_i} \bar{F}^{ik} \cdot (\bar{R}_z^i + \bar{S}_z^{ik})\delta q + \sum_{\ell=1}^{M_i} \bar{T}^{i\ell} \cdot \bar{\theta}_z^i \,\delta q \tag{3.2}$$

The virtual work of a system of n bodies can be written as the sum over all bodies, defining the system generalized force as

$$\delta W = \sum_{i=1}^{n} \delta W_i \equiv Q^T \delta q \tag{3.3}$$

Force Equations From Lagrange's Equations of Motion

Lagrange's equations of motion for a dynamic system can be applied as force equations for kinematic systems. Consider Lagrange's equations for a constrained mechanical system [23],

$$\frac{d}{dt}(T_{\dot{q}}) - T_q + \Phi_q^T \lambda = Q \tag{3.4}$$

where T is kinetic energy of the system and λ is a 3n - vector of time dependent Lagrange multipliers. In this form, Lagrange's equations are a system of 3n second order differential equations in 3n components of q and 3n components of λ.

Denote m_k as the mass of body i for k = 3i - 2 and 3i - 1, i = 1, ..., n, and m_k as the moment of inertia of body i for k = 3i, i = 1, ..., n. Then, the kinetic energy of the system may be written as

$$T = \frac{1}{2} \sum_{k=1}^{3n} m_k \dot{q}_k^2 \equiv \frac{1}{2} \dot{q}^T M \dot{q} \tag{3.5}$$

where

$$M = \mathrm{diag}(m_1, \ldots, m_{3n}) \tag{3.6}$$

Noting that $T_q = 0$, Eq. 3.4 can be rewritten as

$$\Phi_q^T \lambda = Q(q,\dot{q},t) - M\ddot{q} \tag{3.7}$$

For kinematically driven systems, Eqs. 2.16, 2.23, and 2.25 may be considered to have been solved for q(t), $\dot{q}(t)$, and $\ddot{q}(t)$ at any time t. Since the coefficient matrix of λ in Eq. 3.7 is the transpose of the nonsingular constraint Jacobian, Eq. 3.7 uniquely determines the Lagrange multiplier $\lambda(t)$ as a function of time. Since for given design and input history $\alpha(t)$, the kinematic and dynamic equations determine the Lagrange multiplier λ, it plays the role of a state variable in force analysis of a kinematically driven system.

System State Equations

Before proceeding, it is of interest to summarize the state equations as a single composite system. From Eqs. 2.16, 2.23, 2.25, and 3.7, one has

$$\left.\begin{aligned} &\Phi(q,b,\alpha) = 0 \\ &\Phi_q(q,b,\alpha)\dot{q} = -\Phi_\alpha(q,b,\alpha)\dot{\alpha} \\ &\Phi_q(q,b,\alpha)\ddot{q} = -[\Phi_q(q,b,\alpha)\dot{q}]_q\dot{q} - \Omega(q,b,\alpha,\dot{\alpha},\ddot{\alpha}) \\ &\Phi_q^T(q,b,\alpha)\lambda = Q(q,\dot{q},t) - M\ddot{q} \end{aligned}\right\} \tag{3.8}$$

Presume that the Jacobian matrix Φ_q is nonsingular and continuously differentiable in a neighborhood of a position q^0 that satisfies the first equation. Then in a neighborhood of q^0, the Implicit Function Theorem of advanced calculus guarantees existence of a unique solution $q = q(b,\alpha)$ that is continuously differentiable with respect to its arguments. Similarly, the second, third, and fourth equations of Eq. 3.8 have unique solutions for $\dot{q} = \dot{q}(b,\alpha,\dot{\alpha})$, $\ddot{q} = \ddot{q}(b,\alpha,\dot{\alpha},\ddot{\alpha})$, and $\lambda = \lambda(b,\alpha,\dot{\alpha},\ddot{\alpha})$, respectively, that are continuously differentiable with respect to their arguments. In particular, the state of the kinematic system is continuously differentiable with respect to design.

Physical Interpretation of Lagrange Multipliers

One might suspect from Eq. 3.7 that the term $\Phi_q^T\lambda$ would play the role of a generalized force that is associated with the kinematic constraint equations. This interpretation is verified in Ref. 27. For the constraint formulation used here, the Lagrange multipliers λ_x and λ_y, corresponding to Eqs. 2.4 and 2.5 for a revolute joint, are the x and y components of the reaction force in the revolute joint. Similarly, the torque reaction in a translational joint is given by

$$T_\theta = -\left\{ (\xi_{ij} \sin\theta_i + \eta_{ij} \cos\theta_i)(\xi_{ji} \sin\theta_j + \eta_{ji} \cos\theta_j) + (\xi_{ij} \cos\theta_i - \eta_{ij} \sin\theta_i)(\xi_{ji} \cos\theta_j - \eta_{ji} \sin\theta_j)\right\}\lambda_\theta \tag{3.9}$$

where λ_θ is the Lagrange multiplier corresponding to the constraint equation of Eq. 2.11. Finally, the normal reaction force in the translational joint is

$$R_n = |\bar{r}^{ij}|\ \lambda_n \tag{3.10}$$

where $\bar{r}^{ij}$ is the vector defined in Fig. 2.3 and λ_n is the Lagrange multiplier corresponding to the constraint equation of Eq. 2.8.

An important aspect of the formulation used here is that the reaction forces in the constraints are prescribed by the state of the system. Thus, if one wishes to design a system with bounds on joint reaction forces or associated stresses, these quantities are explicitly available.

4. STATEMENT OF THE OPTIMAL DESIGN PROBLEM

Sections 2 and 3 provide the theory necessary to compute the kinematics of a mechanism and forces acting on links of the mechanism. It should, therefore, be possible to put design constraints (bounds) directly on these variables, or on functions of these variables. It should also be possible to extremize a function of the state variables, subject to design constraints.

Continuous Optimization Problem

A general class of optimal design problems can be stated as follows:

Find a design $b \in R^s$ to minimize the cost function

$$\psi_0 \equiv \psi_0(b) \tag{4.1}$$

subject to State Equations of Eqs. 3.8 and the following design constraints:

(B.1) Inequality Constraints,

$$\psi_i(q,\dot{q},\ddot{q},\lambda,b) \leq 0 \ , \qquad i = 1,\ldots,p \quad , \quad \alpha \in A \tag{4.2}$$

(B.2) Equality Constraints,

$$\psi_i(q,\dot{q},\ddot{q},\lambda,b) = 0, \qquad i = p + 1,\ldots,p + q \quad , \quad \alpha \in A \tag{4.3}$$

where the composite design constraints of Eq. 4.2 and 4.3 are required to hold over the entire range A of the input parameter; i.e., for all $\alpha \in A$. Such constraints are called parametric constraints. Techniques for extremizing cost functions subject to parametric constraints are presented in Refs. 20 and 25. Since these techniques require considerable computation, a simpler approximate technique is used here. The range A of input parameters α is discretized into a finite set of grid points α^j, $j = 1,\ldots,\tau$. The composite design constraints are then required to hold at every point on this grid.

Essentially any type of design constraint can be treated in this formulation. Representation of the cost function in the form of Eq. 4.1 does not restrict the technique from being applied to cost functions that involve state variables. An upper bound technique [20] that may be used in such cases is now illustrated for a general function $\hat{\psi}_0(q,\dot{q},\ddot{q},\lambda,b)$, the maximum value of which is to be minimized, over a specified range of input parameters α. Thus,

$$\min_b \hat{\psi}_0 \equiv \min_b \{\max_{\alpha \in A} \hat{\psi}_0(q,\dot{q},\ddot{q},\lambda,b)\} \tag{4.4}$$

The above formulation of the cost function is natural for kinematic optimization. Since state variables take on a range of values over the range of input parameter, it is natural to minimize the maximum value of functions of these variables.

Equation 4.4 represents a min-max problem [20] and is not simple to deal with directly. One scheme that may be employed for such problems is to introduce an artificial design variable b_{s+1} as an upper bound on $\hat{\psi}_0(q,\dot{q},\ddot{q},\lambda,b)$. Then, the minimization problem in Eq. 4.4 can be written as

$$\min \psi_0 \equiv b_{s+1} \tag{4.5}$$

subject to the additional constraints

$$\text{(i)} \quad \psi_1(q,\dot{q},\ddot{q},\lambda,b) \equiv \{\max_{\alpha \in A} \hat{\psi}_0(q,\dot{q},\ddot{q},\lambda,b)\} - b_{s+1} \leq 0 \tag{4.6}$$

(ii) State Equations and other design constraints.

The minimization problem, as stated in Eqs. 4.5 and 4.6, amounts to generating a minimizing sequence of upper bounds of the function $\hat{\psi}_0$.

Discretized Optimal Design Problem

The optimal design problem, with a grid imposed on the range of the input parameters, can be stated as follows:

Find a design $b \in R^s$ to minimize the cost function

$$\psi_0 \equiv \psi_0(b) \tag{4.7}$$

subject to the following state equations:

(A.1) Position state equations of Eq. 2.16

$$\Phi(q^j, b, \alpha^j) = 0 \quad , \qquad j = 1,\ldots,\tau \tag{4.8}$$

where τ is the number of grid points on the range of the input parameters,

(A.2) Velocity state equation of Eq. 2.23,

$$\Phi_q(q^j, b, \alpha^j)\dot{q}^j = -\Phi_\alpha(q^j, b, \alpha^j)\dot{\alpha}^j \quad , \qquad j = 1,\ldots,\tau \tag{4.9}$$

(A.3) Acceleration state equation of Eq. 2.25,

$$\Phi_q(q^j, b, \alpha^j)\ddot{q}^j = -\left[\Phi_q(q^j, b, \alpha^j)\dot{q}^j\right]_q \dot{q}^j - \Omega^j \quad , \qquad j = 1,\ldots,\tau \tag{4.10}$$

(A.4) Force Equation of Eq. 3.8,

$$\Phi_q^T(q^j, b, \alpha^j)\lambda^j = G^j \quad , \qquad j = 1,\ldots,\tau \tag{4.11}$$

and the following composite design constraints:

(B.1) Inequality design constraints

$$\psi_i(q^j, \dot{q}^j, \ddot{q}^j, \lambda^j, b) \leq 0, \; i = 1,\ldots,p, \qquad j = 1,\ldots,\tau \tag{4.12}$$

(B.2) Equality design constraints

$$\psi_i(q^j, \dot{q}^j, \ddot{q}^j, \lambda^j, b) = 0, \quad i = p+1,\ldots,p', \quad j = 1,\ldots,\tau \tag{4.13}$$

5. DESIGN SENSITIVITY ANALYSIS

Most optimization algorithms require that derivatives of the cost and constraint functions with respect to design variables be provided. Computing derivatives of functions that involve only design variables is easy. However, for functions involving state variables, dependence on design arises indirectly through the state equations.

Recall that the state variable is continuously differentiable with respect to design. Thus, the chain rule of differentiation can be used to obtain the design derivative of a typical constraint function at $\alpha = \alpha^j$ as

$$\psi_b^j = \left\{ \psi_q q_b + \psi_{\dot{q}} \dot{q}_b + \psi_{\ddot{q}} \ddot{q}_b + \psi_\lambda \lambda_b + \psi_b \right\}^j \quad , \quad j = 1, \ldots, \tau \qquad (5.1)$$

In vector form, Eq. 5.1 can be rewritten as

$$\psi_b^j = \left\{ \psi_q, \psi_{\dot{q}}, \psi_{\ddot{q}}, \psi_\lambda \right\}_j \begin{bmatrix} q_b \\ \dot{q}_b \\ \ddot{q}_b \\ \lambda_b \end{bmatrix}_j + (\psi_b)^j \ , \ j = 1, \ldots, \tau \qquad (5.2)$$

Since state variables are functions of design, it is required that derivative of the state variables that appear in Eq. 5.2 be written in terms of computable derivatives with respect to design. The objective is then to write Eq. 5.2 as

$$\psi_b^j \equiv \ell^{j\,T} \qquad (5.3)$$

where ℓ^j is the design sensitivity vector of the constraint at the jth grid point α^j.

Observing that the state equations couple the design and state variables, the first derivative of the four state equations of Eq. 3.8 can be written in matrix form as

$$
\begin{bmatrix}
\Phi_q & 0 & 0 & 0 \\
[\Phi_q \dot{q}]_q + [\Phi_\alpha \dot{\alpha}]_q & \Phi_q & 0 & 0 \\
[\Phi_q \ddot{q}]_q + \Omega_q + [[\Phi_q \dot{q}]_q \dot{q}]_q & [[\Phi_q \dot{q}]_q \dot{q}]_{q_z} & \Phi_{q_z} & 0 \\
[\Phi_q^T \lambda]_q - G_q & -G_{\dot{q}} & -G_{q_z} & \Phi_q^T
\end{bmatrix}^j
\begin{bmatrix}
q_b \\ \dot{q}_b \\ \ddot{q}_b \\ \lambda_b
\end{bmatrix}^j
=
\begin{bmatrix}
-\Phi_b \\
-[\Phi_q \dot{q}]_b - [\Phi_\alpha \dot{\alpha}]_b \\
-[[\Phi_q \dot{q}]_q \dot{q}]_b - [\Phi_q \ddot{q}]_b - \Omega_b \\
-[\Phi_q^T \lambda]_b + G_b
\end{bmatrix}^j
\tag{5.4}
$$

Equations 5.4 can be symbolically written in the form

$$A(q^j, \dot{q}^j, \ddot{q}^j, \lambda^j, b)U_b^j = B(q^j, \dot{q}^j, \ddot{q}^j, \lambda^j, b) \tag{5.5}$$

where A represents the matrix on the left side of Eq. 5.4, B represents the matrix on the right, and

$$U^j = [q^T, \dot{q}^T, \ddot{q}^T, \lambda^T]_j^T$$

The design derivative of the composite vector of state variables can be determined from Eq. 5.5 as

$$U_b^j = (A^j)^{-1} B^j \tag{5.6}$$

where $A^j \equiv A(q^j, \dot{q}^j, \ddot{q}^j, \lambda^j, b)$ and $B^j \equiv B(q^j, \dot{q}^j, \ddot{q}^j, \lambda^j, b)$. In deriving Eq. 5.6 from Eq. 5.5, it is required that the matrix A^j be invertible. The existence of an inverse is proved by showing that

$AU = 0$ implies $U = 0$. This follows, with some manipulation, from the fact that Φ_q is nonsingular.

Substituting U_b^j from Eq. 5.6 into Eq. 5.2 gives

$$\psi_b^j = [\psi_q, \ \psi_{\dot q}, \ \psi_{\ddot q}, \ \psi_\lambda]^i \ (A^j)^{-1} B^j + (\psi_b)^j \tag{5.7}$$

To avoid calculation of $(A^j)^{-1}$, the product of the row vector of constraint derivatives and $(A^j)^{-1}$ is denoted by a composite adjoint vector p^{j^T}; i.e.,

$$p^{j^T} \equiv [\psi_q, \ \psi_{\dot q}, \ \psi_{\ddot q}, \ \psi_\lambda]^j \ (A^j)^{-1} \tag{5.8}$$

Equation 5.8 can be rewritten as

$$(A^j)^T \ p^j = [\psi_q, \ \psi_{\dot q}, \ \psi_{\ddot q}, \ \psi_\lambda]^{j^T} \tag{5.9}$$

The composite adjoint vector p^j, defined in Eq. 5.8, is a $12n \times 1$ vector. This vector can be written as a composite vector of four $(3n\times1)$ adjoint vectors μ_1^j to μ_4^j. Equation 5.9 can thus be expanded as

$$\begin{bmatrix} \Phi_q^T & [\Phi_q \dot q]_q^T + [\Phi_\alpha \dot\alpha]_q^T & [\Phi_q \ddot q]_q^T + \Omega_q^T + [[\Phi_q \dot q]_q \dot q]_q^T & [\Phi_q^T \lambda]_q^T - G_q^T \\ 0 & \Phi_q^T & [[\Phi_q \dot q]_q \dot q]_{\dot q}^T & - G_{\dot q}^T \\ 0 & 0 & \Phi_q^T & - G_{\ddot q}^T \\ 0 & 0 & 0 & \Phi_q \end{bmatrix}_j \begin{bmatrix} \mu_1 \\ \mu_2 \\ \mu_3 \\ \mu_4 \end{bmatrix}_j = \begin{bmatrix} \psi_q^T \\ \psi_{\dot q}^T \\ \psi_{\ddot q}^T \\ \psi_\lambda^T \end{bmatrix}_j \tag{5.10}$$

Equation 5.10 is a system of $12n$ linear equations. Rather than solve this as a coupled system, the form of the coefficient matrix makes it possible to solve four separate linear systems of 3n equations each. The last 3n equations of Eq. 5.10 depend on only μ_4^j and have the same coefficient matrix as Eq. 2.19. As noted in Section 2, the solution of linear systems with this coefficient matrix is very efficient, since the LU factored form of the coefficient matrix has already been computed and stored. Equations (6n + 1) to (9n) of Eq. 5.10 involve only the unknown μ_3^j, with coefficient matrix Φ_q^T, since μ_4^j has been determined. Since the coefficient matrix of μ_3^j is the transpose of the coefficient matrix of Eq. 2.19, the remarks made above about solution efficiency are again valid. Continuing this process of backward solution, equations (3n+1) to (6n) of Eq. 5.10 can now be efficiently solved for μ_2^j, since μ_3^j and μ_4^j are known respectively. Finally the first 3n equations of Eq. 5.10 can be efficiently solved for μ_1^j, since μ_2^j, μ_3^j, and μ_4^j are known.

Now that the solution of Eq. 5.9 (equivalently Eq. 5.10) for the composite adjoint vector p^{j^T} is known, Eq. 5.8 can be substituted into Eq. 5.7 to obtain

$$\psi_b^j = \left[p^{j^T} B^j + (\psi_b)^j \right] \tag{5.11}$$

Equation 5.11 expresses the derivative of a composite design constraint with respect to design. Comparing Eqs. 5.3 and 5.11, it can be concluded that the quantity premultiplying δb in Eq. 5.11 is the design sensitivity vector of the constraint function at the jth grid point; i.e.,

$$\ell^{j^T} = p^{j^T} B^j + (\psi_b)^j$$

Substituting for B^j from Eq. 5.4, this can be written explicitly as

$$\begin{aligned} \ell^{j^T} = - \Big\{ & \mu_1^T(\Phi_b)^j + \mu_2^{j^T} [(\Phi_q\dot{q})_b + (\Phi_\alpha\dot{\alpha})_b]^j \\ & + \mu_3^{j^T}\Big[[(\Phi_q\dot{q})_b\dot{q}]_b + (\Phi_q\ddot{q})_b + \Omega_b\Big]^j \\ & + \mu_4^{j^T} [(\Phi_q^T\lambda)_b - G_b]^j - (\psi_b)^j \Big\} \end{aligned} \tag{5.12}$$

Equation 5.12 gives the design sensitivity of a composite design constraint at grid point j, in terms of derivatives of the position state equation and the adjoint variables.

6. DESIGN OPTIMIZATION ALGORITHMS

Active Set Strategy

With the design sensitivity information computed in the previous section, one can proceed to implement the optimization algorithm of his choice. A gradient projection algorithm with constraint error correction [20] has been used in the past in related applications [27]. This rather crude algorithm is used here for design optimization. A more modern and powerful sequential quadratic programming algorithm [26] has been used to solve some of these problems with greater efficiency.

Some of the composite design constraints of Eqs. 4.12 and 4.13 can be put in a simpler form than is indicated in these equations. Constraints that do not involve the state variables do not explicitly or implicitly depend on the input variables α. Such constraints are called "non-parametric" constraints. Design constraints that involve the state variables are called "parametric" constraints. It is necessary to make this distinction, since only the parametric constraints are required to be satisfied over the entire range of input variables α. For a given design, the non-parametric constraints need only be evaluated once. However, parametric constraints must be evaluated at all points on the grid of input variables.

An active set strategy may be adopted to determine a reduced set $\tilde{\psi}$ of active constraints. Since equality constraints, parametric or non-parametric, are always active, they are always included in the active set $\tilde{\psi}$. Parametric inequality constraints, due to their dependence on the input variable α, must be evaluated at all points on the grid of the input variable α. Some computational efficiency can be realized from the fact that the gradient projection algorithm, as stated in the following section, allows only small changes in design, hence leading to only small changes in state. A design constraint with a large violation at a given design iteration may not be fully satisfied at the subsequent design iteration. This is because the optimization algorithm uses only first-order information about the design constraints and design constraints are generally nonlinear. The regions of the input variable grid in which a parametric inequality constraint is active are also not expected to change rapidly from design iteration to iteration. It is thus possible to avoid evaluating, for a few design iterations, a parametric constraint in regions in which it is not ε-active.

Design iterations in which the parametric constraint is evaluated at each point on the α-grid are defined as "sweep" iterations. Iterations in which the parametric constraint is evaluated only in the active region are called "non-sweep" iterations. The interval between two sweep iterations depends on how rapidly the active regions are changing. For non-sweep design iterations, it is not necessary to solve the state equations on the entire range of the input variable. Considerable computational saving can be realized by having a large number of non-sweep iterations between successive sweep iterations. Since new active regions can only be detected during sweep iterations, having a large number of non-sweep iterations separating two sweep iterations could lead to new active regions going undetected for a number of design iterations.

Two alternative definitions could be used to define active regions on the grid of the input variable. The first definition, as illustrated in Fig. 6.1, involves determining the ε-active relative maxima of the constraint function on the α-grid. The active region is then defined to be the set of points at which the relative maxima occur and one grid point on either side of these grid points. The active region can thus be defined by the index set I_R,

$$I_R = I_{R_L} \cup I_{R_M} \cup I_{R_R} \tag{6.1}$$

where

I_{R_M} = {j-1, j, j+1 | $\psi^{ij} > -\varepsilon$, $i = 1,\ldots,p$, and j, for $2 < j < \tau-1$, is a relative maximum point}

I_{R_L} = {j, j+1 | $\psi^{ij} > -\varepsilon$, $i = 1,\ldots,p$, and $j = 1$ is a relative maximum point}

I_{R_R} = {j-1, j, | $\psi^{ij} > -\varepsilon$, $i = 1,\ldots,p$, and $j = \tau$ is a relative maximum point}

The second definition of active constraint region involves determining all the points on the α grid at which the constraint

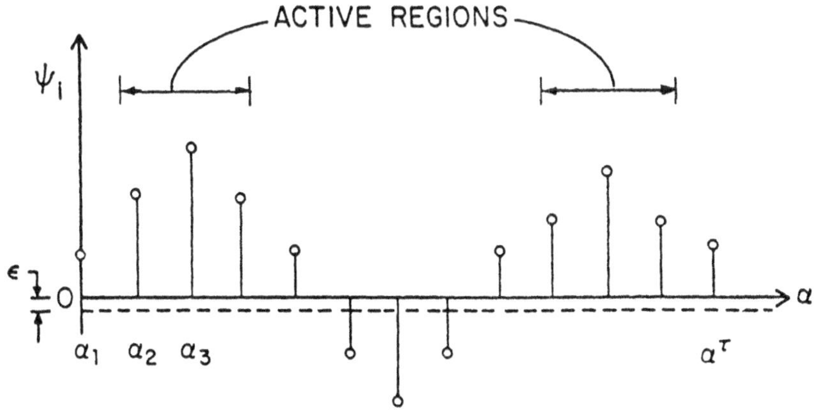

Figure 6.1 Definition of Active Regions on Basis of Relative Maxima of Parametric Constraints

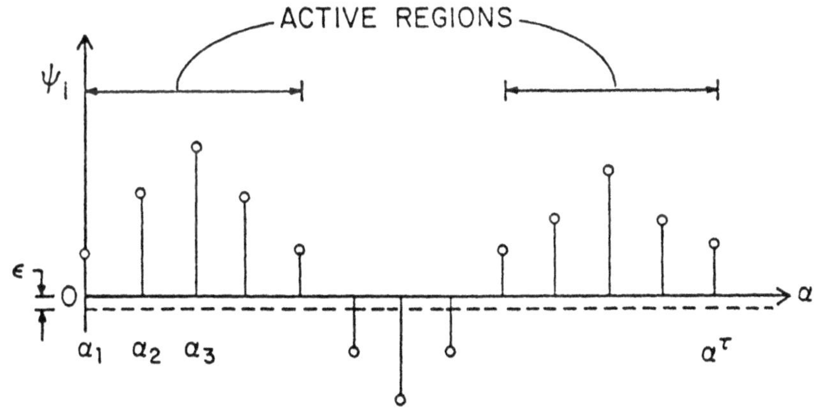

Figure 6.2 Definition of Active Regions on Basis of Epsilon-Active Parametric Constraints

function is ε-active. This set is defined to be the ε-active region. Denoting this set as I_E,

$$I_E = \{j \mid \psi^{ij} > -\varepsilon,\ i = 1,\ldots,p,\ 1 < j < \tau\} \qquad (6.2)$$

Epsilon-Active Parametric Constraints

The relative maximum strategy has been suggested for defining active regions for parametric constraints in Ref. 25. When applied to constraints in mechanism optimization, this strategy many times causes rapid oscillation of the relative maximum point on the α-grid. A switch to the ε-active strategy overcomes this problem. As is evident from Figs. 6.1 and 6.2, there is a penalty to be paid for this switch, since the latter strategy requires a larger number of grid points to be included in the ε-active region.

Gradient Projection Algorithm

The gradient projection algorithm [20] for design optimization can now be stated in the following steps:

Step 1: Estimate a design b^0 and impose a grid on the range of the input parameters.

Step 2: Solve the state equations of Eq. 3.8 for state variables q^j, $\dot{q}^j$, $\ddot{q}^j$, and λ^j, where $j = 1,\ldots,\tau$ if the current iteration is a sweep design iteration or if $j \in I_R$ or I_E. Note, the first design iteration must be a sweep iteration.

Step 3: Determine the active region, depending on the strategy chosen, and form a reduced vector $\tilde{\psi}$ consisting of all ε-active non-parametric inequality and all equality parametric and non-parametric constraints. For inequality parametric constraints, the constraints evaluated in the active regions are included in $\tilde{\psi}$.

Step 4: Compute adjoint variables μ_4^j, μ_3^j, μ_2^j, and μ_1^j from Eq. 5.10 and construct design sensitivity vectors ℓ^j of Eq. 5.12 for the constraint functions in $\tilde{\psi}$. Form the matrix $\tilde{P}=[\ell^j]$, whose columns are the vectors ℓ^j corresponding to constraint functions in $\tilde{\psi}$. Thus, $\tilde{\psi}_b = \tilde{P}^T$.

Step 5: Compute the vector $M_{\psi\psi_0}$ and matrix $M_{\psi\psi}$ from the following relations

$$M_{\psi\psi_0} = \tilde{P}^T W^{-1} \ell^0 \tag{6.3}$$

$$M_{\psi\psi} = \tilde{P}^T W^{-1} \tilde{P} \tag{6.4}$$

Step 6: In the first iteration, compute a parameter γ that is related to step size as

$$\gamma = \frac{\ell^{0^T} W^{-1} \ell^0}{2\beta \psi_0(b^0)} \tag{6.5}$$

where β is the desired fractional reduction in the cost function. The usual range of β is 0.03 to 0.15. In succeeding iterations, the factor γ is adjusted to enhance convergence of the algorithm.

Step 7: Compute $\bar{\mu}^1$ and $\bar{\mu}^2$ from

$$M_{\psi\mu}\bar{\mu}^1 = - M_{\psi\psi 0} \tag{6.6}$$

$$M_{\psi\psi}\bar{\mu}^2 = - \Delta\tilde{\psi} \tag{6.7}$$

where $\Delta\bar{\psi} = C_\psi \bar{\psi}$ and C_ψ is the fraction correction of the constraint desired, usually in the range 0.30 to 1.0.

Step 8: Compute δb^1 and δb^2 from

$$\delta b^1 = W^{-1} \ell^0 + P \bar{\mu}^1 \tag{6.8}$$

$$\delta b^2 = -W^{-1} P \bar{\mu}^2 \tag{6.9}$$

Step 9: Compute an update in design δb from

$$\delta b = - \frac{1}{2\gamma} \delta b^1 + \delta b^2 \tag{6.10}$$

Step 10: Update the estimate of the optimal design using

$$b^1 = b^0 + \delta b \tag{6.11}$$

Step 11: If all constraints are satisfied to within the prescribed tolerance and

$$||\delta b^1|| = \left[\sum_{i=1}^{s} W_i (\delta b_i^1)^2 \right]^{1/2} < \delta \qquad (6.12)$$

terminate the process, where δ is a specified convergence tolerance. If Eq. 6.12 is not satisfied, return to Step 2.

7. NUMERICAL EXAMPLES

An experimental computer program [27], based on the Dynamic Analysis and Design System (DADS) planar code, has been developed to implement the method presented in this paper. To use this code, one need only provide data that define bodies and kinematic elements that make up the system being designed. All derivatives that are required for assembly of the kinematic and dynamic equations and the design sensitivity analysis equations have been precomputed and stored in subroutines. The experimental code then assembles and solves all the equations that are required to analyze the system and calculate design sensitivity. The gradient projection method (or a more refined method [26]) is then used for iterative optimization.

Four numerical examples are presented here to illustrate use of the method. Kinematic synthesis of one and two degree of freedom path generators, stress constrained design, and force balancing design problems are addressed.

Example 1 - Kinematic Synthesis of a Path Generator

Problem Description: A segment of straight line is to be approximated by a point P on the coupler of the 4-bar mechanism shown in Fig. 7.1. In addition to having the lengths of various links as design variables (b_1 to b_3), the orientation of the base link (body 1) is a design variable (b_5). The orientation of the reference line with respect to the base link, about which the input variable α is measured, is also a design variable (b_4). The other two design variables (b_7 and b_8) locate the point P on link 3, as indicated in Fig. 7.1. The length of the base link is kept fixed at 10 units. This problem is the same as Example 3 in Ref. 25.

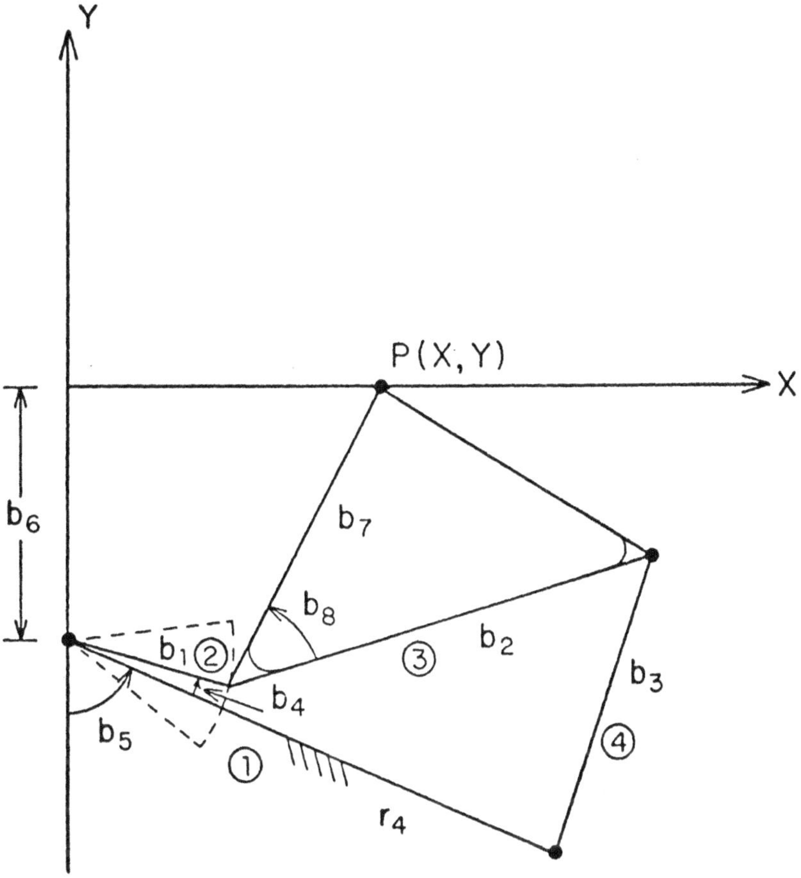

Figure 7.1 Four-Bar Path Generator Mechanism

<u>Problem Formulation</u>: Deviation of the y coordinate of coupler point P from zero in the global coordinate system is to be minimized. The position vector for point P, in terms of design and state variables, can be written as

$$\bar{R}_p = \{x_3 + \cos\theta_3(-\frac{b_2}{2} + b_7 \cos b_8) - b_7 \sin b_8 \sin\theta_3\}\,\bar{I}$$

$$+ \{y_3 + \sin\theta_3(-\frac{b_2}{2} + b_7 \cos b_8) + b_7 \cos\theta_3 \sin b_8\}\,\bar{J} \qquad (7.1)$$

Equation 7.1 can be symbolically expressed in the form

$$\bar{R}_p = (R_{p_x})\,\bar{I} + (R_{p_y})\,\bar{J} \tag{7.2}$$

where R_{p_x} and R_{p_y} represent the x and y coordinates of point p in the global coordinate system.

An artificial design variable b_9 is introduced such that

$$\left| R_{p_y}(\alpha) \right| \leq b_9 \ , \qquad \alpha_{min} \leq \alpha \leq \alpha_{max}$$

where the range of α is given as $\alpha_{min} = -0.3491$ rad. and $\alpha_{max} = 0.3491$ rad. This problem can now be formulated in the standard form of Section 4; i.e., minimize

$$\text{Min } \psi_0 = b_9 \tag{7.3}$$

subject to discretized design constraints

$$\psi_1^j \equiv \left| R_{p_y} \right|^j - b_9 \leq 0, \qquad j = 1, \ldots, \tau \tag{7.4}$$

The constraint that the input link b_1 is a crank is imposed in the form [25]

$$\psi_2 \equiv |r_4 - b_2| - b_3 + b_1 \leq 0 \tag{7.5}$$

$$\psi_3 \equiv b_1 + b_3 - b_2 - r_4 \leq 0 \tag{7.6}$$

Non-negativity constraints on b_7 and b_8 are written in the form

$$\psi_4 \equiv -b_7 \leq 0 \tag{7.7}$$

$$\psi_5 \equiv -b_8 \leq 0 \tag{7.8}$$

The error bound constraint of Eq. 7.4 is imposed over a grid of $\tau = 19$ equally spaced points on the range of α. The driving kinematic constraint for this problem is given as

$$\Phi_1^d \equiv \theta_2 - \{b_5 + b_4 + \alpha\} = 0$$

Verification of Design Sensitivity Analysis: It is necessary to numerically verify accuracy of design sensitivity analysis. The procedure used is briefly explained here.

The problem is set up through user supplied routines and input data and the code is allowed to run for 2 design iterations. On the basis of the design sensitivity vector calculated and the change in design δb obtained in the first iteration, changes in each constraint can be predicted. These predicted values can be compared with the actual changes that are obtained when the constraint functions are evaluated in the second design iteration. If design sensitivity analysis is valid, the predicted and actual changes in constraints should agree within a reasonable tolerance. This procedure is now used to verify accuracy of the design sensitivity analysis for this problem.

Consider grid point 19, where the parametric constraint of Eq. 7.4 is violated during a design iteration. The design sensitivity vector for ψ_1 is obtained as

$$\ell^{1,19^T} = [-0.67893,\ -.04693,\ 0.06083,\ 2.30624,\ 2.85158,\ -1.0,\ 0.99853,\ 0.34923,\ -1.0]$$

The change in design defined by the optimization algorithm at this iteration is $\delta b^T = [0.01663,\ 0.00115,\ -0.00149,\ -0.05650,\ -0.06985,\ 0.02450,\ -0.02446,\ -0.00853,\ -0.2096]$. The predicted change in the constraint value is thus given as

$$\delta\psi^{1,19} \equiv \ell^{1,19^T}\ \delta b = -0.1832$$

The actual change in the constraint function is evaluated by taking the difference in ψ_1 between the modified and original design and is

$$\Delta\psi^{1,19} \equiv \psi_1(q^{19}, b+\delta b, \alpha^{19}) - \psi_1(q^{19}, b, \alpha^{19}) = -0.1709$$

The difference between the predicted and actual change is 6.7%. The design sensitivity analysis is thus considered to be valid.

Optimization Results: Results obtained from the optimization procedure are presented in Table 7.1. None of the non-parametric constraints were active at the optimum. As given in Table 7.1, there are 5 critical points on the grid of the input parameter at the optimum design, where the error in function generation is a relative maximum. The maximum error obtained in Ref. 25 for this problem was 0.0019, which is greater than that obtained by the present technique.

	b_1	b_2	b_3	b_4	b_5	b_6	b_7	b_8	
Initial Design	3.5	9.0	10.0	0.0	-1.0	-2.5	6.5	1.0	1.36
Final Design (38th iter)	3.2463	9.3608	6.8637	-.2455	-0.7762	-2.7825	7.3475	0.6532	0.001175

Critical grid points at Final Design:

-0.3491	-0.2327	0.0	0.2327	0.3491

Error at critical grid points:

0.001092	0.001175	0.001175	0.0011705	0.001096

At the final design $||\delta b^1|| = 5 \times 10^{-3}$

Computing time on IBM 370/168 approximately 0.39 secs/iteration

Table 7.1 Results for Example 1

Example 2 - Two Degree of Freedom Function Generator

Problem Description: The relationship $u = (1 + v)\log_{10}(1 + w)$ is to be mechanized in the input variable region $0 < v < 1$, $0 < w < 1$. This problem is similar to the numerical example given in Ref. 6. The mechanism to be used for function generation is the 7 link mechanism shown in Fig. 7.2. The inputs v and w are the displacements of bodies 5 and 6, respectively. These displacements are measured from reference positions along the global axis, which are taken as design variables b_1 and b_2. The output u of the mechanism is the displacement of body 2, relative to the origin of the global coordinate system. This displacement is measured along the line of translation that makes an angle $\gamma = 27.881°$ with respect to the global x-axis. The lengths of links 3, 4, 6, and 7 are design variables b_4 to b_7. The angle between links 4 and 6 is design variable b_3. The other parameters related to this problem are as indicated in the Fig. 7.2.

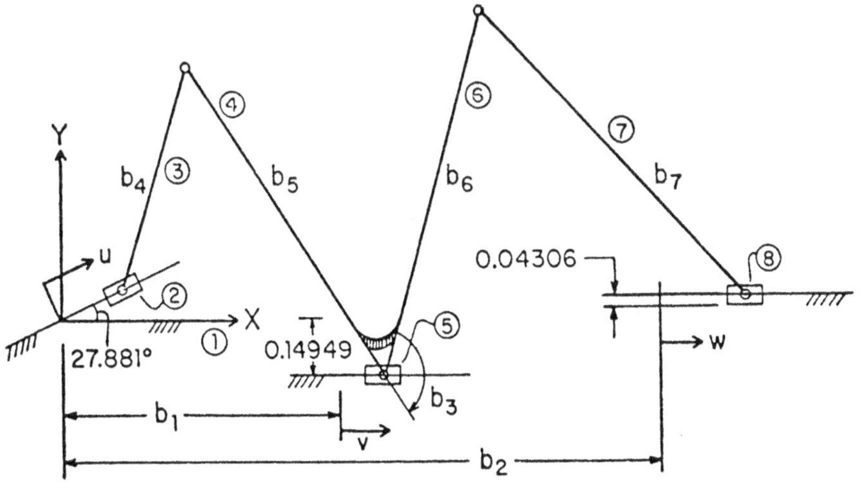

Figure 7.2 Two Degree-of-Freedom Function Generator

Problem Formulation: The displacement of the output slider, body 2, along the line of translation can be written, in terms of the generalized coordinates of body 2, as

$$u = \frac{x_2}{\cos \beta_1} \tag{7.9}$$

where $\beta_1 = 27.881°$.

An artificial design variable b_8 is introduced such that

$$\left| (u_g)^{j,k} - (u_d)^{j,k} \right| \le b_8, \qquad j,k = 1, \ldots, 4$$

where $(u_g)^{j,k}$ represents the generated value of u, for specific values of the input parameters α_1^j and α_2^k, and $(u_d)^{j,k}$ is the desired value of u, for the same values of the input parameters α_1^j and α_2^k. The function to be generated in this example is

$$(u_d)^{j,k} = (1 + \alpha_1^j) \log_{10}(1 + \alpha_2^k)$$

This problem can be formulated in standard form as; minimize

$$\psi_0 \equiv b_8 \tag{7.10}$$

Subject to discretized design constraints

$$\psi^{1,j,k} \equiv \left| (u_g)^{j,k} - (1 + \alpha_1^j) \log_{10}(1 + \alpha_2^k) \right| - b_8 \leq 0,$$

$$j,\ k = 1,\ldots,4 \qquad (7.11)$$

The error bound constraint is imposed on a two dimensional grid of equally spaced points on each of the input variable α_1 and α_2, as shown in Fig. 7.3. A total of 16 grid points are used. No other design constraints are imposed in this problem. The driving kinematic constraints for this problem are given as

$$\Phi_1^d \equiv x_5 - (b_1 + \alpha_1) = 0$$

$$\Phi_2^d \equiv x_8 - (b_2 + \alpha_2) = 0$$

Verification of Design Sensitivity Analysis: The design sensitivity of the upper bound constraint ψ_1 of Eq. 7.11, at design iteration 11 and grid point j = 1, k = 4, was obtained as

$$\ell^{1,1,4^T} = [0.15546,\ 0.23616,\ -0.57531,\ -1.45199,\ 1.29105,\ 0.06958,\ -0.32795,\ -1.0]$$

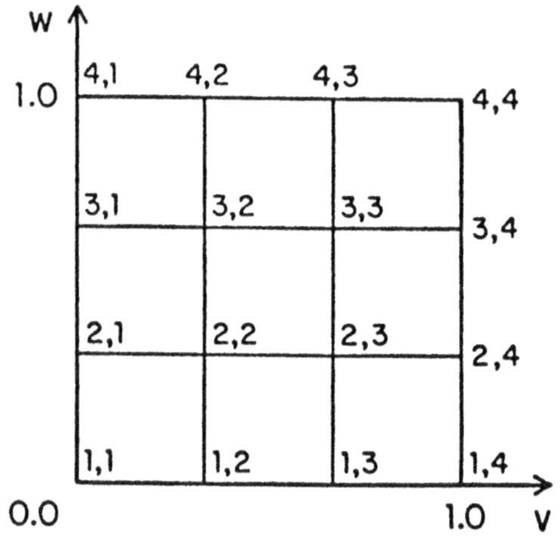

Figure 7.3 Grid Spacing for Example 2

The change δb in design defined by the optimization algorithm was

$$\delta b^T = [-0.01027,\ 0.00197,\ 0.01753,\ 0.00938,\ 0.01017,\ 0.01178,\ -0.00640,\ -0.00070]$$

The predicted change in constraint is thus computed as

$$\delta\psi^{1,1,4} = \ell^{1,1,4^T}\ \delta b = -0.00805$$

The actual change in constraint was obtained as

$$\Delta\psi^{1,1,4} = 0.07259 - 0.08055 = -0.00796$$

The difference between the actual and predicted change in this constraint is 1.0%, which verifies that design sensitivity analysis is valid.

Optimization Results: Results of the optimization procedure for this problem are presented in Table 7.2. The maximum error in function generation that is obtained from the present procedure is 25% higher than that obtained in Ref. 6. This can be attributed to two differences in the formulation. First, the design variables used in the present formulation and in Ref. 6 are not the same. Second, the grid points in the present formulation and in Ref. 6 are not identical. Location of grid points is known to have a significant effect on the error in function generation [14].

	b_1	b_2	b_3	b_4	b_5	b_6	b_7	Max Error
Initial Design	0.750	2.4	-2.792	0.7	0.8	1.5	2.4	0.05
Final Design (30th iter)	0.538	2.45	-2.46	0.942	1.1152	1.687	2.159	0.006

At the final design $||\delta b^1|| = 1 \times 10^{-4}$

Computing time on IBM 370/168 approximately 1.0 Secs/Iteration

Table 7.2 Results for Example 2

Example 3 - Stress Constrained Design of a Four-Bar Mechanism

Problem Description: The four-bar mechanism shown in Fig. 7.4 has its input crank rotating at a constant angular speed of 300 rpm. This mechanism is to be designed for minimum mass, requiring that bending stresses in mobile links 2, 3, and 4 do not exceed an allowable stress. The design variables in the mechanism are the circular cross-sectional areas of the three mobile links. The link lengths are kept fixed at the values specified in Table 7.3. The mass density and allowable stress in the links are as given in Table 7.3. Specifications for this problem are the same as those of Example 1 in Ref. 30 and Example 1 in Ref. 29.

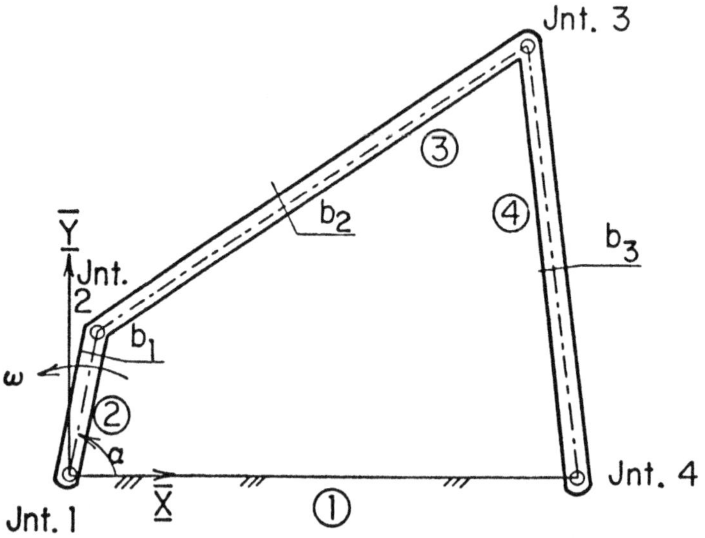

Figure 7.4 Minimum Weight Design of Four-Bar Mechanism

Problem Formulation: As shown in Section 3, the Lagrange multipliers that are obtained as a solution of the force equation are directly related to reaction forces in the joints. As shown in Fig. 7.5, reactions at the ends of a link can be represented by the appropriate Lagrange multipliers. However, to compute bending moments, the end reactions are required to be expressed in a local coordinate system. As shown in Fig. 7.5, the λ_k - λ_{k+1} system of forces at a joint must be transformed to the P_k - P_{k+1} system.

	LINK			
	1	2	3	4
Length (m)	0.9144	0.3048	0.9144	0.762
mass density (kg/m3)	--	2757.25	2757.24	2757.24
Modulus of Elasticity (Pa)	--	6.8948×10^{10}	6.8948×10^{10}	6.8948×10^{10}
Stress upper bound (Pa)	--	2.7579×10^{7}	2.7579×10^{7}	2.7579×10^{7}

Table 7.3 Data for Example 3

The external load on the beam depends on its normal acceleration. To compute the distribution of normal acceleration along the length of the link, a cross section of the link at point c, a distance ν along the x-axis, is considered. The position vector for point c can be written as

$$\bar{R}_{c_i} = x_i \bar{I} + y_i \bar{J} + \nu(\cos \theta_i \bar{I} + \sin \theta \bar{J}) \qquad (7.12)$$

where x_i, y_i, and θ_i are the three generalized coordinates that locate the body in the plane and $\bar{I}$ and $\bar{J}$ are unit vectors in the global $\bar{X}$ and $\bar{Y}$ directions.

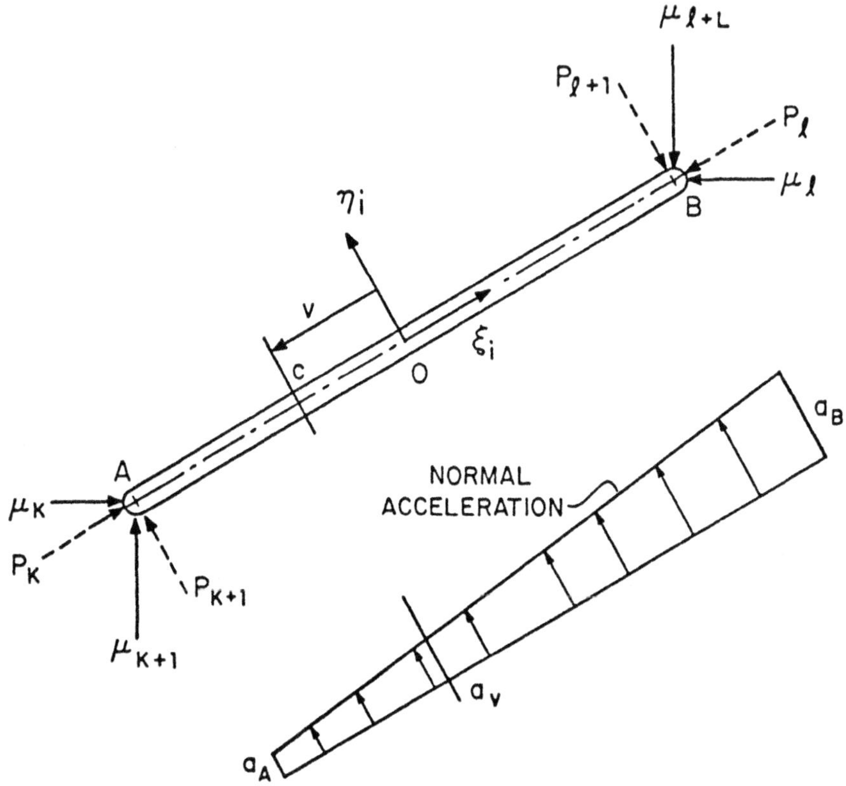

Figure 7.5 Loading System in Example 3

The absolute acceleration of point c can thus be written as

$$\ddot{\bar{R}}_{c_i} = (\ddot{x}_i - \nu \cos \theta_i \, \dot{\theta}_i^2 - \nu \sin \theta_i \ddot{\theta}_i) \, \bar{I}$$

$$+ (\ddot{y}_i - \nu \sin \theta_i \, \dot{\theta}_1^2 + \nu \cos \theta_i \ddot{\theta}_i) \, \bar{J}$$

Transforming the global unit vectors to the local system, acceleration normal to the length of the link can be written as

$$a_{y_i} = (- \ddot{x}_i \sin \theta_i + \ddot{y}_i \cos \theta_i) + \nu \, \ddot{\theta}_i \qquad (7.13)$$

where $\ddot{x}_i$, $\ddot{y}_i$, and $\ddot{\theta}_i$ are the generalized accelerations of body i. From Eq. 7.13, it can be seen that the normal acceleration varies linearly along the length of the link.

The distributed loading on the beam can thus be written as

$$f_i = -m_{\ell_i} a_{\ell_i} + \nu \ddot{\theta}_i \tag{7.14}$$

where $a_{\ell_i} = -\ddot{x}_i \sin\theta_i + \ddot{y}_i \cos\theta_i$.

Since elementary beam theory is used to compute bending stress, it is necessary to first determine the point of maximum bending moment. This is the point at which the shear force changes sign. The shear force at any point D, at a distance X_0 from 0, can be written

$$F_s = P_{k+1} + - m_{\ell_i} \int_{-\ell_i/2}^{X_0} (a_{\ell_i} + \nu \ddot{\theta}_i) d\nu$$

$$= P_{k+1} - m_{\ell_i} \left\{ a_{\ell_i} (x_0 + \frac{\ell_i}{2}) + \frac{\ddot{\theta}_i}{2} (x_0^2 - \frac{\ell_i^2}{4}) \right\} \tag{7.15}$$

At the point of zero shear force, the right side of Eq. 7.15 will be zero, giving rise to the condition

$$\frac{\ddot{\theta}_i}{2} X_0^2 + a_{\ell_i} X_0 + (\frac{a_{\ell_i} \ell_i}{2} - \frac{\ell_i^2 \ddot{\theta}_i}{8} - \frac{P_{k+1}}{m_{\ell_i}}) = 0 \tag{7.16}$$

Solving for X_0 from Eq. 7.16,

$$X_0 = \frac{-a_{\ell_i} \pm \sqrt{a_{\ell_i}^2 - 2\ddot{\theta}_i \left(\frac{a_{\ell_i} \ell_i}{2} - \frac{\ell_i^2 \ddot{\theta}_i}{8} - \frac{P_{k+1}}{m_{\ell_i}} \right)}}{\ddot{\theta}_i} \tag{7.17}$$

The maximum bending moment can thus be written as

$$M_{0_i} = P_{k+1} (\frac{\ell_i}{2} + X_0) + m_\ell \int_{-\ell_i/2}^{X_0} (a_{\ell_i} + \nu \ddot{\theta}_i) (\eta - X_0) d\nu \tag{7.18}$$

Integration of Eq. 7.18 gives

$$M_{0_i} = P_{k+1}\left(\frac{\ell_i}{2} + X_0\right) + m_\ell \quad a_{\ell_i}\left\{\left[-\frac{X_0^2}{2} - \left(\frac{\ell_i^2}{8} + \frac{\ell_i X_0}{2}\right)\right]\right.$$
$$\left. + \ddot{\theta}_i\left[-\frac{X_0^3}{6} + \frac{\ell_i^3}{24} + \frac{X_0\,\ell_i^2}{8}\right]\right\} \tag{7.19}$$

where X_0 is given by Eq. 7.17.

Equation 7.19 is valid only for links 3 and 4. The crank, link 2, is rotating at constant angular velocity and has no normal acceleration. The maximum moment on this link is the torque that is required to drive the mechanism. It occurs at $\eta = -\frac{\ell_1}{2}$ and is given as

$$M_{0_2} = \lambda_{12} \tag{7.20}$$

Since the links are of circular cross section, the area moments of inertia can be written in terms of the design variables as

$$I_i = \frac{b_{i-1}^2}{4\pi} \tag{7.21}$$

where $2 < i < 4$ is the link number. The distance of the extreme fiber from the neutral axis is given as

$$y_i = \sqrt{\frac{b_{i-1}}{\pi}}$$

The absolute values of the maximum bending stresses, from elementary beam theory [31], can now be written as

$$(\sigma_b)_i = \frac{4\quad\pi}{b_{i-1}^{3/2}}|M_0|, \qquad 2 < i < 4 \tag{7.22}$$

where M_0 is given by Eq. 7.20 for $i = 2$ and by Eq. 7.19 for $i = 3$ and 4.

The optimization problem can be written in standard form as; minimize

$$\psi_0 = \rho_2\,\ell_2\,b_1 + \rho_3\,\ell_3\,b_2 + \rho_4\,\ell_4\,b_3 \tag{7.23}$$

where ρ_i is mass density of link i, subject to discretized design constraints

$$\psi_k^j \equiv \frac{(\sigma_b^j)_{k+1}}{\sigma_a} - 1 \leq 0 , \qquad j = 1, \ldots, \tau, \qquad 1 \leq k \leq 3 \qquad (7.24)$$

where σ_a is allowable stress and τ is the total number of grid points. For this problem, 19 input grid points are employed. The driving constraint for this problem is given as

$$\Phi_1^d \equiv \theta_2 - \alpha_1 = 0$$

<u>Verification of Design Sensitivity Analysis</u>: The design sensitivities of the stress constraints at a certain iteration were obtained as

$$\ell^{1,17^T} = [-1801.18, 1928.83, 853.239]^T$$

$$\ell^{2,15^T} = [0.0 , -3481.67, 0.0]^T$$

$$\ell^{3,15^T} = [0.0, 0.0, 2051.6]^T$$

The change in design defined by the optimization algorithm at this iteration was

$$\delta b^T = [2.022\times10^{-5}, \quad 1.76\times10^{-5}, \quad -8.626\times10^{-6}]$$

The predicted changes in the constraint functions can thus be computed at grid point as

$$\delta\psi^{1,17} = \ell^{1,17^T} \delta b = -0.00983$$

$$\delta\psi^{2,15} = \ell^{2,15^T} \delta b = -0.0613$$

$$\delta\psi^{3,15} = \ell^{3,15^T} \delta b = -0.01739$$

The actual changes in the constraint functions were obtained as

$$\Delta\psi^{1,17} = -0.0097$$

$$\Delta\psi^{2,15} = -0.06002$$

$$\Delta\psi^{3,15} = -0.0169$$

Comparison of the actual and predicted changes shows that the two sets of data agree to within 2.5%. The design sensitivity analysis can thus be considered valid.

Optimization Results: Results of the optimization procedure are presented in Table 7.4. Results obtained in Refs. 29 and 30 for this problem are also given. As can be seen, there is a substantial difference between the designs obtained. This difference is attributed to the fact that Refs. 29 and 30 use different methods for stress analysis than is used in the present formulation. The constraints are, therefore, not identical.

Data for the mechanism considered in this problem are given in Table 7.5. This problem is the same as the single counterweight optimization example given in Ref. 4.

Example 4 - Dynamic Balancing of a Four-Bar Mechanism

Problem Description: It is desirable to make the shaking forces of a mechanism vanish by balancing the mechanism. However, doing so

	Design Variables			
	$b_1 (m^2)$	$b_2 (m^2)$	$b_3 (m^2)$	Mass (kg)
Starting Design	1×10^{-1}	1×10^{-1}	1×10^{-1}	546.26
Optimal Design 30th iter	1.03×10^{-3}	0.621×10^{-3}	0.1169×10^{-3}	2.686
Results from Reference 29	0.401×10^{-3}	0.385×10^{-3}	0.385×10^{-3}	2.134
Results from Reference 30	0.953×10^{-3}	0.488×10^{-3}	0.309×10^{-3}	2.703

At the final design $\|\delta b^1\| = 2.5 \times 10^{-6}$

Approximate CPU time (IBM 370/168) = 1.25 Sec/Iteration

Table 7.4 Results for Example 3

introduces additional difficulties. For example, the RMS bearing forces could increase by as much as 100 percent [4]. In the present example, the four bar mechanism shown in Fig. 7.6 is considered. A trade-off is resorted to, in which the RMS shaking forces are minimized, while limiting the increase in ground bearing forces. Minimization of RMS shaking force is accomplished by modifying the inertial properties of the output link, by adding a circular counterweight to it, as shown in Fig. 7.7. The mass and location of this counterweight are design variables.

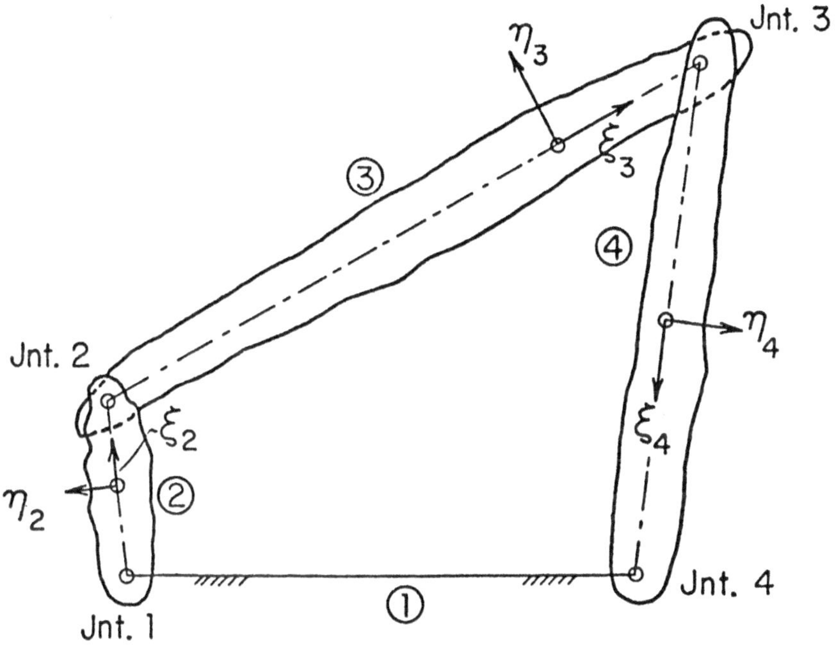

Figure 7.6 Four-Bar Mechanism to be Dynamically Balanced

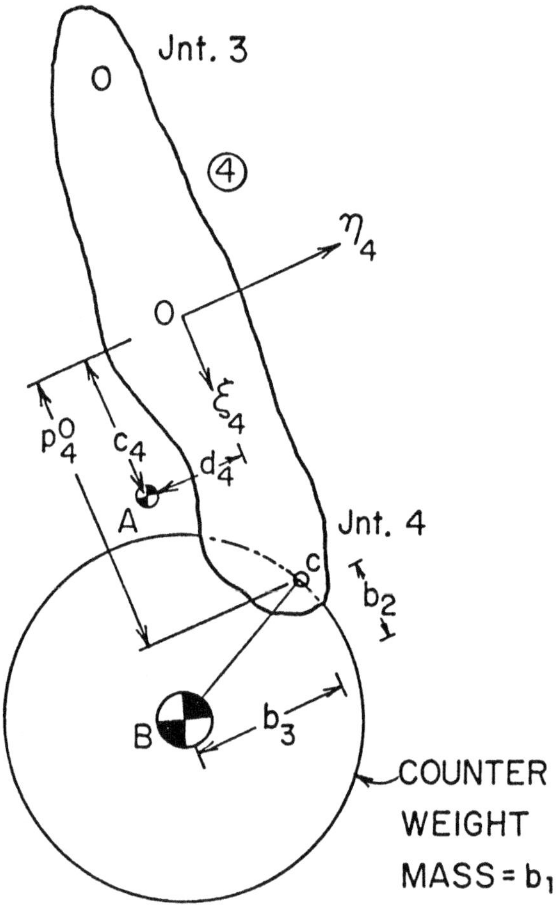

Figure 7.7 Schematic of Counterweight Used in Dynamic Balancing

Problem Formulation: The RMS shaking force for the four-bar mechanism being considered here is given as [4]

$$(R_{M/G})_{RMS} = \sqrt{\frac{1}{2\pi}\int_0^{2\pi} [(F_{14X} + F_{34X})^2 + (F_{14Y} + F_{34Y})^2]d\theta_2} \quad (7.25)$$

where F_{14X} and F_{14Y} are the ground bearing forces at joint 1, in the X and Y directions, respectively; F_{34X} and F_{34Y} are the ground bearing forces at joint 4, in the X and Y directions, respectively; and θ_2 is

	LINK			
	1	2	3	4
Length (m)	7.62×10^{-2}	2.54×10^{-2}	1.016×10^{-1}	7.62×10^{-2}
Mass (kg)	--	4.588×10^{-2}	1.084×10^{-1}	6.602×10^{-2}
Moment of inertia @ C.G kg-m^2	--	--	3.208×10^{-4}	6.779×10^{-5}

Table 7.5 Data for Example 4

the angular orientation of the input crank. The RMS ground bearing forces are given as [4]

$$(F_1)_{RMS} = \sqrt{\frac{1}{2\pi}\int_0^{2\pi}[(F_{41X}^2 + F_{41Y}^2)]d\theta_2} \tag{7.26}$$

$$(F_4)_{RMS} = \sqrt{\frac{1}{2\pi}\int_0^{2\pi}[(F_{43X}^2 + F^2_{43Y})]d\theta_2} \tag{7.27}$$

where $(F_1)_{RMS}$ and $(F_4)_{RMS}$ are the RMS ground bearing forces at joints 1 and 4, respectively.

There is a one-to-one correspondence between the design variables used here and those used in Ref. 4. Location of the center of mass of the combined link, in terms of design variables and given data, can be written, using relationships given in Ref. 4, as

$$C_4 = p_4^0 - \frac{1}{(m_4^0 + b_1)}(m_4^0 p_4^0 - b_1 b_2) \tag{7.28}$$

and

$$d_4 = \frac{1}{(m_4^0 + b_1)} \tag{7.29}$$

where c_4 and d_4 define the location of the combined center of mass, b_1 is the mass of the counterweight, b_2 and b_3 are the distances shown in Fig. 7.7, m_4^0 is the mass of the original output link, and p_4^0 is given data, as defined in Fig. 7.7.

Acceleration of point A relative to point 0 can be computed by the relation

$$\bar{a}_{A0} = \frac{d^2}{dt^2} (\bar{r}_{A0}) \tag{7.30}$$

where $\bar{r}_{A0}$ is the position vector of point A, relative to point 0, given as

$$\bar{r}_{A0} = (c_4 \cos\theta_4 - d_4 \sin\theta_4)\bar{I} + (c_4\sin\theta_4 + d_4\cos\theta_4)\bar{J} \tag{7.31}$$

Substituting $\bar{r}_{A0}$ from Eq. 7.31 into Eq. 7.30 gives

$$\bar{a}_{A0} = \{-(c_4 \cos\theta_4 - d_4 \sin\theta_4)(\dot{\theta}_4)^2 - (c_4 \sin\theta_4 - d_4 \cos\theta_4)\ddot{\theta}_4\}\bar{I}$$
$$\{-(c_4 \sin\theta_4 + d_4 \cos\theta_4)(\dot{\theta}_4)^2 + (c_4 \cos\theta_4 - d_4 \sin\theta_4)\ddot{\theta}_4\}\bar{J} \tag{7.32}$$

The absolute acceleration of point A can thus be written as

$$\bar{a}_A = \bar{a}_0 + \bar{a}_{A0} \tag{7.33}$$

D'Alembert's force on the output link can now be written as

$$\bar{F}_4 = -(m_4^0 + b_1)\,\bar{a}_A \tag{7.34}$$

Computation of D'Alembert moments requires the combined moment of inertia of the original link and the counterweight. Since the counterweight is circular, it's moment of inertia about point B is given as

$$I^* = b_1(b_2^2 + b_3^2)/2 \tag{7.35}$$

Using the parallel axis theorem [23], the combined moment of inertia of the original link and the counterweight can be written as

$$I_A = I_0 + m_4^0(|\bar{r}_{0A}|)^2 + I^* + b_1(|\bar{r}_{AB}|)^2 \tag{7.36}$$

where I_A is the combined moment of inertia about point A and I_0 is the moment of inertia of the original link about point 0. Since moment of inertia of the original link given in Ref. 4 is about point C, the parallel axis theorem must be used to write I_0 in Eq. 7.36 in terms of the given data as

$$I_C = I_0 + m_4^0 (p_4^0)^2 \tag{7.37}$$

Substituting for I_0 in Eq. 7.35 from Eq. 7.37 gives

$$\begin{aligned} I_A = I_C - m_4^0 (p_4^0)^2 + m_4^0 (c_4^2 + d_4^2) \\ + I^* + b_1 \left[(p_4^0 - c_4 + b_2)^2 + (b_3 - d_4)^2 \right] \end{aligned} \tag{7.38}$$

The D'Alembert moment can now be written as

$$\overline{M} = - I_A \ddot{\theta}_4 \tag{7.39}$$

Forces in the ground bearings, used in Eqs. 7.25 to 7.27, can be related, from the results of Section 3, to the Lagrange multipliers by the following relationships:

$$\left. \begin{aligned} F_{14X} &= -\lambda_4 \\ F_{14Y} &= -\lambda_5 \\ F_{34X} &= \lambda_{10} \\ F_{34Y} &= \lambda_{11} \end{aligned} \right\} \tag{7.40}$$

Equations 7.25 to 7.27 can thus be rewritten as

$$(F_{M/G})_{RMS} = \sqrt{\frac{1}{2\pi} \int_0^{2\pi} [(-\lambda_4 + \lambda_{10})^2 + (-\lambda_5 + \lambda_{11})^2] \, d\theta_2} \tag{7.41}$$

$$(F_1)_{RMS} = \sqrt{\frac{1}{2\pi} \int_0^{2\pi} [\lambda_4^2 + \lambda_5^2] \, d\theta_2} \tag{7.42}$$

$$(F_4)_{RMS} = \sqrt{\frac{1}{2\pi} \int_0^{2\pi} [\lambda_{10}^2 + \lambda_{11}^2] \, d\theta_2} \tag{7.43}$$

The optimization problem can now be stated in the standard form of minimizing

$$\psi_0 \equiv (F_{M/G})_{RMS} \tag{7.44}$$

subject to constraints

$$\psi_1 \equiv \frac{(F_1)_{RMS}}{F_1} - 1 \leq 0 \tag{7.45}$$

$$\psi_2 \equiv \frac{(F_4)_{RMS}}{F_4} - 1 \leq 0 \tag{7.46}$$

where $(F_{M/G})_{RMS}$ and $(F_1)_{RMS}$ are given by Eqs. 7.41, 7.42, and 7.43, respectively, and F_1 and F_4 are the upper bounds on the bearing forces at revolute joints 1 and 4, respectively.

The driving kinematic constraint for this problem is

$$\Phi_1^d \equiv \theta_2 - \alpha = 0$$

where fifteen input grid points are chosen.

The rationale used to compute the values of F_1 and F_4 is explained in detail in Ref. 4. Briefly, to arrive at these numbers, the RMS ground bearing forces for the unbalanced mechanism and the fully force balanced mechanism are computed.Since the RMS bearing forces for the fully force balanced mechanism are higher than those for the unbalanced mechanism, the values of F_1 and F_4 are chosen to lie midway between the two extreme values. For the present, these values are taken as $F_1 = 9.136 \times 10\text{-}3$ N and $F_4 = 6.427 \times 10\text{-}3$ N.

The cost function and design constraints considered in example problems thus far do not involve integral quantities, such as those appearing in Eqs. 7.41 to 7.43. Computation of design sensitivities of these functions can also be handled by the formulation developed here. Design sensitivities of the integral cost and constraint functions can be written by taking the derivatives with respect to design of Eqs. 7.44 to 7.46 [32], yielding

$$\ell^{0T} = \frac{d\psi_0}{db} = \frac{1}{2\pi(F_{M/G})_{RMS}} \int_0^{2\pi} [(-\lambda_4 + \lambda_{10}) (- \frac{d\lambda_4}{db} + \frac{d\lambda_{10}}{db}) + (-\lambda_5 + \lambda_{11}) (- \frac{d\lambda_5}{db} + \frac{d\lambda_{11}}{db})] \, d\theta_2 \tag{7.47}$$

$$\ell^{1T} = \frac{d\psi 1}{db} = \frac{F_1}{2\pi(F_1)_{RMS}} \int_0^{2\pi} [\lambda_4 \frac{d\lambda_4}{db} + \lambda_5 \frac{d\lambda_5}{db}] \, d\theta_2 \tag{7.48}$$

$$\ell^{2^T} = \frac{d\psi 2}{db} = \frac{F_4}{2\pi (F_4)_{RMS}} \int_0^{2\pi} [\lambda_{10} \frac{d\lambda_{10}}{db} + \lambda_{11} \frac{d\lambda_{11}}{db}] \, d\theta_2 \tag{7.49}$$

Numerical integration of Eqs. 7.47 to 7.49 requires evaluation of the integrand at specific points in the interval of integration. Computation of the derivatives of Lagrange multipliers appearing in the integrands of Eqs. 7.47 to 7.49 can be routinely done in the present formulation. The computer code treats λ_4, λ_5, λ_{10}, and λ_{11} as constraint functions, purely for the purpose of computing the design sensitivity. The design sensitivity of these variables is evaluated at points on the θ_2 grid corresponding to the nodes of the 15 point Gauss-Legendre integration formula [33]. Since constraints ψ_1 and ψ_2 do not depend on the input parameter, they are nonparametric constraints.

Verification of Design Sensitivity Analysis: The design sensitivity vectors of the cost and constraint functions, at a certain design iteration, where obtained as

$$\ell^{0^T} = [-.000799, .02880, .004828]$$

$$\ell^{1^T} = [4.1542, -12.7106, 3.1470]$$

$$\ell^{2^T} = [3.9492, -14.846, -.3122]$$

The change in design in this iteration was

$$\delta b = [-0.000885, 0.001334, 0.000373]$$

The predicted change in cost and constraints are thus computed as

$$\delta\psi_0 = \ell^{0^T} \delta b = 0.0000409$$

$$\delta\psi_1 = -.01946$$

$$\delta\psi_2 = -.0234$$

The actual changes in the cost and constraints were

$$\Delta\psi_0 = 0.000057$$

$$\Delta\psi_1 = -.011355$$

$$\Delta\psi_2 = -.01406$$

Comparing the predicted and actual changes in cost and constraint functions, it can be seen that the two sets of data do not agree as well as in the preceding examples. This can be attributed to the highly nonlinear nature of the cost and constraint functions in this example. For example, consider constraint 1, for which the design sensitivity vector averaged over the two design iterations is

$$[\ell^{1^T}]_{AVE} = [4.1145, -12.4848, 3.1727]$$

The predicted change in the constraint, on the basis of the averaged design sensitivity vector, is

$$[\Delta\psi_p^1]_{AVE} = -.01183$$

As can be seen, this figure agrees closely with the actual change in the constraint.

Optimization Results: Results obtained from the optimization procedure are presented in Table 7.6. For the purpose of comparison, the design variables used in Ref. 4 are converted to those used in the present formulation. The ratio of the RMS shaking force for the partially balanced mechanism to that for the unbalanced mechanism is given by r_f. As can be seen from Table 7.6, a substantial reduction in RMS shaking force is obtained by using the present formulation. Results obtained from the present formulation and those in Ref. 4 differ substantially. This difference is attributed to the fact that the present formulation imposes the RMS ground bearing force constraints as inequality constraints and Ref. 4 considers these to be equality constraints. Significant improvement in design from the present formulation over that from Ref. 4 is thus to be expected and is evident from Table 7.6.

	Design Variables in present paper			
	b_1(kg)	b_2(m)	b_3(m)	r_f
Starting Design	0.30	0.0	0.0	--
Optimal Design, 15th Iter	0.203	0.040	0.0098	0.25
Optimal Design Ref. 4	0.207	0.0183	-0.017	0.69
At the final design $\|\|\delta b^1\|\| = 1.6 \times 10^{-4}$ Approximate computing time IBM 370/168 = 1.0 Secs/iteration				

Table 7.6 Results in Example 4

REFERENCES

1. Freudenstein, F., "Structural Error Analysis in Plane Kinematic Synthesis," ASME, Journal of Engineering for Industry, Ser. B, Vol. 81, No. 1, Feb. 1959, pp. 15-22.

2. Kramer, S.N. and Sandor, G.N., "Selective Precision Synthesis - A General method of Optimization for Planar Mechanisms," ASME, Journal of Engineering for Industry, Ser. B, Vol. 97, No. 2, 1975.

3. Tepper, F.R. and Lowen, G.G., "General Theorems Concerning Full Force Balancing of Planar Linkages by Internal Mass Redistribution, ASME, Journal of Engineering for Industry, Ser. B, Vol. 94, No. 3, Aug. 1972, pp. 789-796.

4. Tepper, F.R. and Lowen, G.G., "Shaking Force Optimization of Four-Bar Linkage with Adjustable Constraints on Ground Bearing Forces," ASME, Journal of Engineering for Industry, Ser. B, Vol. 97, No. 2, May 1975, pp. 643-651.

5. Berkof, R.S., "Force Balancing of a Six-Bar Linkage," Proceedings of the Fifth World Congress on the Theory of Machines and Mechanism, July 8-13, 1979, Montreal, Canada, pp. 1082-1085.

6. Ramaiyan, G. and Lakshminarayan, K., "Synthesis of Seven Link Two-Freedom Linkage with Sliding Inputs and Outputs using Identical Link Positions," Mechanisms and Machine Theory, Vol. 11, No. 3, 1976, pp. 181-185.

7. Ramaiyan, G., Lakshminarayan, K., and Narayanamurthi, R.G., "Nine-Link Plane Mechanisms for Two-Variable Function Generation - II. Synthesis," Mechanisms and Machine Theory, Vol. 11, No. 3, 1976, pp. 193-199.

8. Mariante, W. and Willmert, K.D., "Optimal Design of a Complex Planar Mechanism," ASME, Journal of Engineering for Industry, Ser. B, Vol. 99, No. 3, Aug. 1977, pp. 539-546.

9. Thronton, W.A., Willmert, K.D., and Khan, M.R., "Mechanism Optimization via Optimality Criterion Techniques," Journal of Mechanical Design, Trans. ASME, Vol. 101, No. 1, July 1979, pp. 392-397.

10. Rubel, A.J. and Kaufman, R.E., "KINSYN III: A New Human Engineered System for Interactive Computer Aided Design of Planar Linkages," ASME, Journal of Engineering for Industry, Vol. 99,No. 2, March 1977, pp. 440-455.

11. Rucci, R.J. "SPACEBAR: Kinematic Design by Computer Graphics," Computer Aided Design, Vol. 8, No. 4, October 1976, pp. 219-226.

12. Erdman, A.G. and Gustavson, J.E. "LINCAGES: Linkage Interactive Computer Analysis and Graphically Enhanced Synthesis Package," ASME paper 77-DET-5.

13. Paul, B., "Analytical Dynamics of Mechanisms - A Computer Related Overview," Mechanism and Machine Theory, Vol. 10, No. 4, 1975, pp. 481-507.

14. Hartenberg, R.S. and Denavit, J., Kinematic Synthesis of Linkages," McGraw-Hill Book Co., 1964.

15. Sheth, P.N. and Uicker, J.J., "IMP Integrated Mechanism Program, A Computer Aided Design System for Mechanisms and Linkages," Journal of Engineering for Industry, Trans. ASME, Vol. 93, No. 4, 1971.

16. Paul, B., Kinematics and Dynamics of Planar Machinery, Prentice-Hall, 1979.

17. Orlandea, N., Chace, M.A., and Calahan, D.A., "A Sparsity-Oriented Approach to the Dynamic Analysis and Design of Mechanical Systems," Part I and II, ASME, Journal of Engineering for Industry, Ser. B., Vol. 99, No. 3, 1977.

18. Wehage, R.A. and Haug, E.J., "Generalized Coordinate Partitioning for Dimension Reduction in Analysis of Constrained Dynamic Systems," ASME paper no. 80-DET-106. To be published in Journal of Mechanical Design.

19. Haug, E.J., Wehage, R., and Barman, N.C., "Design Sensitivity Analysis of Planar Mechanism and Machine Dynamics," ASME, Journal of Mechanical Design, Vol. 103, No. 3, pp. 560-570, 1981.

20. Haug, E.J. and Arora, J.S., Applied Optimal Design, John Wiley & Sons, 1979.

21. Atkinson, K.E., An Introduction to Numerical Analysis,John Wiley & Sons, 1978.

22. United Kingdom Atomic Energy Authority: "Harwell Subroutine Library: A Catalogue of Subroutines," Report AERE R7477, Subroutine Librarian, CSS Division, Atomic Energy Research Establishment, Harwell Didcot, Oxfordshire, England OX11 ORA, 1973.

23. Greenwood, D.T., Principles of Dynamics, Prentice-Hall, 1965.

24. Goldstein, H., Classical Mechanics, 2nd ed., Addison-Wesley Publishing Company, 1980.

25. Kwak, B.M., Parametric Optimal Design, Ph.D. Thesis, University of Iowa, Iowa City, 1975.

26. Choi, K.K., Haug, E.J., Hou, J.W., and Sohoni, V.N., "Pshenichyni's Linearization Method For Mechanical System Optimization," ASME J. of Mechanisms, Transmissions, and Automation In Design, Vol. 105, No. 1, 1983, pp. 97-103.

27. Sohoni, V.N. and Haug, E.J., A State Space Technique for Kinematic Synthesis and Design of Planar Mechanisms and Machines, Report No. 81-5, Center for Computer Aided Design, College of Engineering, The University of Iowa, Iowa City, Iowa, Oct. 1981.

28. Bhatia, D.H. and Bagci, C., "Optimal Synthesis of Multiloop Planar Mechanisms for the Generation of Paths and Rigid Body Positions by the Linear Partition of Design Equations." ASME paper No. 74-DET-14.

29. Khan, M.R., Thornton, W.A., and Willmert, K.D., "Optimality Criterion Techniques Applied to Mechanical Design," Journal of Mechanical Design, ASME, Vol. 100, April 1978, pp. 319-321.

30. Imam, I. and Sandor, G.N., "A General Method of Kineto-Elastodynamic Design of High Speed Mechanisms," Mechanism and Machine Theory, Vol. 8, 1973, pp. 497-516.

31. Ugural, A.C. and Fenster, S.K., Advanced Strength and Applied Elasticity, American Elsevier Publishing Company, 1975.

32. Taylor, A.E. and Mann, W.R., Advanced Calculus, Wiley, 1972.

33. Carnahan, B., Luther, H.A., and James, W.O., Applied Numerical Methods, John Wiley & Sons, 1969.

DESIGN SENSITIVITY ANALYSIS AND OPTIMIZATION OF DYNAMICALLY DRIVEN SYSTEMS*

Edward J. Haug
Neel K. Mani
Prakash Krishnaswami
Center for Computer Aided Design
University of Iowa
Iowa City, Iowa 52242

Abstract. Methods for calculation of first and second design derivatives of measures of response for dynamic systems are presented. Dynamic mechanical systems are considered in which applied forces are prescribed and the dynamic response is determined by equations of motion. Design sensitivity analysis formulations for dynamic systems are presented in two alternate forms; (1) the system of equations of motion are written in terms of independent generalized coordinates and are reduced to first order form and (2) the system equatons of motion are written in terms of a mixed system of second order differential and algebraic equations of motion for constrained dynamic systems. Both first and second order design sensitivity analysis methods are developed, using a theoretically simple direct differentiation approach and a somewhat more subtle, but numerically efficient, adjoint variable method. Detailed derivations are presented and computational algorithms are given and discussed. Examples of first and second order design sensitivity analysis of simple mechanisms and machines and optimization of a five degree-of-freedom vehicle suspension system are presented and analyzed.

1. INTRODUCTION

The general theory of sensitivity analysis has been well developed in the theory of differential equations and for control

* Research supported by Army Research Office Project No. P-18576-M.

NATO ASI Series, Vol. F9
Computer Aided Analysis and Optimization of Mechanical System Dynamics
Edited by E. J. Haug

systems [1,2]. Early developments made primary use of a first order formulation of the initial-value problem that describes dynamics of the system and introduced an adjoint or co-state variable to obtain design derivatives. While a direct design differentiation approach is presented in Ref. 1, it has not received widespread use, due to its computational inefficiency.

Design sensitivity analysis theory that was originally developed in optimal control theory [2] has been adapted to structural and mechanical dynamic systems that are described in first order form in Ref. 3. More recently, second order differential equation formulations have been employed to develop more practical and directly useable formulations for system dynamic design sensitivity analysis [4,5]. In a related development, second order design sensitivity analysis has recently been treated and analyzed [6]. While general formulas for second derivative calculation are quite complex, the advent of symbolic computation techniques may make routine calculation of second design derivatives a practical matter in the future.

Several alternative formulations of initial-value problems that describe system dynamics are employed in this paper, to yield algorithms for computation of design sensitivity. In Section 2, direct differentiation and adjoint variable methods are presented for design sensitivity analysis of systems that are described by nonlinear systems of first order differential equations. This theory is extended in Section 3 to present three different approaches for second order design sensitivity analysis. The first is a direct second order differentiation technique that yields an inordinately large number of equations and is impractical. The second is a pure adjoint variable technique that is computationally feasible, but rather complex. The third is a new method that is based on combining of the direct differentiation and adjoint variable methods, to obtain a more efficient and less complex second order design sensitivity formulation.

Elementary problems are solved in Sections 2 and 3 to illustrate use of the equations derived. Automatic cannon and five degree-of-freedom vehicle examples are treated in Sections 4 and 5, respectively. Design sensitivity calculations are given and the validity of first and second order approximations, using design sensitivity results, are presented and analyzed. In the case of the vehicle example of Section 5, an iterative optimization technique is used with the design sensitivity analysis results to optimize suspension characteristics of the vehicle.

An extension of the methods of Section 2 for first order design sensitivity analysis of large scale systems that are described by mixed differential and algebraic equations is presented in Section 6. As in the preceding, both the direct differentiation and adjoint variable approaches are presented. It is shown that both formulations lead to linear adjoint systems of mixed differential and algebraic equations that have a similar structure to the original system. Examples using this second order formulation with a computer code that automatically generates and solves the system equations of motion is presented to illustrate feasibility of automated design sensitivity analysis, coupled with large scale computer codes in a computer aided design environment.

2. FIRST ORDER DESIGN SENSITIVITY ANALYSIS FOR SYSTEMS DESCRIBED BY FIRST ORDER DIFFERENTIAL EQUATIONS

Alternate Approaches

Design sensitivity analysis of the dynamics of mechanical systems whose equations of motion have been put in first order form has progressed to the point that first derivatives of dynamic response measures with respect to design parameters can be calculated [1-6]. Direct differentiation of equations of motion was initially used to obtain state sensitivity functions [1]. More recently, an adjoint variable method that was borrowed from optimal control theory [2] and developed for mechanical design [3] has been successfully employed for design sensitivity analysis of large scale planar dynamic systems [4] and smaller scale dynamic systems with intermittent motion [5]. These design derivatives give the designer trend information on the effect that design variations will have on system performance and may be used directly in gradient-based iterative optimization algorithms.

First Order Formulation of the Problem

Dynamic systems treated here are described by a design variable vector $b = [b_1, \ldots, b_k]^T$ and a state variable vector

$z(t) = [z_1(t),\ldots,z_n(t)]^T$, which is the solution of an initial-value problem of the form

$$\left.\begin{aligned} \dot{z} &= f(z,b) \quad , \qquad t^1 \leq t^2 \\ z(t^1) &= h(b) \end{aligned}\right\} \tag{2.1}$$

where $\dot{z} = \frac{dz}{dt}$, t^1 is given, and t^2 is determined by the condition

$$\Omega(t^2, z(t^2), b) = 0 \tag{2.2}$$

The function f that appears on the right side of the differential equation of Eq. 2.1, the function h in the initial condition of Eq. 2.1, and the function Ω in Eq. 2.2 are assumed to be twice continuously differentiable in their arguments. Classical results of the theory of ordinary differential equations [7] assure that the solution $z(t;b)$ of Eq. 2.1 exists and is twice continuously differentiable with respect to both t and b.

Consider now a typical functional that may arise in a design formulation,

$$\psi = g(t^2, z(t^2), b) + \int_{t^1}^{t^2} F(t,z,b)dt \tag{2.3}$$

where the first term involves only behavior of the state of the system at the terminal time and the design. The second term involves mean behavior over some period of motion. This form of functional is adequate for treating a large class of dynamic system optimal design problems [2,3].

As noted, dependence on the design variable b in Eq. 2.3 arises both explicitly and through the state variable, which is written in the form $z(t;b)$ to emphasize that it is a function of time that depend on design. In order to obtain the derivative of ψ with respect to b, Leibniz rule of differentiation [8] may be applied to obtain

$$\begin{aligned} \frac{d\psi}{db} = {} & g_{t^2} t_b^2 + g_z [z_b(t^2) + \dot{z}(t^2) t_b^2] + g_b \\ & + \int_{t^1}^{t^2} [F_z z_b + F_b] dt + F(t^2) t_b^2 \end{aligned} \tag{2.4}$$

where $_b$ denotes derivative with respect to b. For a summary of matrix differentiation notation employed in this paper, the reader is referred to Appendix A. It is important to note that z_b is the derivative of a vector function with respect to a vector variable. It is thus a matrix, so the order of terms in matrix products is important.

Equation 2.4 may be reduced to a form that depends only on z_b and not on $t^2{}_b$. Differentiating Eq. 2.2 with respect to b yields

$$\left[\Omega_{t^2} + \Omega_z \dot{z}(t^2)\right] t_b^2 + \Omega_z z_b(t^2) + \Omega_b = 0$$

Since Eq. 2.2 must determine t^2, the coefficient of t_b^2 cannot be zero. Using Eq. 2.1 for $\dot{z}$, this gives

$$t_b^2 = -\left[\frac{\Omega_z}{\Omega_{t^2} + \Omega_z f(t^2)}\right] z_b(t^2) - \left[\frac{\Omega_b}{\Omega_{t^2} + \Omega_z f(t^2)}\right] \tag{2.5}$$

Substituting this result into Eq. 2.4 yields

$$\frac{d\psi}{db} = G^T(t^2, z(t^2), b) z_b(t_2) + g_b - \left[\frac{g_{t^2} + g_z f(t^2) + F(t^2)}{\Omega_{t^2} + \Omega_z f(t^2)}\right]\Omega_b + \int_{t^1}^{t^2} (F_z z_b + F_b) dt \tag{2.6}$$

where

$$G(t^2, z(t^2), b) = \left\{ g_z^T(t^2) - \left[\frac{g_{t^2} + g_z f(t^2) + F(t^2)}{\Omega_{t^2} + \Omega_z f(t^2)}\right] \Omega_z^T \right\} \tag{2.7}$$

In order to make practical use of Eq. 2.6, the terms involving z_b must be calculated or replaced by terms that are written explicitly in terms of computable design derivatives. Two very different methods of achieving this objective are now presented. The first and most classical method [1] uses direct differentiation of Eq. 2.1 to obtain an initial-value problem for z_b, which is solved and the result is substituted in Eq. 2.6. The second method, which has attractive computational features, introduces an adjoint variable [2-6] that is used to write an alternate form of Eq. 2.6 that requires less computation than the direct differentiation method.

First Order Design Sensitivity Analysis

Direct Differentiation Method: Direct differentiation of Eq. 2.1 yields

$$\left.\begin{aligned} \dot{z}_b - f_z z_b &= f_b , \qquad t^1 < t < t^2 \\ z_b(t^1) &= h_b \end{aligned}\right\} \tag{2.8}$$

Note that since z_b is a n × k matrix, Eq. 2.8 is in fact a system of k first order linear differential equations for the k columns of the matrix z_b.

The initial-value problem of Eq. 2.8 can be solved numerically by forward numerical integration, to obtain the solution $z_b(t)$ on the interval $t^1 < t < t^2$. The result may be substituted directly into Eq. 2.6 to obtain the first derivative of the functional ψ of Eq. 2.3 with respect to design. While this computational algorithm is conceptually very simple and in fact can be implemented with a minimum of programming difficulty, if k is large, it requires a massive amount of numerical computation and data storage.

Adjoint Variable Method: It is desirable to find an alternate method of first order design sensitivity analysis that avoids the massive computation associated with the direct differentiation method of the preceeding subsection.

To meet this objective, an adjoint variable λ is introduced by multiplying both sides of Eq. 2.1 by λ^T and integrating from t^1 to t^2 to obtain the identity

$$\int_{t^1}^{t^2} \lambda^T [\dot{z}_b - f_z z_b - f_b] dt = 0 \tag{2.9}$$

Integrating the first term by parts gives

$$\int_{t^1}^{t^2} [(\dot{\lambda}^T + \lambda^T f_z) z_b + \lambda^T f_b] dt - \lambda^T(t^2) z_b(t^2) + \lambda^T(t^1) h_b = 0 \tag{2.10}$$

where $z_b(t^1) = h_b$ has been substituted from the initial condition of Eq. 2.1.

Recall that Eq. 2.10 holds for any function λ. In order to obtain a useful identity, the function λ is chosen so that the coefficient of z_b in the integrands of Eqs. 2.6 and 2.10 are equal; i.e.,

$$\dot{\lambda} + f_z^T \lambda = F_z^T \tag{2.11}$$

Further, if one equates coefficients of $z_b(t^2)$ in Eqs. 2.6 and 2.10; i.e.,

$$\lambda(t^2) = - G(t^2, z(t^2), b) \equiv - g_z^T + \left[\frac{F(t^2) + g_{t^2} + g_z f(t^2)}{\Omega_{t^2} + \Omega_z f(t^2)} \right] \Omega_z^T \tag{2.12}$$

then, all terms in Eqs. 2.6 and 2.10 involving z_b are equal. The terms in Eq. 2.6 that involve z_b may now be replaced by terms in Eq. 2.10 that involve only explicit derivatives with respect to design and the adjoint variable λ. Thus, Eq. 2.6 becomes

$$\frac{d\psi}{db} = g_b - \frac{(g_{t^2} + g_z f(t^2) + F(t^2))\Omega_b}{\Omega_{t^2} + \Omega_z f(t^2)} - \lambda^T(t^1) h_b + \int_{t^1}^{t^2} [F_b - \lambda^T f_b] dt \tag{2.13}$$

Thus, the first derivative of ψ with respect to design can be evaluated. The computational cost associated with evaluating this derivative vector includes backward numerical integration of the terminal-value problem of Eqs. 2.11 and 2.12, to obtain $\lambda(t)$. Numerical integration in Eq. 2.13 then yields the desired design derivative. This method is called the adjoint variable method of design sensitivity analysis. It has been used extensively in optimal control and mechanical design [2,3]. Note that since the differential equation of Eq. 2.11 and the terminal condition of Eq. 2.12 depend on b, $\lambda = \lambda(t;b)$.

Note that the derivative of ψ with respect to design is obtained in Eq. 2.13, with only backward numerical integration of a single terminal-value problem of Eqs. 2.11 and 2.12. If several design variables are involved; i.e. $k >> 1$, then the adjoint variable method is substantially more efficient than the direct differentiation method. If, on the other hand, the design variable is a scalar, then the same number of differential equations must be integrated. There is one practical complication associated with the adjoint variable method. The state variable z(t) must be stored for use in subsequent backward integration.

The conclusion is relatively clear. For large numbers of design variables, the adjoint variable is clearly superior.

Design Sensitivity Analysis of a Simple Oscillator

Consider the simple oscillator of Fig. 2.1 as an example. The objective is to derive first order design derivatives of position at a given time.

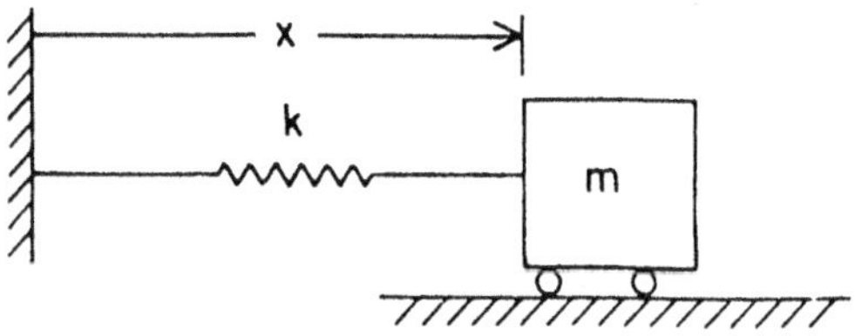

Figure 2.1 Simple Oscillator

The second order equation of motion and initial conditions are

$$\left.\begin{aligned} &m\ddot{x} + kx = 0, \quad 0 < t < \tfrac{\pi}{2} \\ &x(0) = 0 \\ &\dot{x}(0) = v \end{aligned}\right\} \tag{2.14}$$

where v is the initial velocity. Equation 2.14 can be written in the first order form of Eq. 2.1 as

$$\left.\begin{aligned} &\begin{bmatrix} \dot{z}_1 \\ \dot{z}_2 \end{bmatrix} = \begin{bmatrix} z_2 \\ -\dfrac{b_1}{m} z_1 \end{bmatrix} \equiv f(z,b) \;, \quad 0 < t < \tfrac{\pi}{2} \\ &\begin{bmatrix} z_1(0) \\ z_2(0) \end{bmatrix} = \begin{bmatrix} 0 \\ b_2 \end{bmatrix} \equiv h(b) \end{aligned}\right\} \tag{2.15}$$

where $z_1 \equiv x$ and $z_2 \equiv \dot{x} = \dot{z}_1$. The design variable vector is taken as $b = [k,v]^T$.

To simplify the problem, let $m = 1$. Then the solution of Eq. 2.15 is

$$\begin{bmatrix} z_1(t) \\ \\ z_2(t) \end{bmatrix} = \begin{bmatrix} \frac{b_2}{\sqrt{b_1}} \sin\sqrt{b_1}\, t \\ \\ b_2 \cos\sqrt{b_1}\, t \end{bmatrix} \tag{2.16}$$

The functional treated in this simple example is the position of the mass at terminal time $t^2 = \pi/2$; i.e.,

$$\psi = z_1(\tfrac{\pi}{2}) \equiv g(z(\tfrac{\pi}{2}), b) = \frac{b_2}{\sqrt{b_1}} \sin \frac{\pi\sqrt{b_1}}{2} \tag{2.17}$$

From Eq. 2.2, since t^2 has a given fixed value, $\Omega = t^2 - \pi/2$ and

$$\left.\begin{aligned} \Omega_z &= 0 \\ \Omega_{t^2} &= 1 \end{aligned}\right\} \tag{2.18}$$

From Eq. 2.7,

$$G(t^2, z(t^2), b) = \begin{matrix} 1 \\ 0 \end{matrix} \tag{2.19}$$

Direct Differentiation Method: To illustrate use of the direct differentiation method, the initial-value problem of Eq. 2.15 may be differentiated with respect to design, using the notation $y_i^j = (z_i)_{b_j}$, to obtain the matrix differential equation and initial condition

$$\left.\begin{aligned} &\begin{bmatrix} \dot{y}_1^1 & \dot{y}_1^2 \\ \dot{y}_2^1 & \dot{y}_2^2 \end{bmatrix} - \begin{bmatrix} 0 & 1 \\ -b_1 & 0 \end{bmatrix} \begin{bmatrix} y_1^1 & y_1^2 \\ y_2^1 & y_2^2 \end{bmatrix} = \begin{bmatrix} 0 & 0 \\ -z_1 & 0 \end{bmatrix} \\ &\begin{bmatrix} y_1^1 & y_1^2 \\ y_2^1 & y_2^2 \end{bmatrix}_{(t^1)} = \begin{bmatrix} 0 & 0 \\ 0 & 1 \end{bmatrix} \end{aligned}\right\} \tag{2.20}$$

Substituting the solution of Eq. 2.16 into Eq. 2.20 for z_1, one may solve the initial-value problem of Eq. 2.20 to obtain

$$\left.\begin{aligned} y_1^1 &\equiv (z_1)_{b_1} = \left(\frac{b_2 t}{2b_1}\right)\cos\sqrt{b_1}\,t - \left(\frac{b_2}{2(b_1)^{3/2}}\right)\sin\sqrt{b_1}\,t \\ y_2^1 &\equiv (z_2)_{b_1} = \dot{y}_1^1 \\ y_1^2 &\equiv (z_1)_{b_2} = \left(\frac{1}{\sqrt{b_1}}\right)\sin\sqrt{b_1}\,t \\ y_2^2 &\equiv (z_2)_{b_2} = \dot{y}_1^2 \end{aligned}\right\} \tag{2.21}$$

These results may be substituted, with Eq. 2.19, into Eq. 2.6 to obtain

$$\begin{aligned} \frac{d\psi}{db} &= [1,0]z_b(\pi/2) = \left[y_1^1(\pi/2)\ ,\quad y_1^2(\pi/2)\right] \\ &= \left[\left(\frac{\pi b_2}{4b_1}\right)\cos\frac{\pi\sqrt{b_1}}{2} - \left(\frac{b_2}{2(b_1)^{3/2}}\right)\sin\frac{\pi\sqrt{b_1}}{2}\ ,\quad \left(\frac{1}{\sqrt{b_1}}\right)\sin\frac{\pi\sqrt{b_1}}{2}\right] \end{aligned} \tag{2.22}$$

A direct calculation of the partial derivatives of ψ of Eq. 2.17, using the solution of Eq. 2.16, shows that the result obtained by the direct differentiation method is correct.

Adjoint Variable Method: Next, the adjoint variable method is applied. From Eq. 2.13, the first design derivatives of ψ are given by

$$\frac{d\psi}{db} = -\lambda^T(0)h_b - \int_0^{\frac{\pi}{2}} \lambda^T f_b\,dt \tag{2.23}$$

where

$$\left.\begin{aligned} h_b &= \begin{bmatrix} 0 & 0 \\ 0 & 1 \end{bmatrix} \\ f_b &= \begin{bmatrix} 0 & 0 \\ -z_1 & 0 \end{bmatrix} \end{aligned}\right\} \tag{2.24}$$

and λ is the solution of Eqs. 2.11 and 2.12, in this case,

$$\left.\begin{aligned}&\begin{bmatrix}\dot{\lambda}_1\\ \dot{\lambda}_2\end{bmatrix}+\begin{bmatrix}0 & -b_1\\ 1 & 0\end{bmatrix}\begin{bmatrix}\lambda_1\\ \lambda_2\end{bmatrix}=0\\ &\lambda(\tfrac{\pi}{2})=-\begin{bmatrix}1\\ 0\end{bmatrix}\end{aligned}\right\}\tag{2.25}$$

The solution of Eq. 2.25 for λ is obtained in closed form as

$$\begin{bmatrix}\lambda_1\\ \lambda_2\end{bmatrix}=\begin{bmatrix}-\cos\sqrt{b_1}(\frac{\pi}{2}-t)\\ -\frac{1}{\sqrt{b_1}}\sin\sqrt{b_1}(\frac{\pi}{2}-t)\end{bmatrix}\tag{2.26}$$

Substituting Eqs. 2.24 and 2.26 into Eq. 2.23,

$$\frac{d\psi}{db}=\left[\cos\frac{\pi\sqrt{b_1}}{2},\ \frac{1}{\sqrt{b_1}}\sin\frac{\pi\sqrt{b_1}}{2}\right]\begin{bmatrix}0 & 0\\ 0 & 1\end{bmatrix}$$
$$+\int_0^{\frac{\pi}{2}}\left[\cos\sqrt{b_1}(\tfrac{\pi}{2}-t),\ \frac{1}{\sqrt{b_1}}\sin\sqrt{b_1}(\tfrac{\pi}{2}-t)\right]\begin{bmatrix}0 & 0\\ -z_1 & 0\end{bmatrix}dt$$

Using the solution for z_1 in Eq. 2.16 and integrating, one has the result

$$\frac{d\psi}{db}=\left[-\frac{b_2}{2b_1^{\frac{3}{2}}}\sin\frac{\pi\sqrt{b_1}}{2}+\frac{\pi b_2}{4b_1}\cos\frac{\pi\sqrt{b_1}}{2},\ \frac{1}{\sqrt{b_1}}\sin\frac{\pi\sqrt{b_1}}{2}\right]\tag{2.27}$$

which is indeed the result obtained by direct differentiation of Eq. 2.17.

3. SECOND ORDER DESIGN SENSITIVITY ANALYSIS FOR SYSTEMS DESCRIBED BY FIRST ORDER DIFFERENTIAL EQUATIONS

In some situations, first derivatives with respect to design are inadequate. For example, optimization algorithms that use second order derivatives are generally superior to gradient based methods. Of more direct importance in design, design requirements may require that bounds be placed on the design sensitivity of system performance, due to variation in some parameter. In this case, the derivative with respect to the parameter must be bounded. The designer thus needs second derivatives, in order to adjust the design to stay within an acceptable range of first order sensitivity. These and other design requirements motivate the desire to calculate second design derivatives. A recently developed method of second order design sensitivity analysis [6] is presented here and extended to a new hybrid approach that substantially reduces computing requirements.

Direct Differentiation Method

To avoid notational difficulties that are associated with defining the derivative of a matrix with respect to a vector, consider one component of Eq. 2.6; i.e., the derivative of ψ with respect to the i-th component of b,

$$\psi_{b_i} = G^T(t^2, z(t^2), b) z_{b_i}(t^2) + g_{b_i} + \int_{t^1}^{t^2} (F_z z_{b_i} + F_{b_i}) dt \tag{3.1}$$

To further simplify notation, attention will be limited here to the case in which the condition $\Omega^2(t_2, z(t_2)) = 0$ does not depend explicitly on design.

Since ψ_{b_i} is a scalar quantity, it may be differentiated with respect to design, using the chain rule of differentiation and Eqs. 2.1 and 2.5, to obtain

$$\begin{aligned}
\psi_{b_i b} = {} & z_{b_i}^T(t^2) \left\{ -G_t^2 \left[\frac{\Omega_z}{\Omega_{t^2} + \Omega_z f(t^2)} \right] + G_z \right\} z_b(t^2) + z_{b_i}^T(t^2) G_b \\
& - G^T f_{b_i} \left[\frac{\Omega_z}{\Omega_{t^2} + \Omega_z f(t^2)} \right] z_b(t^2) + \left\{ g_{b_i z} - g_{b_i t^2} \left[\frac{\Omega_z}{\Omega_{t^2} + \Omega_z f(t^2)} \right] \right\} z_b(t^2) \\
& + g_{b_i b} - [F_z z_{b_i} + F_{b_i}] \left[\frac{\Omega_z}{\Omega_{t^2} + \Omega_z f(t^2)} \right] z_b(t^2) \\
& + \int_{t^1}^{t^2} [(F_z \tilde{z}_{b_i})_z z_b + F_{b_i z} z_b + F_{b_i b}] dt \\
& + \left\{ G^T(t^2) z_{b_i b}(t^2) + \int_{t^1}^{t^2} F_z z_{b_i b} dt \right\}
\end{aligned} \tag{3.2}$$

where $\tilde{}$ denotes a variable that is to be held fixed for the partial differentiation indicated.

In order to evaluate terms on the right side of Eq. 3.2, the first partial derivatives of z with respect to b may be evaluated using the direct design sensitivity analysis method of Section 2. Second derivatives of z with respect to design, however, arise and must be evaluated. To extend the direct design sensitivity analysis method, one can consider the differentiated state equations of Eq. 2.8 for one component of b and take a second derivative with respect to the vector b, to obtain

$$\left.\begin{aligned} \dot{z}_{b_i b} - f_z z_{b_i b} &= (f_z \tilde{z}_{b_i})_z z_b + (f_z \tilde{z}_{b_i})_b + f_{b_i z} z_b + f_{b_i b} \\ z_{b_i b}(t^1) &= h_{b_i b} \end{aligned}\right\} \qquad (3.3)$$

One might consider solving Eq. 3.3 for $i = 1 \ldots, k$ to obtain all second derivatives of state with respect to design and to substitute the result into Eq. 3.2, to complete calculation of the matrix of second derivatives of ψ with respect to design. While this is mathematically feasible, an exceptionally large number of computations would be required. First, the system of k first order equations of Eq. 2.8 must be solved for z_b, which is then substituted in the right side of Eq. 3.3. Then, the system of k^2 equations of Eq. 3.3 must be solved for the second derivatives of z with respect to design. All of these results would have to be stored and the results substituted into Eq. 3.2, for evaluation of second derivatives of ψ with respect to design. Taken with the original state equations of Eq. 2.1, this constitutes a total of $1 + k + k^2$ systems of differential equations, each being n first order differential equations in n unknowns. This approach is not pursued further, since it is clearly intractable.

Adjoint Variable Method

In order to more efficiently obtain second design derivatives of ψ, note that the functions appearing in Eq. 2.13 involve both the state variable $z(t;b)$ and the solution $\lambda(t;b)$ of Eqs. 2.11 and 2.12. Thus, in differentiating Eq. 2.13, the pair of initial- and terminal-value problems of Eqs. 2.1, 2.11 and 2.12 must be treated as state equations.

Using Leibniz rule [8], Eq. 2.13 yields, with ~ denoting a term that is treated as constant in the differentiation,

$$\frac{d^2\psi}{db^2} = g_{bb} + g_{bz}[\dot{z}(t^2)t_b^2 + z_b(t^2)] + g_{bt^2}t_b^2 - \left(h_b^T\tilde{\lambda}(t^1)\right)_b$$

$$- h_b^T\lambda_b(t^1) + \int_{t^1}^{t^2} \{[F_{bz} - (f_b^T\tilde{\lambda})_z]z_b - f_b^T\lambda b$$

$$+ [F_{bb} - (f_b^T\tilde{\lambda})_b]\}dt + [F_b(t^2) - \lambda^T(t^2)f_b(t^2)]^T t_b^2 \quad (3.4)$$

Using Eq. 2.5, the term t_b^2 may be eliminated and the result may be simplified to obtain

$$\frac{d^2\psi}{db^2} = \left\{- (g_{bz}f(t^2) + g_{bt^2} + [F_b(t^2) - \lambda^T(t^2)\ f_b(t^2)]^T)\left(\frac{\Omega_z}{\Omega_{t^2} + \Omega_z f(t^2)}\right)\right.$$

$$\left. + g_{bz}\right\} z_b(t^2) - h_b^T\lambda_b(t^1) + g_{bb} - (h_b^T\tilde{\lambda}(t^1))_b$$

$$+ \int_{t^1}^{t^2} \left\{[F_{bz} - (f_b^T\tilde{\lambda})_z]z_b - f_b^T\lambda_b + [F_{bb} - (f_b^T\tilde{\lambda})_b]\right\} dt \quad (3.5)$$

In order to explicitly evaluate the second derivatives of Eq. 3.5, terms involving z_b and λ_b must be rewritten explicitly in terms of computable quantities. Multiplying through Eq. 2.11 by an arbitrary matrix function Λ^T and integrating gives the identity

$$\int_{t^1}^{t^2} \Lambda^T[\dot{\lambda} + f_z^T\lambda - F_z^T]dt = 0 \quad (3.6)$$

Taking the derivative of Eq. 3.6 with respect to b gives

$$\int_{t^1}^{t^2} \Lambda^T[\dot{\lambda}_b + (f_z^T\tilde{\lambda})_z z_b + (f_z^T(b)\tilde{\lambda})_b + f_z^T\lambda_b - F_{zz}z_b - F_{zb}]dt = 0$$

Integrating the first term by parts gives the identity

$$\int_{t^1}^{t^2} \left\{[-\dot{\Lambda}^T + \Lambda^T f_z^T]\lambda_b + [\Lambda^T(f_z^T\tilde{\lambda})_z - \Lambda^T F_{zz}]z_b\right.$$

$$\left. + [\Lambda^T(f_z^T\tilde{\lambda})_b - \Lambda^T F_{zb}]\right\} dt + \Lambda^T(t^2)\lambda_b(t^2) - \Lambda^T(t^1)\lambda_b(t^1) = 0 \quad (3.7)$$

Differentiating Eq. 2.12 and using Eqs. 2.1, 2.5, and 2.11 yields

$$\lambda_b(t^2) = G_b + \left\{ Gz(t^2) - [Gt^2 + Gz(t^2)f(t^2) - F_z^T(t^2) + f_z^T(t^2)\lambda(t^2)] \times \left[\frac{\Omega_z}{\Omega_{t^2} + \Omega_z f(t^2)}\right]\right\} z_b(t^2) \equiv G_b + Hz_b(t^2) \tag{3.8}$$

where H is defined to be the extended expression in curved brackets. Substituting this result into Eq. 3.7 gives the following identity in Λ:

$$\int_{t^1}^{t^2} \left\{ [-\dot{\Lambda}^T + \Lambda^T f_z^T]\lambda_b + [\Lambda^T(f_z^T\tilde{\lambda})_z - \Lambda^T F_{zz}]z_b + [\Lambda^T(f_z^T\tilde{\lambda})_b - \Lambda^T F_{zb}] \right\} dt + \Lambda^T(t^2)G_b + \Lambda^T(t^2)Hz_b(t^2) - \Lambda^T(t^1)\lambda_b(t^1) = 0 \tag{3.9}$$

To take advantage of this identity, Λ may be selected to satisfy the initial-value problem (which has a unique solution [7])

$$\left.\begin{aligned} \dot{\Lambda} &= f_z \Lambda = f_b \\ \Lambda(t^1) &= h_b \end{aligned}\right\} \tag{3.10}$$

so that terms under the integral and boundary terms involving λ_b in Eq. 3.9 are equal to those that appear in Eq. 3.4. These terms may then be rewritten, using Eq. 3.9, and substituted into Eq. 3.5 to obtain

$$\frac{d^2\psi}{db^2} = \left\{ -\left(g_{bz} f(t^2) + g_{bt^2} + [F_b(t^2) - \lambda^T(t^2)\ f_b(t^2)]^T\right)\left(\frac{\Omega_z}{\Omega_{t^2} + \Omega_z f(t^2)}\right) + g_{bz} - \Lambda^T(t^2)H \right\} z_b(t^2) + [g_{bb} - (h_b^T\tilde{\lambda}(t^1))_b - \Lambda^T(t^2)G_b] + \int_{t^1}^{t^2} \left\{ [F_{bz} - (f_b^T\tilde{\lambda})_z - \Lambda^T(f_z^T\tilde{\lambda})_z + \Lambda^T F_{zz}]z_b + [F_{bb} - (f_b^T\tilde{\lambda})_b - \Lambda^T(f_z^T\tilde{\lambda})_b + \Lambda^T F_{zb}] \right\} dt \tag{3.11}$$

It now remains only to rewrite terms in Eq. 3.11 involving z_b in terms of computable quantities. To do this, the adjoint variable identity of Eqs. 2.9 and 2.10 may be rewritten in matrix form and a terminal-value problem with unique solution introduced as

$$\left.\begin{aligned}
&\dot{\Theta} + f_z^T\Theta = F_{bz}^T - (f_b^T\bar{\lambda})_z^T - (f_z^T\bar{\lambda})_z^T\Lambda + F_{zz}\Lambda \\
&\Theta(t^2)^T = \left(g_{bz}f(t^2) + g_{bt^2} + [F_b(t^2) - \lambda^T(t^2)f_b(t^2)]^T\right) \\
&\qquad \times \left(\frac{\Omega_z}{\Omega_{t^2} + \Omega_z f(t^2)}\right) - g_{bz} + \Lambda^T(t^2)H
\end{aligned}\right\} \tag{3.12}$$

where Θ is a matrix adjoint variable. Substituting from Eq. 3.12 into Eq. 3.11, terms involving z_b in Eq. 3.11 may be rewritten in terms of computable quantities, yielding

$$\begin{aligned}
\frac{d^2\psi}{db^2} &= g_{bb} - (h_b^T\bar{\lambda}(t_1))_b - \Lambda^T(t_2)G_b - \Theta^T(t_1)h_b \\
&+ \int_{t^1}^{t^2} [F_{bb} - (f_b^T\bar{\lambda})_b - \Lambda^T(f_z^T\bar{\lambda})_b + \Lambda^T F_{zb} - \Theta^T f_b]dt
\end{aligned} \tag{3.13}$$

Equation 3.13 now provides a computable expression for the second derivative of ψ with respect to design. Evaluation of these derivatives requires that the adjoint variables λ, Λ, and Θ be evaluated, as solutions of initial- and terminal-value problems of Eqs. 2.1, 2.11, 2.12, 3.10, and 3.12, respectively. All other terms involved in Eq. 3.13 are computable, so evaluation of the second derivatives involves only integration of known quantities. Note that this process requires numerical solution of 2 + 2k initial-value problems, whereas the direct differentiation method requires solution of $1 + k + k^2$ initial-value problems.

It is important to note that all the adjoint equations arising in this formulation have similar form. The coefficient matrix f_z or its transpose arise in all adjoint equations. In order to solve the state equation and the three adjoint equations numerically, a Runge-Kutta-Fiehlberg method of order 4 and 5 is used.

Integrating the state equation of Eq. 2.1 forward in time, the values of z at each node point in time can be obtained. Since the values of state variables are needed between adjacent node points, to solve the adjoint equations, cubic interpolation coefficients are calculated, using the values of state variables and their derivatives at each node point, and are stored. All data that are needed to integrate the adjoint equation of Eq. 3.10 forward in time are now available. The solution for Λ is obtained and interpolation coefficients are calculated and stored, for use in integrating Eq. 3.12.

Terminal-value problems of Eqs. 2.11, 2.12, and 3.12 may be transformed to initial-value problems by letting $\tau = t^2 + t^2 - t$ and changing the signs of $\dot{\lambda}$ and $\dot{\theta}$. Interpolation coefficients of λ are also calculated for use in Eq. 3.12. In this step, the integrals appearing in Eqs. 2.13 and 3.13 are calculated at each time step. The integration is performed using a Newton-Cotes formula of order 4.

The detailed <u>computational algorithm</u> used is as follows:

Step 1. Read input data, b_i, $i = 1, \ldots, k$, t^1, where k is the number of design variables and t^1 is the initial value of time.

Step 2. Define appropriate time steps Δt^1 to divide interior grid points in time.

Step 3. Set $t = t^1$ and calculate initial values of z_i, $i = 1, \ldots, n$.

Step 4. Integrate the state equation of Eq. 2.1 from t^j to $t^{j+1} = t^j + \Delta t^j$, $j = 0, \ldots, m$, where Δt is the time step defined in each time interval and m is the number of time steps. Store the calculated value of z_i at each time step as $z_i(t^{j+1})$.

Step 5. Calculate the cubic interpolation coefficients for z_i between t^j and t^{j+1}, using

$$z_i(t) = z_i(t^j) + a_{ij}\Delta t + b_{ij}(\Delta t)^2 + c_{ij}(\Delta t)^3,$$

$$i = 1, \ldots, n, \; j = 0, \ldots, m,$$

where

$$\Delta t = (t - t^j)$$

$$a_{ij} = \dot{z}_i(t^j)$$

$$b_{ij} = 3(z_i(t^{j+1}) - z_i(t^j))/(\Delta t^j)^2 - (\dot{z}_i(t^{j+1}) + 2\dot{z}(t^j))/\Delta t^j$$

$$c_{ij} = -2(z_i(t^{j+1}) - z_i(t^j))/(\Delta t^j)^3 + (\dot{z}_i(t^{j+1}) + \dot{z}_i(t^j))/(\Delta t^j)^2$$

Continue until Eq. 2.2 defines time t^2.

Step 6. Integrate the adjoint equation of Eq. 3.10 forward in time, as in steps 3 and 4. In this step, interpolation coefficients of z_i are used to calculate f_z and f_b, which are functions of z_i. Store the calculated value of $\Lambda_{ij}(t^\ell)$, $i = 1,\ldots, n$, $j = 1,\ldots, k$, $\ell = 0,\ldots, m$.

Step 7. Calculate cubic interpolation coefficients for Λ_{ij}, as in step 5, and store these values.

Step 8. For backward integration, define the new time parameter $\tau = t^2 + t^1 - t$. Set $t = t^2$ and calculate terminal value of Eqs. 2.12 and 3.12 for $\lambda(t^2)$ and $\Theta(t^2)$.

Step 9. Increasing τ from τ^{j+1}, integrate Eq. 2.11 by calculating the values of $\dot{\lambda}$ at t^j and changing the sign of these values. In this step, interpolation coefficients of z_i are used to calculate $\dot{\lambda}$.

Step 10. Calculate interpolation coefficients of λ_i, as in step 5.

Step 11. Using the interpolation coefficients of z_i, Λ_{ij}, and λ_i, calculate Θ_{ij} of Eq. 3.12 at time t_j, change the sign, and integrate from τ^j to τ^{j+1}.

Step 12. Calculate the integrands in Eqs. 2.14 and 3.13, using stored values of $z_i(t^j)$ and $\Lambda_{ij}(t^j)$ and the calculated values of λ_i and Θ_{ij}. Integrate using a Newton-Cotes formula of order 4.

Step 13. Repeat steps 9, 10, and 11 until $\tau = t^2$.

Step 14. Complete calculation of the first and second design derivatives of Eq. 2.13 and 3.13.

A Hybrid Method for Second Order Design Sensitivity Analysis

A simple observation allows for coupling of the direct first order design sensitivity analysis method and the adjoint variable technique to more efficiently solve the second order design sensitivity analysis problem. Consider that the direct first order design sensitivity analysis method of Section 2 has been used to obtain all first order derivatives of state with respect to design.

In this situation, the only terms in the second order design sensitivity formula of Eq. 3.2 that are not known are those involving second order derivatives of state with respect to design; i.e.,

$$SOT_i \equiv G^T z_{b_i b} + \int_{t^1}^{t^2} F_z z_{b_i b} dt \tag{3.14}$$

where SOT denotes second order terms that are to be computed. As noted earlier, direct solution of the second order state sensitivity equations of Eq. 3.3 is impractical.

As an alternative, multiply the differential equation of Eq. 3.3 by λ^T and integrate from t^1 to t^2, using an integration by parts, to obtain the following identity in λ:

$$\begin{aligned} -\int_{t^1}^{t^2} [\lambda^T(\dot{z}_{b_i b}) - \lambda^T f_z z_{b_i b}]dt \\ = - \lambda^T(t^2) z_{b_i b}(t^2) + \lambda^T(t^1) h_{b_i b} + \int_{t^1}^{t^2} [\dot{\lambda}^T + \lambda^T f_z] z_{b_i b} dt \\ \int_{t^1}^{t^2} \lambda^T [(f_z \bar{z}_{b_i})_z z_b + (f_z \bar{z}_{b_i})_b + f_{b_i z} z_b + f_{b_i b}]dt \end{aligned} \tag{3.15}$$

Following exactly the same argument as in the adjoint variable method for first order design sensitivity analysis, one may select λ so that the coefficients of $z_{b_i b}$ in the second line of Eq. 3.15 and in Eq. 3.14 are identical; i.e.,

$$\left.\begin{aligned} \dot{\lambda} + f_z^T \lambda = F_z^T \\ \lambda(t^2) = - G(t^2) \end{aligned}\right\} \tag{3.16}$$

Note that the terminal-value problem of Eq. 3.16 is identical to the terminal-value problem of Eqs. 2.11 and 2.12 for first order adjoint design sensitivity analysis. Using this result, one may evaluate the second order terms of Eq. 3.14, using Eq. 3.15 evaluated at the solution of Eq. 3.16, to obtain

$$\begin{aligned} SOT_i = - \lambda(t^1) h_{b_i b} - \int_{t^1}^{t^2} \lambda^T [(f_z \bar{z}_{b_i})_z z_b \\ + (f_z \bar{z}_{b_i})_b + f_{b_i z} z_b + f_{b_i b}]dt \end{aligned} \tag{3.17}$$

This result may be substituted into Eq. 3.2 to obtain an evaluation of all second derivatives of ψ with respect to design.

The remarkable aspect of this approach is that the adjoint equation of Eq. 3.16 does not depend on the index i. Therefore, only a single backward adjoint equation must be solved. Thus, to evaluate the full matrix of second derivatives in Eq. 3.2, one need only solve the single state equation of Eq. 2.1, the system of k first order state sensitivity equations of Eq. 2.8, and the single adjoint equation of Eq. 3.16. This is a system of only 2 + k systems of first order equations, each in n variables. For n >> 1, this is only one half the amount of computation that is required for the second order adjoint method presented in the preceeding subsection.

Second Order Design Sensitivity for the Oscillator Example

Since the oscillator example of Section 2 involves only two design variables, there is not a substantial difference in computational cost between the adjoint variable and the hybrid methods. Therefore, computations are carried out using the adjoint variable method. One first writes the adjoint equation of Eq. 3.10, as

$$\left.\begin{aligned}
&\begin{bmatrix} \dot{\Lambda}_{11} & \dot{\Lambda}_{12} \\ \dot{\Lambda}_{21} & \dot{\Lambda}_{22} \end{bmatrix} - \begin{bmatrix} 0 & 1 \\ -b_1 & 0 \end{bmatrix} \begin{bmatrix} \Lambda_{11} & \Lambda_{12} \\ \Lambda_{21} & \Lambda_{22} \end{bmatrix} = \begin{bmatrix} 0 & 0 \\ -z_1 & 0 \end{bmatrix} \\
&\qquad\qquad \begin{bmatrix} \Lambda_{11} & \Lambda_{12} \\ \Lambda_{21} & \Lambda_{22} \end{bmatrix}_{t=0} = \begin{bmatrix} 0 & 0 \\ 0 & 1 \end{bmatrix}
\end{aligned}\right\} \qquad (3.18)$$

The solution of Eq. 3.18 is

$$\begin{bmatrix} \Lambda_{11} & \Lambda_{12} \\ \Lambda_{21} & \Lambda_{22} \end{bmatrix} = \begin{bmatrix} \dfrac{b_2 t}{2b_1} \cos\sqrt{b_1}t - \dfrac{b_2}{2b_1^{\frac{3}{2}}} \sin\sqrt{b_1}t \;, & \dfrac{1}{\sqrt{b_1}} \sin\sqrt{b_1}t \\ -\dfrac{b_2 t}{2\sqrt{b_1}} \sin\sqrt{b_1}t \;, & \cos\sqrt{b_1}t \end{bmatrix} \qquad (3.19)$$

Similarly, Eq. 3.12 becomes

$$\begin{bmatrix} \dot\Theta_{11} & \dot\Theta_{12} \\ \dot\Theta_{21} & \dot\Theta_{22} \end{bmatrix} + \begin{bmatrix} 0 & -b_1 \\ 1 & 0 \end{bmatrix} \begin{bmatrix} \Theta_{11} & \Theta_{12} \\ \Theta_{21} & \Theta_{22} \end{bmatrix} = \begin{bmatrix} \lambda_2 & 0 \\ 0 & 0 \end{bmatrix}$$

$$\begin{bmatrix} \Theta_{11} & \Theta_{12} \\ \Theta_{21} & \Theta_{22} \end{bmatrix}_{t=\frac{\pi}{2}} = \begin{bmatrix} 0 & 0 \\ 0 & 0 \end{bmatrix} \tag{3.20}$$

Using the solution for λ_2 in Eq. 2.26, the solution of Eq. 3.20 is

$$\begin{bmatrix} \Theta_{11} & \Theta_{12} \\ \Theta_{21} & \Theta_{22} \end{bmatrix} = \begin{bmatrix} \left(\frac{\pi}{4\sqrt{b_1}}\right)\sin\sqrt{b_1}\,(\frac{\pi}{2} - t) - \left(\frac{t}{2\sqrt{b_1}}\right)\sin\sqrt{b_1}\,(\frac{\pi}{2} - t)\, , & 0 \\ \left(\frac{1}{2b_1^{\frac{3}{2}}}\right)\sin\sqrt{b_1}\,(\frac{\pi}{2} - t) - \left(\frac{\pi}{4b_1}\right)\cos\sqrt{b_1}\,(\frac{\pi}{2} - t) + \left(\frac{t}{2b_1}\right)\cos\sqrt{b_1}\,(\frac{\pi}{2} - t), & 0 \end{bmatrix} \tag{3.21}$$

The second design derivative of Eq. 3.13 can now be written as

$$\frac{d^2\psi}{db^2} = -\,\Theta^T(0)h_b + \int_0^{\frac{\pi}{2}} [-\,\Lambda^T(f_z^T\bar\lambda)_b - \Theta^T f_b]dt$$

$$= -\begin{bmatrix} 0 & \Theta_{21}(0) \\ 0 & 0 \end{bmatrix} + \int_0^{\frac{\pi}{2}} \left\{ \begin{bmatrix} \Lambda_{11}\lambda_2 & , & 0 \\ \Lambda_{12}\lambda_2 & , & 0 \end{bmatrix} + \begin{bmatrix} \Theta_{21}z_1 & , & 0 \\ \Theta_{22}z_1 & , & 0 \end{bmatrix} dt \right\} \tag{3.22}$$

Substituting the solutions for z, λ, Λ, and Θ of Eq. 2.16, 2.26, 3.19, and 3.21 into Eq. 3.22, one has the desired result

$$\frac{d^2\psi}{db^2} = \begin{bmatrix} \left\{ \frac{3b_2}{4b_1^{5/2}} - \frac{\pi^2 b_2}{16b_1^{3/2}} \sin\frac{\pi\sqrt{b_1}}{2} - \frac{3\pi b_2}{8b_1^2}\cos\frac{\pi\sqrt{b_1}}{2} \right\} & , & \left\{ -\frac{1}{2b_1^{3/2}}\sin\frac{\pi\sqrt{b_1}}{2} + \frac{\pi}{4b_1}\cos\frac{\pi\sqrt{b_1}}{2} \right\} \\ -\frac{1}{2b_1^{3/2}}\sin\frac{\pi\sqrt{b_1}}{2} + \frac{\pi}{4b_1}\cos\frac{\pi\sqrt{b_1}}{2} & , & 0 \end{bmatrix} \tag{3.23}$$

This is the same result as is obtained by direct differentiation of Eq. 2.17, so the second order design sensitivity analysis method yields precise results. The adjoint variable approach is, of course, not needed for this simple example. It is only presented in detail here to illustrate the method and to verify the theory.

As a numerical example, using the computation algorithm, the first and second design derivatives are calculated with $b = [1.0, 1.0]^T$. The numerical results and exact values of Eqs. 2.27 and 3.23 are given in Table 3.1.

Table 3.1 First and Second Design Derivatives of Simple Oscillator, for $b = [1.0, 1.0]^T$

Design Derivatives*	Numerical Results	Exact Values
ψ_1	-0.49999972724	-0.5
ψ_2	1.0000000000	1.0
ψ_{11}	0.13299722240	0.13314972501
ψ_{21}	-0.49963656519	-0.5
ψ_{22}	-0.49999972723	-0.5
ψ	0.0	0.0

$^*\psi_{ij} = \dfrac{\partial^2 \psi}{\partial b_i \partial b_j}$

Using the first and second design derivatives obtained, first and second order approximations of the functional ψ may be calculated, using Taylors formula; i.e.,

$$\psi(b^i) \simeq \psi(b^0) + \frac{d\psi}{db}(b^0)(b^i - b^0) \tag{3.24}$$

$$\psi(b^i) \simeq \psi(b^0) + \frac{d\psi}{db}(b^0)(b^i - b^0) + \frac{1}{2}(b^i - b^0)^T \frac{d^2\psi}{db^2}(b^0)(b^i - b^0) \tag{3.25}$$

where $b^0 = [1.0, 1.0]^T$, and $b^i = b^0 + i\delta b$, with $\delta b = [0.05, 0.05]^T$. Results are compared with the exact value of $\psi(b^i)$ in Fig. 3.1. It can be seen that the second order approximation is much more accurate than the first order approximation. This illustrates one potential value of second order design derivatives.

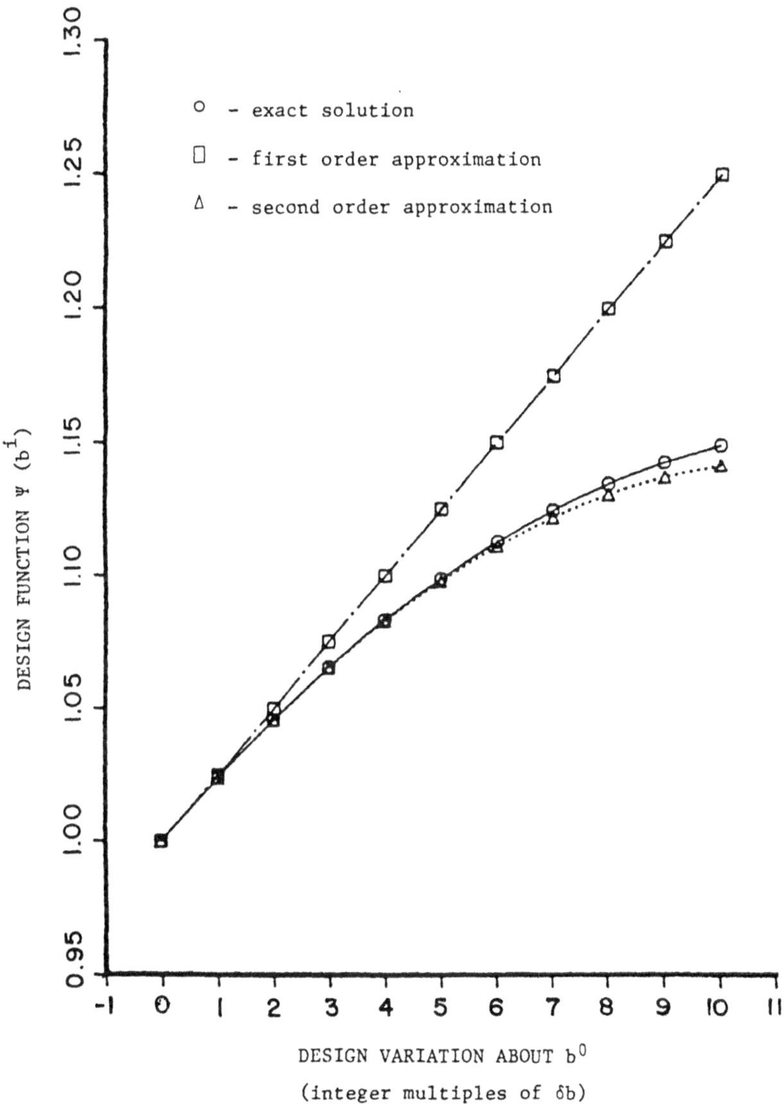

Figure 3.1 Linear and Quadratic Approximations of $\psi = z_1(t_2)$

4. DESIGN SENSITIVITY ANALYSIS OF A BURST FIRE AUTOMATIC CANNON

A burst fire automatic weapon can be modeled as shown in Fig. 4.1. The second order equation of motion of the recoiling mass is

$$m\ddot{x} - f_0 + cd(\dot{x}) + mg\sin\theta + F(t) = 0, \ 0 < t < \tau \tag{4.1}$$

with initial conditions

$$\left.\begin{aligned} x(0) &= 0 \\ \dot{x}(0) &= 0 \end{aligned}\right\} \tag{4.2}$$

where $cd(\dot{x})$ represents a general damping force.

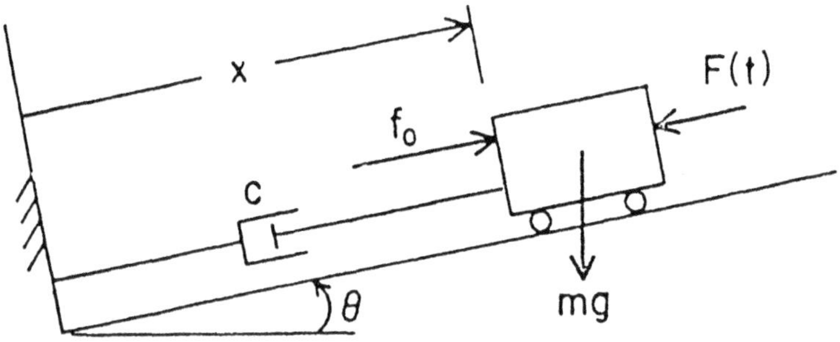

Figure 4.1 Model of Burst Fire Automatic Weapon

At $t = 0$, a latch is released, allowing the recoiling parts to move forward prior to firing, under the action of the actuator force f_0. Three shots are contained in a magazine that is attached to the recoiling parts and are to be fired in rapid succession, with a time Δt between shots. The system is to be designed so that the impulse of the rounds fired brings the recoiling parts to rest at a fixed distance to the rear of the initial position, so that they can be latched again at the position $x = 0$.

The period of feeding and firing a three round burst is τ sec., with each round exiting the tube in a characteristic period ε sec., with an impulse I_0. The ballistic force $F(t)$ acting on the recoiling parts is approximated by

$$F(t) = \begin{cases} \left(\frac{2I_0}{\varepsilon}\right) \sin^2(t - t_i)\frac{\pi}{\varepsilon} & , t_i < t < t_i + \varepsilon \quad , i = 1,2,3 \\ 0 & , \text{otherwise} \end{cases} \tag{4.3}$$

as shown in Fig. 4.2

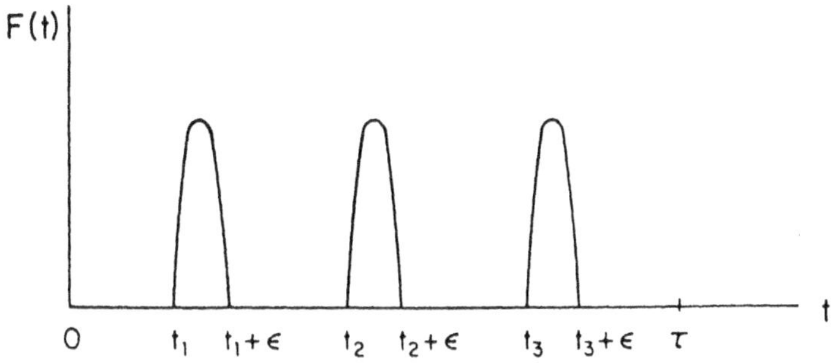

Figure 4.2 Ballistic Force

As a numerical example, let given data be as follows: mg = 1000 lb, m = 2.5892 lb/(in/sec^2), I_0 = 1000 lb - sec, $f_0 = 3.555 \times 10^3$ lb., and $\varepsilon = 0.005$ sec. Calculating the dynamic response with F(t) given in Eq. 4.3, it is found that for the given value of f_0 and with $c = 0$ and $\theta = 0$, $\tau = 8.461 \times 10^{-1}$ sec and $x(\tau) = -1.103494$ in.

It is important to know how $x(\tau)$ varies with variations in the parameters f_0, c, and θ. If variations of these parameters that may occur in application cause $x(\tau)$ to be positive, then a failure of function will occur.

<u>Linear Damping Case:</u> In the case of linear damping, $cd(\dot{x}) = c\dot{x}$ in Eq. 4.1. Using a first order state variable formulation, Eq. 4.1 can be written as

$$\begin{bmatrix} \dot{z}_1 \\ \dot{z}_2 \end{bmatrix} = \begin{bmatrix} z_2 \\ \frac{1}{m}\{b_1 - b_2 z_2 - mg\sin b_3 - F(t)\} \end{bmatrix} \equiv f(z,b) \ , \quad 0 < t < \tau \tag{4.4}$$

with initial conditions

$$\begin{bmatrix} z_1(0) \\ z_2(0) \end{bmatrix} = \begin{bmatrix} 0 \\ 0 \end{bmatrix} \equiv h(b) \tag{4.5}$$

where $b = [f_0, c, \theta]^T$ and F(t) is given in Eq. 4.3. The terminal time τ (playing the role to t^2 in earlier sections) is the time at which the extreme rearward position occurs, so it is determined by

$$\Omega(\tau, z(\tau)) = z_2(\tau) = 0 \tag{4.6}$$

The functional for which design sensitivity is to be calculated is

$$\psi = z_1(\tau) \equiv g(z(\tau), b) \tag{4.7}$$

Writing non-zero terms in Eqs. 2.11, 2.12, 3.10, and 3.12, the adjoint equations for λ are

$$\left.\begin{aligned} \dot{\lambda} + f_z{}^T \lambda &= 0 \\ \lambda(\tau) &= - g_z^T + \left(\frac{g_z f(\tau)}{\Omega_z f(\tau)}\right) \Omega_z^T \end{aligned}\right\} \tag{4.8}$$

where

$$f_z^T = \begin{bmatrix} 0 & 0 \\ 1 & -\frac{b_2}{m} \end{bmatrix}, \qquad f(\tau) = \begin{bmatrix} 0 \\ \dfrac{b_1 - mg\sin b_3 - F(\tau)}{m} \end{bmatrix} \tag{4.9}$$

$$g_z^T = \begin{bmatrix} 1 \\ 0 \end{bmatrix}, \qquad \Omega_z^T = \begin{bmatrix} 0 \\ 1 \end{bmatrix} \tag{4.10}$$

for Λ are

$$\left.\begin{aligned} \dot{\Lambda} - f_z \Lambda &= f_b \\ \Lambda(0) &= 0 \end{aligned}\right\} \tag{4.11}$$

where

$$f_b = \begin{bmatrix} 0 & 0 & 0 \\ \frac{1}{m} & -\frac{z_2}{m} & -g\cos b_3 \end{bmatrix} \tag{4.12}$$

and for θ are

$$\left.\begin{aligned} \dot{\theta} + f_z^T \theta &= - (f_b^T \tilde{\lambda})_z^T \\ \theta(\tau) &= - \frac{\Omega_z^T}{\Omega_z f(\tau)} (\lambda^T(\tau)\ f_b(\tau)) + H^T \Lambda(\tau) \end{aligned}\right\} \quad (4.13)$$

where

$$(f_b^T \tilde{\lambda})_z^T = \begin{bmatrix} 0 & 0 & 0 \\ 0 & -\frac{\lambda_2}{m} & 0 \end{bmatrix} \quad (4.14)$$

and

$$\frac{\Omega_z^T}{\Omega_z f(\tau)} (\lambda^T(\tau) f_b(\tau)) + H^T \Lambda(\tau) = \frac{1}{f_2(\tau)} \begin{bmatrix} 0 & 0 & 0 \\ \Lambda_{21}(\tau) & \Lambda_{22}(\tau) & \Lambda_{23}(\tau) \end{bmatrix}$$

In terms of the solutions of the state and adjoint equations, the first and second design derivatives of Eqs. 2.13 and 3.13 are

$$\frac{d\psi}{db} = - \int_0^\tau [\lambda^T f_b] dt \quad (4.15)$$

$$\frac{d^2\psi}{db^2} = - \int_0^\tau [(f_b^T \tilde{\lambda})_b + \Lambda^T (f_z^T \tilde{\lambda})_b + \theta^T f_b] dt \quad (4.16)$$

Using the nominal design $b^0 = [f_0, 0, 0]^T$, the first and second derivatives of ψ are calculated. From Eqs. 3.24 and 3.25, with the design variation $\delta b = [0.05f_0, 0.05, 0.05]^T$ and multiplies $b^i = b^0 + i\delta b$, $i = 1,2,\ldots, 10$, one and two term approximations of $\psi(b^i)$ are calculated. Results are given in Fig. 4.3. As can be expected, the second order approximation is more accurate than the first order approximation.

Quadratic Damping Case: In case of quadratic damping, $cd(\dot{x}) = c\dot{x}|\dot{x}|$ and the state equation becomes

$$\begin{bmatrix} \dot{z}_1 \\ \dot{z}_2 \end{bmatrix} = \begin{bmatrix} z_2 \\ \frac{1}{m}\ b_1 - b_2 z_2 |z_2| - mg\sin b_3 - F(t) \end{bmatrix} \equiv f(z,b) \qquad 0 < t < \tau \quad (4.17)$$

$$\begin{bmatrix} z_1(0) \\ z_2(0) \end{bmatrix} = 0 \equiv h(b)$$

where the terminal time τ is determined by Eq. 4.6. The response functional is the same as in Eq. 4.7.

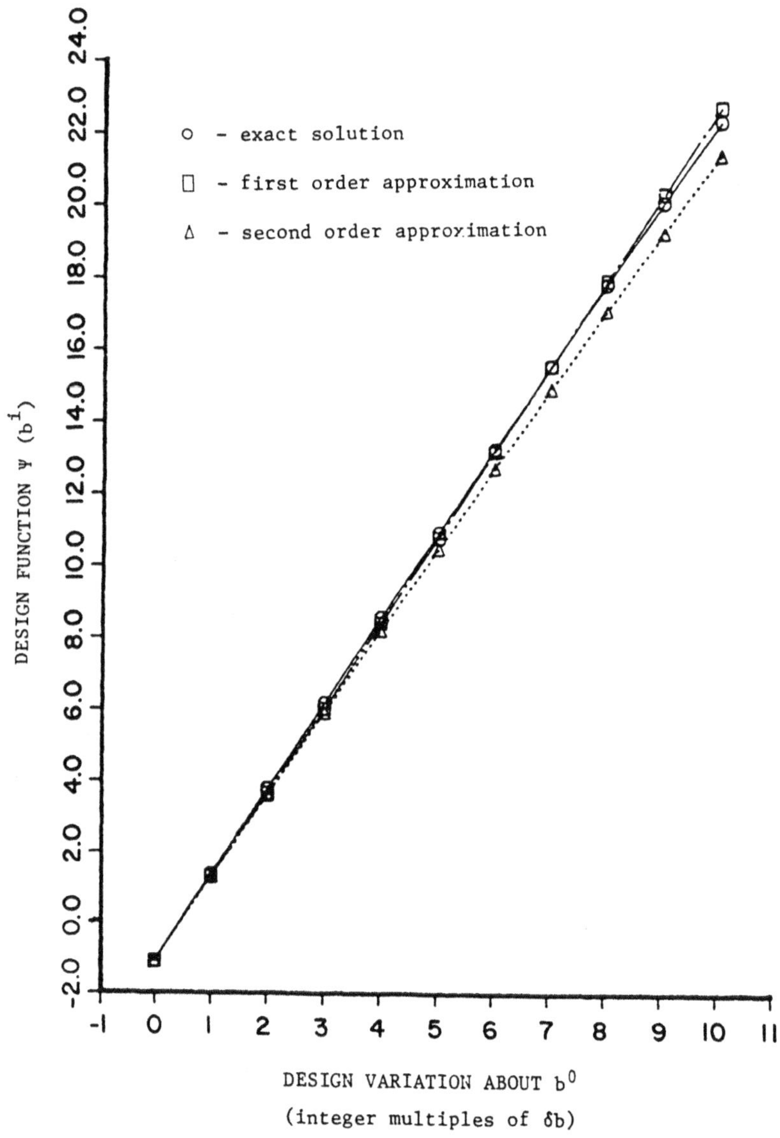

Figure 4.3 First and Second Order Approximation of ψ in Linear Damping Case

The adjoint equations for λ are

$$\left.\begin{aligned} &\dot{\lambda} + f_z^T\lambda = 0 \\ &\lambda(\tau) = - g_z^T + \left(\frac{g_z f(\tau)}{\Omega_z f(\tau)}\right) \Omega_z^T \end{aligned}\right\} \quad (4.18)$$

where

$$f_z^T = \begin{bmatrix} 0 & 0 \\ 1 & -\frac{2b_2}{m}|z_2| \end{bmatrix} \quad , \quad f(\tau) = \begin{bmatrix} 0 \\ \frac{b_1 - mgsinb_3 - F(\tau)}{m} \end{bmatrix} \quad (4.19)$$

$$g_z^T = \begin{bmatrix} 1 \\ 0 \end{bmatrix} \quad , \quad \Omega_z^T = \begin{bmatrix} 0 \\ 1 \end{bmatrix} \quad (4.20)$$

for Λ are

$$\left.\begin{aligned} &\dot{\Lambda} - f_z\Lambda = f_b \\ &\Lambda(0) = 0 \end{aligned}\right\} \quad (4.21)$$

where

$$f_b = \begin{bmatrix} 0 & 0 & 0 \\ \frac{1}{m} & -\frac{1}{m} z_2|z_2| & -gcosb_3 \end{bmatrix} \quad (4.22)$$

and for Θ are

$$\left.\begin{aligned} &\dot{\Theta} + f_z^T \,\Theta = - (f_b^T\tilde{\lambda})_z^T - (f_b^T\tilde{\lambda})_z^T\Lambda \\ &\Theta(\tau) = - \frac{\Omega_z}{\Omega_z f(\tau)}(\lambda^T(\tau) f_b(\tau)) + H^T\Lambda(\tau) \end{aligned}\right\} \quad (4.23)$$

where

$$(f_b^T\tilde{\lambda})_z^T = \begin{bmatrix} 0 & 0 & 0 \\ 0 & -\frac{2\lambda_2}{m}|z_2| & 0 \end{bmatrix} \quad (4.24)$$

$$\frac{\Omega_z^T}{\Omega_z f(\tau)}(f_b^T(\tau)\lambda(\tau)) + H^T\Lambda(\tau) = \frac{1}{f_2(\tau)} \begin{bmatrix} 0 & 0 & 0 \\ \Lambda_{21}(\tau) & \Lambda_{22}(\tau) & \Lambda_{23}(\tau) \end{bmatrix}$$

Using the solutions of the state and adjoint equations, first and second design derivatives are calculated from Eqs. 4.15 and 4.16, for

$b^0 = [f_0, 0, 0]^T$ and $b^3 = b^0 + 3\delta b$, where $\delta b = [0.005f_0, 0.005, 0.05]^T$. The one term and two term approximations of $\psi(b^i)$ are calculated for these two nominal designs with $b^i = b^0 + i\delta b$, $i = 1,2, \ldots, 10$, and $b^i = b^0 + (i+3)\delta b$, $i = -3, -2, \ldots, 7$, respectively. Results are given in Figs. 4.4 and 4.5.

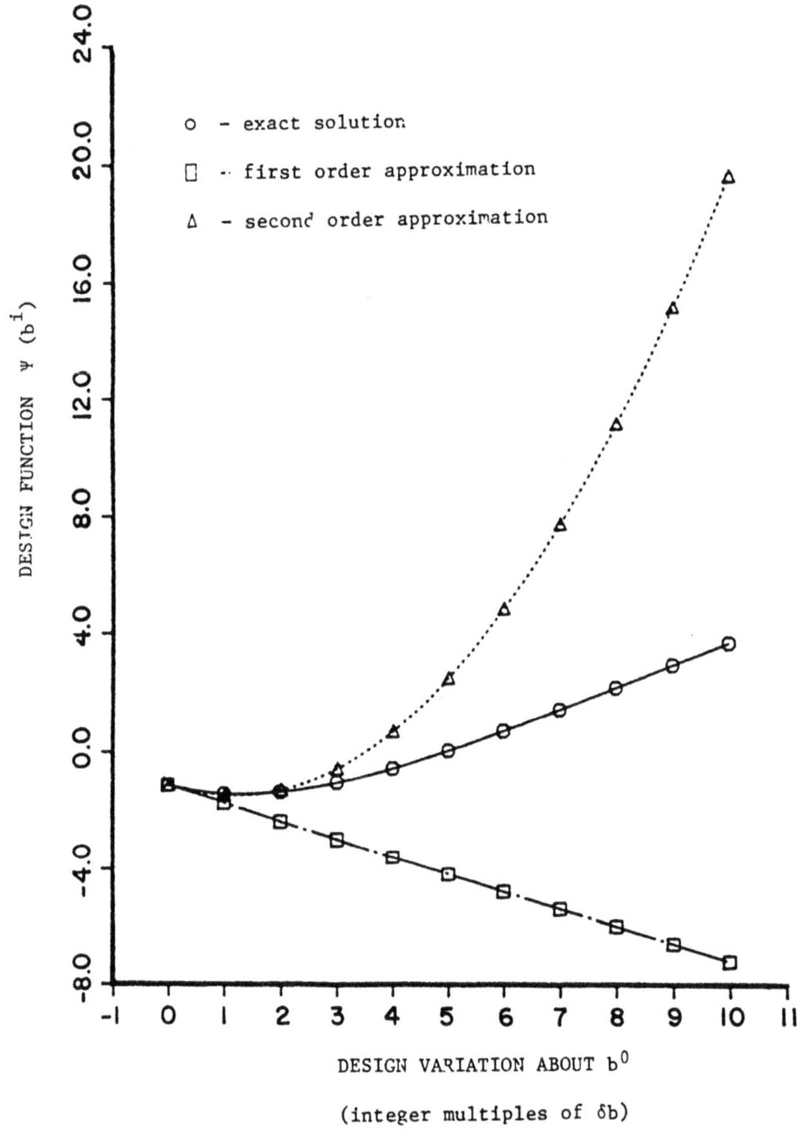

Figure 4.4 First and Second Order Approximations of ψ, with Quadratic Damping about b^0

Note that in both cases the quadratic approximation is far more accurate than the linear approximation, for moderate design variations. For the first nominal design, Fig. 4.4 shows that for very large design perturbations, the error in the quadratic approximation can be of the same order of magnitude as the first order approximation.

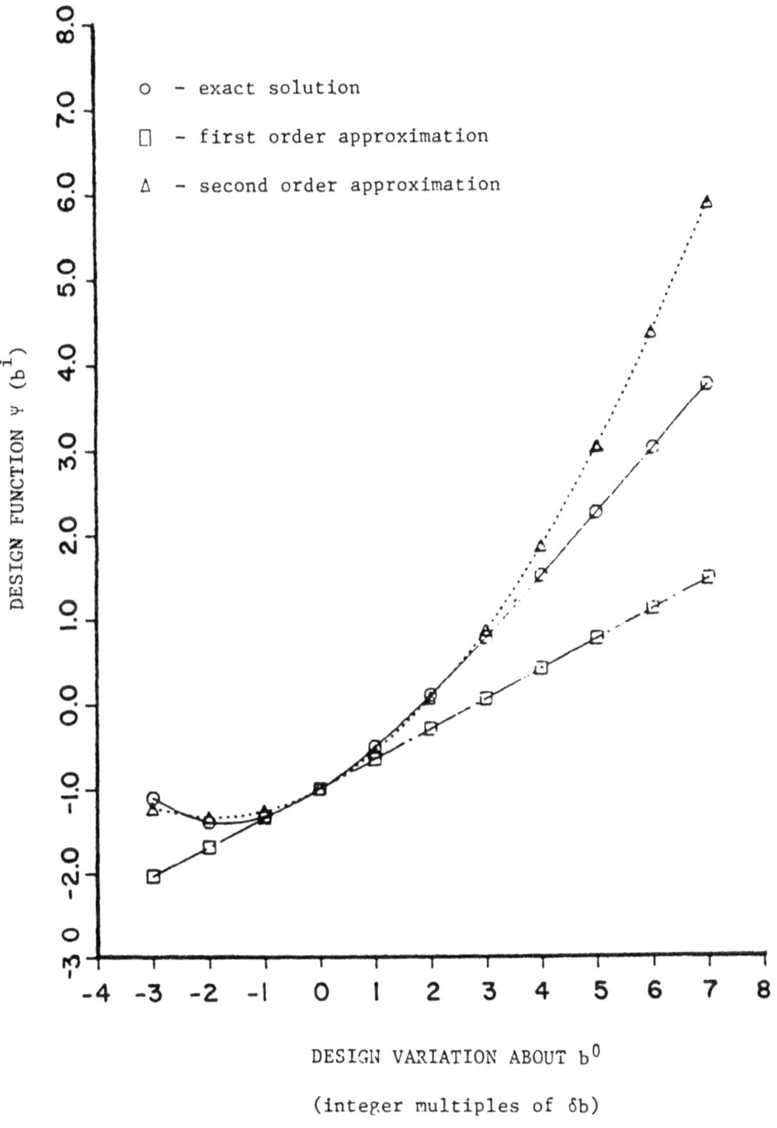

Figure 4.5 First and Second Order Approximations of ψ, with Quadratic Damping about b^3

5. VEHICLE SUSPENSION DYNAMIC OPTIMIZATION

A technique for vehicle suspension dynamic design optimization is presented in this section, using the design sensitivity analysis results of Section 2. Dynamic response measures included in the formulation, for use as the objective function or as constraints, include driver absorbed power [9], driver peak acceleration, and suspension element travel on three distinct road surfaces. Design parameters that are to be selected in the optimization process are suspension spring and damping characteristics.

A vehicle design optimization formulation is presented to minimize driver absorbed power on a nominal road, subject to bounds on absorbed power on a rough road, driver peak acceleration over a discrete obstacle, suspension jounce and rebound travel, wheel hop, and limits on design parameters. The adjoint variable method of Section 2 is used for calculation of first order design derivatives of vehicle dynamic response measures. An iterative optimization algorithm [10] is then used for vehicle design optimization.

Vehicle Model and Road Conditions

The vehicle model used is a five degree-of-freedom, plane, linear model shown in Fig. 5.1. The vehicle generalized coordinates are the passenger seat displacement x_1, the vehicle body vertical displacement x_2 and rotation x_3, and the front and rear wheel vertical displacements x_4 and x_5. The suspension spring stiffnesses are denoted by k_1 to k_5 and damping coefficients are c_1 to c_5. Lengths are denoted by L_1 to L_4. The functions $f_1(t)$ and $f_2(t)$ represent displacements of the front and rear wheels, respectively, due to undulation of the road surface on which the vehicle is traveling.

Equations of motion for the vehicle are derived, using Lagrange's equations. In matrix form, they are

$$M\ddot{x} + C\dot{x} + Kx = q(t) \tag{5.1}$$

where

$$M = \text{diag}(m_1, m_2, \ldots, m_5) \tag{5.2}$$

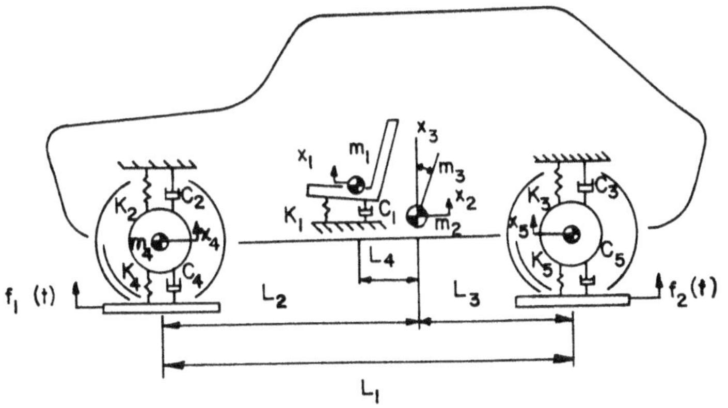

Figure 5.1 Five Degree-of-freedom Vehicle Model

m_3 is the pitch moment of inertia of the chassis and

$$C = \begin{bmatrix} c_1 & -c_1 & -L_4c_1 & 0 & 0 \\ & (c_1+c_2+c_3) & (L_4c_1+L_2c_2-L_3c_3) & -c_2 & -c_3 \\ & & (L_4^2c_1+L_2^2c_2+L_3^2c_3) & -L_2c_2 & L_3c_3 \\ & \text{Symmetric} & & (c_2+c_4) & 0 \\ & & & & (c_3+c_5) \end{bmatrix} \tag{5.3}$$

$$K = \begin{bmatrix} k_1 & -k_1 & -L_4k_1 & 0 & 0 \\ & (k_1+k_2+k_3) & (L_4k_1+L_2k_2-L_3k_3) & -k_2 & -k_3 \\ & & (L_4^2k_1+L_2^2k_2+L_3^2k_3) & -L_2k_2 & L_3k_3 \\ & \text{Symmetric} & & (k_2+k_4) & 0 \\ & & & & (k_3+k_5) \end{bmatrix} \tag{5.4}$$

$$q(t) = \begin{bmatrix} 0 \\ 0 \\ 0 \\ k_4f_1(t) + c_4\dot{f}_1(t) \\ k_5f_2(t) + c_5\dot{f}_2(t) \end{bmatrix} \tag{5.5}$$

If one defines

$$\left.\begin{aligned} z_i &= x_i \\ z_{5+i} &= \dot{x}_i \end{aligned}\right\} \quad i = 1,2,3,4,5 \tag{5.6}$$

then the differential equation of Eq. 5.1 and initial conditions can be transformed to the first-order form

$$\left.\begin{aligned} \dot{z}(t) &= f(t,z,b) \\ z(0) &= 0 \end{aligned}\right\} \tag{5.7}$$

where

$$z(t) = [z_1,z_2,\ldots z_{10}]^T \tag{5.8}$$

$$b = [b_1,b_2,\ldots b_6]^T = [k_1,k_2k_3,c_1,c_2,c_3]^T \tag{5.9}$$

and

$$f(t,z,b) = \begin{bmatrix} z_6 \\ z_7 \\ z_8 \\ z_9 \\ z_{10} \\ \frac{1}{m_1}(-b_1z_1+b_1z_2+L_4b_1z_3-b_4z_6+b_4z_7+L_4b_4z_8) \\ \frac{1}{m_2}\{b_1z_1-(b_1-b_2-b_3)z_2-(L_4b_1+L_2b_2-L_3b_3)z_3+b_2z_4+b_3z_5 \\ +b_4z_6-(b_4+b_5+b_6)z_7-(L_4b_4+L_2b_5-L_3b_6)z_8+b_5z_9+b_6z_{10}\} \\ \frac{1}{m_3}\{L_4b_1z_1-(L_4b_1+L_2b_2-L_3b_3)z_2-(L_4^2b_1+L_2^2b_2+L_3^2b_3)z_3 \\ +L_2b_2z_4-L_3b_3z_5+L_4b_4z_6-(L_4b_4+L_2b_5-L_3b_6)z_7 \\ -(L_4^2b_4+L_2^2b_5+L_3^2b_6)z_8+L_2b_5z_9-L_3b_6z_{10}\} \\ \frac{1}{m_4}\{b_2z_2+L_2b_2z_3-b_2z_4-k_4(z_4-f_1(t)) \\ +b_5z_7+L_2b_6z_8-b_5z_9-c_4(z_9-\dot{f}_1(t))\} \\ \frac{1}{m_5}\{b_3z_2-L_3b_3z_3-b_3z_5-k_5(z_5-f_2(t)) \\ +b_6z_7-L_3b_6z_8-b_6z_{10}-c_5(z_{10}-\dot{f}_2(t))\} \end{bmatrix} \tag{5.10}$$

Numerical values of the model parameters used in this study are given in Table 5.1. The spring and damping coefficients given for the driver's seat and suspension will be the starting values for iterative optimization.

Table 5.1. System Parameters

Index	1	2	3	4	5
Generalized masses m_i (slugs or lb-in-sec^2)	9.01	139.75	41000	3.0	3.0
Spring coefficient k_i (lb/in.)	100	300	300	1500	1500
Damping coefficient c_i (lb-sec/in.)	10	25	25	5	5
Distance, L_i (in)	120	40	80	10	-

In this vehicle model, the tires are modeled as point followers, that are always in contact with the ground. If a tensile force occurs between wheel and ground, wheel hop would actually occur. In the design formulation, a constraint that precludes wheel hop is imposed.

Dynamic response of the vehicle is determined by the vertical displacement history of the wheels on the road surface. A typical road condition is defined as a sinusoidal undulation, with amplitude y_0 and variable half-wavelength ℓ_i. The front tire displacement v(y) at a location y is defined as

$$v(y) = \begin{cases} y_0[1 - \cos \dfrac{\pi(y-y^{i-1})}{\ell_i}], & y^{i-1} < y < y^i, \text{ i is odd} \\ y_0[1 + \cos \dfrac{\pi(y-y^{i-1})}{\ell_i}], & y^{i-1} < y < y^i, \text{ i is even} \end{cases}$$

where y is a coordinate that is measured along the road and $y^i = \sum_{j=1}^{i} \ell_j$.

If the speed of the vehicle is denoted by s, the elapsed time between front and rear tire encounter of the same point on the road surface is $t_\sigma = L_1/s$, where L_1 is the distance between front and rear wheels. Then,

$$v(t) = \begin{cases} y_0[1 - \cos(w_i(t-t^{i-1}))], & t^{i-1} \le t \le t^i, \text{ i is odd} \\ y_0[1 + \cos(w_i(t-t^{i-1}))], & t^{i-1} \le t \le t^i, \text{ i is even} \end{cases} \quad (5.11)$$

where $w_i = \pi s/\ell_i$ and $t^i = y^i/s$. The vertical displacement function for the front wheel can therefore be defined as

$$f_1(t) = \begin{cases} v(t), & 0 \le t \le T_1 \\ 0 \quad , & \text{otherwise} \end{cases} \quad (5.12)$$

where T_1 is the time at which the road undulation ceases. The vertical displacement of the rear wheel has the same value as that of the front wheel, but with a time lag; i.e.,

$$f_2(t) = f_1(t-t_\sigma), \quad t_\sigma \le t \le T_1 + t_\sigma \quad (5.13)$$

Consideration is limited here to the three road profiles shown in Fig. 5.2. Profile 1 is a continous sinusoidal curve with a constant half-wavelength of 480 in. and an amplitude of $y_0 = 2$ in. This profile represents a smooth highway condition. Profile 2 is a continuous sinusoidal curve with a constant half-wavelength of 60 in. and an amplitude of 2 in., which represents a rough highway condition. Profile 3 is a combination of two sinusoidal curves with different half-wavelengths $\ell_1 = 360$ in. and $\ell_2 = 144$ in. and an amplitude of 5 in. This profile represents a severe bump condition. The vehicle speeds for each road profile are as follows:

960 in/sec (54.5 mile/h) for profile 1
616 in/sec (35.0 mile/h) for profile 2
450 in/sec (25.6 mile/h) for profile 3

In actual design of a vehicle, a more extensive set of road profiles and vehicle speeds would be employed, perhaps including random road profiles. The set of road conditions used in this study was selected to illustrate use of the optimization method on multiple road conditions.

Driver Absorbed Power

Driver absorbed power is selected here as a measure of the rate at which energy is absorbed by the human body. It is a quantity that may be used to determine human tolerance to vibration when a vehicle

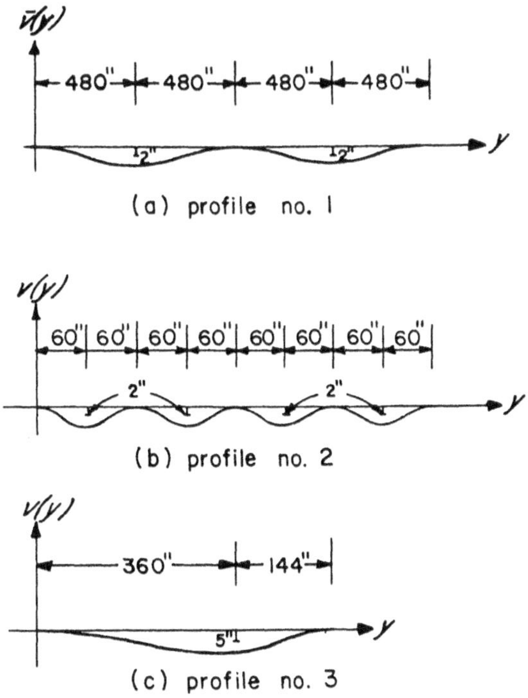

Figure 5.2 Road Surface Profiles

is negotiating rough terrain. Absorbed power equations that have been developed through extensive experimentation are given in the NATO Reference Mobility Model [11] as

$$\left.\begin{aligned} \dot{p}(t) &= h(p,z,b), \quad 0 < t < \tau \\ p(0) &= 0 \end{aligned}\right\} \tag{5.14}$$

where

$$p(t) = [p_1, p_2, \ldots, p_7]^T$$

and

$$h = \begin{bmatrix} -29.8p_1 - 497.49\ddot{z}_1 - 100p_2 \\ 10p_1 \\ 1736.9p_1 - 108p_4 \\ 100p_1 - 35.19p_3 - 39.1p_4 \\ -315.7p_1 + 34.0956p_4 + 171.075p_6 \\ -80.0p_1 - 91.36p_4 - 30.28p_5 \\ p_1 - 0.108p_4 + 0.25p_6 - 6p_7 \end{bmatrix} \qquad (5.15)$$

Here, $\ddot{z}_1$ is the vertical acceleration of the driver's seat of the vehicle, in units of g's. From Eq. 5.10,

$$\ddot{z}_1 = -\frac{1}{m_1 g}\{b_1(z_1 - z_2 - L_4 z_4) + b_4(z_6 - z_7 - L_4 z_8)\} \qquad (5.16)$$

where g is the gravitational acceleration. The average absorbed power is defined, in units of Watts, as

$$AP \equiv \frac{1}{\tau}\int_0^{\tau} (p_1(t) - 0.108p_4(t) + 0.25p_6(t) - p_7(t))^2 dt \qquad (5.17)$$

Using the solution of Eq. 5.14, absorbed power can be calculated numerically.

The Optimal Design Problem

With equations of motion, absorbed power equations, and terrain displacement functions defined, one can now define the optimal design problem. It is desired to make the driver as comfortable as possible, over a range of road conditions and traveling speeds. The design objective selected here is to minimize driver absorbed power on the smooth highway condition, subject to bounds on the following:

(i) absorbed power and wheel hop at the rear wheel on the rough highway condition,

(ii) maximum vertical acceleration of the driver's seat, wheel hop at the rear wheel, and rattle spaces between the seat and body, body and wheels, and wheels and ground on the bump road condition,

(iii) design variables.

The optimal design problem may be summarized as follows: Choose the design variable vector b to minimize

$$\psi_0 = \int_0^{\tau} F(p^1(t))dt \tag{5.18}$$

where the superscript denotes road profile number and

$$F(p) = \{p_1(t) - 0.108p_4(t) + 0.25p_6(t) - p_7(t)\}^2/\tau \tag{5.19}$$

subject to the state equation of Eq. 5.7, the absorbed power equation of Eq. 5.14, and the following constraints:

$$\psi_1 = \int_0^{\tau} G(p^2(t))dt - \theta_1 \le 0 \tag{5.20}$$

$$\psi_2 = z_5^2 - f_2(t) - \theta_2 \le 0, \quad 0 \le t \le \tau \tag{5.21}$$

$$\psi_3 = -(z_5^2 - f_2(t)) - \theta_3 \le 0, \quad 0 \le t \le \tau \tag{5.22}$$

$$\psi_4 = \left| \frac{1}{m_1} \{b_1(z_2^3 + L_4^3 z_3 - z_1) + b_4(z_7^3 + L_4 z_8^3 - z_6^3)\} \right|$$

$$- \theta_4 \le 0, \quad 0 \le t \le \tau \tag{5.23}$$

$$\psi_5 = \left| z_2^3 + L_4 z_3^3 - z_1^3 \right| - \theta_5 \le 0, \quad 0 \le t \le \tau \tag{5.24}$$

$$\psi_6 = \left| z_4^3 - z_2^3 - L_2 z_3^3 \right| - \theta_6 \le 0, \quad 0 \le t \le \tau \tag{5.25}$$

$$\psi_7 = \left| z_5^3 - z_2^3 - L_3 z_3^3 \right| - \theta_7 \le 0, \quad 0 \le t \le \tau \tag{5.26}$$

$$\psi_8 = \left| z_4^3 - f_1(t) \right| - \theta_8 \le 0, \quad 0 \le t \le \tau \tag{5.27}$$

$$\psi_9 = z_5^3 - f_2(t) - \theta_9 \le 0, \quad 0 \le t \le \tau \tag{5.28}$$

$$\psi_{10} = -(z_5^3 - f_2(t)) - \theta_{10} \le 0, \quad 0 \le t \le \tau \tag{5.29}$$

where θ_i, i = 1,2,...,10 are constraint bounds, z^j and p^j are state and power solutions, and j is the road profile number, defined as j=1 for the smooth highway, j=2 for the rough highway, and j=3 for the bump. Constraints on design variables are imposed as

$$b_i^L < b_i < b_i^U, \quad i=1,2,3,4,5,6 \tag{5.30}$$

where

θ_1 = maximum allowable absorbed power, on rough highway.

θ_2, θ_9 = maximum allowable upward distance between rear wheel and road surfaces. Since the tire should be always in contact with the road, this value may be static deflection of rear tire, neglecting dynamic effect of tire. Road surfaces are rough highway and bump road, respectively.

θ_3, θ_{10} = maximum allowable downward distance between rear wheel and rough highway and bump road, respectively.

θ_4 = maximum allowable acceleration at driver's seat, on bump road.

θ_5 = maximum allowable distance between driver's seat and chassis, on bump road.

θ_6, θ_7 = maximum allowable distance between body, and front wheel and rear wheel, respectively, for bump road.

θ_8 = maximum allowable distance between front wheel and road, on bump road.

The values of θ_i and design bounds for this study are given in Tables 5.2 and 5.3.

Table 5.2. θ_i Values

θ_1	3.5 Watts
θ_2	1.1127 in
θ_3	3.0 in
θ_4	350 in/sec^2
θ_5	2.0 in
θ_6	5.5 in
u_7	5.5 in
θ_8	2.0 in
θ_9	1.1127 in
θ_{10}	3.0 in

Table 5.3. Design Bounds

i	b_i^L	b_i^U
1	25 lb/in	500 lb/in
2	100 lb/in	1000 lb/in
3	100 lb/in	1000 lb/in
4	1.0 lb-sec/in	50 lb-sec/in
5	2.5 lb-sec/in	80 lb-sec/in
6	2.5 lb-sec/in	80 lb-sec/in

Design Sensitivity Analysis

The dynamic system treated here is described by a design variable vector $b = [b_1, b_2, \ldots b_6]^T$ and state variable vectors $z(t) = [z_1(t), z_2(t), \ldots, z_{10}(t)]^T$ and $p(t) = [p_1(t), p_2(t), \ldots p_7(t)]^T$, which are the solutions of initial value problems of Eqs. 5.7 and 5.14, rewritten here as

$$\begin{bmatrix} \dot{z} \\ p \end{bmatrix} = \begin{bmatrix} f(t,z,b) \\ h(p,z,b) \end{bmatrix} \tag{5.31}$$

$$\begin{bmatrix} z(0) \\ p(0) \end{bmatrix} = \begin{bmatrix} 0 \\ 0 \end{bmatrix} \tag{5.32}$$

The design problem may be summarized in terms of functionals of the form

$$\psi_i = \int_0^\tau F(p^i(t))dt - \theta_i, \quad i=0,1 \tag{5.33}$$

where $\theta_0 = 0$ and

$$\psi_i = g_i(t^i,z,b) - \theta_i, \quad i=2,3,...10 \tag{5.34}$$

The functions g_i are defined by the right sides of Eqs. 5.21-29 and t^i is determined by the conditions

$$\Omega^i(t^i,z,b) \equiv \frac{d}{dt} g_i(t,z,b)\Big|_{t=t^i} = 0, \quad i=2,3,..., 10 \tag{5.35}$$

That is, Eq. 5.35 defines times t^i at which the functions on the right sides of Eqs. 5.21-29 are maximum. In this way, the constraints of Eq. 5.34 are imposed only at isolated times t^i, whereas Eqs. 5.21-29 must hold for $0 \leq t \leq \tau$.

The absorbed power functionals of Eq. 5.33 involve both p and z, so the composite state equations of Eqs. 5.31 and 5.32 must be used. Denoting the composite adjoint variable as $[\lambda^T, \gamma^T]^T$, the adjoint equations of Eqs. 2.12 and 2.13 are

$$\begin{bmatrix} \dot{\lambda} \\ \dot{\gamma} \end{bmatrix} + \begin{bmatrix} f_z & 0 \\ h_z & h_p \end{bmatrix}^T \begin{bmatrix} \lambda \\ \gamma \end{bmatrix} = \begin{bmatrix} 0 \\ F_p \end{bmatrix} \tag{5.36}$$

$$\begin{bmatrix} \lambda(\tau) \\ \gamma(\tau) \end{bmatrix} = 0 \tag{5.37}$$

since $f_p = 0$, $g = 0$, and $\Omega = t_2 - \tau = 0$.

The differential equation of Eq. 5.36 may be expanded as

$$\dot{\gamma} + h_p^T\gamma = F_p \tag{5.38}$$

$$\dot{\lambda} + f_z^T\lambda = - h_z^T\gamma \tag{5.39}$$

Note that one can first integrate Eq. 5.38 backward in time for $\gamma(t)$ and then use the result to integrate Eq. 5.39 backward in time for $\lambda(t)$. Uncoupling the equations in this way substantially reduces the computational burdon. These computations are repeated for i = 0 and 1, to obtain design derivatives of ψ_0 and ψ_1.

Since $\Omega^i = t^i - \tau = 0$ and $g_i = 0$, i = 0 and 1, h = 0, and $F_b = 0$, Eq. 2.14 yields

$$\frac{d\psi_i}{db} = -\int_0^T [\lambda^T f_b + \gamma^T h_b]dt \tag{5.40}$$

For constraint functionals of Eq. 5.34, t^i plays the role of t^2 in Section 2 and is determined by Eq. 5.35. Since the absorbed power state variable does not arise in these functionals, only the physical state equation of Eq. 5.7 need be considered. From Eqs. 2.11 and 2.12, the adjoint problem is

$$\dot{\lambda} + f_z^T \lambda = 0 \tag{5.41}$$

$$\lambda(t^i) = -(g_i)_z^T + \beta_i \Omega_z^{iT} \tag{5.42}$$

where

$$\beta_i = \frac{(g_i)_z f(t^i) + (g_i)_{t^i}}{\Omega^i_{t^i} + \Omega^i_z f(t^i)} \tag{5.43}$$

Equation 2.14 yields the desired result,

$$\frac{d\psi_i}{db} = (g_i)_b - \beta_i \Omega_b^i - \int_0^{t^i} \lambda^{i^T} f_b dt$$

where λ^i is the solution of Eqs. 5.41 and 5.42.

Computation of design sensitivities for a given design is summarized as follows:

For ψ_i, i=0,1 (absorbed power functional);

(1) Integrate state equations of Eqs. 5.7 and 5.14 from t=0 to τ, where τ is the time when dynamic response of the vehicle is near the steady state.

(2) Integrate adjoint equations of Eqs. 5.38 and 5.39 backward from τ to 0, using known state variables.

(3) Calculate design derivatives of Eq. 5.40.

For ψ_i, i=2,3,4,...10;

(1) Integrate the state equation of Eq. 5.7 from t=0 to some large time τ for the given road condition, to include the time t^i of maximum g_i.

(2) Find t^i satisfying Eq. 5.35 by using a root finding algorithm.

(3) Integrate the adjoint equations of Eqs. 5.41 and 5.42 backward from $t=t^i$ to 0.

(4) Calculate design derivatives of Eq. 5.44.

The cost and constraint values and their design derivatives for the initial design $b = [100, 200, 230, 2.8, 70, 22]^T$ are calculated. In that design, only ψ_1 and ψ_5 are violated. The state and absorbed power equations and the corresponding adjoint equations are integrated using a Runge-Kutta-Fiehlberg method of order 4 and 5, with a relative error of 10^{-9} and an absolute error of 10^{-6}. The error tolerance in satisfying $\Omega_i = 0$ to find t_i is 10^{-9}.

To check the validity of design derivatives, design variations are created by making a change of 5% in each design variable, one variable at at time. Predicted variations are calculated using the design derivatives derived. They agree with the actual variation, with a maximum deviation of 9%. With design variation $\delta b = 0.05\ [b_1, b_2, \ldots b_6]^T$ and multiples $b^i = b^0 + i\delta b$, $i=1,2,3,4$; i.e., from 5% to 20% design variation, the actual and predicted values of cost and selected constraints are given in Figs. 5.3 to 5.5, where the solid line is the actual value and the dotted line is the predicted value. Predicted values show good correspondence with actual values, for substantial variations in design.

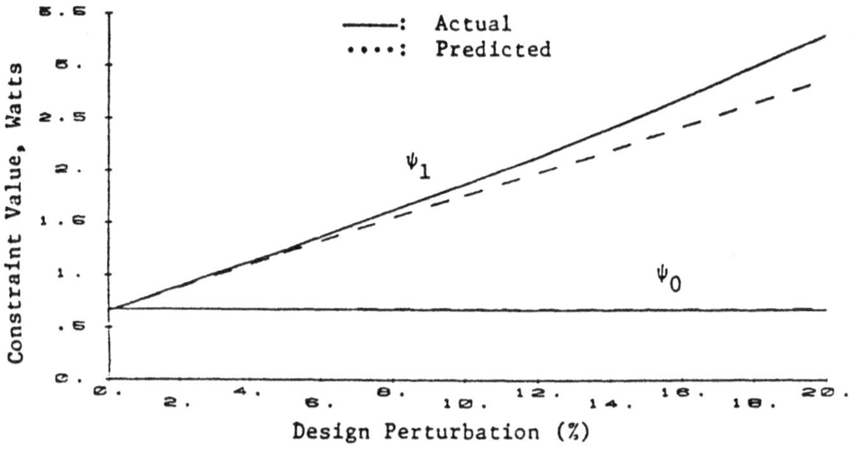

Figure 5.3 Accuracy Test of Design Sensitivity of Cost (ψ_0) and Absorbed Power on Rough Highway (ψ_1)

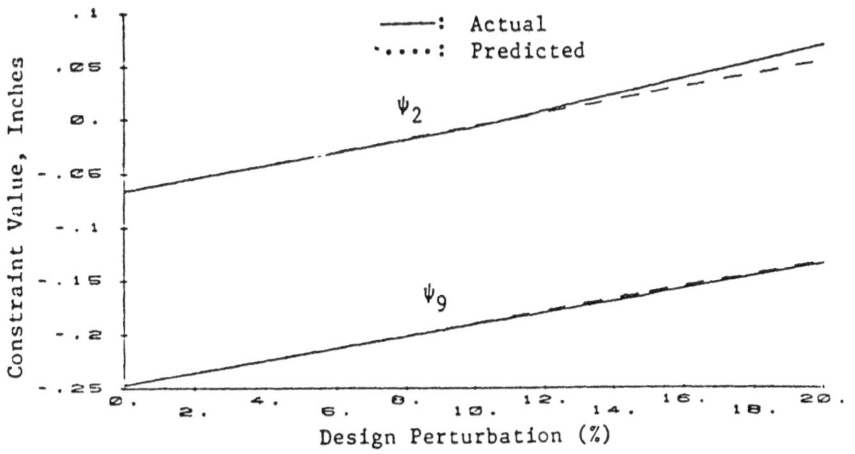

Figure 5.4 Accuracy Test of Design Sensitivity of Rear Wheel Hop Constraint on Rough Highway (ψ_2) and Bump (ψ_9)

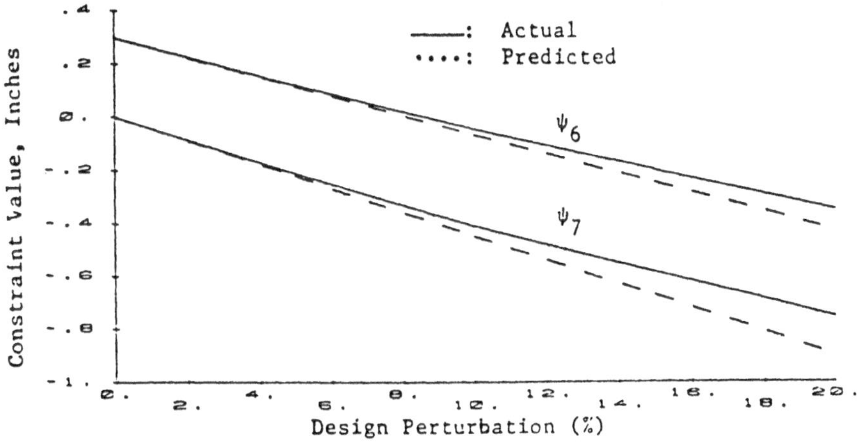

Figure 5.5 Accuracy Test of Design Sensitivity of Rattle Space Constraint Between Chassis and Front Wheel (ψ_6) and Chassis and Rear Wheel (ψ_7), Over Bump Road Conditions

Iterative Optimization Algorithm

Many mathematical programming algorithms for solving optimization problem have been developed for engineering design optimization. In this paper, a sequential quadratic programming approach, called the linearization method of Pshenichny [10], is used. This algorithm was originally presented in the Russian literature, but has apparently only recently come to the attention of workers in the West. Pshenichny has proved global convergence of the algorithm, using an active-set strategy that is essential in large scale mechanical optimization problems. Details of the algorithm are summarized in Appendix B of this paper.

Numerical Results

The initial design shown in Table 5.4 is chosen from Ref. 3. With the given initial design, the peak acceleration is 331.8 in/sec^2 and the cost is 0.847 watts. For this design, constraints on absorbed power on the rough highway and rear wheel-hop and rattle space constraints between chassis and wheel assemblies on the bump condition are violated. This initial design, from the viewpoint of ride comfort and safety, is poor.

After the 7th design iteration, the rear wheels are in contact with the road surface; i.e., the wheel hop constraint is satisfied. Constraints on absorbed power and rattle space between chassis and front wheel assembly are satisfied after the 21st iteration. The optimum design, given in Table 5.4, is obtained at the 23rd iteration. The peak acceleration is 342 in/sec^2 and the cost is 1.08 watts. However, comfort and safety on each road condition are

Table 5.4. Initial and Optimum Designs

Design Variable Description			Initial Design	Optimum Design
Driver seat spring constant	[lb/in]	b_1	100.	126.26
Spring constant of front suspension	[lb/in]	b_2	300.	356.35
Spring constant of rear suspension	[lb/in]	b_3	300.	274.08
Driver seat damping coefficient	[lb-sec/in]	b_4	10.	2.47
Damping coefficient of front suspension	[lb-sec/in]	b_5	25.	49.87
Damping coefficient of rear suspension	[lb-sec/in]	b_6	25.	14.90

much improved. As a measure of comfort, the history of absorbed power on the rough highway is shown in Fig. 5.6. Cost and maximum violation are plotted in Fig. 5.7.

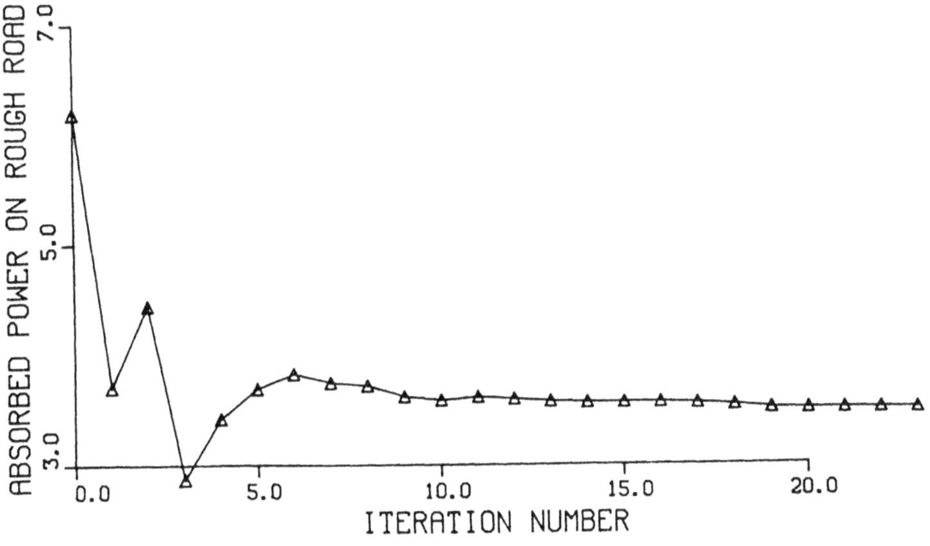

Figure 5.6 History of Absorbed Power Constraint (ψ_1)

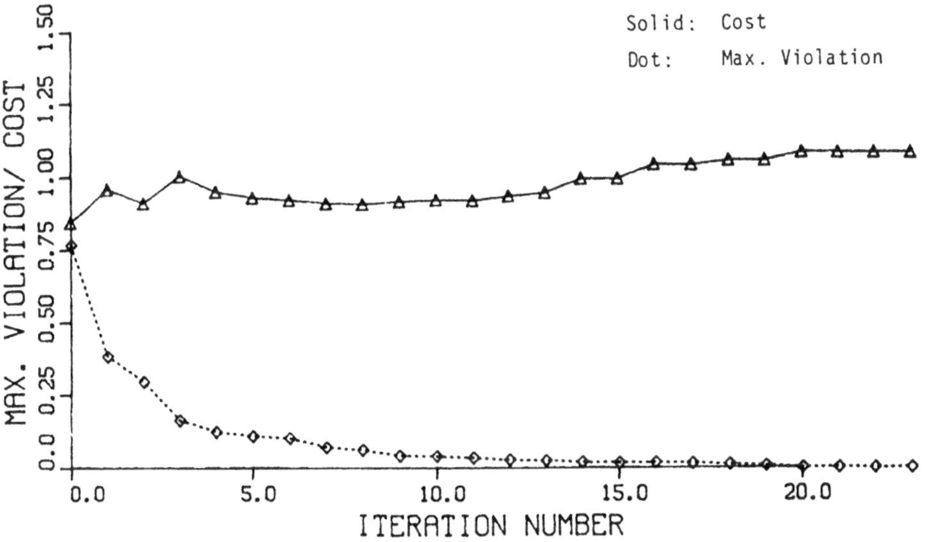

Figure 5.7 History of Cost and Maximum Violation

Results of the optimization process, as illustrated in Fig. 5.7, show that absorbed power over the smooth highway has been increased slightly, but that potential hazardous characteristics of the initial design have been brought under control by eliminating violations in constraints. The general tendency in this design optimization example is

(1) to increase stiffness and damping in the front wheel suspension and

(2) to reduce stiffness and damping in the rear wheel suspension.

This indicates that more energy should be dissipated by the front wheel suspension system to satisfy the given constraints.

No general suspension design optimization trends can be drawn based on this study, since results obtained here reflect the cost function and constraints selected. The method presented, however, is shown to correct constraint violations (design deficiencies) of the trial design and to converge to an optimum design. The approach presented can be used to account for a broad range of dynamic performance factors and ride quality measures.

6. FIRST ORDER DESIGN SENSITIVITY ANALYSIS FOR SYSTEMS DESCRIBED BY SECOND ORDER DIFFERENTIAL AND ALGEBRAIC EQUATIONS

The first order design sensitivity analysis methods presented in Section 2 are based on the first order form of the system of differential equations. Most mechanical systems involve numerous bodies that are connected by kinematic joints, which may be described by mixed systems of differential and algebraic equations. Automated formulation techniques are now available that provide computer generation of system equations of motion and numerical algorithms for their direct solution. The purpose of this section is to present a formulation that extends design sensitivity analysis methods to treat such classes of problems, with the ultimate objective of both computer generation and solution of equations of design sensitivity analysis.

A general formulation for constrained equations of mechanical system dynamics is presented in this section. Both the direct differentiation and adjoint variable methods of Section 2 are extended to treat these problems and examples that have been solved with a general purpose computer code are presented.

Problem Formulation

Dynamic systems under consideration are presumed to involve constrained rigid body motion, under the influence of time varying forcing functions. Design of such systems is defined by a vector of k design parameters, denoted

$$b = [b_1, \ldots, b_k]^T \tag{6.1}$$

These parameters are at the disposal of the designer and represent physical properties that prescribe the system, such as dimensions, spring constants, damping coefficients, masses, or force magnitudes. In contrast to parameters that are specified by the designer, dynamic response of the system is described by a vector of n generalized coordinates q(t),

$$q(t) = [q_1(t), \ldots, q_n(t)]^T \tag{6.2}$$

The generalized coordinates are determined by the governing equations of motion for the system, under the action of applied loads. Forces applied to the system are transformed, using the principle of virtual work [12], to obtain generalized forces that correspond to the generalized coordinates of Eq. 2. The generalized force vector treated here is of the form

$$Q = [Q_1(t,q(t),\dot{q}(t),b), \ldots, Q_n(t,q(t),\dot{q}(t),b)]^T \tag{6.3}$$

The class of systems under consideration is subject to holonomic kinematic constraints of the form

$$\Phi(t,q,b) = 0 \tag{6.4}$$

where Φ is a vector of constraint functions,

$$\Phi = [\Phi_1(t,q,b), \ldots, \Phi_m(t,q,b)]^T \tag{6.5}$$

For details concerning the form of such equations, see the companion kinematics paper of Ref. 13. These algebraic, time dependent constraints define dependencies among the state variables that must be accounted for in the equations of motion of the system. Note also that the constraints are design dependent.

Since dynamic systems under consideration are nonlinear, the kinetic energy of the system is written in the form

$$T = \frac{1}{2}\,\dot{q}^T M(q,b)\dot{q} \tag{6.6}$$

where M is a mass matrix that depends on the position of the system and design. Using the kinetic energy expression of Eq. 6.6 and the constraints of Eq. 6.4, the Lagrange multiplier formulation of the constrained dynamic equations of motion [12] may be written as

$$\frac{d}{dt}\,(M\dot{q}) - T_q^T - Q + \Phi_q^T\lambda = 0 \tag{6.7}$$

where λ is the Lagrange multiplier vector that is associated with the constraints of Eq. 6.4 and a subscript denotes partial derivative. The reader is referred to Appendix A for definition of matrix calculus notation that is employed here. The differential equations of Eq. 6.7 and the algebraic constraint equations of Eq. 6.4 constitute the system equations of motion. To simplify notation, the values of generalized coordinates at specific times t^i are denoted as

$$q^i \equiv q(t^i) \quad , \qquad i = 1,2 \tag{6.8}$$

It is presumed that initial and final times t^1 and t^2 of the dynamic event are determined by relations of the form

$$\Omega^i(t^i,q^i,b) = 0 \quad , \qquad i = 1,2 \tag{6.9}$$

which are defined by the engineer. The generalized coordinates must satisfy Eq. 6.4 at t^1 and t^2, so

$$\Phi^i \equiv \Phi(t^i,q^i,b) = 0 \quad , \qquad i = 1,2 \tag{6.10}$$

A complete characterization of the motion of the system requires definition of initial conditions on position and velocity. The initial conditions are specified by relations of the form

$$A^1q^1 = a(b) \tag{6.11}$$

$$A^2\dot{q}^1 = c(b) \tag{6.12}$$

where A^1 and A^2 are matrices that define initial conditions of position and velocity, respectively. Since Eqs. 6.4 and 6.11 are to determine the full set of initial generalized coordinates q^1, the matrix

$$\begin{bmatrix} A^1 \\ \Phi_q(t^1) \end{bmatrix} \tag{6.13}$$

must be nonsingular.

In addition to the initial conditions, the generalized velocities must satisfy

$$\Phi_t^i + \Phi_q^i \dot{q}^i = 0 \quad , \qquad i = 1,2 \tag{6.14}$$

Equation 6.15 is obtained by taking the total time derivative of Eq. 6.4. Equation 6.12 and Equation 6.15 with i = 1 must determine the initial velocity, so the matrix

$$\begin{bmatrix} A^2 \\ \Phi_q(t^1) \end{bmatrix} \tag{6.15}$$

must also be nonsingular.

A necessary and sufficient condition for Eqs. 6.4 and 6.7 to uniquely determine the motion of the system is that the following matrix be nonsingular:

$$\begin{bmatrix} M & \Phi_q^T \\ \Phi_q & 0 \end{bmatrix} \tag{6.16}$$

Under the foregoing assumptions, once the design b is specified, dynamic response of the system is uniquely predicted by the system constraints and equations of motion of Eqs. 6.4 and 6.7 and the initial conditions of Eqs. 6.11 and 6.12. An efficient numerical method of integrating these equations of motion is presented in Ref. 14, employing a generalized coordinate partitioning technique for automatic formulation and integration of the equations of motion.

Well-known results from the theory of initial-value problems [7] show that the dynamic response of this system is continuously differentiable with respect to the design variables, as long as the matrix of Eq. 6.16 retains full rank. It is therefore of interest to consider a typical functional that may arise in an optimal design problem of selecting the design b to minimize some cost, subject to performance constraints,

$$\psi = g(t^i, q^i, \dot{q}^i, b) + \int_{t^1}^{t^2} F(t, q, \dot{q}, \lambda, b)dt \tag{6.17}$$

where variables associated with both i = 1 and 2 appear in the function g. The function ψ is allowed to depend on all variables of the problem. Since the Lagrange multiplier λ uniquely determines reaction forces in the constrained system, such as reaction forces in bearings, bounds on force transmitted are included in the integral of the second term of Eq. 6.17.

Design Derivative of ψ

Derivatives of the functional of Eq. 6.17 with respect to design are to be calculated. Since t^i, q^i, $\dot{q}^i$, $q(t)$, $\dot{q}(t)$, and $\lambda(t)$ depend on design, one may use Leibniz rule [8] for derivative of an integral and the chain rule of differentiation to obtain

$$\psi_b = \sum_{i=1}^{2} g_{t^i} t_b^i + \sum_{i=1}^{2} g_{q^i}(q_b^i + \dot{q}^i t_b^i) + \sum_{i=1}^{2} g_{\dot{q}^i}(\dot{q}_b^i + \ddot{q}^i t_b^i) + g_b + F^2 t_b^2 - F^1 t_b^1$$
$$+ \int_{t^1}^{t^2} (F_q q_b + F_{\dot{q}} \dot{q}_b + F_\lambda \lambda_b + F_b)dt \tag{6.18}$$

where $F^i = F(t^i, q^i, \dot{q}^i, b)$ and the following relations are employed

$$\frac{dq^i}{db} = q_b^i + \dot{q}^i t_b^i \quad , \qquad i = 1,2 \tag{6.19}$$

$$\frac{d\dot{q}^i}{db} = \dot{q}_b^i + \ddot{q}^i t_b^i \quad , \qquad i = 1,2 \tag{6.20}$$

Integration by parts of the second term in the integral of Eq. 6.18 and rearranging terms yields

$$\psi_b = (g_{t^1} + g_{q^1}\dot{q}^1 + g_{\dot{q}^1}\ddot{q}^1 - F^1)t_b^1 + (g_{t^2} + g_{q^2}\dot{q}^2 + g_{\dot{q}^2}\ddot{q}^2 + F^2)t_b^2$$

$$+ (g_{q^1} - F^1)q_b^1 + (g_{q^2} + F^2)q_b^2 + g_{\dot{q}^1}\dot{q}_b^1 + g_{\dot{q}^2}\dot{q}_b^2$$

$$+\int_{t^1}^{t^2}[(F_q - \frac{d}{dt}F_{\dot{q}})q_b + F_\lambda\lambda_b]dt$$

$$+ g_b + \int_{t^1}^{t^2} F_b dt \tag{6.21}$$

In order to make use of Eq. 6.21, partial derivatives of q, $\dot{q}$, t^1, and t^2 with respect to b must be evaluated or rewritten in terms of computable quantities. This is done in the following two subsections, using direct differentiation and adjoint variable methods, respectively.

Direct Differentiation Method

For direct evaluation of Eq. 6.21, partial derivatives of all state related terms with respect to design must be calculated. This can be done by direct differentiation of the state equations. Beginning with the differential equations of motion of Eq. 6.7, one may take a total derivative with respect to design to obtain

$$\left.\begin{aligned} &\frac{d}{dt}(M\dot{q}_b) - T_{q\dot{q}}\dot{q}_b - T_{qq}\dot{q}_b - Q_{\dot{q}}\dot{q}_b - Q_q q_b \\ &\qquad + (\Phi_q^T\lambda)_q q_b + \Phi_q^T\lambda_b \\ &\quad = -\frac{d}{dt}(M\dot{q})_b + T_{qb} + Q_b - (\Phi_q^T\lambda)_b \\ &\Phi_q q_b = -\Phi_b \end{aligned}\right\} \tag{6.22}$$

This is a system of linear second order differential equation in the variable q_b. In order to solve these differential equations, a set of initial conditions for $q_b(t^1)$ and $\dot{q}_b(t^1)$ must be calculated.

To complicate matters, the time t^1 at which the initial condition is imposed depends on design, through Eq. 6.9. Taking the total design derivative of Eq. 6.9, one has

$$\Omega^i_{t^i} t^i_b + \Omega^i_{q^i}(q^i_b + \dot{q}^i t^i_b) + \Omega^i_b = 0, \; i=1,2 \tag{6.23}$$

Since Eq. 6.9 is to determine t^i, the coefficient of its derivative with respect to the t^i must be nonzero. Therefore, one may solve Eq. 6.23 for t^i_b to obtain

$$t^i_b = -\left[\Omega^i_{q^i} q^i_b + \Omega^i_b\right] \Big/ \left[\Omega^i_{t^i} + \Omega^1_{q^i} \dot{q}^i\right], \; i=1,2 \tag{6.24}$$

In order to obtain initial conditions for q_b, one may differentiate Eq. 6.11 to obtain

$$A^1 q^1_b - A^1 \dot{q}^1 \left[\frac{\Omega^1_{q^1}}{\Omega^1_{t^1} + \Omega^1_{q^1} \dot{q}^1}\right] q^1_b = a_b + A\dot{q}^1 \left[\frac{\Omega^1_b}{\Omega^1_{t^1} + \Omega^1_{q^1} \dot{q}^1}\right] \tag{6.25}$$

Since q^1 must satisfy the kinematic equations of Eq. 6.10 at t^1, one may take the total design derivative of Eq. 6.10 to obtain

$$\Phi^1_q q^1_b - \left[\Phi^1_q \dot{q}^1 + \Phi^1_t\right] \left[\frac{\Omega^1_{q^1}}{\Omega^1_{t^1} + \Omega^1_{q^1} \dot{q}^1}\right] q^1_b = -\Phi^1_b + \Phi^1_q \dot{q}^1 \left[\frac{\Omega^1_b}{\Omega^1_{t^1} + \Omega^1_{q^1} \dot{q}^1}\right] \tag{6.26}$$

Equations 6.25 and 6.26 constitute a system of n equations to determine the n initial value q^1_b.

In order to obtain initial conditions for $\dot{q}^1_b$, one may differentiate Eq. 6.12 to obtain

$$A^2 \dot{q}^1_b - A^2 \ddot{q}^1 \left[\frac{\Omega^1_{q^1}}{\Omega^1_{t^1} + \Omega^1_{q^1} \dot{q}^1}\right] \dot{q}^1_b = c_b + A\ddot{q}^1 \left[\frac{\Omega^1_b}{\Omega^1_{t^1} + \Omega^1_{q^1} \dot{q}^1}\right] \tag{6.27}$$

Since $\dot{q}^1$ must satisfy Eq. 6.15, one may take its total design derivative to obtain

$$
\begin{aligned}
\Phi_q \dot{q}_b^1 = &- \Phi_{tb}^1 - (\Phi_q^1 \dot{q}^1)_b + \left[\Phi_t^1 + 2\Phi_{qt}^1 \dot{q}^1 + (\Phi_q^1 \dot{q}^1)_q \dot{q}^1 + \Phi_q^1 \ddot{q}^1\right]\left[\frac{\Omega_b^1}{\Omega_{t^1}^1 + \Omega_{q^1}^1 \dot{q}^1}\right] \\
&+ \left\{ - \Phi_{tq}^1 - (\Phi_q^1 \dot{q}^1)_q + \left[\Phi_t^1 + 2\Phi_{qt}^1 \dot{q}^1 + (\Phi_q^1 \dot{q}^1)_q \dot{q}^1 + \Phi_q^1 \ddot{q}^1\right] \right. \\
&\times \left. \left[\frac{\Omega_{q^1}^1}{\Omega_{t^1}^1 + \Omega_{q^1}^1 \dot{q}^1}\right] \right\} q_b^1
\end{aligned}
\tag{6.28}
$$

Note that the right side of Eq. 6.28 involves q_b^1, which was determined from Eqs. 6.25 and 6.26 and may, therefore, be treated as known. Equations 6.27 and 6.28 constitute the proper number of equations to determine the initial condition $\dot{q}_b^1$.

Having calculated the initial conditions from Eqs. 6.25-28, one may now solve the mixed system of differential-algebraic equations of Eq. 6.22 for $q_b(t)$ and λ_b. Having calculated these quantities, Eq. 6.24 may be used to obtain t_b^i, i=1,2, yielding all terms that are required to evaluate the total derivitave of ψ with respect to design in Eq. 6.21.

Numerical Examples by Direct Differentiation

The direct differentiation algorithm presented here was coded in FORTRAN and implemented on a PRIME 750 supermini computer, as presented in Ref. 15. The program was tested on several problems, results of which are summarized here.

Verification Procedure: Results of design sensitivity calculations can be checked by methods based on perturbation theory. Two such methods are used to verify the design sensitivity calculations by the present method, as follows:

(a) Check on Functional Design Sensitivity: The matrix ψ_b of functional design sensitivity coefficients is checked by calculating state design sensitivity and constraint functional at a nominal design b. The design is then given a small perturbation δb, so that the new design becomes

$$
b^* = b + \delta b \tag{6.29}
$$

The system is now solved at the new design b^* and the constraint functions are re-evaluated. Let the value of constraint function at the original design be $\psi(b)$ and its value at the perturbed design be $\psi(b^*)$. The actual change in the value of the constraint function is given by

$$\Delta\psi = \psi(b^*) - \psi(b) \tag{6.30}$$

The design sensitivity prediction of the change in functional value for a design change of δb is given by

$$\delta\psi = \psi_b \, \delta b \tag{6.31}$$

If design sensitivity analysis is correct, then the value of $\Delta\psi_i$ obtained from Eq. 6.30 should be approximately equal to the value of $\delta\psi_i$ obtained from Eq. 6.31.

(b) Check on State Design Sensitivity: As a check on state design sensitivity q_b, the system is solved at the original design b and the perturbed design b^*. Consider the variation in acceleration of generalized coordinate i at time t when the design is changed from b to b^*. This variation may be written as

$$\Delta\ddot{q}^i(t) = \ddot{q}^i(t, b^*) - \ddot{q}^i(t, b) \tag{6.32}$$

For a design change δb, the change in acceleration at time t is predicted by design sensitivity theory as

$$\delta\ddot{q}^i(t) = \ddot{q}^i_b(t)\,\delta b \tag{6.33}$$

The perturbations of Eqs. 6.32 and 6.33 should be approximately equal. Acceleration design sensitivity accuracy is checked, since if it is accurate, then position and velocity sensitivities will be even more accurate.

For each example, the perturbation in design is chosen to be

$$\delta b_i = \begin{cases} +\,0.001 & \text{for } b_i > 0 \\ -\,0.001 & \text{for } b_i < 0 \end{cases} \tag{6.34}$$

Example 1; Four-Bar Linkage under Self-Weight: The initial configuration of a four-bar linkage is shown in Fig. 6.1. The design variables are the coordinates of the revolute joints, as indicated in Fig. 6.1. The only loading is self-weight of the members. The simulation time is from 0 to 1 sec. Input data are as follows, with in slug-inch units:

Masses;

$$m_1 = 0,\ m_2 = 8.0,\ m_3 = 8.0,\ m_4 = 8.0$$

Moments of Inertia;

$$J_1 = 0,\ J_2 = 8.0,\ J_3 = 8.0,\ J_4 = 8.0$$

The nominal design is

$$b \equiv [-70.9107,\ 70.7107,\ -50.,\ -55.]^T$$

The functional is chosen to be

$$\psi \equiv \int_0^1 (\sin\phi_2 - 0.2389)^2 dt$$

The design sensitivity vector for ψ is found to be

$$\psi_b \equiv [-5.841,\ 1.616,\ -2.356,\ 2.601] \times 10^{-3}$$

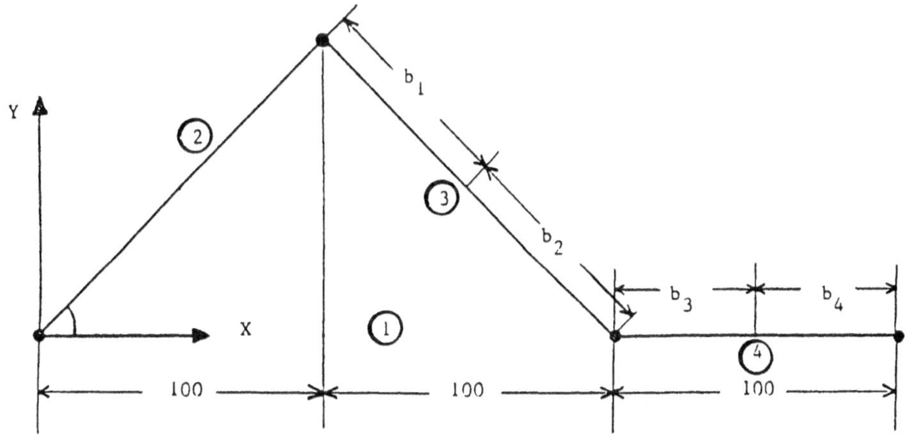

Figure 6.1 Four-bar Linkage under Self Weight

Using this design sensitivity vector and the design perturbation of Eq. 6.34, the predicted change in cost is

$$\delta\psi = 1.24 \times 10^{-5}$$

Reanalysis with the perturbed design yields

$$\Delta\psi = 1.28 \times 10^{-5}$$

which is quite close.

The check on state sensitivity was carried out for $\ddot{C}_3 \equiv \frac{d^2}{dt^2} \cos\phi_3$, where ϕ_3 is the angle of link 3 in Fig. 6.1 with the x-axis. Plots of $\Delta\ddot{C}_3$ and $\delta\ddot{C}_3$ for this test are shown in Fig. 6.2. The curve obtained from state design sensitivity and the curve obtained by perturbation coincide to be within numerical accuracy. The graph of $\ddot{C}_3$ is shown in Fig. 6.3., which indicates a substantial acceleration variation.

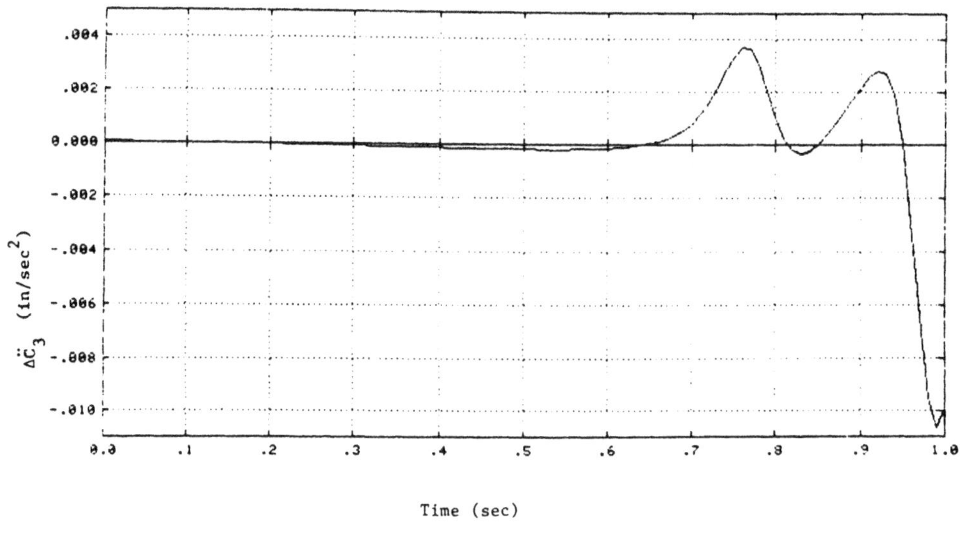

Figure 6.2 State Sensitivity Check for Example No. 1

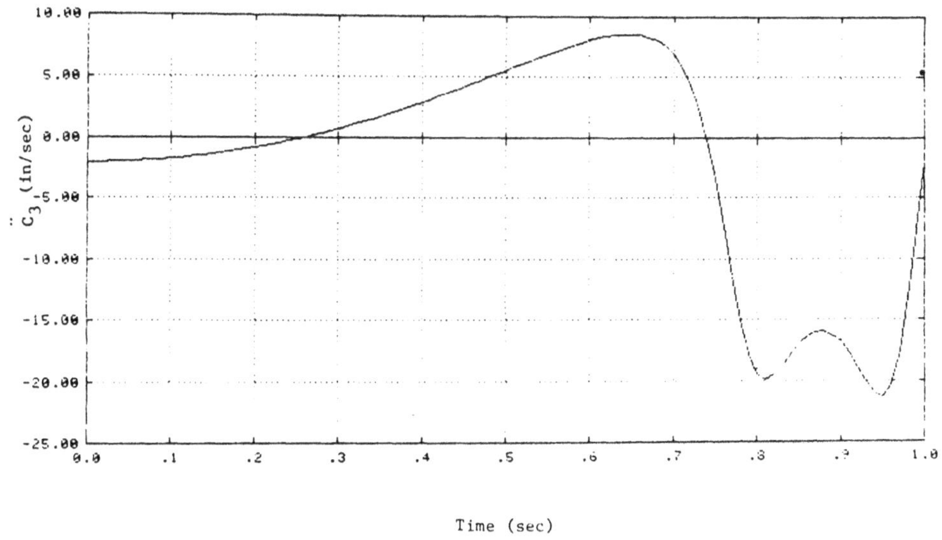

Figure 6.3 $\ddot{C}_3$ Versus Time Curve for Example 1

Example 2; Four Bar Linkage with Applied Torque: The initial position of a four-bar linkage is shown in Fig. 6.4. In addition to the weights of the members, a constant torque T is applied to member 2. This torque and the coordinates of the revolute joints shown in Fig. 6.4 are the design variables. The simulation time is from 0 to 0.84 seconds. The input data, in MKS units, are as follows:

Masses;

$m_1 = 0,\ m_2 = 6.0,\ m_3 = 1.0,\ m_4 = 4.0$

Moments of Inertia;

$J_1 = 0,\ J_2 = 0.125,\ J_3 = 0.6667,\ J_4 = 0.0833$

The nominal design is

$$b \equiv [0.,\ 0.25,\ -0.25,\ -0.5,\ 0.5,\ 0.25,\ -0.25]^T$$

The functional selected for analysis is

$$\psi \equiv \int_0^{0.84} b_1\ (\cos\phi_2(\dot{\mathrm{Sin}}\phi_2) - \sin\phi_2(\dot{\mathrm{Cos}}\phi_2))dt$$

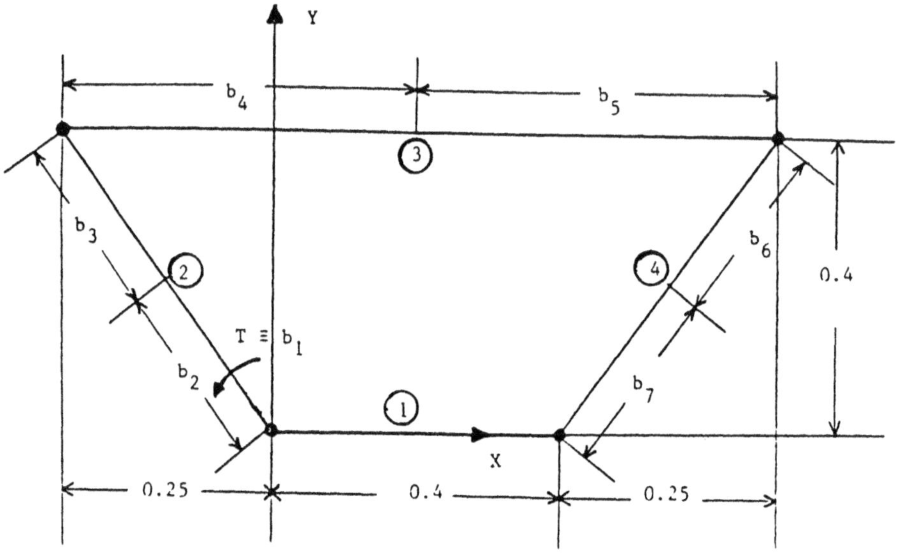

Figure 6.4 Four-bar Linkage with Applied Torque

The design sensitivity vector was calculated as

$$\psi_b = [2.649, 0, 0, 0, 0, 0, 0]$$

This predicts a change

$$\delta\psi = .00265$$

Perturbation analysis yields a value of

$$\Delta\psi = .00263$$

which is quite close.

State design sensitivity verification was done for $\ddot{S}_4 = \frac{d^2}{dt^2}(\sin\phi_4)$. The graphs of $\Delta\ddot{S}_4$ and $\delta\ddot{S}_4$ are shown in Fig. 6.5. The curve obtained from design sensitivity analysis again coincides with that obtained by perturbation. The plot of $\ddot{S}_4$ versus time is shown in Fig. 6.6.

Example 3; Slider-Crank Mechanism: Figure 6.7 shows a slider-crank mechanism in its initial position. The loading in this case is the weight of the members and a constant force F = 125 lbs, which acts on the piston in the direction indicated in the figure. The

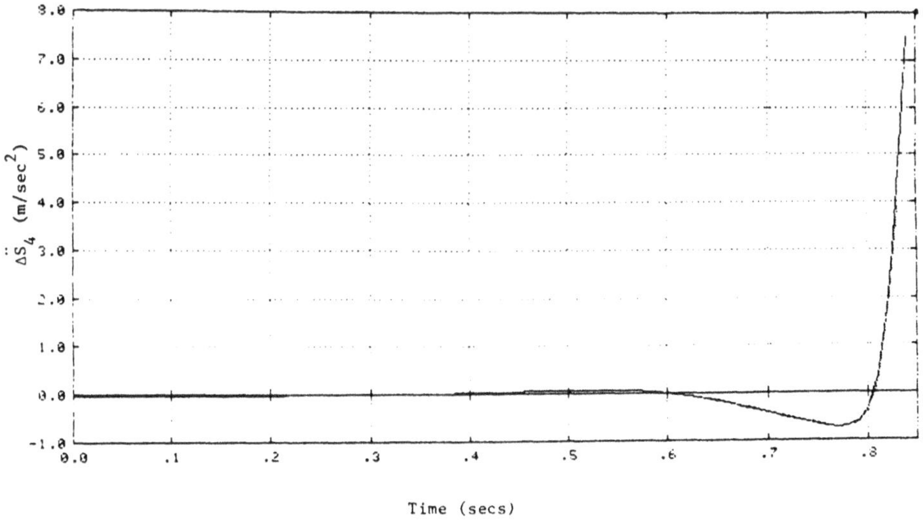

Figure 6.5 State Sensitivity Check for Example 2

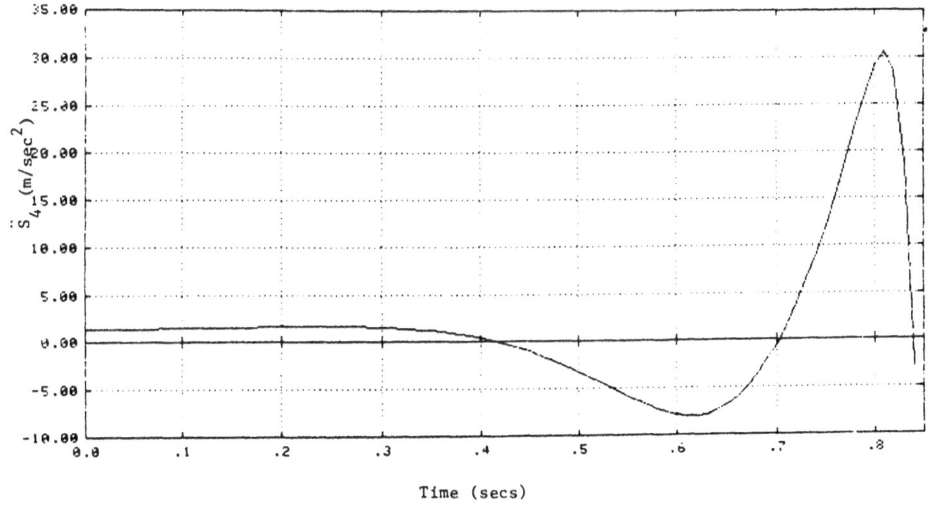

Figure 6.6 $\ddot{S}_4$ Versus Time Curve for Example 2

simulation time is from 0 to 1 sec. The input data are given, in foot-pound units, as follows:

Masses;

$m_1 = 0,\ m_2 = 5.0,\ m_3 = 15.0,\ m_4 = 8.0$

Moments of Inertia;

$J_1 = 0,\ J_2 = 0.5,\ J_3 = 2.5,\ J_4 = 8.0$

The nominal design is

$$b = [-0.25,\ 0.25]^T$$

and the functional selected for analysis is

$$\psi = \int_0^1 (x_4 - 20)^2 dt$$

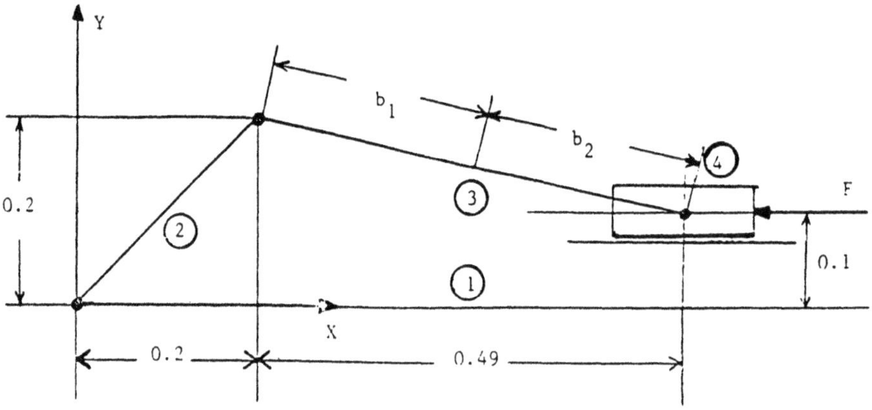

Figure 6.7 Slider-crank Mechanism

For this problem, the functional design sensitivity vector was found to be

$$\psi_b = [46.48,\ -34.73]$$

The predicted cost function variation is

$$\delta\psi = -0.0812$$

Perturbation and reanalysis yielded the comparable result

$\Delta\psi = -0.0814$

which again shows good agreement.

The state design sensitivity verification was done on $\ddot{x}_4$ and, again, coincident curves of Fig. 6.8 were obtained. Figure 6.9 is the plot of $\ddot{x}_4$ versus time.

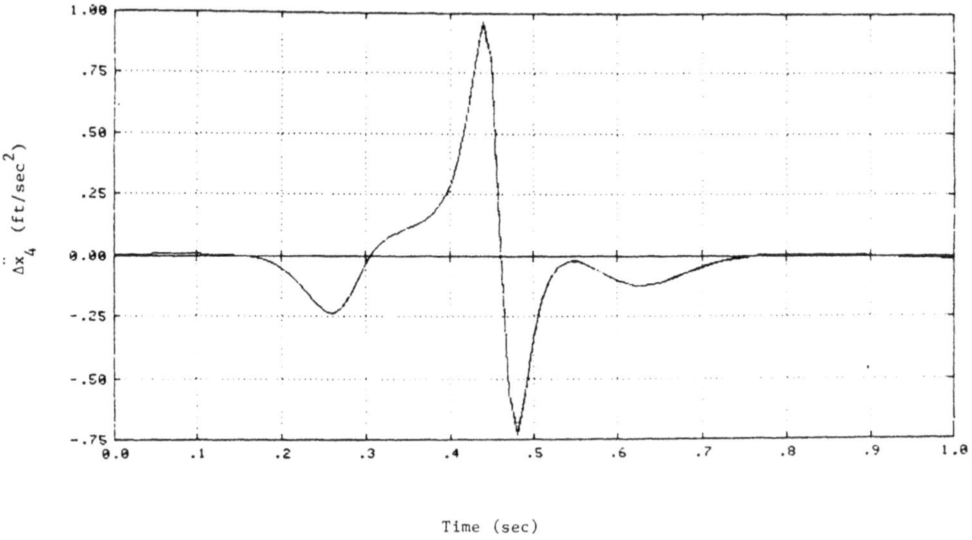

Figure 6.8 State Sensitivity Check of Example 3

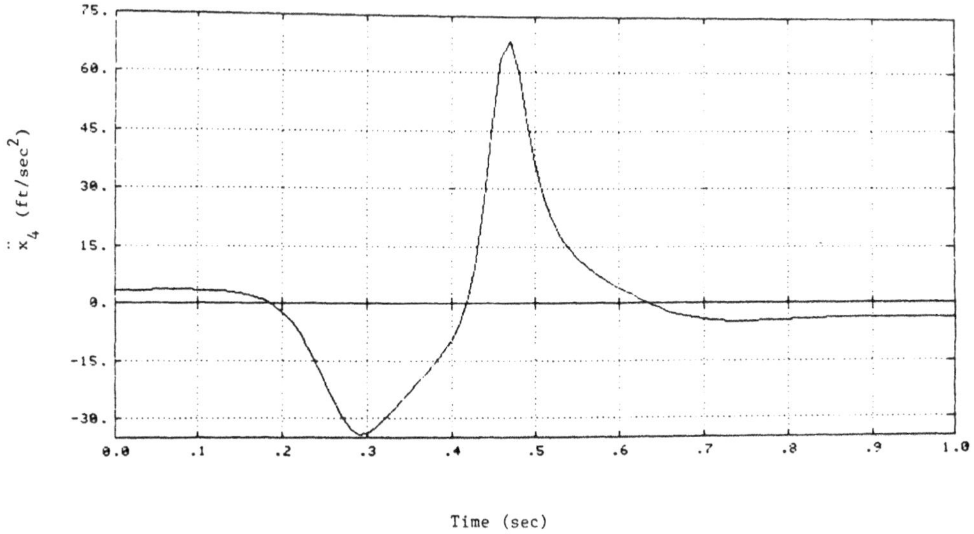

Figure 6.9 $\ddot{x}_4$ Versus Time Curve for Example 3

Adjoint Variable Method

Variations in state variables and initial and final times must be consistent with the system constraints and equations of motion. To implicitly account for this dependence and to avoid explicit computation of derivatives of state with respect to design, a sequence of adjoint relationships is now developed, using a method that has been widely applied in mechanical and structural design [3]. First, both sides of the equations of motion of Eq. 6.7 and the constraints of Eq. 6.4 are multiplied by arbitrary multiplier vector functions $\mu(t)$ and $\nu(t)$, to obtain the identities

$$\int_{t^1}^{t^2} \mu^T\left(\frac{d}{dt}(M\dot{q}) - T_q^T - Q + \Phi_q^T \lambda\right)dt = 0 \tag{6.35}$$

and

$$\int_{t^1}^{t^2} \nu^T \Phi dt = 0 \tag{6.36}$$

These equations hold for all values of design, so the total design derivative of both sides of Eqs. 6.26 and 6.27 may be taken, yielding

$$\int_{t^1}^{t^2} \mu^T\Big[\frac{d}{dt}(M\dot{q}_b + (M\dot{q})_q q_b) - T_{qq}q_b - (M\dot{q})_q\dot{q}_b - Q_q q_b - Q_{\dot{q}}\dot{q}_b$$

$$+ (\Phi_q^T\lambda)_q q_b + \Phi_q^T\lambda_b + \frac{d}{dt}(M\dot{q})_b - T_{qb} - Q_b + (\Phi_q^T\lambda)_b\Big]dt = 0 \tag{6.37}$$

and

$$\int_{t^1}^{t^2} \nu^T(\Phi_q q_b + \Phi_b)dt \tag{6.38}$$

Integrating terms in the integrals of Eq. 6.37 involving $\dot{q}_b$ by parts and using Eqs. 6.19 and 6.20 yields the identity

$$\int_{t^1}^{t^2} \Big\{ [\tfrac{d}{dt}(\dot{\mu}^T M) - \dot{\mu}^T (M\dot{q})_q - \mu^T T_{qq} + \tfrac{d}{dt}(\mu^T (M\dot{q})_q) - \mu^T Q_q$$

$$+ \tfrac{d}{dt}(\mu^T Q_{\dot{q}}) + \mu^T (\Phi_q^T \lambda)_q] q_b + \mu^T \Phi_q^T \lambda_b \Big\} dt$$

$$+ [\mu^T M \dot{q}_b - \dot{\mu}^T M q_b - \mu^T Q_{\dot{q}} q_b] \Big|_{t^1}^{t^2}$$

$$+ \int_{t^1}^{t^2} [\mu^T (\Phi_q^T \lambda)_b - \dot{\mu}^T (M\dot{q})_b - \mu^T T_{qb} - \mu^T Q_b] dt + \mu^T (M\dot{q})_b \Big|_{t^1}^{t^2} \quad (6.39)$$

Equation 6.9 may similarly be multiplied by an arbitrary multiplier ξ^i to obtain the identity

$$\xi^i \Omega^i (t_i, q^i, b) = 0 \quad , \qquad i = 1,2 \quad (6.40)$$

Since this equation must hold for all design, the total design derivative of both sides yields the identity

$$\xi^i \Omega^i_{t^i} t^i_b + \xi^i \Omega^i_{q^i} (q^i_b + \dot{q}^i t^i_b) + \xi^i \Omega^i_b = 0 \quad , \qquad i = 1,2 \quad (6.41)$$

Multiplying the kinematic constraint equation of Eq. 6.10 by an arbitrary multiplier vector γ^i yields the identity

$$\gamma^{i^T} \Phi^i (t^i, q^i, b) = 0 \quad , \qquad i = 1,2 \quad (6.42)$$

Taking the total design derivative of both sides of Eq. 6.42 yields the identity

$$\gamma^{i^T} \Phi^i_t t^i_b + \gamma^{i^T} \Phi^i_q (q^i_b + \dot{q}^i t^i_b) + \gamma^{i^T} \Phi^i_b = 0 \quad , \qquad i = 1,2 \quad (6.43)$$

Multiplying the initial condition of Eq. 6.11 by an arbitrary multiplier vector α yields the identity

$$\alpha^T A^1 q^1 = \alpha^T a = 0 \quad (6.44)$$

Taking the total design derivative of both sides of Eq. 6.44 yields the identity

$$\alpha^T A^1 (q^1_b + \dot{q}^1 t^1_b) - \alpha^T a_b = 0 \quad (6.45)$$

Similarly, multiplying the velocity initial condition of Eq. 6.12 by an arbitrary multiplier vector β, yields the identity

$$\beta^T A^2 \dot{q}^0 - \beta^T c = 0 \tag{6.46}$$

Taking the total design derivative of both sides yields the identity

$$\beta^T A^2 (\dot{q}_b^1 + \ddot{q}^1 t_b^1) - \beta^T c_b = 0 \tag{6.47}$$

Velocity variations must satisfy Eq. 6.15 at t^1 and t^2. Hence, Eq. 6.15 may be multiplied by an arbitrary multiplier vector η^i to obtain the identity

$$\eta^{i^T} [\Phi_t^i + \Phi_q^i \dot{q}^i] = 0 \quad , \qquad i = 1,2 \tag{6.48}$$

Taking the total design derivative of both sides yields the identity

$$\eta^{i^T} \Phi_{tt}^i t_b^i + \eta^{i^T} [\Phi_{tq}^i + (\Phi_q^i \dot{q}^i)_q][q_b^i + \dot{q}^i t_b^i]$$

$$+ \eta^{i^T} \Phi_q^i (\dot{q}_b^i + \ddot{q}^i t_b^i) + \eta^{i^T} \Phi_{tb}^i + (\Phi_q^i \dot{q}^i)_b] = 0 \quad , \qquad i = 1,2 \tag{6.49}$$

The identities of Eqs. 6.38, 6.39, 6.41, 6.43, 6.45, 6.47, and 6.49 are relationships among design derivatives of state, Lagrange multiplier, and initial and final times. All of these identities are valid for arbitrary multiplier functions $\mu(t)$ and $\nu(t)$ and multiplier parameters ξ^i, γ^i, α, β, and η^i. The objective in use of these identities is to select the arbitrary multipliers in such a way that terms involving state derivatives with respect to design in Eq. 6.21 may be written explicitly in terms of computable quantities. The technique employed here is an extension of the adjoint variable method that has been used extensively in optimal control and optimal design literature [2,3]. The idea is simply to sum identities of Eqs. 6.38, 6.39, 6.41, 6.43, 6.45, 6.47, and 6.49, to obtain a linear expression in all the derivatives involved. The coefficients of each design derivative of state, Lagrange multiplier, and initial and final times are equated to corresponding coefficients on the right of Eq. 6.21. This process yields the following set of adjoint relations:

Equating coefficients of q_b in the integrals yields

$$\frac{d}{dt}(M\dot{\mu}) - T_{qq}\mu + \left(\frac{d}{dt}(M\dot{q})_q^T\right)\mu - Q_q^T\mu + \left(\frac{d}{dt} Q_{\dot{q}}\right)^T\mu$$

$$+ (\Phi_q^T\lambda)_q^T\mu + \Phi_q^T\nu = F_q^T - \frac{d}{dt}(F_{\dot{q}})^T \tag{6.50}$$

Equating coefficients of λ_b in the integrals yields

$$\Phi_q\mu = F_\lambda^T \tag{6.51}$$

Equating coefficients of q_b^1 yields

$$M^1\dot{\mu}^1 + Q_{\dot{q}}^{1^T}\mu^1 + \Omega_{q^1}^{1^T}\xi^1 + \Phi_q^{1^T}\gamma^1 + A^{1^T}\alpha + [\Phi_{tq}^1 + (\Phi_q^1\dot{q}^1)_q]^T\eta^1$$

$$= g_{q^1}^T - F_{\dot{q}}^{1^T} \tag{6.52}$$

Equating coefficients of q_b^2 yields

$$- M^2\dot{\mu}^2 - Q_{\dot{q}}^{2^T}\mu^2 + \Omega_{q^2}^{2^T}\xi^2 + \Phi_q^{2^T}\gamma^2 + [\Phi_{tq}^2 + (\Phi_q^2\dot{q}^2)_q]^T\eta^2$$

$$= g_{q^2}^T + F_{\dot{q}}^2 \tag{6.53}$$

Equating coefficients of $\dot{q}_b^1$ yields

$$- M^1\mu^1 + A^{2^T}\beta + \Phi_q^{1^T}\eta^1 = g_{\dot{q}^1}^T \tag{6.54}$$

Equating coefficients of $\dot{q}_b^2$ yields

$$M^2\mu^2 + \Phi_q^{2^T}\eta^2 = g_{\dot{q}^2}^T \tag{6.55}$$

Equating coefficients of t_b^1 yields

$$\Omega_{t^1}^1\xi^1 + \Omega_{q^1}^1\dot{q}^1\xi^1 + \Phi_t^{1^T}\gamma^1 + \dot{q}^{1^T}\Phi_q^{1^T}\gamma^1$$

$$+ \dot{q}^{1^T}A^{1^T}\alpha + \ddot{q}^{1^T}A^{2^T}\beta + \Phi^{1^T}\eta_{tt}^1 + \dot{q}^{1^T}\,(\Phi_q^1\dot{q}^1)_q + \Phi_{tq}^1{}^T\eta^1$$

$$+ \ddot{q}^{1^T}\Phi_q^{1^T}\eta^1 = g_{t^1} + g_{q^1}\dot{q}^1 + g_{\dot{q}^1}\ddot{q}^1 - F^1 \tag{6.56}$$

Equating coefficients of t_b^2 yields

$$(\Omega^2_{t^2} + \Omega^2_{q^2}\dot{q}^2)\xi^2 + \Phi_t^{2^T}\gamma^2 + \dot{q}^{2^T}\Phi_q^{2^T}\gamma^2 + \Phi_{tt}^{2^T}\eta^2 + \dot{q}^{2^T}[(\Phi^2_q\dot{q}^2)_q + \Phi^2_{tq}]^T\eta^2$$

$$+ \ddot{q}^{2^T}\Phi_q^{2^T}\eta^2 = g_{t^2} + g_{q^2}\dot{q}^2 + g_{\dot{q}^2}\ddot{q}^2 + F^2 \tag{6.57}$$

Presuming that Eqs. 6.50 through 6.57 determine all multipliers, the resulting terms in the summation of Eqs. 6.38, 6.39, 6.41, 6.43, 6.45, 6.47, and 6.49 yield a formula for the sum of all terms in Eq. 6.21 involving design derivatives of state, Lagrange multiplier, and initial and final times. This identity is substituted into Eq. 6.21, yielding an explicit expression for the derivative of ψ with respect to b

$$\psi_b = g_b + \mu^{1^T}(M\dot{q}^1)_b - \mu^{2^T}(M\dot{q}^2)_b - \sum_{i=1}^{2}(\gamma^i\Phi^i_b + \xi^i\Omega^i_b]$$

$$+ \alpha^T a_b + \beta^T c_b \quad - \sum_{i=1}^{2}\eta^{i^T}(\Phi^i_{tb} + (\Phi_q\dot{q}^i)_b) - \xi^i\Omega_b^{i^T}$$

$$+\int_{t_1}^{t_2}[F_b + \dot{\mu}^T(M\dot{q})_b - \mu^T(\Phi_q^T\lambda)_b - T_{qb} - Q_b \quad - \nu^T\Phi_b]dt \tag{6.58}$$

The design sensitivity analysis vector of Eq. 6.58 may be evaluated numerically once the state and adjoint equations have been solved.

Adjoint Variable Design Sensitivity Analysis Algorithm

In order to evaluate the design sensitivity vector ψ_b in Eq. 6.58, it is important to define a practical computational algorithm that determines all variables that arise. For a nominal design b, the following sequence of computations yields all variables that are required for evaluating the sensitivity vector in Eq. 6.58:

Step 1: Integrate the equations of motion Eqs. 6.4 and 6.7, with initial conditions of Eqs. 6.11, 6.12, and 6.15. Store $q(t)$, $\dot{q}(t)$, $\ddot{q}(t)$, and $\lambda(t)$. The numerical method of Ref. 14 can be employed for this calculation.

Step 2: Equations 6.55 and 6.51 are

$$\begin{bmatrix} M^2 & \Phi_q^{2^T} \\ \Phi_q^2 & 0 \end{bmatrix} \begin{bmatrix} \mu^2 \\ \eta^2 \end{bmatrix} = \text{known terms} \tag{6.59}$$

Since the coefficient matrix is the nonsingular matrix of Eq. 6.16, μ^2 and η^2 are uniquely determined.

Step 3: Equation 6.53 and $\Phi_q \dot{\mu} = -(\frac{d}{dt} \Phi_q)\mu + \frac{d}{dt} F_\lambda^T$ are

$$\begin{bmatrix} M^2 & \Phi_q^{2^T} \\ \Phi_q^2 & 0 \end{bmatrix} \begin{bmatrix} \dot{\mu}^2 \\ -\gamma^2 \end{bmatrix} = \begin{bmatrix} \Omega_{q^2}^{2^T} \\ 0 \end{bmatrix} \xi^2 + \text{known terms} \tag{6.60}$$

Since the coefficient matrix is nonsingular, $\dot{\mu}^2$ and γ^2 are uniquely determined as functions of ξ^2.

Step 4: Equation 6.57, with the solution of Eq. 6.60, determines ξ^2, hence $\dot{\mu}^2$ and γ^2.

Step 5: Backward integration of Eqs. 6.50 and 6.51, employing the numerical methods of Ref. 14, yields $\mu(t)$, $\dot{\mu}(t)$, and $\nu(t)$, hence μ^1 and $\dot{\mu}^1$. Uniqueness follows since these equations are linearizations of the constraints and equations of motion of Eqs. 6.4 and 6.7.

Step 6: Equation 6.54 is

$$\begin{bmatrix} A^{2^T} & \Phi_q^T \end{bmatrix} \begin{bmatrix} \beta \\ \eta_1 \end{bmatrix} = \text{known terms} \tag{6.61}$$

Since the coefficient matrix is the transpose of the nonsingular matrix of Eq. 6.14, it is nonsingular and Eq. 6.61 uniquely determines β and η^1.

Step 7: Equation 6.52 is

$$\begin{bmatrix} A^{1^T} & \Phi_q^T \end{bmatrix} \begin{bmatrix} \alpha \\ \gamma^1 \end{bmatrix} = -\Omega_{q^1}^{1^T} \xi^1 + \text{known terms} \tag{6.62}$$

Since the coefficient matrix is the transpose of the nonsingular matrix of Eq. 6.13, it is nonsingular and Eq. 6.62 determines α and γ^1 in terms of ξ^1.

Step 8: Equation 6.56, with the solution of Eq. 6.62, determines ξ^1, hence α and γ^1.

Step 9: The design sensitivity vector ψ_b of Eq. 6.58 is evaluated, using results of Steps 1 through 8.

This algorithm has been implemented in the dynamic analysis program DADS [14]. The program automatically assembles the constraint equations and the governing differential equations for the problem from user supplied data. The user identifies design variables and all derivatives that are needed in the adjoint equations are assembled. Integration forward in time is carried out for the state and backward integration in time is carried out for the adjoint variables. Backward integration requires that the state, Lagrange multipliers, mass matrix, and Jacobian matrix be stored on disk during forward integration, to be retrieved at the appropriate times during backward integration. When a successful time step is taken in forward integration, q, $\dot{q}$, $\ddot{q}$, λ, the Jacobian, and the mass matrix are written on disk. Polynomials that interpolate for $\ddot{q}$ and λ are also generated and stored on disk. During the backward integration, accelerations are obtained by interpolation and $\dot{q}$ and q are obtained by integrating the polynomial for $\ddot{q}$.

Numerical Examples by the Adjoint Variable Method

Several example problems have been analyzed with the software developed, two of which are discussed here.

Example 4; Two Degree Freedom Spring Mass System: As the first example, a simple two degree freedom spring mass system shown in Fig. 6.10 is considered. Two masses of 20 Kg. and moment of inertia 125 Kgm^2 are connected by springs and dampers, as shown in Fig. 6.10. Body 1 is excited by a force F=1000 sin20t. The spring constants and damping coefficients are the design variables, as shown in Fig. 6.10. Analysis is carried out for a period of 1 sec. and a nominal design $b^0 = [3920, 10, 3920, 10]^T$.

The dynamic response functional for this problem is taken as

$$\psi = \int_0^1 (y_1 - y_1^1)^2 dt$$

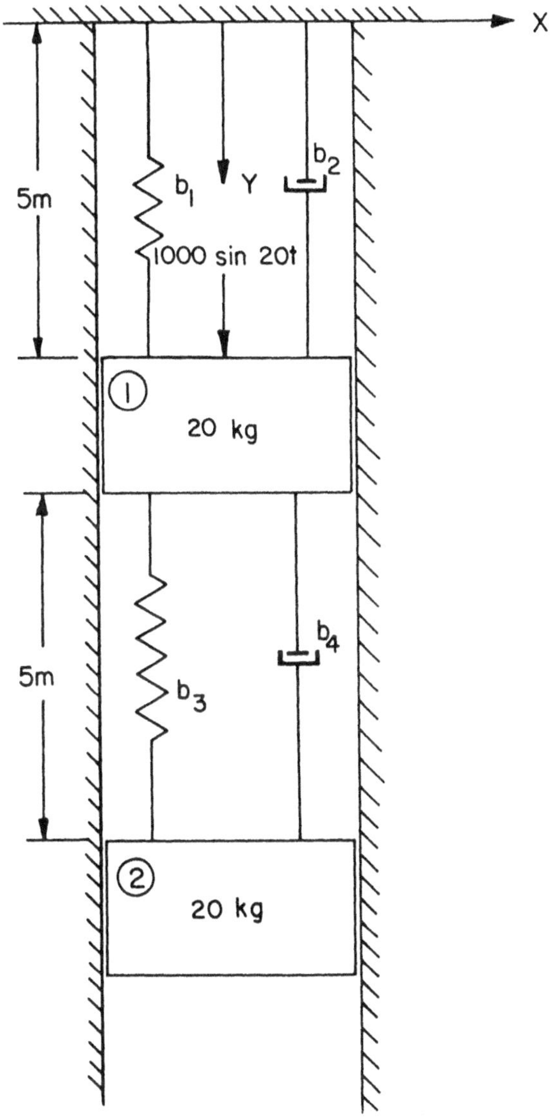

Figure 6.10 Elementary Two Mass Example

where $y_1^1 = 5$ is the initial position of body 1. For this system, the vector of constraints is simply

$$\Phi = [x_1, \theta_1, x_2, \theta_2]^T = [0]$$

and the Jacobian matrix is

$$J = \Phi_q = \begin{bmatrix} 1 & 0 & 0 & 0 & 0 & 0 \\ 0 & 0 & 1 & 0 & 0 & 0 \\ 0 & 0 & 0 & 1 & 0 & 0 \\ 0 & 0 & 0 & 0 & 0 & 1 \end{bmatrix}$$

The mass matrix M is

$$M = \text{diag}\ (20,\ 20,\ 125,\ 20,\ 20,\ 125)$$

Bodies 1 and 2 are initially at distances 5m and 10m from the global X-axis respectively. The terminal values of μ and η are obtained (Step 2) from

$$\begin{bmatrix} M & \Phi_q^T \\ \Phi_q & 0 \end{bmatrix} \begin{bmatrix} \mu^2 \\ \eta^2 \end{bmatrix} = \begin{bmatrix} 0 \\ 0 \end{bmatrix} \tag{6.63}$$

which gives $[\mu^2, \eta^2]^T = [0]$ and $\dot{\mu}^2$ and γ^2 are obtained (Step 3) from

$$\begin{bmatrix} M & \Phi_q^T \\ \Phi_q & 0 \end{bmatrix} \begin{bmatrix} \dot{\mu}^2 \\ \gamma^2 \end{bmatrix} = \begin{bmatrix} 0 \\ 0 \end{bmatrix} \tag{6.64}$$

which gives $[\dot{\mu}^2, \gamma^2]^T = [0]$. Using these terminal conditions, the equations for μ and ν in Eqs. 6.50 and 6.51 are solved to provide $\dot{\mu}$ and $\ddot{\mu}$ (Step 5). Substitution of these quantities in Eq. 6.58 gives the vector of design sensitivities. The design sensitivity vector was calculated from the program as

$$\psi_b = [-0.1681\times10^{-4},\ -0.5707\times10^{-3},\ -0.1049\times10^{-3},\ 0.608\times10^{-3}]^T$$

The design sensitivity vector obtained approximately by perturbation (finite difference) analysis is

$$(\psi_b)_{pert} = [-0.1788\times10^{-4},\ -0.6259\times10^{-3},\ -0.119\times10^{-3},\ 0.608\times10^{-3}]^T$$

Good agreement is seen between the design sensitivities computed by the program and by differencing.

Example 5; Four Bar Mechanism: As a second example, a four bar mechanism (Fig. 6.11) that falls from rest under its own weight is considered. Links 1 and 2 are initially at rest at angles of 45° and -45°, with respect to the global X-axis. The design variables are locations of revolute joints, as shown in Fig. 6.11. The simulation was carried out for a period of one second.

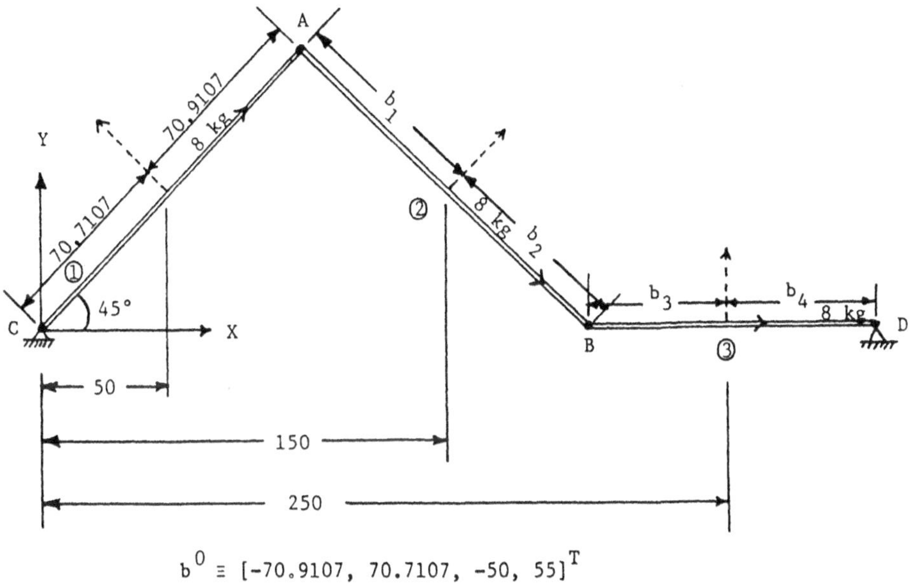

Figure 6.11 Four-Bar Linkage Example

In this problem, the dynamic response functional is taken as

$$\psi = \int_0^1 [\sin\theta_3 + 0.23893455]^2 dt$$

For this system, the constraint equations are the constraint equations for the four revolute joints,

$$\Phi = \begin{bmatrix} x_1 + 70.9107 \cos\theta_1 - x_2 - b_1\cos\theta_2 \\ y_1 + 70.9107 \sin\theta_1 - y_2 + b_1\sin\theta_2 \\ x_2 + b_2\cos\theta_2 - x_3 - b_3\cos\theta_3 \\ y_2 + b_2\sin\theta_2 - y_3 - b_3\sin\theta_3 \\ x_1 - 70.7107 \cos\theta_1 \\ y_1 - 70.7107 \sin\theta_1 \\ x_3 + b_4\cos\theta_3 \\ y_3 + b_4\sin\theta_3 \end{bmatrix} = [0]$$

and the Jacobian matrix is

$$\Phi_q = \begin{bmatrix} 1 & 0 & -70.9107\sin\theta_1 & -1 & 0 & b_1\sin\theta_2 & 0 & 0 & 0 \\ 0 & 1 & 70.9107\cos\theta_1 & 0 & -1 & -b_1\cos\theta_2 & 0 & 0 & 0 \\ 0 & 0 & 0 & 1 & 0 & -b_2\sin\theta_2 & -1 & 0 & b_3\sin\theta_3 \\ 0 & 0 & 0 & 0 & 1 & b_2\cos\theta_2 & 0 & -1 & -b_3\cos\theta_3 \\ 1 & 0 & 70.71707\sin\theta_1 & 0 & 0 & 0 & 0 & 0 & 0 \\ 0 & 1 & -70.7107\cos\theta_1 & 0 & 0 & 0 & 0 & 0 & 0 \\ 0 & 0 & 0 & 0 & 0 & 0 & 1 & 0 & -b_4\sin\theta_3 \\ 0 & 0 & 0 & 0 & 0 & 0 & 0 & 1 & b_4\cos\theta_3 \end{bmatrix}$$

The mass matrix is a 9×9 matrix. The terminal values of the adjoint variables and their derivatives are computed as in Eqs. 6.63 and 6.64 of Example 4.

The initial design was selected as

$$b^0 = [-70.9107,\ 70.7109,\ -50,\ 55]^T$$

The design sensitivity vector obtained from the program is

$$\psi_b = [-0.9731\times10^{-2}, 0.272\times10^{-2}, -0.3897\times10^{-2}, 0.4125\times10^{-2}]$$

The design sensitivity vector obtained by perturbation (finite difference) is

$$(\psi_b)_{pert} = [-0.96\times10^{-2},\ 0.24\times10^{-2},\ -0.35\times10^{-2},\ 0.42\times10^{-2}]$$

yielding good agreement between results obtained by the program and result obtained by differencing.

REFERENCES

1. Tomovic, R., and Vukobratonic, M., General Sensitivity Theory, American Elsevier, NY, 1972.

2. Bryson, A.E., and Ho, Y.C., Applied Optimal Control, Wiley, NY, 1975.

3. Haug, E.J., and Arora, J.S., Applied Optimal Design, Wiley, NY, 1979.

4. Haug, E.J., Wehage, R.A., and Barman, N.C., "Design Sensitivity Analysis of Planar Mechanism and Machine Dynamics", ASME Journal of Mechanical Design, Vol. 103, No. 3, July 1981, pp. 560-570.

5. Ehle, P.E., and Haug, E.J., "A Logical Function Method for Dynamic and Design Sensitivity Analysis of Mechanical Systems with Intermittent Motion", ASME Journal of Mechanical Design, to appear.

6. Haug, E.J., and Ehle, P.E., "Second Order Design Sensitivity Analysis of Mechanical System Dynamics", International Journal for Numerical Methods in Engineering, Vol. 18, 1982, pp. 1699-1717.

7. Coddington, E.A., and Levinson, N., Theory of Ordinary Differential Equations, McGraw-Hill, NY, 1955.

8. Hildebrand, F.B., Advanced Calculus for Applications, Prentice-Hall, Englewood Cliffs, NJ, 1976.

9. Lins, W.F., Human Vibration Response Measurement, Technical Report No. 11551, US Army Tank-Automotive Command, Warren, Michigan, 1972.

10. Choi, K.K., Haug, E.J., Hou, J.W. and Sohoni, V.N., "Pshenichny's Linearization Method for Mechanical System Optimization," ASME Journal of Mechanisms, Transmission, and Automation in Design, No. 1, Vol. 105, 1983, pp. 97-103.

11. Murphy, N.R. and Ahlvin, R.B., AMC-74 Vehicle Dynamics Module, Technical Report No. M-76-1, Waterways Experiment Station, Vicksburg, Mississippi, 1976.

12. Greenwood, D.T., Classical Dynamics, Prentice-Hall, Englewood Cliffs, NJ, 1977.

13. Haug, E.J., and Sohoni, V.N., "Design Sensitivity Analysis and Optimization of Kinematically Driven Systems", these proceedings.

14. Wehage, R.A., and Haug, E.J., "Generalized Coordinate Partitioning for Dimension Reduction in Analysis of Constrained Dynamic Systems," ASME Journal of Mechanical Design, Vol. 104, No. 1, 1982, pp. 247-255.

15. Krishnaswami, P., Wehage, R.A., and Haug, E.J., Design Sensitivity Analysis of Constrained Dynamic Systems by Direct Differentiation, Technical Report No. 83-5, Center for Computer Aided Design, The University of Iowa, 1983.

APPENDIX A

Matrix Calculus Notation

For $x \in R^k$, $y \in R^m$, $a(x,y) \in R^1$, A an m×n constant matrix, $g(x) \in R^n$, and $h(x) \in R^n$, using i as row index and j as column index, define

$$a_x \equiv \frac{\partial a}{\partial x} \equiv \left[\frac{\partial a}{\partial x_j}\right]_{1\times k} \tag{A.1}$$

$$g_x \equiv \left[\frac{\partial g_i}{\partial x_j}\right]_{n\times k} \tag{A.2}$$

$$a_{xy} \equiv \left[\frac{\partial^2 a}{\partial x_i \partial y_j}\right]_{k\times m} = \frac{\partial}{\partial y}\left(\frac{\partial a}{\partial x}^T\right) \equiv (a_x^T)_y \tag{A.3}$$

Using this notation, the following formulas are obtained:

$$\frac{\partial}{\partial x}(Ag) = \left[\frac{\partial}{\partial x_j}(A_{i\ell}g_\ell)\right] = \left[A_{i\ell}\frac{\partial g_\ell}{\partial x_j}\right] = Ag_x \tag{A.4}$$

$$\frac{\partial}{\partial x}(g^T h) = \left[\frac{\partial}{\partial x_j}(g_\ell h_\ell)\right] = \left[\frac{\partial g_\ell}{\partial x_j}h_\ell + g_\ell\frac{\partial h_\ell}{\partial x_j}\right]$$

$$= h^T g_x + g^T h_x \tag{A.5}$$

where summation notation is used with repeated indices in the same term.

APPENDIX B

The Mathematical Programming Problem

The general mathematical programming problem is to find $b \in R^n$ to minimize $f_0(b)$, with constraints

$$\left.\begin{array}{ll} f_i(b) \le 0, & i=1,2,\ldots,m' \\ f_i(b) = 0, & i=m'+1,\ldots,m \end{array}\right\} \tag{B.1}$$

where $f_i(b)$, $i=0,1,\ldots,m$, are continuously differentiable functions. Note that $f_i(b) = 0$ is equivalent to the inequalities $f_i(b) \le 0$, and $-f_i(b) \le 0$. Hence, one can limit considerations to the case with inequality constraints. Thus, one wishes to minimize $f_0(b)$, subject to the constraints

$$f_i(b) \le 0, \quad i=1,2,\ldots,m \tag{B.2}$$

Basic Assumptions

Let

$$F(b) = \max\,\{0, f_1(b), \ldots, f_m(b)\} \tag{B.3}$$

Note that $F(b) \ge 0$ for all $b \in R^n$. Given $\varepsilon \ge 0$, define the ε-active constraint set

$$A(b,\varepsilon) = \{i : f_i(b) \ge F(b) - \varepsilon,\ i=1,2,\ldots,m\} \tag{B.4}$$

(a) Suppose there is an integer $N > 0$ such that the set

$$\Omega_N = \{b : \Phi_N(b) \le \Phi_N(b^0)\} \tag{B.5}$$

is bounded, where b^0 is an initial design and

$$\Phi_N(b) = f_0(b) + NF(b).$$

(b) Suppose gradients of the functions $f_i(b)$, $i=0,1,2,\ldots,m$, satisfy Lipschitz conditions in Ω_N; i.e., there exists $L > 0$ such that

$$||f_i'(b^1) - f_i'(b^2)|| \leq L||b^1 - b^2|| \qquad (B.6)$$

where $f_i' = [\partial f_i/\partial b_1, \ldots, \partial f_i/\partial b_n]^T$. This condition is satisfied if f_i has piecewise continuous first derivatives.

(c) Suppose the problem of quadratic programming; find $p \in R^n$ to minimize

$$(f_0'(b), p) + \tfrac{1}{2} ||p||^2 \qquad (B.7)$$

subject to the linearized constraints

$$(f_i'(b), p) + f_i(b) \leq 0, \quad i \quad A(b, \varepsilon) \qquad (B.8)$$

is solvable with any $b \in \Omega_N$ and there are Lagrange multipliers $u_i(b)$, $i \in A(b, \varepsilon)$, such that

$$\sum_{i \in A(b,\varepsilon)} u_i(b) \leq N \qquad (B.9)$$

Theoretical Algorithm

Under the above hypotheses, one may state the following theoretical Linearization Algorithm of Pshenichny [10]:

Let b^0 be an initial approximation and $0 < \delta < 1$. For the kth iteration,

(1) Solve the quadratic programming problem of Eqs. B.7 and B.8, with $b = b^k$ and solution $p^k = p(b^k)$.

(2) Find the smallest integer i such that

$$\Phi(b^k + \frac{1}{2^i} p^k) \leq \Phi(b^k) - \frac{1}{2^i} \delta ||p^k||^2 \qquad (B.10)$$

If this inequality is satisfied with $i = i_0$, let
$\alpha_k = 2^{-i_0}$ $b^{k+1} = b^k + \alpha_k p^k$

Under the basic assumptions, Pshenichny has proved convergence criteria for the algorithm, which are given in Ref. 10.

Numerical Algorithm

The following numerical algorithm is intended for solving the problem of minimizing $f_0(b)$, subject to the constraints of Eq. B.1. Define

$$F(b) = \max\{0, f_1(b), \ldots, f_{m'}(b), |f_{m'+1}(b)|, \ldots, |f_m(b)|\}$$

$$A(b,\varepsilon) = \{i : f_i(b) > F(b) - \varepsilon,\ i = 1,2,\ldots,m'\}$$

$$B(b,\varepsilon) = \{i : |f_i(b)| > F(b) - \varepsilon,\ i = m'+1,\ldots,m\}$$

$$\phi_N(b) = f_0(b) + NF(b)$$

Select the initial approximation b^0, N_0 sufficiently large, $\varepsilon_0 > 0$, and $0 < \delta < 1$.

Step 1. In the kth iteration, solve the problem of finding u to minimize

$$\phi(u) = \tfrac{1}{2}\,\Big|\Big|f_0'(b^k) + \sum_{i \in A(b^k,\varepsilon) \cup B(b^k,\varepsilon)} u_i f_i'(b^k)\Big|\Big|^2 - \sum_{i \in A(b^k,\varepsilon) \cup B(b^k,\varepsilon)} u_i f_i(b^k)$$

subject to $u_i > 0$, $i \in A(b^k,\varepsilon)$, and u_i arbitrary for

$i \in B(b^k,\varepsilon)$, where $f_i' = [\frac{\partial f_i}{\partial b_1}, \ldots, \frac{\partial f_i}{\partial b_n}]^T$.

If the solution u^k is such that $\phi(u^k) = -\infty$, then set $b^{k+1} = b^k$, $\varepsilon_{k+1} = (\tfrac{1}{2})\varepsilon_k$, and $N_{k+1} = N_k$ and return to Step 1. Otherwise, let

$$p^k = -f_0'(b^k) - \sum_{i \in A(b^k,\varepsilon) \cup B(b^k,\varepsilon)} u_i^k f_i'(b^k)$$

and go to Step 2.

Step 2. Set

$$b^{k+1} = b^k + \alpha_k p^k$$

$$\varepsilon_{k+1} = \varepsilon_k$$

where α_k is chosen equal to $\frac{1}{2^{q_0}}$ and q_0 is the smallest integer for which

$$\Phi_{N_k}\left(b^k + \frac{1}{2^q} p^k\right) < \Phi_{N_k}(b^k) - \frac{1}{2^q}\delta ||p^k||^2$$

Step 3. If

$$2 \sum_{i \in A(b^k,\varepsilon)} u_i^k + \sum_{i \in B(b^k,\varepsilon)} |u_i^k| > N_k > \sum_{i \in A(b^k,\varepsilon)} u_i^k + \sum_{i \in B(b^k,\varepsilon)} |u_i^k|$$

then let $N_{k+1} = N_k$. Otherwise, let

$$N_{k+1} = 2 \sum_{i \in A(b^k,\varepsilon)} u_i^k + \sum_{i \in B(b^k,\varepsilon)} |u_i^k|$$

Step 4. If $||p^k||$ is sufficiently small, terminate. Otherwise, return to Step 1.

In implementing the algorithm, the derivatives f_i' are calculated using the adjoint variable method, which requires that the state and adjoint equations be solved forward and backward in time, respectively. This is a substantial amount of computation that must be carried out during each optimization iteration.

OPTIMIZATION METHODS

C. FLEURY — Research Associate, NFSR
V. BRAIBANT — Research Assistant, NFSR
Aerospace Laboratory
University of Liege
21, Rue E. Solvay
B-4000 Liege, Belgium

Table of contents.

1. INTRODUCTION

In this lecture a review is presented of numerical methods that can efficiently solve optimization problems related to the synthesis of mechanical systems. These methods usually perform iteratively, which means, for our purpose, that the system design will be modified step by step (redesign). Any idealized engineering system can be described by a finite set of quantities. For example, an elastic structure modelled by finite elements is characterized by the node coordinates, the types of element, their thicknesses and material properties, etc ... Some of these quantities are fixed in advance and they will not be changed by the redesign process (they are often called prescribed parameters). The others are the design variables ; they will be

NATO ASI Series, Vol. F9
Computer Aided Analysis and Optimization of Mechanical System Dynamics
Edited by E. J. Haug

modified during each redesign process in order to gradually optimize the mechanicsl system. A function of the design variables must be defined, whose value permits selecting different feasible designs ; this is the objective function (e.g. the weight of an aerospace structure). A design is said to be feasible if it satisfies all the requirements that are imposed to the mechanical system when performing its tasks. Usually, requiring that a design is feasible amounts to assigning upper or lower limits to quantities characterizing the system behaviour (inequality constraints). Sometimes given values, rather than lower or upper bounds, are imposed to these quantities (equality constraints). Taking again the case of structural optimization, the bahaviour constraints are placed on stresses, displacements, frequencies, buckling loads, etc ...

Therefore the optimization problem consists in minimizing an objective function, which represents a cost associated with the mechanical system, subject to equality and inequality constraints which insure the design feasibility.

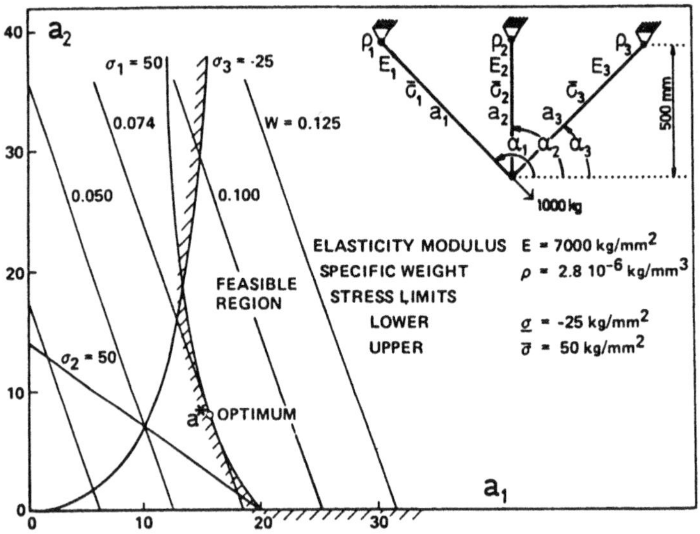

DESIGN SPACE

FIG. 1

To help fix ideas let us consider the minimum weight design of the 3-bar truss shown in Fig. 1. Several types of design variables could be chosen for optimizing this structure : the bar cross-sectional areas [a_1, b_2, a_3], their material properties [(ρ_1, E_1, $\bar{\sigma}_1$); (ρ_2, E_2, $\bar{\sigma}_2$); (ρ_3, E_3, $\bar{\sigma}_3$)] the coordinates of the free node (or, equivalently, the angles [α_1, α_2, α_3]). To simplify the problem, we restrict ourselves to the first class of variables (optimal sizing problem). Adopting the same material for each bar and fixing the prescribed geometrical parameters to the values $\alpha_1 = 45°$, $\alpha_2 = 90°$, $\alpha_3 = 135°$, the minimum weight design problem for maximum allowable compressive and tensile stresses $\underline{\sigma}$ and $\bar{\sigma}$ can be expressed as follows :

$$\text{minimize} \quad W(a) = \rho \ell (2\sqrt{2}\, a_1 + a_2) \tag{1}$$

$$\text{subject to} \quad \sigma_1(a) \equiv \frac{P}{\sqrt{2}\, a_1} \, \frac{a_2 + \sqrt{2}\, a_1}{a_1 + \sqrt{2}\, a_2} \leqslant \bar{\sigma} \tag{2}$$

$$\sigma_2(a) \equiv \frac{P}{a_1 + \sqrt{2}\, a_2} \leqslant \bar{\sigma} \tag{3}$$

$$-\sigma_3(a) \equiv \frac{P}{\sqrt{2}\, a_1} \, \frac{a_2}{a_1 + \sqrt{2}\, a_2} \leqslant \underline{\sigma} \tag{4}$$

$$a_1, \; a_2 \geqslant 0 \tag{5}$$

Note that, for symmetry reasons, only two design variables a_1 and a_2 define the problem, which admits the geometrical interpretation given in the design space of Fig. 1. Each point in the design space corresponds to a possible structural design. The objective function (1) is represented by a set of constant weight planes, and the stress limitations (2-4), by restraint surfaces that permits defining the feasible domain. Clearly the optimal design a^* corresponds to the point where a constant weight plane is tangent to the boundary of the feasible domain. At the optimum, only one constraint is satisfied as an equality (active constraint). The others are satisfied as inequalities (inactive constraints), which means that the stress level in bars 2 and 3 is below the allowable upper limit.

In this simple example, it is quite easy to detect that only one constraint is really meaningful. Realistic optimization problems, however, involve many design variables and constraints. Their solution can no longer be obtained analytically or graphically, but they require efficient mathematical tools. The question of finding how many and which constraints are active is especially crucial.

2. MATHEMATICAL PROGRAMMING PROBLEM

The optimal design problem can be expressed as finding a point

$x^* = (x_1^*, \ldots, x_n^*)^T \in E^n$

in a n-dimensional Euclidian space E^n, solution of the mathematical programming problem

$$\begin{array}{llll} \text{minimize} & f(x) & & \text{objective function (6)} \\ \text{subject to} & h_j(x) \geq 0 & j = 1,\dots,m & \text{inequality constraints (7)} \\ & \bar{x}_i \geq x_i \geq \underline{x}_i & i = 1,\dots,n & \text{side constraints (8)} \end{array}$$

This problem is formulated in terms of inequality constraints only. The reason is that, for algorithmic purposes, any equality constraint can be written as two opposed inequalities. Note also that the lower and upper bound constraints (8) are considered apart from the general constraints (7), because these side constraints are very simple explicit functions and they can be treated separately in most mathematical programming algorithms. In particular there is no need to associate Lagrangian multipliers with the side constraints in order to establish optimality conditions. The well known KUHN-TUCKER conditions take the form :

$$\frac{\partial f}{\partial x_i} - \sum_{j=1}^{m} \lambda_j \frac{\partial h_j}{\partial x_i} \quad \begin{array}{lll} > 0 & \text{if} & x_i = \underline{x}_i \\ = 0 & \text{if} & \bar{x}_i > x_i > \underline{x}_i \\ < 0 & \text{if} & x_i = \bar{x}_i \end{array} \qquad (9)$$

where the multipliers λ_j must be non negative :

$$\begin{array}{llll} \lambda_j = 0 & \text{if} & h_j > 0 & \text{(inactive constraint)} \\ \lambda_j > 0 & \text{if} & h_j = 0 & \text{(active constraint)} \end{array} \qquad (10)$$

It is important to point out that these conditions are only necessary for local optimality. As a result, mathematical programming algorithms do not generally generate the global solution to problem (6-8), but only a local solution which depends upon the starting point in the design space.

The optimality conditions (9,10) involve the first derivatives of the objective function and constraints. Computing these derivatives is an essential ingredient of any optimization scheme (see the chapters in this book devoted to the sensitivity analysis of mechanical systems).

<u>Classification of mathematical programming problems</u>

When the objective function f(x) and all the constraint functions

$\{h_j(x),\ j = 1,m\}$ are linear, the problem (6-7) is called a <u>linear programming problem</u>. This special case has been treated extensively in the literature and we shall assume that we are provided with a standard linear programming problem solver. If any of the functions $\{f,\ h_j\}$ is nonlinear, the problem is called a <u>nonlinear programming problem</u>. Solution methods are then much less standard. The <u>unconstrained minimization problem</u>

$$\text{minimize} \quad f(x) \quad \text{for} \quad x \in E^n \tag{11}$$

is a very important special case, because it is at the origin of the basic theory underlying nonlinear programming methods. Another interesting special case is that of a <u>linearly constrained minimization problem</u>. Indeed methods for unconstrained problems can be readily adapted when linear constraints are considered. Moreover some effective general purpose optimization techniques proceed by transforming the original problem into a sequence of linearly constrained subproblems. The problem (6-8) is called a <u>convex programming problem</u> when $f(x)$ is a convex function and each $h_j(x)$ is a concave function. The feasible region is then a convex set and any local solution is also global. In addition, the KUHN-TUCKER conditions becomes sufficient for global optimality, and many of the duality results, that will be used later in this chapter are strictly valid for convex problems only. For a convex separable problem, these duality results can be implemented into efficient algorithms. A <u>separable programming problem</u> is one that can be written

$$\begin{aligned} &\text{minimize} && f(x) \equiv \sum_{i=1}^{n} f_i(x_i) && \\ &\text{subject to} && h_j(x) \equiv \sum_{i=1}^{n} h_{ji}(x_i) \geq 0 && j = 1,m \\ & && \bar{x}_i \geq x_i \geq \underline{x}_i && i = 1,n \end{aligned} \tag{12}$$

where each function $\{f_i(x_i),\ h_{ji}(x_i)\}$ depends only on the single variable x_i.

<u>Primal and dual problem statement</u>

As indicated previously, only the main constraints (7) have to be associated with lagrangian multipliers, or <u>dual variables</u>, $(\lambda_j,\ j=1,m)$, while the side constraints can be treated separately. Let therefore X define the set of all primal points satisfying the side constraints (8) that is,

$$X = \{x : \underline{x}_i \leq x_i \leq \bar{x}_i \ ; \quad i = 1,n\} \tag{13}$$

and let Λ denote the set of all dual points satisfying the nonnegativity conditions, that is,

$$\Lambda = \{\lambda : \lambda_j \geq 0 \ ; \quad j = 1,m\} \tag{14}$$

Corresponding to the <u>primal problem</u>

$$\begin{aligned} &\text{minimize} \quad f(x) \quad \text{for} \quad x \in X \\ &\text{s.t.} \qquad h_j(x) \geq 0 \qquad\qquad j = 1,m \end{aligned} \tag{15}$$

there exists a unique dual problem if the function $f(x)$ is strictly convex and if the functions $h_j(x)$ are concave. Formally the solution of the dual problem can be obtained through a two phase procedure as follows :

$$\max_{\lambda \in \Lambda} \ \min_{x \in X} \ L(x,\lambda)$$

where

$$L(x,\lambda) = f(x) - \sum_{j=1}^{m} \lambda_j \ h_j(x) \tag{16}$$

is the lagrangian function. Therefore the <u>dual problem</u> can be written

$$\begin{aligned} &\text{maximize} \quad \ell(\lambda) \\ &\text{s.t.} \qquad \lambda_j \geq 0 \qquad j = 1,m \end{aligned} \tag{17}$$

where

$$\ell(\lambda) = \min_{x \in X} L(x,\lambda) \tag{18}$$

is defined as the dual function.

While the primal problem involves n variables, m general constraints and 2n side constraints, the dual problem involves m variables and m nonnegativity constraints. The dual problem is thus quasi-unconstrained, and it can be readily solved by slightly modifying any unconstrained minimization algorithm. This requires the gradient of the dual function to be known. Fortunately, $\nabla\ell(\lambda)$ is extremely simple to compute, because it is given by the primal constraints :

$$\frac{\partial \ell}{\partial \lambda_j} = - h_j \ [x(\lambda)] \tag{19}$$

where $x(\lambda)$ denotes the primal point that minimizes $L(x,\lambda)$ over X for given λ (see Eq. 18). In terms of this $x(\lambda)$, the dual function can also be written

$$\ell(\lambda) = f[x(\lambda)] - \sum_{j=1}^{m} \lambda_j h_j [x(\lambda)] \tag{20}$$

When a numerical maximization scheme is employed to solve the dual problem, the evaluation of the dual function calls for the determination of the primal constraint values $h_j[x(\lambda)]$, so that the first derivatives (19) are available without additional computation. To obtain $\ell(\lambda)$ it is necessary to find the x that minimizes the Lagrangian, as formally stated in Eq. (18). For certain problems, this is not very difficult. For the separable programming problem (12), the dual function takes the form :

$$\ell(\lambda) = \sum_{i=1}^{n} \{ \min_{\underline{x}_i \leq x_i \leq \bar{x}_i} [f_i(x_i) - \sum_{j=1}^{m} \lambda_j h_{ji}(x_i)] \} \tag{21}$$

So, a one-dimensional search in each of n components is all that is required. In some problems, the simplicity of each single variable minimization problem appearing in (21) is such that it can be solved in closed form, yielding thus an explicit dual function.

3. UNCONSTRAINED MINIMIZATION

Most algorithms for solving a minimization problem are iterative. They require an initial estimate of the solution, $x^{(o)}$, and then, for $k = 0, 1, 2, \ldots$, the kth iteration replaces $x^{(k)}$ by $x^{(k+1)}$, which should be a better estimate of the solution. Furthermore, nearly all of the unconstrained minimization methods described in this section are descent methods, that is :

$$f(x^{(k+1)}) < f(x^{(k)}) \tag{22}$$

They usually involve sequential minimization of $f(x)$ along successive search directions $s^{(k)}$, so that

$$x^{(k+1)} = x^{(k)} + \alpha^{(k)} s^{(k)} \tag{23}$$

Clearly $s^{(k)}$ must be a downhill direction, which means that for sufficiently small $\alpha > 0$, the inequality

$$f(x^{(k)} + \alpha s^{(k)}) < f(x^{(k)})$$

should hold. Assuming differentiability, an equivalent requirement is that

$$s^{(k)^T} g^{(k)} < 0 \tag{24}$$

where $g^{(k)} = \nabla f(x^{(k)})$ denotes the gradient of $f(x)$ at $x^{(k)}$.

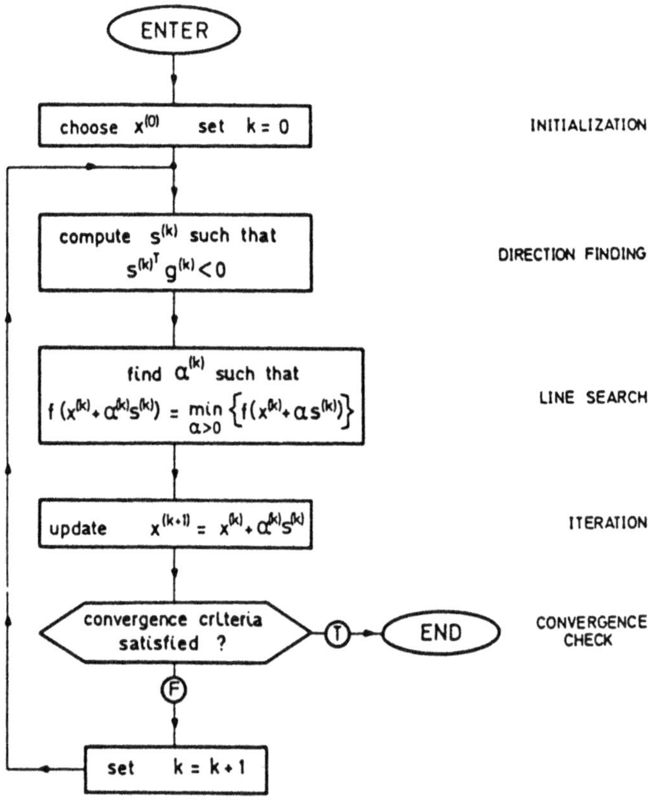

FIG. 2 BASIC DESCENT ALGORITHM

A basic descent algorithm for minimizing $f(x)$ is shown in Fig. 2. Each iteration involves two parts. The first part calculates a downhill direction, $s^{(k)}$, at $x^{(k)}$, and the second part evaluates a steplength $\alpha^{(k)}$ from which the new point $x^{(k+1)}$ is computed by using Eq. (23). Most often the steplength $\alpha^{(k)}$ is estimated so as to minimize the objective function along the search direction $s^{(k)}$. Then $\alpha^{(k)}$ satisfies at least approximately, the requirement

$$f(x^{(k+1)}) = \min_{\alpha} \phi(\alpha) \tag{25}$$

where $\phi(\alpha)$ represents the objective function along the line $s^{(k)}$, regarded as a function of α alone :

$$\phi(\alpha) = f(x^{(k)} + \alpha\, s^{(k)}) \tag{26}$$

Therefore each iteration in the basic descent algorithm implies solving the equation

$$\phi'(\alpha) = 0 \tag{27}$$

From (26), it is apparent that

$$\begin{aligned}\phi'(\alpha) &= \sum_{i=1}^{n} \frac{\partial f}{\partial x_i}(x^{(k)} + \alpha\, s^{(k)}) \frac{\partial}{\partial \alpha}(x_i^{(k)} + \alpha\, s_i^{(k)}) \\ &= g(x^{(k)} + \alpha\, s^{(k)})^T s^{(k)}\end{aligned} \tag{28}$$

Condition (27) and the definition (23) of $x^{(k+1)}$ show that the gradient $g^{(k+1)}$ of $f(x)$ at $x^{(k+1)}$ is orthogonal to the search direction $s^{(k)}$

$$g^{(k+1)^T} s^{(k)} = 0 \tag{29}$$

This property is illustrated in Fig. 3. Note that because of Eq. (29) $s^{(k)}$ is tangent to the contour of f on which $f(x) = f(x^{(k+1)})$. Note also that

$$\begin{aligned}\phi'(\alpha) &< 0 \quad \text{if} \quad \alpha < \alpha^{(k)} \\ \phi'(\alpha) &> 0 \quad \text{if} \quad \alpha > \alpha^{(k)}\end{aligned} \tag{30}$$

which are useful inequalities for trapping the optimal step length $\alpha^{(k)}$.

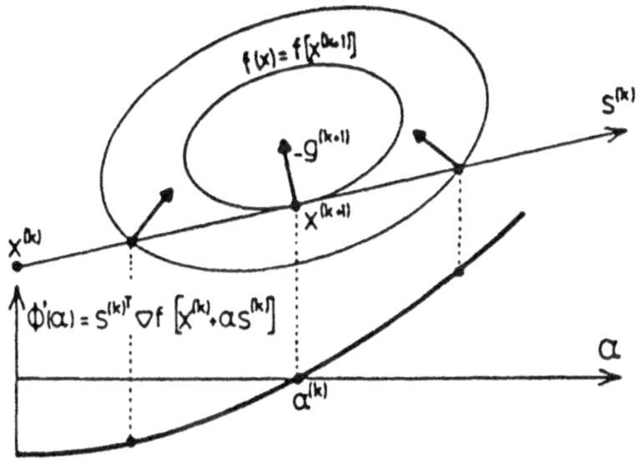

ONE - DIMENSIONAL MINIMIZATION (line search)

FIG. 3

Line search techniques

Solving the one-dimensional minimization problem (25), or, equivalently, the single variable nonlinear equation (27), is an important part of many descent methods. This process of finding $\alpha^{(k)}$ by estimating a minimum of $\phi(\alpha)$ is called line search. Most often it is performed by using iterative numerical procedures which are terminated when some convergence criteria are satisfied. Line searches are therefore usually not exact and their accuracy must be adapted to the type of descent method employed. In any event the computational efficiency of the line search is very important for the success of the whole minimization method, because repeated function and gradient evaluations are required.

Many line search techniques are based on curve fitting procedures. Depending upon whether or not derivatives of the function can be measured, one or several points must be used to determine the fit, and a variety of line search techniques can be devised. By applying the Newton-Raphson method to Eq. (27), a one-point pattern is obtained, which demands to compute both $\phi'(\alpha)$ and $\phi''(\alpha)$ at each iteration in the line search. In most problems, however, only first derivative informations are available. Two measurement points are then necessary to genérate a polynomial fit. Considering an initial bracket $[\alpha_1, \alpha_2]$ in which the minimum α^* is known to lie, the idea is to gradually reduce the bracket - sometimes called interval of uncertainty - until the minimum is trapped within sufficient accuracy. The general equation yielding a new estimate of the minimum is

$$\alpha_3 = \alpha_2 - \beta(\alpha_2 - \alpha_1) \quad \text{with} \quad 0 \leqslant \beta \leqslant 1 \tag{31}$$

Clearly this equation is such that $\alpha_1 \leqslant \alpha_3 \leqslant \alpha_2$, so that each refit will narrow the interval of uncertainty. In view of Eq. (30), the new bracket for the next iteration can be determined according to the sign of $\phi'(\alpha_3)$ (see Fig. 3).

The value of β in Eq. (31) depends upon the interpolating formula being used. A very crude technique is to take $\beta = \frac{1}{2}$, which amounts to halving the interval of uncertainty at each iteration (bisection iteration). It is much better, however, to exploit the derivative informations on $\phi(\alpha)$ in order to approximate it by a polynomial in α of degree two (method of false position) or three (cubic interpolation), and finding analytically the minimum of the polynomial fit. Finally, it is worth noticing that, in order to avoid first derivative computations, a three point pattern can be used, that is based on the objective function values only (quadratic interpolation).

The method of steepest descent

The fundamental method in nonlinear programming is probably the method of steepest descent, in which the downhill directions are taken as the negative gradient vectors. This method is not in itself a very efficient one, but it provides the basis for all gradient methods. Furthermore convergence properties can be established, which serve as a reference situation for other minimization algorithms. In view of the descent condition (24), a natural choice for a downhill direction in the basic descent algorithm schematized in Fig. 2 is $s^{(k)} = -g^{(k)}$, yielding $s^{(k)T}g^{(k)} < 0$ if $g^{(k)} \neq 0$. So, from the point $x^{(k)}$, we search along the direction of negative gradient to a minimum point $x^{(k+1)}$ on this line. This is the method of steepest descent, and Fig. 4 shows how it works.

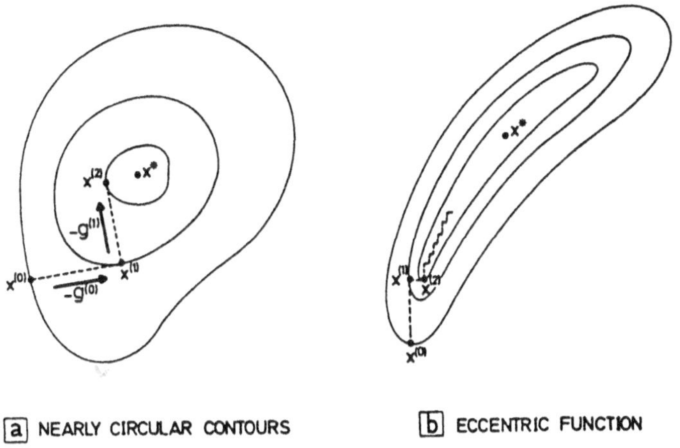

FIG. 4 THE METHOD OF STEEPEST DESCENT

From the line search condition (29) it is apparent that the successive search directions are orthogonal. Also, it can be seen from Fig. 4 that the method should well perform on an objective function with nearly circular contours. On the other hand convergence will be extremely slow if the contours are very elongated. These intuitive results are confirmed by studying the rate of convergence of the steepest descent method when applied to quadratic problems.

Consider the quadratic problem

$$\text{minimize} \quad q(x) = \frac{1}{2} x^T A x - b^T x \tag{32}$$

where A is a symmetric positive definite matrix with eigenvalues $0 < e = e_1 \leqslant e_2 \leqslant \ldots \leqslant e_n = E$. Clearly q is strictly convex and has a unique minimum point $x^* = A^{-1}b$. It can be shown that the method of steepest descent converges to x^* for any starting point $x^{(o)}$ (global convergence). The rate of convergence is linear and the convergence ratio is

$$\rho = \left(\frac{E - e}{E + e}\right)^2 = \left(\frac{r - 1}{r + 1}\right)^2 \tag{33}$$

where r is the condition number of the matrix A, that is, the ratio E/e of its largest to its smallest eigenvalue. The meaning of Eq.(33) is that each iteration will reduce the error in the objective function by at least a factor ρ. Note that $0 < \rho < 1$. The smaller is ρ, the more rapid is the convergence. Therefore convergence is slowed as r increases, and is very rapid when r is close to unity, which can only happen if all the eigenvalues of A are close to each other. A geometrical interpretation is that the speed of convergence depends upon the ratio of the longest to the shortest principal axes of the elliptical contours of q, that is, it is primarily governed by the eccentricity of the ellipsoïds. For example, in the limiting case of unit condition number (E = e), corresponding to circular contours, convergence occurs in a single step. Conversely the rate of convergence is reduced as the contours of q become more eccentric (see Fig. 4).

The conjugate gradient algorithm

The vectors $s^{(i)}$ ($i = 1,2,\ldots,n$) are said to be conjugate with respect to the symmetric matrix A (or A-orthogonal) if

$$s^{(i)T} A s^{(j)} = 0 \quad \text{for } i \neq j \tag{34}$$

If A is positive definite, then the vectors $s^{(i)}$ are linearly independent. In this case, it is easily shown that an arbitrary vector v can be expressed in the form

$$v = \sum_{i=0}^{n-1} \frac{s^{(i)^T} A\, v}{s^{(i)^T} A\, s^{(i)}}\, s^{(i)} \tag{35}$$

Applying this expansion to the solution $x^* = A^{-1}b$ of the quadratic problem (32), it comes

$$x^* = \sum_{i=0}^{n-1} \gamma_i\, s^{(i)} \quad \text{with} \quad \gamma_i = \frac{s^{(i)^T} b}{s^{(i)^T} A\, s^{(i)}} \tag{36}$$

This result suggests that problem (32) can be solved by using an iterative process of n steps in which the successive coefficients γ_i are evaluated without knowing x^*. This forms the basis of the class of conjugate direction methods, by which a quadratic function can be minimized in at most n iterations, starting from an arbitrary initial estimate of x^*. Such methods are said to be quadrically convergent, or to have the property of quadratic termination. A stronger result is that if a line search is performed successively along a set of mutually conjugate directions, the (quadratic) function is minimized in the space spanned by those directions (Expanding Subspace Theorem). This is easily shown from the fact that the gradient vector or $x^{(k+1)}$ is orthogonal to all of the preceding descent directions $s^{(i)}$:

$$g^{(k+1)^T} s^{(i)} = 0 \qquad i = 0, \ldots, k \tag{37}$$

The conjugate gradient algorithm is a special case of these methods, in which the set of conjugate directions is generated from A-orthogonalization of the successive gradients. The first search direction is the steepest descent vector and then, $s^{(k+1)}$ is determined as a linear combination of $- g^{(k+1)}$ and $s^{(k)}$ (see Fig. 5).

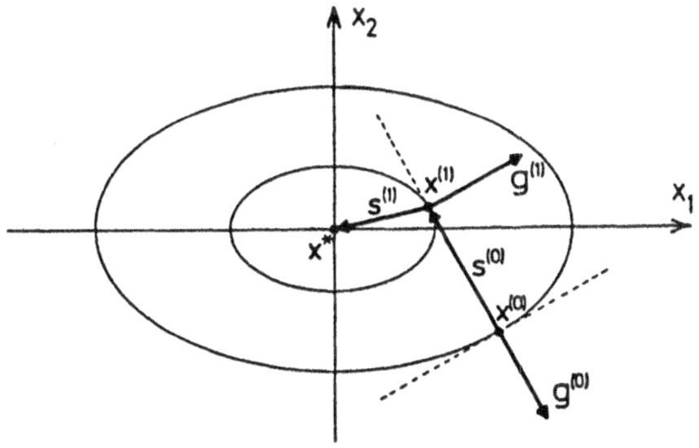

FIG. 5 THE CONJUGATE GRADIENT METHOD

There exists several implementation of the conjugate gradient method, depending upon the way the conjugacy coefficients are evaluated. The FLETCHER-REEVES algorithm is probably the best known :

initialization (i) choose $x^{(o)}$ arbitrary ; compute $g^{(o)} = \nabla f(x^{(o)})$; set $s^{(o)} = -g^{(o)}$

iterative process (ii) $x^{(k+1)} = x^{(k)} + \alpha^{(k)} s^{(k)}$; $\alpha^{(k)}$ optimal

(iii) $g^{(k+1)} = \nabla f(x^{(k+1)})$

$$\beta^{(k)} = \frac{g^{(k+1)^T} g^{(k+1)}}{g^{(k)^T} g^{(k)}} \tag{38}$$

(iv) $s^{(k+1)} = -g^{(k+1)} + \beta^{(k)} s^{(k)}$

(v) set $k = k + 1$ and return to (ii)

It is important to realize that the successive search directions are mutually conjugate only if the line search implied in step (ii) is exact. Note also that, because the vectors $[s^{(1)}, \ldots, s^{(i)}]$ and $[g^{(1)}, \ldots, g^{(i)}]$ span the same subspace, equation (37) is equivalent to

$$g^{(k+1)^T} g^{(i)} = 0 \qquad i = 0, \ldots, k \tag{39}$$

which means that the gradient at $x^{(k+1)}$ is orthogonal to all the previously computed gradients.

The above algorithm needs only function and gradient evaluations and it is therefore applicable, with minor modifications, to non quadratic objective functions. Clearly, if $x^{(o)}$ is sufficiently close to x^*, then $f(x)$ is sufficiently well approximated by a quadratic function and the directions $s^{(k)}$ constructed in the FLETCHER-REEVES algorithm are nearly conjugate in the basis of the local Hessian. We expect $x^{(n)}$ generated by n steps of the algorithm to be much closer to x^* than $x^{(o)}$, so that the quadratic approximation becomes better in the vicinity of $x^{(n)}$. This suggests a strategy called resetting, which consists of restarting the algorithm after n steps with $s^{(o)} = -g^{(o)}$. Another straighforward modification is that a test of convergence has to be included to stop the algorithm. Also, if the line search is not accurate enough, it might happen that $g^{(k+1)^T} s^{(k)} < 0$, in which case the algorithm must be restarted.

Conjugate gradient methods are much better than the steepest descent method because they take into account, in a simple manner, some informations about the curvature of the objective function. In fact, superlinear convergence can be proved. They are especially useful when the number of variables is large, because of the relatively small amount of machine storage requirement, in opposition with Newton-type and quasi-Newton methods to be described next.

The method of Newton

Newton's method is very efficient to solve an unconstrained minimization problem, because it has order 2 convergence, as opposed to gradient methods characterized by order 1 convergence. It consists in taking $s^{(k)} = -[H^{(k)}]^{-1} g^{(k)}$ and $\alpha^{(k)} = 1$ in the basic iterative scheme (23), where H denotes the Hessian of the objective function, i.e., the matrix of its second partial derivatives. Applied to the quadratic function (32), Newton's method generates the minimum in one step. For a general non linear function f(x), it proceeds iteratively as follows

$$x^{(k+1)} = x^{(k)} - [H^{(k)}]^{-1} g^{(k)} \qquad (40)$$

where $x^{(k+1)}$ is the minimum of the local quadratic approximation to f(x) at $x^{(k)}$ (second order Taylor series expansion) :

$$\tilde{f}(x) = f(x^{(k)}) + (x-x^{(k)})^T \nabla f(x^{(k)}) + \frac{1}{2}(x-x^{(k)})\nabla^2 f(x^{(k)})(x-x^{(k)}) \qquad (41)$$

For a highly eccentric function, the Newton search direction tends to be parallel to the eigenvectors of $H(x^{(k)})$ corresponding to the lowest eigenvalues. In other words, even for a steep-sided valley, the Newton direction roughly points toward the minimum (see Fig. 6).

Provided that the local Hessian is positive definite, the Newton direction is obviously downhill, since $s^{(k)T} g^{(k)} = -s^{(k)T} H^{(k)} s^{(k)} < 0$. However, because a unit step length is adopted, the descent condition $f(x^{(k+1)}) < f(x^{(k)})$ may not be fulfilled. Hence the idea of the generalized Newton's method, which consists in adding a line search in the iterative scheme (40). If $H^{(k)}$ remains positive definite, then the descent condition (22) can always be satisfied by choosing the step length α sufficiently small. Several remedies against failure of Newton's method can be devised when the Hessian happens to be **indefinite** (non convex functions).

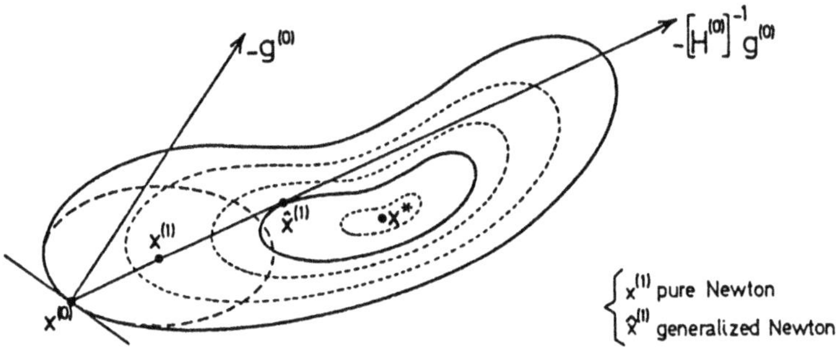

FIG. 6 THE METHOD OF NEWTON

For example, the Newton step can be replaced by a steepest descent iteration or, if the Newton direction is uphill, it can simply be reversed. More sophisticated procedures for dealing with positive undefiniteness of the Hessian exist. Most often they consist in adding appropriate diagonal terms to the Hessian matrix (Newton-type methods):

$$\tilde{H}^{(k)} = H^{(k)} + \varepsilon^{(k)} I \qquad \varepsilon^{(k)} > 0 \tag{42}$$

where $\tilde{H}^{(k)}$ is a modified definite positive matrix to be used in the iterative process.

Quasi-Newton methods

The principal objection to Newton's method is that each iteration requires the direct evaluation of the Hessian matrix $H^{(k)}$ and the solution of the associated linear system (40). Whence the idea of constructing minimization algorithms in which the Hessian is approximated from available quantities (first derivatives only) rather than calculated directly. The basis for such approximations results from a first order expansion of the gradient of f(x) in the vicinity of $x^{(k+1)}$:

$$g^{(k)} = g^{(k+1)} + H^{(k+1)} (x^{(k)} - x^{(k+1)}) + \Delta \tag{43}$$

where $\Delta \rightarrow 0$ as $x^{(k)} \rightarrow x^{(k+1)}$. Setting

$$y^{(k)} = g^{(k+1)} - g^{(k)} \tag{44}$$

$$s^{(k)} = x^{(k+1)} - x^{(k)} \tag{45}$$

and neglecting the second order term Δ, it comes

$$y^{(k)} = H^{(k+1)} s^{(k)} \tag{46}$$

Therefore, defining $S^{(k+1)}$ as an approximation to the inverse Hessian $[H^{(k+1)}]^{-1}$, it should be constructed so as to verify the so-called quasi-Newton equation

$$S^{(k+1)} y^{(k)} = s^{(k)} \qquad k \geqslant 0 \tag{47}$$

The matrix $S^{(k+1)}$ is easily computable from $S^{(k)}$ if it is obtained by adding to $S^{(k)}$ a correction term $C^{(k)}$ which depends upon $S^{(k)}$, $y^{(k)}$ and $s^{(k)}$ only :

$$S^{(k+1)} = S^{(k)} + C^{(k)} \tag{48}$$

Furthermore the correction term $C^{(k)}$ should preserve the symmetry and the positive definiteness of $S^{(k)}$ in order to obtain a downhill search direction at $x^{(k)}$:

$$s^{(k)} = - S^{(k)} g^{(k)} \tag{49}$$

Various quasi-Newton algorithms have been devised, which only differ by the choice of the matrix updating formula (48). The Davidon-Powell (DFP) method (also called variable metric method) is probably the most widely used. It consists of a rank two correction procedure :

$$S^{(k+1)} = S^{(k)} + \frac{s^{(k)} s^{(k)T}}{s^{(k)T} y^{(k)}} - \frac{S^{(k)} y^{(k)} y^{(k)T} S^{(k)}}{y^{(k)T} S^{(k)} y^{(k)}} \tag{50}$$

It can be proven that if $S^{(k)}$ is positive definite, then so is $S^{(k+1)}$ provided that an exact line search is accomplished to calculate $x^{(k+1)}$. Applied to the quadratic function (32), the DFP method has two fundamental properties : it is quadratically convergent and it generates conjugate directions. Furthermore, if the initial approximation $S^{(o)}$ is taken as the identity matrix, the method becomes the conjugate gradient algorithm. An important corollary is that the vectors $s^{(o)}$, $s^{(1)}$, ..., $s^{(k)}$ are eigenvectors corresponding to unity eigenvalues for the matrix $S^{(k+1)}A$. These eigenvectors being A-orthogonal, they are linearly independent and therefore, $S^{(n)} = A^{-1}$. So the DFP update formula (50) generates the inverse Hessian after n iterations.

Another interesting approach is to use the rank one update formula

$$S^{(k+1)} = S^{(k)} + \frac{[s^{(k)} - S^{(k)} y^{(k)}][s^{(k)} - S^{(k)} y^{(k)}]^T}{y^{(k)T} [s^{(k)} - S^{(k)} y^{(k)}]} \qquad (51)$$

Convergence is still achieved within n steps, yielding the inverse Hessian $S^{(n)} = A^{-1}$. The advantage of using this rank-one formula is that it does not require any line search. It is thus expected that its application to nonquadratic cases dispenses also with an accurate line search, which makes it very attractive. Unfortunately the positive definite character of $S^{(k)}$ may be lost in the iterative process, which might lead to a breakdown of the algorithm.

When compared to the conjugate gradient methods, quasi-Newton methods exhibit a stronger stability of convergence, specially in the case of highly eccentric functions. On the other hand they have the drawback of requiring a significantly larger machine storage ($n^2/2$ additional positions).

4. LINEARLY CONSTRAINED MINIMIZATION

The problem addressed in this section consists in minimization a nonlinear objective function subject to linear constraints:

$$\text{minimize} \quad f(x) \qquad (52)$$

$$\text{s.t.} \quad \sum_{i=1}^{n} c_{ij}\, x_i \geq b_j \qquad j = 1,m \qquad (53)$$

$$\bar{x}_i \geq x_i \geq \underline{x}_i \qquad i = 1,n \qquad (54)$$

The so-called projection methods are very efficient to solve this problem. They are characterized by a move along downhill directions that are constrained to reside on the polyhedral boundary of the feasible domain. Therefore each point in the process is feasible and the objective function value constantly decreases. Furthermore an estimate of the Lagrangian multipliers associated with the active constraints is generated at each iteration.

The gradient projection method

Let $x^{(k)}$ be a feasible point at which q linear constraints are active (i.e. $c_j^T x^{(k)} - b_j = 0$; $j = 1,q$), the others being inactive (i.e. $c_j^T x^{(k)} - b_j > 0$; $j > q$). Let N_q denotes the matrix composed of the gradients of the active constraints :

$$N_q = [\, c_1 \; \ldots \; c_q] \tag{55}$$

For simplicity, the side constraints (54) will be treated like regular linear constraints, so that the columns of N_q may contain vectors c_j that are simple base vectors [e.g. $(0, \ldots, 0, 1, 0, \ldots, 0)^T$]. However it can be shown that the side constraints can be handled in a more efficient way. Assuming regularity of the active constraints, N_q is a n x q matrix of rank $q < n$. The kth iteration must lead from $x^{(k)}$ to another feasible point $x^{(k+1)}$ according to the usual descent iteration (23). For notational convenience we shall henceforth omit the iteration index k and simply use the upperscript $^+$ to indicate a new point (at iteration k+1). Therefore the current iteration (23) becomes

$$x^+ = x + \alpha\, s \tag{56}$$

where α is the steplength made along the search direction s. We want s to be a downhill direction, so we require $s^T g < 0$, where $g = \nabla f(x)$. In addition we wish s to lie in the intersection of the active constraint hyperplanes, so we require

$$N_q^T s = 0 \tag{57}$$

so that all currently active constraints remain active at x^+. The particular search direction that we shall use is the projection of the negative gradient into the constraints intersection :

$$s = -\, P_q g \tag{58}$$

where P_q is an <u>orthogonal projection operator</u>. To find the form of the matrix P_q we notice that any vector v can be written as the difference between the projected vector $P_q v$ and a vector $N_q \lambda$ orthogonal to the constraint intersection :

$$v = P_q v - N_q \lambda \tag{59}$$

where $\lambda \in E^q$. Taking v as the negative gradient and using condition (57), it follows that

$$N_q^T \, [P_q g] \equiv N_q^T (g - N_q \lambda) = 0 \tag{60}$$

Because N_q has rank q, we can solve this equation for λ and obtain

$$\lambda = (N_q^T N_q)^{-1} N_q^T g \tag{61}$$

Now Eq. (58) can be written

$$s = - P_q g = - g + N_q \lambda \tag{62}$$

which shows that

$$P_q = I - N_q (N_q^T N_q)^{-1} N_q^T \tag{63}$$

The direction s given by (62) is the "projected gradient". By construction the feasibility requirement (57) is satisfied. Furthermore the descent condition is also fulfilled, because $s^T g = - s^T s < 0$ if $s \neq 0$. On the other hand, if $s = 0$, then from (62) :

$$g - N_q \lambda = 0 \tag{64}$$

Since N_q is made up of the active constraint gradients, (64) implies that the KUHN-TUCKER conditions are satisfied provided that the components of λ are all nonnegative. The process then is terminated. Suppose now that $s = 0$ and at least one component of λ is negative, say $\lambda_r < 0$. It is thus possible to find a new feasible search direction $\tilde{s}$ by relaxing the corresponding inequality $c_r^T x - b_r \geqslant 0$ and projecting the negative gradient onto the intersection of the remaining q-1 active constraints :

$$\tilde{s} = - P_{q-1} g \tag{65}$$

where P_{q-1} is the new projection matrix, which is computed from Eq. (63) with N_q replaced by N_{q-1} (N_{q-1} is simply the matrix N_q with column c_r deleted). The new vector $\tilde{s}$ is a feasible downhill direction. Indeed the constraint just left cannot be violated, because it can be shown that $c_r^T \tilde{s} = - \frac{1}{\lambda_r} \tilde{s}^T \tilde{s} > 0$. This process of dropping a binding constraint at a point satisfying (64) is illustrated in Fig. 7.a.

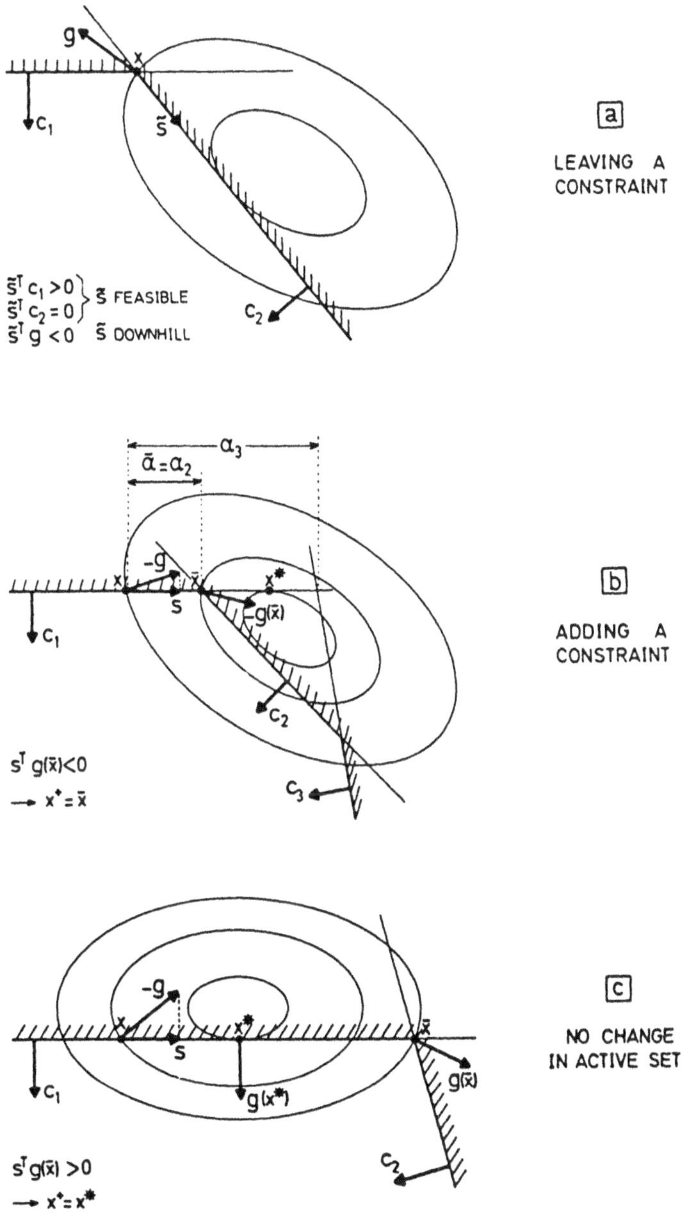

THE GRADIENT PROJECTION METHOD (linear constraints)

FIG. 7

From the foregoing developments, it appears that λ satisfying (64) can be identified as the vector of Lagrangian multipliers associated with the active constraints. When condition (64) is not fulfilled, then the λ_j's computed from (61) are no longer the true Lagrangian multipliers. They are called first order estimates of the multipliers.

We next consider selection of the step size α in Eq. (56), which leads to eventually adding a constraint to the active set. Iteration (56) must be performed so as to minimize the objective function along the direction s, with the additional requirement that x^+ must still be a feasible point. Therefore there exists a maximum allowable step length along s, which can be computed as follows :

$$\bar{\alpha} = \min_{j=q+1,m} \{\alpha_j > 0 : x + \alpha_j s \quad \text{is feasible}\} \tag{66}$$

It is easily seen that

$$\alpha_j = - \frac{c_j^T x - b_j}{c_j^T s} \qquad j = q+1,m \tag{67}$$

which implies $\alpha_j > 0$ only if $c_j^T s < 0$. As indicated in Fig. 7.b., the α_j's are the intercept distances to the constraint hyperplanes corresponding to previously inactive constraints. With $\bar{\alpha}$ known from Eq.(66), compute $\bar{x} = x + \bar{\alpha}\, s$. If $s^T g(\bar{x}) \leq 0$, $\bar{x}$ is the minimum of $f(x)$ along s, because for $\alpha > \bar{\alpha}$, at least one constraint would be violated. Thus set $x^+ = \bar{x}$ and add the newly encountered constraint to the active set. This means that the corresponding constraint gradient is added to the matrix N_q to form the n x (q+1) matrix N_{q+1}. The associated projection matrix P_{q+1} is given by Eq. (63) with N_{q+1} replacing N_q. The iterative process can now be repeated at x^+. On the other hand, if $s^T g(\bar{x}) > 0$, then there exists $\alpha^* \in [0, \bar{\alpha}]$ such that $x^* = x + \alpha^* s$ is the minimum of $f(x)$ along s (see Fig. 7.c.). This value α^* can be determined by using a line search technique (see section 3.1). In this case, no new constraint has been added to the active set. Thus simply set $x^+ = x^*$ and repeat the iteration at x^+.

First and second order projection methods

Just as the method of steepest descent, the gradient projection method is largely inefficient because of its slow convergence speed. It is therefore natural to consider Newton-type, quasi-Newton and conjugate gradient methods. This implies defining adequate projection operators. If H is the objective function Hessian matrix,

$$\hat{P}_q = I - H^{-1} N_q (N_q^T H^{-1} N_q)^{-1} N_q^T \tag{68}$$

is an <u>oblique projection operator</u>, weighed by H^{-1}, which projects vectors of E^n so that they are orthogonal to the space spanned by the colunms of E^n. $\hat{P}_q$ can be thought of as a projection operator in the metric of the Hessian matrix, rather than the Euclidian metric. The matrix $\hat{P}_q H^{-1}$ can be considered as a projected inverse Hessian and consequently the vector $s = - \hat{P}_q H^{-1} g$ is the direction of Newton within the intersection of the active constraint hyperplanes. It is easily verified that for a quadratic function with linear equality constraints, the optimum is then generated in one iteration, with a step length equal to unity, exactly like Newton's method for unconstrained minimization.

When applied to the general linearly constrained problem(52-54), the second order projection algorithm proceeds iteratively, just as the first order projection algorithm of section 4.1. The main difference is that we are now provided with second order estimates for the lagrangian multipliers,

$$\hat{\lambda} = (N_q^T H^{-1} N_q)^{-1} N_q^T H^{-1} g \tag{69}$$

in terms of which the search direction is written as

$$s = - H^{-1} (g - N_q \hat{\lambda}) \tag{70}$$

Also the line search can be done approximately by adopting a unit step length. This "projected" Newton method is probably the most appropriate for solving linearly constrained problems, provided that the Hessian can be readily evaluated and inverted at each new point in the iterative process. However, because the Hessian is not necessarily positive definite at a constrained minimum, techniques for enforcing positive definiteness are more likely to be required than in the unconstrained case.

Another approach is to resort to quasi-Newton methods for approximating the Hessian matrix or its inverse on the basis of first order informations accumulated from the preceding steps. For example, the rank-one formula (51) can be employed, because it avoids the necessity of an exact one-dimensional minimization. In such a quasi-Newton method, H^{-1} is replaced by $S^{(k)}$ in Eqs. (69-70), where the matrix $S^{(k)}$ is the approximation to the inverse Hessian at the kth iteration. A similar strategy is to mix the DFP update procedure and the gradient projection method, yielding search directions

$$s^{(k)} = - S_q^{(k)} g^{(k)} \tag{71}$$

as long as the set of active constraints remains unchanged. Indeed, if the initial approximation to the inverse Hessian is orthogonal to the constraint gradients (e.g. $S_q^{(o)} = P_q$ as given by Eq. 63), then so are all subsequent approximations if $S_q^{(k)}$ is updated by using the DFP formula (50). Consequently all search directions will reside within the initial subspace of linear constraints. Of course, when a constraint is added to, or dropped from the active set, the update formula has to be modified. In this first order projection method S_q is an approximation to the matrix $\hat{P}_q H^{-1}$ where $\hat{P}_q$ is defined in Eq. (68). If the objective function is a positive definite quadratic form and if n-q successive iterations are performed with a constant set of q active constraints, then, starting from $S_q^{(o)} = P_q$, $S_q^{(n-q)}$ will be equal to $\hat{P}_q H^{-1}$. In addition the search directions generated in these n-q iterations are H-conjugate.

Finally, it is worth giving a simple but effective way of improving the convergence speed of the gradient projection method. It merely consists of conjugating the projected gradient vectors according to the FLETCHER-REEVES formula (38) with $g^{(k)}$ replaced by $P_q g^{(k)}$. For a quadratic problem with q linear equality constraints, convergence is then achieved within n-q iterations. Of course, in the general case, the conjugacy procedure must be reinitiated whenever the projection matrix is modified because of a change in the set of active constraints. This corresponds to loosing useful informations and constitutes thus a disadvantage with respect to the methods which approximate inverse Hessians.

The active set strategy

An essential ingredient in any projection algorithm is the determination at each iteration of the set of active constraints whose intersection forms the basis for projection. The strategy for adding constraints to the basis is rather straightforward and is fixed by computing the step to the nearest constraint (see Eq. 66). Much less apparent are the strategies for determining which constraints to delete from the basis. The simplest strategy consists in retaining all the currently active constraints until a minimum is found with respect to the corresponding subspace. Only at this point is the sign of the Lagrangian multipliers examined, which eventually may lead to deleting constraints from the basis. It can be shown that this scheme will terminate at a strong local minimum after a finite number of basis changes. Unfortunately this strategy suffers from a major drawback, in that it requires finding the minimum on each successive subspace. If the initial basis differs substantially from the optimal one, it is obvious that a prohibitive computational labour will be needed before reaching the final solution.

An alternative strategy is to compute estimates of the Lagrangian multipliers every iteration and move off the constraint with the most negative multiplier, if any. This is essentially the rule used in the simplex method of linear programming, in which case the estimates of the multipliers are exact. This feature is no longer true when a general function is minimized subject to linear constraints, and in fact, the multipliers tend to be very inaccurate when computed far from the minimum in the current subspace. The error is $O(h)$ for the first order estimates (Eq. 61) and $O(h^2)$ for the second order ones (Eq. 69), where h is the distance to the minimum. This inaccuracy can be responsible for the phenomenon of zigzagging, in which a given constraint may be dropped from the basis and then reintroduced later, repeatedly until the sign of the corresponding multiplier is stabilized. Consequently progress to the solution is considerably slowed and jamming can even occur. In fact global convergence cannot be proved, that is, this type of strategy is not guaranteed to terminate after a finite number of basis changes.
Note however that there exists several efficient anti-zigzagging rules, e.g., "if a constraint previously dropped from an active set if added again, it is retained in the active set until a constrained stationary point is attained".

Yet another approach consists in deleting a constraint only if its removal will result in a sufficient decrease in the objective function. The additional benefit that is gained when deleting the jth constraint can be computed approximately as $\frac{1}{2}\hat{\lambda}_j^2/u_j$, where $\hat{\lambda}_j$ is the second order multiplier estimate and u_j is the jth diagonal element of $(N_q^T H^{-1} N_q)^{-1}$. Therefore a suitable active set strategy is as follows : find the constraint for which $\frac{1}{2} \hat{\lambda}_j^2/u_j$ is maximum over the constraints that are candidate for deletion (i.e. $\hat{\lambda}_j < 0$) and drop the corresponding constraint only if the additional reduction in f(x) exceeds γ times the reduction that would be obtained by keeping it in the active basis. Mathematically this constraint (say the qth) is deleted if

$$- s^T g = s^T H s \leqslant \frac{1}{\gamma} \frac{\hat{\lambda}_q^2}{u_q} \tag{72}$$

where γ is a selected positive constant. In the gradient projection method the Hessian is not computed and a substitute test must be used, e.g.

$$- s^T g = s^T s \leqslant \frac{1}{\gamma} \frac{\lambda_q^2}{v_q} \tag{73}$$

where λ_q is the first order estimate of the Lagrangian multiplier and v_q the qth diagonal element of $(N_q^T N_q)^{-1}$. The validity of the foregoing strategy is strongly dependent on the quality of the Lagrangian multiplier estimates. Therefore it is interesting to apply them only it

$$|s^T g| \leqslant \varepsilon \tag{74}$$

where ε is a weak tolerance. This additional condition guarantees that the estimates of the multipliers are sufficiently accurate, because (74) can only be satisfied in the neighbourhood of a stationary point, where they are known to be exact.

5. GENERAL NONLINEAR PROGRAMMING METHODS

In this section general methods for solving a constrained nonlinear programming problem (6-8) are examined. We first look at conventional techniques that we classify as follows : primal (or direct) methods, which attack directly the problem by using sequential one-dimensional minimizations along usable feasible directions ; linearization methods, where the problem is replaced with a sequence a linear programming problems and transformation methods, in which the constrained problem is approximated by a sequence of unconstrained problems. Finally some recent and promising techniques for general purpose minimization are briefly described.

Primal methods

The gradient projection method of section 4.1 can be extended to deal with nonlinear constraints. It consists of projecting the objective function gradient onto the tangent plane to the active constraints. Let the active nonlinear constraints ($h_j(x) = 0$; $j = 1, \ldots, q$) be approximated linearly in the form :

$$\tilde{h}_j(x) = h_j(\tilde{x}) + (x - \tilde{x})^T \nabla h_j(\tilde{x}) \qquad (75)$$

where $\tilde{x}$ is a feasible boundary point. In an analogous manner to the case of linear constraints, the search direction (projected gradient) is given by (58). The orthogonal operator P_q can still be computed from (63), where the matrix N_q is built up from the active constraint gradients :

$$N_q = [\nabla h_1(\tilde{x}), \ldots, \nabla h_q(\tilde{x})] \qquad (76)$$

However the nonlinearity of the constraints leads to substantial changes in the algorithm. First the matrix $(N_q^T N_q)^{-1}$ involved in the projection scheme has to be reevaluated at each iteration step, even if the set of active constraints is not modified. In addition, the minimization phase involves the solution of nonlinear equations to locate the maximal allowable step length (intercept distances with the constraint surfaces). Finally, after completing the line search along the projected gradient, the resulting point is not necessarily feasible. Hence the need for a restoration phase, which leads back to the boundary of the feasible domain before calculating the next descent direction.

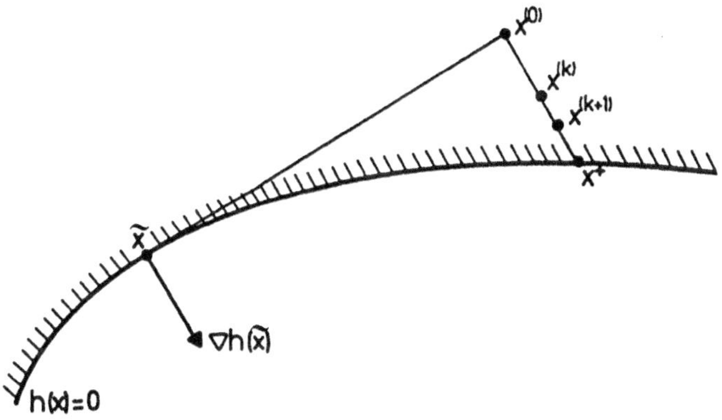

FIG. 8 THE GRADIENT PROJECTION METHOD (non linear constraints)

To perform the restoration step the linear approximation to the constraints (75) may be used : a progression is made along a direction normal to the constraint intersection at the initial point (see Fig. 8). According to Eqs. (75) and (76), the new point x^+, such that $h_j(x^+) = 0$ $(j = 1, \ldots, q)$ can be generated by the iterative process

$$x^{(k+1)} = x^{(k)} - N_q (N_q^T N_q)^{-1} h(x^{(k)}) \qquad (77)$$

The convergence of this restoration step clearly depends upon the distance of $x^{(o)}$ to the constraint surface, and therefore upon the step length accomplished in the minimization phase. It is of course necessary that the descent condition $f(x^+) < f(\tilde{x})$ remain verified. It is worthile noticing that in the restoration phase, advantage is taken of the availability of the matrix $(N_q^T N_q)^{-1}$ which was needed in the minimization step. More powerful minimization algorithms based on conjugate gradient, Newton or quasi-Newton iterations can be devised, just as in the case of projection methods for linear constraints (section 4.2).

Another type of primal method uses the concept of <u>feasible directions</u>. According to it, each standard iteration (56) consists of a line search in a direction s which does not immediately leave the feasible domain. The feasibility condition at a boundary point where q constraints are simultaneously active is written :

$$s^T \nabla h_j \geqslant 0 \qquad j = 1, \ldots, q \qquad (78)$$

with strict inequality for nonlinear constraints. Any vector s verifying Eq. (78) lies at least partly in the feasible domain, as indicated in Fig. 9.a. Moreover the feasible direction s is said to be <u>usable</u> if it is a dowhill direction, that is, if it satisfies the descent condition

$$s^T \nabla f \leqslant 0 \qquad (79)$$

If the current point is not a local minimum, both inequalities (78) and (79) define a cone of feasible-usable directions as shown in Fig. 9.a. The direction finding problem can be formulated as a linear programming problem : the feasible-usable direction which leads to the maximum decrease in the objective function is solution of

$$\begin{aligned} &\text{maximize} && \beta \\ &\text{subject to} && s^T \nabla f + \beta \leqslant 0 \\ & && s^T \nabla h_j - \theta_j \beta \geqslant 0 \qquad j = 1, q \\ & && -1 \leqslant s_i \leqslant 1 \end{aligned} \tag{80}$$

where the θ_j's are arbitrary positive constants. Note that the normalization condition on s might be replaced by another bound to the length of s. Clearly if $\beta_{max} > 0$, the strict inequalities (78, 79) hold and the selected vector s is a feasible-usable direction. If $\beta_{max} = 0$, then the initial point is a local minimum.

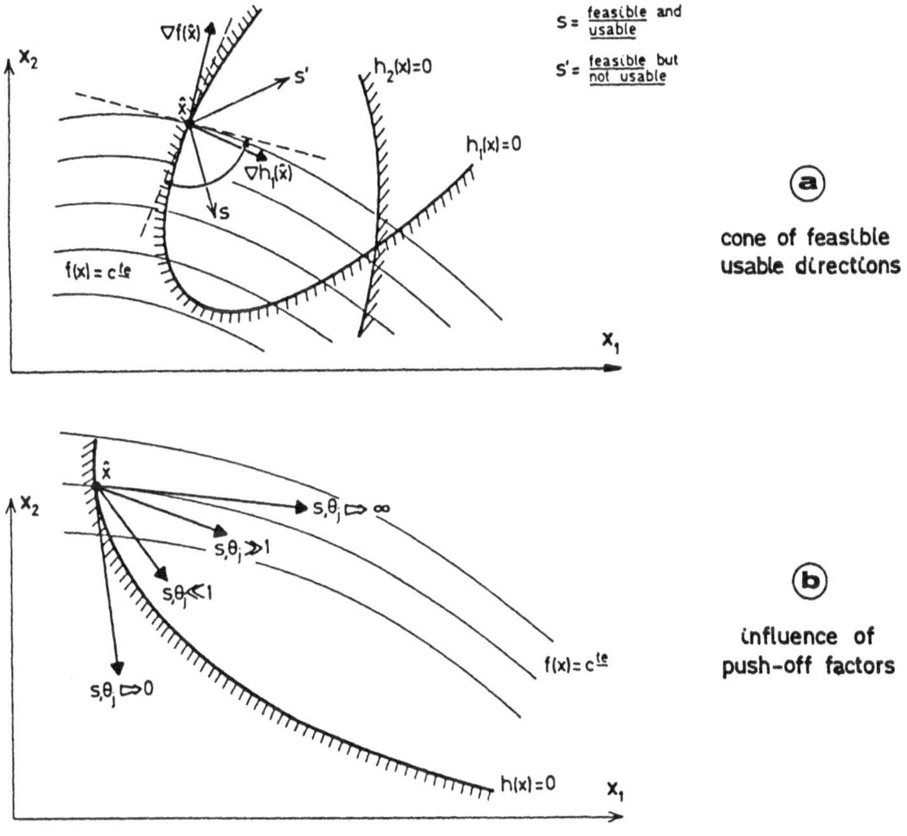

FIG. 9 THE METHOD OF FEASIBLE DIRECTIONS

The positive constants θ_j are called <u>push-off factors</u>. They measure to what extent the direction s is rejected from the boundary of the feasible domain. The influence of the push-off factors is illustrated in Fig. 9.b. If they are taken close to zero, the feasible direction is essentially chosen so that $s^T \nabla f + \beta = 0$. Therefore the objective function is decreased rapidly but with a search direction that almost follows the boundary of the feasible domain. On the other hand, if the θ_j's are taken very large, the direction s tends to follow the contour of the objective function. The risk of running out the feasible domain is reduced, but is paid by a lower decrease in the objective function. Intermediate values of θ_j (usually $\theta_j = 1$) furnish search directions for which, in the vicinity of the initial point, the constraints and the objective function are decreased at similar rates. There remains the problem of selecting the step size along the feasible-usable direction, which is similar as in the gradient projection method (line search). An additional feature is that one of the previously active constraints, at the point where s has been evaluated, may become active again. Moreover the zigzagging phenomena due to instability in the selection of active constraints may be very acute in the feasible direction methods. A suitable active set strategy is thus an important ingredient.

Linearization methods

Perhaps the most natural approach to a nonlinear problem is to transform it into a sequence of linear programming problems as follows:

$$\begin{aligned} &\text{minimize} \quad f(\tilde{x}) + (x - \tilde{x})^T \nabla f(\tilde{x}) \\ &\text{s.t.} \quad h_j(\tilde{x}) + (x - \tilde{x})^T \nabla h_j(\tilde{x}) \geq 0 \qquad j = 1,m \end{aligned} \tag{81}$$

where $\tilde{x}$ is the point where the objective function and the constraints are linearized. The solution of this problem is a starting point to the next linearization. Because it corresponds to a rather natural process, this technique is appealing to the engineer. Also the availability of standard linear programming packages facilitates its practical implementation. Unfortunately this very attractive recursive method suffers from severe limitations. It does not converge to a local minimum unless the latter occurs at a vertex of the feasible domain. Otherwise the process either converges to a non optimal vertex, or it oscillates indefinitely between two vertices. Note also that the problem (81) admits a solution only if the number of constraints exceeds the number of variables.

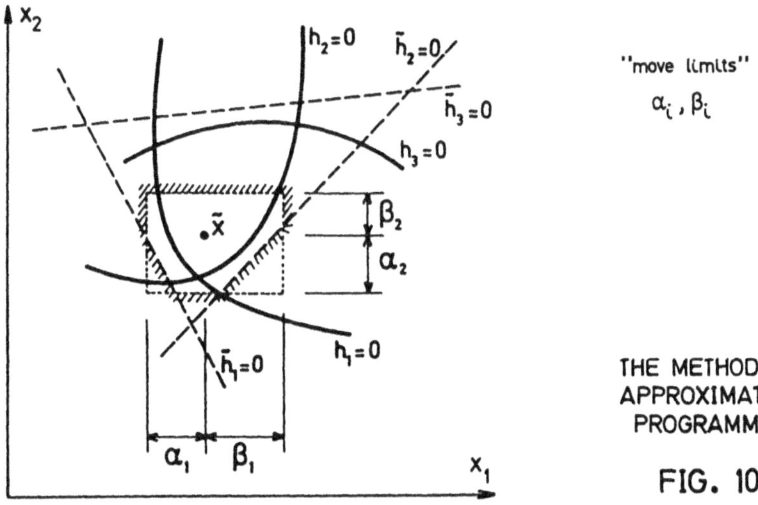

THE METHOD OF APPROXIMATION PROGRAMMING

FIG. 10

The method of approximation programming is an interesting variant on the foregoing recursive linear programming method. It solves the linearized problem (81), but with artificial side constraints :

$$\tilde{x}_i - \alpha_i \leqslant x_i \leqslant \tilde{x}_i + \beta_i \tag{82}$$

where α_i and β_i are properly chosen positive constants, called <u>move limits</u>. After solution of the problem (81) with the additional constraints (82), the objective function and the main constraints are again linearized, and the move limits are possibly modified. The technique is illustrated in Fig. 10.

Transformation methods

The <u>barrier function approach</u> proceeds by forming an auxiliary function whose minima are unconstrained inside the feasible region. The auxiliary function is defined so as to construct a barrier at the boundary of the feasible region, preventing thus violation of the constraints. By gradually removing the effect of the constraints in the auxiliary function through controlled changes in the value of a parameter, a sequence of unconstrained problems is generated, whose solutions are interior points converging to a minimum of the original constrained problem. Therefore barrier methods are often referred to as <u>interior point unconstrained minimization techniques</u>. The methods are especially attractive for mechanical system design applications, since they exhibit the reassuring feature that, should the algorithm be terminated prematurely, a feasible solution is nevertheless returned, which usually corresponds to a better design than the initial one (same philosophy as the primal methods). The barrier function transformation is stated as

$$\phi(x, r) = f(x) + r\,B(x) \qquad r > 0 \tag{83}$$

where B(x), the barrier function, is positive in the interior of the feasible region and $B(x) \to \infty$ as x approaches the boundary of the feasible region (note that the barrier function approach is not a valid transformation for problems involving equality constraints). Frequently used barrier functions are the logarithmic function

$$B(x) = - \sum_{j=1}^{m} \ell n\,[\,h_j(x)\,] \tag{84}$$

and the inverse function

$$B(x) = \sum_{j=1}^{m} \frac{1}{h_j(x)} \tag{85}$$

It can be proved under mild conditions that if $\{r^{(k)}\}$ is a monotonic decreasing sequence with $r^{(k)} \to 0$ as $k \to \infty$, then the solutions of the unconstrained problems

$$\underset{x}{\text{minimize}} \quad \phi(x, r^{(k)}) \tag{86}$$

initiated at an interior point, are interior point $\{x^{(k)} = x\,[\,r^{(k)}]\,\}$ converging to x^*, a solution to the constrained problem. Furthermore

$$\lim_{k\to\infty} \phi(x^{(k)}, r^{(k)}) = \lim_{k\to\infty} f(x^{(k)}) = f(x^*) \tag{87}$$

and $f(x^{(k)})$ is monotonic decreasing. An intuitive understanding of these theoritical results is as follows. If, for example, the jth constraint is active at x^*, then as $x \to x^*$, $h_j(x) \to 0$ and, from Eqs. (84) or (85), it is apparent that $B(x) \to \infty$. However, if $r \to 0$ the growth of B(x) is canceled, so making possible the constraint value to be reduced, and x to approach x^*.

The basic algorithm in the barrier function approach is therefore as follows :

(i) select a monotonic decreasing sequence $\{r^{(k)}\} \to 0$ as $k \to \infty$; find an interior point $x^{(o)}$ and set $k = 0$;

(ii) with $x^{(k)}$ as a starting point, minimize $\phi(x, r^{(k)})$ to find $x^{(k+1)} = x[r^{(k)}]$;

(iii) if convergence criteria are not satisfied, set $k = k+1$ and return to (ii).

The barrier function transformation provides a powerful way of solving general constrained minimization problems. Unfortunately this approach suffers from an essential difficulty, in that when the controlling parameter $r^{(k)}$ becomes small, the auxiliary function $\phi(x, r^{(k)})$ is more and more difficult to minimize. This undesirable behaviour can be related to the illconditioned nature of the Hessian matrix. As explained in section 3.2., it is important, in order to evaluate the difficulty of an unconstrained minimization problem, to determine the eigenvalue structure of the Hessian. For the auxiliary function (83) this structure becomes increasingly unfavorable as r decreases. Assuming that there are q active constraints at the solution x^* to the original constrained problem, then the Hessian of the auxiliary function, $\nabla^2\phi(x, r)$, has q eigenvalues that vary with r^{-1} and thus tend to infinity as $r \to 0$. The other n-q eigenvalues tend to finite positive limits. This implies that as r decreases the condition number of $\nabla^2\phi(x, r)$ varies with r^{-1} and so, problem (86) becomes less and less manageable. In other words the function $\phi(x,r)$ is more and more eccentric as $r \to 0$, therebly slowing considerably the speed of convergence of any first order minimization algorithm. One idea that may be used for avoiding slow convergence is to resort to Newton's method, since its order two convergence is unaffected by the poor eigenvalue structure. In applying the method, however, attention should be paid to the manner by which the ill-conditioned Hessian is inverted. Also Newton's method requires the second derivatives of the problem functions to be readily available, which is not often the case in practical problems.

The question of knowing how the sequence $\{r^{(k)}\}$ should be chosen is thus very important, because it can seriously affect the computational effort required to find a solution. The conflict is clear : ultimately $r^{(k)}$ must be small enough to force the minimum $x(r^{(k)})$ to approach the boundary of the feasible region, but large enough to enable the auxiliary function $\phi(x, r^{(k)})$ to be minimized without excessive difficulty. In practice a convenient choice is

$$r^{(k+1)} = r^{(k)} f \tag{88}$$

with f ranging from 0.1 to 0.5 depending upon the nature of the problem and the unconstrained minimization algorithm employed.

In contrast to the barrier function transformation, the <u>penalty function transformation</u> is defined so as to prescribe a high cost for violation of the constraints :

$$\psi(x,r) = f(x) + \frac{1}{r} P(x) \qquad r > 0 \tag{89}$$

where $P(x) \geqslant 0$ for all $x \in E^n$ and $P(x) = 0$ if and only if x is an exterior point. $\psi(x,r)$ is defined on E^n and $\psi(x,r) \rightarrow \infty$ as constraint violation increases. Frequently used penalty functions are the quadratic loss function

$$P(x) = \sum_{j=1}^{m} [\min(0, h_j(x))]^2 \tag{90}$$

and the Zangwill loss function

$$P(x) = -\sum_{j=1}^{m} \min(0, h_j(x)) \tag{91}$$

It should be emphasized that this type of method can handle equality constrained problems. An important penalty function transformation for such a problem is (assuming each h_j represents an equality constraint) :

$$\psi(x,r) = f(x) + \frac{1}{r} \sum_{j=1}^{m} [h_j(x)]^2 \tag{92}$$

The controlling parameter r is used effectively to increase the magnitude of the penalty, i.e., constraint violation is gradually weighed as $r \rightarrow 0$. For small r, it is clear that the minimum point of the unconstrained problem

$$\underset{x}{\text{minimize}} \quad \psi(x,r) \tag{93}$$

will be in a region where P(x) is small. Thus, for a decreasing sequence $\{r^{(k)}\}$, it is expected that the corresponding solution points of Eq. (93) will approach the feasible region and will minimize f(x). Ideally then, as $r^{(k)} \to 0$, the solution points of the penalty problems (93) will converge to a solution of the original constrained problem.

Therefore the behaviour of the penalty function transformation is similar to that given for barrier functions. Convergence can still be ensured under mild conditions. The only difference in the basic algorithm is in step (i), where the initial point $x^{(o)}$ does no longer need to be feasible. Again, computational difficulties result from $\psi(x,r)$ forming an increasingly steep-sided valley as the controlling parameter r decreases. It can be shown that the Hessian $\nabla^2\psi(x,r)$ becomes ill-conditioned as $r \to 0$. As a final observation, note that in general the sequence $x^{(k)}$ approaches x^* from outside the feasible region. Therefore the penalty function transformation methods are also called exterior point unconstrained minimization techniques.

Recent general purpose optimization methods

The use of a Lagrangian function is probably the basis of the most powerful minimization methods with nonlinear constraints. Their success relies on the fact that account of the curvature of the constraints is taken through their quadratic approximation contained in the Lagrangian function. Let us assume for sake of simplicity that our minimum problem is subjected only to equality constraints :

$$\begin{aligned} &\text{minimize} \quad f(x) \\ &\text{s.t.} \quad h_j(x) = 0 \qquad j = 1, \ldots, q \end{aligned} \tag{94}$$

The extension to inequality constraints is then made using an active constraint set strategy based on the magnitude of the Lagrangian multipliers.

The basis for recursive quadratic programming methods is to write down the stationarity conditions of the Lagrangian function

$$L(x,\lambda) = f(x) - \sum_{j=1}^{q} \lambda_j h_j(x) \tag{95}$$

which gives rise to the system of nonlinear equations

$$\begin{cases} \nabla f(x) - \sum_{j=1}^{q} \lambda_j \nabla h_j(x) = 0 \\ h_j(x) = 0 \qquad\qquad j = 1,q \end{cases} \tag{96}$$

If Newton's method is applied to this system of equations, a better approximation (x,λ) to the solution is obtained from an estimate $(\hat{x},\hat{\lambda})$ by solving the linear system of equations

$$\begin{bmatrix} G(\hat{x},\hat{\lambda}) & -N(\hat{x}) \\ -N^T(\hat{x}) & 0 \end{bmatrix} \begin{bmatrix} x - \hat{x} \\ \lambda - \hat{\lambda} \end{bmatrix} = \begin{bmatrix} -\nabla f(\hat{x}) + N(\hat{x})\hat{\lambda} \\ h(\hat{x}) \end{bmatrix} \tag{97}$$

where $G(x,\lambda)$ is the Hessian matrix of the Lagrangian function (95) and $N(x)$ is a matrix collecting the constraint gradients (see Eq. 76). The quadratic approximation is obtained by noticing that the correction vector $\delta = (\hat{x} - x)$ that solves Eq. (97) can also be found by solving the quadratic problem

$$\min_{\delta} \frac{1}{2} \delta^T G(\hat{x},\hat{\lambda})\delta + \delta^T \nabla f(\hat{x}) \tag{98}$$

subject to the linear approximations to the constraints

$$N^T(\hat{x})\delta + h(\hat{x}) = 0 \tag{99}$$

Equations (98,99) suggest thus to transform the initial problem (94) into a succession of quadratic programs in which the objective function is a second order approximation of the Lagrangian calculated with the active constraints and where the constraints are replaced by their linear approximation (99). A further improvement to the method is obtained if the second derivative matrix $G(\hat{x},\hat{\lambda})$ is no longer evaluated at each step of the method, but simply obtained from successive quasi-Newton updates (see the chapter by GILL for more details).

In the <u>multiplier method</u>, the Lagrangian function (95) again plays an important role. It is well known that if there exists some λ^* for which x^* solves the unconstrained problem $\min_x L(x,\lambda^*)$, while satisfying the equality constraints, then x^* is a solution to the original problem. Therefore the problem

$$\begin{aligned} &\min_x L(x,\lambda) \quad \text{given suitable } \lambda \\ &\text{s.t. } h_j(x) = 0 \qquad\qquad j = 1,q \end{aligned} \tag{100}$$

is equivalent to the original problem (94) in the sense that both problems have x^* as local minimum and λ^* as associated Lagrangian multiplier vector. Now consider solving the equivalent problem (100) by resorting to the penalty function transformation (92). This leads us to add the penalty term not to be objective function f(x), but rather to the Lagrangian function $L(x,\lambda)$, thus forming the Augmented Lagrangian Function for the equality constrained problem :

$$\chi(x,\lambda,r) = f(x) - \sum_{j=1}^{q} \lambda_j \, h_j(x) + \frac{1}{r} \sum_{j=1}^{q} [h_j(x)]^2 \tag{101}$$

Given values of λ and r, the multiplier method consists in applying an algorithm for unconstrained minimization to the function (101), yielding a point $x(\lambda,r)$. Then $\lambda$and r are adjusted for the next iteration so that $x(\lambda,r)$ converges to x^* as the iterations proceed. Consideration of two extreme cases is instructive. If $\lambda = 0$, then (101) reduces to the standard penalty function transformation (92), with the associated ill-conditioning troubles when r is decreased. If $\lambda = \lambda^*$, then minimizing $\chi(x,\lambda^*,r)$ gives the solution to the original problem for any value of $r > 0$. These observations suggest that, by updating the Lagrangian multipliers λ so that they approach λ^*, convergence may occur without the need for r to be very small. Thus the ill-conditioning associated with penalty methods can be avoided. Usually the update formula for the λ_j's is as follows :

$$\lambda_j^{(k+1)} = \lambda_j^{(k)} - \frac{2}{r^{(k)}} \, h_j(x^{(k)}) \tag{102}$$

where k is the iteration index. The problem of adjusting λ can also be viewed as the problem of maximizing the auxiliary dual function

$$\ell_r(\lambda) = \min_x \chi(x,\lambda,r) \tag{103}$$

by using a steepest ascent move with step size $2/r^{(k)}$ in the dual space. Indeed the first derivatives of the dual function $\ell_r(\lambda)$ are given by minus the constraint values (see Eq. 19). Therefore the multiplier method can be interpreted as a kind of primal-dual optimization method with very limited search in the dual space for optimum Lagrangian multipliers.

Sequence of conservative subproblems

In this last section a new and rather general mathematical programming method is briefly described. It is based upon an approximation concepts approach that has been successfully applied to many various problems of structural optimization. The method consists of transforming the primary nonlinear programming problem (6-8) into a sequence of "linearized" subproblems having each a simple explicit algebraic structure. Each subproblem is strictly convex and separable and it constitutes a first order conservative approximation of the primary problem.

The key idea is to perform the linearization process with respect to mixed variables, either direct (x_i) or reciprocal ($\frac{1}{x_i}$), independently for each function involved in the problem. By normalizing the variables x_i so that they become equal to unity at the current point $\tilde{x}$ where the problem is linearized, the following convex, separable subproblem is generated

$$\min \quad \sum_{+} f_i\, x_i - \sum_{-} \frac{f_i}{x_i} \tag{104}$$

$$\text{s.t.} \quad \sum_{+} h_{ij}\, x_i - \sum_{-} \frac{h_{ij}}{x_i} \leq \bar{h}_j \tag{105}$$

$$\underline{x}_i \leq x_i \leq \bar{x}_i \tag{106}$$

where $f_i = \dfrac{\partial f}{\partial x_i}\Big|_{\tilde{x}}$

$h_{ij} = \dfrac{\partial h_j}{\partial x_i}\Big|_{\tilde{x}}$

and $\bar{h}_j = h_j(\tilde{x}) - \sum_{+} h_{ij}\, \tilde{x}_i + \sum_{-} \dfrac{h_{ij}}{\tilde{x}_i}$

In these expressions the symbol $\sum_{+}$ ($\sum_{-}$) means "summation over all positive (negative) terms".

It can be proved that the first order approximation of the objective function (Eq. 104) and of the constraint functions (Eq. 105) are locally conservative, which means that they overestimate the value of the true functions. As a result the "linearized" feasible domain corresponding to the explicit subproblem (104-106) is generally located inside the true feasible domain corresponding to the primary problem (6-8) (see Fig. 11). In other words, the method tends to generate a sequence of design points that "funnel down the middle" of the feasible region. This is a very attractive property from a practical viewpoint, since the designer may stop the optimization process any time, and still get a better solution than its initial estimate.

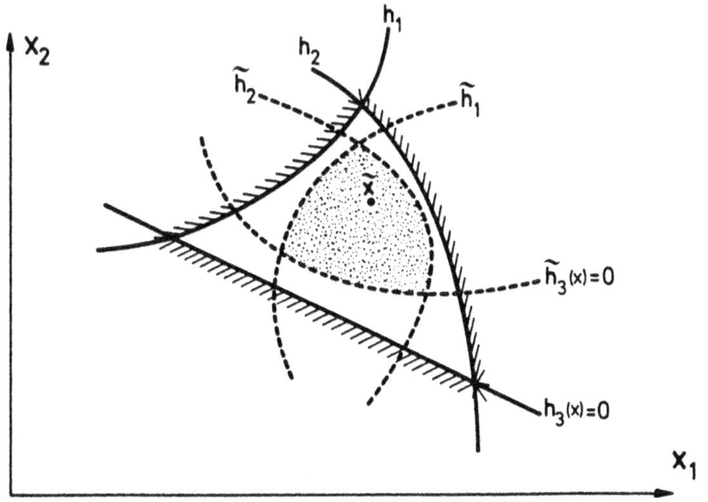

FIG. 11 CONSERVATIVE APPROXIMATIONS

Because of its properties of convexity and separability, the explicit problem (104-106) can be efficiently solved by dual methods of mathematical programming (see section 2). The dual function has the form of Eq. 21, where each single variable minimization problem admits a closed form solution. This yields the following explicit relations giving the primal variables x_i in terms of the dual variables λ_j :

$$x_i(\lambda) = \left(\frac{\sum_+ h_{ij}\lambda_j}{f_i - \sum_- h_{ij}\lambda_j}\right)^{1/2} \qquad \text{if} \quad f_i > 0 \tag{107}$$

$$x_i(\lambda) = \left(\frac{f_i - \sum_+ h_{ij}\lambda_j}{\sum_- h_{ij}\lambda_j}\right)^{1/2} \qquad \text{if} \quad f_i < 0 \tag{108}$$

Note that these equations must be employed in connection with the side constraints (106), which have always to be enforced. The dual problem (17) can now be written explicitly :

$$\max \quad \ell(\lambda) = \sum_+ f_i x_i(\lambda) - \sum_- \frac{f_i}{x_i(\lambda)} - \sum_j \lambda_j \left[\bar{h}_j - \sum_+ h_{ij} x_i(\lambda) + \sum_- \frac{h_{ij}}{x_i(\lambda)}\right]$$

$$\text{s.t.} \quad \lambda_j \geqslant 0$$

where $x_i(\lambda)$ is given by Eqs. (107,108), or is fixed to its lower or upper bound. To solve this problem a second-order Newton-type algorithm is specially recommended, because the Hessian matrix of the dual function is easily available.

6. CONCLUDING REMARKS

After reading the foregoing text it is apparent that there exists a broad variety of methods for solving an optimization problem. One might then get the deceiving feeling that the choice of a method is very problem dependent. This statement is only partially correct. Although some general purpose methods are beginning to emerge, it is the opinion of the authors that no such method is available now, that is capable of solving efficiently any optimization problem. This is especially true in engineering sciences, where most often large scale problems have to be dealt with, involving many design variables and many constraints.

In fact two extreme situations should be considered. On one hand, if an occasional use of optimization techniques is needed for a single specific problem, then resorting to a standard general purpose package is obviously the best choice (e.g. recursive quadratic programming). On the other hand, if an engineer has to make an everyday use of optimization methods for always solving the same class of problems, then he should better develop its own system, by taking advantage of its deep knowlegde of the subject, its physical insight, its previous experience, etc ...

Because the idea of optimizing the dynamics of mechanical system is relatively new, it will probably be valuable to study in details the essential nature and the algebraic structure of the functions involved in the nonlinear problem to be dealt with. For this purpose, solving analytically simple but representative problems is quite instructive. Such a fundamental research work has been accomplished in the field of structural optimization with considerable success. It has given rise (after 20 years !) to a unified approach that can solve sizing as well as shape optimal design problems in usually less than ten iterations. This unified approach is based on generating a sequence of approximate subproblems (e.g. convex, separable problems). It is hoped that similar results will be gained in the next few years in the emerging field considered in this book.

BIBLIOGRAPHY

A.V. FIACCO and G.P. Mc CORMICK, Non-Linear Programming : Sequential Unconstrained Minimization Techniques, Wiley, New York, 1968

L.S. LASDON, Optimization Theory for Large Systems, Macmillan, New York, 1970

F.A. LOOTSMA (èd), Numerical Methods for Non-Linear Optimization, Academic Press, London, 1972

W. MURRAY (ed), Numerical Methods for Unconstrained Optimization, Academic Press, London, 1972

D.M. HIMMELBLAU, Applied Nonlinear Programming, Mc Graw Hill, New York, 1972

D.G. LUENBERGER, Introduction to Linear and Nonlinear Programming, Addison-Wesley, Reading, 1973

P.E. GILL and W. MURRAY (eds), Numerical Methods for Constrained Optimization, Academic Press, London, 1974

M. AVRIEL, Nonlinear Programming - Analysis and Methods, Prentice-Hall, Englewood Cliffs, 1976

M.A. WOLFE, Numerical Methods for Unconstrained Optimization - An Introduction, Van Nostrand Reinhold, Wokingham, 1978

R.L. FOX, Optimization Methods for Engineering Design, Addison-Wesley, 1971

R.H. GALLAGHER and O.C. ZIENKIEWICZ (eds), Optimum Structural Design - Theory and Applications, John Wiley, London, 1973

C. FLEURY, Le Dimensionnement Automatique des Structures Elastiques, Doctoral Dissertation, LTAS Report SF-72, University of Liège, Belgium 1978

L.A. SCHMIT and H. MIURA, Approximation Concepts for Efficient Structural Synthesis, NASA CR-2552, 1976

E.J. HAUG and J. ARORA, Applied Optimal Design : Mechanical and Structural Systems, Wiley-Interscience, 1979

C. FLEURY and L.A. SCHMIT, Dual Methods and Approximation Concepts in Structural Synthesis, NASA CR-3226, 1980

A.J. MORRIS (ed), Foundations of Structural Optimization : A Unified Approach, John Wiley, London, 1982

SEQUENTIAL QUADRATIC PROGRAMMING METHODS FOR NONLINEAR PROGRAMMING

Philip E. Gill, Walter Murray,
Michael A. Saunders and Margaret H. Wright
Department of Operations Research
Stanford University
Stanford, California 94305

Abstract. Sequential quadratic programming (SQP) methods are among the most effective techniques known today for solving nonlinearly constrained optimization problems. This paper presents an overview of SQP methods based on a quasi-Newton approximation to the Hessian of the Lagrangian function (or an augmented Lagrangian function). We briefly describe some of the issues in the formulation of SQP methods, including the form of the subproblem and the choice of merit function. We conclude with a list of available SQP software.

1. INTRODUCTION

In this paper we consider the implementation of quasi-Newton methods for the solution of the nonlinear programming problem:

NIP	$\underset{x\in\Re^n}{\text{minimize}}$	$F(x)$
	subject to	$c_i(x) \geq 0, \quad i = 1, \ldots, m.$

The set of functions $\{c_i\}$ will be called the *constraint functions.* The objective function F and the constraint functions taken together comprise the *problem functions.* Unless otherwise stated, the problem functions will be assumed to be at least twice-continuously differentiable. Let $g(x)$ denote the gradient of $F(x)$, and $a_i(x)$ the gradient of $c_i(x)$. Quasi-Newton methods utilize the values of the problem functions and their gradients at trial iterates, but do not assume the availability of higher derivative information. When explicit first derivatives are not available, quasi-Newton methods can be implemented using finite-difference approximations to the gradient (see, e.g., Gill, Murray and Wright, 1981).

NATO ASI Series, Vol. F9
Computer Aided Analysis and Optimization of Mechanical System Dynamics
Edited by E. J. Haug

The solution of NIP will be denoted by x^*. All the methods of interest are *iterative*, and generate a sequence $\{x_k\}$ that is intended to converge to x^*. At the k-th iteration, the new iterate is defined by

$$x_{k+1} = x_k + \alpha_k p_k, \tag{1.1}$$

where α_k is a non-negative scalar called the *step length*, and p_k is an n-vector called the *search direction*.

2. QUASI-NEWTON METHODS FOR UNCONSTRAINED OPTIMIZATION

2.1. Basic Theory: Before turning to the constrained problem, we briefly review some important features of quasi-Newton methods applied to *unconstrained* minimization. In many unconstrained methods (including quasi-Newton methods), the search direction p_k is defined through an approximating *quadratic model* of the objective function. The most obvious candidate for such a model is the quadratic function defined by the first three terms of the Taylor-series expansion of F about the current iterate, i.e.,

$$F(x_k + p) \approx F(x_k) + g(x_k)^T p + \frac{1}{2} p^T \nabla^2 F(x_k) p. \tag{2.1}$$

The class of *Newton-like* methods is based on choosing p_k in (1.1) to minimize the quadratic function (2.1). However, (2.1) involves the Hessian matrix of F, which may not be available. Rather than using the exact Hessian, quasi-Newton methods use a changing approximation to the Hessian matrix to develop a quadratic model, and build up second-order information about F as the iterations proceed. (This feature is emphasized by the term "variable metric", which is used by some authors instead of "quasi-Newton".)

Let B_k denote the approximate Hessian at the k-th iteration. The search direction p_k is taken as the minimum of the quadratic model function

$$g_k^T p + \frac{1}{2} p^T B_k p, \tag{2.2}$$

where g_k denotes $g(x_k)$. Note that (2.1) and (2.2) differ only in their Hessian matrices. If B_k is positive definite, (2.2) has a unique minimum that satisfies

$$B_k p_k = -g_k. \tag{2.3}$$

In addition, the solution of (2.3) is a *descent direction* for F at x_k, i.e., $g_k^T p_k < 0$.

An essential feature of a quasi-Newton method is the incorporation in B_k of new information about F acquired during the k-th iteration. The matrix B_{k+1} is typically defined by adding a matrix of low rank to B_k such that B_{k+1} will satisfy the *quasi-Newton condition*

$$B_{k+1} s_k = y_k, \tag{2.4}$$

where $s_k = x_{k+1} - x_k$ and $y_k = g_{k+1} - g_k$. If F is quadratic, the quasi-Newton condition is always satisfied by the exact (constant) Hessian.

In most quasi-Newton methods, B_{k+1} is defined by

$$B_{k+1} = B_k + U_k, \tag{2.5}$$

where U_k is a matrix of rank *two*. For given vectors s_k and y_k, an infinite number of matrices U_k exist that would satisfy (2.4). (For a general characterization, see Dennis and Moré, 1977.) This limitless choice of formulae has, not surprisingly, led to much research devoted to finding the "best" method. (The term "best" has generally come to mean the definition of U_k that tends to solve a large set of test problems with the smallest number of evaluations of F and g.) It is widely accepted today that the best available quasi-Newton method uses the *BFGS update*, in which B_{k+1} is given by

$$B_{k+1} = B_k - \frac{1}{s_k^T B_k s_k} B_k s_k s_k^T B_k + \frac{1}{y_k^T s_k} y_k y_k^T. \tag{2.6}$$

If B_k is positive definite, B_{k+1} as defined by (2.6) will be positive definite if and only if

$$y_k^T s_k > 0. \tag{2.7}$$

An important feature of the BFGS method is that the iterates generally exhibit *superlinear convergence* in practice, i.e., if $\alpha_k = 1$, the sequence $\{x_k\}$ defined by (1.1) and (2.3) satisfies

$$\lim_{k\to\infty} \frac{\|x_{k+1} - x^*\|}{\|x_k - x^*\|} = 0. \tag{2.8}$$

(See Dennis and Schnabel, 1983, for a summary of convergence results for quasi-Newton methods.)

2.2. Computation of the step length: For any iterative method, it is helpful to have some way of deciding whether a new point is "better" than the old point. In the case of unconstrained minimization, the value of the objective function provides a "natural" measure of progress. This leads to a requirement that the step length α_k in (1.1) should be chosen to satisfy the *descent condition* $F(x_k + \alpha_k p_k) < F(x_k)$. However, the descent condition alone is not sufficient to prove convergence of $\{x_k\}$ to x^*, and stronger conditions must be imposed on α_k. Various conditions that ensure convergence are given by Wolfe (1969), Goldstein and Price (1967) and Ortega and Rheinboldt (1970). In general, α_k is obtained through an iterative process called the *line search*.

First, the value of α_k cannot be "too large" relative to the reduction achieved in F. A measure that relates α_k and the expected reduction in F is provided by $g_k^T p_k$, which gives the directional derivative of F along p_k at x_k. A condition that ensures a large enough decrease in F relative to α_k is

$$F(x_k) - F(x_k + \alpha_k p_k) \geq -\mu \alpha_k g_k^T p_k, \tag{2.9}$$

where $0 < \mu \le \frac{1}{2}$.

On the other hand, convergence will not occur if α_k is "too small". To avoid this possibility, α_k can be forced to approximate the step to the first minimum of F along p_k, by requiring that the magnitude of the directional derivative at $x_k + \alpha_k p_k$ should be "sufficiently" reduced from that at x_k. If the gradient of F is available at trial step lengths, a suitable condition is

$$|g(x_k + \alpha_k p_k)^T p_k| \le -\eta g_k^T p_k, \tag{2.10}$$

where $0 \le \eta < 1$. If the gradient of F is *not* available at trial points during the line search, we require instead that

$$|F(x_k + \alpha_k p_k) - F(x_k + \nu p_k)| \le -\eta(\alpha_k - \nu) g_k^T p_k, \tag{2.11}$$

where ν is any scalar such that $0 \le \nu < \alpha_k$. Under mild conditions on p_k, the combination of condition (2.9) with either (2.10) or (2.11) is said to guarantee a *sufficient decrease* in F (provided that $\mu \le \eta$), and allows a proof of global convergence to $\overset{*}{x}$. A further benefit of the criterion (2.10) for quasi-Newton methods is that it guarantees satisfaction of (2.7), and hence ensures positive-definiteness of the BFGS update.

The acceptance criteria (2.9) and (2.10) (or (2.9) and (2.11)) are computationally feasible. In particular, methods of safeguarded polynomial interpolation (see, e.g., Brent, 1973; Gill, Murray and Wright, 1981) can find a suitable α_k efficiently. These methods assume that values δ_k and Δ_k are known at x_k such that

$$0 < \delta_k \le \|\alpha_k p_k\| \le \Delta_k. \tag{2.12}$$

In general, δ_k and Δ_k depend on x_k and on F. The value of δ_k defines the *minimum allowed distance* between x_k and x_{k+1}, and reflects the accuracy to which F can be computed; Δ_k is an upper bound on the change in x, and is usually taken as a very large number in the unconstrained case.

A popular alternative technique for computing α_k is known as *backtracking* (see, e.g., Dennis and Schnabel, 1983). In this case, given a fixed $0 < \rho < 1$, a sequence $\{\beta_j\}$ is generated that satisfies $\beta_0 = 1$ and $\beta_j > \beta_{j+1} \ge \rho\beta_j$. The value of α_k is taken as the first element in the sequence $\{\beta_j\}$ satisfying

$$F(x_k) - F(x_k + \beta_j p_k) \ge -\mu\beta_j g_k^T p_k,$$

where $0 < \mu < 1$. The simplicity of backtracking algorithms and their utility in convergence proofs have led to their frequent appearance in the literature. However, since a backtracking method can never generate a value of α_k greater than unity, *it cannot guarantee that* $y_k^T s_k > 0$. Most implementations of quasi-Newton methods with a backtracking line search simply skip the update in this case. (As the iterates converge to a local minimum, it can be shown that $y_k^T s_k > 0$ when $\alpha_k = 1$, and hence the difficulty does not arise in the limit.)

2.3. Computation of the update: A computational benefit of quasi-Newton methods is that a factorized form of B_k can be *updated* (rather than recomputed) following a low-rank change, so that (2.3) can be solved efficiently. In the earliest quasi-Newton methods, an explicit inverse of B_k was updated. However, from the viewpoint of numerical stability, it is preferable to recur the Cholesky factorization of B_k (see Gill and Murray, 1972). Suppose that $B_k = R_k^T R_k$, where R_k is an upper-triangular matrix. The vector p_k that satisfies (2.3) may be computed from the two triangular systems

$$R_k^T q = -g_k \quad \text{and} \quad R_k p_k = q. \tag{2.13}$$

Dennis and Schnabel (1981) have shown that the BFGS update (2.6) may be expressed as a *rank-one* update to R_k. Let β and γ denote the scalars $(s_k^T B_k s_k)^{\frac{1}{2}}$ and $(y_k^T s_k)^{\frac{1}{2}}$ respectively. The BFGS update may then be written as $B_{k+1} = \bar{R}_{k+1}^T \bar{R}_{k+1}$, where

$$\bar{R}_{k+1} = R_k + vw^T, \quad \text{with} \quad v = \frac{1}{\beta} R_k s_k, \quad w = \frac{1}{\gamma} y_k - \frac{1}{\beta} B_k s_k.$$

The matrix $\bar{R}_k$, which is not upper-triangular, may be restored to upper-triangular form by finding an orthogonal matrix P such that

$$P\bar{R}_{k+1} = R_{k+1},$$

where R_{k+1} is upper triangular. Then $B_{k+1} = \bar{R}_{k+1}^T \bar{R}_{k+1} = \bar{R}_{k+1}^T P^T P \bar{R}_{k+1} = R_{k+1}^T R_{k+1}$, as required. A suitable matrix P can be constructed from two sweeps of plane rotations; for more details, see Gill *et al.* (1974).

If p_k satisfies (2.3), two matrix-vector multiplications may be avoided in the implementation of the BFGS update (2.6). Substituting from (2.13), we obtain

$$v = \frac{1}{\sigma} q \quad \text{and} \quad w = \frac{1}{\gamma} y_k + \frac{1}{\sigma} g_k,$$

where $\sigma = |g_k^T p_k|^{\frac{1}{2}}$ and $\gamma = (y_k^T s_k)^{\frac{1}{2}}$.

The use of the Cholesky factorization avoids a serious problem that would otherwise arise in quasi-Newton methods: *the loss (through rounding errors) of positive-definiteness in the Hessian (or inverse Hessian) approximation.* With exact arithmetic, satisfaction of (2.7) should ensure that the BFGS update generates strictly positive-definite Hessian approximations. However, in practice the formula (2.6) can lead to a singular or indefinite matrix B_{k+1}. When B_k is represented by its Cholesky factorization and updates are performed directly to the factorization, every B_k will be *numerically* positive definite.

3. METHODS FOR NONLINEAR EQUALITY CONSTRAINTS

3.1. Basic theory: In this section, we consider methods for problems that contain only nonlinear *equality* constraints, i.e.

$$\text{NEP} \qquad \underset{x \in \Re^n}{\text{minimize}} \quad F(x) \qquad \text{subject to} \quad c_i(x) = 0, \quad i = 1, \ldots, m.$$

The gradient vector of the function $c_i(x)$ will be denoted by the n-vector $a_i(x)$. The $m \times n$ matrix $A(x)$ whose i-th row is $a_i(x)^T$ is termed the *Jacobian matrix* of the constraints. For simplicity of exposition, we shall assume that $A(x)$ has full row rank for all x.

The Kuhn-Tucker conditions for NEP state the existence of an m-vector λ^* (the *Lagrange multiplier vector*) such that

$$g(x^*) = A(x^*)^T \lambda^*. \tag{3.1}$$

(For a detailed discussion of first- and second-order Kuhn-Tucker conditions for optimality, see, for example, Fiacco and McCormick, 1968, and Powell, 1974.)

Let $Z(x)$ denote a matrix whose columns form a basis for the set of vectors orthogonal to the rows of $A(x)$, i.e., $A(x)Z(x) = 0$. An equivalent statement of (3.1) in terms of Z is

$$Z(x^*)^T g(x^*) = 0.$$

The vector $Z(x)^T g(x)$ is termed the *projected gradient* of F at x.

The *Lagrangian function*

$$L(x, \mu) = F(x) - \mu^T c(x),$$

where μ is an m-vector of Lagrange-multiplier estimates, plays an important role in understanding and solving constrained problems. Condition (3.1) is a statement that x^* is a *stationary point* (with respect to x) of the Lagrangian function when $\mu = \lambda^*$.

Unfortunately, x^* is not necessarily a local *minimum* of the Lagrangian function. However, the second-order sufficiency conditions for optimality imply that x^* must be a minimum (with respect to x) of $L(x, \lambda^*)$ when x is restricted to lie in the linear subspace $A(x^*)(x - x^*) = 0$, i.e., x^* is a solution of the minimization problem

$$\begin{aligned} &\underset{x \in \Re^n}{\text{minimize}} \quad L(x, \lambda^*) \\ &\text{subject to} \quad A(x^*)(x - x^*) = 0. \end{aligned} \tag{3.2}$$

This property is equivalent to a requirement that the Lagrangian function should have *non-negative curvature* along vectors orthogonal to $A(x^*)$. Let $W(x, \mu)$ denote the Hessian (with respect to x) of the Lagrangian function

$$W(x, \mu) \equiv \nabla^2 F(x) - \sum_{i=1}^{m} \mu_i \nabla^2 c_i(x). \tag{3.3}$$

Since x^* solves (3.2), the *projected Hessian of the Lagrangian function*, $Z(x)^T W(x, \mu) Z(x)$, must be positive semi-definite at $x = x^*$, $\mu = \lambda^*$.

In the following, we describe the class of *sequential quadratic programming* (SQP) methods, in which the search direction p_k in (1.1) is the solution of a *quadratic program*, i.e., the minimization of a quadratic form subject to linear constraints. (SQP methods are also known as QP-based methods and recursive quadratic programming methods.)

The QP subproblem in an SQP method is based on approximating the "ideal" problem (3.2). Let x_k be an estimate of x^*. A set of linear constraints is suggested by the usual Taylor-series linearization of c about x_k:

$$c(x_k + p) \approx c_k + A_k p, \tag{3.4}$$

where c_k and A_k denote $c(x_k)$ and $A(x_k)$. Since $c(x^*) = 0$, we simply impose the requirement that the linearized constraints (3.4) vanish at $x_k + p_k$, i.e., p_k must satisfy the linear constraints

$$A_k p = -c_k. \tag{3.5}$$

Note that (3.5) is analogous to defining p_k as a Newton step to the solution of the nonlinear equations $c(x) = 0$. If $x_k = x^*$, (3.5) defines the same subspace as the constraints in the "ideal" problem (3.2).

The derivation of (3.2) indicates that the quadratic objective function should be an approximation to the Lagrangian function. Since the optimal multiplier λ^* is unknown, some procedure is necessary to obtain Lagrange-multiplier *estimates*. An obvious strategy is to construct the quadratic function so that the Lagrange multipliers of the subproblem approach the optimal multipliers as x_k converges to x^*.

Based on all these considerations, the search direction p_k solves the following QP subproblem:

$$\underset{p \in \Re^n}{\text{minimize}} \quad g_k^T p + \frac{1}{2} p^T B_k p \tag{3.6a}$$

$$\text{subject to} \quad A_k p = -c_k. \tag{3.6b}$$

The Lagrange multiplier vector of (3.6) (denoted by μ_k) satisfies

$$B_k p_k + g_k = A_k^T \mu_k. \tag{3.7}$$

Comparing (3.1) and (3.7), we see that μ_k must approach λ^* as x_k approaches x^*.

One might assume that the *first-order* term of the quadratic model (3.6a) should use the current gradient of the Lagrangian function, $g_k - A_k^T \lambda$, where λ is the best available multiplier estimate. However, using g_k alone in (3.6a) allows the multiplier μ_k (3.7) of the QP subproblem to be taken as an estimate of λ^*. If the gradient of $L(x, \mu)$ were used instead and $x_k = x^*$, μ_k would be zero rather than λ^*.

The matrix B_k is almost always taken as a symmetric positive-definite approximation to the Hessian of the Lagrangian function. Hence, any update to B_k will require the definition of a multiplier estimate. The choice of multiplier estimate will be considered in Section 3.4.

3.2. Equality-constraint quadratic programming: In this section we shall consider methods for the solution of quadratic programs with only equality constraints, of the form (3.6). For simplicity of notation, we drop the subscript k. Nearly all methods for solving (3.6) are based on the *augmented system* of equations for p and μ

$$\begin{pmatrix} B & A^T \\ A & 0 \end{pmatrix}\begin{pmatrix} p \\ -\mu \end{pmatrix} = -\begin{pmatrix} g \\ c \end{pmatrix}, \tag{3.8}$$

which express the optimality and feasibility conditions.

Methods for solving (3.8) are often based upon constructing an equivalent, but simpler system. Let S_1 and S_2 be nonsingular $(n+m)\times(n+m)$ matrices. The solution of (3.8) is equivalent to the solution of

$$S_1\begin{pmatrix} B & A^T \\ A & 0 \end{pmatrix}S_2\begin{pmatrix} \bar{p} \\ -\bar{\mu} \end{pmatrix} = -S_1\begin{pmatrix} g \\ c \end{pmatrix} \quad \text{with} \quad \begin{pmatrix} p \\ -\mu \end{pmatrix} = S_2\begin{pmatrix} \bar{p} \\ -\bar{\mu} \end{pmatrix}. \tag{3.9}$$

For example, if S_1 is given by

$$S_1 = \begin{pmatrix} AB^{-1} & -I \\ I & 0 \end{pmatrix}$$

and S_2 is the identity, we obtain the following equations for p and μ:

$$AB^{-1}A^T\mu = AB^{-1}g - c \tag{3.10a}$$

$$Bp = -g + A^T\mu. \tag{3.10b}$$

In order to solve (3.10), factorizations of B and $AB^{-1}A^T$ are required.

A less obvious choice utilizes the LQ factorization of A:

$$AQ = \begin{pmatrix} L & 0 \end{pmatrix}, \tag{3.11}$$

where Q is an $n\times n$ orthonormal matrix and L is an $m\times m$ lower-triangular matrix: Assume that the columns of Q are partitioned so that

$$Q = \begin{pmatrix} Y & Z \end{pmatrix},$$

where Z is an $n\times(n-m)$ matrix and Y is an $n\times m$ matrix. Then let

$$S_2 = \begin{pmatrix} Y & Z & 0 \\ 0 & 0 & I \end{pmatrix}, \tag{3.12}$$

and choose S_1 as S_2^T. Let p_Y and p_Z denote the first m and last $n-m$ elements of $\bar{p}$, respectively. Substituting from (3.12) into (3.9), we obtain

$$\begin{pmatrix} Y^TBY & Y^TBZ & L^T \\ Z^TBY & Z^TBZ & 0 \\ L & 0 & 0 \end{pmatrix}\begin{pmatrix} p_Y \\ p_Z \\ -\mu \end{pmatrix} = -\begin{pmatrix} Y^Tg \\ Z^Tg \\ c \end{pmatrix}.$$

Thus, p and μ may be found by solving the equations

$$Lp_Y = -c \tag{3.13a}$$

$$Z^TBZp_Z = -Z^Tg - Z^TBYp_Y \tag{3.13b}$$

$$p = Yp_Y + Zp_Z \tag{3.13c}$$

$$L^T\mu = Y^T(g + Bp). \tag{3.13d}$$

In addition to the LQ factorization (3.11), a factorization of Z^TBZ is required in ordet to solve (3.13).

A third method for solving (3.8) can be effective when the Cholesky factorization of B is available, i.e., $B = R^T R$, where R is upper-triangular. In this case, if we define

$$S_2 = \begin{pmatrix} R^{-1} & \\ & I \end{pmatrix} = S_1^T,$$

the linear system associated with the solution $\bar{p}$ of (3.9) is

$$\begin{pmatrix} I & \bar{A}^T \\ \bar{A} & 0 \end{pmatrix} \begin{pmatrix} \bar{p} \\ -\mu \end{pmatrix} = -\begin{pmatrix} \bar{g} \\ c \end{pmatrix}, \tag{3.14}$$

where $R\bar{A} = A$ and $R^T\bar{g} = g$.

As with (3.8), the solution of (3.14) can be computed using the LQ factorization of $\bar{A}$:

$$\bar{A}\bar{Q} = (\,\bar{L} \quad 0\,),$$

where $\bar{Q}$ is orthonormal and $\bar{L}$ is lower-triangular. Let $\bar{Q} = (\,\bar{Y} \quad \bar{Z}\,)$, and partition $\bar{Q}^T\bar{p}$ into $\bar{p}_Y$ and $\bar{p}_Z$. Putting $B = I$ in (3.13), we see that

$$\begin{aligned} \bar{L}\bar{p}_Y &= -c \\ \bar{p}_Z &= -\bar{Z}^T\bar{g} \\ \bar{p} &= \bar{Y}\bar{p}_Y + \bar{Z}\bar{p}_Z \\ \bar{L}^T\mu &= \bar{Y}^T\bar{g} + \bar{p}_Y. \end{aligned}$$

The desired vector p is then obtained by solving $Rp = \bar{p}$.

3.3. The choice of merit function: In the unconstrained case, the value of F provides a natural measure of progress to guide the choice of the steplength α_k in (1.1) (see Section 2.2). By contrast, the definition of a suitable step length for a nonlinearly constrained problem is extremely complicated. For example, it might seem at first that α_k could be chosen simply to produce a sufficient decrease in F. However, if the constraints are violated, F may already be below the true optimal value. Since it is impossible (except in a few special cases) to generate a sequence of iterates that satisfy a set of nonlinear equality constraints, the step length must be chosen to balance the (usually) conflicting aims of reducing the objective function and satisfying the constraints. A standard approach for problem NEP is to define a *merit function* Φ_M that measures progress toward x^*, and then to choose α_k to yield a "sufficient decrease" in Φ_M (using criteria like (2.9) and (2.10)).

It is desirable for Φ_M to have certain properties. First, Φ_M should provide a sensible measure of progress toward the solution — e.g., at least one of the quantities F, $\|c\|$ and $\|Z^Tg\|$ should decrease at every iteration. Second, it must be possible to achieve a sufficient decrease in Φ_M at each iteration — i.e., p_k must be a descent direction for Φ_M. Third, if the steplength enforces a sufficient decrease in Φ_M, it should also guarantee that the quasi-Newton update is positive definite. (This property will be satisfied in the unconstrained case if the "merit function" $F(x)$ is reduced according to the criteria (2.9) and (2.10).) Finally, the merit function should not restrict the rate of convergence of the SQP method. For example, if B_k is taken as the exact Hessian of the Lagrangian function, a step of unity should produce a sufficient decrease in Φ_M near the solution. Unfortunately, no known merit function has *all* these desirable properties. Therefore, existing methods tend to be *hybrid* in the sense that a combination of different techniques is applied.

Many different merit functions have been suggested. For reasons of brevity, we shall consider only three choices. A common theme is the assignment of a positive "penalty" for constraint violation. Let ρ denote a non-negative *penalty parameter*. (It is possible to assign a different penalty parameter to each constraint, but for simplicity we consider only a single ρ.)

One possible merit function is the *quadratic penalty function* P_Q, which is simply F plus a multiple of the squared constraint violations:

$$P_Q(x,\rho) = F(x) + \frac{\rho}{2}\sum_{i=1}^{m} c_i(x)^2 = F(x) + \frac{\rho}{2}\|c(x)\|_2^2,$$

where $\frac{1}{2}\rho\|c(x)\|_2^2$ is called the *penalty term*. The effect of this transformation is to create a local minimum of P_Q which, for sufficiently large ρ, is "near" x^*. It can be shown that the unconstrained minimum of $P_Q(x,\rho)$ approaches x^* as ρ approaches infinity; however, x^* is not an unconstrained minimum of P_Q for any finite ρ, and as ρ increases, the function becomes more and more difficult to minimize.

A pcpular non-smooth merit function is the ℓ_1 *penalty function*:

$$P_1(x,\rho) = F(x) + \rho\sum_{i=1}^{m}|c_i(x)| = F(x) + \rho\|c(x)\|_1.$$

A crucial distinction between P_Q and P_1 is that, under mild conditions, there is a *finite* threshold value $\bar{\rho}$ such that x^* is a local minimum of P_1 for $\rho > \bar{\rho}$. (For this reason, penalty functions like P_1 are sometimes termed *exact penalty functions*.) Note that P_1 is not differentiable at points where a constraint function vanishes — in particular, at x^*. The ℓ_1 merit function provides a simple means of enforcing steady progress towards a solution (see, e.g., Powell, 1983). Unfortunately, requiring a decrease in P_1 at every iteration may inhibit superlinear convergence (see Maratos, 1979). One method of overcoming this difficulty is to allow a limited number of iterations in which another criterion is used to choose the step length if the unit step does not give a sufficient decrease in P_1 (see Chamberlain et *al*., 1982; Mayne and Polak, 1982).

A smooth merit function that includes a quadratic penalty term, but does not require an infinite penalty parameter, is the *augmented Lagrangian function*

$$L_A(x,\lambda,\rho) = F(x) - \sum_{i=1}^{m}\lambda_i c_i(x) + \frac{\rho}{2}\sum_{i=1}^{m} c_i(x)^2 = F(x) - \lambda^T c(x) + \frac{\rho}{2}\|c(x)\|_2^2, \qquad (3.15)$$

where λ is a Lagrange-multiplier estimate. Since $c(x^*) = 0$, both the quadratic penalty term of (3.15) and its gradient vanish at x^*. If $\lambda = \lambda^*$, it follows from (3.1) that x^* is a stationary point (with respect to x) of (3.15). At x^*, the Hessian of the penalty term is simply $\rho A(x^*)^T A(x^*)$, which is a positive semi-definite matrix with strictly positive eigenvalues corresponding to eigenvectors in the range of $A(x^*)^T$. Thus, the penalty term in L_A does not alter the stationary property of x^*, but adds positive curvature in the range space of $A(x^*)^T$. It can be shown that there exists a *finite* $\bar{\rho}$ such that x^* is an *unconstrained minimum* of $L_A(x, \lambda^*, \rho)$ for all $\rho > \bar{\rho}$. If μ denotes the QP multiplier vector, the QP search direction is a descent direction for $L_A(x, \mu, \rho)$ for all non-negative values of ρ. Note that this implies that the QP search direction is also a descent direction for the Lagrangian function $L(x, \mu)$.

The properties of the augmented Lagrangian as a merit function are complementary to those of P_1 — i.e., the augmented Lagrangian will not inhibit superlinear convergence, but cannot be guaranteed to ensure steady progress to a solution. For example, a step length that decreases L_A may *increase* both F and $\|c\|$. One means of attempting to achieve a consistent measure of progress is to define intermediate iterations in which the augmented Lagrangian function is reduced while keeping the multipliers fixed. During the intermediate steps, progress towards a solution is enforced by choosing the penalty parameter to guarantee some reduction in the constraint violations. For further details of this type of technique, see Bertsekas (1982) and Coope and Fletcher (1980).

Another way of measuring progress with an augmented Lagrangian merit function is to compute a vector $\delta\lambda$ that serves as the *change* in the multipliers (analogous to computing p as a change to x) and perform the linesearch with respect to p and $\delta\lambda$ simultaneously. In this case, on completion of the linesearch we have a new pair $(\bar{x}, \bar{\lambda})$ such that

$$\begin{pmatrix} \bar{x} \\ \bar{\lambda} \end{pmatrix} = \begin{pmatrix} x \\ \lambda \end{pmatrix} + \alpha \begin{pmatrix} p \\ \delta\lambda \end{pmatrix}.$$

There are many choices for the multiplier "search direction" $\delta\lambda$. For example, if λ is the best estimate available before the QP subproblem is solved, $\delta\lambda$ may be defined as $\mu - \lambda$, where μ are the QP multipliers (3.7). Under certain conditions, it can be shown that there exists a *finite* value of ρ such that $(p, \delta\lambda)$ will be a descent direction for $L_A(x, \lambda, \rho)$. Numerical experience on standard sets of test problems indicates that the augmented Lagrangian can be an effective merit function (see, e.g., Schittkowski, 1982).

It is important to note that any of the merit functions discussed can be guaranteed to have *at best* a local minimum along the search direction. For example, there are simple cases where all three merit functions are unbounded below *for any value of the penalty parameter*. The difficulties caused by an unbounded merit function will vary from one merit function to another. For example, with a given value of ρ, unboundedness is much more likely for the ℓ_1 penalty function than the quadratic penalty function because constraint violations are penalized less severely. Moreover, larger values of ρ may be needed for the ℓ_1 merit function in order to force convergence from poor starting points.

Other smooth merit functions have been proposed that treat the multiplier vector as a continuous function of x; some of these ensure global convergence and permit local superlinear convergence (see, e.g., Fletcher, 1970; DiPillo and Grippo, 1979, 1982; Dixon, 1979; Bertsekas,

1980; Boggs and Tolle, 1980, 1981). With these functions, the descent property of the SQP direction depends on the value of a single parameter, but the first derivatives of these merit functions involve *second* derivatives of the problem functions. Hence, the determination of whether the SQP direction is a descent direction is complicated, and these functions would be impractical if first derivatives of the objective and constraints are expensive or unavailable.

For other merit functions and their properties, see Bartholomew-Biggs (1981), Dixon (1979), Fletcher (1983) and Tapia (1977).

3.4. Approximating the Hessian of the Lagrangian: Based on the unconstrained case (see Section 2.1), the BFGS formula (2.6) seems a logical choice for updating an approximation to the Hessian of the Lagrangian function. For convenience, we rewrite the BFGS formula in terms of a general vector v:

$$\bar{B} = B - \frac{1}{s^T B s} B s s^T B + \frac{1}{v^T s} v v^T, \tag{3.16}$$

where $s = \bar{x} - x$, the change in x. With this definition, $\bar{B}$ satisfies $\bar{B}s = v$; if B is positive definite, $\bar{B}$ is positive definite if and only if

$$v^T s > 0. \tag{3.17}$$

In the unconstrained case, the matrix B is intended to approximate the Hessian of F at x^*, and v is taken as $\bar{g} - g$, the difference in gradients of F. As noted in Section 2.2, (3.17) can always be satisfied with a line search based on (2.10); for a backtracking line search, the update can be skipped if (3.17) is not satisfied. Furthermore, (3.17) is satisfied by the *unit step* in the neighborhood of an unconstrained minimum, and hence the updated matrices will eventually remain positive definite for either line search strategy.

For the nonlinear-constraint case, we seek a quasi-Newton approximation to the Hessian of the *Lagrangian function*. Hence, it might seem that the standard BFGS update (3.16) could be applied with v taken as y_L, the difference in gradients of the Lagrangian function, i.e.,

$$y_L = \bar{g} - g - (\bar{A}^T - A^T)\lambda,$$

where λ is the best available multiplier estimate. Most authors have recommended taking λ as μ, the QP multiplier vector (3.7). An alternative is $\bar{\lambda}_L$, the vector of *least-squares multipliers* that is the solution of the least-squares problem $\min_\lambda \|\bar{g} - \bar{A}^T\lambda\|_2^2$.

Unfortunately, taking v as y_L leads to serious difficulties. For numerous reasons, it is desirable for B to remain positive definite (see, e.g., Powell, 1977; Murray and Wright, 1982). However, since x^* is not an unconstrained minimum of the Lagrangian function, it may be *impossible*, with *any* line search, to find a step length for which (3.17) holds. Hence, skipping the update in this case could destroy the local convergence properties, since no update might ever be performed.

A widely used strategy to avoid this difficulty is given by Powell (1977). At the end of the line search, the following condition is tested:

$$y_L^T s \geq \sigma s^T B s, \tag{3.18}$$

where $0 < \sigma < 1$. If (3.18) holds, $y_L^T s$ is considered "sufficiently positive", and the standard BFGS update is performed with $v = y_L$. If not, the BFGS update is carried out with v taken as a linear combination of y_L and Bs:

$$v = \theta y_L + (1-\theta)Bs, \quad \text{where} \quad \theta = \frac{(\sigma - 1)s^T B s}{s^T(y_L - Bs)}. \tag{3.19}$$

With this choice of θ, $v^T s = \sigma s^T B s$ and $\det(\bar{B}) = \sigma \det(B)$, where $\det(\cdot)$ denotes the determinant of a matrix. With the Powell modification (3.19), the determinant of the updated Hessian approximation is bounded away from zero. Powell (1977) has shown that the BFGS update with the modification (3.19) is two-step superlinearly convergent (i.e., formula (2.8) holds in the limit with x_{k+2} replacing x_{k+1}).

A second alternative is to consider B as an approximation to the Hessian of the *augmented* Lagrangian function L_A (3.20) (see, e.g., Tapia, 1977; Han, 1977; Bertsekas, 1982). (Recall from Section 3.3 that x^* is an unconstrained minimum of L_A if $\mu = \lambda^*$ and ρ is large enough.) With this approach, v in (3.16) is taken as y_A, the difference in gradients of L_A:

$$y_A \equiv y_L + \rho(\bar{A}^T \bar{c} - A^T c). \tag{3.20}$$

Note that $\bar{B}$ will then satisfy a quasi-Newton condition with respect to L_A. Boggs, Tolle and Wang (1982) have shown that, under certain conditions, one-step superlinear convergence can be achieved with the definition (3.20). In order to define v from (3.20), a suitable penalty parameter ρ must be found. One idea is to alter ρ only when an iteration occurs in which $v^T s$ is not sufficiently positive. At this point, ρ is increased by an amount $\delta\rho$ that gives a sufficiently positive value of $v^T s$ when v is defined as

$$y_L + (\rho + \delta\rho)(\bar{A}^T \bar{c} - A^T c).$$

Fortunately, a suitable value of $\delta\rho$ can always be found *if the step length is chosen based on an augmented Lagrangian merit function* (as described below in Section 3.5). In this case, any $\delta\rho$ greater than the scalar

$$\frac{-y_A^T s}{s^T(\bar{A}^T \bar{c} - A^T c)}$$

will suffice, where y_A is given by (3.20).

Yet another possibility is based on the property that the projected Hessian of the Lagrangian function must be positive semi-definite at x^* (with $\mu = \lambda^*$) (see Section 3.1). This suggests developing a positive-definite quasi-Newton approximation to the *projected* Hessian $Z^T W Z$, thereby avoiding the need to modify the BFGS update or to augment the Lagrangian function. A closely related approach is to update an unsymmetric approximation to the *one-sided projection* $Z^T W$, as in quasi-Newton methods for solving nonlinear equations. Both types of "projected" quasi-Newton methods are currently the subject of active research; see, e.g., Wright (1976), Murray and Wright (1978), Coleman and Conn (1982), Gabay (1982), and Nocedal and Overton (1983).

3.5. Further comments on the line search: Since any merit function may be unbounded along the search direction, the upper bound Δ on the step length (2.12) must be selected with care. If Δ is "too large", $|F|$ and $\|c\|$ may increase without limit. Clearly, a "safe" strategy would be to choose $\Delta = 1$, as in a backtracking linesearch. However, this restriction may lead to some inefficiency. Away from the solution, $\|p\|$ may be *poorly scaled* in the sense that a local minimum of the merit function may occur for step lengths that are substantially larger than one. Moreover, allowing more freedom in the choice of α may improve the performance of an algorithm when the merit function is bounded below along the search direction, yet has local negative curvature. For example, there may exist positive-definite quasi-Newton updates for the augmented Lagrangian at values of α larger than one. We suggest a compromise strategy in which steps greater than unity are permitted as long as both the objective function and the Euclidean norm of the constraints are decreasing. In practice, this may be achieved by imposing an *additional termination condition* on α when $\alpha \geq 1$ and $\|c(x+\alpha p)\| > 0$. Specifically, the line search terminates with the first value of α that satisfies (2.2) *or* either of the conditions

$$g(x+\alpha p)^T p \geq 0 \quad \text{or} \quad p^T A(x+\alpha p)^T c(x+\alpha p) \geq 0. \tag{3.21}$$

With this scheme, any negative curvature will be treated as local unless there is the potential for the merit function to decrease without limit.

A second point of interest about the line search is the relationship between the definition of an acceptable step length (in terms of the merit function) and the Hessian update. If B represents an approximation to the Hessian of the augmented Lagrangian, it is convenient to use the augmented Lagrangian as a merit function, since (3.17) will hold automatically if the steplength satisfies the usual conditions (2.9) and (2.10). Hence, the BFGS update will be well defined and positive definite. If (2.10) cannot be satisfied (in which case the augmented Lagrangian must have local negative curvature), satisfaction of the additional termination criteria (3.21) ensures the existence of a penalty parameter for which (3.17) will hold.

4. METHODS FOR NONLINEAR INEQUALITY CONSTRAINTS

4.1. Background: In the final problem to be considered, all the constraints are nonlinear *inequalities*:

$$\text{NIP} \qquad \begin{array}{ll} \underset{x\in\Re^n}{\text{minimize}} & F(x) \\ \text{subject to} & c_i(x) \geq 0, \quad i = 1,\ldots,m. \end{array}$$

The matrix $A(x)$ will denote the Jacobian of $c(x)$.

The constraint c_i is said to be *active* at x if $c_i(x) = 0$, and *violated* if $c_i(x) < 0$. The optimality conditions for NIP are similar to those for the equality-constraint case, except that they involve only constraints active at x^*, and impose a sign restriction on the Lagrange multipliers. In particular, the following conditions are necessary for x^* to be a local minimum of NIP:

(i) The point x^* is feasible, i.e., $c(x^*) \geq 0$. Let c denote the vector of constraint functions active at x^*, and let t^* denote the number of active constraints at x^*;

(ii) There exists a t^*-vector $\hat{\lambda}^*$ such that

$$g(x^*) = A(x^*)^T \hat{\lambda}^* \quad \text{with} \quad \hat{\lambda}^* \geq 0, \tag{4.1}$$

where A denotes the Jacobian of c;

(iii) Let $W(x, \mu)$ denote the matrix (3.3), and let $Z(x)$ denote a matrix such that $A(x)Z(x) = 0$; then $Z(x^*)^T W(x^*, \lambda^*) Z(x^*)$ is positive semi-definite.

The major difference between inequality- and equality-constrained problems is that the set of constraints active at the solution is *unknown* in the inequality case. Therefore, algorithms for NIP must include some procedure — usually termed an *active-set strategy* — for determining the correct active set. In this section we discuss the additional algorithmic complexity in SQP methods that arises specifically from the presence of inequality constraints.

4.2. Formulation and solution of the QP subproblem: Broadly speaking, two extreme types of QP subproblems might be posed when solving inequality-constrained problems. The first extreme is represented by a subproblem in which *all* the nonlinear inequality constraints are included *as inequalities* in the QP subproblem; this has been by far the most widely used formulation in published SQP methods. With this *IQP subproblem*, the search direction p is the solution of

$$\underset{p \in \Re^n}{\text{minimize}} \quad d^T p + \frac{1}{2} p^T B p \tag{4.2a}$$

$$\text{subject to} \quad Ap \geq -b, \tag{4.2b}$$

where B is an approximation to the Hessian of the Lagrangian function and A is the Jacobian of $c(x)$ evaluated at the current iterate.

In general, the solution of (4.2) must be found by iteration. An essential feature of most QP algorithms is a systematic search for the correct active set. To determine the active set, QP methods maintain a *working set* that serves as an approximation to the active set, and solve a sequence of QP subproblems in which the constraints in the working set are treated as *equalities*. The major differences among QP methods arise from the numerical methods that solve the equations in the subproblems (see Section 3.3), and the strategies that control changes in the working set. Modern QP methods are surveyed in Fletcher (1981) and Gill, Murray and Wright (1981).

As noted in Section 4.1, methods for nonlinear inequality-constrained problems must determine the set of constraints active at x^*. Because (4.2) is an optimization problem in which all the constraints of NIP are represented, it is convenient to take *the active set of the QP* as a prediction of the active set of the nonlinearly constrained problem. The theoretical justification for this strategy is that, if $d = g(x)$ and $b = -c(x)$, the QP (4.2) will make a correct prediction of the active set of the original problem in a neighborhood of x^* for *any* bounded positive-definite matrix B (see Robinson, 1974, for a precise specification of the required conditions). The solution of (4.2) with these choices for d and b will be called the *standard IQP search direction*. With this formulation, the QP multipliers will approach λ^* as the iterates converge to x^*, and hence it is common to take the QP multipliers as a Lagrange multiplier estimate for NIP.

One complication with choosing $-c(x)$ as the right-hand side of the constraints in (4.2) is that there may be no feasible point for the linearized constraints $Ap \geq -c$, even when feasible points exist for the *nonlinear* constraints (this can happen only when at least one nonlinear constraint is violated at x). Various strategies have been proposed to overcome this difficulty.

One possibility (Powell, 1977) is to define a damped right-hand side for the linearized version of a violated constraint in (4.2*b*). The damping is controlled by a scalar δ ($0 \leq \delta \leq 1$), and the modified constraint is given by

$$a_i^T p \geq -(1-\delta)c_i \quad \text{if} \quad c_i(x) < 0. \tag{4.3}$$

If constraints like (4.3) are included in (4.2*b*) for a sequence of decreasing values of δ, at least one feasible point must eventually exist ($p = 0$ and $\delta = 1$). Rather than repeatedly solving (4.2), Schittkowski (1982) proposed that δ be computed by adding a quadratic penalty term $\frac{1}{2}\rho\delta^2$ to the quadratic objective function of (4.2) and including the constraint $0 \leq \delta \leq 1$. Unfortunately, even with the modified constraints (4.3), problems exist for which the only feasible point is the zero vector. (For example, consider the constraints $x_1^2 + x_2^2 \geq 4$, $x_1 \geq 0$ and $x_2 \geq 0$ at the point $x_1 = 0$, $x_2 = 0$.)

Several other strategies have been suggested that can be viewed as a more general form of constraint relaxation. The idea is to minimize some measure of the violations of the linearized constraints $Ap \geq -c$, or a weighted combination of the linearized constraint violations and the quadratic objective function. Methods of this type have been suggested by Fletcher (1981), Tone (1983) and Gill *et al.* (1983). With such an approach, it is desirable for p to be a descent direction for an appropriate merit function (see Section 4.3).

The other extreme form of QP subproblem used in SQP methods involves a QP with only *equality* constraints. In order to pose an equality-constrained QP (EQP), some determination must be made before posing the QP as to which constraints are to be included; this is sometimes called a *pre-assigned active-set strategy*. The ideal choice would clearly be the set of constraints active at x^*. Therefore, pre-assigned active-set strategies tend to choose constraints that satisfy properties of the active constraints in a neighborhood of the solution — e.g., are "small" in magnitude. The selection may also be based on Lagrange multiplier estimates, or on the behavior of the merit function. Any pre-assigned active-set strategy must have the property that it will identify the correct active set in some neighborhood of x^*.

A benefit of posing an EQP subproblem is that, in general, the subproblem will be *easier to solve* than one with inequality constraints. Moreover, the computation can be arranged so that only the gradients of the constraints in the working set need be computed. The difficulties with infeasible subproblems that occur with an IQP formulation can be avoided by ensuring that constraints in the working set are linearly independent, or by redefining the subproblem (see Murray and Wright, 1982).

4.3. Definition of the merit function: Most merit functions used in inequality problems are generalizations of those given in Section 3.3. As the iterates approach the solution, only constraints active at x^* should affect the merit function and its derivatives.

The standard definitions of the quadratic and ℓ_1 penalty functions for the inequality problem are

$$P_Q(x,\rho) = F(x) + \frac{\rho}{2}\sum_{i=1}^{m}\left(\min(0,c_i(x))\right)^2;$$

$$P_1(x,\rho) = F(x) + \rho\sum_{i=1}^{m}|\min(0,c_i(x))|.$$

Since constraints that are strictly satisfied do not appear in these definitions, only constraints active at x^* will affect penalty functions in a neighborhood of the solution. Under mild conditions it can be shown that the standard IQP search direction is a direction of descent for P_1 for sufficiently large (finite) ρ (Han, 1976).

When working with an augmented Lagrangian merit function, the associated Lagrange multiplier estimate λ implies a prediction of the active set (i.e., the active constraints have non-zero multipliers). Even if a violated constraint has a zero multiplier estimate, it should be included in the *penalty* term of an augmented Lagrangian in order to provide a safeguard if the predicted active set is incorrect (otherwise, there would be no control on the violations of constraints not in the predicted active set).

Another possibility for the merit function is an augmented Lagrangian function that involves *only a subset of the constraints*. The most common such definition is

$$L_A(x,\lambda,\rho) = F(x) + \sum_{i=1}^{m}\begin{cases} -\lambda_i c_i(x) + \dfrac{\rho_i}{2}c_i(x)^2, & \text{if} \quad c_i \le \dfrac{\lambda_i}{\rho}; \\ -\dfrac{\lambda_i^2}{2\rho_i} & \text{if} \quad c_i > \dfrac{\lambda_i}{\rho}, \end{cases} \tag{4.4}$$

where the term involving λ_i^2 is included to make the function continuous (Rockafellar, 1974). In effect, the i-th constraint is considered "significant" if $c_i \le \lambda_i/\rho$. (Note that violated constraints are always significant if $\lambda \ge 0$.) If λ converges to the optimal multiplier vector, only the constraints active at x^* will be included as the iterates converge to the solution.

4.4. Solving QP Subproblems Within an SQP Method: The original conception of an SQP algorithm based on an inequality QP was that the subproblem could be solved using a "black box" quadratic programming algorithm. However, there are significant advantages in using a QP algorithm that takes advantage of certain features of SQP algorithms.

In particular, it is beneficial for a QP algorithm to allow the initial working set and part (or all) of its factorization to be specified for the first iteration of the QP method. This feature increases the speed with which each QP subproblem can be solved, since the active set from one QP subproblem can be taken as the initial working set for the next QP. Eventually, the active set of the QP will become the correct active set for the nonlinear problem (see Section 4.2), and thus *QP subproblems near the solution will reach optimality in only one iteration.* Furthermore, the work associated with factorizing the working set in order to compute a least-squares multiplier estimate at each SQP iteration need not be repeated within the QP (which would almost certainly be the case with a "black box" QP code).

5. AVAILABLE SOFTWARE

Although there has been substantial interest in SQP methods for several years, very little high-quality SQP software has been made available. This delay in software production is typical in all areas of numerical mathematics, and stems largely from the need to cater for difficulties that can be ignored (or assumed away) in a theoretical context. As the algorithmic details of SQP methods become more refined, we expect a variety of further implementations to be produced.

In the meantime, we are aware of four SQP codes: ORQP/XROP (Bartholomew-Biggs, 1979), which is available from the Numerical Optimization Centre, Hatfield Polytechnic, Hertfordshire, England; VMCWD (Powell, 1982, 1983), which is part of the Harwell Subroutine Library, AERE Harwell, Oxfordshire, England; NLPQL (Schittkowski, 1983), which is available from the Institut für Angewandte Mathematik und Statistik, Universität Würzburg, Germany; and NPSOL (Gill *et al.*, 1983), which is available from the Office of Technology Licensing, Stanford University, Stanford, California 94305.

ORQP/XROP are the latest versions of one of the first SQP codes that was made widely available; they solve an EQP subproblem that includes the linearizations of the violated constraints. At each iteration a BFGS approximation to the Hessian of the Lagrangian is updated; indefinite updates are avoided using the Powell modification (3.19). The quadratic penalty function is used as merit function.

At each iteration, VMCWD solves a full IQP subproblem using a "black box" QP algorithm. Progress to the solution is maintained using the ℓ_1 merit function (with safeguards to ensure that superlinear convergence is not impeded). The matrix B_k is a positive-definite BFGS approximation to the Hessian of the Lagrangian function.

NLPQL solves an inequality constrained QP that involves linearizations of a subset of the original constraints. This means that only the constraints predicted to be active are evaluated at each iteration. An augmented Lagrangian merit function is used in a line search with respect to x and λ.

Both VMCWD and NLPQL treat infeasible subproblems using (4.3), and use the Powell modification (3.19) to ensure positive-definite updates.

The method of NPSOL is an IQP method, in which the active set of the QP is used as a prediction of the active set of the nonlinear problem. The code treats bounds, linear constraints and nonlinear constraints separately. The matrix B_k is a positive-definite approximation to the Hessian of an augmented Lagrangian function, and a sufficient decrease in an augmented Lagrangian merit function is required at each iteration. Infeasible subproblems are treated by finding the least-infeasible point for (4.2), and then (temporarily) relaxing the violated linearized constraints. Each QP subproblem is solved using a quadratic programming code that has the features mentioned in Section 4.4.

ACKNOWLEDGEMENTS

This research was supported by the U.S. Department of Energy Contract DE-AC03-76SF00326, PA No. DE-AT03-76ER72018; National Science Foundation Grants MCS-7926009 and ECS-8012974; the Office of Naval Research Contract N00014-75-C-0267; and the U.S. Army Research Office Contract DAAG29-79-C-0110.

REFERENCES

Bartholomew-Biggs, M. C. (1979). An improved implementation of the recursive quadratic programming method for constrained minimization, Report 105, Numerical Optimisation Centre, Hatfield Polytechnic, Hatfield, England.

Bartholomew-Biggs, M. C. (1981). Line search procedures for nonlinear programming algorithms with quadratic programming subproblems, Report 116, Numerical Optimisation Centre, Hatfield Polytechnic, Hatfield, England.

Bertsekas, D. P. (1980). "Variable metric methods for constrained optimization based on differentiable exact penalty functions", in *Proceedings of 18th Allerton Conference on Communication, Control and Computing*, pp. 584–593, Allerton Park, Illinois.

Bertsekas, D. P. (1982). *Constrained Optimization and Lagrange Multiplier Methods*, Academic Press, New York and London.

Biggs, M. C. (1972). "Constrained minimization using recursive equality quadratic programming", in *Numerical Methods for Non-Linear Optimization* (F. A. Lootsma, ed.), pp. 411–428, Academic Press, London and New York.

Boggs, P. T. and Tolle, J. W. (1980). Augmented Lagrangians which are quadratic in the multiplier, *J. Opt. Th. Applics.* **31**, pp. 17–26.

Boggs, P. T. and Tolle, J. W. (1981). A family of descent functions for constrained optimization, Report 81-3, Department of Mathematics, University of North Carolina, Chapel Hill, North Carolina.

Boggs, P. T., Tolle, J. W. and Wang, P. (1982). On the local convergence of quasi-Newton methods for constrained optimization, *SIAM J. Control and Optimization* **20**, pp. 161–171.

Brent, R. P. (1973). *Algorithms for Minimization without Derivatives*, Prentice-Hall, Inc., Englewood Cliffs, New Jersey.

Chamberlain, R. W., Lemaréchal, C., Pedersen, H., and Powell, M. J. D. (1982). The watchdog technique for forcing convergence in algorithms for constrained optimization, *Math. Prog. Study* **16**, pp. 1–17.

Coleman, T. F. and Conn, A. R. (1982). Nonlinear programming via an exact penalty function, *Math. Prog.* **24**, pp. 123–161.

Coope, I. D. and Fletcher, R. (1980). Some numerical experience with a globally convergent algorithm for nonlinearly constrained optimization, *J. Opt. Th. Applics.* **32**, pp. 1–16.

Dennis, J. E., Jr. and Moré, J. J. (1977). Quasi-Newton methods, motivation and theory, *SIAM Review* **19**, pp. 46–89.

Dennis, J. E., Jr. and Schnabel, R. E. (1981). "A new derivation of symmetric positive definite secant updates", *Nonlinear Programming 4* (O. L. Mangasarian, R. R. Meyer and S. M. Robinson, eds.), pp. 167–199, Academic Press, London and New York.

Dennis, J. E., Jr. and Schnabel, R. B. (1983). *Numerical Methods for Unconstrained Optimization and Nonlinear Equations*, Prentice-Hall, Inc., Englewood Cliffs, New Jersey.

DiPillo, G. and Grippo, L. (1979). A new class of augmented Lagrangians in nonlinear programming, *SIAM J. Control and Optimization* **17**, pp. 618–628.

DiPillo, G. and Grippo, L. (1982). A new augmented Lagrangian function for inequality constraints in nonlinear programming problems, *J. Opt. Th. Applics.* **36**, pp. 495–519.

Dixon, L. C. W. (1979). Exact penalty functions in nonlinear programming, Report 103, Numerical Optimisation Centre, Hatfield Polytechnic, Hatfield, England.

Fiacco, A. V. and McCormick, G. P. (1968). *Nonlinear Programming: Sequential Unconstrained Minimization Techniques*, John Wiley and Sons, New York and Toronto.

Fletcher, R. (1970). "A class of methods for nonlinear programming with termination and convergence properties", in *Integer and Nonlinear Programming* (J. Abadie, ed.), pp. 157–175, North-Holland, Amsterdam.

Fletcher, R. (1981). *Practical Methods of Optimization, Volume 2, Constrained Optimization*, John Wiley and Sons, New York and Toronto.

Fletcher, R. (1983). "Penalty functions", in *Mathematical Programming: The State of the Art*, (A. Bachem, M. Grötschel and B. Korte, eds.), pp. 87–114, Springer-Verlag, Berlin, Heidelberg, New York and Tokyo.

Gabay, D. (1982). Reduced quasi-Newton methods with feasibility improvement for nonlinearly constrained optimization, *Math. Prog. Study* **16**, pp. 18–44.

Gill, P. E., Golub, G. H., Murray, W. and Saunders, M. A. (1974). Methods for modifying matrix factorizations, *Math. Comp.* **28**, pp. 505–535.

Gill, P. E. and Murray, W. (1972). Quasi-Newton methods for unconstrained optimization, *J. Inst. Maths. Applics.* **9**, pp. 91–108.

Gill, P. E., Murray, W., Saunders, M. A. and Wright, M. H. (1983). User's guide for SOL/NPSOL: a Fortran package for nonlinear programming, Report SOL 83-12, Department of Operations Research, Stanford University, California.

Gill, P. E., Murray, W. and Wright, M. H. (1981). *Practical Optimization*, Academic Press, London and New York.

Goldstein, A. and Price, J. (1967). An effective algorithm for minimization, *Numer. Math.* **10**, pp. 184–189.

Han, S.-P. (1976). Superlinearly convergent variable metric algorithms for general nonlinear programming problems, *Math. Prog.* **11**, pp. 263–282.

Han, S.-P. (1976). Superlinearly convergent variable metric algorithms for general nonlinear programming problems,

Han, S.-P. (1977). Dual variable metric algorithms for constrained optimization, *SIAM J. Control and Optimization* **15**, pp. 546–565.

Maratos, N. (1978). *Exact Penalty Function Algorithms for Finite-Dimensional and Control Optimization Problems*, Ph. D. Thesis, University of London.

Mayne, D. Q. and Polak, E. (1982). A superlinearly convergent algorithm for constrained optimization problems, *Math. Prog. Study* **16**, pp. 45–61.

Murray, W. and Wright, M. H. (1978). Methods for nonlinearly constrained optimization based on the trajectories of penalty and barrier functions, Report SOL 78–23, Department of Operations Research, Stanford University.

Murray, W. and Wright, M. H. (1982). Computation of the search direction in constrained optimization algorithms, *Math. Prog. Study* **16**, pp. 63–83.

Nocedal, J. and Overton, M. (1983). Projected Hessian updating algorithms for nonlinearly constrained optimization, Report 95, Department of Computer Science, Courant Institute of Mathematical Sciences, New York University, New York.

Ortega, J. M. and Rheinboldt, W. C. (1970). *Iterative Solution of Nonlinear Equations in Several Variables*, Academic Press, London and New York.

Powell, M. J. D. (1974). "Introduction to constrained optimization", in *Numerical Methods for Constrained Optimization* (P. E. Gill and W. Murray, eds.), pp. 1–28, Academic Press, London and New York.

Powell, M. J. D. (1977). A fast algorithm for nonlinearly constrained optimization calculations, Report DAMTP 77/NA 2, University of Cambridge, England.

Powell, M. J. D. (1982). VMCWD: a Fortran subroutine for constrained optimization, Report DAMTP 82/NA 4, University of Cambridge, England.

Powell, M. J. D. (1983). "Variable metric methods for constrained optimization", in *Mathematical Programming: The State of the Art*, (A. Bachem, M. Grötschel and B. Korte, eds.), pp. 288–311, Springer-Verlag, Berlin, Heidelberg, New York and Tokyo.

Robinson, S. M. (1974). Perturbed Kuhn-Tucker points and rates of convergence for a class of nonlinear programming algorithms, *Math. Prog.* **7**, pp. 1–16.

Rockafellar, R. T. (1974). Augmented Lagrange multiplier functions and duality in nonconvex programming, *SIAM J. Control and Optimization* **12**, pp. 268–285.

Schittkowski, K. (1981). The nonlinear programming method of Wilson, Han, and Powell with an augmented Lagrangian type line search function, *Numerische Mathematik* **38**, pp. 83–114.

Schittkowski, K. (1982). On the convergence of a sequential quadratic programming method with an augmented Lagrangian line search function, Report SOL 82-4, Department of Operations Research, Stanford University.

Schittkowski, K. (1983). Some implementation details for a nonlinear programming algorithm, *Mathematical Programming Society Committee on Algorithms Newsletter*, March 1983.

Tapia, R. A. (1977). Diagonalized multiplier methods and quasi-Newton methods for constrained optimization, *J. Opt. Th. Applics.* **22**, pp. 135–194.

Tone, K. (1983). Revisions of constraint approximations in the successive QP method for nonlinear programming, *Math. Prog.* **26**, pp. 144–152.

Wilson, R. B. (1963). *A Simplicial Algorithm for Concave Programming*, Ph.D. Thesis, Harvard University.

Wolfe, P. (1969). Convergence conditions for ascent methods, *SIAM Review,* **11**, pp. 226–235.

Wright, M. H. (1976). Numerical Methods for Nonlinearly Constrained Optimization, Ph. D. Thesis, Stanford University.

NATO ASI Series

Series F: Computer and Systems Sciences

No. 5

International Calibration Study of Traffic Conflict Techniques

Editor: **E. Asmussen**

Published in cooperation with NATO Scientific Affairs Division

1984. VII, 229 pages. ISBN 3-540-12716-X

Contents: Introduction speech. - General papers on conflict technique development and application. - Technical papers on conflict techniques. - Background paper for the international calibration study. - International calibration study of traffic conflict techniques: General design. - Summary of discussions and conclusions. - Closing remarks. - List of abstracts. - List of participants.

No. 6

Information Technology and the Computer Network

Editor: K. G. Beauchamp

Published in cooperation with NATO Scientific Affairs Division

1984. VIII, 271 pages. ISBN 3-540-12883-2

Contents: Information Technology. - Standardisation. - Information Services. - Network Development. - Network Management. - Protocols and Secure Systems. - Delegates "Short" Papers. - Participants.

No. 7

High-Speed Computation

Editor: **J. S. Kowalik**

Published in cooperation with NATO Scientific Affairs Division

1984. IX, 441 pages. ISBN 3-540-12885-9

Contents: Opening Address. - Keynote Address. - Hardware Systems. - Performance and Capabilities of Computer Systems. - Algorithms and Applications. - Names and Address of Participants.

No. 8

Program Transformation and Programming Environments

Editor: **P. Pepper**

Directed by F. L. Bauer, H. Remus

Published in cooperation with NATO Scientific Affairs Division

1984. 24 figures. XIV, 378 pages. ISBN 3-540-12932-4

In preparation

Contents: Part I: Digest of the Discussions: Introduction. Life Cycle Models and Programming Environments. Management & Organization. Formal Methods in Program Development. Software Specification. Program Development by Transformations. Acceptance of Formal Methods. Outlook. Conclusion. - Part II: Position Statements and Papers. - Appendix: List of Participants. Glossary.

NATO ASI Series

Series F: Computer and Systems Sciences

No. 1

Issues in Acoustic Signal - Image Processing and Recognition

Editor: **C.H.Chen**

Published in cooperation with NATO Scientific Affairs Division

1983. VIII, 333 pages. ISBN 3-540-12192-7

Contents: Overview. - Pattern Recognition Processing. - Artificial Intelligence Approach. - Issues in Array Processing and Target Motion Analysis. - Underwater Channel Characterization. - Issues in Seismic Signal Processing. - Image Processing. - Report of Discussion Session on Unresolved Issues and Future Directions. - List of Participants.

No. 2

Image Sequence Processing and Dynamic Scene Analysis

Editor: **T.S.Huang**

Published in cooperation with NATO Scientific Affairs Division

1983. IX, 749 pages. ISBN 3-540-11997-3

Contents: Overview. - Image Sequence Coding. - Scene Analysis and Industrial Applications. - Biomedical Applications. - Subject Index.

No. 3

Electronic Systems Effectiveness and Life Cycle Costing

Editor: **J.K.Skwirzynski**

Published in cooperation with NATO Scientific Affairs Division

1983. XVII, 732 pages. ISBN 3-540-12287-7

Contents: Mathematical Background and Techniques. - Reliability: Hardware. Software. - Life Cycle Costing and Spares Allocation. - Two Panel Discussions. - List of Lectures and Delegates.

No. 4

Pictorial Data Analysis

Editor: **R.M.Haralick**

Published in cooperation with NATO Scientific Affairs Division

1983. VIII, 468 pages. ISBN 3-540-12288-5

Contents: Neighborhood Operators: An Outlook. - Linear Approximation of Quantized Thin Lines. - Quadtrees and Pyramids: Hierarchical Representation of Images. - Fast In-Place Processing of Pictorial Data. - C-Matrix, C-Filter: Applications to Human Chromosomes. - The Application of Gödel Numbers to Image Analysis and Pattern Recognition. - Segmentation of Digital Images Using a Priori Information About the Expected Image Contents. - A Syntactic-Semantic Approach to Pictorial Pattern Analysis. - Relational Matching. - Representation and Control in Vision. - Computer Vision Systems: Past, Present, and Future. - Artificial Intelligence: Making Computers More Useable. - Automation of Pap Smear Analysis: A Review and Status Report. - Medical Image Processing. - 2-D Fitting and Interpolation Applied to Image Distortion Analysis. - Pictorial Pattern Recognition for Industrial Inspection. - Pattern Recognition of Remotely Sensed Data. - Satellite Image Understanding Through Synthetic Images. - A Diffusion Model to Correct Multi-Spectral Images for the Path-Radiance Atmospheric Effect. - Analysis of Seasat-Synthetic Aperture Radar (SAR) Imagery of the Ocean Using Spatial Frequency Restoration Techniques (SFRT). - Adjacency Relationships in Aggregates of Crystal Profiles.

Springer-Verlag
Berlin
Heidelberg
New York
Tokyo

GPSR Compliance

The European Union's (EU) General Product Safety Regulation (GPSR) is a set of rules that requires consumer products to be safe and our obligations to ensure this.

If you have any concerns about our products, you can contact us on ProductSafety@springernature.com

In case Publisher is established outside the EU, the EU authorized representative is:

Springer Nature Customer Service Center GmbH
Europaplatz 3
69115 Heidelberg, Germany

Batch number: 10762209

Printed by Printforce, the Netherlands